The
Science of
Footwear

Human Factors and Ergonomics

Series Editor

Gavriel Salvendy

Professor Emeritus
School of Industrial Engineering
Purdue University

Chair Professor & Head
Dept. of Industrial Engineering
Tsinghua Univ., P.R. China

The
Science of
Footwear

Ravindra S. Goonetilleke

CRC Press
Taylor & Francis Group
Boca Raton London New York

CRC Press is an imprint of the
Taylor & Francis Group, an **Informa** business

CRC Press
Taylor & Francis Group
6000 Broken Sound Parkway NW, Suite 300
Boca Raton, FL 33487-2742

First issued in paperback 2017

© 2013 by Taylor & Francis Group, LLC
CRC Press is an imprint of Taylor & Francis Group, an Informa business

Library of Congress Cataloging-in-Publication Data

The science of footwear / editor, Ravindra S. Goonetilleke.
     p. cm. -- (Advances in human factors and ergonomics series)
    Includes bibliographical references and index.
    ISBN 978-1-4398-3568-5 (hardback)
    1. Footwear. 2. Footwear industry. I. Goonetilleke, Ravindra S.

TS990.S334 2012
685'.3--dc23
                                        2012030465

Visit the Taylor & Francis Web site at
http://www.taylorandfrancis.com

and the CRC Press Web site at
http://www.crcpress.com

To the memory of my parents,
Reggie and Catherine Goonetilleke

Contents

PART I   The Human Foot

PART II   Scanning and Processing

PART III   Footwear Design

PART IV   Testing

PART V Footwear Effects

PART VI Activity-Specific Footwear

PART VII   Customization

Preface

The global footwear market in 2011 was \$241,294.3 million, and it grew by 5% from the year before. In the next five years, the footwear market is expected to grow by 29% (MarketLine, 2012) with most footwear manufactured in South-East Asia. Footwear suppliers have maintained their position in the market by differentiating the products, by offering specialized items, and by manufacturing high-end designer footwear, all of which require careful thought and research. All footwear go through the same process — from eliciting market needs to design, from manufacture to testing, and finally to market even though how each aspect is executed changes every day. Footwear-related information and research are spread across different disciplines. I have focused on bringing together high-quality contributions from designers, biomechanists, ergonomists, engineers, podiatrists, and scientists from industry and academia into one book. The differing perspectives give a holistic view of the science behind footwear. The book contains very useful data from past research and the state-of-the-art methodologies that are now used in the design and manufacture of footwear. The first four chapters are related to the foot. Feet among people are quite varied, and knowing the various characteristics helps with the design as well as the selection of footwear.

In the early days, lasts were made using simple measurements taken from people's feet. Even today, foot length is the common measurement to determine the size of a shoe that one can wear. With improvements in technology, we now have sophisticated devices that can give the complete three-dimensional shape of an object. These scanning technologies are ideal for determining the shapes and sizes of feet and for matching the fit for shoes and feet using differing algorithms and techniques. The three chapters on scanning and processing give a very broad view of the various processes and highlight the challenges of using the scanning technologies. Even though shoes were primarily "carved" by artisans, modern-day shoemaking uses various technologies so that multiples of the same product can be made reliably knowing the basic needs of the consumers. The use of computers, scanners, numerically controlled machines, and rapid prototyping machines is quite common in footwear manufacture. The chapters on design provide an exclusive coverage of these computer-aided technologies. A number of chapters outline the design aspects of the various components of a shoe from mass manufacturing to personalized shoes and components. Footwear testing may be performed in vivo or in vitro. Both types have been covered. Footwear is meant to enhance safety, improve performance, and reduce stress on the human body. However, not all footwear can make a positive impact on a person, and some could be detrimental, if not carefully selected. The various effects of footwear are highlighted, and many chapters discuss activity-specific footwear as well. In today's world, high variability, injury, or being "outsized" often makes one search for a more personalized and customized product. The last few chapters discuss the issues of customization with a view to minimizing waste, enhancing overall performance, and delighting customers.

The figures in the book are all in black and white. Color figures are on the CRC website related to the book, *Science of Footwear*. Each chapter in the book has excellent content, written by an expert. Anyone interested in footwear has all the information he or she needs, from eliciting customer needs to testing footwear. The book will undoubtedly be useful to academicians, practitioners, designers, researchers, and all others working in the area of footwear who strive to incorporate the biomechanical and ergonomics principles into footwear to enhance performance, safety, and comfort and to reduce injury.

REFERENCE

MarketLine. *Global Footwear*. London, U.K., February 2012.

Ravindra S. Goonetilleke
Hong Kong University of Science and Technology
Clear Water Bay, Hong Kong

MATLAB® is a registered trademark of The MathWorks, Inc. For product information, please contact:

The MathWorks, Inc.
3 Apple Hill Drive
Natick, MA, 01760-2098 USA
Tel: 508-647-7000
Fax: 508-647-7001
E-mail: info@mathworks.com
Web: www.mathworks.com

Acknowledgments

I would like to express my sincere appreciation to the many people who were directly and indirectly involved with this book. First and foremost, I am very grateful to Professor Gavriel Salvendy for inviting me to edit this book. His unending perseverance and support made this work possible. Special thanks to my wife, Sharmala, and children, Reshanga, Shehani, and Eshani, for their patience, support, and encouragement throughout the development process. The editorial work took quite some time, and their question, "Still not finished?" gave me the motivation to complete the work a little more quickly. It is unfortunate that neither of my parents were able to see this work. Their love, guidance, encouragement, and unending support were truly amazing and a source of strength. They were always there when I needed them—every day, all the time. I have learnt very much from my sister, Nirmala, and my brothers, Gamini and Anslem, and their advice is truly valued. I thank my academic advisor, Professor Colin Drury, for his mentoring; Professor Mitchel M. Tseng, our former department head, for recruiting me to the Hong Kong University of Science and Technology and for supporting my research; our present head, Professor Fugee Tsung, for approving my sabbatical leave during which most of the editorial work was completed; and Professors Uwe Reischl and Sarah Toevs, for hosting my sabbatical leave. Much of my research would not have been possible without the assistance of my colleagues in the Department of Industrial Engineering and Logistics Management, and past and present students. Most importantly, this book would not have been a reality without the high-quality work of colleagues and friends. It has been a wonderful experience, and I value this opportunity to compile the outstanding work of all the authors. I apologize and thank all those whose names I may have inadvertently missed.

Editor

Ravindra S. Goonetilleke is a professor in the Department of Industrial Engineering and Logistics Management at the Hong Kong University of Science and Technology. He has been actively involved in footwear research in the United States, Hong Kong, and China for over 20 years. He was a project manager and consultant at the Biomechanics Corporation of America and the human factors manager at the Nike Sport Research Laboratory. He set up a fully operational state-of-the-art footwear research laboratory in Wenzhou for the Zhejiang Advanced Manufacturing Institute. He has numerous patents, has worked with local and international footwear companies, and is an active researcher in the field with many publications. His primary research interests are mathematical modeling in human factors and ergonomics and the development of culture- and user-friendly products and systems.

Contributors

Sandra Alemany
R&D Area
Instituto de Biomecánica de Valencia
Valencia, Spain

Katharina Althoff
Biomechanics Laboratory
Institute for Sports and Movement
  Sciences
University Duisburg-Essen
Essen, Germany

Emily Y.L. Au
Human Performance Laboratory
Department of Industrial Engineering
  and Logistics Management
Hong Kong University of Science
  and Technology
Clear Water Bay, Hong Kong

Philip Azariadis
Department of Product and Systems
  Design Engineering
University of the Aegean
Ermoupolis, Syros, Greece

Lorilynn Bloomer
Nike Sports Research Lab
Nike, Inc.
Beaverton, Oregon

Claudio R. Boër
Institute of Computer Integrated
  Manufacturing for Sustainable
  Innovation
University of Applied Science
  of Southern Switzerland
Manno, Switzerland

Georgeanne Botek
Orthopaedic and Rheumatologic
  Institute
Cleveland Clinic
Cleveland, Ohio

Helen Branthwaite
Faculty of Health
Centre for Sport, Health and Exercise
  Research
Staffordshire University
Stoke on Trent, United Kingdom

Kai-Ming Chan
Department of Orthopaedics
  and Traumatology
Prince of Wales Hospital
and
Faculty of Medicine
The Chinese University of Hong Kong
Sha Tin, New Territories, Hong Kong

Jeffrey Chang
Vorum Research Corporation
Vancouver, British Columbia,
  Canada

Jason Tak-Man Cheung
Li Ning Sports Science
  Research Center
Li Ning Company Limited
Beijing, China

Nachiappan Chockalingam
Faculty of Health
Centre for Sport, Health and Exercise
  Research
Staffordshire University
Stoke-on-Trent, United Kingdom

Sharna M. Clark-Donovan
Nike Sports Research Lab
Nike, Inc.
Beaverton, Oregon

Brian D. Corner
United States Army Natick Soldier
Research, Development,
    and Engineering Center
Natick, Massachusetts

Yihua Ding
Computer School
Wuhan University
Wuhan, People's Republic of China

Daniel Tik-Pui Fong
Department of Orthopaedics
    and Traumatology
Prince of Wales Hospital
and
Faculty of Medicine
The Chinese University of Hong Kong
Sha Tin, New Territories, Hong Kong

Juan Carlos González
R&D Area
Instituto de Biomecánica de Valencia
Valencia, Spain

Ravindra S. Goonetilleke
Human Performance Laboratory
Department of Industrial Engineering
    and Logistics Management
Hong Kong University of Science
    and Technology
Clear Water Bay, Hong Kong

William Gordon
Faculty of Engineering and Science
Design and Manufacturing Centre
Queensland University of Technology
Brisbane, Queensland, Australia

Allison H. Gruber
Biomechanics Laboratory
University of Massachusetts
Amherst, Massachusetts

Joseph Hamill
Biomechanics Laboratory
University of Massachusetts
Amherst, Massachusetts

Anna Lucy Hatton
Centre for Functioning and Health
    Research
Queensland Health
and
School of Health and Rehabilitation
    Sciences
The University of Queensland
Brisbane, Queensland, Australia

Aoife Healy
Faculty of Health
Centre for Sport, Health and Exercise
    Research
Staffordshire University
Stoke-on-Trent, United Kingdom

Ewald M. Hennig
Biomechanics Laboratory
Institute for Sports and Movement
    Sciences
University Duisburg-Essen
Essen, Germany

Youlian Hong
Department of Sports Medicine
Chengdu Sports University
Chengdu, People's Republic of China

Ajay Joneja
Department of Industrial
    Engineering and Logistics
    Management
Hong Kong University of Science
    and Technology
Clear Water Bay, Hong Kong

Noël Keijsers
Department of Research, Development
   and Education
Sint Maartenskliniek
Nijmegen, the Netherlands

Claudia Kieserling
selve—The Shoe Individualizer
Munich, Germany

Fan Sai Kit
Department of Industrial Engineering
   and Logistics Management
Hong Kong University of Science
   and Technology
Clear Water Bay, Hong Kong

Makiko Kouchi
Digital Human Research Center
National Institute of Advanced
   Industrial Science and Technology
Tokyo, Japan

Inga Krauss
Department of Sports Medicine
Medical Clinic
University of Tuebingen
Tuebingen, Germany

Wing Kai Lam
Li Ning Sports Science
   Research Center
Li Ning Company Limited
Beijing, China

Karl Landorf
Department of Podiatry
La Trobe University
Melbourne, Victoria, Australia

Jing Xian Li
School of Human Kinetics
University of Ottawa
Ottawa, Ontario, Canada

Evalotte Lindgens
School of Business and Economics
Technology and Innovation
   Management Group
RWTH Aachen University
Aachen, Germany

Stephen R. Lord
Falls and Balance Research Group
Neuroscience Research Australia
and
School of Public Health and
   Community Medicine
University of New South Wales
Sydney, New South Wales, Australia

Ameersing Luximon
Institute of Textiles and Clothing
The Hong Kong Polytechnic University
Hung Hom, Hong Kong

Yan Luximon
School of Design
The Hong Kong
   Polytechnic University
Hung Hom, Hong Kong

Marlene Mauch
Praxisklinik Rennbahn AG
Muttenz, Switzerland

Allison R. Medellin
Nike Sports Research Lab
Nike, Inc.
Beaverton, Oregon

Jasmine C. Menant
Falls and Balance Research Group
Neuroscience Research Australia
and
School of Public Health and
   Community Medicine
University of New South Wales
Sydney, New South Wales, Australia

Ross H. Miller
Biomechanics Laboratory
University of Massachusetts
Amherst, Massachusetts

Kam-Ming Mok
Department of Orthopaedics
    and Traumatology
Prince of Wales Hospital
and
Faculty of Medicine
The Chinese University of Hong Kong
Sha Tin, New Territories, Hong Kong

George S. Murley
Department of Podiatry
La Trobe University
Melbourne, Victoria, Australia

José Olaso
R&D Area
Instituto de Biomecánica de Valencia
Valencia, Spain

Tammy M. Owings
Department of Biomedical Engineering
Cleveland Clinic
Cleveland, Ohio

Paolo Pedrazzoli
Institute of Computer Integrated
    Manufacturing for Sustainable
    Innovation
University of Applied Science
    of Southern Switzerland
Manno, Switzerland

Frank Piller
School of Business and Economics
Technology and Innovation
    Management Group
RWTH Aachen University
Aachen, Germany

Sergio Puigcerver
R&D Area
Instituto de Biomecánica de Valencia
Valencia, Spain

Sudhakar Rajulu
Anthropometry and Biomechanics
    Facility
Johnson Space Center
National Aeronautics and Space
    Administration
Houston, Texas

Anthony C. Redmond
Division of Musculoskeletal Disease
Leeds Institute of Molecular Medicine
University of Leeds
and
Biomedical Research Unit
Chapel Allerton Hospital
National Institute for Health Research
Leeds, United Kingdom

Lloyd Reed
School of Clinical Sciences
Queensland University of Technology
Brisbane, Queensland, Australia

Asanka S. Rodrigo
Department of Electrical Engineering
University of Moratuwa
Katubedda, Sri Lanka

Keith Rome
Health and Rehabilitation Research
    Institute
and
School of Podiatry
Auckland University of Technology
Auckland, New Zealand

Carl G. Saunders
Vorum Research Corporation
Vancouver, British Columbia, Canada

Johan Steenwyk
Steenwyk Custom Shoes & Orthotics
Red Deer, Alberta, Canada

Frank Steiner
School of Business and Economics
Technology and Innovation
  Management Group
RWTH Aachen University
Aachen, Germany

Thorsten Sterzing
Li Ning Sports Science
  Research Center
Li Ning Company Limited
Beijing, China

Mitchell M. Tseng
Advanced Manufacturing Institute
Hong Kong University of Science and
  Technology
Clear Water Bay, Hong Kong

Stephen Urry
School of Clinical Sciences
Queensland University of Technology
Brisbane, Queensland, Australia

Gordon A. Valiant
Nike Sports Research Lab
Nike, Inc.
Beaverton, Oregon

Chenjie Wang
Advanced Manufacturing Institute
Hong Kong University of Science
  and Technology
Clear Water Bay, Hong Kong

Thilina W. Weerasinghe
Human Performance Laboratory
Department of Industrial Engineering
  and Logistics Management
Hong Kong University of Science
  and Technology
Clear Water Bay, Hong Kong

Ma Xiao
Institute of Textiles and Clothing
The Hong Kong Polytechnic
  University
Hung Hom, Hong Kong

Shuping Xiong
School of Design and Human
  Engineering
Ulsan National Institute of Science
  and Technology
Ulsan, South Korea

Jianhui Zhao
Computer School
Wuhan University
Wuhan, People's Republic of China

Part I

The Human Foot

1 Foot Structure and Anatomy

Ma Xiao, Yan Luximon, and Ameersing Luximon

CONTENTS

The human feet are very complex. Each foot consists of 26 bones; 33 joints; and muscles, tendons, and ligaments; and a network of blood vessels, nerves, skin, and other surrounding soft tissues. These components work together to create a complex flexible structure to provide the body with support, balance, and mobility. The human feet combine mechanical complexity and structural strength. The feet can sustain large pressure and provide flexibility and resiliency.

In order to have a good understanding of the foot, directional terms are very useful. The inner side or big toe side of the foot is called the medial side, while the outer side or little toe side is called the lateral side of the foot. Distal is away from the trunk or origin—in this case away from the center of foot. Proximal is the nearest to the trunk or origin—in this case it means near the center of foot. Anterior is the front side while posterior is the back or rear side.

1.1 SKELETAL SYSTEM

There are 26 bones in each human foot (Figure 1.1). Considering that the number of bones in the body is 206, both feet make up one quarter of the bones in the entire human body. The bones can be divided into three groups according to their locations and functions (Drake et al., 2010). Seven tarsals or tarsus form the ankle as the connection of the foot and leg. Five metatarsals or metatarsus form the medial side to the lateral side. They are numbered I–V. Fourteen phalanges are the bones of toes. There are three for each toe, except the big toe, which has two (Logan et al., 2004).

1.1.1 Tarsus

The tarsus consists of calcaneus (1), talus (2), navicular (3), cuboid (7), and three cuneiforms (4–6) (Figure 1.1). The calcaneus, also called heel bone, is the largest bone of the foot. The talus, which sits on the calcaneus, forms the ankle joint together with the calcaneus by articulating above with the fibula and tibia (the two leg bones). It also articulates forward with the navicular in the medial side. The pressure and impact generated by body weight or walking are transmitted to the ground mostly through these two bones.

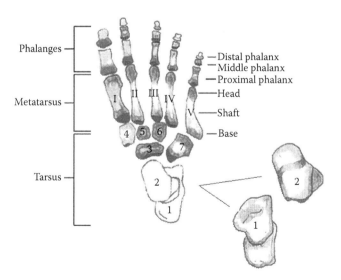

FIGURE 1.1 Top view of the skeletal foot. Note: 1, calcaneus; 2, talus; 3, navicular; 4, the medial cuneiform; 5, the intermediate cuneiform; 6, the lateral cuneiform; 7, cuboid.

The navicular is boat shaped when seen on the medial side of the foot. It is the connection between the talus and cuneiforms. On its medial surface, there is a prominent tuberosity for the attachment of the tibialis posterior tendon, which may abrade with the inside surface of the shoe and cause foot pain if it is too large.

The cuboid is located between the metatarsal IV and V and the calcaneus, which also articulates with the navicular and the lateral cuneiform on its medial side. There are three cuneiforms, named the medial cuneiform (4), the intermediate cuneiform (5), and the lateral cuneiform (6), respectively (Figure 1.1). They are also called the first cuneiform, the second cuneiform, and the third cuneiform sometimes. The three cuneiforms articulate with the navicular behind and with the bases of the medial three metatarsals in front.

1.1.2 METATARSUS

The five metatarsals, which are numbered I–V from the medial side to the lateral side, are the connection of the tarsus and phalanges (Figure 1.1). Each metatarsal has a proximal base connected to the tarsus, a slim shaft and a distal head near the digits. The sides of the bases of metatarsals II–V also articulate with each other. The lateral side of the base of metatarsal V has a prominent tuberosity, which projects posteriorly and is the attachment site for the tendon of the fibularis brevis muscle.

1.1.3 PHALANGES

The 14 phalanges form the toes. The great toe (first toe) has just two phalanges, while the other four have three each: the proximal phalanx, the middle phalanx, and the distal phalanx. Each phalanx has a base, a shaft, and a head (Figure 1.1). As discussed the distal side is away from the center of foot; hence, the distal phalanx is at the toe tip, while the proximal phalanx is the one near the metatarsals.

1.2 ARCHES OF FOOT

The foot has two types of arches: longitudinal arches and transverse arches. The longitudinal arches consist of medial and lateral parts, which distribute the body weight and pressure in different directions with the transverse arches together (Jenkins, 2009). The longitudinal arches are essential for support and reducing the cost of walking. There are several transverse arches, which are all essential for the foot functions.

The foot bones are not arranged in a horizontal plane, but in longitudinal and transverse arches supported and controlled by tendons, which absorb and transmit forces and pressure from the body to the ground when standing or moving. When the longitudinal arches are higher than normal, the foot is classified as high arch foot. When the longitudinal arches are low, it is called flat foot. Both flat foot and high arch foot do not transmit forces efficiently and hence can lead to foot pains. In addition, these also affect the pressure distribution causing irregular pressure in other parts of the body with long-term problems such as back pains.

1.2.1 LONGITUDINAL ARCH

The foot has two longitudinal arches: medial and lateral. The medial longitudinal arch is composed of the calcaneus, the talus, the navicular, the three cuneiforms, and the metatarsal I–III, which is higher than the lateral longitudinal arch and called foot arch normally (Figure 1.2). The bones of lateral longitudinal arch include the calcaneus, the cuboid, and the metatarsal IV and V (Figure 1.3). Under the medial longitudinal arch, soft tissues (such as the plantar calcaneonavicular ligament) with elastic properties act as springs.

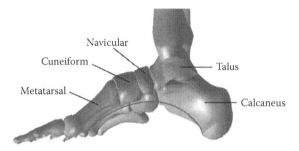

FIGURE 1.2   The medial longitudinal arch.

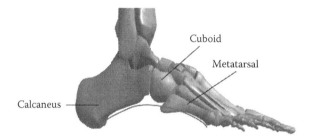

FIGURE 1.3   The lateral longitudinal arch.

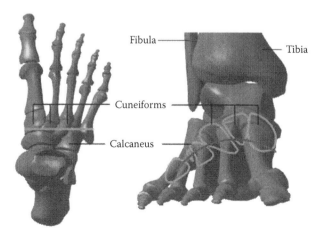

FIGURE 1.4   Transverse arch.

1.2.2 Transverse Arch

The transverse arch is the arch across the foot from the medial to the lateral side. The shape of the arch can be different at different locations. For illustration, we have shown a cross section of the foot at the cuneiform region (Figure 1.4). The transverse arches together with the longitudinal arches enable the foot in its function of support and locomotion.

1.3 FOOT JOINTS

Between the foot bones, there are joints that enable foot dynamics. There are 33 joints in each foot, which are constructed by two or more bones to allow movement, providing mechanical support and absorbing the shock.

1.3.1 Ankle Joint

The ankle joint consists of the talus of the foot and the tibia and fibula of the leg (Figure 1.4). It is a synovial hinge joint, which mainly enables dorsiflexion and plantar flexion of the foot. There are the medial (deltoid) and lateral ligaments that stabilize the ankle joints. The lateral ligament is weaker than the medial ligament, which consists of three slender bands: the anterior talofibular ligament (front ligament between the talus and fibular), the calcaneofibular ligament (between the calcaneus and fibular), and the posterior talofibular ligament (back side ligament between talus and fibular). The dorsiflexion (moving the toe upward away from ground) involves in the medial, calcaneofibular, and posterior talofibular ligaments, while the plantar flexion (moving the toe downward) involves the anterior talofibular ligament and the anterior part of the medial ligament. The ankle is used frequently when walking or running, especially in sports. Hence, it is the most frequently injured joint of the foot. The common ankle injuries include sprains (torn ligaments), fracture-dislocation when the foot is extremely everted (look for Eversion, Figure 1.5), and fractures of relative bones.

1.3.2 Intertarsal Joints

The intertarsal joints are the joints between the tarsals. These are mainly the subtalar (talocalcaneal) joint and the transverse tarsal joint (including the talocalcaneonavicular and the calcaneocuboid joints) (Figure 1.6). The main movements involving these joints are inversion, eversion, pronation, and supination (Figure 1.5). Table 1.1 illustrates the main information of the intertarsal joints.

1.3.3 Other Joints of Foot

Table 1.2 shows the articulation, ligaments, and corresponding movements of other foot joints (Figure 1.6). The cuneonavicular joints and the tarsometatarsal joints have limited movements. Apart from the ankle joint, the metatarsophalangeal joints are important as most motions take place during walking and running here.

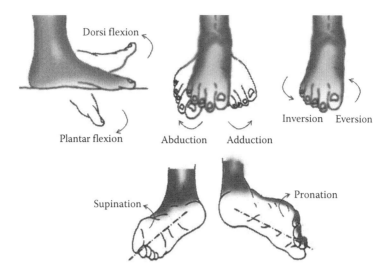

FIGURE 1.5 Foot motions.

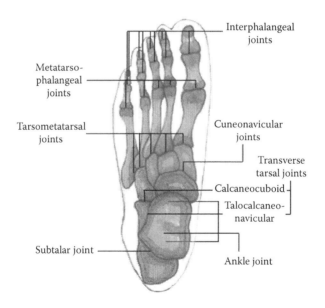

FIGURE 1.6 The joints of the foot.

1.4 MUSCLES AND TENDONS

1.4.1 MUSCLES OF THE DORSUM ASPECT

Most of the foot muscles exist on the plantar aspect, except for the extensor digitorum brevis and the extensor halluces brevis on the dorsum of the foot. The extensor halluces brevis is considered as the extension of the extensor digitorum brevis. Extensor digitorum brevis extends the metatarsophalangeal joint of the great toe, and the three

TABLE 1.1
Intertarsal Joints

Joint	Articulation	Ligaments	Movements
Talocalcaneal	Inferior surface of the talus (large posterior calcaneal facet) articulates with superior surface of the calcaneus (posterior talar facet)	Lateral, medial, posterior, and interosseous talocalcaneal ligaments	Gliding and rotation, also involved in inversion and eversion
Talocalcaneonavicular	Head of the talus articulates with the calcaneus and the navicular in front	Interosseous talocalcaneal ligament, talonavicular ligament, and plantar calcaneonavicular ligament	Gliding and rotation, also involved in inversion, eversion, pronation, and supination
Calcaneocuboid	Anterior end of the calcaneus articulates with posterior surface of the cuboid	Bifurcate ligament, long plantar ligament, and the plantar calcaneocuboid ligament (short plantar ligament)	Gliding and rotation, also involved in inversion, eversion, pronation, and supination

TABLE 1.2
Other Joints of Foot

Joint	Articulation	Ligaments	Movements
Cuneonavicular	Anterior navicular articulates with bases of cuneiforms	Dorsal and plantar cuneonavicular ligaments	Limited movement
Tarsometatarsal	Metatarsals articulate with adjacent tarsal bones	Dorsal, plantar, and interosseous tarsometatarsal ligaments	Limited sliding movements
Metatarsophalangeal	Heads of metatarsals articulate heads of corresponding proximal phalanxes	Bifurcate ligament, long plantar ligament, and the plantar calcaneocuboid ligament (short plantar ligament)	Gliding and rotation, also involved in inversion, eversion, pronation, and supination
Interphalangeal	Phalanges articulate with each other	Medial and lateral collateral ligaments and by plantar ligaments	Flexion and extension

middle toes through attachments to the long extensor tendons and extensor hoods. The muscles of the dorsal foot help the long extensors to extend the toes.

1.4.2 Muscles of the Plantar Aspect

Muscles of the plantar foot can be divided into four layers from superficial to deep (Moore and Agur, 2007). The plantar muscles are able to support the stance, maintain the foot arches and also refine further the efforts of the long muscles, producing supination and pronation in enabling the foot to adjust to rough ground.

There are three muscles in the first layer, which is the most superficial layer. From medial to lateral, they are abductor hallucis, flexor digitorum brevis, and abductor

TABLE 1.3
First Muscle Layer of the Plantar Aspect

Muscle	Function	Innervation	Corresponding Tendon
Abductor hallucis	Abducts and flexes great toe at metatarsophalangeal joint	Innervated by the medial plantar branch of the tibial nerve	The tendon inserting on the medial side of the base of the proximal phalanx of the great toe
Flexor digitorum brevis	Flexes lateral four toes at proximal interphalangeal joint	Innervated by the medial plantar branch of the tibial nerve	Flexor digitorum brevis tendons on the lateral four toes
Abductor digiti minimi	Abducts little toe at the metatarsophalangeal joint	Innervated by the lateral plantar branch of tibial nerve	The tendon inserting on the lateral side of base of proximal phalanx of little toe

TABLE 1.4
Second Muscle Layer of the Plantar Aspect

Muscle	Function	Innervation	Corresponding Tendon
Quadratus plantae	Assists flexor digitorum longus tendon in flexing toes II–V	Innervated by the lateral plantar branch of tibial nerve	Flexor digitorum longus tendons
Lumbricals	Flexes metatarsophalangeal joint and extend interphalangeal joints	First lumbrical is innervated by the medial plantar nerve from the tibial nerve; second, third, and fourth lumbricals are innervated by the lateral plantar nerve from the tibial nerve	Flexor digitorum longus tendons

TABLE 1.5
Third Muscle Layer of the Plantar Aspect

Muscle	Function	Innervation	Corresponding Tendon
Flexor hallucis brevis	Flexes metatarsophalangeal joint of the great toe	Innervated by the lateral plantar branch of tibial nerve	The tendon originating from plantar surface of cuboid and lateral cuneiform; tendon of tibialis posterior muscle
Adductor hallucis	Adducts great toe at metatarsophalangeal joint	Innervated by the lateral plantar branch of tibial nerve	Fibularis longus tendon
Flexor digiti minimi brevis	Flexes little toe at metatarsophalangeal joint	Innervated by the lateral plantar branch of tibial nerve	Fibularis longus tendon

TABLE 1.6
Fourth Muscle Layer of the Plantar Aspect

Muscle	Function	Innervation
Dorsal interossei	Abducts toes II–IV at metatarsophalangeal joints; resists extension of metatarsophalangeal joints and flexion of interphalangeal joints	Innervated by the lateral plantar branch of tibial nerve; first and second dorsal interossei also innervated by deep fibular nerve
Plantar interossei	Adducts great toe at metatarsophalangeal joint	Innervated by the lateral plantar branch of tibial nerve

digiti minimi (Table 1.3). The second muscle layer consists of quadratus plantae muscle and lumbrical muscles (Table 1.4). There are three muscles in the third muscle layer. The flexor hallucis brevis and adductor hallucis are associated with the great toe, while flexor digiti minimi brevis is associated with the little toe (Table 1.5). The fourth muscle layer, the deepest layer, has two groups of muscles: the dorsal and plantar interossei. These bipennate muscles are located on the sides of metatarsals (Table 1.6).

1.5 VASCULAR SYSTEM

1.5.1 ARTERIES

The blood supply of the plantar foot is mainly provided by the Posterior Tibial Artery (PTA), which bifurcates into two primary channels: the medial plantar artery and the lateral plantar artery. The PTA enters the sole of foot on the medial side of the ankle and the posterior to the medial malleolus. The medial branch goes deep

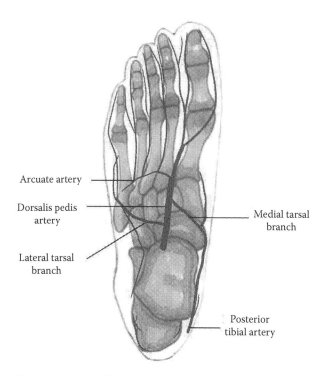

Arcuate artery

Dorsalis pedis
artery

Lateral tarsal
branch

Medial tarsal
branch

Posterior
tibial artery

FIGURE 1.7   The dorsal artery distribution.

through the medial sole to supply adjacent muscles and communicate with digital
branches of the lateral plantar artery on the medial side of the great toe. The lateral
plantar artery passes the sole diagonally to the base of the metatarsal V, and then
turns medially to form the deep plantar arch, which communicate with the branch
of the dorsalis pedis artery at the bases between the metatarsals I and II finally. The
deep plantar arch has several branches as well, which supply the digital and adjacent
muscles and also form circulation by joining with the vessels on the dorsal aspect of
the foot (Gray and Goss, 1973).

The dorsalis pedis artery is the continuation of the anterior tibial artery, which
enters the dorsal aspect of the foot from the anterior malleolus (Figure 1.7). It passes
anteriorly over the dorsal aspect of the talus, the navicular, and the intermediate
cuneiforms, and then passes inferiorly to form the deep plantar artery. Another main
branch goes laterally to form the arcuate artery at the bases of metatarsals I and II,
which branches to supply the metatarsal and digital arteries (Gray and Goss, 1973).

1.5.2   Veins

The veins in the foot vary in its position. They can be divided into two groups: the
superficial system and the deep systems. The superficial venous group includes the
saphenous veins. The great saphenous vein is the largest vein on the dorsum of the foot
and it lies anterior to the medial malleolus at the ankle. The small saphenous vein lies
posterior and lateral to the lateral malleolus. It is smaller than the great saphenous vein

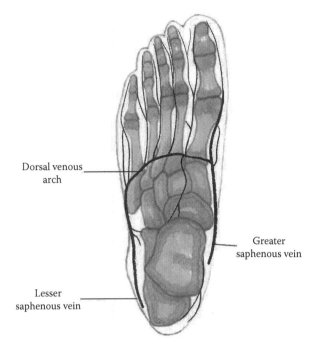

Dorsal venous arch

Greater saphenous vein

Lesser saphenous vein

FIGURE 1.8   The dorsal vein distribution.

and drains blood from the lateral side of the foot, while the great saphenous vein drains from the medial side. The dorsal venous arch connects the small saphenous vein with the great saphenous vein (Figure 1.8; Snell, 2004).

1.6   NERVES

1.6.1   PLANTAR NERVES

The distribution of the nerves (Figure 1.9) on the plantar aspect of the foot is similar with the plantar arteries mostly. The tibial nerve enters the sole on the medial side of the ankle and the posterior to the medial malleolus as well (Drake et al., 2010). It bifurcates into two major branches—a smaller lateral plantar nerve and a larger medial plantar nerve that is the major nerve of the sole. The medial plantar nerve is mainly a sensory nerve in the sole of the foot. It innervates skin on the most of the sole of the forefoot and skin of the first to third toes and the medial side of the fourth toe. Moreover, it innervates four muscles, which are abductor hallucis, flexor digitorum brevis, flexor hallucis brevis, and the first lumbrical. The lateral plantar nerve is a motor nerve of the plantar aspect of the foot. It innervates all major muscles in the sole except the four innervated by the medial plantar nerve. It also innervates the skin of the lateral two toes and the lateral area of the sole where it passes through. Another small branch of the tibial nerve is the sural nerve. It is in the lateral superficial area of the foot and mainly innervates skin on the lateral side of the foot and dorsolateral surface of the little toe.

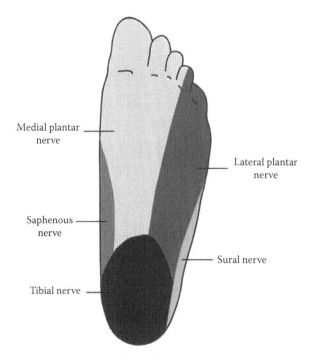

FIGURE 1.9   The plantar cutaneous distribution.

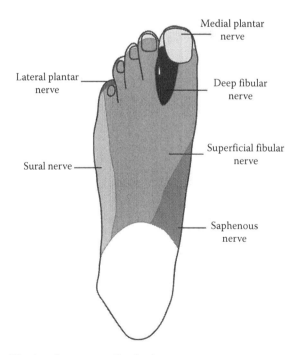

FIGURE 1.10   The dorsal cutaneous distribution.

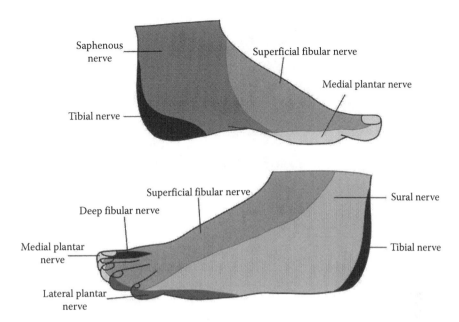

FIGURE 1.11   The cutaneous distribution on the medial side (top) and lateral side (bottom).

1.6.2 DORSAL NERVES

There are two major nerves on the dorsal aspect of the foot—the deep fibular nerve and the superficial fibular nerve. The deep fibular nerve goes along the same route with dorsalis pedis artery on the lateral side. It mainly supplies skin over adjacent surfaces of the medial two toes (Figure 1.10). The superficial fibular nerve enters the dorsal aspect of the foot from the anterolateral side of the lower leg and generates the cutaneous branches and digital branches of the dorsal surface. In addition, skin in superficial fascia on the medial side of the ankle is innervated by the saphenous nerve, which is a cutaneous branch of the femoral nerve that originates in the thigh. The side view of the nerve distribution is shown in Figure 1.11.

1.7 SURFACE ANATOMY

In many cases we start with surface anatomy, but we preferred to discuss the surface anatomy after we have understood the location of bones, tendons, ligaments, and the blood vessels. Depending on different foot, the foot surface provides guides for tendons, muscles, bony landmarks, even arteries, veins, and nerves. These can be observed or palpated by hand. We have listed 12 landmarks (Figure 1.12 and Table 1.7), but other landmarks can be found by flexion, extension, eversion, or inversion of the foot (Tixa, 1999).

1.8 MRI AND DIGITAL FOOT MODEL

Magnetic resonance imaging (MRI) machine uses powerful magnetic field to visualize detailed internal structures. Using MRI scanning technique, the internal structures of the foot can be obtained. The MRI creates a gray color coded

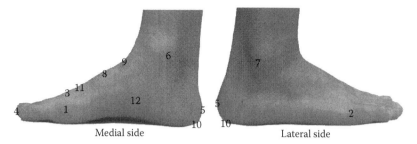

FIGURE 1.12 Landmarks on the surface of the foot.

TABLE 1.7
The Definitions of Landmarks

	Definition or How to Locate It
1. The first metatarsophalangeal joints (medial MPJ)	The salient point at the head of the first metatarsal on the medial side of the foot
2. The fifth metatarsophalangeal joints (lateral MPJ)	The salient point at the head of the fifth metatarsal on the medial side of the foot. It is the boundary point of the definition of foot width as well as the first MPJ
3. Top of the first MPJ	On the dorsal aspect of the head of the first metatarsal. The ball girth is defined by this point, the medial, and lateral MPJ
4. Toe tip	The extremity of the great toe
5. Pternion	The extremity of the heel
6. Medial malleolus	The salient point of tibia
7. Lateral malleolus	The salient point of fibula
8. Instep point	The middle cuneiform prominence
9. Junction point	The junction of the leg and foot on the dorsal aspect
10. Landing point	The extremity of the foot print
11. Waist point	At the approximate center of the metatarsal
12. Arch point	The tuberosity of navicular

image at a given section. In order to create the complete foot, continuous section at regular interval is obtained. For a given foot of length 250 mm, in order to generate the complete foot shape at 1 mm interval, 250 images are obtained. Example cross sections are shown in Figure 1.13. From each picture, the edge of the bones and other soft tissue can be located. Some software (e.g., MIMICS, or MATLAB® programs) can locate the edges automatically (Figure 1.14a), but in most cases manual verification and modification are required to obtain the desired structure such as bone outline (Figure 1.14b). This is a tedious process considering the different structures in each of the 250 cross-sectional MRI images. Once the edges have been obtained, 3D geometric shape can be created by a method called lofting. Figure 1.15 shows an example of a digital foot including bones, muscles, and other tissues. Nowadays, with the extensive use of technology, digital anatomy is becoming possible. The 3D digital foot model can be used for learning foot anatomy, design or shoe-last, and shoes.

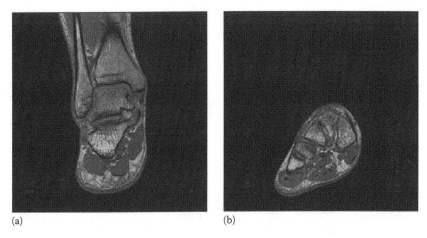

(a)   (b)

FIGURE 1.13   Example of MRI images: (a) section around fibula and (b) section around instep.

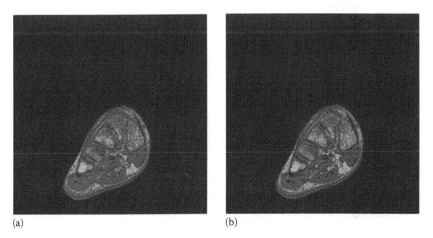

(a)   (b)

FIGURE 1.14   Edges extraction from MRI images: (a) automatic edge detection using MATLAB® program and (b) selection of only the edges of the bones.

1.9   CONCLUSION

The human foot has evolved for its function of locomotion and support. The foot has 26 bones, muscles, veins, arteries, nerves, connecting tissues, and other soft tissues. The human foot is very complex, and a very detailed explanation is out of scope in this book. This chapter has provided basic information that will enable knowledge of foot when designing shoe-last and shoes. The knowledge of the location of the joints, especially the metatarsophalangeal joint, is essential to design footwear that flexed at the right location. Knowledge of the ankle is also important since many foot injuries occurs at the ankle joint. With advancement of technology, detailed 3D digital foot anatomical model will be available to shoe designers. Then the footwear designers will be able to consider the foot internal structures and foot shape appropriately in their design.

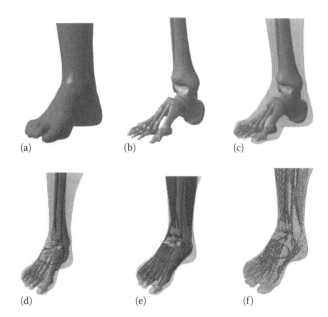

FIGURE 1.15 Digital model of foot: (a) surface model, (b) skeletal model, (c) visualization of bones and surface, (d) blood vessels and connective tissues, (e) muscles model, and (f) mesh model.

QUESTIONS

1.1 How many bones do we have in the foot? List them.
1.2 Describe the ankle joint making reference to the bones and the ligaments at the ankle joint?
1.3 Name the veins and the arteries of the foot.
1.4 Draw sketches to show the arches of the foot and describe the function of the foot arches.

REFERENCES

Drake, R. L., Gray, H., Mitchell, A. W. M., and Vogl, W. (2010). *Gray's Anatomy for Students*, Vol. 2. Philadelphia, PA: Churchill Livingstone.

Gray, H. and Goss, C. M. (1973). *Anatomy of the Human Body*. Philadelphia, PA: Lea & Febiger.

Jenkins, D. B. (2009). *Hollinshead's Functional Anatomy of the Limbs and Back*, Vol. 9. St. Louis, MO: Saunders.

Logan, B. M., McMinn, R. M. H., Singh, D., and Hutchings, R. T. (2004). *McMinn's Color Atlas of Foot and Ankle Anatomy*, Vol. 3. Edinburgh, New York: Mosby.

Moore, K. L. and Agur, A. M. R. (2007). *Essential Clinical Anatomy*. Philadelphia, PA: Lippincott Williams & Wilkins.

Snell, R. S. (2004). *Clinical Anatomy*, Vol. 7. Philadelphia, PA: Lippincott Williams & Wilkins.

Tixa, S. (1999). *Atlas of Palpatory Anatomy of the Lower Extremities: A Manual Inspection of the Surface*. New York: McGraw-Hill, Health Professions Division.

Van De Graaff, K. M. (2002). *Human Anatomy*, Vol. 6. New York: McGraw-Hill.

2 Foot Morphology

Inga Krauss and Marlene Mauch

CONTENTS

Knowledge of foot morphology is important in several domains. Psychological issues such as physical attractiveness are related to anthropometric dimensions and its proportions (Fessler et al. 2005a,b, Voracek et al. 2007). Anthropometrics are also important in forensic research if identification of a dead person is hindered. In that case, knowledge of foot morphology may help to identify sex and population group of the victim or suspect that is tried to be defined (Smith 1997, Ozden et al. 2005). Another very important field of application is footwear industry. Footwear should be based on foot shape (Janisse 1992, Hawes and Sovak 1994, Hawes et al. 1994, Kouchi and Mochimaru 2003); hence, knowledge of foot anthropometrics is an important prerequisite to optimize last design (Mauch et al. 2009, Krauss et al. 2010). Footwear should account for general aspects of sex-specific differences in foot shape and differences being related to different age classes, races, and body compositions. Beyond that, it is well known that foot shape is manifold and can differ from one individual to another (Jung et al. 2001, Mauch et al. 2009, Krauss et al. 2010). Various foot types should therefore be considered in the manufacturing of shoes (Krauss et al. 2008, 2010, Mauch et al. 2009).

The following sections aim for providing an insight into methods and measures to quantify foot shape. Distinctive foot measures can be analyzed in different ways to describe the nature of foot morphology. It is important to know specific qualities of each way to analyze the data in order to be able to interpret them properly. Subsequent to these more technically oriented paragraphs, the focus is on distinctive characteristics of foot morphology in different population groups such as age groups, different body compositions, male and female feet, and differences being related to ethnic or regional origin.

2.1   METHODS TO QUANTIFY FOOT MORPHOLOGY

Various methods are available to quantify foot morphology. This paragraph will not comment on methods based on clinical assessment, radiographs, and magnetic resonance imaging, as they are prone to subjectivity (Hawes et al. 1992) or have major drawbacks such as harmful x-rays, difficult access, or great costs.

Aside from basic tools such as caliper rulers, tape measures, and blueprints, high-tech measurement instruments such as plantar pressure distribution mats or three-dimensional (3D) foot scanners are available on the market. Selection of the suitable method depends on the field of application. In everyday routine of an orthopedic shoemaker, easy-to-use and cost-effective procedures such as caliper, ruler, and plaster cast may be sufficient, although new technologies will likely get more adapted as soon as systems are offered at affordable costs (Huppin 2009). Though large studies were conducted using basic measurement instruments (Parham et al. 1992, Mauch et al. 2008b), today data collection of large population groups mostly relies on more sophisticated systems such as 3D foot scanners (Krauss et al. 2008, Mauch et al. 2008b, 2009, Yu and Tu 2009).

2.1.1   CALIPER RULE AND TAPE MEASURES

Caliper rule and tape measures are classic approaches to quantify foot anthropometrics. Tapes are used to measure circumferences of the foot, whereas caliper rules measure height, width, and length dimensions. A common instrument to measure foot dimensions is the Brannock Foot-Measuring Device®. A caliper for foot length (heel to toe) and arch length (heel to ball) is combined with a caliper for foot breadth at the first metatarsophalangeal joint. Foot length and width dimensions can easily be read off the scaling of the instrument.

2.1.2 TWO- AND THREE-DIMENSIONAL SYSTEMS

Several technologies to capture the plantar surface of the foot or 3D information of the foot and leg are available on the market (Telfer and Woodburn 2010). Two-dimensional (2D) scanners are comparable to classic flatbed scanners. They only give information of the plantar curvature of the foot. Three-dimensional representations of the human foot can be produced with digitizers or scanners. For digitizers, discrete points of the foot surface are stored as digital codes on a computer one at a time. In contrast, optical- or video-equipped scanning systems capture 3D images of the foot that are converted to digital form at once (Telfer and Woodburn 2010). For some of these systems, anatomical landmarks are identified and marked prior to the scanning process. The corresponding software then automatically calculates specific foot measures based on this anatomical landmark (Nácher et al. 2006, Witana et al. 2006). Other foot scan systems automatically derive specific width, height, and circumference measures at different percentage lengths of the foot. Additional anatomical-based measures can be quantified by editing digital points of the 3D image of the foot after the scanning process (Mauch et al. 2009, Krauss et al. 2011).

2.1.3 FOUR-DIMENSIONAL SYSTEMS

Newsworthy innovative scanner concepts are developing, using four-dimensional (4D) information of the roll-over process with the time domain as the fourth dimension of the human foot shape in dynamic situations (Kimura et al. 2005, Kouchi et al. 2009, Schmeltzpfenning et al. 2009). These systems give further information on deformation of foot structures while walking and prolongation of foot length during roll-over process.

2.1.4 BLUEPRINTS

Blueprints (pedographs) are an easy and inexpensive option to capture static weight-bearing footprints. They display foot outlines and allow the assessment of foot structure (Riddiford-Harland et al. 2000). One foot is placed on the measurement device while standing with equally distributed weight on both feet. The foot is placed on a membrane that has an ink pad on the reverse side, printing the footprint onto a pedograph paper placed below the membrane. Aside from its use by orthopedic shoemakers, blueprints are applied to quantify foot outline and foot indices such as the Staheli Index (SI), Chippaux-Smirak Index (CSI), or the Foot Print Angle as described in Section 2.2.2.

2.1.5 PLANTAR PRESSURE MEASUREMENT SYSTEMS

Systems quantifying pressure distribution patterns under the feet are long-established instruments to quantify foot function in dynamic situations (Hennig and Rosenbaum 1991). They provide information about actual forces and pressures applied to each region of contact during standing and walking (Rosenbaum and Becker 1997, Mickle et al. 2006a). Pressure is the result of the vertical force per unit area of the sensors that are loaded. Pressure distribution patterns can be quantified with pressure mat systems or in-shoe-systems. Whereas in-shoe systems primarily reveal interaction between foot and footwear in walking and running, pressure mats are mostly used to capture barefoot plantar pressures in static and dynamic conditions.

2.2 MEASURES DESCRIBING FOOT MORPHOLOGY

In the description of foot morphology, a comprehensive consideration on all dimensions is necessary. Particularly, the anatomic characteristics including the bony, muscular, and ligament structures and their resulting anthropometric measures have an important influence in forming foot shape. These measures are commonly used in orthopedic questions, but are also relevant in last and shoe construction. A various number of different foot measures have been reported by several authors (Parham et al. 1992, Kouchi 1995, Witana et al. 2006, Krauss et al. 2008, Mauch et al. 2009). The following sections describe a common selection of foot measures describing foot morphology.

2.2.1 LENGTHS, WIDTHS, CIRCUMFERENCES, AND FOOT HEIGHT

As foot measures are based on anatomical structures of the foot, different landmarks are used to define these measures. Length and width measures are dependent on the longitudinal axis of the foot. There are two methods commonly used:

1. Medial axis: Medial tangent of the foot touching the heel outline as well as the first metatarsophalangeal protrusion (Krauss et al. 2008, 2010, 2011, Mauch et al. 2008a, 2009).
2. Brannock axis: Central tangent from the posterior point of the heel to a point 38.1 mm (1.5 in.) from the medial side of the first metatarsal bone (Brannock) or as a variation to the tip of the second toe (Kouchi 2003).

An overview of foot measures are shown in Table 2.1.

TABLE 2.1
Foot Measures Describing Foot Morphology

Measure	Description	
Length		
Foot length (FE – FT$_1$)	Distance between foot end (FE) (heel) and foot tip (FT$_1$) (anterior point of the most protruding toe) along the medial tangent of the foot (foot measuring line; *y*-axis)	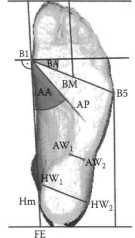
Ball-of-foot length (FE – B1)	Distance between FE (heel) and the first metatarsophalangeal protrusion (MTP) (B1)	
Outside ball-of-foot length (FE – B5)	Distance between FE (heel) and the fifth metatarsophalangeal protrusion (MTP) (B5) parallel to foot measuring line (FE – FT$_1$)	
Toe length (BM – FT$_2$)	Distance between the bisected ball line (B1 – B5) (BM) and foot tip (FT$_2$) (anterior point of the most protruding toe)	
Heel to medial/lateral malleolus	Length from FE to the most medially/laterally protruding point on the medial/lateral malleolus measured along the foot axis (Kouchi 2003)	See Kouchi (2003)
Widths		
Ball-of-foot width (B1 – B5)	Connection line between first MTP joint (B1) and fifth MTP joint protrusion (B5) (ball line)	
Ball-of-foot width (orthogonal)	Orthogonal connection line starting at the first MTP joint (B1) to the outside curvature of the foot	
Heel width (HW$_1$ – HW$_2$)	Widest part of the heel (plantar print) (HW$_1$ – HW$_2$) parallel to the ball line (B1 – B5)	
Heel width (orthogonal)	Orthogonal connection line starting on the medial side of the heel (Hm) to the outside curvature of the heel	
Plantar arch width (AW$_1$ – AW$_2$)	Narrowest section of the plantar medial longitudinal arch (AW$_1$ – AW$_2$)	

(continued)

TABLE 2.1 (continued)
Foot Measures Describing Foot Morphology

Measure	Description	
Bimalleolar width	Distance between the most medially protruding point on the medial malleolus and the most laterally protruding point on the lateral malleolus measured along a line perpendicular to the brannock axis (Kouchi 2003)	See Kouchi (2003)
Circumferences		
Ball girth (B1 – B5)	Maximum circumference over the first (B1) and fifth MTP joint protrusion (B5)	
Minimum arch girth	The vertical plane of minimum instep circumference, orthogonal to the foot measuring line, approx. 50%–60% of foot length	See Hawes and Sovak (1994)
Heel girth	Circumference passing through the point of distal heel curvature to the dorsal junction of the foot and leg (Kouchi 2003)	See Kouchi (2003)
Heights		
Medial/lateral malleolus height (H2)	Height of the most medially/ laterally protruding point of the medial/lateral malleolus (H2) (Kouchi 2003)	See Kouchi (2003)
Dorsal arch height (DA$_1$ – DA$_2$)	Height of medial dorsal junction of the foot and leg to the floor	
Angles		
Ball angle (BA)	Angle between the horizontal (90° to foot measuring line) and orthogonal ball line (B1 – B5) in B1	See the preceding figure

---

TABLE 2.1 (continued)
Foot Measures Describing Foot Morphology

Measure	Description	
Hallux angle (HA)	Angle between the orthogonal ball line (B1 – B5) and the medial definition of the hallux in B1	
Digitus minimus angle (DMA)	Angle between the orthogonal ball line (B1 – B5) and the lateral definition of the digitus minimus in B5	

Source: With kind permission from Macmillan Publishers Ltd., *Int. J. Obes.* (*Lond.*), Mauch, M. et al., Foot morphology of normal, underweight and overweight children, 32(7), 1068–1075. Copyright 2008a.

Tables 2.2 and 2.3 summarizes results of different studies investigating foot anthropometrics. Differences in absolute and relative foot measures are presented regarding age, gender, and ethnic origin. Comparison of sexual dimorphism within the same foot length was not included for lack of space (Table 2.3).

2.2.2 Measures and Indices for Foot Posture and Arch Structure

Foot posture and arch structure is considered to be an important determinant in the evaluation of the function of the foot and lower limb. Measures are often used to classify feet into different foot types. Applied methods are visual nonquantitative inspections, anthropometric values, footprint parameters, plantar pressure values, or radiographic evaluation (Gould et al. 1989, Hennig and Rosenbaum 1991, Hawes et al. 1992, Staheli 1994, Razeghi and Batt 2002, Echarri and Forriol 2003).

The most common measures based on footprints and anthropometric measures are described in Table 2.4.

Next to the previously described measures, plantar pressure measurement devices are used to assess loadings of the feet not only in standing but also in walking (e.g., in patients with obesity or diabetes) by quantifying the pressure distributed on the plantar surface of the feet. The pressure data is observed for the total foot on the one hand, but also for divisions of the foot, which are commonly referred to masks. There are different approaches regarding the determination and the used number of masks. It ranges from a division of two masks, that is, the forefoot and rearfoot mask, divided by 50% foot length (Dowling et al. 2001), to a set of 10 masks based on anatomical landmarks, for example, medial and lateral heel, medial and lateral midfoot, metatarsal head (MTH) 1, MTH 2, MTH 3-5, hallux, second toe, and toes 3–5 (Cavanagh et al. 1987).

TABLE 2.2
Data on Absolute Foot Measures in Different Populations (mm/deg), (Mean ± SD)

Source	Mauch (2007)[a]	Kouchi (2003)[b]		Wunderlich and Cavanagh (2001)[c]	
Subjects Measure	$n = 2963$	♂ ($n = 217$)	♀ ($n = 206$)	♂ ($n = 293$)	♀ ($n = 491$)
Foot length (FE − FT$_1$)	145.8 ± 10.3 − 249.7 ± 16.1	254.0 ± 10.9	233.1 ± 9.8	269.8	243.8
Ball-of-foot length (FE − B1)	106.9 ± 8.3 − 184.7 ± 12.2	185.6 ± 8.4	172.0 ± 7.6	196.7	177.6
Outside ball-of-foot length (FE − B5)	85.9 ± 7.1 − 151.5 ± 11.2	163.1 ± 7.7	148.9 ± 7.1	166.6	149.2
Toe length (BM − FT$_2$)	49.8 ± 3.7 − 82.3 ± 5.5	/	/	/	/
Heel to medial/ lateral malleolus	/	66.9 ± 6.3/ 53.0 ± 5.8	55.6 ±4.7/ 49.3 ± 5.2	/	/
Ball-of-foot width (B1 − B5)	60.0 ± 4.8 − 95.8 ± 6.5	100.7 ± 5.4	95.3 ± 4.5	105.4	95.0
Heel width (HW$_1$ − HW$_2$)	35.0 ± 2.7 − 52.8 ± 3.7	/	/	/	/
Heel width (orthogonal)	/	63.8 ± 3.7	60.8 ± 3.2	70.2	63.3
Bimalleolar width	/	73.4 ± 3.5	66.8 ± 3.3	73.1	65.2
Ball girth (B1 − B5)	153.0 ± 9.5 − 234.0 ± 13.6	249.5 ± 12.1	231.9 ± 10.2	251.8	226.7
Minimum arch girth	/	247.9 ± 11.7	227.2 ± 10.3	/	/
Heel girth	/	/	/	342.8	308.8
Medial/lateral malleolus height (H2)	/	82.6 ±5.3/ 72.0 ± 5.5	75.5 ± 4.8/ 63.4 ± 4.5	81.2/72.6	71.7/65.9
Dorsal arch height (DA$_1$ − DA$_2$)	43.6 ± 3.1 − 67.7 ± 4.1	61.9 ± 5.4	53.2 ± 3.2	88.8	80.5
Ball angle (BA)	19.5 ± 2.7	77.3 ± 2.6[d]	76.1 ± 2.5[d]	/	/
Hallux angle (HA)	103.9 ± 4.6 − 103.9 ± 4.6	8.6 ± 4.6[d]	11.9 ± 4.7[d]	/	/

TABLE 2.2 (continued)
Data on Absolute Foot Measures in Different Populations (mm/deg),
(Mean ± SD)

Source	Mauch (2007)[a]	Kouchi (2003)[b]		Wunderlich and Cavanagh (2001)[c]	
Subjects Measure	n = 2963	♂ (n = 217)	♀ (n = 206)	♂ (n = 293)	♀ (n = 491)
Digitus minimus angle (DMA)	71.5 ± 3.7 − 79.0 ± 4.8	12.4 ± 4.3[d]	11.7 ± 4.5[d]	/	/

[a] 2963 children of a European population (Germany), 1499 male, 1464 female, measured in 2003, age 2–14 years.

[b] Young adults living in and around Tsukuba City, Japan: 217 male, 206 female, measured in 1991/1992, age 18–29 years.

[c] Men and women of the U.S. Army reflecting the racial balance of the U.S. Army at the time of data collection.

[d] Differing measurement.

The most established determinants in plantar pressure measurements are force, pressure, contact area, and contact time, which are further used to calculate force–time integrals or pressure–time integrals.

Peak force (F_{max}, N): Forces over all sensors of the whole foot or in a discrete mask are added and the maximum is displayed.

Peak pressure (P_{max}, kPa): Forces of each sensor within the whole foot or within one mask is divided by each sensor area. The maximum over all sensors will be determined.

Contact area (CA, mm²): Total area or subarea of a mask, which is loaded during the roll-over process.

Contact time (CT, s): Time of contact over all or in specific masks. It represents the temporal proportion of a mask in the roll-over process.

2.3 STATISTICAL TECHNIQUES ANALYZING FOOT MORPHOLOGY

This chapter gives an overview on different options to analyze foot anthropometrics. To ease the understanding of these techniques, basic examples are given (i.e., for sex-specific differences). These examples solely aim for understanding the idea of different types of analysis.

2.3.1 Absolute Foot Measures

2.3.1.1 Comparison of Absolute Data

Foot measures of different population groups such as gender can be compared by using mean values of each population. An example is given in Table 2.5 where average

TABLE 2.3

Data on Relative Foot Measures in Different Populations (% Foot Length), (Mean ± SD)

Author	Mauch (2007)[a]	Wunderlich et al. (2001)[b]		Krauss et al. (2008)[c]	
Subjects — Measure	$n = 2963$	♂ $(n = 293)$	♀ $(n = 491)$	♂ $(n = 423)$	♀ $(n = 424)$
Foot length (FE − FT$_1$)	/	100	100	100	100
Ball-of-foot length (FE − B1)	73.1 ± 1.6 − 74.2 ± 1.1	70.7	72.8	73.2 ± 1.0	73.6 ± 1.0
Outside ball-of-foot length (FE − B5)	58.9 ± 1.8 − 61.1 ± 1.8	61.7	61.2	61.8 ± 1.7	61.5 ± 1.7
Toe length (BM − FT$_2$)	33.0 ± 1.1 − 34.5 ± 1.5	/	/	32.6 ± 1.1	32.6 ± 1.0
Heel to medial/ lateral malleolus	/	/	/	/	/
Ball-of-foot width (B1 − B5)	37.9 ± 1.7 − 41.2 ± 2.1	39.1	39.0	38.8 ± 1.7	38.9 ± 1.7
Heel width (HW$_1$ − HW$_2$)	21.1 ± 1.3 − 24.1 ± 1.3	/	/	22.2 ± 1.1	22.0 ± 1.1
Heel width (orthogonal)	/	26.0	26.0	/	/
Bimalleolar width	/	27.1	26.8	/	/
Ball girth (B1 − B5)	93.5 ± 4.1 − 105.1 ± 5.1	93.4	93.1	/	/
Minimum arch girth	/	/	/	/	/
Heel girth	/	127.0	126.8	/	/
Medial/lateral malleolus height (H2)	/	30.1/27.0	29.5/27.1	/	/
Dorsal arch height (DA$_1$ − DA$_2$)	27.2 ± 1.7 − 30.0 ± 1.6	32.9	33.1	28.6 ± 1.6	28.6 ± 1.6
Ball angle (BA)	/	/	/	/	/
Hallux angle (HA)	/	/	/	/	/
Digitus minimus angle (DMA)	/	/	/	/	/

[a] 2963 children of a European population (Germany), 1499 male, 1464 female, measured in 2003, age 2–14 years.

[b] Men and women of the U.S. Army reflecting the racial balance of the U.S. Army at the time of data collection.

[c] 918 European subjects of a predominantly physically active population with an age between 18 and 60 years.

TABLE 2.4
Foot Measures Describing Foot Posture and Arch Structure

Anthropometric Values

Arch height	Highest point of the MLA (mostly represented by the navicular bone)
Plantar arch width ($AW_1 - AW_2$)	Narrowest section of the plantar medial longitudinal arch ($AW_1 - AW_2$)
Rearfoot angle (Longitudinal)	Angle between a longitudinal line bisecting the rearfoot (calcaneus) and the bisecting line of the distal one-third of the lower leg or to the floor
Arch angle Also, Clarke's angle, footprint angle	Angle between line connecting the first MTP joint (B1) and the medial heel (Hm) with a second line drawn to the apex of the medial longitudinal arch concavity (AP)

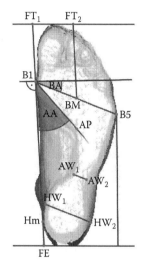

Footprint Indices

Arch index (Cavanagh and Rodgers 1987)	$$\frac{\text{Midfoot area}}{\text{Rearfoot area} + \text{Midfoot area} + \text{Forefoot area}}$$	See Cavanagh and Rodgers (1987)
Chippaux-Smirak Index (Forriol and Pascual 1990)	Relation between arch width ($AW_1 - AW_2$) and ball width ($B1 - B5$) A greater index indicates a high width in the arch area and thus a lowered medial arch of the foot	
Staheli Index (Staheli et al., 1987)	Relation between the arch width ($AW_1 - AW_2$) and heel width ($HW_1 - HW_2$) High arch index is the result of a descending of the arch of the foot	

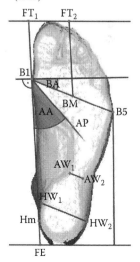

TABLE 2.5
Comparison of Ball-of-Foot Length between Gender Using Averaged Data (mm)

Subject	Men	Women
1	212	177
2	202	162
3	178	192
4	187	177
5	203	167
6	195	160
Mean value	196.17	172.5
Difference between groups	**23.7**	

ball-of-foot length of six male subjects is compared with average ball-of-foot length of six female subjects. In this example, this measure differs by 23.7 mm between genders.

This measure is especially valuable in forensic research (Ozden et al. 2005, Agnihotri et al. 2007). Drawbacks are related to its global perspective disallowing statements on distinctive length categories.

2.3.1.2 Comparison of Absolute Data within the Same Foot Length Category

This type of analysis is frequently used to describe the influence of factors such as gender, BMI, racial and/or environmental factors, or age on foot morphology (Anil et al. 1997, Wunderlich and Cavanagh 1999, Krauss et al. 2008, Mauch et al. 2008b, 2009, Krauss et al. 2011). Length categories (i.e., European Stitch, Mondopoint, or foot length in different age groups in children) have to be defined prior comparison. This is important because foot proportions change according to foot length and therefore influence foot anthropometrics (Mochimaru et al. 2000, Krauss et al. 2008). Average values are calculated within each length category and subsequently compared between population groups of interest. The example given in Table 2.6 compares average values of a given male and female population in predefined foot length categories defined in millimeter. Sex-specific differences of 4–6 mm are apparent for width and height measures in both foot length categories, whereas only minor differences are observable for medial ball length.

This kind of analysis is valuable for footwear industry as the last's shape is based on foot shape (Janisse 1992, Hawes et al. 1994). Foot proportions change according to size and footwear recommendations should be based on size-specific information (Krauss et al. 2008).

2.3.2 RELATIVE FOOT MEASURES

Normalization of foot measures to stature, foot length, or ball-of-foot length allows comparison of foot anthropometrics although absolute values of the given measures

TABLE 2.6

Comparison of Foot Measures between Men (M) and Women (W) in Different Foot Length (FL) Categories (mm)

Foot Length	250–256 mm FL			257–262 mm FL		
Variable	Men	Women	M – W	Men	Women	M – W
Ball-of-foot length	187	186	1	191	191	0
Ball-of-foot width	102	96	6	102	98	4
Dorsal arch height	75	71	4	76	72	4

TABLE 2.7

Foot Measures in % FL Averaged across All Foot Lengths ± Standard Deviation (StdDev)

Variable (% FL)	Men		Women		M – W
	Mean	StdDev	Mean	StdDev	MD
Ball-of-foot length	73.2	1.0	73.6	1.0	−0.3
Ball-of-foot width	38.8	1.7	38.9	1.7	−0.1
Dorsal arch height	28.6	1.6	28.6	1.6	0.0

Mean difference (MD) of the two population groups in % FL.

differ remarkably as a result of different size dimensions of the population groups to be compared. Equation 2.1 exemplarily displays the formula for calculating normalized foot measures relative to foot length.

Calculation of foot measures relative to foot length in %:

$$\text{Foot Measure (\% FL)} = \frac{\text{Foot measure}}{\text{Foot length}} \times 100 \qquad (2.1)$$

The example given in Table 2.7 compares foot measures of a given male and female population with different foot lengths. Measures are normalized to foot length. Sex-specific differences are between 0.0% and 0.3% FL.

Foot measures relative to stature can be used in the context of ethnological and psychological research (Fessler et al. 2005b, O'Connor et al. 2006). They have also been used for footwear recommendation (Wunderlich et al. 2001). However, the use of relative measures averaged across sizes has to be questioned as foot anthropometrics change according to size. Therefore, the use of relative data may cause size-specific information to be overlooked (Krauss et al. 2008).

2.3.3 MULTIVARIATE ANALYSIS

In the last decade, multivariate analyses have been increasingly used to characterize foot morphology and to classify feet into different foot types.

2.3.3.1 Principal Component Analysis

As described earlier, investigators consider a great many of variables describing foot shape. To classify feet into different foot types, for example, it is necessary to reduce these manifold variables into a simplified number of variables but still maintain much of the information of the original data. Principal component analysis (PCA) transforms items that are highly correlated to each other and generate a set of widely independent variables (principal components) (Jolliffe 2002).

Basically, the extraction of principal components amounts to a *variance maximizing* (*varimax*) *rotation* of the original variable space. For example, in a scatterplot we can think of the regression line as the original *x*-axis, rotated so that it approximates the regression line. This type of rotation is called *variance maximizing* because the goal of the rotation is to maximize the variance (variability) of the "new" variable (principal component) relative to the previous one, while minimizing the variance around the new variable.

Several authors (Bataller et al. 2001, Mauch et al. 2008a, 2009) used PCA prior to clustering for generating uncorrelated and standardized variables that are requisites for the cluster analysis (see Section 2.3.3.2). Therefore, Bataller et al. (2001) reduced the number of 39 parameters measured from Spanish feet to 13 principal components. Mauch et al. (2008a) reduced the number of 11 foot dimensions applying varimax rotation and allowing for eigenvalues after rotation being greater than one. Four principal components were identified which accounted for 88% of the variance, describing the "arch" of the foot (including arch angle, SI, and CSI), the "volume" of the foot (including girths, widths, heights), "angle" (characterizing the forefoot shape), and "length" describing the forefoot–hindfoot proportion.

2.3.3.2 Cluster Analysis

To classify foot types with similar foot dimensions, clustering techniques have been applied by several investigators (Bataller et al. 2001, Krauss et al. 2008, 2011, Mauch et al. 2009). Cluster analysis is a method to classify objects (i.e., feet) on the base of several criteria (i.e., foot measures). As foot measures are continuous data, the distance between foot data is calculated using the squared euclidean distance (Hair et al. 1998). For this kind of calculation, it is necessary that all involved variables are standardized, for example, as principal components or *z*-values.

The Ward's method is starting with a partition having each foot in a separate "group." Then those feet are merged into one group, which display the smallest distance to each other. The decision on the number of groups has to be done by the operator based on statistical aspects as well as with regard to content. Compared to this method, the partition method (k-means method) restructures feet based on a preassigned partition. It is advisable to determine the adequate number of groups by using the Ward's method and then optimizing this solution using the k-means method.

For example, Mauch et al. (2008a) used three independent factors, *arch*, *volume*, and *length*, for conducting cluster analysis. As a result, five clusters representing different foot types were identified. They can be briefly described as follows, taking into account that other factors do also have an influence on the characteristic of the single foot type:

1. *Flat feet*—mainly characterized by a lowered longitudinal arch
2. *Slender feet*—characterized by a small volume, that is, narrow widths
3. *Robust feet*—characterized by a rather large volume, that is, wide and high
4. *Short feet*—characterized by a short hindfoot and a long forefoot proportion
5. *Long feet*—characterized by a long hindfoot and a short forefoot proportion

2.3.4 MODELING FOOT MORPHOLOGY

Principal component or cluster analyses as described in the previous sections are generally based on 2D foot dimensions. In contrast to this, approaches modeling foot forms capture the 3D foot shape of the foot mainly for the use in product design (i.e., shoe construction). For example, Mochimaru et al. (2000) describe the Free Form Deformation (FFD) method—one of the computer graphics technologies to model and classify feet. The method transforms the shape of the feet by moving lattice points around the object. Next to this polygon-based object, it is feasible to produce mathematical surface-based objects. One application of this method is to design shoe lasts. This can be done by transforming a foot model into a last model using the distortion of control lattice points that convert a standard body form into other body forms (Mochimaru et al. 2000).

Luximon and Goonetilleke (2004) used methods to determine a 3D foot shape from 2D information.

They tried to model a standard foot using foot length, foot width, foot height, and measures of the foot curvature. First, polynomial regression was applied to determine the metatarsaophalangeal joint (MPJ curve). Second, a "standard" foot and foot shape prediction model was generated using 3D foot shape data. The complete procedure is described in Luximon and Goonetilleke (2004).

2.4 FOOT MORPHOLOGY IN DIFFERENT POPULATIONS

2.4.1 FOOT MORPHOLOGY IN THE COURSE OF LIFE

2.4.1.1 Children's Feet

Foot morphology is changing in the course of childhood and adolescence until it is full-grown in adulthood. The foot matures, it changes in its shape and function, and it grows, which means it changes in its lengths, widths, and proportions. In the following, the development of the main foot measures will be described.

2.4.1.1.1 Lengths

Foot length changes from 145.8 mm (±10.3) in 2 year olds to 249.7 mm (±16.1) in 14 year olds. About 98% of the feet are full-grown in the age of 12–13 years in girls and in the age of 15 years in boys (Andersen et al. 1956, Cheng et al. 1997). Growth decelerates with increasing age. While in a 3 year old the growth lies between 5.4 mm (girls) and 5.1 mm (boys) per half-year, the growth in a 12 year old is 2.2 mm (girls) and 4.7 mm (boys) per half-year (Mauch 2007). Other authors mention a yearly growth in the first 3 years in girls and boys with up to 2 mm per month, decreasing to approximately 1 mm per month until the age of 5 years (Wenger et al. 1983, Gould et al. 1990).

Comparably to foot length, the absolute characteristics of all other lengths, widths, circumferences, and height measures show a linear growth across the ages. Due to the fact that these measures are strongly dependent on foot length, relative measures (normalized to foot length) were also considered. *Ball-of-foot length* changes from 106.9 mm (±8.3) in 2 year olds to 184.7 mm (±12.2) in 14 year olds with relative values at an average of 73.8% (±1.4) over all ages (Mauch 2007).

Toe length changes from 49.8 mm (±3.7) in 2 year olds to 82.3 mm in 14 year olds. Relative to foot length, toe length changes its proportions from 34.2% in 2 year olds to 33.0% in 14 year olds (Mauch 2007).

2.4.1.1.2 Widths, Heights, and Circumferences

Widths measures show—relative to foot length—a parabolic decrease of the respective measures, that is, with an increasing foot length, width, height, and circumference measures decrease in the first years stronger and then more and more to a more weakly degree.

Ball-of-foot width changes from 60 mm (±4.8) in 2 year olds to 95.8 mm (±6.5) in 14 year olds, accompanied with a continuous relative change (normalized to foot length) from 41.2% in 2 year olds to 38.4% in 14 year olds (Mauch 2007). Other authors confirm the (relative) wider foot in younger ages to a (relative) smaller foot in older ages (Gould et al. 1990, Kouchi 1998). Similar values are evident for *ball-of-foot circumference*, which changes from 153 mm (±9.5) and 105.1% (relative) in 2 year olds to 234.0 mm (±13.6) and 93.9% (relative) in 14 year olds.

For *heel width*, absolute and relative changes are recognizable: 35.0 mm (±2.7), that is, 41.1%, in 2 year olds to 52.8 mm (±3.7), that is, 21.2%, in 14 year olds (Kouchi 1998).

Dorsal arch height is similar to the other width measures; it changes its proportions from 29.9% in 2 year olds (43.6 ± 3.1 mm) to 27.2% (67.7 ± 4.1 mm) in 14 year olds (Mauch 2007).

2.4.1.1.3 Angles

In contrast to the length and width measures, for the *ball angle* across all ages, no decrease or increase is observable. Nevertheless, a wide variance within the age groups is evident with a range of 11°–29° and a mean value of 20.1° (±2.6°) (Mauch 2007).

For all measures, it has to be mentioned that there are differing measuring practices to assess foot measures, and thereby, the results of different studies have to be interpreted with caution.

2.4.1.1.4 Medial Longitudinal Arch

The development of the medial longitudinal arch in children is a controversially discussed issue. Few authors have the opinion that the bony concavity already exists at the fetal stage, but will be visible at the age of 2 years and will get higher with the child maintaining an upright posture and walk. Other authors describe a fat pad that hides the medial longitudinal arch until the age of 4–6 years. This pad is supposed to cushion the forces underneath the talonavicular joint, and therefore protect against overload and damage of the still-maturing cartilage structures. The characteristic of the medial longitudinal arch is indirectly measured by footprint angle or footprint indices (see Section 2.2.2). Medial arch height increases considerably until the age of 5–6 years, and after this age the erection reduces until the age of 12 years. In general, girls show constantly higher arch values compared to their equally aged counterparts—boys tend to have more flat feet than girls (Cheng et al. 1997, Bordin et al. 2001, Echarri and Forriol 2003, Mauch 2007).

2.4.1.1.5 Development of Comprehensive Foot Shape (Foot Types)

Not only for single foot measures but also for the comprehensive foot shape in terms of foot types, a change of proportion in a child's foot is apparent. There is a high proportion of *flat* (low longitudinal arch), *robust* (wide and high feet), and *short feet* (long toes and slender) in the 2–4 year olds, with a decreasing proportion until the age of 14 years. The proportional course for the *slender* and *long feet* run in the opposite direction; there is a high proportion in the 9–14 year olds—in younger ages the proportion of these foot types is quite small (see Figure 2.1; Mauch 2007).

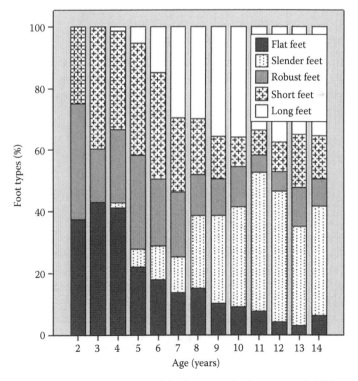

FIGURE 2.1 Distribution (percentage) of the foot types in the course of childhood.

2.4.1.2 Senior's Feet

Kouchi (1998) investigated intergeneration differences and found larger values for foot length measurements of the younger generations, except toe length. Ball girth, ball width, and heel width were smaller in younger generation females, but not in males. Younger generations tended to have a smaller relative ball girth. Older people have also been shown to have flatter, longer, and wider feet than younger adults (Frey et al. 1993, Frey 2000, Scott et al. 2007). These findings can explain why more than two-thirds of elderly feet are consistently wider than the available footwear, and why up to 80% of elderly people may wear shoes being too narrow and too short for their feet (Chantelau and Gede 2002, Menz and Morris 2005, Chaiwanichsiri et al. 2008). The aforementioned intergeneration differences may be caused by secular changes such as socioeconomic circumstances or differential bone growth during adolescence or by aging after the end of linear growth. Results of Kouchi (1998) indicate that the secular change in foot length accelerated for those born after approximately 1950. According to Kouchi (2003), changes in foot length and longitudinal arches due to aging are very small, thus negligible, and factors affecting growth of foot bones and foot structure (secular change) are more important than changes at the end of growth (aging) to explain intergeneration differences in foot morphology.

As exposure to footwear increases with age, it should be mentioned that footwear can be responsible for foot deformities such as hallux valgus, hammer toes, claw toes, and corns (Frey 2000, Manna et al. 2001, Menz and Morris 2005, O'Connor et al. 2006, Mickle et al. 2010).

2.4.2 Body Mass and Foot Morphology

The prevalence of obesity is rising throughout the world. In Europe, adiposity is the most common childhood disease. There are several studies investigating the influence of overweight on foot morphology and function. Results of these studies show an effect on foot morphology of obese children, who displayed significantly larger foot dimensions in terms of broader, taller, and thicker feet compared to nonobese children (Mauch 2007, Riddiford-Harland et al. 2010). Additionally, obese children show flatter feet compared to nonobese children quantified by an increased CSI and a decreased arch angle in obese children (Dowling et al. 2001, Riddiford-Harland et al. 2000, Mickle et al. 2006b, Villarroya et al. 2008). Generally, it was concluded that structural differences, in terms of a flattened medial arch in obese children, are due to the additional mass and therefore the increased load.

Additionally to the influence of excessive mass of overweight children on single foot dimensions, others investigated the influence of body mass (BMI, kg m^{-2}) on the comprehensive foot shape in terms of the prevalence of different foot types. This was observed not only for overweight but also for underweight and a normal population of children (Mauch et al. 2008a). Thereby, significant differences between the foot types with respect to the BMI of the children were detected: Children with a robust foot type had the highest BMI values, whereas children with a slender foot type displayed the lowest BMI values. As foot shape alters within the course of childhood (see Section 2.4.1.1), the prevalence of foot type differentiation for under- and overweight children was also observed. Figure 2.2 shows differences in the foot

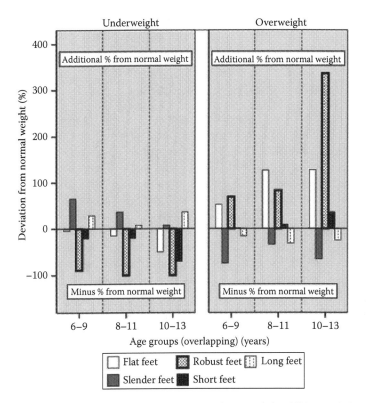

FIGURE 2.2 Proportion of foot types of under and overweight children relative to normal weight children. 0 = proportion of normal weight set to 100% within the three age groups. (With kind permission from Macmillan Publishers Ltd., *Int. J. Obes. (Lond.)*, Mauch, M. et al., Foot morphology of normal, underweight and overweight children, 32(7), 1068–1075. Copyright 2008a.)

types, thereby the proportion of normal weight children is presented by the zero line, set to 100%. The most pronounced finding is the higher incidence of robust feet and flat feet in overweight children with a higher proportion of 69%–337% and 53%–128%, respectively, compared to normal weights over all age groups. The underweight children display a higher proportion of slender feet (65%–7%), especially in the age groups from 6 to 9 years and a lower proportion of robust (−89% to 100%), flat (−4% to −50%), and short feet (−21% to −70%), especially in the age from 10 to 13 years.

Differences regarding body weight were observable not only for children but also for adults. There are biomechanical studies that investigated the influence of additional weight on plantar pressure measurements in adults resulting in an increased plantar pressure under the heel, midfoot, and metatarsals in standing and walking. The highest values were found underneath the longitudinal arch and the metatarsal bones (Hills et al. 2001). Additionally, total plantar force, as well as total contact area, was found to be larger in obese adults (Birtane and Tuna 2004).

Overall, the influence of the children's BMI on foot morphology and foot shape is established. Nevertheless, uncertainties exist regarding the long-term effects of additional load on the musculoskeletal systems and structural changes on the foot anatomy. With the increasing incidence of the obese population, altered foot morphology should be further considered, for example, in shoe construction.

2.4.3 FOOT MORPHOLOGY OF MEN AND WOMEN

2.4.3.1 Gender Differences in Absolute Foot Measures

Male and female feet differ (to varying degrees) with respect to length, height, and width measures with larger values for the male population (Manna et al. 2001, Wunderlich et al. 2001, Luo et al. 2009). For example, Manna et al. (2001) reported foot length differences between genders of up to 10%. Luo et al. (2009) described mean differences of 17 mm for ball length (165 versus 182 mm), and Wunderlich and Cavanagh (2001) reported differences of 19 mm for ball length (177 versus 196 mm). Differences of 23 mm foot length were also reported for an elderly population (Mickle et al. 2010). A marked difference between men and women was noted for foot volume, quantified with a water displacement method. On average, foot volume showed approximately 50% higher values for male subjects (Manna et al. 2001).

2.4.3.2 Gender Differences in Foot Measures Relative to Stature

Fessler et al. (2005a) examined data from three previous anthropometric studies of different population groups (United States, Turkey, Native North and Central American, and historical foot traces from different indigenous American populations). They concluded that men have longer feet than women for a given stature. This is somewhat surprising as—from a biomechanical standpoint of view—women could profit from a larger area of support during pregnancy—foot length affects dorsoventral stability. Therefore, differences between men and women may better be explained by an intersexual selection for small female foot size in the evolutional process (Fessler et al. 2005a), indexing youth, and nulliparity—a reasonable prerequisite to maximize returns on male reproductive investment (Fessler et al. 2005b).

2.4.3.3 Gender Differences in Foot Measures Relative to Foot Length

Comparisons of foot measures relative to foot or ball-of-foot length showed inconsistent results across studies (Wunderlich et al. 2001, Krauss et al. 2008, 2011, Luo et al. 2009, Mickle et al. 2010). For ball-of-foot length, Wunderlich and Cavanagh (2001) reported no sex-specific differences whereas recent studies reported significantly larger values for the female population. However, these differences were quite small (0.3%–1% foot length) (Krauss et al. 2008, 2011, Mickle et al. 2010). Because of a lack of consensus, results of previous studies related to other foot measures are not described in detail. Furthermore, most of the reported differences were quite small (<1% foot length) and their practical relevance has to be questioned (Wunderlich et al. 2001, Krauss et al. 2008, 2011).

2.4.3.4 Gender Differences for Hallux Angle and Heel Bone Angle

Hallux angle can be used as a measure of hallux adduction in the forefoot region. Measurements of articular surfaces of foot bones suggest that the female's first metatarsal bones have more potential to move in the direction of adduction in comparison with their male counterparts (Ferrari et al. 2004). This finding may underline the predisposition of the female foot to develop hallux valgus deformity. According to O'Connor et al. (2006), development of hallux valgus is two to four times more likely in women. Frey even reported a rate of 9:1 and directly related this ratio to the style of shoes that are worn by women (Frey 2000). A link between flattened and more pronated feet and hallux valgus seems reasonable as this link has been demonstrated in a previously published study on elderly people (Mickle et al. 2010).

2.4.3.5 Gender Differences within the Same Foot Length Categories

Comparison of foot anthropometrics in the same foot length are particularly important for footwear industry as many females' lasts are simply downscaled versions of males' last (Frey 2000, Wunderlich et al. 2001, Krauss et al. 2010). However, only few studies focused on a comparison of foot morphology within the same foot length. This may be put down to the fact that the number of overlapping foot length categories is limited.

Differences between genders are not uniformly described for foot length measures such as ball-of-foot length, dorsal arch length, and outside ball-of-foot length. In contrast, dorsal arch height, width, and circumference measures are smaller in women in comparison with men when looking at the same foot length. Differences are relevant in a practical manner and account for 2–8 mm for width measures and up to 5 mm for height measures (Anil et al. 1997, Chaiwanichsiri et al. 2008, Krauss et al. 2008, 2011, Mickle et al. 2010).

2.4.3.6 Gender and Foot Types

Several authors used multivariate statistics to account for interindividual differences in foot shape. One recent study especially focused on gender-specific differences in foot types (Krauss et al. 2008, 2011). Dorsal arch height, ball-of-foot width, heel width, ball-of-foot length, and ball angle were used as input variables. Aside from ball angle, measures were normalized to foot length to allow a discrimination of relative foot proportions in the subsequent cluster analysis. Three foot types (clusters) were obtained: (C1) a voluminous foot type with large values for dorsal arch height, heel width, and ball-of-foot width; (C2) a flat and pointed foot type particularly characterized by its low dorsal arch height, long ball-of-foot length, large ball angle, and small heel width; and (C3) a slender foot type especially characterized by small values for ball-of-foot width and heel width.

Figure 2.3 demonstrates that the voluminous foot type is more frequent in smaller sizes, whereas Clusters 2 and 3 are more frequent in larger sizes. These two foot types mainly differ with respect to forefoot shape. Both are narrower in height and width than the voluminous foot type in cluster 1. The proportional change of foot dimensions is described as allometry; increasing foot length is associated with a relative decrease in width and height dimensions of the foot (Mochimaru et al. 2000, Krauss et al. 2008). The distribution of the predominant foot type is similar for men and women. However, there is a size-specific offset of approximately three shoe sizes. The picture is important to highlight previously described results of sex-specific comparisons in relative (Section 2.4.3.3) and absolute (Section 2.4.3.1) foot measures: Only small and mostly inconsistent differences were observed for relative

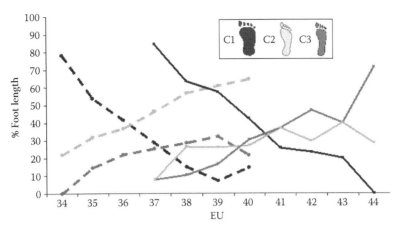

FIGURE 2.3 Cluster distribution within different sizes (European Stitch EU) for women (dashed lines) and men (solid lines). (From Krauss, I. et al., *Ergonomics*, 51(11), 1693, 2008. With kind permission of the publisher Taylor & Francis Group, http://www.informaworld.com)

measures. By contrast, comparison of absolute foot measures within the same foot length show sex-specific differences for width and height measures with larger values for the male population. The distribution of foot types for men and women in Figure 2.3 depicts the discrepancies between results based on comparisons of averaged data across all sizes and comparisons within the same size. Although the proportional change of foot dimensions exists in both genders, foot proportions of men and women are not the same within the same shoe size. The following example illustrates this fact. The incidence of C1 is comparable between genders (39% in men, 35% in women). It is most common in the small foot length categories of each group. Yet it is important to know that the definition "small" is sex-dependent, since small women's feet are about three sizes shorter than small men's feet. When comparing feet in EU 39, most of the women's feet belong to foot type C2 or C3 (more narrow and flat feet), whereas most of the men's feet belong to C1 (voluminous feet).

According to the previous remarks, it is important to keep in mind that allometric changes of foot shape are size and sex specific. This is especially important for the manufacturing of footwear.

2.4.4 ETHNICAL AND REGIONAL DIFFERENCES IN FOOT MORPHOLOGY

Anthropometrics of the human body and the feet vary between different races, ethnics, regions, and continents due to genetic, environmental, socioeconomic, life-style, and shoe-wearing differences (Hawes et al. 1994, Sachithanandam and Joseph 1995, Kusumoto et al. 1996, Ashizawa et al. 1997, Razeghi and Batt 2002, Fessler et al. 2005b). People residing in warmer climates, such as Australia, are found to have broader feet with straight toes (Cheskin 1987), whereas Caucasian women have narrow feet, possibly influenced by fashioned shoes (Frey 2000). Hawes et al. (1994) found differences regarding the location and angularity of the metatarsophalangeal joint axis and the shape of the anterior margin of the foot between Oriental males (Japanese and Korean) and North Americans, revealing a more squared foot shape for the Oriental population. Kouchi (1998) analyzed differences in foot shape characteristics due to ethnic backgrounds and reported Japanese having wider feet for foot length compared to Caucasoid or Australoids and smaller foot length for height compared to Southeast Asians and Africans. It was observed that these differences may be caused by genetic but also environmental factors, for example, shoe-wearing habits, like wearing open footwear such as sandals or toe thongs, walking frequently barefooted, or wearing closed, fashioned shoes. Mauch et al. (2008b) found differences between the feet of German and Australian children. Thereby, the German children displayed significantly longer and flatter feet compared to their Australian counterparts, whereas Australian children showed a squarer forefoot shape relative to the German children. It is speculated that due to the mixture of racial backgrounds in all continents and the difficulty of determining "race" that environmental factors, rather than race alone, may explain foot shape differences among different populations. Independent of possible causal factors, ethnic and regional specifics in foot anthropometrics should be considered in the manufacturing of footwear.

ACKNOWLEDGMENTS

The authors thank the publishers Taylor & Francis Group (http://www.informaworld.com) for their kind permission to reuse some material already published in the *Journal of Ergonomics* and the Nature Publishing Group (www.nature.com) for permission to use material from the *International Journal of Obesity*.

QUESTIONS

2.1 What foot dimensions can be measured with a Brannock Device?

2.2 Please name the four dimensions that are captured with a 4D scanner. What advantage with respect to footwear has the new technology in contrast to 3D scanners?

2.3 What are common foot dimensions and measures to quantify foot morphology? Which measures are used to quantify arch structure?

2.4 Which problems occur when comparing different studies on foot morphology?

2.5 Why is the comparison of foot measures within the same foot length category important for the manufacturing of shoes?

2.6 Which advantages and disadvantages hold multivariate statistical methods, like cluster analysis?

2.7 What is a specific characteristic in the development of a child's foot shape? What consequences are affected by this development regarding the science of footwear?

2.8 How does a foot of overweights differ regarding their foot morphology and plantar pressure?

2.9 Are there any regional differences in foot morphology? What are the possible reasons for differences? Which consequences occur out of possible differences for the manufacturing of shoes?

2.10 What problem can occur if a men's last is used for the manufacturing of women's shoes?

REFERENCES

Agnihotri, A. K., Shukla, S., and Purwar, B. 2007, Determination of sex from the foot measurements, *Internet J. Forensic Sci.*, 2(1), DOI: 10.5580/1283.

Andersen, M., Blais, M., and Greene, W. T. 1956, Growth of the normal foot during childhood and adolescence; length of the foot and interrelations of foot, stature, and lower extremity as seen in serial records of children between 1–18 years of age, *Am. J. Phys. Anthropol.*, 14(2), 287–308.

Anil, A., Peker, T., Turgut, H. B., and Ulukent, S. C. 1997, An examination of the relationship between foot length, foot breath, ball girth, height and weight of Turkish university students aged between 17 and 25, *Anthropol. Anz.*, 55(1), 79–87.

Ashizawa, K., Kumakura, C., Kusumoto, A., and Narasaki, S. 1997, Relative foot size and shape to general body size in Javanese, Filipinas and Japanese with special reference to habitual footwear types, *Ann. Hum. Biol.*, 24(2), 117–129.

Bataller, A., Alcantara, E., Gonzales, J. C., Garcia, A. C., and Alemany, S. 2001, Morphological grouping of Spanish feet using clustering techniques. In *5th Symposium on Footwear Biomechanics*, July 5, 2001, Zürich, Switzerland, E. Henning and A. Stacoff, eds., pp. 12–13.

Birtane, M. and Tuna, H. 2004, The evaluation of plantar pressure distribution in obese and non-obese adults, *Clin. Biomech. (Bristol, Avon.)*, 19(10), 1055–1059.

Bordin, D., De, G. G., Mazzocco, G., and Rigon, F. 2001, Flat and cavus foot, indexes of obesity and overweight in a population of primary-school children, *Minerva Pediatr.*, 53(1), 7–13.

Cavanagh, P. R. and Rodgers, M. M. 1987, The arch index: A useful measure from footprints, *J. Biomech.*, 20, 547–551.

Cavanagh, P. R., Rodgers, M. M., and Iiboshi, A. 1987, Pressure distribution under symptom-free feet during barefoot standing, *Foot Ankle*, 7(5), 262–276.

Chaiwanichsiri, D., Tantisiriwat, N., and Janchai, S. 2008, Proper shoe sizes for Thai elderly, *Foot (Edinb.)*, 18(4), 186–191.

Chantelau, E. and Gede, A. 2002, Foot dimensions of elderly people with and without diabetes mellitus—A data basis for shoe design, *Gerontology*, 48(4), 241–244.

Cheng, J. C., Leung, S. S., Leung, A. K., Guo, X., Sher, A., and Mak, A. F. 1997, Change of foot size with weightbearing. A study of 2829 children 3 to 18 years of age, *Clin. Orthop. Relat. Res.*, (342), 123–131.

Cheskin, M. P. 1987, Foot types, in *The Complete Handbook of Athletic Footwear*, M. P. Cheskin, K. J. Sherkin, and B. T. Bates, eds., Fairchild Publications, New York, pp. 234–236.

Dowling, A. M., Steele, J. R., and Baur, L. A. 2001, Does obesity influence foot structure and plantar pressure patterns in prepubescent children? *Int. J. Obes. Relat. Metab. Disord.*, 25(6), 845–852.

Echarri, J. J. and Forriol, F. 2003, The development in footprint morphology in 1851 Congolese children from urban and rural areas, and the relationship between this and wearing shoes, *J. Pediatr. Orthop. B*, 12(2), 141–146.

Ferrari, J., Hopkinson, D. A., and Linney, A. D. 2004, Size and shape differences between male and female foot bones: Is the female foot predisposed to hallux abducto valgus deformity? *J. Am. Podiatr. Med. Assoc.*, 94(5), 434–452.

Fessler, D. M., Haley, K. J., and Lal, R. D. 2005a, Sexual dimorphism in foot length proportionate to stature, *Ann. Hum. Biol.*, 32(1), 44–59.

Fessler, D. M., Nettle, D., Afshar, Y., Pinheiro, I. A., Bolyanatz, A., Mulder, M. B., Cravalho, M. et al. 2005b, A cross-cultural investigation of the role of foot size in physical attractiveness, *Arch. Sex Behav.*, 34(3), 267–276.

Forriol, F. and Pascual, J. 1990, Footprint analysis between three and seventeen years of age, *Foot Ankle*, 11, 101–104.

Frey, C. 2000, Foot health and shoewear for women, *Clin. Orthop.*, 372, 32–44.

Frey, C., Thompson, F., Smith, J., Sanders, M., and Horstman, H. 1993, American orthopaedic foot and ankle society women's shoe survey, *Foot Ankle*, 14(2), 78–81.

Gould, N., Moreland, M., Alvarez, R., Trevino, S., and Fenwick, J. 1989, Development of the child's arch, *Foot Ankle*, 9(5), 241–245.

Gould, N., Moreland, M., Trevino, S., Alvarez, R., Fenwick, J., and Bach, N. 1990, Foot growth in children age one to five years, *Foot Ankle*, 10(4), 211–213.

Hair, J. F., Anderson, R. E., Tatham, R. L., and Black, W. C. 1998, *Multivariate Data Analysis*, 5 edn., 730p., Prentice-Hall International, Inc., Upper Saddle River, NJ.

Hawes, M. R., Nachbauer, W., Sovak, D., and Nigg, B. M. 1992, Footprint parameters as a measure of arch height, *Foot Ankle*, 13(1), 22–26.

Hawes, M. R. and Sovak, D. 1994, Quantitative morphology of the human foot in a North American population, *Ergonomics*, 37(7), 1213–1226.

Hawes, M. R., Sovak, D., Miyashita, M., Kang, S. J., Yoshihuku, Y., and Tanaka, S. 1994, Ethnic differences in forefoot shape and the determination of shoe comfort, *Ergonomics*, 37(1), 187–196.

Hennig, E. M. and Rosenbaum, D. 1991, Pressure distribution patterns under the feet of children in comparison with adults, *Foot Ankle*, 11(5), 306–311.

Hills, A. P., Hennig, E. M., Mcdonald, M., and Bar-or, O. 2001, Plantar pressure differences between obese and non-obese adults: A biomechanical analysis, *Int. J. Obes. Relat. Metab. Disord.*, 25(11), 1674–1679.

Huppin, L. 2009, Technology: Choosing a digital foot scanner, *Low. Extremity Rev.* (August).

Janisse, D. J. 1992, The art and science of fitting shoes, *Foot Ankle*, 13(5), 257–262.

Jolliffe, I. T. 2002, *Principal Component Analysis*, 2 edn., Springer, New York.

Jung, S., Lee, S., Boo, J., and Park, J. 2001, A classification of foot types for designing footwear of the Korean elderly. In *Proceedings of the 5th Symposium on Footwear Biomechanics*, Zurich, Switzerland, E. Hennig and A. Stacoff, eds., pp. 48–49.

Kimura, M., Mochimaru, M., Kouchi, M., Saito, H., and Kanade, T. 2005, 3-D Cross-sectional shape measurement of the foot while walking. In *Proceedings of the 7th Symposium on Footwear Biomechanics*, Hamill, C.L., Harding, E., and Williams, K.R. (eds.) Cleveland, p. 34.

Kouchi, M. 1995, Analysis of foot shape variation based on the medial axis of foot outline, *Ergonomics*, 38(9), 1911–1920.

Kouchi, M. 1998, Foot dimensions and foot shape: Differences due to growth, generation and ethnic origin, *Anthropol. Sci.*, 106(Suppl.), 161–188.

Kouchi, M. 2003, Inter-generation differences in foot morphology: Aging or secular change?, *J. Hum. Ergol. (Tokyo)*, 32(1), 23–48.

Kouchi, M., Kimura, M., and Mochimaru, M. 2009, Deformation of foot cross-section shapes during walking, *Gait Posture*, 30(4), 482–486.

Kouchi, M. and Mochimaru, M. 2003, The feet of runners: Do they differ from non-runners feet. August 9, 1931, P. Milburn, ed., In *Proceedings of the 6th Symposium on Footwear Biomechanics*, Queenstown, New Zealand, pp. 49–51.

Krauss, I., Grau, S., Mauch, M., Maiwald, C., and Horstmann, T. 2008, Sex-related differences in foot shape, *Ergonomics*, 51(11), 1693–1709.

Krauss, I., Langbein, C., Horstmann, T., and Grau, S. 2011, Sex-related differences in foot shape of adult Caucasians—A follow-up study focusing on long and short feet, *Ergonomics*, 54(3), 294–300.

Krauss, I., Valiant, G., Horstmann, T., and Grau, S. 2010, Comparison of female foot morphology and last design in athletic footwear—Are men's lasts appropriate for women? *Res. Sports Med.*, 18(2), 140–156.

Kusumoto, A., Suzuki, T., Kumakura, C., and Ashizawa, K. 1996, A comparative study of foot morphology between Filipino and Japanese women, with reference to the significance of a deformity like hallux valgus as a normal variation, *Ann. Hum. Biol.*, 23(5), 373–385.

Luo, G., Houston, V. L., Mussman, M., Garbarini, M., Beattie, A. C., and Thongpop, C. 2009, Comparison of male and female foot shape, *J. Am. Podiatr. Med. Assoc.*, 99(5), 383–390.

Luximon, A. and Goonetilleke, R. S. 2004, Foot shape modeling, *Hum. Fact.*, 46(2), 304–315.

Manna, I., Pradhan, D., Ghosh, S., Kar, S. K., and Dhara, P. 2001, A comparative study of foot dimension between adult male and female and evaluation of foot hazards due to using of footwear, *J. Physiol. Anthropol. Appl. Hum. Sci.*, 20(4), 241–246.

Mauch, M. 2007, *Kindliche Fussmorphologie*, 1 edn., pp. 1–228, VDM Verlag Dr. Müller, Saarbruecken, Germany.

Mauch, M., Grau, S., Krauss, I., Maiwald, C., and Horstmann, T. 2008a, Foot morphology of normal, underweight and overweight children, *Int. J. Obes. (Lond.)*, 32(7), 1068–1075.

Mauch, M., Grau, S., Krauss, I., Maiwald, C., and Horstmann, T. 2009, A new approach to children's footwear based on foot type classification, *Ergonomics*, 52(8), 999–1008.

Mauch, M., Mickle, K. J., Munro, B. J., Dowling, A. M., Grau, S., and Steele, J. R. 2008b, Do the feet of German and Australian children differ in structure? Implications for children's shoe design, *Ergonomics*, 51(4), 527–539.

Menz, H. B. and Morris, M. E. 2005, Footwear characteristics and foot problems in older people, *Gerontology*, 51(5), 346–351.

Mickle, K. J., Munro, B. J., Lord, S. R., Menz, H. B., and Eele, J. R. 2010, Foot shape of older people: Implications for shoe design, *Footwear Sci.*, 2(3), 131–139.

Mickle, K. J., Steele, J. R., and Munro, B. J. 2006a, Does excess mass affect plantar pressure in young children? *Int. J. Pediatr. Obes.*, 1(3), 183–188.

Mickle, K. J., Steele, J. R., and Munro, B. J. 2006b, The feet of overweight and obese young children: Are they flat or fat? *Obesity (Silver Spring)*, 14(11), 1949–1953.

Mochimaru, M., Kouchi, M., and Dohi, M. 2000, Analysis of 3-D human foot forms using the free form deformation method and its application in grading shoe lasts, *Ergonomics*, 43(9), 1301–1313.

Nácher, B., Alemany, S., Gonzáles, J. C., Alcántara, E., Garcia-Hernández, J., Heras, S., and Juan, A. 2006, A footwear fit classification model based on anthropometric data. In *2006 Digital Human Modeling for Design and Engineering Conference*, SAE Technical Paper 2006-01-2356, Lyon, France, DOI: 10.4271/2006-01-2356.

O'connor, K., Bragdon, G., and Baumhauer, J. F. 2006, Sexual dimorphism of the foot and ankle, *Orthop. Clin. North Am.*, 37(4), 569–574.

Ozden, H., Balci, Y., Demirustu, C., Turgut, A., and Ertugrul, M. 2005, Stature and sex estimate using foot and shoe dimensions, *Forensic Sci. Int.*, 147(2–3), 181–184.

Parham, K. R., Gordon, C. C., and Bensel, C. K. 1992, Anthropometry of the foot and lower leg of U.S. army soldiers: Fort Jackson, SC-1985, Technical Report United States Army Natick Research, Development and Engineering Center, pp. 1–341.

Razeghi, M. and Batt, M. E. 2002, Foot type classification: A critical review of current methods, *Gait Posture*, 15(3), 282–291.

Riddiford-Harland, D. L., Steele, J. R., and Baur, L. A. 2010, Are the feet of obese children fat or flat? Revisiting the debate, *Int. J. Obes. (Lond.)*, 35(1), 115–120.

Riddiford-Harland, D. L., Steele, J. R., and Storlien, L. H. 2000, Does obesity influence foot structure in prepubescent children? *Int. J. Obes. Relat. Metab. Disord.*, 24(5), 541–544.

Rosenbaum, D. and Becker, H.-P. 1997, Plantar pressure distribution measurements. Technical background and clinical applications, *Foot Ankle Surg.*, 3(1), 1–14.

Sachithanandam, V. and Joseph, B. 1995, The influence of footwear on the prevalence of flat foot. A survey of 1846 skeletally mature persons, *J. Bone Joint Surg. Br.*, 77(2), 254–257.

Schmeltzpfenning, T., Plank, C., Krauss, I., Aswendt, P., and Grau, S. 2009, Dynamic foot scanning: A new approach for measurement of the human foot shape while walking, *Footwear Sci.*, 1(S1), 28–30.

Scott, G., Menz, H. B., and Newcombe, L. 2007, Age-related differences in foot structure and function, *Gait Posture*, 26(1), 68–75.

Smith, S. L. 1997, Attribution of foot bones to sex and population groups, *J. Forensic Sci.*, 42(2), 186–195.

Staheli, L. T. 1994, Footwear for children, Instructional Course Lectures, 43, pp. 193–197.

Staheli, L. T., Chew, D. E., and Corbett, M. 1987, The longitudinal arch: A survey of eight hundred and eighty-two feet in normal children and adults, *J. Bone Joint Surg. Am.*, 69, 426–428.

Telfer, S. and Woodburn, J. 2010, The use of 3D surface scanning for the measurement and assessment of the human foot, *J. Foot Ankle Res.*, 3(19), 1–9.

Villarroya, M. A., Esquivel, J. M., Tomas, C., Buenafe, A., and Moreno, L. 2008, Foot structure in overweight and obese children, *Int. J. Pediatr. Obes.*, 3(1), 39–45.

Voracek, M., Fisher, M. L., Rupp, B., Lucas, D., and Fessler, D. M. 2007, Sex differences in relative foot length and perceived attractiveness of female feet: Relationships among anthropometry, physique, and preference ratings, *Percept. Mot. Skills*, 104(3 Pt 2), 1123–1138.

Wenger, D. R., Mauldin, D., Morgan, D., Sobol, M. G., Pennebaker, M., and Thaler, R. 1983, Foot growth rate in children age one to six years, *Foot Ankle*, 3(4), 207–210.

Witana, C. P., Xiong, S., Zhao, J., and Goonetilleke, R. S. 2006, Foot measurements from three-dimensional scans: A comparison and evaluation of different methods, *Int. J. Ind. Ergon.*, 36(9), 789–807.

Wunderlich, R. E. and Cavanagh, P. R. 1999, Sexual dimorphism in foot shape. In *4th Symposium of the Technical Group of Footwear Biomechanics*, 1999, Canmore, Alberta, Canada.

Wunderlich, R. E. and Cavanagh, P. R. 2001, Gender differences in adult foot shape: Implications for shoe design, *Med. Sci. Sports Exerc.*, 33(4), 605–611.

Yu, C. Y. and Tu, H. H. 2009, Foot surface area database and estimation formula, *Appl. Ergon.*, 40(4), 767–774.

3 Foot Characteristics and Related Empirical Models

Shuping Xiong, Asanka S. Rodrigo,
and Ravindra S. Goonetilleke

CONTENTS

3.1 INTRODUCTION

Foot characteristics such as its shape and structure have an influence on foot function (Razeghi and Batt, 2002). Thus, understanding these characteristics is useful to design footwear that are comfortable to wear (Dohi et al., 2001; Goonetilleke, 2001; Goonetilleke and Luximon 2001; Witana et al., 2006).

The foot can be described with dimensions such as lengths, widths, heights, and girths. Some of the dimensions are related to each other and can be expressed as a percentage of foot length (Krauss and Mauch, 2012). For example, foot length has a high correlation with arch length (for Hong Kong males, arch length = 0.7369 * foot length − 2.317; $R^2 = 0.86$) and so does foot circumference with foot width (for Hong Kong males, foot circumference at the metatarsophalangeal joint = 2.711 * foot width − 8.995; $R^2 = 0.86$) (Goonetilleke et al., 1997). However, foot width does not increase proportionately with foot length (Cabrera et al., 2004; Kouchi and Yamazaki, 1992). Even though foot width can be found as a proportion of foot length, the explained variance is relatively low ($R^2 = 0.43$, for Hong Kong males) (Goonetilleke et al., 1997). Allometry is a means to investigate these disproportionate differences in one or more of the dimensions when there is a change in size.

3.2 FOOT SHAPE

3.2.1 MIDFOOT SHAPE CHARACTERISTICS

Xiong et al. (2008) found no direct relationship between foot heights (at 35%, 40%, 45%, and 50% of foot length) and foot length of Hong Kong Chinese feet, thereby indicating that a longer foot is not necessarily higher or vice versa. After eliminating the toe region, a normalized ball-to-section length was defined as the percentage of foot length (NBL = 100 * BL/FL) (Figure 3.1). This variable, NBL, has a significant allometric relationship ($R^2 > 0.97$ for both males and females) with the ball-to-section height (BH) (Equation 3.1). A validation of the model resulted in a mean absolute prediction error of 1.3 mm (SD = 0.7 mm) and 1.0 mm (SD = 0.7 mm) for males and females, respectively, indicating reasonable accuracy.

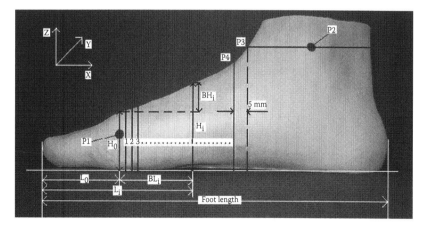

FIGURE 3.1 Dimensions to characterize the foot. P1 is the most medial prominence of the first MPJ; P2 is on the medial malleolus; H_0 is foot height from the floor and L_0 is the length from the tip of longest toe at the first MPJ section; H_i is foot height from the floor and L_i is the length from the tip of the toe at the ith section. The relative heights and lengths are $BH_i = H_i − H_0$; $BL_i = L_i − L_0$.

$$\text{For males, BH} = 1.068 * \text{NBL}^{1.038} \quad\quad (3.1)$$

$$\text{For females, BH} = 1.074 * \text{NBL}^{1.007}$$

Thus, beyond the toes the foot height has a proportional relationship with length. The foot heights among people can then be accounted by considering the big toe heights of each person. Hence, the toe height is a dimension that should be measured to generate a good fit. The relationship can be shown to be valid across different sizes of feet as well.

3.2.2 EFFECTS OF LOAD BEARING

Load bearing can significantly change the foot shape. Increased loading due to weight-bearing will generally result in increases in length and width and a reduction in foot height with a rotation to the medial side (eversion). From a no-weight bearing state to full body weight on each foot, increased length by 3.2 mm (1.3%) in males and 3.4 mm (1.5%) in females; arch length increases were 2.3 mm (1.3%) and 2.2 mm (1.3%) for Hong Kong Chinese males and females, respectively (Xiong et al., 2009). The width increases for the male participants were 3.0 mm (3.0%) for foot width, 4.0 mm (4.5%) for midfoot width, and 1.1 mm (1.7%) for heel width. The changes in the widths of female feet were 1.8 mm (2.1%) for foot width, 3.3 mm (4.3%) for midfoot width, and 0.8 mm (1.3%) for heel width. Xiong et al. (2009) also showed changes in three height dimensions: for males, the height reductions were 4.3 mm (6.3%) at midfoot, 1.3 mm (1.6%) decrease of medial malleolus height, and a lateral malleolus height reduction of 1.3 mm (1.9%). The height reductions for females were 3.8 mm (6.0%) at midfoot and 1.1 mm (1.7%) at lateral malleolus. There were no changes in the medial malleolus height of females. The mean ball girth increases were 3.5 mm (1.5%) in males and 1.6 mm (0.8%) in females. These data show that the percentage dimensional changes due to loading are small and are generally less than 3% except midfoot height and width, primarily due to the foot arch. The data among the various studies are comparable. In 2829 children (3–18 years old), Cheng et al. (1997) found a foot length increase of 3.14 mm (1.2%) for males and a 3.26 mm (1.4%) increase for females from no load to half body weight on the foot. Rossi (1983) reported a 4.2 mm (1.7%) increase in foot length in 85% of American women. The largest increase in foot length of 6.3 mm (2.7%) was reported by Tsung et al. (2003). The differences among the various studies can be attributed to the differing sample sizes, populations tested, measurement protocols and devices, and so on. The important point is to account for these dimensional changes when designing footwear.

3.3 FOOT CLASSIFICATION

3.3.1 TOE TYPES

The toes help to take up load when the heel is raised. They are in contact with the ground or footwear for approximately three-fourths of the walking cycle. The pressures on the toes can be similar to those under the metatarsal heads (Hughes et al., 1990;

Egyptian type   Greek type   Square type

FIGURE 3.2   Three toe types of foot classified by the relative lengths of digits.

Lambrinudi, 1932). Thus, they are quite important for various activities such as walking, providing balance and weight-bearing, and also to generate push-off.

The toes are called hallux (big toe), index toe (second toe), middle toe, fourth toe, and little toe. The hallux has two phalanx bones while the other four toes consist of three phalanx bones. The hallux accounts for about 60%–80% of the loading on the toes during walking, indicating that the others play a minor role (Hughes et al., 1990; Hutton and Dhanendran, 1979).

The common form of foot is the "Egyptian type" that has a longer hallux (1st > 2nd > 3rd > 4th > 5th). Those of the "Greek type" have a longer second toe than the hallux (2nd > 1st > 3rd > 4th > 5th), and few others have a "Square type" with similar lengths for most toes (Figure 3.2). A survey conducted by Hawes and Sovak (1994) with 708 American and 513 Japanese and Korean male subjects found that 76.09% of Americans and only 50.8% of Japanese and Korean male subjects had the Egyptian type of foot.

Even though predominant patterns of toe lengths may be different among differing ethnicities, the ratio of arch length to foot length appears to be in the range of 71.8%–73.0% in both male and female adults (Table 3.1). Thus, locating the flex line of a shoe is somewhat easy for a given foot length (Mochimaru and Kouchi, 2003).

3.3.2   Arch Types

The sole of the foot generally does not make full contact with the ground (Figure 3.3) due to the two longitudinal arches (medial longitudinal arch, MLA, and lateral longitudinal arch, LMA) and the transverse arch in each foot. The flexibility of these arches makes walking and running easier, as they provide the necessary shock absorption for the foot (Chu et al., 1995; Shiang et al., 1998; Williams and McClay, 2000). A reliable, valid, and easy way to classify the foot arch type can help discover possible risk factors or potential causes of foot injuries so that appropriate orthotics can be prescribed to prevent them (Howard and Briggs, 2006; Lin et al., 2004; McCrory et al., 1997; Williams and McClay, 2000). A common categorization is high-arch, normal-arch, or low-arch (Figure 3.4).

1. *The low-arched or flat foot* has an imprint that looks like the sole of the foot. This type of foot is an overpronated foot that strikes on the lateral side of the heel and has an excessive roll inward (Özgü, 2005) with a greater risk of soft-tissue damage on the medial side of the foot. Thus, motion control and stability are important for a flexible, low-arched foot when selecting a shoe.

TABLE 3.1
Arch Length to Foot Length Ratio for Adults in Different Ethnic Groups (Balanced Standing)

Ethnic Groups	Number of Test Participants	Mean Age (Years)	Mean Foot Length (FL) (mm)	Mean Arch Length (AL) (mm)	AL/FL Ratio (%)	Source
Males						
Japanese	478	34.9	247.9	180.8	72.9	Kouchi (1998)
Australian Aborigines	33	N/A	259.1	188.6	72.8	Kouchi (1998)
French	31	36.4	264.2	191.3	72.4	Kouchi (1998)
Indonesians	50	33.7	248.6	179.1	72.0	Kouchi (1998)
Hong Kong Chinese	31	22.1	252.0	183.4	72.7	Goonetilleke et al. (1997)
Mainland Chinese	49205	N/A	251.3	181.8	72.3	Xing et al. (2000)
American	5574	N/A	268.4	192.6	71.8	U.S. Armored Medical Research Lab (1946)
Females						
Japanese	410	33.7	227.4	165.9	73.0	Kouchi (1998)
Australian Aborigines	32	N/A	238.6	173.2	72.6	Kouchi (1998)
French	31	36.9	237.4	171.6	72.3	Kouchi (1998)
Indonesians	32	32.3	230.0	166.6	72.4	Kouchi (1998)
Hong Kong Chinese	15	22.1	234.8	171.5	73.0	Xiong et al. (2009)
Mainland Chinese	34425	N/A	232.1	168.3	72.5	Xing et al. (2000)
American	N/A	N/A	N/A	N/A	N/A	N/A

N/A, Not applicable.

Often a straight- or mild semi-curve-lasted shoe will be most appropriate. Low-arched feet are common in infants, children, and about 25% of adults (Cavanagh and Rodgers, 1987; Staheli et al., 1978).

2. *The normal-arched foot* has an imprint with a flare (Goonetilleke and Luximon, 1999), and the forefoot and heel are connected by a wide band. The normal foot would land on the outside of the heel and would roll slightly inward to absorb shock.

3. *The high-arched foot* has an imprint without a band or with a very narrow band connecting the forefoot and heel regions. This type of foot is at an increased risk of injuring the bony structures on the lateral side of the foot (over-supinated). High-arched feet tend to have smaller areas for

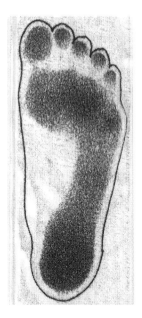

FIGURE 3.3 Footprint collected during balanced standing.

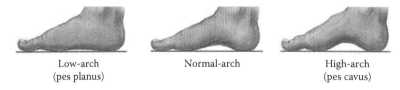

| Low-arch (pes planus) | Normal-arch | High-arch (pes cavus) |

FIGURE 3.4 Low-, normal-, and high-arched type feet.

weight-bearing and also tend to be more rigid, thereby transmitting higher stresses to the foot and leg (Özgü, 2005). Thus, shock absorption is a prime concern for a rigid, high-arched foot when selecting a shoe. Often a curve-lasted shoe with a high level of cushioning and shock absorption (Goonetilleke, 1999) will be most appropriate for those with a high-arch.

3.4 FOOT TYPE CLASSIFICATION TECHNIQUES AND MEASURES

Many different methods are available to classify the arches even though there is no agreement on which is best (Razeghi and Batt, 2002; Xiong et al., 2010). Most of the available metrics belong to one of three types (Table 3.2) (Razeghi and Batt, 2002):

- Anthropometric indices
- Radiographic indices
- Footprint indices

TABLE 3.2

Foot Type Classification Techniques and Representative Indices

Technique Category	Representative Indices for Foot Type Classification	Related References
Anthropometric	NH	Nigg et al. (1993), Williams and McClay (2000)
	NH/FL	Mall et al. (2007), Williams and McClay (2000)
	ND	Billis et al. (2007), Vinicombe et al. (2001)
	Navicular drift	Billis et al. (2007), Vinicombe et al. (2001)
	Midfoot dorsum angle	Xiong et al. (2010)
Radiographic	Calcaneal-first metatarsal angle	Razeghi and Batt (2002), Simkin et al. (1989), Smith et al. (1986)
	Calcaneal inclination angle	Menz et al. (2008), Younger et al. (2005)
	Rearfoot–forefoot angle	Freychat et al. (1996), Razeghi and Batt (2002)
Footprint	AI	Cavanagh and Rodgers (1987), Shiang et al. (1998)
	FI	Hawes et al. (1992), Shiang et al. (1998)
	MAI	Chu et al. (1995), Xiong et al. (2010)

3.4.1 ANTHROPOMETRIC INDICES

Anthropometric approaches involve measuring the locations of anatomical land-marks or feature points within the foot. These include the navicular height (NH), the ratio of the NH to foot length (NH/foot length), the navicular drop (ND), the navicular drift, the midfoot dorsal angle, the longitudinal arch angle, the dorsum height/foot length, and so on. A few representative indices are introduced below and their values are summarized in Table 3.3.

- *Navicular height*: The height of the prominent navicular bone measured from the supporting surface is the NH. This measure is probably the most popular anthropometric index, possibly because it has very direct physical meaning. Its reliability tends to be high (ICC > 0.95 for both intra-rater and inter-rater reliabilities) when using digital calipers (Hawes et al., 1992). Even though this measure has excellent face validity, a shortcoming is that it does not account for differences in foot length (Xiong et al., 2010).
- *Navicular height/foot length*: A larger foot may have a larger NH. Hence, Williams and McClay (2000) normalized the NH with respect to foot length. They showed that the normalized metric has improved reliability and validity. Saltzman et al. (1995) too found that this ratio is strongly cor-related with the radiographic measurements of the MLA structure.
- *Navicular drop*: The sagittal displacement of the navicular tuberosity from a non-weight-bearing to half-body-weight-bearing state of the foot. This metric has gained increasing acceptance in recent years due to its high face validity. However, the reliability of the variable has only been mod-erate due to difficulties in controlling half body weight, assessing small

TABLE 3.3
Previously Reported Data (Mean [SD]) on Anthropometric Indices of Foot Arch (Bipedal Stance; BW, Body Weight)

Anthropometric Indices	Williams and McClay (2000)	Vinicombe et al. (2001)	Mall et al. (2007)	Billis et al. (2007)	Shrader et al. (2005)	Jonely et al. (2011)	Queen et al. (2007)
NH[a] (cm)	4.0 (0.6) and 3.5 (0.6) at 10% and 90% BW	N/A[b]	3.8 (0.73) and 3.8 (0.74) for sessions 1 and 2 (at 90% BW)	N/A	3.6 (0.77) and 3.6 (0.74) for sessions 1 and 2	N/A	3.7 (0.7)
NH/FL (%)	16.4% (2.5%) and 14.2% (2.6%) 10% and 90% BW	N/A	15.1% (2.7%) and 14.9% (2.8%) for sessions 1 and 2 (at 90% BW)	N/A	N/A	N/A	14.4%[c]
ND (mm)	N/A	9.5 (2.5)	N/A	11.0 (3.4)	8.4 (5.3) and 8.3 (5.2) for sessions 1 and 2	7.0 (5.0)	N/A
Navicular drift (mm)	N/A	7.0 (3.0)	N/A	10.1 (3.2)	N/A	6.6 (3.8)	N/A

a NH is measured at half body weight unless specified otherwise.
b N/A: Not applicable.
c Calculated based on the available NH and foot length data.

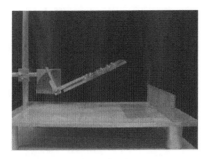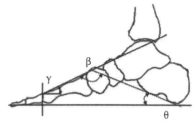

FIGURE 3.5 Jig for measuring midfoot dorsal angle (γ) and its representation. (Adapted from Xiong, S. et al., *J. Am. Podiatr. Med. Assoc.*, 100, 14, 2010. With permission.)

displacements, and the difficulty in judging the subtalar joint objectively. (Brody, 1982; Gross et al., 1991).

- *Navicular drift*: Navicular drift, which quantifies frontal plane movements and complements the ND (Billis et al., 2007). This measure is not considered to be too useful as the displacement is too subtle to be detected and is measurement error prone (Billis et al., 2007; Razeghi and Batt, 2002).
- *Midfoot dorsum angle*: γ is the angle between the midfoot dorsal plane from the head of first metatasophalangeal joint (MPJ) and the foot supporting plane (Figure 3.5) (Xiong et al., 2010). This angle is quite easy to measure and can be very useful in clinical settings. The intra-rater reliability of this index is relatively high (ICC = 0.94). The data of 48 subjects (24 males, 24 females) have shown that mean midfoot dorsum angle is 25.1° (SD = 2.6°) and 24.6° (SD = 2.4°) for males and females, respectively. An angle larger than 26° can be categorized as high arched and low arched if γ is smaller than 22°.

3.4.2 RADIOGRAPHIC INDICES

Radiographic imaging and measurement is a useful validation tool to evaluate anthropometric or footprint indices (Razeghi and Batt, 2002; Smith et al., 1986). The calcaneal-first metatarsal angle (β) and calcaneal inclination angle (θ) are widely used for this purpose (Figure 3.5). Their representative values are summarized in Table 3.4.

- *Calcaneal-first metatarsal angle*: Menz et al. (2008) defined the calcaneal-first metatarsal angle (β) as the "angle subtended between the tangent to the inferior surface of the calcaneus and the line drawn along the dorsum of the midshaft of the first metatarsal" (Figure 3.5). A larger calcaneal-first metatarsal angle would indicate a flatter foot. Using radiographs of 100 patients with foot problems, Saltzman et al. (1995) reported high intra-rater and inter-rater reliabilities (correlation coefficients are larger than 0.9); even higher intra-rater reliability (ICC = 0.99) have been reported by Wearing et al. (2004).

TABLE 3.4
Previously Reported Data (Mean [SD]) on Radiographic Angle Parameters of Foot Arch (Half-Body-Weight-Bearing)

Radiographic Angle Indices	Menz et al. (2008)	Younger et al. (2005)	Villarroya et al. (2009)	Murley et al. (2009)	Yalcin et al. (2010)
Calcaneal-first metatarsal angle (β,°)	132.5 (7.5) and 133.8 (8.9) for Calcaneal spur absent and present groups	N/A	N/A	141.7 (6.7), 132.8 (4.0), and 129.0 (7.7) for flat-, normal-, and high-arched groups	144[a]
Calcaneal inclination angle (θ,°)	21.2 (5.1) and 20.4 (6.1) for Calcaneal spur absent and present groups	20.8 (4.1) and 15.4 (8.4) for control and flatfoot group	14.2 (4.8) and 15.5 (4.9) for obese Spanish boys and girls (right foot)	16.1 (5.0), 20.9 (3.4), and 24.9 (4.9) for flat-, normal-, and high-arched groups	41 (6.9)

[a] This angle was derived from the geometry and the available angle data in the paper.

- *Calcaneal inclination angle*: The calcaneal inclination angle (θ in Figure 3.5) is the angle between the tangent to the inferior aspect of the calcaneus and the surface on which the foot rests as seen in the lateral view (Simkin et al., 1989). This angle determines the alignment of the hindfoot; it is high in a high-arched foot and low with flat feet (Yalcin et al., 2010). The angle is generally measured together with the calcaneal-first metatarsal angle (β) from radiographs (Table 3.4). Menz and Munteanu (2005) reported high reliability (ICC $\geq$ 0.98) for this measurement.
- *Rearfoot–forefoot angle*: The angle between the forefoot and rearfoot axes is defined as the rearfoot–forefoot angle α (Figure 3.6) (Freychat et al., 1996) and measured on an x-ray. The rearfoot axis is the line joining the posterior calcaneus (REC) and the calcaneocuboid joint (CCJ). The forefoot axis is formed by CCJ and the center of the second and third metatarsal bones (MM). The angle in static and dynamic conditions is denoted as α_S and α_R, respectively (Freychat et al., 1996; Razeghi and Batt, 2002). A positive α_R is a medial rotation of the forefoot relative to the rearfoot (closed foot), and a negative α_R is a lateral rotation or an "open foot" type (Razeghi and Batt, 2002). The difference between α_S and α_R, that is, ($\alpha_R - \alpha_S$), can represent the arch deformation (Razeghi and Batt, 2002).

3.4.3 FOOTPRINT INDICES

There is no doubt that radiographic measurements have high face validity, high reliability, and potentially high correlation with injury. The downside is the high cost and the amount of radiation exposure associated with its measurement. An alternative to

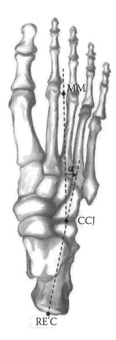

FIGURE 3.6 Forefoot–rearfoot angle from foot radiograph. (Reproduced and adapted from *Gait Posture*, 15, Razeghi, M. and Batt, M.E., Foot type classification: A critical review of current methods, Figure 4A, 282–291, Copyright 2002, with permission from Elsevier.)

radiographic measurements is the footprint technique, which has been used since the early 1920s (Chu et al., 1995). Recent developments in pressure transducer technology have made this technique more ubiquitous. In its simplest form, an ink or digital footprint is obtained when standing (Figure 3.7). The footprint is then processed to obtain a range of measures to characterize the arch. These are arch index (AI), modified arch index (MAI), Staheli index, Chippaux-Smirak Index, arch or footprint angle, footprint index (FI), the truncated AI, and Brucken index. A fundamental assumption in all these measures is that the structure of the arch is related to the footprint.

The three common measures are the AI, FI, and MAI. Their representative values are summarized in Table 3.5.

- *Arch index*: Cavanagh and Rodgers (1987) proposed this measure and it is based on the division of the footprint excluding the toe area. It is calculated as the middle area to the total area of a footprint when excluding the toes: AI = B2/(B1 + B2 + B3) (Figure 3.7). AI is one of the most commonly used measures for foot classification. Cavanagh and Rodgers (1987) reported that the arch index from an inked footprint (AIF) acquired with half body weight has a high within-day and between-day measurement reliability. Based on the distribution of AI values of 107 American participants, they classified the foot as high arch: AI ≤ 0.21; low arch: AI ≥ 0.26; and normal arch: 0.21 < AI < 0.26. Xiong et al. (2010) also reported a relatively high intra-rater reliability (ICC = 0.96) for 48 Hong Kong Chinese

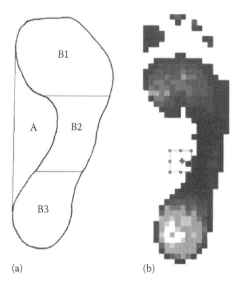

(a)   (b)

FIGURE 3.7 Illustration of arch-related parameters (arch index = B2/(B1 + B2 + B3), FI = A/(B1 + B2 + B3)) from a traditional inked footprint (a) and a sample of the electronic footprint (b).

TABLE 3.5
Previously Reported Data (Mean [SD]) on Footprint Parameters for Foot Arch

Footprint Indices[a]	Cavanagh and Rodgers (1987)	Chu et al. (1995)	Shiang et al. (1998)	Murley et al. (2009)	Xiong et al. (2010)
AI	0.230 (0.046)	0.213 (0.066)	0.210 (0.067)	0.240 (0.04) for normal-arched group	Males: 0.241 (0.056); females: 0.235 (0.035)
FI	N/A	N/A	0.262 (0.086)	N/A	Males: 0.308 (0.110); females: 0.327 (0.084)
MAI	N/A	0.144 (0.072)	0.138 (0.062)	N/A	Males: 0.101 (0.063); females: 0.105 (0.059)

[a] All from traditional ink footprints except modified arch index, which is from digital footprints.

subjects. Interestingly, the first (Q1) and third quartiles (Q3) of the AI distribution (Figure 3.8) from the Hong Kong participants were 0.217 and 0.261, respectively. These values are quite comparable with the range of 0.21–0.26 proposed by Cavanagh and Rodgers, indicating that the AI values are possibly not dependent on the population tested. However, it should be noted that even though AI from the digital footprint was highly correlated (Pearson correlation coefficient R = 0.84) with AIF, the AI of the

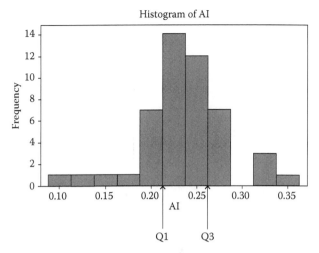

FIGURE 3.8 The distribution of AIF with the first quartile (Q1) and the third quartile (Q3) to classify the foot arch types (AI < Q1 is high arched, Q1 ≤ AI < Q3 is normal arched, AI ≥ Q3 is low arched). (Adapted from Xiong, S. et al., *J. Am. Podiatr. Med. Assoc.,* 100, 14, 2010. With permission.)

digital print was significantly smaller (Xiong et al., 2010). Hence, the criterion of using AI derived from a digital footprint should be different. Thus, Chu et al. (1995) proposed AI < 0.178 for high arch, AI > 0.250 for low arch, and 0.178 ≤ AI ≤ 0.250 for normal arch.

- *Footprint index*: Initially designed to study flat-footedness of children, the footprint index is defined as the ratio of the noncontact area to the contact area of a footprint when excluding the toe areas (Figure 3.7), that is, FI = A/(B1 + B2 + B3). Hawes et al. (1992), Shiang et al. (1998), and Xiong et al. (2010) all reported a significant and moderate to high (R = −0.55 to −0.89) negative correlation between FI and AI. The negative correlation is because a high-arched foot would have less contact area with the ground, especially in the midfoot region; thus, AI will be smaller and FI will be larger.
- *Modified arch index*: Chu et al. (1995) proposed the MAI that incorporates the force values rather than areas (Figure 3.7). So MAI accounts for the force supported in the middle third of the foot relative to that on the foot. Feet are then classified as follows: MAI < 0.093 for high arch, MAI > 0.178 for low arch, and 0.093 ≤ MAI ≤ 0.178 for normal arch (Chu et al., 1995). MAI has been shown to be correlated with AIF (R = 0.76) and with AI from a digital footprint (R = 0.89) (Xiong et al., 2010).

3.4.4 CORRELATION ANALYSIS AND FACTOR ANALYSIS OF DIFFERENT FOOT ARCH INDICES

The relations among arch height index (AHI), the NH to arch length ratio, the AI, the FI, the subjective ranking (SR), the MAI, the malleolar valgus index (MVI), and the midfoot dorsal angle have been reported by Chu et al. (1995), Queen et al. (2007),

TABLE 3.6
Pearson Correlation Coefficient among the Different Metrics
from Various Studies

	AIF	FI	MAI	NH	ND
AIF	1				
Inked FI	−0.55 (Shiang et al., 1998); −0.89 (Xiong et al., 2010)	1			
MAI	0.94 (Shiang et al., 1998) 0.76 (Xiong et al., 2010)	−0.59 (Shiang et al., 1998) −0.67 (Xiong et al., 2010)	1		
NH	−0.70 (Chu et al., 1995)	0.44 (Queen et al., 2007)	−0.71 (Chu et al., 1995)	1	
	−0.42 (Queen et al., 2007)	0.60 (Queen et al., 2007)[a]	−0.34 (Xiong et al., 2010)[b]		
	−0.61 (Queen et al., 2007)[a] −0.66 (Xiong et al., 2010)[b]	0.69 (Xiong et al., 2010)[b]			
ND	0.32 (Billis et al., 2007)	N/A	N/A	N/A	1

[a] NNH (= NH/FL) was used in the cited study.
[b] NNH (= NH/arch length) was used in the cited study.

Shiang et al. (1998), and Xiong et al. (2010) (Table 3.6). Xiong et al. (2010) found that correlations among most parameters (Table 3.7) were significant ($p < 0.05$), except all measures with MVI and between MAI and AHI. In particular, the AI from the traditional ink-footprint method (AIF) shows moderate-to-strong correlations (ranging from 0.52 to 0.89) with all parameters except MVI.

A factor analysis using the principal component method with varimax rotation has shown three dominant groups that explain 87.8% of the variance (Table 3.8) (Xiong et al., 2010). The first group is dominated by the area-related measures of AIF, arch index from Fscan pressure (AIP), FI, and the force-related parameter, MAI; the second group includes the foot dimension-related measures of AHI, normalized navicular height (NNH), and the midfoot dorsal angle (α); the third group includes the foot posture–related parameter and MVI. The grouping clearly explains the significance and the potential similarities and differences among the indices and can form the basis for clinical evaluations.

3.5 FOOT SENSATIONS

3.5.1 Tactile Sensation

Sensation of touch spans from feeling to pain. The sensitivity to touch or the tactile detection threshold (TDT) has been measured at various locations of the foot (Figure 3.9) (Hodge et al., 1999; Rosenbaum et al., 2006). Generally, TDTs are obtained using the Semmes–Weinstein monofilaments (SWMF) having different force levels (Figure 3.10).

TABLE 3.7

Pearson's Correlation Coefficient, R, among the Different Metrics

	AHI	NNH	AIF	AIP	FI	MAI	SR	MVI	Angle α
AHI	1								
NNH	**0.767**	1							
AIF	-0.520	**-0.664**	1						
AIP	-0.323	-0.472	**0.841**	1					
Inked FI	0.436	**0.688**	**-0.892**	**-0.760**	1				
MAI	-0.273[a]	-0.342	**0.759**	**0.893**	**-0.674**	1			
SR	-0.633	**-0.740**	**0.713**	0.517	**-0.716**	0.516	1		
MVI	-0.427	-0.422	0.290[a]	-0.033[a]	-0.221[a]	-0.131[a]	0.465	1	
Midfoot dorsal angle (α)	**0.814**	**0.682**	**-0.723**	-0.550	0.628	-0.522	**-0.723**	-0.353	1

Source: Xiong, S. et al. *J. Am. Podiatr. Med. Assoc.*, 100, 14, 2010. With permission.

R > 0.65 is shown in bold.

[a] Correlation is not significant at significance level of 0.05.

TABLE 3.8

Factor Analysis of Nine Metrics Using the Principal Component Method with Varimax Rotation

	Factor								
	1	2	3	4	5	6	7	8	9
AHI	−0.13	**0.92**	−0.20	0.26	−0.15	0.06	−0.10	−0.02	0.00
NNH	−0.24	**0.50**	−0.21	**0.74**	−0.24	0.22	0.03	−0.03	0.01
AIF	**0.74**	−0.30	0.21	−0.21	0.19	−0.38	−0.08	0.31	−0.01
AIP	**0.94**	−0.16	−0.04	−0.17	0.09	−0.15	−0.04	0.00	−0.20
Inked FI	**−0.60**	0.20	−0.12	0.29	−0.26	**0.66**	0.05	−0.01	0.00
MAI	**0.94**	−0.15	−0.13	−0.02	0.19	−0.07	−0.03	−0.04	0.20
SR	0.36	−0.38	0.28	−0.25	**0.74**	−0.20	−0.05	0.03	0.00
MVI	−0.08	−0.21	**0.96**	−0.12	0.14	−0.06	−0.02	0.02	0.00
Midfoot dorsal angle (α)	−0.38	**0.76**	−0.17	0.10	−0.24	0.18	0.39	−0.04	0.06
Variance explained by each factor	5.59	1.68	0.63	0.39	0.28	0.20	0.09	0.07	0.06
% Variance explained by each factor	62.10%	18.68%	7.04%	4.33%	3.12%	2.24%	1.03%	0.77%	0.69%
Cumulative % variance	62.10%	80.78%	87.82%	92.15%	95.3%	97.51%	98.54%	99.31%	100.0%

Source: Xiong, S. et al., *J. Am. Podiatr. Med. Assoc.*, 100, 14, 2010. With permission.
Factor loadings ≥0.5 are shown in bold.

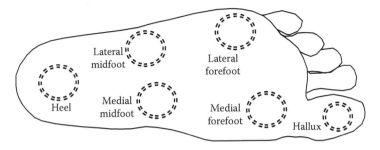

FIGURE 3.9 Locations at which tactile sensitivity has been measured.

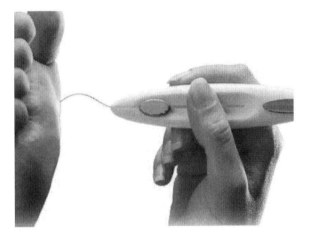

FIGURE 3.10 Procedure of measuring TDT using SWMF.

Dohi et al. (2003) reported the TDT of 17 test locations on feet of healthy adults and found that the foot is least sensitive in the plantar region except for the plantar arch. They reported that the dorsal region was the most sensitive. McPoil and Cornwall (2006) found that males generally have a higher TDT than females except on the hallux (Table 3.9).

3.5.2 RECEPTORS AND GATE CONTROL THEORY

There has been little research exploring the transformation of touch to discomfort or pain. Pain is a means to warn us of potentially damaging situations to help us avoid cuts, burns, broken bones, and so on. When the pressure that acts on the body is excessive, people experience pain and discomfort (Gonzalez et al., 1999). Perl (1985) stated that "Pain has attributes of a sensation, yet its usual capacity to make us uncomfortable or to suffer distinguishes it from other sensations." Pain is defined as "an unpleasant sensory and emotional experience associated with actual or potential tissue damage, or described in terms of such damage" (Merskey, 1986).

TABLE 3.9

Average Monofilament Tactile Detection Thresholds of the Plantar Foot for Males and Females

	Male (N1 = 20)			Female (N2 = 20)		
	Tactile Force Threshold[a] (g)	Filament Diameter (mm)[b]	Tactile Pressure Threshold (kPa)	Tactile Force Threshold (g)	Filament Diameter (mm)	Tactile Pressure Threshold (kPa)
Heel	2.052	0.305	27.5	0.696	0.203	21.1
Lateral midfoot	0.408	0.178	16.1	0.166	0.158	8.3
Medial midfoot	0.408	0.178	16.1	0.166	0.158	8.3
Lateral forefoot	0.696	0.203	21.1	0.408	0.178	16.1
Medial forefoot	0.696	0.203	21.1	0.408	0.178	16.1
Hallux	0.408	0.178	16.1	0.408	0.178	16.1

Source: McPoil, T.G. and Cornwall, M.W., *Foot*, 16, 192, 2006.

[a] Tactile force threshold is from Tables 1 and 3 in McPoil and Cornwall (2006).

[b] Tactile pressure threshold = Tactile force threshold/Area of filament.

Varying stimuli are sensed through the activation of differing receptors: mechanoreceptors for sensing mechanical displacement, thermoreceptors for temperature changes, and nociceptors for sensing pain. The mechanoreceptors that respond to mechanical stimulation of pressure, stretching, and vibration lie on the external layer (epidermis) and the layer below (dermis) it. Four types of mechanoreceptors are responsible for tactile perception:

1. The Merkel receptor (for light touch) is located near the border between the epidermis and the dermis.
2. The Meissner corpuscle (for light touch) is in the dermis.
3. The Ruffini cylinder (for sensing heat) is located in the dermis.
4. The Pacinian corpuscle (for deep pressure) is located deep in the skin.

The mechanoreceptors differ in the way they respond to stimulation. For example, the Merkel discs and the Ruffini cylinders respond when a stimulus is present without much decrease in its firing rates and are called slowly adapting (SA) fibers. The fibers associated with Meissner corpuscles and Pacinian corpuscles respond with a sudden burst at the onset of a stimulus and are thus called rapidly adapting (RA) fibers. An important property of these mechanoreceptors is the receptive field, which is the area of skin when stimulated directly affects the firing rate of neurons. The Merkel (SA1) and Meissner (RA1) receptors have small receptive fields, while the Ruffini endings (SA2) and the Pacinian corpuscles (RA2) respond over a larger area and thus have larger receptive fields. The receptors with the small receptive areas are located close to the surface of the skin, whereas the fibers with the larger receptive areas (SA2 and RA2) are located deeper in the skin. As each mechanoreceptor has its

TABLE 3.10
Details of Different Nerve Fibers

Type of Fiber	Myelinated/ Insulation	Size (μm)	Transmission Speed (m/s)	Specific Stimuli
C-fiber	Unmyelinated	0.25–1.5	0.25–1.25	Mechanical, thermal, chemical
Aδ	Thinly myelinated	1–5	6–30	Mechanical, thermal, chemical
Aβ	Heavily myelinated	5–15	30–100	Touch

Source: Britton, N.F. and Skevington, S.M., *J. Theor. Biol.*, 137, 91, 1989.

specialized response, any given stimulation on the skin will generally activate many of these types, some stronger than others, depending on the stimulus. For example, if someone grasps a glass, the SA1 (Merkel) will fire to feel small details and the texture of the surface, while the SA2 (Ruffini) will fire as the hand stretches around the glass. The Meissner receptors also may fire depending on the texture. The overall perception will be determined by all of them (Roland, 1992). Each of these receptors and the free endings are connected to the spinal cord through fibers: small (Aδ and C-fiber) and large (Aβ) nerve fibers (Britton and Skevington, 1989). The properties of these nerve fibers are given in Table 3.10. It is also known that more stimulation on the small nerve fibers of the skin results in more pain (Hallin and Torebjork, 1973; van Hees and Gybels, 1972) and more stimulation of the large fibers, results in a temporal increase of pain, but may have a relieving effect in the long term (Chapman et al., 1976; Wall and Sweet, 1967). Melzack and Wall (1965) proposed the gate theory of pain to explain these phenomena. According to that theory, the large fibers close the "gate" or inhibit the transmission while the small fibers open the "gate" or facilitate the transmission.

Britton and Skevington (1989) have mathematically modeled the neurotransmission of pain from a molecular biology point of view (Figure 3.11).

3.5.3 PSYCHOPHYSICS OF PRESSURE SENSATION

Pressure-related pain is generally measured as the pressure pain threshold (PPT) and/or pressure discomfort threshold (PDT), where PPT is defined as the minimum pressure that induces pain. PDT is the minimum pressure that induces some discomfort (Fischer, 1987). It is known that discomfort precedes pain, and hence, PDT is lower than PPT (Gonzalez et al., 1999; Goonetilleke, 2001). Pressure thresholds are generally measured using a pressure algometer (Gonzalez et al., 1999; Goonetilleke and Eng, 1994; Kosek et al., 1993; Messing and Kilbom, 2001); however, some researchers have used custom equipment to have better control on some variables (Dohi et al., 2003; Fransson-Hall and Kilbom, 1993; Xiong et al., 2011). The thresholds tend to be different at varying sites with nerve tissues or sites over bones having a lower PPT than nearby muscles (Dohi et al., 2003; Gerecz-Simon et al., 1989; Kosek et al., 1993; Xiong et al., 2011). Pressure thresholds depend on the effective contact area. They decrease with an increase of stimulus size (Defrin et al., 2006; Goonetilleke and Eng, 1994; (Greenspan and McGills, 1991, 1997;

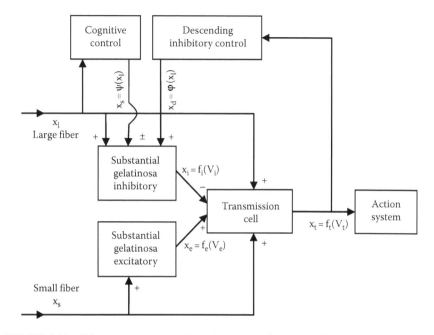

FIGURE 3.11 Schematic representation of conceptual models of gate control theory of pain perception mechanisms; + and − represent the excitation and inhibition, respectively. (Adapted from Britton, N.F. and Skevington, S.M., *J. Theor. Biol.*, 137, 91, 1989.) x is the firing frequency in each path; V is the potential of substantial gelatinosa (SG) or transmission cell (T-cell); φ and ψ are strictly monotonically increasing functions; and f is strictly a monotonically increasing function with zero for values below a certain potential.

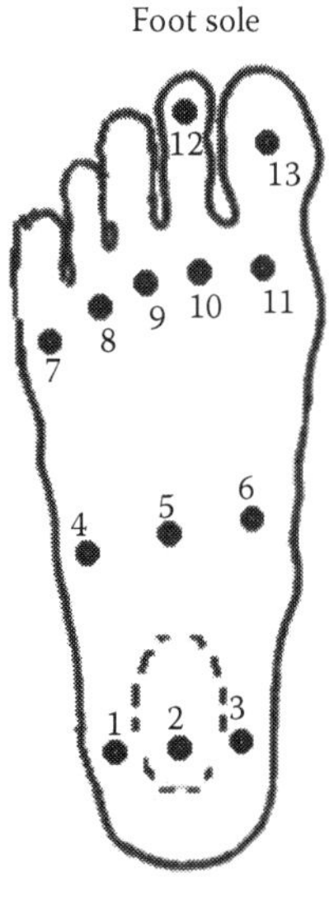

Location	PDT (kPa)
1	613
2	630
3	635
4	543
5	463
6	440
7	463
8	453
9	470
10	455
11	430
12	325
13	415

Foot sole

FIGURE 3.12 PDT variation on plantar foot.

Rodrigo et al., 2010; Xiong et al., 2011) and increase with an increase in the stimulus rate of change (Defrin et al., 2006; Rodrigo et al., 2010; Xiong et al., 2011) (Figure 3.12). Thresholds such as PPT and PDT can be measured using various sized probes. However, an important issue is whether the threshold would change when a nearby site is stimulated. Rodrigo (2010) has shown that an uneven load distribution on two nearby points does have an effect on the thresholds.

3.6 PRESSURE THRESHOLDS ON FOOT

Much of the literature related to PPT and PDT has shown that there are no differences between genders (Clausen and King, 1950; Fillingim et al., 1998; Gonzalez et al., 1999, Kenshalo, 1986; Rodrigo, 2010). In contrast, some of the pain-related studies report that there is a significant gender effect on pain thresholds, with females having lower thresholds than males (Brennum et al., 1989; Chesterton et al., 2003; Fillingim and Maixner, 1996; Fillingim and Ness, 2000). The difference may be attributed to varying locations, assessment methods, and the populations tested in the different studies. Several researches have studied the age-related changes in pain perception (Gagliese et al., 1999; Gibson and Helme, 2001) and most studies demonstrate that there is an age-related increase in the pain threshold. It was also reported that PPT increase is more noticeable in females than males (Jensen et al., 1992).

Gonzalez et al. (1999) tested the PDT variations on 13 locations of the plantar foot using a manual pressure algometer with an aluminum probe that had a cylindrical rounded edge of area of $1.3\,\text{cm}^2$. The highest PDT was in the heel area followed by the midfoot. The lowest was on the toes (Figure 3.12).

Xiong et al. (2011) investigated seven locations on the plantar foot and reported that the PPT and PDT decreased when the probe area increased, but increased when the indentation speed increased (Figure 3.13).

The highest to lowest PPT as reported by Xiong et al. (2011) was foot heel (P7), followed by the three locations under the metatarsal heads (P1, P2, P3), then the test location under the lateral plantar arch (P6), and finally, the foot center (P4) and medial plantar arch (P5). The PPT across the six test locations on the dorsum of foot showed that the dorsal point (P11) had the higher PPT when compared with the other five locations (Figure 3.14).

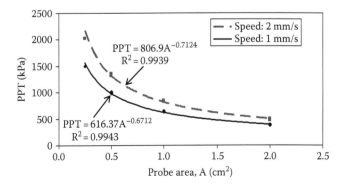

FIGURE 3.13 Variation of PPT on heel with probe area and indentation speed.

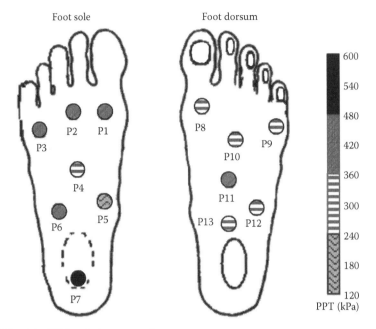

FIGURE 3.14 PPT variation across foot.

Rodrigo (2010) used dimensional analysis to show that

$$\text{PPT} = c\left(\frac{(V \cdot t)^2}{A}\right)^{\alpha} \quad \text{or} \quad \text{PPT} = c\left(\frac{d^2}{A}\right)^{\alpha}, \quad (r^2 \geq 0.9547),$$

where
  A is the stimulated area
  d is the tissue deflection
  V is the speed of indentation
  "c" is the coefficient related to the tissue and exponent
  α is possibly a representation of the spatial summation effect of pressure-related
    pain

It was found that the coefficient c for heel, medial arch, and third MPJ are 3115.8, 634.8, and 3441.5 kPa, respectively, and exponent α for these three locations are 0.63, 0.62, and 0.61, respectively. The model indicates that PPT increases with V and decreases with the increase of the probe area (A) as shown in Figure 3.12.

Xiong et al. (2012) evaluated the effects of area, A, and speed, V, using a different approach: PPT = $[a + b\ln(v)]A^{\beta}$, where β depends on location and gender (-0.55 for male and -0.63 for female on the heel), b depends on indentation speed v (103 and 70 for the hard and soft foot tissues of males and 79 and 60 for the hard and soft foot tissues of females for speeds of 1 and 2 mm/s, respectively), and a is highly correlated with foot tissue stiffness.

The differing phenomena related to area of stimulation, rate of stimulation, and multiple stimulation have been explained with theories such as spatial summation (Greenspan et al., 1997), temporal summation (Price et al., 1977; Sarlani and Greenspan, 2002), and spatial discrimination (Cain, 1973; Defrin et al., 2006).

3.6.1 SPATIAL SUMMATION OF PAIN

The decrease in pain threshold with increase in the area stimulated is referred to as the spatial summation of pain (Goonetilleke and Eng, 1994; Greenspan et al., 1997) and is applicable to heat stimuli as well (Nielsen and Arendt-Nielsen, 1997; Price et al., 1989). This behavior has been explained with the following two neural nociceptive hypotheses:

1. *Local integration*, in which individual neurons integrate nociceptive information from varying receptors (Figure 3.15)
2. *Neuron recruitment*, in which the number of neurons recruited increase with an increase in the stimulation area (Figure 3.15; Price et al., 1989)

It has been reported that the receptive field of a human muscle nociceptor is ~5 cm² (Marchettini et al., 1996), and thus, local integration mainly involves a smaller stimulus area whereas neuron recruitment might occur with a larger stimulus area (Nie et al., 2009). These two mechanisms are not mutually exclusive and both are probably necessary for spatial summation of pain (Price et al., 1989).

3.6.2 SPATIAL DISCRIMINATION OF PAIN

Spatial discrimination is the ability to discriminate and separate nociceptive input (Defrin et al., 2006). In contrast to spatial summation of pain, spatial discrimination has received less attention, and thus, there is a lack of literature on spatial discrimination in pressure pain. Nielsen and Arendt-Nielsen (1997), and Defrin et al. (2006) reported that the ability to perform two-point discrimination gradually improves with an increase of heat stimuli separation (Figure 3.16).

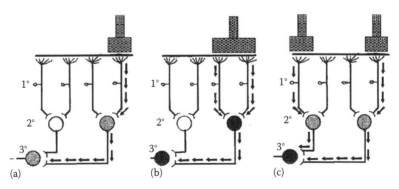

FIGURE 3.15 Neural mechanism of spatial summation of pain: (a) nociceptor stimulation, (b) local integration, and (c) neuron recruitment. (Data from Price, DD, et al., *Journal of Neurophysiol*, 62, 1270, 1989.)

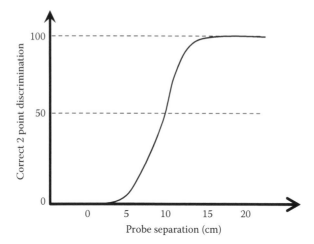

FIGURE 3.16 Effect of probe separation on the correct identification of the number of probes. (Data from Defrin, R et al., *Pain*, 126, 123, 2006.)

3.7 FOOT TISSUE DEFORMATION

Foot tissue deformation is important in the development of athletic and orthopedic shoes (Klaesner et al., 2001). A parameter that characterizes tissue is its stiffness, which is defined as the relationship between the applied force and the resultant tissue deformation. Typically, the force–deformation curve is nonlinear (Figure 3.17) with a low initial stiffness and a high stiffness at large forces.

Soft tissue parameters are different at differing locations, but no differences have been reported between genders (Dohi et al., 2003; Rodrigo, 2010). Typical force–deformation curves under heel, third MPJ, and the medial arch are shown in Figure 3.18.

The nonlinear pattern of tissue deformation (loading or unloading) can be approximated as two piecewise linear segments (Makhsous et al., 2007) of stiffness, K^I and K^{II} (Figure 3.19).

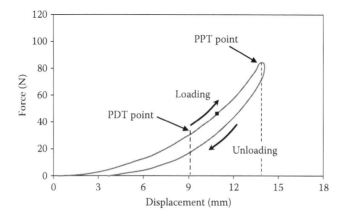

FIGURE 3.17 Typical force–displacement curve on the plantar foot.

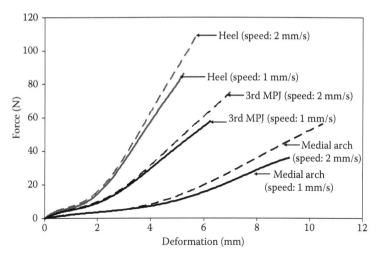

FIGURE 3.18 Illustration of typical tissue deformation pattern on plantar foot.

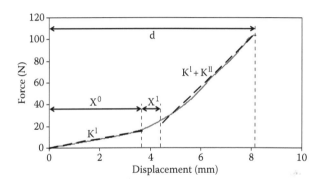

FIGURE 3.19 Representation of revised model parameters on force–deformation curve.

TABLE 3.11

Tissue Deformation Model Parameters under Heel, Medial Arch, and Third MPJ

Location	Speed (mm/s)	d (mm)	X^0 (mm)	X^1 (mm)	K^1 (N/mm)	K^{II} (N/mm)
Heel	1	7.372261	2.351277	0.325075	5.346888	18.71787
	2	7.729851	2.566141	0.681481	5.021233	19.16146
Medial arch	1	12.35822	1.935981	0.220773	1.729678	6.51748
	2	12.88778	1.801367	0.783226	1.490309	7.09558
Third MPJ	1	8.305468	3.257849	0.384768	1.998927	11.96125
	2	8.978034	2.737982	1.321071	1.984804	12.89224

Source: Rodrigo, A.S, Pressure related spatial summation effects on plantar foot, PhD thesis, Industrial Engineering and Logistic Management, Hong Kong University of Science and Technology, Hong Kong, 2010.

This piecewise linear model can be mathematically represented as

$$F = K^1 x \qquad\qquad \text{for } 0 \le x \le X^0$$

$$F = K^1 x + K^{11}(x - X^0 - X^1) \quad \text{for } (X^0 + X^1) \le x$$

where
X^0 is the displacement in the initial relatively low stiffness region
X^1 is the displacement in the transition region
d is the total deformation at PPT

The model parameters for the heel, medial arch, and third MPJ are shown in Table 3.11.

3.8 CONCLUSIONS

Each person's foot is unique and numerous parameters allow the quantification of shape, changes in shape in critical regions, and how pressure and heat are sensed. The various metrics allow the foot to be evaluated so that tissue-compliant footwear can be designed and generated to match the various characteristics.

ACKNOWLEDGMENTS

The authors would like to thank the Research Grants council for supporting the numerous studies under the General Research Fund of the Research Grants Council of Hong Kong through grants 613205, 613406, 613008, and 612711. The support from the National Research Foundation of Korea (NRF 2011-0022185), the Shanghai Pujiang Program (PJ [2009] 00861) and the Ministry of Education of China (20100073120046) is also appreciated.

QUESTIONS

3.1 What are the implications of the allometric model related to midfoot shape? Explain how the model can be used in footwear design.

3.2 How does foot shape change with the increased weight-bearing? Explain how to address those changes during footwear design and development.

3.3 What is the normal range for the ratio of arch length to foot length and what is its implication on footwear design?

3.4 Why is it important to reliably classify the feet into high, normal, and low arched? What techniques are available for classification and what are their advantages and disadvantages?

3.5 There are variations in the foot arch measures among the various studies. What are possible sources contributing to these variations?

3.6 How can the sensation characteristics of the foot be used in designing footwear?

3.7 What are the possible reasons that foot tissue have two distinct regions in the force deformation curve? Identify the differences in the tissue deformation characteristics in soft tissue and hard tissue.

3.8 How does PPT correlate with the tissue characteristics? What possibilities exist to artificially raise PPT at a given location?

3.9 Why are there differences in foot sensation and deformation characteristics in different groups?

REFERENCES

Billis E, Katsakiori E, Kapodistrias C, and Kapreli E (2007). Assessment of foot posture: Correlation between different clinical techniques. *The Foot*, 17: 65–72.

Brennum J, Kjeldsen M, Jensen K, and Jensen T (1989). Measurements of human pain-pressure thresholds on fingers and toes. *Pain*, 38(2): 211–217.

Britton, NF and Skevington SM (1989). A mathematical model of the gate control theory of pain. *Journal of Theoretical Biology*, 137: 91–105.

Brody DM (1982). Techniques in the evaluation and treatment of the injured runner. *Orthopedic Clinics of North America*, 13(3): 541–558.

Cabrera J, Tsui KL, and Goonetilleke RS (2004). A scale model for fitting object shapes from fixed location data. *IIE Transactions*, 36: 1099–1105.

Cain WS (1973). Spatial discrimination of cutaneous warmth. *American Journal of Psychology*, 89(1): 169–181.

Cavanagh PR and Rodgers MM (1987). The arch index: A useful measure from footprint. *Journal of Biomechanics*, 20: 547–551.

Chapman CR, Wilson ME, and Gehrig JD (1976). Comparative effects of acupuncture and transcutaneous stimulation on the perception of painful dental stimuli. *Pain*, 2(3): 265–283.

Cheng JCY, Leung SSF, Leung AKL, Guo X, Sher A, and Mak AFK (1997). Change of foot size with weightbearing: A study of 2829 children 3 to 18 years of age. *Clinical Orthopaedics and Related Research*, 342: 123–131.

Chesterton LS, Barlas P, Foster NE, Baxter GD, and Wright CC (2003). Gender differences in pressure pain threshold in healthy humans. *Pain*, 101(3): 259–266.

Chu WC, Lee SH, Chu W et al. (1995). The use of arch index to characterize arch height: A digital image processing approach. *IEEE Transactions on Biomedical Engineering*, 42: 1088–1093.

Clausen J and King HE (1950). Determination of the pain threshold on untrained subjects. *Journal of Psychology*, 30: 299–306.

Defrin R, Givon R, Raz N, and Urca G (2006). Spatial summation and spatial discrimination of pain sensation. *Pain*, 126(1–3): 123–131.

Dohi M, Mochimaru M, and Kouchi M (2001). Foot shape and shoe fitting comfort for elderly Japanese women. *Japanese Journal of Ergonomics*, 37(5): 228–237.

Dohi M, Mochimaru M, and Kouchi M (2003). The tactile sensitivity and elasticity of the sole of the foot as factors of shoe comfort. *Proceeding of IEA2003*, Seoul, South Korea, pp. 46–49.

Fillingim RB and Maixner W (1996). Gender differences in the responses to noxious stimuli. *Pain Forum*, 4(4): 209–221.

Fillingim RB, Maixner W, Kincaid S, and Silva S (1998). Sex differences in temporal summation but not sensory-discriminative processing of thermal pain. *Pain*, 75(1): 121–127.

Fillingim RB and Ness TJ (2000). Sex-related hormonal influences on pain and analgesic responses. *Neuroscience and Biobehavioral Reviews*, 24(4): 485–501.

Fischer AA (1987). Tissue compliance meter for objective, quantitative documentation of soft tissue consistency and pathology. *Archives of Physical Medicine and Rehabilitation*, 68(2): 122–125.

Fransson-Hall C and Kilbom A (1993). Sensitivity of the hand to surface pressure. *Applied Ergonomics*, 24(3): 181–189.

Freychat P, Belli A, Carret JP and Lacour JR (1996). Relationship between rearfoot and fore-foot orientation and ground reaction forces during running. *Medicine and Science in Sports and Exercise*, 28(2): 225–232.

Gagliese L, Katz J, and Melzack R (1999). Pain in the elderly. In *Textbook of Pain*, Wall P, Melzack R. (Eds.), Churchill Livingstone, New York.

Gerecz-Simon EM, Tunks ER, Heale JA, Kean WF, and Buchanan WW (1989). Measurement of pain threshold in patients with rheumatoid arthritis, osteoarthritis, ankylosing spondylitis, and healthy controls. *Clinical Rheumatology*, 8(4): 467–474.

Gibson S and Helme R (2001). Age-related differences in pain perception and report. *Clinics in Geriatric Medicine*, 17: 433–456.

Gonzalez JC, Garcia AC, Vivas MJ, Ferrus E, Alcantara E, and Forner A (1999). A new portable method for the measurement of pressure discomfort threshold on the foot plant. *Fourth Symposium of the Technical Group on Footwear Biomechanics*, August 5–7, Canmore, Canada, pp. 48–49.

Goonetilleke RS (1999). Footwear cushioning: Relating objective and subjective measurements. *Human Factors*, 41(2): 241–256.

Goonetilleke RS (2001). The comfort-discomfort phase change. In *International Encyclopedia of Ergonomics and Human Factors*, Karwowski W (Ed.), Taylor & Francis Group, Boca Raton, FL, pp. 399–402.

Goonetilleke RS and Eng T (1994). Contact area effects on discomfort. *Proceedings of the 38th Human Factors and Ergonomics Society Conference*, October 24–28, 1994, Nashville, TN, pp. 688–690.

Goonetilleke RS, Ho ECF, and So RHY (1997). Foot anthropometry in Hong Kong. *Proceedings of the ASEAN 97 Conference*, Kuala Lumpur, Malaysia, 1997, pp. 81–88.

Goonetilleke RS and Luximon A (1999). Foot flare and foot axis. *Human Factors*, 41(4): 596–607.

Goonetilleke RS and Luximon A (2001). Designing for comfort: A footwear application. *Proceedings of Computer-Aided Ergonomics and Safety Conference*, Maui, Hawaii (Plenary Session).

Greenspan JD, McGillis SLB. (1991) Stimulus features relevant to the preception of sharpness and mechanically evoked cutaneous pain. *Somatosensory and Motor Research*, 8: 137–147.

Greenspan JD, Thomadaki M, and McGillis SLB (1997). Spatial summation of perceived pressure, sharpness and mechanically evoked cutaneous pain. *Somatosensory and Motor Research*, 14(2): 107–112.

Gross ML, Davlin LB, and Evanski PM (1991). Effectiveness of orthotic shoe inserts in the long-distance runner. *American Journal of Sports Medicine*, 19: 409–412.

Hallin RG and Torebjork HE (1973). Electrically induced A and C fibre responses in intact human skin nerves. *Experimental Brain Research*, 16: 309–320.

Hawes MR, Nachbauer W, Sovak D et al. (1992). Footprint parameters as a measure of arch height. *Foot and Ankle*, 13: 22–26.

Hawes MR and Sovak D (1994). Quantitative morphology of the human foot in a north American population. *Ergonomics*, 37(7): 1213–1226.

Hawes MR, Sovak D, Miyashita M, Kang SJ, Yoshihuku Y, and Tanaka S (1994). Ethnic differences in forefoot shape and the determination of shoe comfort. *Ergonomics*, 37(1): 187–196.

van Hees J and Gybels JM (1972). Pain related to single afferent C fibers from human skin. *Brain Research*, 48: 397–400.

Hodge MC, Bach TM, and Carter GM (1999). Orthotic management of plantar pressure and pain in rheumatoid arthritis. *Clinical Biomechanics*, 14: 567–575.

Howard JS and Briggs D (2006). The arch-height-index measurement system: A new method of foot classification. *International Journal of Athletic Therapy and Training*, 11(5): 56–57.

Hughes J, Clark P, and Klenerman L (1990). The importance of the toes in walking. *Journal of Bone and Joint Surgery*, 72-B: 245–251.

Hutton WC and Dhanendran M (1979). A study of the distribution of load under the normal foot during walking. *International Orthopaedics*, 3: 130–141.

Jensen R, Rasmussen B, Pedersen B, Lous I, and Olesen J (1992). Cephalic muscle tenderness and pressure pain threshold in a general population. *Pain*, 48: 197.

Jonely H, Brismée J, Sizer P, and James C (2011). Relationships between clinical measures of static foot posture and plantar pressure during static standing and walking. *Clinical Biomechanics,* 26: 873–879.

Kenshalo, DR (1986). Somesthetic sensitivity in young and elderly humans. *The Journals of Gerontology*, 41(6): 732–742.

Klaesner JW, Hastings MK, Zou DQ, Lewis C, and Mueller MJ (2001). Plantar tissue stiffness in patients with diabetes mellitus and peripheral neuropathy. *Archives of Physical Medicine and Rehabilitation*, 83: 1796–1801.

Kosek E, Ekholm J, and Nordemar R (1993). A comparison of pressure pain thresholds in different tissues and body regions. Long-term reliability of pressure algometry in healthy volunteers. *Scandinavian Journal of Rehabilitation Medicine*, 25(3): 117–124.

Kouchi M (1998). Foot dimensions and foot shape: Differences due to growth, generation and ethnic origin. *Anthropological Science*, 106 (suppl): 161–188.

Kouchi M and Yamazaki N (1992). Allometry of the foot and the shoe last. *Journal of the Anthropological Society of Nippon*, 100(1): 101–118 (in Japanese).

Krauss I and Mauch M (2012). Foot morphology. In *Science of Footwear*, Goonetilleke RS (Ed.), CRC Press, Boca Raton, FL.

Lambrinudi C (1932). Use and abuse of toes. *Postgraduate Medical Journal*, 8: 459–464.

Lin CH, Chen JJ, Wu CH et al. (2004). Image analysis system for acquiring three-dimensional contour of foot arch. *Computer Methods and Programs in Biomedicine*, 75: 147–157.

Makhsous M, Rowles DM, Rymer WZ, Bankard J, Nam EK, and Chen D (2007). Periodically relieving ischial sitting load to decrease the risk of pressure ulcers. *Archives of Physical Medicine and Rehabilitation*, 88: 862–870.

Mall N, Hardaker WM, Nunley J, and Queen R (2007). The reliability and reproducibility of foot type measurements using a mirrored foot photo box and digital photography compared to caliper measurements. *Journal of Biomechanics*, 40: 1171–1176.

Marchettini P, Simone DA, Caputi G, and Ochoa JL (1996). Pain from excitation of identified muscle nociceptors in humans. *Brain Research,* 740: 109–116.

McCrory JL, Young MJ, Boulton AJM et al. (1997). Arch index as a predictor of arch height. *The Foot*, 7: 79–81.

McPoil TG and Cornwall MW (2006). Plantar tactile sensory thresholds in healthy men and women. *The Foot*, 16(4): 192–197.

Melzack R and Wall PD (1965). Pain mechanisms: A new theory. *Science*, 150: 971–979.

Menz HB and Munteanu SE (2005). Validity of 3 clinical techniques for the measurement of static foot posture in older people. *Journal of Orthopaedics & Sports Physical Therapy*, 35: 479–486.

Menz HB, Zammit GV, Landorf KB, and Munteanu SE (2008). Plantar calcaneal spurs in older people: Longitudinal traction or vertical compression? *Journal of Foot and Ankle Research*, 1: 7.

Merskey H (1986). Psychiatry and pain. In *The Psychology of Pain*, Sternback R. (Ed.), Raven, New York.

Messing K and Kilbom A (2001). Standing and very slow walking: Foot pain-pressure threshold, subjective pain experience and work activity. *Applied Ergonomics*, 32(1): 81–90.

Mochimaru M and Kouchi M (2003). Last customization from an individual foot form and design dimensions. *Sixth ISB Footwear Biomechanics,* Queenstown, New Zealand, pp. 62–63.

Murley GS, Menz HB, and Landorf KB (2009). A protocol for classifying normal- and flat-arched foot posture for research studies using clinical and radiographic measurements. *Journal of Foot and Ankle Research*, 2: 22.

Nie HL, Graven-Nielsen T, and Arendt-Nielsen L (2009). Spatial and temporal summation of pain evoked by mechanical pressure stimulation. *European Journal of Pain*, 13(6): 592–599.

Nielsen J and Arendt-Nielsen L (1997). Spatial summation of heat induced pain within and between dermatomes. *Somatosensory Motor Research*, 14(2): 119–125.

Nigg BM, Cole GK, and Nachbauer W (1993). Effects of arch height of the foot on angular motion of the lower extremities in running. *Journal of Biomechanics*, 26(8): 909–916.

Özgü HÖ (2005). A research on footwear and foot interaction through anatomy and human engineering, Master thesis, Izmir Institute of Technology, Izmir, Turkey.

Perl E (1985). Pain and pain management. In: *Research Briefings, National Academy of Sciences*, National Academy Press, Washington, DC, pp. 19–32.

Price DD, Hu JW, Dubner R, and Gracely RH (1977). Peripheral suppression of first pain and central summation of second pain evoked by noxious heat pulses. *Pain*, 3(1): 57–68.

Price DD, McHaffie JG, and Larson MA (1989). Spatial summation of heat induced pain: Influence of stimulus area and spatial separation of stimuli on perceived pain sensation intensity and unpleasantness. *Journal of Neurophysiology*, 62(6): 1270–1279.

Queen RM, Mall NA, Hardaker WM, and Nunley JA (2007). Describing the medial longitudinal arch using footprint indices and a clinical grading system. *Foot and Ankle International*, 28(4): 456–462.

Razeghi M and Batt ME (2002). Foot type classification: A critical review of current methods. *Gait and Posture*, 15: 282–291.

Rodrigo AS, Goonetilleke RS, and Xiong S (2010). Perception of pressure on foot. In *Advances in Ergonomics Modeling and Usability Evaluation*, Tareq ZA (Ed.), CRC Press, Boca Raton, FL, pp. 80–88.

Roland P (1992). Cortical representation of pain. *Trends in Neuroscience*, 15: 3–5.

Rosenbaum D, Schmiegel A, Meermeier M, and Gaubitz M (2006). Plantar sensitivity, foot loading and walking pain in rheumatoid arthritis. *Rheumatology*, 45: 212–214.

Rossi WA (1983). The high incidence of mismatched feet in population. *Foot and Ankle*, 4: 105–112.

Saltzman CL, Nawoczenski DA, and Talbot KD (1995). Measurement of the medial longitudinal arch. *Archives of Physical Medicine and Rehabilitation*, 76(1): 45–49.

Sarlani E and Greenspan JD (2002). Gender differences in temporal summation of mechanically evoked pain. *Pain*, 97(1–2): 163–169.

Shiang TY, Lee SH, Lee SJ et al. (1998). Evaluating different footprint parameters as a predictor of arch height. *Engineering in Medicine and Biology Magazine, IEEE*, 17(6): 62–66.

Shrader J, Popovich J, Gracey GC, and Danoff JV (2005). Navicular drop measurement in people with Rheumatoid Arthritis: Interrater and intrarater reliability. *Physical Therapy*, 85(7): 656–664.

Simkin A, Leichter I, Giladi M, Stein M, and Milgrom C (1989). Combined effect of foot arch structure and an orthotic device on stress fractures. *Foot and Ankle*, 19(11): 738–742.

Smith LS, Clarke TE, Hamill CL et al. (1986). The effects of soft and semi-rigid orthoses upon rearfoot movement in running. *Journal of the American Pediatric Medical Association*, 76(4): 227–233.

Staheli L, Chew D, and Corbett M (1987). The longitudinal arch. *Journal of Bone and Joint Surgery*, 69a: 71–78.

Tsung BYS, Zhang M, Fan YB, and Boone DA (2003). Quantitative comparison of plantar foot shapes under three weighting-bearing conditions. *Journal of Rehabilitation Research and Development*, 40(6): 517–526.

U.S. Armored Medical Research Laboratory (1946). Foot dimensions of soldiers. Project No. T-13: Survey of foot measurements of proper fit of army shoes, Fort Knox, KY.

Villarroya M, Esquivel J, Tomás C, Moreno L, Buenafé A, and Bueno G (2009). Assessment of the medial longitudinal arch in children and adolescents with obesity: Footprints and radiographic study. *European Journal of Pediatrics*, 168: 559–567.

Vinicombe A, Raspovic A, and Menz HB (2001). Reliability of navicular displacement measurement as a clinical indicator of foot posture. *Journal of the American Pediatric Medical Association*, 91(5): 262–268.

Wall PD (1978). The gate control theory of pain mechanisms: A re-examination and re-statement. *Brain*, 101(1): 1–18.

Wall PD and Sweet WH (1967). Temporary abolition of pain in man. *Science*, 155: 108–109.

Wearing SC, Hills AP, Byrne NM et al. (2004). The arch index: A measure of flat or fat feet? *Foot and Ankle International*, 25(8): 575–581.

Williams DS and McClay IS (2000). Measurements used to characterize the foot and the medial longitudinal arch. *Physical Therapy*, 80(9): 864–871.

Witana CP, Xiong S, Zhao J, and Goonetilleke RS (2006). Foot measurements from three-dimensional scans: A comparison and evaluation of different methods. *International Journal of Industrial Ergonomics*, 36(9): 789–807.

Xiong S, Goonetilleke RS, Rodrigo AS, and Zhao J (2012). A model for the perception of surface pressure on human foot. To appear in *Applied Ergonomics*.

Xing DH, Deng QM, Ling SL, Chen WL, and Shen DL (2000). *Handbook of Chinese Shoe Making: Design, Technique and Equipment*. Press of Chemical Industry, Beijing, China (in Chinese).

Xiong S, Goonetilleke RS, Witana CP, and Au E (2008). Modelling foot height and foot shape-related dimensions. *Ergonomics*, 51(8): 1272–1289.

Xiong S, Goonetilleke RS, Witana CP, and Au EYL (2010). Foot-arch characterization: A review, a new metric and a comparison. *Journal of the American Podiatric Medical Association*, 100(1): 14–24.

Xiong S, Goonetilleke RS, Zhao J, Li W and Witana CP (2009). Foot deformations under different load-bearing conditions and their relationships to stature and body weight. *Anthropological Science*, 117(2):77–88.

Xiong S, Goonetilleke RS, and Jiang Z (2011) Pressure thresholds of the human foot: Measurement reliability and effects of stimulus characteristics. *Ergonomics*, 54(3): 282–293; *Erratum*, 54(4): 412.

Yalcin N, Esen E, Kanatli U, and Yetkin H (2010). Evaluation of the medial longitudinal arch: A comparison between the dynamic plantar pressure measurement system and radiographic analysis. *Acta Orthopaedica et Traumatologica Turcica*, 44(3): 241–245.

Younger AS, Sawatzky B, and Dryden P (2005) Radiographic assessment of adult flatfoot. *Foot and Ankle International*, 26: 820–825.

4 Foot Posture Index and Its Implications for Footwear Selection

Anthony C. Redmond

CONTENTS

4.1 FOOT POSTURE MEASUREMENT: AN INTRODUCTION

The extremes of foot posture (e.g., frank pes planus and pes cavus) are easily rec-ognizable clinically and without recourse to sophisticated technology, but such foot evaluations undertaken subjectively are often criticized for a lack of subtlety. Where pathology is accompanied by clear neurological and histological change, even sub-jective classifications have good face validity. Consequently, simple dichotomous clinical classifications are in common use, with most descriptors indicating simply whether "pathology" is present or absent. Such discrete classifications (for instance, "cavus" versus "normal" or "pes planus") are consequently used widely (Figure 4.1; Redmond, 2012).

Problems arise when there is a need for an evaluation of foot posture or function where the intention is to capture more subtle variations from the "normal" or "ideal" foot (McPoil and Hunt, 1995, Staheli, 1987). In these situations, simple dichotomous classifications are generally inadequate. Tailoring footwear to specific foot types is one example of this dilemma, where a more sophisticated classification system may be more meaningful in the real world setting.

In clinical settings and particularly in retail footwear fitting settings the ideal of objective, dynamic assessment of foot posture is often not possible due to the time, complexity, and costs associated with detailed assessment (Toro et al., 2003). Most compromise systems for quantifying variations in foot types lack reliability and accuracy (LaPointe et al., 2001, McDowell et al., 2000, Pierrynowski and Smith, 1996, Weiner-Ogilvie and Rome, 1998, Williams et al., 1999). Difficulties arise primarily because the foot is such a complex structure, with motions occur-ring about composite and dynamically mobile axes with many interdependencies (Engsberg, 1987, Engsberg and Andrews, 1987, Kessler, 1983, Kitaoka et al., 1998, Nester, 1997).

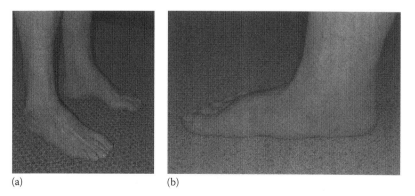

(a)   (b)

FIGURE 4.1   (a, b) Frank cavus and pes planus foot deformities, in both cases associated with systemic conditions (Charcot–Marie–Tooth disease and systemic joint hypermobility). (After Redmond, A.K. and Keenan A.M., Foot problems, in Warburton, L. (ed.), *The New MRCGP Handbook of Musculoskeletal Problems*, Royal College of General Practitioners, London, U.K., 2012. Reproduced with permission.)

FIGURE 4.2 Laboratory assessment of lower limb structure and function. The subject is wearing the marker set for evaluation using the Vicon Oxford multisegment foot model.

The ideal assessment remains a full dynamic evaluation of gait, essentially a laboratory-based exercise still, although in practice the most common compromise is to use static measurements to infer dynamic function indirectly, based on the assumption that the static foot posture is a valid indicator of the function of that same foot when walking (McPoil and Cornwall, 1996a, Menz, 1998, Razeghi and Batt, 2002). While associated to some degree, the relationship between static and dynamic measurements is by no means absolute, and any use of static measures should include some consideration of the degree to which the static measure predicts the true motions in the foot occurring during walking (Figure 4.2).

Even as static measures improve, the extent of the relationship between the static measures and the subsequent dynamic function will always be limited (Razeghi and Batt, 2002), as the dynamic component is substantially more complex. In summary, while dynamic measures remain an ideal, the immediate utility of static measures means that they have a valid role in the assessment of foot postures for footwear choice.

4.2 METHODS FOR FOOT POSTURE MEASUREMENT

Radiographic approaches were commonplace in the retail environment until the middle of the twentieth century and have a limited role still in clinical applications. Radiographic measures are reliable, although still subject to some measurement error due to parallax and the requirement for an observer to measure lines and angles manually (Johnson and Timins, 1998). Advanced techniques such as computed tomography and magnetic resonance imaging address many of these technical deficits and may offer some exciting opportunities in the future although again they remain impractical for most applications of foot posture assessment, especially in footwear assessment (Woodburn et al., 2002).

Many alternative techniques have been proposed that are supposed to produce better data than from clinical observation, while not requiring the expensive and time-consuming methods outlined in the previous section. Without exception though, these are compromise solutions and while they may be more suited to the various demands of the clinical setting, all have significant technical limitations (Razeghi and Batt, 2002).

Some techniques use inked foot prints or more sophisticated electronic capture to quantify foot type and estimate function, according to angular relationships of anatomical landmarks, or the calculation of relative area of different parts of the print. These can be relatively reliable (Cavanagh and Rodgers, 1987, Freychat et al., 1996) and can provide useful clinical information although are often limited to relationships between single planes of motion of simple estimates of foot segment weightbearing (Freychat et al., 1996, Kouchi and Tsutsumi, 1996). Although the angular relationships are useful, this approach, in common with other measures based on foot prints (Atkinson-Smith and Betts, 1992, Cavanagh and Rodgers, 1987, Forriol and Pascual, 1990, Hawes et al., 1992, Volpon, 1994, Weiner-Ogilvie and Rome, 1998), pressure readings (Alexander et al., 1990, Atkinson-Smith and Betts, 1992, Cavanagh et al., 1997, Hennig et al., 1994, Morag and Cavanangh, 1999, Scherer and Sobiesk, 1994), and measures such as the Arch Index (Cavanagh and Rodgers, 1987), fails to take into consideration the many other factors affecting dynamic gait. Hawes reported that in 115 subjects it was found that apart from a very general association at the extremes, the footprint parameters were not related to directly measured arch height. Hawes summarized the study with a strongly worded conclusion, "The categorisation of the human foot according to the footprint measures evaluated in this paper represent no more than indices and angles of the plantar surface of the foot itself" (Hawes et al., 1992).

There is also high between-subject variability in these measures. For one set of angular measures derived from footprints from a normal population of 17 year olds (Forriol Campos et al., 1990), the mean angles of 42.9°–46.3° were associated with 95% CIs of 23.5°–65.3°. Area-based measures such as the Arch Index fare worse still, with the 95% CI for the normal range equal to −0.1 to 0.99 for males and 0.13 to 0.93 for females, almost the entire possible range (Forriol Campos et al., 1990). While changes in these measures can provide some indication of group effects in large group studies, individual comparisons must be made with care.

Measurement of the angles and positions of anatomical landmarks directly on the foot are probably the most common approaches. In turn, the most common of these are measurement of the angle of the calcaneus, relative to the leg or the floor, and measures of arch height. Calcaneal position has been reported widely in the literature although studies suggest poor reliability (LaPointe et al., 2001, McDowell et al., 2000, Pierrynowski and Smith, 1996, Weiner-Ogilvie and Rome, 1998, Williams et al., 1999). Typical figures indicate that practitioners can place the rearfoot within ±3° of the true position on 90% of occasions although this range of 6° error represents about 50% of the total variation encountered in clinical practice (Pierrynowski et al., 1996). The direct measurement of the height of medial arch of the standing foot has been reported several times and employing a number of different protocols.

In its simplest forms, this involves direct measurement of either arch height, the height of the navicular tuberosity, or the height of the dorsum of the foot at a specific point. More complex techniques normalize these measures against various interpretations of foot length. Finally, a quasi-functional component is introduced by the use of composite measures such as "navicular drop," in which the foot is measured in its resting state and again in an artificially induced "neutral" position. If the raw height of an anatomical landmark is used, the intra-subject reliability is high, but the value of the measure tends to be limited to repeat measures within individual patients, as inter-subject variation makes between-subject comparisons inappropriate. The reliability of double measure techniques such as navicular drop is low for inexperienced practitioners (Picciano et al., 1993), although they have been reported to be acceptable in intrarater evaluations of more experienced practitioners (Mueller et al., 1993, Sell et al., 1994).

The coefficient of determination (R^2) for navicular drop in predicting composite forefoot/rearfoot position ranges from 0.24 to 0.33, and for the explanation of variability in transverse plane shank motion by single plane measurement of arch height is 0.27 (Mueller et al., 1993, Nigg et al., 1993). This supports the assertion that measuring motion in only one plane results in poor overall prediction and usefulness restricted to carefully selected, paired measures.

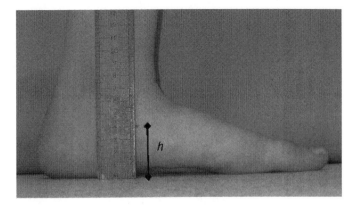

FIGURE 4.3 Measurement of navicular height. (After from Redmond, A.C. et al., *J. Foot Ankle Res.*, 1, 6, 2008. Reproduced with permission of Churchill Livingstone.)

Various measures of arch angle have been also described, employing differing combinations of anatomical landmarks, with the apex usually acting as the origin of the angle between proximal and distal landmarks (Coplan, 1989, Franco, 1987, Giallonardo, 1988, Jonson and Gross, 1997). The arch angle is thus a variation of the other measures of arch or navicular height with the angular changes reflecting anatomical movement as the arch or navicular moves either dorsally or plantar-ward with change in foot posture. Arch angle measures have been reported to be adequately reliable (Jonson and Gross, 1997), although they are not in widespread use clinically (Figure 4.3).

4.3 FOOT POSTURE INDEX

4.3.1 HISTORY

About 15 years ago, organizations such as the Research Council of the American Orthopedic Foot and Ankle Society had highlighted as a priority the need for better measures of foot pathology (Saltzman et al., 1997). In 2001, the American Physical Therapy Association produced consensus statement summarizing priorities for foot outcomes research including the following (McClay, 2001):

1. The foot is a complex structure, and classification systems should be reflective of this complexity.
2. Both dynamic and static methods of describing foot function have identifiable strengths, and both methods have shortcomings that cannot be resolved, given the limitations of current technology.
3. Some means of standardization of approaches is desirable.
4. Where possible, the description of foot structure and function should be based on continuous measures rather than grouping into distinct "foot type" categories.
5. Multifactorial variations in structure and function must be recognized and incorporated into assessment. Measures need to reflect the complexity of the interrelationships occurring within the leg and foot.
6. The development of valid and reliable functional assessment tools should be a priority.

Kitaoka et al. (1995) had previously described the following key features in a desirable foot measure, and these informed the development of the new measure:

- The scale should be reproducible.
- It should be comprehensive enough to reflect the complexity of the condition yet simple enough to be user friendly.
- It should include subjective and objective clinical factors graded to allow for quantitation.
- It should be sensitive enough to detect changes.

- The scale should facilitate statistical analysis.
- It should not be time-consuming to use.
- It should not require sophisticated equipment such as a gait laboratory.
- The measures should be taken in a weightbearing posture, as joint responses to input forces are dependent on foot loading.

Kelly, as early as 1947, had recognized the inadequacy of relying on single measures, and classified the foot using a range of some 32 measures derived from footprints and other indicators obtained with the subject standing, sitting, or semi-weightbearing (Kelly, 1947).

Dahle et al., using a criterion-based, multifactorial rating schema, further found that clinicians were able to classify foot type reliably using multiple, predefined criteria (Dahle et al., 1991).

An ongoing difficulty with foot type assessment is that of defining an appropriate reference point or joint "neutral position." This is highlighted by the range of examples in which definitions of neutral are based on anatomical convenience rather than functionally meaningful positions. (Elveru et al., 1988, McPoil and Hunt, 1995, Sobel and Levitz, 1997). Consequently, whatever compromise definition is accepted for the reference position, the problem remains that an inherently subjective measure acts as the starting point for all subsequent objective (or quasi-objective) measures (Ball and Johnson, 1993). Experienced practitioners have been shown to reach some consensus in finding a foot "neutral" position, but only after as many as six measures on the same subject (Pierrynowski and Smith, 1996). Astrom and Arvidson proposed that the entire paradigm of an "ideal" foot is based on an invalid theoretical concept, arguing that a proper reference should be based on clinical observations, rather than on unreliable and theoretically driven but unsubstantiated considerations (Astrom and Arvidson, 1995). It is certainly possible to identify characteristics associated with various types of foot morphology (Schwend and Drennan, 2003), and these are so well documented as to be taken for granted.

Where pragmatic clinical assessments are required, measurements using clinician-generated criteria are widely used in medicine. In the evaluation of benign joint hypermobility syndrome, for instance, hypermobility is almost universally defined according to a nine-point scale known as the Beighton–Carter–Wilkinson (BCW) scale (Beighton et al., 1973). In using the BCW scale, the presence or absence of a range of specific criteria are simply noted and scored by the clinician. Although the Beighton scale has been criticized for its simplicity and potential subjectivity (all criticisms which could equally be leveled at a criterion-based measure of foot posture), in the absence of a definitive laboratory or biochemical assay which can provide a definitive quantification of joint mobility, the BCW scale remains the almost unanimous choice of clinicians (Biro et al., 1983). Modifications to increase objectivity have been suggested over the years (Dijkstra et al., 1994, Fairbank et al., 1984) but without exception have fallen rapidly into disuse. The simplicity and clinical utility of such a measure remain its most compelling features (Biro et al., 1983).

4.4 DEVELOPMENT AND VALIDATION OF THE FOOT POSTURE INDEX

The aim of the body of work was to derive an instrument for better assessing foot posture from these principles. A four phase process was used including derivation of the measures, development of a scoring system, proof of concept, and component selection and reduction. The formal validation of the final instrument is described separately in this chapter although the body of work is captured in three published papers (Keenan et al., 2007, Redmond et al., 2006, 2008).

4.4.1 DEVELOPING THE MEASURE

In a comprehensive review of the literature, more than 100 papers were identified as describing in adequate detail, the clinical evaluation of foot posture. From these, 36 discrete clinical measures were identified and a matrix was used to map the candidate measures based on their capacity to measure postural changes in each of the three body planes, and according to the anatomical segment to be evaluated (hindfoot/midfoot/forefoot/multiple). Measures requiring either a formal measurement or special equipment were excluded as were measures requiring the drawing of bisections based on anatomical features, or necessitating time-consuming measurements. Sixteen items were short-listed for inclusion in the draft instrument against the criteria described by Kitaoka and McClay noted earlier, and eight were selected for a draft version. In line with previous recommendations, the measure was to be used weightbearing (Kitaoka et al., 1995, McClay, 2001) and while consideration was given to the assertion that single limb standing may be a better indicator of dynamic function than double limb quiet stance, double limb standing was preferred because (a) the relaxed stance position has been shown to be repeatable (McPoil and Cornwall, 1996b), (b) this position is well known to practitioners, and (c) the populations in which the new measure will be used includes patients with balance and postural control problems.

Next, a common scoring system was required across all criteria that would reflect variability within the population, was clear and understandable, and allowed varied measures to be compared and aggregated. A five-point Likert-type scale was used in the first draft, as a five-point scale provides a reasonable compromise between sensitivity, reliability, and ease of use (Bennett et al., 2001, Gadbury-Amyot et al., 1999). Likert-type scales can use between 3 and 20 or more graduations, but while increasing the number of graduations increases the sensitivity of the scale, reliability is reduced. In this instance, it was desirable to have specific criteria matching each point on the scale, and so a smaller number of intervals was considered preferable (Bennett et al., 2001, Hernandez et al., 2000). The scale was anchored such that the central response was "zero," with the sign of the deviation from this central response indicating the direction of postural change. A deviation of one interval (±1 point) on the scoring scale for each component was used to indicate small variations from normal, and two intervals (±2 points) to indicate marked deviations from normal posture. The scoring criteria were individualized to each of the component measures and described in an accompanying manual.

Aggregated scores for foot postures tending toward a more supinated position are thus denoted by overall negative scores, with higher positive scores indicating aggregation of features associated with a pronated foot posture. With eight component measures, each scored on a symmetrical five-point scale, the possible aggregate scores for the initial eight-item version of the foot posture index (FPI) ranged from −16 to +16.

Following a small study to investigate empiric reliability, ease of use, and perceived usefulness, two further studies investigated more thoroughly the validity of the components and informed the removal of components to create a final version of the instrument. Firstly, the eight-item FPI scores were compared in a field trial to concurrently derived Valgus Index (VI) scores (Rose, 1991). In a second stage, a three-dimensional static lower limb model was reconstructed from data obtained from an electromagnetic motion tracking system. Ordinal regression modeling was used to quantify the strength of the relationship between the EMT variables and each of the FPI components.

4.4.2 FIELD TRIALS

Rose's VI was first employed as the benchmark in a field trial of the new measure. Ratings of the 8 clinical components making up the draft FPI were undertaken for each of 131 subjects, while they stood on a "pedograph," ink and paper mat. The VI was calculated later from the inked footprint. An ordinal regression model was constructed to evaluate the capacity of the total FPI-8 score to predict the VI scores, and the inter-item reliability for each of the eight components was evaluated using Cronbach's alpha coefficient. A principal components analysis was also conducted to identify latent factors that might explain the variance in the FPI scores, but which were not apparent a priori (Figure 4.4).

A sample of 91 (69.5%) male and 40 (30.5%) female volunteers aged 18–65 (Mean = 33.7 years). yielded VI scores ranging from −3.6 to 33.61 (Mean = 10.28, SD = 6.52). FPI scores ranged from −7.0 to 15.0 (Mean = 4.9, SD = 3.9). An ordinal regression model was constructed to assess the degree to which the FPI total score predicted the VI score. The ordinal regression modeling indicated good agreement between the VI and FPI-8 scores, and the FPI-8 total scores predicted 59% of the variance in VI values (Cox and Snell $R^2 = 0.590$, $B = 0.551$, $P < 0.001$, $N = 131$).

The inter-item reliability of the eight-item set using Cronbach's α was 0.834, indicating good inter-item reliability. The individual coefficients were high or very high for six of the eight FPI components although the components measuring Helbing's sign and the congruence of the lateral border of the foot showed poor inter-item reliability (Cronbach's $\alpha < 0.40$). Subsequent principal components analysis also highlighted limitations in the lateral border measurement and these two items were subsequently removed from the final version of the FPI.

To measure FPI ratings concurrently with a detailed objective measure of static foot posture, a FasTrak™ system (Polhemus Inc, Colchester, VT) was used to reconstruct a three-dimensional lower limb model for the right leg of 20 healthy volunteers in each of three positions. An electromagnetic tracking system with a digital stylus was used to derive the Cartesian coordinates of landmarks and to define rigid

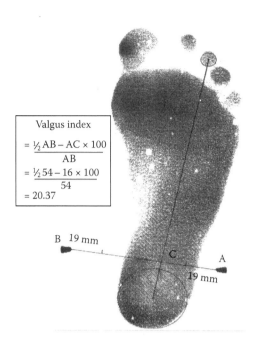

Valgus index

$$= \frac{\frac{1}{2} AB - AC \times 100}{AB}$$

$$= \frac{\frac{1}{2} 54 - 16 \times 100}{54}$$

$$= 20.37$$

B 19 mm

C   A

19 mm

FIGURE 4.4 Valgus index calculation derived from an inked footprint. (From Burns, J. et al., *J. Am. Podiatr. Med. Assoc.*, 95, 235, 2005b. Reproduced with permission.)

models of the shank, hindfoot, and forefoot in each of three postures, which were then reconstructed in software. Subjects were measured in (1) a functionally neutral position corresponding to a "0" score using the FPI, (2) a maximally pronated stance position, and (3) a maximally supinated stance position.

The eight FPI components were entered in turn as the dependent variable into a series of ordinal regression models (Walters et al., 2001), and the contribution of the EMT measures to the explanation of variance in each FPI component score was established. Six components performed well but the FPI component measuring lateral border congruence and Helbing's sign again performed poorly and did not fit the ordinal regression model.

As a consequence of the early field trials and EMT study, these two criteria were removed, yielding a six-item final version of FPI-6 for final validation.

4.5 FPI-6 EXAMINATION

A summary is provided here for the purposes of this chapter, but a full set of documentation including a fully illustrated manual (Figure 4.5) reference sheets and data sheets is available for download at http://www.leeds.ac.uk/medicine/FASTER/fpi.htm

Note: The FPI and associated materials are made freely available for clinical and noncommercial use. The author reserves all rights however and permission must be sought to use the FPI or its materials for commercial purposes or to reproduce documentation in any publically available form.

FIGURE 4.5 The cover of the FPI-6 Manual. (Reproduced with permission http://www. leeds.ac.uk/medicine/FASTER/fpi.htm, all rights reserved.)

To conduct the FPI-6 examination, the patient stands in their relaxed stance position with double limb support. It can be helpful to ask the patient to take several steps, marching on the spot, prior to settling into a comfortable stance position. The patient will need to stand still for approximately 2 min in total in order for the assessment to be conducted. The assessor should be able to move around the patient during the assessment and have an uninterrupted access to the posterior aspect of the leg and foot.

If observation of a single criterion cannot be made (e.g., because of soft tissue swelling), simply miss it out and indicate on the datasheet that the item was not scored.

If there is genuine doubt about how high or low to score an item, it is best always to use the more conservative score.

Item 1. Talar head palpation
This is the only scoring criterion that relies on palpation rather than observation. The head of the talus is palpated on the medial and lateral side of the anterior aspect of the ankle, according to the standard method described variously by Root, Elveru, and many others. Scores are awarded for the observation of the position as follows:

Score	−2	−1	0	+1	+2
	Talar head palpable on lateral side/ but not on medial side	Talar head palpable on lateral side/ slightly palpable on medial side	Talar head equally palpable on lateral and medial side	Talar head slightly palpable on lateral side/ palpable on medial side	Talar head not palpable on lateral side/ but palpable on medial side

Item 2. Supra and infra lateral malleolar curvature
In the neutral foot it has been suggested that the curves should be approximately equal. In the pronated foot, the curve BELOW the malleolus will be more acute than the curve above due to the abduction of the foot and eversion of the calcaneus. The opposite is true in the supinated foot.

Score	−2	−1	0	+1	+2
	Curve below the malleolus either straight or convex	Curve below the malleolus concave, but flatter/more shallow than the curve above the malleolus	Both infra and supra malleolar curves roughly equal	Curve below malleolus more concave than curve above malleolus	Curve below malleolus markedly more concave than curve above malleolus

Item 3. Calcaneal frontal plane position
This is an observational equivalent of the measurements often employed in quantifying the relaxed and neutral calcaneal stance positions. With the patient standing in the relaxed stance position, the posterior aspect of the calcaneus is visualized with the observer in line with the long axis of the foot.

Score	−2	−1	0	+1	+2
	More than an estimated	Between vertical and an estimated	Vertical	Between vertical and an estimated	More than an estimated
	5° inverted (varus)	5° inverted (varus)	—	5° everted (valgus)	5° everted (valgus)

Item 4. Bulging in the region of the talo-navicular joint (TNJ)
In the neutral foot, the area of skin immediately superficial to the TNJ will be flat. The TNJ becomes more prominent if the head of the talus is adducted in rearfoot pronation. Bulging in this area is thus associated with a pronating foot. In the supinated foot this area may be indented.

Score	−2	−1	0	+1	+2
	Area of TNJ markedly concave	Area of TNJ slightly, but definitely concave	Area of TNJ flat	Area of TNJ bulging slightly	Area of TNJ bulging markedly

Item 5. Height and congruence of the medial longitudinal arch
While arch height is a strong indicator of foot function, the shape of the arch can also be equally important. In a neutral foot, the curvature of the arch should be relatively uniform, similar to a segment of the circumference of a circle. When a foot

is supinated, the curve of the medial longitudinal arch (MLA) becomes more acute at the posterior end of the arch. In the excessively pronated foot, the MLA becomes flattened in the center as the midtarsal and Lisfranc's joints open up.

Score	−2	−1	0	1	2
	Arch high and acutely angled toward the posterior end of the medial arch	Arch moderately high and slightly acute posteriorly	Arch height normal and concentrically curved	Arch lowered with some flattening in the central portion	Arch very low with severe flattening in the central portion—arch making ground contact

Item 6. Abduction/adduction of the forefoot on the rearfoot

When viewed from directly behind, and in line with the long axis of the heel (not the long axis of the whole foot), the neutral foot will allow the observer to see the forefoot equally on the medial and lateral sides. In the supinated foot, the forefoot will adduct on the rearfoot resulting more of the forefoot being visible on the medial side. Conversely, pronation of the foot causes the forefoot to abduct resulting in more of the forefoot being visible on the lateral side.

Score	−2	−1	0	1	2
	No lateral toes visible. Medial toes clearly visible	Medial toes clearly more visible than lateral	Medial and lateral toes equally visible	Lateral toes clearly more visible than medial	No medial toes visible. Lateral toes clearly visible

The final FPI score will be a whole number between −12 and +12.

Scores can be collated in the reader's own notes or entered in a datasheet downloadable from http://www.leeds.ac.uk/medicine/FASTER/fpi.htm

All materials are provided for interested parties to download and use for clinical purposes. Anyone considering using the FPI-6 is strongly urged to download and familiarize themselves with the manual, however, and to undertake simple training. Initial training should be undertaken with a partner to allow cross referencing of scores. Conducting at least 30 sets of ratings, in as wide a range of foot types as can be sourced, is usually sufficient to prepare the new user to use the FPI clinically with adequate reliability.

4.6 VALIDATION

The final version of the FPI-6 has been subjected to extensive validation, both by the author and by various independent researchers (Barton et al., 2010, Cain et al., 2007, Keenan et al., 2007, Redmond et al., 2006). This section presents firstly the results of the capacity of the static FPI to predicted concurrent static posture and subsequent dynamic functions (Redmond et al., 2006), followed by an analysis of the performance of the measure using modern item response theory (Rasch analysis) to evaluate individual criteria and the measure as a whole (Keenan et al., 2007).

The process for deriving a set of normative values from more than 1400 observations is then outlined, with the section closing with a summary of independent studies conducted since the publication of the FPI-6 (Redmond et al., 2008).

Concurrent validity of FPI-6 scores was assessed against a static EMT model of the ankle joint complex (AJC), followed by a predictive validation to determine the extent to which static FPI-6 scores predict systematic variations in AJC position during normal walking.

Ankle joint complex positions and motions were again captured using a FasTrak™ electromagnetic motion tracking system (Polhemus Inc., Colchester, VT, USA), although this study uses surface-mounted sensors to describe segmental motions rather than stylus-based digitization of landmarks (Figure 4.6). In the static studies, participants stood still in the center of a calibrated capture volume, while in the dynamic studies, the participants started walking at one end of a 9 m walkway and passed uninterrupted through a calibrated capture volume at a self-selected walking pace, continuing to the far end of the walkway. FPI total scores were regressed against AJC excursions using linear regression modeling with static AJC entered as the dependent variable and FPI-6 total score entered as the independent variable. The FPI-6 scores during quiet double limb standing predicted 64% of the variation in the concurrent static AJC position derived through technically demanding EMT studies (adjusted $R^2 = 0.64$, $F = 73.529$, $P < 0.001$, $N = 14$).

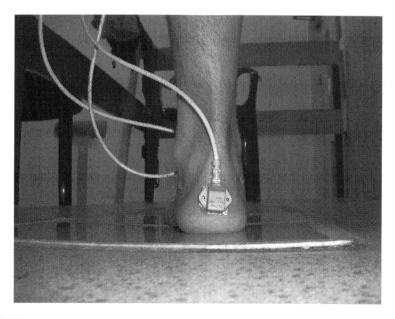

FIGURE 4.6 The setup used for capturing posture and function using electromagnetic motion tracking (note that the sensors were covered tightly in conforming tape for the trials to minimize skin movement artifact).

For the dynamic studies, the within-subject coefficient of multiple correlations for AJC motions derived from EMT was 0.93. Midstance was identified as the point at which the EMT derived static and dynamic AJC_β rotations are most closely related ($R = 0.864$). The instant of midstance was entered, therefore, as the dependent variable in the FPI predictive regression modeling, with the FPI-6 total score entered as the sole independent variable. The resulting linear regression model yielded an adjusted R^2 of 0.41 ($F = 31.786$, $P < 0.001$) indicating that the static FPI total score predicted 41% of the subsequent dynamic variation in midstance foot position.

Laboratory gait analysis remains the gold-standard, but the facilities to produce high-quality objective data are expensive, and the process of acquiring the data can be overly time-consuming for routine patient assessment. The strength of the relationship between the FPI-6 as a static clinical measure and the EMT data is weaker for the dynamic studies than for the concurrent static measures, in common with previous studies (Cavanagh et al., 1997, McPoil and Cornwall, 1996b). Nevertheless, the FPI-6 score predicted over 40% of the variance in dynamic AJC motion, a stronger association than in most reports in the literature for pairs of static and dynamic measures (Cashmere et al., 1999, McPoil and Cornwall, 1996a). In common with previous studies of composite measures (Cavanagh et al., 1997), this suggests that the composite nature of the FPI, accounting for variation in all three body planes, may provide a more complete description of foot posture than most currently used static clinical measures.

The final in-house evaluation explored the fit of the FPI-6 data to a standard model of psychometric validity, in this case the Rasch model. Rasch analysis is a useful technique for assessing properties of outcome measures including dimensionality (the extent to which items measure a single construct), item difficulty (the relative difficulty of the items when compared to one another), and person separation (the extent to which items distinguish between distinct levels of functioning). Rasch analysis (Rasch, 1960) has been widely used in the development and validation of outcome measures since its introduction. Data collected from questionnaires or scales that are summated into an overall score are tested against the fit expectations of the measurement model. The model defines the ideal item response characteristics if measurement (at the interval level) is to be achieved. The observed response patterns achieved are tested against expected patterns and a variety of fit statistics determine whether fit is adequate to the model (Hunt et al., 2000).

Importantly, Rasch modeling is also capable of transforming ordinal scores obtained by summation of criteria into interval measures. Transformed scores (as logit values) can then be used for parametric statistical analysis which improves their use in research and practice.

FPI scores from 143 people (98 male and 45 female) with a range of foot types were assessed using all eight items described in the original FPI, and total scores were derived for both FPI-8 and FPI-6 version. Using the RUMM2020 software package, the original FPI-8 indicated misfit to the Rasch model due to Helbing's sign displaying disordered thresholds, and the measure of congruence of the lateral border of the foot not measuring the same dimensions as the other items. This confirmed the appropriateness of the previous decision to reduce the original eight-item version to the final FPI-6.

TABLE 4.1

Table for Converting Raw FPI-6 Scores to Rasch Transformed Scores

FPI-6 Raw Score	Transformed Score
−12	−10.47
−11	−7.96
−10	−6.45
−9	−5.54
−8	−4.84
−7	−4.25
−6	−3.71
−5	−3.2
−4	−2.67
−3	−2.12
−2	−1.54
−1	−0.91
0	−0.21
1	0.5
2	1.16
3	1.75
4	2.33
5	2.98
6	3.81
7	4.83
8	5.68
9	6.36
10	7.01
11	7.77
12	8.65

Source: Keenan, A.M. et al., *Arch. Phys. Med. Rehabil.*, 88, 88, 2007. Reproduced with permission.

The FPI-6 demonstrated good overall fit to the model with no disordered thresholds and evidence of unidimensionality. The person frequency distribution demonstrated that there was a large measurement range of the item thresholds retained in the FPI-6, indicating that the measure was able to discriminate across different foot positions.

A table of logit transformed values, suitable for use in parametric analyses, has been generated and is provided as Table 4.1.

4.7   NORMATIVE VALUES

A set of normative values have also been compiled using complete anonymized datasets originating from 16 studies undertaken in 9 centers and representing 1648 individual participants' FPI-6 scores. Data covers age, gender, pathology

TABLE 4.2

Logit Scores and Back-Transformed FPI-6 Raw Scores for the Normal Adult Population

	Pathological	Potentially Abnormal	Normal Range			Potentially Abnormal	Pathological
	<−2 SD	−2 SD	−1 SD	Mean	+1 SD	+2 SD	>+2 SD
FPI logit		−2.2	+0.1	+2.4	+4.7	+7.0	
FPI raw score	<−3	−3	+1	+4	+7	+10	>+10

(where relevant), and body mass index (BMI) values, where available. As noted in the previous section, logit transformed scores are more valid for fuller analysis, and so logit transformed scores were used throughout.

Reference ranges were defined using the cut points employed previously for similar studies (Moseley et al., 2001), namely,

1. *Normal*: values lying in the range, mean ±1 standard deviation (SD)
2. *Potentially abnormal*: values 1–2 SDs from the mean
3. *Pathological*: values lying outside 2 SDs from the mean

The normal adult sample comprised 619 observations of a single limb from each participant. There were no systematic differences between observations derived from left or right feet and between the scores derived from males and females nor evidence of any relationship with BMI. The means and SDs of the logit scores were back-transformed into FPI raw scores and normal, potentially abnormal and truly pathological ranges defined. These are presented in Table 4.2.

4.8 FOOT POSTURE AND ANTHROPOMETRICS

4.8.1 AGE-RELATED DIFFERENCES

In our own normative dataset, the normative data were explored for age-related trends and initial graphical plotting suggested that FPI scores may indicate more pronated postures with extremes of age (Figure 4.7). Within the adult group, those over 60 years appeared to represent a potentially different population, as did a group of minors ($n = 388$, mean age 8.5 years, range 3–17 years) who had been omitted from the analysis outlined in the previous section. The dataset was expanded to include the minors, and data were recorded by age group into normal minors (<18 years), normal adults (18–59 years), and normal older adults (≥60 years).

Graphical output suggested some systematic difference by age group (Figure 4.8), with both minors (mean FPI logit score = 3.7, SD = 2.5) and older adults (mean = 2.9, SD = 2.6) showing higher mean scores than seen in the general population (ANOVA, $F = 51.07$, $P < 0.001$). The differences between groups were all also confirmed as

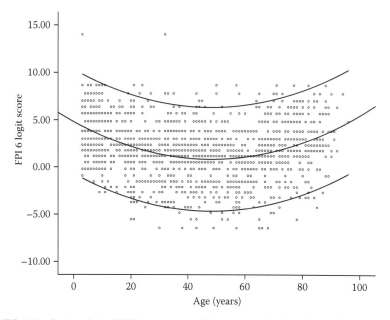

FIGURE 4.7 Scatterplot of FPI scores according to age showing the tendency toward a more pronated foot posture in children and older adults. (After Redmond, A.C. et al., *J. Foot Ankle Res.*, 1, 6, 2008. Reproduced under the terms of the BiomedCentral Creative Commons Attribution License.)

individually significant by Tukey's post hoc test ($P < 0.001$). Separate reference ranges have therefore been defined for the minor and older adult groups (Table 4.3).

There was some evidence of age-related variation in mean foot posture scores and this is in agreement with previous studies. In the study by Scott et al. (2007), a sample of older adults (mean age $80.2 \pm$ SD 5.7) had more pronated foot postures than a group of younger adults (mean age $20.9 \pm$ SD 2.6). A tendency toward more pronated foot postures in younger children is also well documented. A flatter, more pronated foot has been reported in young children as a consequence of the process of development of the longitudinal arch (Staheli et al., 1987). The values reported in this study of FPI normative values support the notion of a U-shaped relationship between age and foot posture reported by Staheli et al. (1987).

4.8.2 Body Mass

Interestingly, our data uncovered no relationship between the BMI and the FPI score. Previous studies undertaken using measures such as the footprint angle (FA) and the Chippaux-Smirak Index had reported lowered longitudinal arches, a broader midfoot area and subsequently flatter feet in people with high BMI values (Riddiford-Harland et al., 2000). However, the studies reporting BMI-related differences, especially those prior to the introduction of the FPI-6 mostly used footprint measures, and the postural data may be confounded by the effect of body fat on the interpretation of

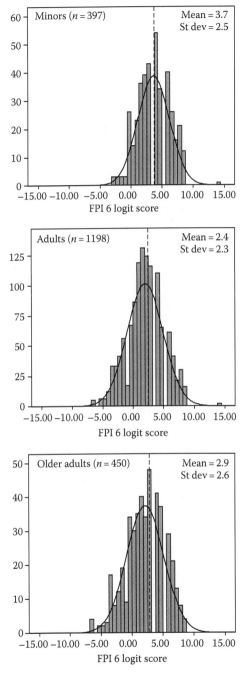

FIGURE 4.8 Histograms of FPI scores for minors, adults, and older adults. Dashed lines represent means. (After Redmond, A.C. et al., *J. Foot Ankle Res.*, 1, 6, 2008. Reproduced under the terms of the BiomedCentral Creative Commons Attribution License.)

TABLE 4.3

Logit Scores and Back-Transformed FPI-6 Raw Scores for Minors and Older Adults

		Pathological	Potentially Abnormal	Normal Range			Potentially Abnormal	Pathological
		<−2 SD	−2 SD	−1 SD	Mean	+1 SD	+2 SD	>+2 SD
Minors (<18 years)	FPI logit		−1.3	+1.2	+3.7	+6.2	+8.7	
	FPI raw score	<−2	−2	+2	+6	+9	+12	+12
Older adults (>60 years)	FPI logit		−2.3	+0.3	+2.9	+5.4	+8.1	
	FPI raw score	<−3	−3	+1	+5	+8	+11	+12

arch height based on these footprint estimates (Aurichio et al., 2011). Indeed it has been suggested previously that footprint parameters are consequently a measure of "fat feet" rather than "flat feet" (Wearing et al., 2004). In a more recent concurrent study employing both footprint measures and the FPI-6 scores in nearly 400 older people, the Arch Index and FPI-6 score were both found to correlate weakly with overall BMI ($r = 0.15–0.2$) (Aurichio et al., 2011). Assigning broader categories, such as normal ($<25 \, kg/m^2$), overweight ($25–30 \, kg/m^2$), and obese ($>30 \, kg/m^2$) did also, as expected, improve the capacity to differentiate by foot posture. In summary, there does appear to be some tendency toward a more pronated foot posture with increasing obesity, but footprint measures are frankly unreliable in this group of people and FPI data are weakly correlated and must be treated with caution.

4.8.3 BALANCE

Only one study has explored the extent to which measures of foot and ankle strength, range of motion, FPI score, and deformity are associated with performance of balance and functional ability tests in older people (Spink et al., 2011). In this cross-sectional study of 305 participants aged 65–93 years, the FPI-6 score, among other factors, was found to be an independent predictor of postural sway on foam, with participants exhibiting more pronated foot postures also achieving poorer performance in sway control. This has some implications for identifying falls risk, particularly in the elderly, although foot posture should be considered alongside factors such as knee extensor strength and ankle strength.

4.9 FOOT POSTURE AND PATHOLOGY: DIFFERENCES BETWEEN NORMAL AND PATHOLOGICAL GROUPS

The FPI-6 has now been employed in more than 50 studies and median FPI raw scores for normal samples have been reported to lie consistently around +4 to +5, that is, slightly pronated rather than strictly "neutral" (Burns et al., 2005b, Cain et al., 2007, Menz and Munteanu, 2005, Yates and White, 2004). Our own study employed a large sample and used the best statistical models to confirm that in the normal adult population the mean (back-transformed) FPI score is +4, confirming that a slightly pronated foot posture is the normal position at rest. Statistically determined reference ranges for postural variations such as standing foot position are inherently wide, so must be used as a general guide only in interpreting FPI scores in a clinical context. It is recognized that clinically, relatively minor variations from the mean may increase risk of mechanically induced pathology, although the strength of these relationships have not been confirmed scientifically and certainly vary for different pathological groups. Except for foot postures falling clearly outside the normal range, the reference ranges alone are probably not adequate for clinical decision making.

It is known, however, from both empirical observation and previous studies that foot posture differences may be encountered in association with underlying disease processes or functional pathology. In our own data, we have quantified the capacity of the FPI-6 to differentiate between pathologies through comparison of the normative data with data from participants in the same 16 studies who had identified pathology.

Comparison of the FPI scores from the normal with those from participants known to have identified pathology revealed variations consistent with those predicted by theory. Four groups were identified: (1) miscellaneous local musculoskeletal symptoms, (2) diabetic neuropathy, (3) neurogenic pes cavus, and (4) idiopathic pes cavus.

The group with neurogenic pes cavus (mean FPI logit score = −2.78, SD = 2.32) and idiopathic pes cavus (mean = −2.63, SD = 1.25) had FPI scores significantly different from the normal population (mean logit score = +2.4) indicating that the FPI data was sensitive to disease-related postural changes. Data have also been reported elsewhere indicating the sensitivity of the FPI to postural change associated with pathological pes planovalgus (median FPI raw score = +12) (Burns and Crosbie, 2005). Conversely, the otherwise healthy group with minor musculoskeletal symptoms (mean FPI logit score = 2.23, SD = 2.35) was not systematically different from the normal population (mean FPI logit score = 2.4, SD = 2.3), nor was a sample of patients with diabetes (mean = 2.14, SD = 2.96).

In summary, values for the miscellaneous musculoskeletal symptoms group and the diabetic neuropathic group were comparable with the normal population, as would be expected for conditions not normally associated with significant structural change. Conversely, the neurogenic cavus and the idiopathic cavus groups were confirmed as representing a clearly pathological population (F = 216.981, P < 0.001) (Figure 4.9).

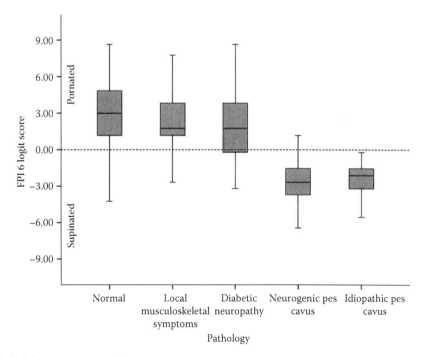

FIGURE 4.9 Boxplots of FPI scores according to presence of pathology. Error bars are 95% confidence intervals and horizontal lines represent medians. (After Redmond, A.C. et al., *J. Foot Ankle Res.*, 1, 6, 2008; Reproduced under the terms of the BiomedCentral Creative Commons Attribution License.)

There appears therefore to be scope for using FPI scores and associated normative values to help identify groups with structural pathology and to assist in the clinical decision-making process.

This relationship between FPI score and symptoms has been confirmed independently since in relation to a range of conditions. In relation to foot pathology, increased risk for plantar heel pain was found in one study of 80 adults with chronic plantar heel pain and 80 controls, to be associated with a combination of both BMI and FPI-6 score, leading to the conclusion that obesity and pronated foot posture are associated with chronic heel pain (Irving et al., 2007). In the forefoot, a very high correlation has been reported ($r = 0.84$, $P = 0.05$) between painful hallux rigidus and FPI-6 score in an independent study of demographics and clinical parameters associated with a group of subjects with hallux functional limitation. In this study of 110 people, footwear was found to be an aggravating factor only in females while the FPI score demonstrated a strong correlation across the sample (Beeson et al., 2009).

Burns and the group at the University of Sydney have undertaken a series of studies exploring the impact of high arched (pes cavus) foot types and resulting pathology. A strong correlation was shown between FPI-6 score and Achilles tendon length ($r = 0.757$, $P < 0.001$) (Burns and Crosbie, 2005) and this is known to contribute to pes cavus severity in a number of neurogenic and idiopathic causes of structural foot pathology. Pes cavus postures of either idiopathic or neurogenic aetiology, as defined by FPI-6 scores, have been shown to result in a higher prevalence of foot pain (~60%) compared to subjects with a normal foot type (23%), and pressures under the foot and rearfoot and forefoot regions in pes cavus are also higher than in the normal foot type (Burns et al., 2005a).

More proximally in the leg, Barton indentified that patients with a more pronated foot type, again as determined by FPI-6 score, demonstrated grater forefoot abduction and an earlier time of peak rearfoot eversion during walking in a group of patients with patellofemoral pain, a finding which did not occur in a control group. This lead the authors to conclude that this was consistent with established biomechanical models of patellofemoral pain syndrome development, linking foot function with patellar tracking (Barton et al., 2011). Also in the knee, people with medial compartment knee OA have been demonstrated to exhibit significantly lower FPI-6 scores compared to controls (Levinger et al., 2010). It has subsequently been recommended that the assessment of patients with knee OA in clinical practice should include simple foot measures, and that the potential influence of foot structure and function on the efficacy of foot orthoses in the management of medial compartment knee OA be further investigated (Levinger et al., 2010).

4.10   FOOT POSTURE IN SHOE DESIGN

4.10.1   Sports Shoes

We have reported previously on the increased risk of overuse injury in athletes with cavus type foot postures, particularly in running sports. We initially used FPI scoring to monitor injury risk over the course of a season in 131 elite triathletes and found that the odds of an injury occurring in those with a supinated foot type were 4.3 times greater than in those who did not have a supinated foot type (Burns et al., 2005c). In soccer players, a low FPI-6 score (high-arched feet) was found to be predictive of greater technical ability but again was associated with increased risk of overuse injury (Cain et al., 2007).

For sporting participants with high-arched foot postures (i.e., those with low-FPI sores), there is some theoretical merit in careful matching of footwear design characteristics to foot type. The injury risk in this group appears elevated across a range of sports and it does seem as though individuals with a supinated foot type may not tolerate linear, repetitive running as well as those with a more neutral or even pronated foot posture. As noted previously, there are reports indicating that greater plantar loading in those with high-arched foot postures, poor shock absorbency in the proximal musculoskeletal system, and Achilles tightness may also play a part in injury risk, although the link between any of these variables is informal and as yet unclear. Despite the absence of quality evidence regarding specific therapeutic interventions either in the form of matched athletic footwear or even shoe modifications or in shoe additions such as orthoses, this area warrants thought. Customized foot orthoses directed at reducing plantar pressures have been shown to be more effective in patients with painful pes cavus–related presentations than non-cavus controls (Burns et al., 2006), but again, high-quality evidence is scant. In the retail sphere, there are many designs of footwear on the market for the high-arched sports person incorporating features such as increased heel shock absorbency, reinforced central zones for midfoot strikers, and various cutouts, additions, and modifications to ameliorate increased forefoot pressure. Of note, however, is the relatively small proportion of athletes across the various studies who exhibited a supinated foot posture.

Far more common in the population at large and therefore more commonly noted in the anecdotal literature is the link between pronated foot postures and injury risk, although the formal evidence remains inconclusive across a range of studies. Of papers reporting the use of the FPI-6 to explore injury risk, two have indicated an increased risk of medial tibial stress syndrome; in runners (Tweed et al., 2008) and military recruits (Yates and White, 2004) but there is much contradictory literature. Countless variations in sports shoe design exist to counter the effects of excessive foot pronation and certainly too many and with too flimsy an evidence base to list in detail here. In principle however, desirable features can be said to include the following:

- A firm heel counter to limit heel eversion
- Support through additions or variations in midsole density in the region of the talo-navicular joint
- Some reinforcement to limit midfoot longitudinal twisting
- An adjustable fastening to allow for slim or unusually proportioned midfoot regions
- A cushioned forefoot midsole
- As required, cutouts or "dynamic" features aimed at improving the function of the first (or lateral) ray and associated metatarsophalangeal joints

4.10.2   STANDARD RETAIL FOOTWEAR

Although sports footwear design receives a good deal of attention in relation to postural tailoring, most people will spend more time in day-to-day retail footwear than in sports shoes. Most people with mild to moderate variations in foot posture and without frank pathology should be able to continue wearing standard retail footwear throughout the lifespan. It can be helpful, though, to have an understanding of the features of retail footwear that will work best with the feet of people with significant variations in foot posture.

Generally speaking, the following characteristics are considered:

- A moderate or low heel (<2 in.)
- A cushioned midsole
- Breathable, conformable uppers
- An adjustable fastening (e.g., laces, straps)
- A stable heel counter

As long as the person with foot problems is adequately informed, then the extent to which these criteria are balanced against style and cosmesis becomes entirely personal. As with sports footwear, there is little hard evidence but there is a general acceptance that pronated foot postures will do best in retail footwear that is designed to be supportive in the heel counter, adjustable in the midfoot region if the foot is slim, and has adequate plantar cushioning to reduce localized areas of high pressure. Conversely, people with a supinated foot type might do better in shoes with a generous upper and adjustable fastening to accommodate the arch height, a medially curved last shape, and a cushioned sole to minimize peaks of plantar pressure and reduce shock transients.

4.10.3   NICHE RETAIL FOOTWEAR

A range of manufacturers and suppliers provide footwear using a niche high street or mail-order business model to sell footwear design-optimized for people who have foot conditions and postures that render standard retail footwear inadequate. Niche retail retains much of the freedom of choice associated with the standard retail model, combined with a more explicit acknowledgment of the specific design needs of people with foot problems.

The types of features typically addressed by niche footwear include increased focus on the five characteristics defined earlier, provision of extra width or depth to accommodate deformity or the need to fit in extra insoles, additional panels to increase flexibility in the uppers and novel fasteners.

4.10.4   SPECIALIST THERAPEUTIC FOOTWEAR—STANDARD
OR "STOCK" AND MODULAR FOOTWEAR

One growth area in the field of musculoskeletal foot pathology is that of specialist therapeutic footwear supplied through a clinic rather than a purely commercial provider and which prioritizes the therapeutic aspects of the shoe. Specialist therapeutic footwear typically comes in either "stock" form, that is, provided by a clinician but is dispensed as is (albeit with enhanced therapeutic design features) or in modular form that is, with the capacity to mix and match features according to the precise needs of the patient/client.

Stock shoes usually have design features such as extra depth to accommodate an additional therapeutic insole, a wider fit, and a range of fastenings. Manufacturers will offer a range of styles often based on lasts with differing fundamental features which can be matched to the wearer as discussed in Section 4.11.

The introduction of stock or readymade therapeutic footwear in the 1960s transformed the provision of prescription footwear for people with the most severe foot problems, as the cost of stock shoes is about one-fifth to one-third that of fully made-to-measure shoes. As well as being cost-effective, stock shoes can often be supplied immediately reducing the number of attendances and waiting times. The immediately availability also increases choice, allowing trialing of a variety of fitting options and providing the patient with an opportunity to see the differing styles and accompanying features.

Stock footwear can be offered in a wide range of styles, and new materials such as injection molded soles have improved appearance and performance.

4.11   SHOE FITTING USING FOOT POSTURE ASSESSMENT

When consulted, clinicians should have sufficient knowledge of basic foot assessment and shoe fitting and styling to enable provision of clear information. General shoe fitting principles include an understanding of posture, overall length, hell-to-ball length and flex angle, ball joint-to-toe length, heel and seat fit, instep, waist and arch fit, joint width and vamp, top line, throat and entry, patterns and styles, and measurement and size systems. Ideally, they should also have some knowledge of the local retail footwear industry for practical advice.

Interactions between foot and shoe are complex and variable. The clinician is usually best served, therefore, to provide the client with a list of footwear characteristics matched generally to their particular foot classification or problem, rather than endorsing a specific model or brand of shoe (McPoil, 2000). As discussed previously, various factors can affect foot posture including age and disease processes. Also, while it has been sown consistently that foot postures in males and females do not appear to be systematically different, there are of course significant differences between the genders in their demands of foot wear, and these do not necessarily reconcile readily with the limitations imposed by varying foot postures. The sizes, shapes, and postures of elderly feet differ from those of young adults, and given that footwear for the elderly is typically designed and made using the same general lasts last as for young persons, fitting shoes for the older person can be difficult for more than aesthetic reasons (Davis et al. 2006). In the pronated foot, typical of the older person, medial bulge and lateral concavity of the lateral foot create "outflare," factors asserted to relate to the fit and comfort of the shoe (Kouchi, 1995) and which require careful matching of a shoe design to the client. In one study of shoe fit in an elderly population, most of the subjects wore shoes narrower than their feet. Unsurprisingly, women in particular wore shoes that were shorter and narrower and had a reduced total area compared to their feet than did the men in the study. Wearing narrow shoes was further associated with corns on the toes, hallux valgus deformity and foot pain, and wearing shoes shorter than the foot was associated with lesser toe deformity. Wearing shoes with heel elevation greater than 25 mm was associated with hallux valgus and plantar calluses in women, highlighting the importance of foot posture and footwear matching in older people (Menz and Morris, 2005).

In sports people, risk factors that predispose people to lower extremity overuse injuries have been described previously in this chapter and include pes planus, pes cavus, restricted ankle dorsiflexion, and increased hindfoot inversion. Many athletes are less than ideal structurally and will exhibit either frank underlying pathology or a range of subtle range of compensations, many of which are theoretically amenable to intervention and possible correction through matching of footwear to the presenting foot type (Kaufman et al., 1999, Yamashita, 2005).

In a prospective study of the effect of the appropriateness of foot–shoe fit and training shoe type on the incidence of overuse injuries among infantry recruits, it was found that trade-offs between shoe size and width were commonplace but less important than shoe type. Notably, recruits who trained in basketball shoes, that is, shoes with many of the characteristics described earlier in this section, had a lower incidence of overuse injuries affecting the feet specifically than recruits who trained in more rigid infantry boots although the incidence of overuse injuries overall was the same in both groups (Finestone et al., 1992).

Ultimately, and especially for foot types at clearly identifiable increased risk of pathology, the goal is to match the footwear to the foot type explicitly. Customized footwear derived from extensive databases of individual customer information and industry templates is in the pipeline although progress remains slow due to technical limitations and the inherent subjectivity of issues such as defining and quantifying "fit." Although both two-dimensional and three-dimensional objective approaches have been outlined in search of the perfect match between foot and shoe (Luximon et al., 2003), in the short term at least, simple measures such as the FPI-6, in combination with an experienced eye are likely to remain central to good shoe fitting.

ACKNOWLEDGMENTS

The author is grateful for the assistance of Professors RA Ouvrier, J Crosbie, and J Burns and Drs AM Keenan, AM Evans, and R Scharfbillig for their assistance in the program of work leading to the finished FPI, to all the others who have contributed to validation studies and the various translations, and finally to Richard Wilkins and Lorraine Loughrey for their assistance in bringing this chapter up to date.

QUESTIONS

4.1 What are the limitations of using single plane measures to infer total foot posture or function?

4.2 What are the main limitations of attempting to infer dynamic foot function from static assessment of foot posture?

4.3 How much practice and experience is needed to be able to use the FPI-6 reliably in a clinical/retail setting?

4.4 Why is it useful to have a table to transform ordinal data to an interval scale?

4.5 Do all adults exhibit similar ranges of foot postures?

4.6 What are the factors affecting fitting of footwear to feet of different postures and functions in (a) males and (b) females?

4.7 To what extent does assessment of foot posture predict pathology or injury?

REFERENCES

Alexander, I. J., Chao, E. Y., and Johnson, K. A. 1990. The assessment of dynamic foot-to-ground contact forces and plantar pressure distribution: A review of the evolution of current techniques and clinical applications. *Foot and Ankle,* 11, 152–167.

Astrom, M. and Arvidson, T. 1995. Alignment and joint motion in the normal foot. *Journal of Orthopaedic and Sports Physical Therapy,* 22, 216–222.

Atkinson-Smith, C. and Betts, R. P. 1992. The relationship between footprints, foot pressure distributions, rearfoot motion and foot function in runners. *The Foot,* 2, 148–154.

Aurichio, T. R., Rebelatto, J. R., and De Castro, A. P. 2011. The relationship between the body mass index (BMI) and foot posture in elderly people. *Archives of Gerontology and Geriatrics,* 52, e89–e92.

Ball, P. and Johnson, G. R. 1993. Reliability of hindfoot goniometry when using a flexible electrogoniometer. *Clinical Biomechanics,* 8, 13–19.

Barton, C., Bonanno, D., Levinger, P., and Menz, H. 2010. Foot and ankle characteristics in patellofemoral pain syndrome: A case control and reliability study. *Journal of Orthopaedic and Sports Physical Therapy,* 40, 286–296.

Barton, C., Menz, H., and Crossley, K. 2011. The immediate effects of foot orthoses on functional performance in individuals with patellofemoral pain syndrome. *British Journal of Sports Medicine,* 45, 193.

Beeson, P., Phillips, C., Corr, S., and Ribbans, W. J. 2009. Hallux rigidus: A cross-sectional study to evaluate clinical parameters. *The Foot,* 19, 80–92.

Beighton, P., Solomon, L., and Soskolne, C. L. 1973. Articular mobility in an African population. *Annals of the Rheumatic Diseases,* 32, 413–418.

Bennett, M. E., Tulloch, J. F., Vig, K. W., and Phillips, C. L. 2001. Measuring orthodontic treatment satisfaction: Questionnaire development and preliminary validation. *Journal of Public Health Dentistry,* 61, 155–160.

Biro, F., Gewanter, H. L., and Baum, J. 1983. The hypermobility syndrome. *Pediatrics,* 72, 701–706.

Burns, J. and Crosbie, J. 2005. Weight bearing ankle dorsiflexion range of motion in idiopathic pes cavus compared to normal and pes planus feet. *The Foot,* 15, 91–94.

Burns, J., Crosbie, J., Hunt, A., and Ouvrier, R. 2005a. The effect of pes cavus on foot pain and plantar pressure. *Clinical biomechanics,* 20, 877–882.

Burns, J., Crosbie, J., Ouvrier, R., and Hunt, A. 2006. Effective orthotic therapy for the painful cavus foot: A randomized controlled trial. *Journal of the American Podiatric Medical Association,* 96, 205.

Burns, J., Keenan, A.-M., and Redmond, A. 2005b. Foot type and overuse injury in triathletes. *Journal of the American Podiatric Medical Association,* 95, 235–241.

Cain, L. E., Nicholson, L. L., Adams, R. D., and Burns, J. 2007. Foot morphology and foot/ankle injury in indoor football. *Journal of Science and Medicine in Sport,* 10, 311–319.

Cashmere, T., Smith, R., and Hunt, A. 1999. Medial longitudinal arch of the foot: Stationary versus walking measures. *Foot and Ankle International,* 20, 112–118.

Cavanagh, P. R., Morag, E., Boulton, A. J., Young, M. J., Deffner, K. T., and Pammer, S. E. 1997. The relationship of static foot structure to dynamic foot function. *Journal of Biomechanics,* 30, 243–250.

Cavanagh, P. R. and Rodgers, M. M. 1987. The arch index: A useful measure from footprints. *Journal of Biomechanics,* 20, 547–551.

Coplan, J. A. 1989. Rotational motion of the knee: A comparison of normal and pronating subjects. *Journal of Orthopaedic and Sports Physical Therapy,* 10, 366–369.

Dahle, L. K., Mueller, M., Delitto, A., and Diamond, J. E. 1991. Visual assessment of foot type and relationship of foot type to lower extremity injury. *Journal of Orthopaedic and Sports Physical Therapy,* 14, 70–74.

Dijkstra, P. U., De Bont, L. G., Van der Weele, L. T., and Boering, G. 1994. Joint mobility measurements: Reliability of a standardized method. *Cranio,* 12, 52–57.

Elveru, R. A., Rothstein, J. M., Lamb, R. L., and Riddle, D. L. 1988. Methods for taking subtalar joint measurements: A clinical report. *Physical Therapy,* 68, 678–682.

Engsberg, J. R. 1987. A biomechanical analysis of the talocalcaneal joint—In vitro. *Journal of Biomechanics,* 20, 429–442.

Engsberg, J. R. and Andrews, J. G. 1987. Kinematic analysis of the talocalcaneal/talocrural joint during running support. *Medicine and Science in Sports and Exercise,* 19, 275–284.

Fairbank, J. C., Pynsent, P. B., and Phillips, H. 1984. Quantitative measurements of joint mobility in adolescents. *Annals of the Rheumatic Diseases,* 43, 288–294.

Finestone, A., Shlamkovitch, N., Eldad, A., Karp, A., and Milgrom, C. 1992. A prospective study of the effect of the appropriateness of foot-shoe fit and training shoe type on the incidence of overuse injuries among infantry recruits. *Military Medicine,* 157, 489–490.

Forriol C. F., Maiques, J. P., Dankloff, C., and Gomez Pellico, L. 1990. Foot morphology development with age. *Gegenbaurs Morphologisches Jahrbuch,* 136, 669–676.

Forriol, F. and Pascual, J. 1990. Footprint analysis between three and seventeen years of age. *Foot and Ankle,* 11, 101–104.

Franco, A. H. 1987. Pes cavus and pes planus. Analyses and treatment. *Physical Therapy,* 67, 688–694.

Freychat, P., Belli, A., Carret, J. P., and Lacour, J. R. 1996. Relationship between rearfoot and forefoot orientation and ground reaction forces during running. *Medicine and Science in Sports and Exercise,* 28, 225–232.

Gadbury-Amyot, C. C., Williams, K. B., Krust-Bray, K., Manne, D., and Collins, P. 1999. Validity and reliability of the oral health-related quality of life instrument for dental hygiene. *Journal of Dental Hygiene,* 73, 126–134.

Giallonardo, L. M. 1988. Clinical evaluation of foot and ankle dysfunction. *Physical Therapy,* 68, 1850–1856.

Hawes, M. R., Nachbauer, W., Sovak, D., and Nigg, B. M. 1992. Footprint parameters as a measure of arch height. *Foot and Ankle,* 13, 22–26.

Hennig, E. M., Staats, A., and Rosenbaum, D. 1994. Plantar pressure distribution patterns of young school children in comparison to adults. *Foot and Ankle International,* 15, 35–40.

Hernandez, L., Chang, C. H., Cella, D., Corona, M., Shiomoto, G., and Mcguire, D. B. 2000. Development and validation of the satisfaction with pharmacist scale. *Pharmacotherapy,* 20, 837–843.

Hunt, A. E., Fahey, A. J., and Smith, R. M. 2000. Static measures of calcaneal deviation and arch angle as predictors of rearfoot motion during walking. *Australian Journal of Physiotherapy,* 46, 9–16.

Irving, D., Cook, J., Young, M., and Menz, H. 2007. Obesity and pronated foot type may increase the risk of chronic plantar heel pain: A matched case-control study. *BMC Musculoskeletal Disorders,* 8, 41.

Johnson, J. E. and Timins, M. E. 1998. Optimal computed tomography imaging of the midfoot: An improved technique. *Foot and Ankle International,* 19, 825–829.

Jonson, S. R. and Gross, M. T. 1997. Intraexaminer reliability, interexaminer reliability, and mean values for nine lower extremity skeletal measures in healthy naval midshipmen. *Journal of Orthopaedic and Sports Physical Therapy,* 25, 253–263.

Kaufman, K. R., Brodine, S. K., Shaffer, R. A., Johnson, C. W., and Cullison, T. R. 1999. The effect of foot structure and range of motion on musculoskeletal overuse injuries. *AmericanJournal of Sports Medicine,* 27, 585–593.

Keenan, A. M., Redmond, A. C., Horton, M., Conaghan, P. G., and Tennant, A. 2007. The foot posture index: Rasch analysis of a novel, foot-specific outcome measure. *Archives of Physical Medicine and Rehabilitation,* 88, 88–93.

Kelly, E. D. 1947. A comparative study of the structure and function of normal, pronated and painful feet among children. *Research Quarterly,* 18, 291–312.

Kessler, R. M. 1983. The ankle and hindfoot. In Kessler, R. M. and Hertling, D. (eds.), *Management of Common Muscular Disorders.* Philadelphia, PA: Harper Row, pp. 267–269.

Kitaoka, H. B., Lundberg, A., Zong Ping, L., and An, K. N. 1995. Kinematics of the normal arch of the foot and ankle under physiologic loading. *Foot and Ankle International,* 16, 492–499.

Kitaoka, H. B., Luo, Z. P., and An, K. N. 1998. Three-dimensional analysis of flatfoot deformity: Cadaver study. *Foot and Ankle International,* 19, 447–451.

Kouchi, M. 1995. Analysis of foot shape variation based on the medial axis of foot outline. *Ergonomics,* 38, 1911–1920.

Kouchi, M. and Tsutsumi, E. 1996. Relation between the medial axis of the foot outline and 3-D foot shape. *Ergonomics,* 39, 853–861.

LaPointe, S. J., Peebles, C., Nakra, A., and Hillstrom, H. 2001. The reliability of clinical and caliper-based calcaneal bisection measurements. *Journal of the American Podiatric Medical Association,* 91, 121–126.

Levinger, P., Menz, H. B., Fotoohabadi, M., Feller, J., Bartlett, J., and Bergman, N. 2010. Foot posture in people with medial compartment knee osteoarthritis. *Journal of Foot and Ankle Research,* 3, 29.

Luximon, A., Goonetilleke, R. S., and Tsui, K. L. 2003. Foot landmarking for footwear customization. *Ergonomics,* 46, 364–383.

McClay, I. S. 2001. Report of "Static and dynamic classification of the foot" meeting, Annapolis, Maryland, May 2000. *Journal of Orthopedic and Sports Physical Therapy,* 31, 158.

McDowell, B. C., Hewitt, V., Nurse, A., Weston, T., and Baker, R. 2000. The variability of goniometric measurements in ambulatory children with spastic cerebral palsy. *Gait & Posture,* 12, 114–121.

McPoil, T. G. 2000. Athletic footwear: Design, performance and selection issues. *Journal of Science and Medicine in Sport,* 3, 260–267.

McPoil, T. G. and Cornwall, M. W. 1996a. The relationship between static lower extremity measurements and rearfoot motion during walking. *Journal of Orthopaedic and Sports Physical Therapy,* 24, 309–314.

McPoil, T. G. and Cornwall, M. W. 1996b. Relationship between three static angles of the rearfoot and the pattern of rearfoot motion during walking. *Journal of Orthopaedic and Sports Physical Therapy,* 23, 370–375.

McPoil, T. G. and Hunt, G. C. 1995. Evaluation and management of foot and ankle disorders: Present problems and future directions. *Journal of Orthopaedic and Sports Physical Therapy,* 21, 381–388.

Menz, H. B. 1998. Alternative techniques for the clinical assessment of foot pronation. *Journal of the American Podiatric Medical Association,* 88, 119–129.

Menz, H. B. and Morris, M. E. 2005. Footwear characteristics and foot problems in older people. *Gerontology,* 51, 346–351.

Menz, H. B. and Munteanu, S. E. 2005. Validity of 3 clinical techniques for the measurement of static foot posture in older people. *Journal of Orthopaedic and Sports Physical Therapy,* 35, 479–486.

Morag, E. and Cavanangh, P. R. 1999. Structural and functional predictors of regional peak pressures under the foot during walking. *Journal of Biomechanics,* 32, 359–370.

Moseley, A. M., Crosbie, J., and Adams, R. 2001. Normative data for passive ankle plantarflexion-dorsiflexion flexibility. *Clinical Biomechanics,* 16, 514–521.

Mueller, M. J., Host, J. V., and Norton, B. J. 1993. Navicular drop as a composite measure of excessive pronation. *Journal of the American Podiatric Medical Association,* 83, 198–202.

Nester, C. J. 1997. Rearfoot complex: A review of its interdependent components, axis orientation and functional model. *Foot,* 7, 86–96.

Nigg, B. M., Cole, G. K., and Nachbauer, W. 1993. Effects of arch height of the foot on angular motion of the lower extremities in running. *Journal of Biomechanics,* 26, 909–916.

Picciano, A. M., Rowlands, M. S., and Worrell, T. 1993. Reliability of open and closed kinetic chain subtalar joint neutral positions and navicular drop test. *Journal of Orthopaedic and Sports Physical Therapy,* 18, 553–558.

Pierrynowski, M. R. and Smith, S. B. 1996. Rear foot inversion/eversion during gait relative to the subtalar joint neutral position. *Foot and Ankle International,* 17, 406–412.

Pierrynowski, M. R., Smith, S. B., and Mlynarczyk, J. H. 1996. Proficiency of foot care specialists to place the rearfoot at subtalar neutral. *Journal of the American Podiatric Medical Association,* 86, 217–223.

Rasch, G. 1960. *Probabilistic Models for Some Intelligence and Attainment Tests.* Chicago, IL: University of Chicago Press.

Razeghi, M. and Batt, M. E. 2002. Foot type classification: A critical review of current methods. *Gait & Posture,* 15, 282–291.

Redmond, A. C., Crane, Y. Z., and Menz, H. B. 2008. Normative values for the foot posture index. *Journal of Foot and Ankle Research,* 1, 6.

Redmond, A. C., Crosbie, J., and Ouvrier, R. A. 2006. Development and validation of a novel rating system for scoring standing foot posture: The foot posture index. *Clinical Biomechanics (Bristol, Avon),* 21, 89–98.

Redmond, A. C., Keenan A. M. 2012. Foot problems. In Warburton, L. (ed.), *The New MRCGP Handbook of Musculoskeletal Problems.* London, U.K.: Royal College of General Practitioners, pp. 121–155.

Riddiford-Harland, D. L., Steele, J. R., and Storlien, L. H. 2000. Does obesity influence foot structure in prepubescent children? *International Journal of Obesity and Related Metabolic Disorders,* 24, 541–544.

Rose, G. K. 1991. Pes planus. In Jahss, M. (ed.), *Disorders of the Foot and Ankle: Medical and Surgical Management.* 2nd edn. Philadelphia, PA: WB Saunders, pp. 892–920.

Saltzman, C. L., Domsic, R. T., Baumhauer, J. F., Deland, J. T., Gill, L. H., Hurwitz, S. R., Kitaoka, H. B., Mcclouskey, L. C., and Porter, D. 1997. Foot and ankle research priority: Report from the Research Council of the American Orthopaedic Foot and Ankle Society. *Foot and Ankle International,* 18, 447–448.

Scherer, P. R. and Sobiesk, G. A. 1994. The center of pressure index in the evaluation of foot orthoses in shoes. *Clinics in Podiatric Medicine and Surgery,* 11, 355–363.

Schwend, R. M. and Drennan, J. C. 2003. Cavus foot deformity in children. *Journal of the American Academy of Orthopaedic Surgeons,* 11, 201–211.

Scott, G., Menz, H. B., and Newcombe, L. 2007. Age-related differences in foot structure and function. *Gait & Posture,* 26, 68–75.

Sell, K. E., Verity, T. M., Worrell, T. W., Pease, B. J., and Wigglesworth, J. 1994. Two measurement techniques for assessing subtalar joint position: A reliability study. *Journal of Orthopaedic and Sports Physical Therapy,* 19, 162–167.

Sobel, E. and Levitz, S. J. 1997. Reappraisal of the negative impression cast and the subtalar joint neutral position. *Journal of the American Podiatric Medical Association,* 87, 32–33.

Spink, M. J., Fotoohabadi, M. R., Wee, E., Hill, K. D., Lord, S. R., and Menz, H. B. 2011. Foot and ankle strength, range of motion, posture, and deformity are associated with balance and functional ability in older adults. *Archives of Physical Medicine and Rehabilitation,* 92, 68–75.

Staheli, L. T. 1987. Evaluation of planovalgus foot deformities with special reference to the natural history. *Journal of the American Podiatric Medical Association,* 77, 2–6.

Staheli, L. T., Chew, D. E., and Corbett, M. 1987. The longitudinal arch. A survey of eight hundred and eighty-two feet in normal children and adults. *Journal of Bone and Joint Surgery,* 69A, 426–428.

Toro, B., Nester, C. J., and Farren, P. C. 2003. The status of gait assessment among physiotherapists in the United Kingdom. *Archives of Physical Medicine and Rehabilitation,* 84, 1878–1884.

Tweed, J. L., Campbell, J. A., and Avil, S. J. 2008. Biomechanical risk factors in the development of medial tibial stress syndrome in distance runners. *Journal of the American Podiatric Medical Association,* 98, 436.

Volpon, J. B. 1994. Footprint analysis during the growth period. *Journal of Pediatric Orthopedics,* 14, 83–85.

Walters, S. J., Campbell, M. J., and Lall, R. 2001. Design and analysis of trials with quality of life as an outcome: A practical guide. *Journal of Biopharmaceutical Statistics,* 11, 155–176.

Wearing, S. C., Hills, A. P., Byrne, N. M., Hennig, E. M., and Mcdonald, M. 2004. The arch index: A measure of flat or fat feet? *Foot and Ankle International,* 25, 575–581.

Weiner-Ogilvie, S. and Rome, K. 1998. The reliability of three techniques for measuring foot position. *Journal of the American Podiatric Medical Association,* 88, 381–386.

Williams, D., McClay, I., and Laughton, C. 1999. A comparison of between day reliability of different types of lower extremity kinematic variables in runners. *23rd Annual Meeting of the American Society of Biomechanics,* October 21–23, 1999, University of Pittsburgh, Pittsburgh, PA.

Woodburn, J., Udupa, J. K., Hirsch, B. E., Wakefield, R. J., Helliwell, P. S., Reay, N., O'Connor, P., Budgen, A., and Emery, P. 2002. The geometric architecture of the subtalar and midtarsal joints in rheumatoid arthritis based on magnetic resonance imaging. *Arthritis and Rheumatism,* 46, 3168–3177.

Yamashita, M. H. 2005. Evaluation and selection of shoe wear and orthoses for the runner. *Physical Medicine and Rehabilitation Clinics of North America,* 16, 801–829.

Yates, B. and White, S. 2004. The incidence and risk factors in the development of medial tibial stress syndrome among naval recruits. *American Journal of Sports Medicine,* 32, 772–780.

Part II

Scanning and Processing

5 3D Shape Capture of Human Feet and Shoe Lasts

Carl G. Saunders and Jeffrey Chang

CONTENTS

5.1 INTRODUCTION

One of the key requirements of appropriate footwear is "good fit." For the user/wearer, this goes beyond considerations of just being too big or too small. To ensure a good fit, shoes must provide support for surfaces of the foot that can tolerate pressure and accommodate regions which cannot. Shoes must also protect the foot from harsh elements in the environment and, at the same time, work in concert with the biomechanics required to ambulate efficiently.

There are many turn-key systems available today that map the complete three-dimensional (3D) shape of a human foot. These provide valuable insights into the 3D shape requirements associated with the design of a shoe to fit a foot for optimal comfort and performance. That said, selecting a scanning system to meet the needs of a particular footwear application requires careful consideration of many issues.

For the construction of shoes, one needs to consider measurement specifications for both human feet and shoe lasts (these are the molds from which the shoes are manufactured). For feet, all surfaces (plantar aspect, dorsal aspect, heel, Achilles tendon, malleoli, and tops and end of toes) are important, with the possible exception of the areas in between individual toes. For shoe lasts, not only are all surfaces important, but there are also two other shape elements—the primary and secondary featherlines—which are unique to shoe lasts in comparison to feet. The primary and secondary featherlines each consist of a sharp edge where two different 3D surfaces intersect. While it can be debated as to whether accurate data describing the geometry of the secondary featherline is essential, the primary featherline (where the upper meets the sole of the shoe) is critical to subsequent processes in shoe manufacturing. This type of edge is not present in human feet, where the entire 3D surface is anatomical and therefore smoothly curved. Consequently, for both feet and shoe lasts, it is important to capture the full 3D surface and, with shoe lasts, there is an added requirement of accurately capturing the 3D space curve corresponding to the primary featherline.

This chapter reviews several 3D shape capture technologies. It is not intended to be an exhaustive analysis of all available systems nor an endorsement of particular brands, but rather an overview of the different kinds of 3D capture technologies that are used in the footwear field. To assist in the selection of an appropriate system for a particular footwear application, this chapter presents criteria that should be investigated prior to acquisition. After the overview, a method for evaluating the measurement performance of candidate systems is presented.

5.2 OVERVIEW OF CURRENT TECHNOLOGIES

The scanning systems described in this section are practical examples of specific 3D capture technologies. Descriptions include how the underlying technology works, one or more sample implementations of that technology, typical 3D measurement performance, and advantages and disadvantages for measuring the shape of feet and shoe lasts.

5.2.1 PROJECTED LASER LINE, LINES, OR PLANE

5.2.1.1 How It Works

A laser projects a line or series of lines onto a surface. A camera, or series of cameras, is tuned to that wavelength, often with filters. The computer system then takes the curve that the camera sees, along with the known position and orientation in space of the camera, and creates a 3D curved line in XYZ coordinates. The system continues acquiring pieces of data in this manner and then assembles all the pieces together to form the 3D representation of the entire object.

5.2.1.2   Individual Implementations

5.2.1.2.1   *Yeti (Vorum Research Corporation, Canada)*
The Yeti is a noncontact laser scanner system designed to collect external 3D surface data on a number of parallel slices through a given foot or shoe last shape. The foot or shoe last rests on a Plexiglas plate and the scanner performs refraction correction on the views from the opposite side of the plate.

Four laser sources combine to shine a flat beam, which results in a circumferential line of reflected light around the surface of interest. Eight cameras record positions along this bright line of light; this is repeated at each successive slice position. From these data, the Yeti software calculates the 3D shape of the foot or the shoe last.

The cameras and lasers are housed in two shielded camera carriages, one on each side of the Yeti. The two camera carriages are mechanically interconnected and move together along the scanning range via an electromechanical drive system. High-quality measurement performance can be achieved over a wide range of object colors and surface textures by using different scanning scenarios (optimized system parameters and camera settings).

The default data produced are a proprietary (structured) point cloud and a proprietary slice-based data set. Optional software is available to yield various industry standard file formats (Figure 5.1).

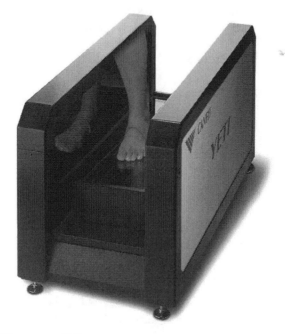

FIGURE 5.1   3D scanner—Yeti.

Typical performance:

1. Scan time is 4 s or less (depends on foot length).
2. Reconstruction time is fast, virtually real-time (<0.5 s).
3. Accuracy reported as ±0.5 mm radially (although ±0.25 mm is achievable).

Advantages/disadvantages:

1. Quick scan time reduces the likelihood of object movement during the scan.
2. Extremely simple single button operation.
3. Scans non-weight-bearing through to full weight-bearing.
4. Due to the form factor of the scanner, the maximum height of the scan is restricted to 18 cm. Therefore, shins and calves cannot be scanned for high boot applications.
5. Transportable, as distinct from portable. For example, it can be moved to another location, but it cannot be carried with you.
6. Several file formats are supported, including the ubiquitous STL format.

5.2.1.2.2 scanGogh II (Vorum Research Corporation, Canada)

This device is a handheld, wand-type scanner with a single camera and a single laser source. A special transmitter that emits a small magnetic field is used for position/orientation tracking. This interacts with a small electromagnetic sensor mounted on the wand. Based on the field strength (direction and magnitude), the system can determine the position and orientation of the wand in space and thus reconstruct the 3D space curves as they are acquired. Proprietary software post-processes the raw data, removes unwanted noise, fills holes, smooths bumps, and removes redundant data (Figure 5.2).

Typical performance:

1. Scan time varies widely with the size of the shape and with operator experience. Typical scan time for a foot or shoe last is less than 1 min.
2. Reconstruction time ranges from 5 to 30 s depending on the size of shape being scanned, the resolution used, and amount of data collected.
3. Accuracy is ±0.5 mm radially.

Advantages/disadvantages:

1. Can scan non-weight-bearing through to full weight-bearing (through a Plexiglas plate).
2. Portable.
3. Larger scan volumes are achievable; hence, this scanner can be used for applications other than feet and shoe lasts.

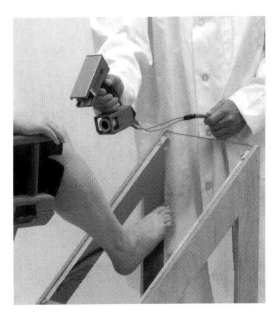

FIGURE 5.2 3D scanner—scanGogh II.

4. The presence of metal or electric fields in the immediate scanning environment can introduce errors in the 3D data obtained.
5. Numerous file formats are supported, including the STL format.

5.2.2 PROJECTED PATTERNS (PROJECTED WHITE LIGHT)

5.2.2.1 How It Works

Projectors cast a pattern or a series of patterns onto the target. Cameras acquire the images (with and without the pattern) and the system identifies surface points to create 3D shell surfaces. Proprietary software then assembles and blends this shell data to yield a 3D shape.

5.2.2.2 Individual Implementations

5.2.2.2.1 FotoScan 3D (Precision 3D, United Kingdom)

This is a fixed frame scanner with five projector/camera pairs that capture the entire foot. The foot or shoe last rests on a Plexiglas plate and the system compensates for the refraction (Figure 5.3).

Typical performance:

1. Scan time is approximately 3–4 s.
2. Reconstruction time is approximately 30 s.
3. Stated accuracy is ±0.5 mm radially.
4. File formats supported include STL, DXF, and VRML.

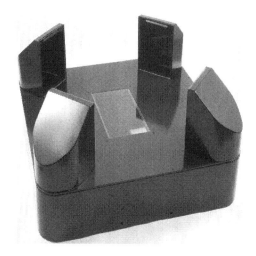

FIGURE 5.3   3D scanner—FotoScan 3D.

Advantages/disadvantages:

1. Simple single button operation.
2. Scans non-weight-bearing through to full weight-bearing (through a Plexiglas plate).
3. Due to the form factor of the scanner, the scan height is restricted to a maximum of 17 cm. Therefore, shins and calves cannot be scanned for high boot applications.

5.2.2.2.2   Artec M-Series (Artec 3D Scanners, Russia)

This handheld wand-type scanner incorporates a camera and a white light projector. The projector pulses a single pattern from which the system captures 3D surface information. Using various pattern matching algorithms, the system finds matching

FIGURE 5.4   3D scanner—Artec.

features between acquisitions and then "stitches" the surfaces together. Software exists to post-process the raw data, align shells, remove unwanted noise, fill holes, smooth bumps, and remove redundant data (Figure 5.4).

Typical performance:

1. Scan time varies widely with the size of the shape and with operator experience. Typical scan time for a foot or shoe last is less than 1 min.
2. Reconstruction time varies widely with the size of the shape being scanned; typical time for foot or shoe last shape range from 20 to 60 s.
3. Typical accuracy is ±0.5 mm radially.

Advantages/disadvantages:

1. Portable.
2. Larger scan volumes are possible; it can therefore be used for applications other than feet and shoe lasts.
3. Only a rough alignment of individual scan surfaces occurs in real time. Fine alignment must be performed after scanning. If there is a problem, it may not be apparent for several minutes.
4. Heavier and bulkier than other wand-type scanners.
5. Numerous file formats are supported, including the STL format.

5.2.2.2.3 FootScanner FTS-4 (Ideas, Belgium)

This is a fixed frame scanner with three projector/camera pairs. The foot is pressed into a foam impression box (such as from Biofoam), then the three subsystems respectively acquire the front/top, left, and right surfaces. The target is removed and

FIGURE 5.5   3D scanner—FootScanner FTS-4.

the top/front subsystem captures the negative of the bottom surface. This bottom surface is inverted and then stitched and blended to the other shell surfaces (Figure 5.5).

Typical performance:

1. Total acquisition time is approximately 10 s.
2. Stated accuracy is ±0.5 mm.
3. Reconstruction time varies due to interactive process required to stitch views together.

Advantages/disadvantages:

1. Semi-portable (collapses to be put into a case).
2. Simple push-button operation.
3. Due to the form factor of the scanner, maximum scan height is restricted to 25 cm.
4. Can only scan weight-bearing objects.
5. Plantar 3D shape can vary due to the use of foam box impressions.
6. Data output is a simple ASCII file (stated to be an open format).

5.2.3   PATTERNS PRE-APPLIED TO OBJECT BEING SCANNED

5.2.3.1   How It Works

Instead of projecting a pattern from the scanner onto the object, this method works by preparing the target object's surface with one or more patterns prior to the actual scan being done. Target patterns are preprinted onto various sizes of sock or tube, and an appropriate sized sock is then donned by the client and multiple digital images of the foot are acquired. Ultimately, this method works the same way as those in Section 5.2.2, but the origin of the pattern is different; projectors are not required.

5.2.3.2   Individual Implementations

5.2.3.2.1   Lightbeam Scanner (corpus.e, Germany)

This scanner is a fixed frame device with a round platform upon which the target foot shape rests. The foot is covered with a form fitting yellow and green sock called "Magical Skin" which has a unique pattern printed on it. A single arm with a camera and light source swings around the target, illuminating the pattern and capturing the images of the surfaces circumferentially around the foot. The shape of the underside of the foot is not measured. However, a pressure-sensitive mat can be used to collect pressure data from the plantar surface. All the pattern and pressure data are then sent away (via the Internet) to be analyzed and reconstructed on the scanner supplier's servers (Figure 5.6).

Typical performance:

1. Scan time is approximately 15 s.
2. Reconstruction time is variable based on Internet speeds and server availability.
3. Accuracy is ±1 mm.

FIGURE 5.6   3D scanner—Lightbeam and Magical Skin.

Advantages/disadvantages:

1. Simple push-button operation.
2. Potential to measure both feet in the same scanning operation.
3. Designed to scan feet in weight-bearing only.
4. Must have an Internet connection.
5. Shapes are reconstructed elsewhere; hence rely on the availability of the supplier's server computers.
6. Requires a specific patterned sock to be worn; new sock is needed for each client, or sock must be washed after each client use (for hygienic reasons).
7. Not suitable for shoe last measurement.
8. Supports many industry-standard file formats.

5.2.4 Mechanical Tracing/Digitizing—Instrumented Linkages

5.2.4.1 How It Works

This measuring device consists of an articulated arm with a pen-like probe. The end of the probe is traced manually across the surface of the target. Based on the angles at each of the articulations and the known arm lengths, the three dimensional position of the end of the probe is calculated. During the tracing process, many of these positions are acquired to yield an XYZ coordinate point cloud, effectively digitizing the surface of the shape.

5.2.4.2   Individual Implementation

5.2.4.2.1   MicroScribe (Solution Technologies Inc., United States)

Several models exist with varying working volumes (spheres with diameters from 127 to 167 cm) and precision (0.005–0.023 cm) specifications. The MicroScribe device has a base that can be mounted to a ceiling, a wall, or a table. As the probe is moved from one point to another, a button is clicked at each location where a measurement is desired (Figure 5.7).

Typical performance:

1. Acquisition time varies with the complexity of the surface being digitized and with operator skill. Shoe lasts can require tens of minutes or a few hours depending on the desired data density.
2. Reconstruction time varies with amount and quality of data obtained and smoothness requirements for final 3D surface data.
3. Accuracy is quoted as 0.023–0.43 mm, depending on the model.

Advantages/disadvantages:

1. Most accurate/precise of all scanners described herein.
2. The system is compact and, when not in use, can be folded up for storage
3. With a careful process which ensures good surface coverage and careful tracing of featherlines, this technology can be used for digitizing the 3D shape of shoe lasts.
4. As a contact-based system, this device is not practical for use on human feet (they will deform under contact and the tracing process takes too long).
5. Longest scan time (an acquisition can take hours) of all scanners described herein.
6. Can be used for objects other than shoe lasts.
7. Files can be exported in many industry standard formats.

FIGURE 5.7   3D scanner—MicroScribe.

5.3 SELECTING AN APPROPRIATE 3D SHAPE SCANNING SYSTEM FOR A FOOTWEAR APPLICATION

All of the scanners described in the previous sections, as well as many others not reviewed here, have applications in the footwear industry. A key question then is, how does one go about choosing an appropriate 3D capture system for a specific footwear application? The answer lies in determining how a particular machine's quoted performance relates to the user's (practitioner's) needs.

A logical approach to determining these "performance needs" is to list the shape measurement criteria, prioritize them, and establish acceptable minimum and maximum values for each. The criteria should include all of the ones discussed in the following, plus any others that are specific to the user's particular situation.

For foot scanning, there are many commercially available systems available. Hence, it should be possible to find a 3D scanner which meets most users' technical requirements/criteria.

For the 3D measurement of shoe lasts, the first step is to decide if accurate capture of the 3D geometry of the primary featherline is critically important. If it is, then the 3D capture options are limited to either an instrumented linkage system, where an operator can carefully trace along the featherline directly, or an optical system with high-resolution capabilities and specific software which allows the 3D featherline edge to be extracted automatically from the dense 3D point cloud obtained. With the featherline capture requirement defined, the focus should then be on accuracy and surface coverage.

For the 3D measurement of human feet, the main criteria to consider are speed of data capture, speed of data reconstruction, accuracy, surface coverage, and usability.

5.3.1 Speed of Data Capture

For capture speed, the axiom "faster is better" applies. This just makes the entire 3D measurement experience more acceptable for both the person (client) being scanned and the person doing the scanning. It also helps to reduce the possibility of distortion in the resulting shape data due to client movement.

5.3.2 Speed of Data Reconstruction

Data reconstruction speed relates to how long the operator and the client need to wait before being sure that the 3D capture worked properly. The same "faster is better" rule applies because if there is a problem in the data reconstruction phase, the 3D scan will often have to be repeated. If this occurs, the client may need to receive and understand new instructions, he may need to be repositioned, and he will definitely have to wait through another data capture and data reconstruction phase. The longer it takes to reconstruct the data to verify the 3D shape result, the more of an inconvenience it is for all involved and the less viable the scanning device.

5.3.3 Accuracy

The measurement accuracy requirements for human feet are typically less than those for shoe lasts. For practical purposes, diameters (or any other linear dimension) of the foot should be accurate to within ±1 mm. For most sizing purposes, the overall length of the foot needs to be accurate to within ±2 mm. Circumferences should be accurate to within ±3 mm. Shoe last measurements need to be at least twice as accurate as their equivalent foot measurements.

5.3.4 SURFACE COVERAGE

The importance of surface coverage varies depending on how the measurement data are used. For example, in a shoe-sizing application, data corresponding to the complete 3D surface of the foot are likely to have little value. One other item to consider is, whether or not the 3D surface of the plantar aspect of the foot is critical to the application. For example, if 3D foot shape data are captured while the client is weight-bearing, then the majority of the bottom of the foot will be flat with very little 3D curvature. On the other hand, in certain custom shoe applications (where a custom shoe last is a key objective of the design process), the complete surface of the foot including the instep, heel, plantar aspect, Achilles tendon, toes, and bunions (if present) needs to be well-captured.

5.3.5 USABILITY

While somewhat subjective, the usability of the scanner is often critical to its successful adoption for everyday use. "Easy to use" is the mantra here and nowhere is this more important than in a retail setting. One-button operation that results in a first-try, quick, and successful 3D scan is the ideal. That said, technical users working in a research or industry laboratory setting may tolerate more complexity, provided it gives them flexibility in terms of configuring the 3D measurement device for their changing needs.

Using the aforementioned criteria, it should be possible to identify some candidate 3D scanning systems. The next step is then to evaluate and confirm the performance of these systems for the user's specific measurement application. Key aspects of the performance evaluation include objective measurement tests, repeatability assessment, and post-processing effort.

5.3.6 MEASUREMENT TESTS

Assuming one can access a candidate system for evaluation, the next step is to conduct objective measurement tests on representative 3D sample shapes of known dimensions. At least one of these samples should be a machined geometric shape whose surface geometry is accurate to an order of magnitude better than the desired accuracy. Once the measurements are obtained, then these must be compared against the actual measurements of the test shapes to determine whether the desired measurement performance is achievable.

5.3.7 REPEATABILITY TESTING

Though often overlooked at this stage, repeatability testing is extremely important. Simply place one or more of the test objects in the 3D capture device in several nominally different positions and orientations and scan them multiple times. Then verify if the resulting 3D data, linear measurements, and accuracy are comparable to what was achieved in the first measurement tests given earlier.

5.3.8   Post-Processing Effort

Further, be sure to consider how much post-processing effort is required on the 3D data produced by the device. If the footwear application requires the 3D data to be used immediately in a time-critical task, then running tests in the proposed measurement environment will confirm if the 3D surface data resulting from the scan are both clean and complete (free of spurious data or holes in the shape) or if it can be quickly and easily edited (ideally within seconds). On the other hand, if the 3D data is the first step in an involved design process (e.g., designing a master sample shoe last that will ultimately be graded into various sizes), then it may be acceptable for the shoe last designer to spend time post-processing the data obtained from the 3D capture device. In this case, the post-processing effort would be much less of an issue.

The benefit of conducting the preceding tests is that it will help avoid surprises during both the introduction and subsequent operation of any new 3D shape capture device.

5.4   SUMMARY

This chapter has described some of the important issues and technologies related to the 3D measurement of human feet and shoe lasts.

For complete 3D measurement of the human foot, optical 3D capture systems are preferred. To determine which system best suits the needs of any given footwear application, list the measurement criteria for the application, being sure to include speed of data capture, time to reconstruct the 3D data, accuracy, surface coverage, and usability. Objective measurement and repeatability tests, prior to purchase, are absolutely critical to the successful implementation of a candidate 3D capture system.

For shoe last measurement, decide if accurate definition of the primary featherline is important. If so, instrumented linkages have been shown to be more accurate. However, instrumented linkages can be prone to human error and are time consuming to use. Optical 3D scanning systems are much faster, yet very few offer a simple and accurate means of extracting the featherline contours from the resulting 3D point cloud data.

By being aware of the important criteria and performance tests of candidate systems, prospective 3D shape capture users should find a "good fit," not only of their shoes and shoe lasts, but also of a 3D scanning system to their footwear application needs.

6 3D Surface Scanning

Sudhakar Rajulu and Brian D. Corner

CONTENTS

6.1 INTRODUCTION

The benefits of 3D surface scanning include its ability to save a surface scan of an object for future processing and verification, reduction of human error associated with manual anthropometric data collection, allowing for measurements to be obtained from them at a later time, and reduction in experiment time so as not to impede on the subject's time or time with the hardware/object. The scans also allow for surface area or volumetric data to be obtained, via post-processing. The benefits of surface area and volumetric anthropometric data to represent the variations in shape and size of human body can eventually enable designers and developers to enhance the overall accommodation, fit, and comfort for a wide range of population as well as to improve the ability to customize for individual needs. In contrast to traditional anthropometric data and databases, 3D scan information can benefit better analytical capabilities; for instance, with traditional anthropometric information as in a tabular form, it is difficult to create a specific individual based on his or her own anthropometry. The reason for this is that in a tabular form, anthropometric data become generic or percentile based and thus it is not possible to recreate a specific individual from a generic tabular data. This is not the case with 3D scan data. It is possible to create shapes and sizes that are representative of the specific individual. Even if one were to extract mere linear dimensions from the scanned image, it is quite possible to retrieve the scanned image of that specific individual and use the shape of that individual as the need arises. There are basic issues associated with extracting data from the 3D scan data. In this chapter, we will discuss the basics of 3D scanning and the ways to process 3D scan data.

6.2 BACKGROUND OF 3D SCANNING

Take a snapshot of your friend with your camera. In doing so, the radiation reflected from your friend is coded onto a radiation-sensitive medium such as photographic film or an electronic sensor. The radiation typically is in the ultraviolet (UV), visible, and near-infrared (IR) spectrum. At those energy levels radiation does not penetrate into your friend's body as do, for example, x-rays, and thus, the photograph is of the object surface. 3D scanners meld photography, electronics, and computation to record an image. As with the child of photographic technology, 3D scanning captures the radiation reflected off an object through the use of electronic sensors. These scanners operate in the near-IR through to UV spectra, although recently the development of millimeter and terahertz scanners has pushed the boundaries to higher energy levels in recent years (Corner et al. 2010, Rodriguez-Quiñonez et al. 2011).

Scanning your friend, like taking a photo, records the body surface. Unlike a snapshot, 3D surface scanners provide depth to an otherwise flat surface. Depth, or range, is generated through three techniques—triangulation of a light point or stripe usually from a laser source, the deformation of structured light patterns projected onto a surface, and stereophotogrammetry (including silhouettes). In all cases, the result is a set of geometric points in space; the point cloud may be used to reconstruct the object's surface. The main difference between the scanning technologies may be summarized by contrasting acquisition time and time to render the point cloud or mesh. Laser scanners have a relatively slow acquisition time, on the order of 15–20 s depending on scanning volume size and point-cloud density, but render in a few seconds. Structured light and stereo-photo systems have very fast capture times, a few seconds, but take longer to display the point cloud, a minute or two is common. Time to capture and time to render become important when large numbers of scans are being taken such as in anthropometric surveys (e.g., CAESAR, SizeUK, SizeKorea, SizeUSA, MC-ANSUR, and ANSUR-II). Subjects may not have time to wait while the scan is processed enough to be evaluated by the scanner operator. Too slow a capture time, however, opens the possibility of subject movement (Corner and Hu 1998, Daanen et al. 1997). Irrespective of the data capture method, all surface scanners record what is in their line-of-sight; thus, surfaces tangent to the

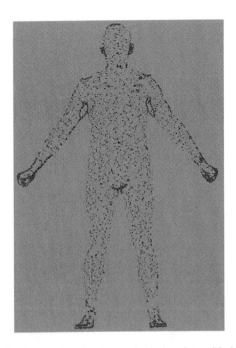

FIGURE 6.1   Whole-body scan showing areas of missing data with dark outline.

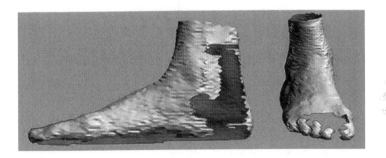

FIGURE 6.2   Feet segmented from a 3D whole-body scan showing holes in scan image.

camera view such as top-of-head and top of shoulders, for example, and occluded surfaces are not recorded and would register as holes or voids in the point-cloud data (Figure 6.1). Unless cameras are positioned to look down at their dorsal surface, feet are often poorly represented in most whole-body scanner system scans (Figure 6.2).

6.3   3D SCANNING SYSTEMS

The majority of human body scanners are stand-alone systems that record a whole-body, head only, or foot only. Torso and single-limb scanners are available as well. Single-limb scanners are most often used in the creation of prosthetics. Scanners specifically for the hand are relatively rare and reflect the difficulty of capturing the multiple surfaces presented (contrast hand and foot surfaces). Stand-alone systems contain multiple cameras fixed to a support structure. In laser-based scanners, cameras are attached to a motion system that moves the laser line over (movement of a horizontal laser line) or around the body (movement of a vertical laser line). In stereo photo and patterned light systems, enough cameras to cover the scan volume are attached to a tripod or rigid frame. To reduce interference, individual cameras are synchronized to fire in rapid succession. All systems have their own method for calibration, and all must be calibrated relatively frequently for optimal performance.

6.3.1 Types of Scanning Systems

In this section, we will discuss the most common types of scanning systems.

6.3.1.1 Stereophotogrammetry

Stereophotogrammetry is a technique that estimates the 3D coordinates of an object using multiple photographic images taken from different positions, front and side, for example, that have similar points in each camera's field of view (FOV). The photographic images from each camera pair must be captured at the same exact time, and must have known distance from the camera to the object. The pros for this technique are that it captures data very fast and thus may be used in cases where the subject is unstable, such as standing on one leg or reaching. Stereo-photo systems are particularly suited for scanning children and for capturing facial movements. The systems are relatively lightweight and portable, and some systems may even be reconfigured to support changes in scan volumes (e.g., 3dMD whole-body system, www.3dmd.com). A limitation with the technique is the relatively longer time to process all camera views into a coherent mesh. Also, the accuracy of the measurements depends on the quality of the photographs which is dependent on the properties of the camera, the required distance between the camera and the object, and the FOV of the camera.

6.3.1.2 White-Light Scanning

White-light scanning projects a pattern of white light, stripes are typical, onto the subject, and then the CCD cameras record the distortion of the light pattern to produce the 3D image. [TC]² scanners use this type of technology ([TC]², Cary, NC). Like the stereo-photo systems, white-light scanning has a fast capture time and is relatively easy to package into a small working footprint. The cons for white-light systems are relatively slow time-to-render and are sensitive to calibration. They also do not capture hair or dark colors.

6.3.1.3 Laser Scanning

Laser scanning is based on a triangulation method, using cameras and lasers. The laser is used as a light source and shined on the object. The cameras record the distortion or reflection of the laser using a coupled-charged device (CCD). The pros for these types of scanners are that they are quite robust to environmental conditions, are easy to calibrate, have fast time-to-render speeds, and tend to have a smaller overall system footprint. Some of the limitations are that they have a relatively slow capture time compared to photometric systems, hair and dark colors are not captured by laser scanner that use visible light (not a problem with infrared systems), they are rather bulky, and typically the camera systems are fixed and may not be reconfigured. Foot scanners tend to be built around laser systems because of the laser cameras may be placed closer to the foot and thereby minimize system size.

6.3.1.4   Millimeter-Wave Scanners

Millimeter-wave scanners use linear and radio wave technology to pass through clothing to scan a subject's body surface. The radio waves pass through the subject's clothing and reflect of the surface of the skin and are captured by an antenna array and are then processed to produce the 3D image (Corner et al. 2010). This is a similar technology to the scanners being implemented in airport security checks. The obvious pros for this system is its ability to scan through clothing, thus removing the need for subjects to don tight-fitting and revealing scan wear, a main impediment to large-scale commercial adaptation of 3D scanning technology. On the con side, the scan image is of relatively low resolution compared to the other systems (Corner et al. 2010).

6.4   SELECTING 3D SCANNERS

Depending on the technology being used for a scanner, a scanner can be small or large, handheld or stand-alone, portable or static, and sensitive to colors or metals. Also depending on how the device determines the location of the scanner head may affect the scan; for example, if the scanner uses a magnetic field to locate the scanner head in 3D space to the internal coordinate system, any metal around the object will distort the scan (Figure 6.3). Examples of handheld scanners are Faro Laser ScanArm and FastScan Polhemus. Processing capabilities for each scanner would vary from minimal to canned capabilities to customer-dependent capabilities.

Hence, selection of appropriate scanners would definitely hinge on the processing needs and capabilities.

Stand-alone scanners are built to be relatively easy to use push-button operations. Examples of stand-alone scanners are the Vitus Smart 3D whole-body scanner (Vitus/ Smart™, Vitronic GmbH, Wiesbaden, Germany), Pedus foot scanner (PEDUS, Vitronic GmbH, Wiesbaden, Germany), and [TC]^2 (TC-squared). Once a subject is prepared, that is, in appropriate clothing, with markers or other fiducials added, and positioned correctly, the operator starts the acquisition cycle and an image is quickly produced for review.

Although it is simple enough to operate, the ability for quick data capture and ease of use do come at a cost. Scan volume is defined by the support structure and motion mechanism. As such there is no way to fill in surfaces missed by the camera through occlusion or areas tangent to the sensor view. More often than not, the missing areas are relatively small compared to the overall surface and the trade-off is worth the cost.

In contrast to stand-alone systems, handheld scanners (Handyscan, FastSCAN, Artec, Faro) are versatile but sensitive to motion because they capture small areas at a time. For example, the laser line produced by the Faro line scanner (Faro Edge) is 4 in. (101 mm) wide (http://www.faro.com/edge/us/scanner). Handhelds are best used on

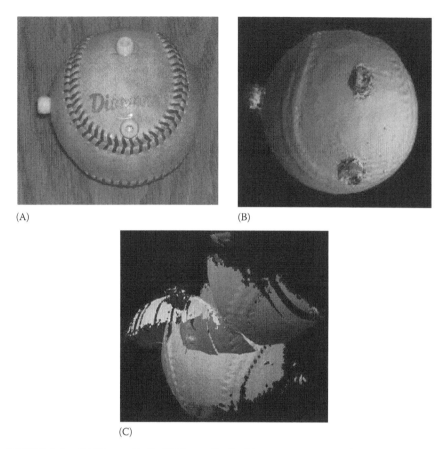

(A)   (B)

(C)

FIGURE 6.3 (A) Photo of ball. (B) Scan of ball with no metal around. (C) Scan of ball with metal 10 ft from receiver and transmitter.

more stable objects such as feet or someone who is in a braced pose (sitting down), for example. A handheld scanner may be used together with a stand-alone system to fill in areas missed by the larger scanner. Handheld scanners are smaller in size, less costly, and are more portable. Many of these devices can be used in confined spaces, and allow the operator to move about the capture space utilizing the device's coordinate system. Disadvantages of handheld scanners are that they require pre-evaluation planning, setup time, and space necessary to operate and move the hardware in order to reach all the desired points of interest. Not all scanners have the same limitations; therefore, it is important to research the scanner technology and application thoroughly before purchasing an investment such as a 3D scanner.

6.5 BASIC PROCESSING

A wide variety of 3D surface scanners are available that employ a number of imaging methods and technologies. In the end, however, they all produce the same thing, a point cloud. Saving a point cloud is the first step in the process of obtaining a useful surface scan. Post-processing sensor data starts with aligning and editing the point cloud then moves to polygonal mesh creation from the point cloud. Once a mesh is made, other editing options are available including mesh regularization, hole (void) filling, surface smoothing, and mesh decimation or subdivision.

6.5.1 CALIBRATION

Calibration handles arranging scans from multiple sensors. Additional adjustment may be done in overlapped regions using iterative closest point (ICP) or another registration method (Besl and McKay 1992, Blais et al. 1995, Rusinkiewicz and Levoy 2001). Once all images are aligned, the overlap between adjacent patches is removed. Which points are kept and which are deleted is not often obvious. Computing a confidence weight helps identify which points to save. In general, points closer to the center of a sensor are kept over those near an edge (Rusinkiewicz and Levoy 2001). Some noise is always present in the point cloud. For example, specularity and sharp transitions between texture, color, or geometry may cause a flare or spike in the image where points are projected from the surface. The edited point cloud is next turned into a polygonal mesh composed of triangles typically, but a quad mesh may also be built (Hoppe 1996, Lee and Lee 2003). Meshing provides a substrate for surface rendering and later is the basis for computing measurements (length, area, or volume), shape analysis, and animation, to name a few applications.

6.5.2 MESH REGULARIZATION

Meshes behave the best in applications when they are regular; that is, they are composed of equilateral triangles or squares that are distributed evenly over the object surface (Figure 6.4). Long thin mesh elements and intersecting faces may cause problems later. Post-processing steps to smooth a mesh and make it more regular are applied in most cases (Huang and Ascher 2007, Ohtake et al. 2000, Taubin 1995).

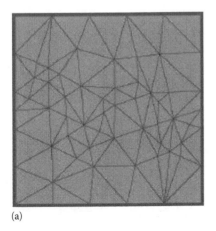
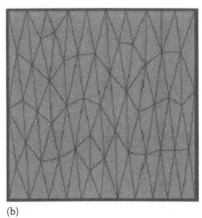

(a)   (b)

FIGURE 6.4 Original mesh (a) and smoothed mesh (b).

6.5.3 HOLE FILLING

One of the most important and difficult post-processing steps is filling holes (voids) in the mesh. A successful hole filling algorithm reconstructs the mesh such that the new surface matches the curvature of the adjacent known surface, that new and original surfaces blend together, and the surface is manifold (Carr et al. 2001). A number of approaches have been described (Bernardini et al. 1999). B-spline patches use the neighboring vertices to identify the four corners of a surface patch along with other vertices to determine the curvature of the patch (Douros et al. 1999). The triangulation method connects each point of the point-cloud data to create triangles that result in a polygon mesh surface. This technique works best with simple voids or nearly flat surfaces. It works less well with more complex shapes that include curves and creases due to the fact that it wants to connect all the data points with a straight edge, which may result in the fill not representing that actual geometry of the object (Figure 6.5).

New methods have improved hole filling in curved and other more complex surfaces (Brunton et al. 2009, Vollmer et al. 2009, Zhao et al. 2007). Figure 6.6 shows hole filling technique based on Hermite splines (Vollmer et al. 2009).

Most of the hole filling algorithms previously noted run in automatic mode. Software packages such as Polyworks, Geomagic, and Rapidform allow the user to fill all holes, all holes of a certain size (defined by, e.g., number or edges), or to step through and fill holes individually. Manual hole filling has its advantages when there are relatively few holes, the holes tend to be small and more-or-less flat, and there are relatively few scans to process. In this case, the user may fill simple holes quickly than spend more time editing and filling holes in more complex surfaces.

Typically, in cases of holes in more complex surfaces, the user segments the hole into more manageable parts by creating a bridge across a gap. One of the best

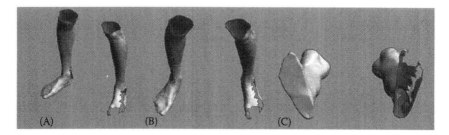

FIGURE 6.5 Comparison of filled and unfilled feet from a whole-body scan. (A) Unfilled holes, holes filled in one foot. (B) Dorsal view. (C) Plantar view.

FIGURE 6.6 Examples of a hole filling algorithm using Hermite splines.

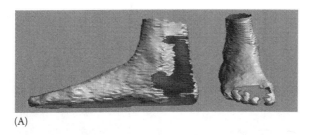

(A)

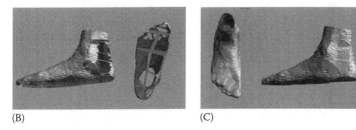

(B)   (C)

FIGURE 6.7 Filling holes in feet from a whole-body scan (with CySlice 2.0). (A) Unedited scan. (B) Bridging large gaps to approximate foot surface. (C) Reconstructed surface.

illustrations of interactive hole filling is reconstructing the feet in a human whole-body scan. In Figure 6.7, bridges are constructed across large gap thereby creating more-or-less flat holes from the initial curved hole. After bridging, the holes are filled with a triangulated surface. Processing scans will always be a combination of automatic and manual tasks. As hole filling algorithms improve, users will spend less time checking and editing scans.

6.5.4 BODY SEGMENTATION

With the surface scan captured, the occlusions filled, and the polygonal mesh water-tight, the scan is now capable of being segmented for isolating a specific body part. Take for example the design of a protective helmet for astronauts. A designer may need to isolate the heads of representative subject scans to determine if a computer-aided design (CAD) interfaces and accommodates these users. In order to isolate the segment of interest from the surface scan, a segmentation technique needs to be applied.

Unique tools are provided in most post-processing software packages that allow a research to segment a part of the body of interest by using cutting planes defined by the user. Figure 6.8 shows a surface scan with multiple horizontal and vertical planes used to segment the body using specific landmarks. Once again the investigator is at the liberty to segment the scan according to his or her liking and the segment can then be used for further analysis such as anthropometric measurement determination, surface area, or its volume.

Manual placement of segmentation planes allows the researcher to segment according to his or her needs, but it is often the case that a large number of scans require segmentation or that segmentation is one part of an application such as

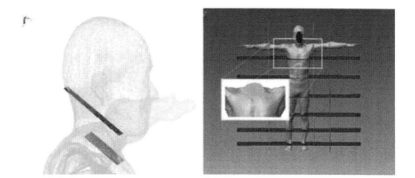

FIGURE 6.8 Segmenting the whole-body scan image.

extracting measuring body regions for clothing size prediction. Automating segmentation is possible through a standardized process across the entire sample. Nurre (1997) and Nurre et al. (2000) developed a robust function for identifying arm, leg, and neck cut planes automatically in whole-body scans. Others have provided automated segmentation functions based on 2D projection (Li et al. 2000) or body contours (Leong et al. 2007).

6.6   ANTHROPOMETRY

Anthropometric measurements are collected using standardized locations on the body established from select skeletal or soft tissue points (Ulijaszek and Mascie-Taylor 2005). In addition to these standard linear measurements, it is possible to obtain surface area and volumetric values, as well as segmental representations from the scanned data.

There are inherently sources of error in the collection of anthropometric data. In traditional anthropometry, these error sources are based on posture and human error. Posture-based error represents shifts in subject position during measurement, and human error can be from errors in locating landmarks, taking the measurement, recording the measurement, and entering the data. In scanner-based measurement extractions, the potential for measurement error includes that of traditional anthropometry as well as differences between the two techniques. For example, circumferential measurements using a tape measure compress the soft tissue around the site of measurement whereas the scanner does not incorporate compression. Past studies have shown that the differences between the two techniques are minimal as long as appropriate care has been taken to match measurement techniques and landmark locations between traditional and scanner-based extractions (Ben Azouz et al. 2006, Bradtmiller and Gross, 2003, Morency et al. 2005, Perkins et al. 2000).

In addition to capturing the overall shape of an object using 3D scanning for anthropometric consideration, there are other advantages to using 3D scanning instead of traditional anthropometry. The ability to save a surface scan as well as the speed of data collection are distinct benefits to 3D scanning. Saving a scanned image allows for future extraction of additional anthropometric data without requiring the scanned individual at a later date. The speed of data collection minimizes the scanned individual's test time. The 3D scans can also be used for additional analysis. However, experience has shown that it may be necessary to scan the subject in various poses to overcome some of the problems of occlusion encountered during scanning. This may increase the amount of subject's time; yet, the ability to extract both linear as well as other non-linear measurements makes the scanning method preferable in the long run.

6.6.1 LINEAR MEASUREMENT

The majority of linear anthropometric dimensions involve the point-to-point distance between two standard landmarks, point-to-point distance along the surface, point-to-plane (e.g., heights from the floor), circumferences, and arcs. Thus, two pieces of information are required for a measurement: (1) location of landmarks and (2) the type of measurement—point-to-point, point-to-plane, along the surface, etc.

Companies that specialize in 3D body scanning generally have measurement software bundled with their scanners. For example, Human Solutions uses Anthroscan™ software to measure a wide range of standard body dimensions from the International Standards Organization (ISO), the clothing industry, and military organizations (www.human-solutions.com/apparel/products_anthroscan_en.php). Similarly, [TC]² bundles their software, [TC]² Body Measurement System, with their whole-body scanners and it is available as a stand-alone program for use with other scanners (http://www.tc2.com/index_3dbodyscan.html). Body measurement is automatic for predefined dimensions and most software solutions allow the user to create and save their own measurements. It is worth observing that most of the automated measurements are obtained from minimally post-processed point clouds. For example, there may be some noise reduction but images from multicamera systems are not merged and holes are not filled. This saves computation time; image merging may take several minutes even on the fastest computer, and provides results quicker.

There are other software sources for body measurement. NatickMeasure was developed in-house by the U.S. Army anthropologists to obtain measurements from any scanner (Paquette et al. 2005), and mesh editing software like Polyworks, Geomagic, Rapidform, and MeshLab may also be used to extract body dimensions from 3D scans.

6.6.2 SURFACE CURVE MEASUREMENT

The main purpose of anthropometric data is to ensure workers/users with different body shapes and sizes can be accommodated in workplaces, such as crew cockpit, in a space suit, as well as have fitting shoes and clothes. While it is true that linear anthropometric measurements have been extensively used for these purposes, in reality, bulk of the fitting issues actually require accommodating the volumetric shape and size of individuals. While there are stringent requirements to measuring the linear measurements (such as standardized posture, proper landmarking, and repeated measurements), there are many relevant but important nonlinear dimensions that are more important and necessary for ensuring proper fit. Unfortunately, these nonlinear type measurements such as vertical trunk diameter (the diagonal distance from the mid shoulder to the crotch area) or spinal curvature are very difficult to standardize while using the anthropometer or tape measure. Instead, they can be measured repeatedly by using the 3D scan and this increases the possibility of taking relevant measurement from subjects who may exhibit different contours than compared to other subjects. Also, suit fitting and cloth fitting such as fire-fighter suit or jump suit could be better improved if we take advantage of the shape information provided by the 3D scanners.

6.6.3 SHAPE ANALYSIS

The adoption of 3D laser scanning has led to an increased ability to provide information on the overall shape of the human body as well as the ability to compare shape across the user population. Traditional anthropometric measurements are linear in nature and cannot describe the complex shape of the human body in its entirety. Placement and comparison of subjects within the context of the population using just a single dimension is fairly simple; however, as more and more dimensions are added into the analysis, it becomes difficult to define and differentiate the individuals within the population as a whole. A 3D scan allows for that immediate relation of linear to multivariate dimensions of the body using the volumetric aspects of the scans. This section details several ways that scanner technology can be used to compare and evaluate the human body.

6.6.4 OVERLAY (SUPERIMPOSING)

Volumetric overlay compares multiple scans (Figure 6.9); however, the goal is to evaluate each scan against each other. Superimposing scans provides a new way of visualizing clothing and equipment and is a powerful tool for evaluating fit, coverage, and other critical measurements. For example, if you have three representative individuals (or mannequins) that all wear a size medium shirt or pants, how will the different anthropometries of the individuals impact the fit of the shirt? Extensive studies have been done in the textile industry to allow folks to select an appropriate

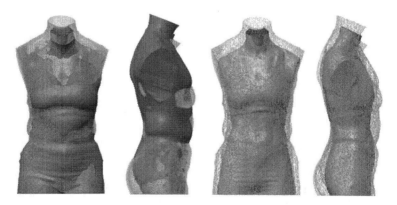

FIGURE 6.9 Superimposition of multiple subjects.

FIGURE 6.10 3D Scan overlay of multiple subjects to determine the overall contour of the seat pan.

size that matches their volumetric profile. For instance, studies done by Ashdown and others (Ashdown and Locker, 2010, Petrova and Ashdown, 2008, Song and Ashdown, 2010) have demonstrated the capability to even change the current pattern in order to make the apparel follow the curvature of the persons wearing them. At NASA, the unique problem of using a unisex suit requires an understanding of how the overall shape and size of both genders vary so that appropriate changes can be made to the interiors of the suit to allow for comfort fit. Efforts have been made at NASA to also determine the seat pan curvature so as to encompass a wide range of user population (Figure 6.10).

It is also possible to perform 3D analysis of whether or not clothing would fit a person or a range of intended users. Figure 6.11 shows how one can determine

FIGURE 6.11 Merging of CAD models with the 3D scan images.

whether a particular suit component, hard upper torso (HUT), would fit a person's torso region by way of collision detection techniques.

6.6.5 Mass Properties

One of the major advantages of 3D scan data is the ability to extract other valuable information besides the linear anthropometric dimensional data. For design purposes such as fighter jet aircraft or space shuttle cockpit seats, designers need to know the effect of the crewmembers' body mass properties while designing the seat and the seat layout. These properties are very essential to make sure that the cockpit module is well balanced and can handle the vehicle during off-nominal takeoff and landing scenarios. Prior to the advent of laser scanning, body mass properties such as whole-body center of gravity (CG), radius of rotation, segment mass CG, segment radius of rotation, as well as segment and whole-body moments of inertia were difficult to obtain and required cumbersome measurement techniques. Due to these hindrances, researchers sought the use of cadaver studies to obtain these hard-to-measure information (Chandler, et al. 1975, Clauser, et al. 1969). Even then, approximations had to be made since all these data were taken from older subjects and until now, questions lingered among designers whether or not the approximations to adjust for younger pilot populations would be accurate and representative of their body mass properties. Now with the use of 3D scan data, it is possible to treat the point clouds of either the entire body or a specific body segment to obtain the surface area and volume and from which it is possible to obtain a rough estimate of the body/segment mass with the acceptable density assumption of 1.006 gm/cc. Studies have begun using this method to improve the ability to provide individualized pilot's data so that structural engineers and designers can quantitatively assess the impact of the cockpit module due to changes in seat layout and crew mass distribution.

6.7 DYNAMICS AND ANIMATION

Digital human modeling is a growing area where the interest is to provide virtual humans in a virtual environment so that designers and developers can simulate movement of virtual humans while interacting with a virtual prototype environment or interfaces. Currently, the virtual humans are computer generated and strides are being made it to look and operate like humans. Modeling software such as Jack, Delmia, Ramsis, and Santos are some examples that are engaged in the development of virtual humans. Besides the need to depict humans virtually in terms of body shape and size, it is also necessary that these virtual humans can represent the physical performance of an individual as well as the user population. Stereo-photo-based scanners have fast capture times and thus are able to approach true dynamic 3D scanning, but scanners capable of capturing images at close to video speed are coming on the market (http://www.di3d.com/index.php). As with other stereo-photo systems, dynamic scanners require longer post-processing times to create 3D meshes.

A good example of 3D scanning in a dynamic setting is scanning the foot during weight-bearing. Models of the weight-bearing foot are important in medial and biomechanical contexts, and scanning has been done through direct imaging or by scanning casts taken while weight was applied (Houston et al. 2006, Oladipo et al. 2009, Tsung et al. 2003, Xiong et al. 2009). These studies are more concerned with deformation of the plantar surface for orthotics development. Capturing whole foot deformation during normal walking or running requires high-speed cameras arranged in an array (Coudert et al. 2006, Jezerek et al. 2009, Kimura et al. 2011, Mochimaru and Kouchi, 2011). Telfer and Woodburn (2010) point out that the more complicated systems have yet demonstrated an improvement over standard methods.

Animation also requires considerable interactive work. Improvements in automated feature extraction and template matching will reduce processing times in the near future. Transforming a static 3D scan into a form that can be animated has proven to be a considerable challenge. A great deal of interactive work is required to get the rigging and skin weights correct so that the model moves correctly. Although animation software has progressed considerably, reliance on interactive manipulation of the mesh makes it more difficult to apply animation to population-level 3D scan databases. However, see James and Twigg (2005) and Baran and Popović (2007) for possible approaches.

While it is still ways to go in terms of using 3D scans for human animation for mathematical predictions during dynamic and functional performance tasks, it is possible to show 3D scans of static humans in various postural and clothing conditions for visual information (Figure 6.12). It should be pointed out, however, that efforts are already underway to show dynamic human movement or postural aspects in a dynamic manner.

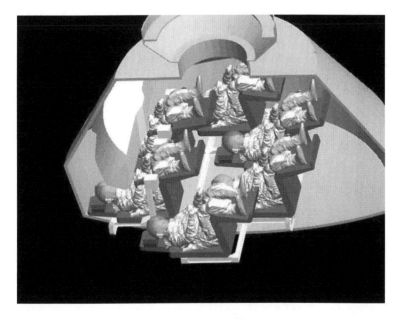

FIGURE 6.12   Suited crew in a seated posture with 3D scan data.

6.8 SUMMARY

Although 3D scanners have been available for many years, it's only with the recent large-scale surveys that samples became large enough to consider what statistical methods are appropriate for analysis of scans themselves rather than the derived 2D measurements. Besides being able to replicate standard traditional anthropometric measurements, scan data also are providing valuable information that is difficult to obtain from the manual, traditional methods. Volumetric representation of body shape and size via 3D scanning would eventually steer many researchers from seeking new methods to utilize scan data that are almost impossible to envision with even a robust linear anthropometric dataset. Unlike traditional anthropometric data, scan data do require extensive processing and may even require post-processing of the scan information before using the data for appropriate use. While these may be difficult in the beginning, software enhancements are happening at a faster rate to make most of these difficulties less painful and eventually may lead to better utilization of scan data for efficient use of it. This chapter hopefully provided ample information and steps that one needs to undertake while dealing with scan data.

QUESTIONS

6.1 What are the benefits and disadvantages of 3D scanning as compared to traditional manual measurement?

6.2 How does the surface scanning work?

6.3 Describe different types of technologies that are currently available.

6.4 Describe one of the post-processing techniques (hole filling, segmentation, and extraction).

6.5 How would you estimate mass properties from body scanning? How does it differ from the previous methods?

6.6 What are the assumptions you would make if you were to use scanning data to calculate mass properties?

6.7 Describe a situation where you see 3D scanning could be effectively used based on the information in this chapter?

REFERENCES

Ashdown SP and Loker S (2010) Mass customized target market sizing: Extending the sizing paradigm for improved apparel fit. *Design Practices*, **2**(2):147–173.

Baran I and Popović J (2007) Automatic rigging and animation of 3D characters. *ACM Transactions on Graphics*, **26**(3):72–78.

Ben Azouz Z, Shu C, and Mantel A (2006) Automatic locating of anthropometric landmarks on 3D human models, 3dpvt. *Third International Symposium on 3D Data Processing, Visualization, and Transmission (3DPVT'06)*, University of North Carolina, Chapel Hill, NC, pp. 750–757.

Bernardini F, Bajaj CL, Chen J, and Schikore DR (1999) Automatic reconstruction of 3D CAD models from digital scans. *International Journal of Computational Geometry and Applications*, **9**(4–5):327.

Besl PJ and McKay ND (1992) A method for registration of 3-D shapes. *IEEE Transactions on Pattern Analysis and Machine Intelligence*, **14**:239–256.

Blais G and Levine MD (1995) Registering multiview range data to create 3D computer objects. *IEEE Transactions on Pattern Analysis and Machine Intelligence,* **17**(8):820–824.

Bradtmiller B and Gross M (2003) Anthropometric dimensions extracted from surface scans and measured directly: Issues for standardization. *Proceedings of the XVth Triennial Congress of the International Ergonomics Association and 7th Joint Conference of Ergonomics Society of Korea/Japan Ergonomics Society,* Seoul, Republic of Korea, 4p.

Brunton A, Wuhrer S, Shu C, Bose P, and Demaine ED (2009) Filling holes in triangular meshes by curve unfolding. *IEEE International Conference on Shape Modeling and Applications (SMI 2009),* Beijing, China, pp. 66–72. http://dx.doi.org/10.1109/SMI.2009.5170165

Carr JC, Beatson RK, Cherrie JB, Mitchell TJ, Fright WR, McCallum BC, and Evans TR (2001) Reconstruction and representation of 3D objects with radial basis functions. *Proceedings of the 28th Annual Conference on Computer Graphics and Interactive Techniques SIGGRAPH'01,* Los Angeles, CA, pp. 67–76.

Chandler RF, Clauser CE, McConville JP, Reynolds HM, and Young JW (1975) *Investigation of Inertial Properties of the Human Body,* AMRL-TR-74-137. Dayton, OH: Aerospace Medical Research Laboratory.

Clauser CE, McConville JT, and Young JW (1969) *Weight, Volume, and Center of Mass of Segments of the Human Body,* AMRL-TR-69-70. Dayton, OH: Aerospace Medical Research Laboratory.

Corner BD and Hu A (1998) The effect of sway on image fidelity in whole body digitizing. In J Nurre and BD Corner (eds.), *Three-Dimensional Image Capture and Applications,* Seattle, WA. *Proceedings of IS&T/SPIE Symposium on Electronic Imaging: Science and Technology,* San Jose, CA, Vol. 3227, pp. 36–45.

Corner BD, Li P, Coyne M, Paquette S, and McMakin DL (2010) Comparison of anthropometry obtained form a first production millimeter wave three dimensional whole body scanner to standard direct body measurements. In V Duffy (ed.), *Advances in Applied Digital Human Modeling.* New York: CRC Press.

Coudert T, Vacher P, Smits C, and Van Der Zande M (2006) A method to obtain 3D foot shape deformation during the gait cycle. *Ninth International Symposium on the 3D Analysis of Human Movement,* June 28–30, 2006, Valenciennes, France. http://www.univ-valenciennes.fr/congres/3D2006/Abstracts/117-Coudert.pdf

Daanen H, Brunsmen M, and Taylor S (1997) Absolute accuracy of the cyberware WB4 whole body scanner. Technical Report AL/CF-TR-1997-0046. Dayton, OH: U.S. Air Force Armstrong Laboratory.

Douros I, Dekker L, and Buxton BF (1999) An improved algorithm for reconstruction of the surface of the human body from 3D scanner data using local B-spline patches. *ICCV99,* Corfu, Greece, pp. 29–36.

Hoppe H (1996) Progressive meshes. *Proceedings of ACM SIGGRAPH 1996,* New Orleans, LA, pp. 99–108.

Houston VL, Luo G, Mason CP, Mussman M, Garbarini M, and Beattie AC (2006) Changes in male foot shape and size with weight bearing. *Journal of the American Podiatric Medical Association,* **96**:330–343.

Huang H and Ascher U (2007) Fast surface mesh denoising with regularization and edge preservation. *Proceedings in Applied Mathematics and Mechanics,* **7**, pp. 2010001–2010002.

James DL and Twigg CD (2005) Skinning mesh animations. *ACM Transactions on Graphics,* **24**(3):399–407.

Jezerek M and Mozina J (2009) High-speed measurement of foot shape based on multiple-laser-plane triangulation. *Optical Engineering,* **48**:1133604.

Kimura M, Mochimaru M, and Kanade T (2011) 3D measurement of feature cross-sections of foot while walking. *Machine Vision and Application.* **22**(2):377–388.

Lee YK and Lee CK (2003) A new indirect anisotropic quadrilateral mesh generation scheme with enhanced local mesh smoothing procedures. *International Journal for Numerical Methods in Engineering*, **58**:277–300.

Leong I-F, Fang J-J, and Tsai M-J (2007) Automatic body feature extraction from a markerless scanned human body. *Computer-Aided Design*, **39**:568–582.

Li P, Corner B, and Paquette S (2000) Segmenting 3D surface scan data of the human body by 2D projection. In BD Corner and JH Nurre (eds.), *Three-Dimensional Image Capture and Applications III, Proceedings of SPIE*, Seattle, WA,Vol. 3958, pp. 172–177.

Mochimaru M and Kouchi M (2011) 4D measurement and analysis of plantar deformation during walking and running. *Footwear Science* **3** (Suppl. 1); *Proceedings Tenth Footwear Biomechanics Symposium*, Tubingen, Germany, pp. S109–S112.

Montague A (1960) *A Handbook of Anthropometry*. Thomas Publishing: New York.

Morency R, Ferrer M, Jaramillo M, Gonzalez L et al. (2005) Evaluation of a full-body scanning technique for the purpose of extracting anthropometrical measurements. *SAE Technical Paper*, doi:10.4271/2005-01-3016.

Norton K and Olds T (1996) *Anthropometrica*. University New South Wales Press: Adelaide, New South Wales, Australia.

Nurre JH (1997) Locating landmarks on human body scan data. *Proceedings of the International Conference on 3D Digital Imaging and Modeling*, Ottawa, Canada, pp. 289–295

Nurre JH, Connor J, Lewark EA, and Collier JS (2000) On segmenting the three dimensional scan data of a human body. *IEEE Transactions on Medical Imaging*, **19**(8):787–797.

Ohtake Y, Belyaev A, and Bogaevski IA (2000) Polyhedral surface smoothing with simultaneous mesh regularization. *Proceedings of Geometric Modeling and Processing*, Hong Kong, pp. 229–237.

Oladipo G, Bob-Manuel I, and Ezenatein G. (2009) Quantitative comparison of foot anthropometry under different weight bearing conditions amongst Nigerians. *Internet Journal of Biological Anthropology*, **3**:1.

Paquette S, Corner BD, Carson JM, Li P, Fisher S, and McEwen S (2005) *NatickMeasure: A 3-D Visualization and Anthropometric Measurement Tool (User's Guide, Version 1.0)*. Technical Report Natick/TR-05/021L. Natick, MA: U.S. Army Research, Development and Engineering Command, Natick Soldier Center.

Perkins T, Burnsides DB, Robinette KM, and Naishadham D (2000) Comparative consistency of univariate measures from traditional and 3-D scan anthropometry. *Proceedings of the SAE International Digital Human Modeling for Design and Engineering International Conference and Exposition*, Dearborn, MI.

Petrova A and Ashdown SP (2008) 3-D body Scan data analysis: Body size and shape dependence of ease values for pants fit. *Clothing and Textiles Research Journal*, **26**(3):227–252.

Rodríguez-Quiñonez JC, Sergiyenko O, Tyrsa V, Básaca-Preciado LC, Rivas-Lopez M, Hernández-Balbuena D, and Peña-Cabrera M (2011) 3D body and medical scanners' technologies: Methodology and spatial discriminations. In O Sergiyenko (ed.), *Optoelectronic Devices and Properties*. ISBN: 978-953-307-204-3, InTech, Available from://www.intechopen.com/articles/show/title/3d-body-medical-scanners-technologies-methodology-and-spatial-discriminations

Rusinkiewicz S and Levoy M (2001) Efficient variants of the ICP algorithm. In *Proceedings Third International Conference on 3D Digital Imaging and Modeling (3DIM 2001)*, Quebec City, Canada, pp. 145–152.

Song HK and Ashdown SP (2010) An exploratory study of the validity of visual fit assessment from three dimensional scans. *Clothing and Textiles Research Journal*, **28**(4):263–278.

Taubin G (1995) A signal processing approach to fair surface design. *Proceedings of SIGGRAPH*, Los Angeles, pp. 351–358.

Telfer S and Woodburn J (2010) The use of 3D surface scanning for the measurement and assessment of the human foot. *Journal of Foot and Ankle Research*, **3**:19–28.

Tsung BY, Zhang M, Fan YB, and Boone DA (2003) Quantitative comparison of plantar foot shapes under different weight bearing conditions. *Journal of Rehabilitation Research and Development*, **40**:517–526.

Ulijaszek S and Mascie-Taylor CG (2005) *Anthropometry: The Individual and the Population*. Cambridge, U.K.: Cambridge University Press.

Vollmer J, Mencl R, and Muller H (2009) Improved Laplacian smoothing of noisy surface meshes. **18**(3):131–138; *Eurographics* '99, Milano, Italy.

Xiong S, Goonetlleke RS, Zhao J, Li W, and Witana CP (2009) Foot deformations under different load-bearing conditions and their relationships to stature and body weight. *Anthropological Science*, **117**:77–88.

Zhao W, Gao S, and Hongwei (2007) A robust hole-filling algorithm for triangular mesh. *Visual Comput*, **23**:987–997. DOI:10.1007s00371-007-0167-y.

7 Three-Dimensional Data Processing Techniques

Jianhui Zhao, Yihua Ding, and Ravindra S. Goonetilleke

CONTENTS

7.1 CURVES AND SURFACES

As one of the most important research topics in computer graphics, curves and surfaces have been applied in footwear manufacture for a long time. With the deepening application of curves and surfaces, the related theories and algorithms have been more and more mature. According to the definition of mathematical expressions of curves and surfaces, generally speaking, there are two kinds of representations: parametric and nonparametric (Foley et al. 1993). Furthermore, nonparametric representation is divided into two subcategories: explicit and implicit equations.

7.1.1 Nonparametric Representation of Curves and Surfaces

In a Cartesian coordinate system, assuming that the coordinate variables of points on a given curve or surface satisfy a definite relation, we are able to describe this relation by an equation, which is the so-called nonparametric representation equation for curves or surfaces. If the relation among the coordinate variables of points is relatively simple, and one coordinate variable can be represented explicitly by other variables, the corresponding representation method is defined as explicit equation. For example, we can use the following equation to represent a straight line in the two-dimensional (2D) space:

$$y = kx + b \tag{7.1}$$

However, we should pay special attention to the limitation of explicit representation: it cannot represent closed or multiple valued curves or surfaces, such as circle and sphere. On the contrary, the implicit equation can fully represent these kinds of curves or surfaces in mathematics. For example, we can use the following equation to describe a circle, where r is its radius:

$$x^2 + y^2 - r^2 = 0 \tag{7.2}$$

7.1.2 Parametric Representation of Curves and Surfaces

The parametric representation is to utilize parametric equation to describe the curves and surfaces. The simplest type of parametric curves in 2D space is given as

$$\begin{cases} x = x(t) \\ y = y(t) \end{cases} \tag{7.3}$$

where t is a parameter and $t \in [a,b]$. Giving a value to t, we can get the coordinates of the corresponding point on this curve. When t changes continuously in the closed interval $[a,b]$, the entire curve is obtained. For convenience, we usually normalize the closed interval $[a,b]$ to $[0,1]$ and $t' = (t - a)/(b - a)$. On the other hand, if we represent the relation among coordinate variables by an equation with two parameters u and v, we get the parametric representation of surfaces.

7.1.3 Comparison of Parametric and Nonparametric Representations

The curves and surfaces represented by parametric equations are called parametric curves and surfaces. Here, we are mainly talking about the advantages of parametric curves and surfaces compared with the nonparametric ones.

1. The normalized parametric interval [0,1], defines the boundary of geometric object, and it is unnecessary to use other parameters to define the boundary.
2. The form of parametric equation is independent of the coordinate system.
3. It is convenient to transform curves and surfaces geometrically.
4. When it comes to nonparametric representation, the rate of change is described by slope, which sometimes is infinity. On the contrary, parametric representation uses tangent vector to represent the rate of change, and the infinity is avoided in this way.
5. The parametric curves and surfaces are more suitable to be generated discretely, easier to be controlled and programmed using computer programming language.

However, we must realize that there are some deficiencies for parametric curves and surfaces that cannot be ignored.

1. Continuity constraints are difficult to maintain.
2. It is hard to find intersections among different curves or surfaces.
3. When it comes to parametric surfaces, control mesh must be quadrilaterals.

7.1.4 Applications

Up to now, curves and surfaces have been applied in various areas. The often used parametric curves include parametric polynomial curve, Bezier curve, B-spline curve, and so on. In 1962, P.E. Bezier in Renault Automobile Company of France constructed a kind of parametric curve based on approximation theory, which is the so-called Bezier curve now. It has become one of the most basic methods for modeling curves and surfaces. It combines functional approximation and geometric representation, and makes it more convenient for the designer to control curves and surfaces using a computer. Besides, bilinear parametric surface, Bezier surface, B-spline surface, and Coons surface are also usually used in the design of products, such as shoes, cars, ships, appliances, and toys. At the same time, they play a significant role in animation and movie productions.

7.2 REVERSE ENGINEERING

As a novel approach for rapid prototype and computer numerical control manufacture, reverse engineering has been developing rapidly in recent years. When it comes to footwear manufacture, reverse engineering creates computer-aided design (CAD) models of feet and lasts from existing physical entities, and further uses these models to perform secondary developments. This process involves acquisition of digital information of feet and lasts, recovery of 3D model, and analysis and inference of intended design. Generally speaking, the reverse engineering procedure in footwear manufacture can be divided into the following four basic steps.

1. *Data acquisition*: There are several different digitalization methods for data acquisition of feet and lasts, as shown in Figure 7.1. Basically, they are classified as tactile methods and noncontact methods. Tactile methods

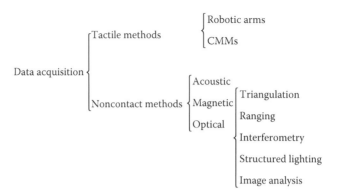

FIGURE 7.1   Classification of data acquisition methods.

are characterized by high accuracy and low speed. Its vital deficiencies are disability of measuring ultrathin and soft materials such as feet and necessity of manual operation all the time. On the contrary, the noncontact methods have wider market owing to the high speed and no necessity for contacting feet and lasts. In practice, the most common instrument, laser scanner, is one kind of noncontact method. It can get the 3D coordinates of the points on the surfaces of feet and lasts quickly and relative completely.

2. *Data preprocessing*: In the practical data acquisition of feet and lasts, incomplete or inaccurate geometrical information from measurement instruments would greatly lower the quality of reconstructed CAD model. Noise, isolated islands, and burrs all need special attention. So, after acquisition of point cloud data, we should do some preprocesses on these raw data, including transformation, registration, noise elimination, and so on.

3. *CAD model creation*: In this step, the core work is to fit CAD model from the scattered 3D point cloud. There are many different kinds of methods for surface recovery: parametric surface, implicit surface, Voronoi method, surface evolution, subdivision surface, and so on. We need to choose the most suitable one from the methods mentioned earlier.

4. *Inspection and amendment of CAD model*: After the creation of CAD model of feet and lasts, it is quite necessary for us to check whether the reconstructed CAD model satisfies the accuracy and other performance requirements. For unqualified models, we need to do amendments until they meet the design requests.

7.3   PROCESSING OF POINT CLOUDS

7.3.1   DENOISING OF NOISY POINT CLOUDS

As previously described, advances in digital scanning devices make 3D models of feet and lasts, typically point cloud models, widely available in footwear manufacture. However, the acquired data inevitably have noise from various sources (Motavalli 1998, Levoy et al. 2000, Rusinkiewicz et al. 2002). Such noise can degrade the visual quality and usability of point cloud models. Therefore, effective denoising algorithms

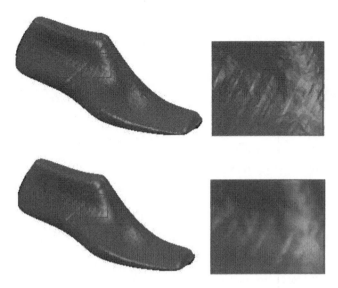

FIGURE 7.2   Original point cloud model with noise and denoised model.

must be achieved to improve the quality of the models for further processing. Up to now, due to the shortage of topological connectivity, eliminating noise from point clouds is a difficult research topic in computer graphics. However, many researchers have proposed inspiring methods for noise removal.

As shown in Figure 7.2, the top image is point cloud model with noise, and the bottom one is the corresponding denoised model (Schall et al. 2005). Roughly speaking, the existing denoising algorithms are classified into the following five categories (Schall et al. 2008):

1. *PDE approaches*: partial differential equations (PDEs) for denoising are first unveiled by Perona and Malik (1990) in image processing. They eliminated the noise in the image via anisotropic diffusion. Later, this method was extended in order to denoise point clouds by introducing theory of anisotropic heat conduction, anisotropic mean curvature flow, dynamic balanced flow, and so on (Lange and Polthier 2005, Xiao et al. 2006, Zhang et al. 2006). Those algorithms all need to solve partial differential equations.

2. *Spectral techniques*: The techniques of spectral filters in image setting have been generalized to 3D models for a long time. Taubin (1995) first introduced signal processing on meshes based on the definition of the Laplacian operator on surfaces. In a later work, Pauly and Gross (2001) denoised point cloud models by spectral analysis and filtering. They first divided the surface of point cloud model into patches, then resampled each patch via local height field, and then blended the patches together. In this way, Fourier-based spectral decomposition was created and performed on point cloud models. The data flow diagram of the spectral processing pipeline is illustrated in Figure 7.3.

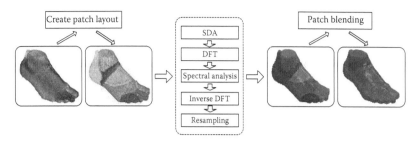

FIGURE 7.3 Spectral processing pipeline (SDA and DFT stand for scattered data approximation and discrete Fourier transform, respectively).

3. *Neighborhood filtering*: Neighborhood filters were early addressed in Yaroslavsky (1985). Later, Fleishman et al. (2003) presented an anisotropic mesh denoising algorithm motivated by the bilateral filtering for image denoising. This method is independent of the topological connectivity, so it is also suitable for denoising of point cloud models. The previous work of neighborhood filters is to determine proper local neighborhood on the point cloud. Based on this, different strategies are made to adjust or eliminate noise.

   We take the work of Huang et al. (2009a,b) as an example to illustrate the core idea of neighborhood filtering. When it comes to point $P(x_0, y_0, z_0)$ in the point cloud, first of all, its neighborhood needs to be found. Then, the best approximation plane, $Ax + By + Cz + D = 0$, is fitted as shown in Figure 7.4. The distance between point P and this plane is defined as

$$d = \frac{\left|Ax_0 + By_0 + Cz_0 + D\right|}{\sqrt{A^2 + B^2 + C^2}} \tag{7.4}$$

Furthermore, we need to find out all distances between every point of neighborhood to the best approximation plane, and then calculate the average of distance $\bar{d}$. If $\bar{d}$ is greater than a given threshold value, we mark point P as noise and remove it.

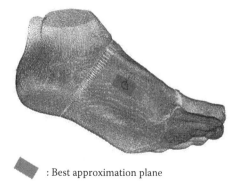

: Best approximation plane

FIGURE 7.4 Neighborhood and best approximation plane.

4. *Projection-based approaches*: Moving least squares (MLS) algorithm is the most common projection-based approach. Many researchers have utilized this method to denoise the point cloud model (Mederos et al. 2003, Amenta and Kil 2004, Dey et al. 2004, Weyrich et al. 2004, Fleishman et al. 2005, Daniels et al. 2007). Its key idea is to use a projection operator to scatter points in the vicinity of a surface onto the surface itself, or specify a user-adjustable blending parameter, and then shifting point positions toward the corresponding MLS surface. However, the main problem of MLS-based methods is that prominent shape features are blurred during smoothing. Therefore, improvement on this algorithm has been made in order to preserve features during denoising (Li 2009).

5. *Statistical techniques*: The processing techniques for point cloud models using statistics were introduced by Pauly et al. (2004). Soon after, various statistical techniques were applied to filter noisy point cloud. Schall et al. (2005) used a mean shift–based clustering procedure to robustly filter a given noisy point cloud model. The smoothed models can be obtained by moving every sample to the maximum likelihood positions. Jenke et al. (2006) presented an algorithm to denoise a given noisy point cloud using Bayesian statistics. Recently, Yoon et al. (2009) used Bayesian technique to estimate the noise in a scanned point cloud. This algorithm provided more robust parameter estimation and assessment of the final model, and further can be used to remove noise on point cloud models.

7.3.2 Point Cloud Transformation

In reverse engineering, coordinate transformation of point clouds is both important and necessary in many cases. For instance, when it comes to scanning feet or lasts using digital scanning device, if they are far away from the origin of coordinate system of digital scanning device, obtained coordinate values of point clouds might be very huge, which makes the following processing-like surface reconstruction difficult. Here, we can use coordinate transformation techniques to solve this problem, that is, transform point clouds according to the transformation matrix that makes the coordinate systems of both point cloud and digital scanning device identical. In order to make this process clear, we consider a specific example.

Giving a point cloud as shown in Figure 7.5, its coordinate system is $P(X, Y, Z)$ and the coordinate system of digital scanning device is $D(X', Y', Z')$, while $O = (x_o, y_o, z_o)$ and $O' = (x'_o, y'_o, z'_o)$ are the origins of P and D, respectively. The transformation can be described as a rotation followed by a translation. Taking D as the reference coordinate system, we first translate O to O', and then rotate P about three orthogonal axes of D with proper angles to make both of them identical. Therefore, the transformation matrix M is defined as

$$M = R_z(\eta) \times R_y(\beta) \times R_x(\alpha) \times T \qquad (7.5)$$

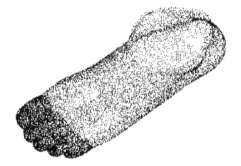

FIGURE 7.5 Original foot point cloud from scanner.

The translation matrix

$$
T = \begin{bmatrix} 1 & 0 & 0 & x'_o - x_o \\ 0 & 1 & 0 & y'_o - y_o \\ 0 & 0 & 1 & z'_o - z_o \\ 0 & 0 & 0 & 1 \end{bmatrix}
$$

(7.6)

The x-axis rotation matrix

$$
R_x(\alpha) = \begin{bmatrix} 1 & 0 & 0 & 0 \\ 0 & \cos\alpha & -\sin\alpha & 0 \\ 0 & \sin\alpha & \cos\alpha & 0 \\ 0 & 0 & 0 & 1 \end{bmatrix}
$$

(7.7)

The y-axis rotation matrix

$$
R_y(\beta) = \begin{bmatrix} \cos\beta & 0 & \sin\beta & 0 \\ 0 & 1 & 0 & 0 \\ -\sin\beta & 0 & \cos\beta & 0 \\ 0 & 0 & 0 & 1 \end{bmatrix}
$$

(7.8)

The z-axis rotation matrix

$$
R_z(\eta) = \begin{bmatrix} \cos\eta & -\sin\eta & 0 & 0 \\ \sin\eta & \cos\eta & 0 & 0 \\ 0 & 0 & 1 & 0 \\ 0 & 0 & 0 & 1 \end{bmatrix}
$$

(7.9)

After determining the transformation matrix M, we can obtain the transformed point cloud by performing M on original point cloud. The entire process is described in Figure 7.6.

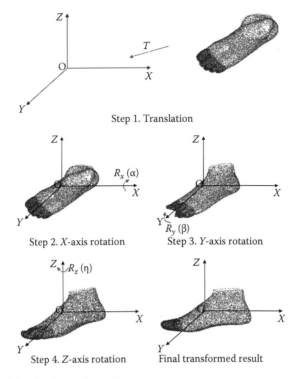

Step 1. Translation

Step 2. X-axis rotation   Step 3. Y-axis rotation

Step 4. Z-axis rotation   Final transformed result

FIGURE 7.6   Point cloud transformation.

Here, the coordinate transformation we are talking about is rigid body transformation, because the transformed point cloud should be in accordance with the measured object in size. However, we are able to scale the point cloud by performing the following matrix on it if needed:

$$
S = \begin{bmatrix} s_x & 0 & 0 & 0 \\ 0 & s_y & 0 & 0 \\ 0 & 0 & s_z & 0 \\ 0 & 0 & 0 & 1 \end{bmatrix}
\tag{7.10}
$$

where s_x, s_y, and s_z are scale factors regarding x, y, and z coordinates, respectively.

7.3.3   REGISTRATION OF POINT CLOUDS

Due to physical constraints of digital scanning devices or limited environment, it is often difficult or even impossible to capture feet and lasts at the same time. The collected point clouds usually cannot be guaranteed to have the same position and posture in the world coordinate system. However, in practical applications such as automatic measurement of human feet, making various point clouds have identical position and posture is both necessary and important. Since automatic determination of different dimensions closely depends on the measuring axis, the consistency of different foot point clouds is a prerequisite to highly accurate measurement. Therefore, registration of point clouds plays a vital role in footwear manufacture.

Roughly speaking, the existing approaches for registration can be classified into two subcategories: coarse registration and fine registration (Xie et al. 2010). Up to now, many techniques have been applied on coarse registration in order to estimate the approximate transformations of different point clouds (Feldmar and Ayache 1996, Várady et al. 1997, Farin et al. 2002, Li et al. 2002, Long et al. 2002, Wyngaerd and Gool 2002, Han et al. 2003). However, it is not enough to get final transformation relationship only in these ways. Therefore, we need to do further processing, that is, fine registration. In this section, we choose principal component analysis (PCA) to determine initial estimation of the transformation between two feet point clouds, and then use interactive closest point (ICP) algorithm (Besl and McKay 1992, Chen and Medioni 1992) to implement more accurate registration.

7.3.3.1 Coarse Registration Using PCA

Supposing there are N points in a point cloud, each point is represented by $p_i(x_i, y_i, z_i)$, $i \in [1, N]$, and the center $O(x, y, z)$ is calculated as the average of the N points. Covariance matrix for the point cloud is defined as

$$
COV = \frac{1}{N}
\begin{bmatrix}
\sum_{i=1}^{N}(x_i - x)^2 & \sum_{i=1}^{N}(x_i - x)(y_i - y) & \sum_{i=1}^{N}(x_i - x)(z_i - z) \\
\sum_{i=1}^{N}(y_i - y)(x_i - x) & \sum_{i=1}^{N}(y_i - y)^2 & \sum_{i=1}^{N}(y_i - y)(z_i - z) \\
\sum_{i=1}^{N}(z_i - z)(x_i - x) & \sum_{i=1}^{N}(z_i - z)(y_i - y) & \sum_{i=1}^{N}(z_i - z)^2
\end{bmatrix}
$$

$$(7.11)$$

Eigenvalues of the covariance matrix are sorted in descending order, and the eigenvector corresponding to the highest eigenvalue is taken as the principal component of the data set. The ordered eigenvectors are used to set up a PCA coordinate system with point O as the coordinate origin. Then the PCA coordinate system is transformed to be coincident with the world coordinate system. In this way, the transformed point cloud is more significant along both X- and Y-dimensions while less significant along Z-dimension of the global coordinate system.

Based on the theory described earlier, we can transform the point clouds of 2 ft, and make them have identical local coordinate system. As shown in Figure 7.7, the left image presents the original two point clouds, and the right image shows the transformed results.

7.3.3.2 Fine Registration Using ICP

As one of the most popular techniques for fine registration, the ICP algorithm (Besl and McKay 1992) has been studied for many years. Its numerous extensions are also widely used to solve this problem. Here, we illustrate how to use ICP algorithm to register point clouds obtained by PCA transformation.

FIGURE 7.7 Coarse registration using PCA.

FIGURE 7.8 Fine registration using ICP.

Suppose the red point cloud R contains N_R points, and the black point cloud B has N_B points, respectively. The iteration is initialized by setting $P_0 = R$, $\vec{q}_0 = [1,0,0,0,0,0,0]^T$, and $k = 0$. The registration vectors are defined relative to P_0 so that the final registration represents the complete transformation. The following four steps are applied until convergence within a given tolerance value, as shown in Figure 7.8:

Step 1: Compute the closest points $Y_k = C(P_k, B)$.
Step 2: Compute the registration $\left(\vec{q}_k, d_k\right) = Q\left(P_0, Y_k\right)$.
Step 3: Apply the registration $P_{k+1} = \vec{q}_k\left(P_0\right)$.
Step 4: $R' = \vec{q}_k\left(R\right)$ and terminate the iteration if $d_k - d_{k+1}$ is less than a given threshold $\tau > 0$, otherwise, go back to Step 1.

Although ICP algorithm has been applied widely, we have to realize that the original ICP algorithm is computationally expensive (Zhang 1994), and can only handle rigid registration. So the extensions of ICP mainly focus on how to complete quick, robust, and fully automatic registration, and how to achieve nonrigid registration (Chua and Jarvis 1996, Wyngaerd and Gool 1999, Granger and Pennec 2002, Sharp et al. 2002, Fitzgibbon 2003, Jost and Heinz 2003, Stamos and Leordeanu 2003, Bendels et al. 2004, Jian and Vemuri 2005, Amberg et al. 2007, Du et al. 2007, 2010, Brenner and Dold 2007, Ho et al. 2007, Wang and Brenner 2008, Jiang et al. 2009, Qiu et al. 2009, Thapa et al. 2009). Besides, apart from ICP algorithm and its extensions, many researchers have been actively searching alternative algorithms to solve 3D registration of point clouds, and can generate promising registration results (Pottmann et al. 2004, Boughorbel et al. 2010, Flory and Hofer 2010).

7.4 SURFACE RECONSTRUCTION

As one of the core contents of CAD/CAM, surface reconstruction from the scattered 3D point cloud is an important research subject in reverse engineering all the time. Starting from a point cloud with or without normal vectors, the purpose of surface

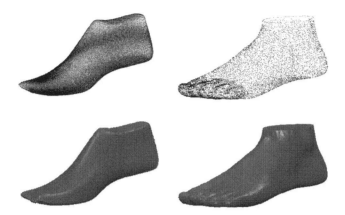

FIGURE 7.9 Last and foot point clouds and the recovered surfaces.

reconstruction is to recover a 3D watertight surface that approximates the original point cloud as closely as possible. As shown in Figure 7.9, the first row illustrates point clouds of last and human foot, and the second row shows the corresponding reconstructed surfaces.

Owing to scarce information, including unknown topology of the original surface, unknown connections among the points, and so forth, the reconstruction procedure is an ill-posed problem, that is, there is no unique solution. However, this topic has inspired constant research over the past years. In 1992, Hoppe et al. (1992) first unveiled the seminal work regarding surface reconstruction. After that, various related algorithms have been proposed. As for all of these approaches, according to different representation of surfaces, they can be divided into the following categories as shown in Figure 7.10.

1. *Parametric surface method*: The parametric surface method is to convert a point cloud representation of an object into a collection of parametric patches (illustrated in Figure 7.11), such as Bezier patch and B-spline patch. An early example is the work of Bajaj et al. (1995), which first constructed a C^1 smooth piecewise algebraic surface (called domain surface), and then calculated the signed distance function. After generating the piecewise linear approximation using alpha shape computation, the continuous Bernstein–Bezier patch reconstructed on the domain surface was taken as the final reconstructed surface. Later, some kinds of B-spline surface reconstruction have been studied extensively (Ma and Kruth 1995, Lee et al. 1997).

$$\text{Surface reconstruction}\begin{cases}\text{Parametric surface method}\\\text{Implicit surface method}\\\text{Voronoi method}\\\text{Surface evolution method}\\\text{Subdivision surface method}\\\text{Other methods}\end{cases}$$

FIGURE 7.10 Classification of surface reconstruction methods.

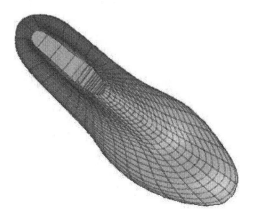

FIGURE 7.11   Constructed parametric surface.

2. *Implicit surface method*: Implicit surface reconstruction can be interpreted
   as the following problem (Hoppe et al. 1992). Suppose there are n points
   $\{(x_i, y_i, z_i)\}_{i=1}^{n}$ on a given surface M. The core job of implicit surface method
   is to find out a surface M' that is a reasonable approximation to M. For this
   purpose, a smooth implicit function $f(\ )$ is modeled. If any point (x, y, z) on
   surface M satisfies the equation

$$f(x, y, z) = 0 \tag{7.12}$$

   we can say $f(\ )$ implicitly defines M, and the reconstructed surface can be
   extracted as the zero set of $f(\ )$. Besides, the function $f(\ )$ is designed to
   be positive outside the object and negative inside the object. Up to now,
   various implicit functions have been applied to surface reconstruction, such
   as signed distance function, radial basis functions, and so on (Carr et al.
   2001, Ohtake et al. 2003, 2004, Amenta and Kil 2004). Recently, as the
   new choice for $f(\ )$, the indicator function, which is one inside the object and
   zero outside, attracted lots of attention. Based on this, Kazhdan et al. (2006)
   used a Poisson system to define a surface, and then obtained recovered surface by extracting the isosurface.
3. *Voronoi method*: Voronoi diagram and its dual Delaunay triangulation have
   been ubiquitous in computational geometry community (Okabe et al. 2000). The
   related methods often provide provable guarantees under prescribed sampling
   criteria (Amenta et al. 1998) or noise models (Dey and Goswami 2006) that are
   generally not realizable by real-world data. The main differences among these
   methods in existing literature are the ways of removing the tetrahedron and
   building the external triangular mesh. However, they have the same purpose:
   constructing a mesh that represents the original surface as accurately as possible (Gopi et al. 2000, Amenta et al. 2002, Cohen and Da 2004). Figure 7.12
   (left) shows the surface representation of reconstructed foot and (right) its wire
   frame representation.

FIGURE 7.12   Reconstructed mesh surface using Voronoi method.

4. *Surface evolution method*: Sometimes, certain complicated surfaces consist of a set of surfaces. It is difficult or even impossible for us to reconstruct them only using a piece of parametric surface or implicit surface. Fortunately, some researcher proposed a new kind of surface reconstruction method—surface evolution—and finally this problem was solved to some extent. The surface evolution method is following the idea of deforming a simple given surface according to certain rules until it reaches the input object. An early work of this idea is due to Miller et al. (1991). Their paper placed a "seed" model in the volume data set. The model is then deformed by a relaxation process that minimizes a set of constraints that provides a measure of how well the model fits the features in the data. In this way, the ultimate surface is obtained. Apart from this, the level-set method has also been used in surface evolution by Zhao et al. (2000), during which implicit and nonparametric shape reconstruction from unorganized points was completed.

5. *Subdivision surface method*: Another class of methods is based on subdivision surface. The basic idea is to define a smooth surface by subdividing a polyhedral mesh recursively. This method is known for its simplicity and ability to describe complex surfaces. In 1994, Hoppe et al. (1994) presented subdivision method to reconstruct surface. This method has three steps. First of all, as the approximation of original point cloud, a triangular mesh is defined. Then optimization is performed in order to reduce vertex and improve accuracy. At last, the reconstructed surface is generated by controlling the number, position, and link relation of vertex. Up to now, related algorithms or variants have been widely applied in various fields, especially in animation design and surface reconnection of complicated statues.

What we have discussed earlier are mainstream algorithms. Actually, there are other methods we also need to pay attention to, such as the surface reconstruction using primitive shapes or from nonparallel curve networks (Liu et al. 2008, Schnabel et al. 2009). Besides, with the development of reverse engineering, more and more new surface reconstruction techniques will appear in the future.

7.5   SHAPE DEFORMATION

Deformation and animation of 3D objects are widely utilized in the film and game industries and virtual reality applications (Chang and Rockwood 1994, DeBoer et al. 2007, Huang et al. 2009a,b). In the meanwhile, we have to realize that deformation also plays an indispensable role in footwear manufacture, especially in interactive last design. As we all know, foot characteristics differ by individuals. For certain shoe-last style, it is almost impossible to match with all feet. Therefore, under this situation, deformation of last is quite necessary in order to make it meet the requirement of fitting evaluation.

Up to now, lots of related algorithms have been proposed to deal with this problem, such as dimension-driven surface deformation, free-form deformation based on lattice, or mesh controlling (Huang et al. 2009a,b, Luximon and Luxinmon 2009). Recently, Wang et al. proposed an approach to achieve interactive deformation, which also can be used in last design (Wang et al. 2011). In this algorithm, piecewise reconstruction is first completed, which divides the whole shoe-last into many regions (gray color) as shown in Figure 7.13, and then contour curves (black color) are generated from the boundaries of every piecewise region. After that, parametric constraints of contour curves are defined. The local surfaces enveloped by contour curves in this paper are quad-boundary surfaces. The deformation region of interest (ROI) is based on its corresponding control point. The moving of control point affects its adjacent quad-boundary surfaces. After the adjusting of contour curves, vertices on the boundary of piecewise surfaces are actually updated. With the help of constraints, it can rationally transfer the deformation magnitude from the contour curves to the surfaces in its ROI.

There are two approaches to make deformation happen: Directly drag a control point, and in real time solve constraints on the point to determine its coordinate values, and then update the relative curves (Figure 7.14a and b). Figure 7.14c shows the ultimate deformation result.

FIGURE 7.13   Piecewise reconstruction.

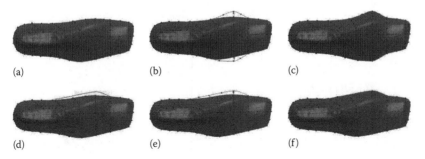

FIGURE 7.14   Process of draft-driven deformation. (a) piecewise surfaces with control points, (b) drag points and update curves, (c) deformed surface, (d) draft a curve, (e) update contour curves, and (f) deformation result.

Draft a curve by hand, map relative control points to the drafted curve, and then update the contour curves (Figure 7.14d and e). Figure 7.14f is the corresponding deformation result.

Although various deformation algorithms are performed on rigid last deformation in footwear manufacture, they do not work well in simulation for real deformation of a 3D human body, since the application has different requirements: There is no reference 3D shape to instruct deformation, and there should be no or only little difference between simulated and real deformation.

The nonrigid limbs of the human body, such as the foot, have flexible structures. However, this does not imply that the foot can deform in any and every way. There are many constraints and allowable degrees of freedom due to the 26 bones, ligaments, and soft tissue present in a foot. Thus, replicating and simulating foot deformation to biomechanically realistic and accurate postures is a real challenge. Deformation is important as feet tend to be scanned with the foot on a flat plate or surface. However, the majority of shoes have a heel and hence prior to any matching procedure, the foot has to be geometrically deformed to a posture corresponding to the footbed surface.

It is a practical way to perform 3D foot deformation using a 2D last line to correspond to the plantar (bottom) surface of the foot. Even though the method has shown some promise, there are issues with regard to the corresponding modeling on the dorsal (top) surface of the foot. Techniques need to be studied and adopted to reduce and eliminate such errors. One possible method is finding relationships between local and global deformation of a 3D object. It involves extensive controlled experimentation with people where empirical relationships can be derived and set up between heel heights and foot shapes. Based on the mappings between local deformation and global deformation, the shape of the 3D object at any time can be computed or interpolated. Another possibility is building a complete 3D model for the object that includes surface as well as volume so that volumetric deformation can account for shape changes as well. With state-of-the-art equipment such as 3D scanners and 3D motion analysis systems, it is possible to compare model-based virtual deformation with the actual deformation of human feet. The models to be developed should have the same geometric structure, exact number of segments and nonlinear materials, and correct interactions among the elements. For every segment, degrees of freedom are defined in the local coordinate system. The model should be adjustable for different objects. A multilayered model is a good method of modeling a hierarchical skeleton with deformable primitives attached to it to simulate soft tissues and skin. Once the multilayered character is constructed, only the underlying skeleton needs to be scripted for movement, while consistent yet expensive shape deformations will be generated automatically.

7.6 DIMENSIONAL MEASUREMENTS

Lengths, widths, heights, and girths are very important for the manufacture of custom footwear and hence it is no surprise that many researchers have developed algorithms to automatically determine them. Bunch (1988) used a 3D digitizer to obtain the coordinates of 34 landmarks on the foot surface from which a series of foot dimensions were calculated using a computer program. Similarly, Liu et al. (1999) obtained the coordinates of 26 points on the surface of the foot and leg using an electromagnetic digitizing device and thereafter computed 23 variables that comprised heights, lengths, widths, and angles. Many others (Butdee 2002, Klassen et al. 2004, Xu et al. 2004, Yahara et al. 2005, Witana et al. 2006, Hu et al. 2007, Zhao et al. 2008, Ding et al. 2010) have used different techniques to obtain dimensions such as foot length, foot width, and foot girths. In this section, the definitions of different kinds of dimensions are presented in detail, and corresponding measuring methods are introduced.

7.6.1 FOOT DIMENSION DEFINITIONS

To the best of our knowledge, there are 18 dimensions on the human foot as shown in Figure 7.15. Out of these foot dimensions, only 10 (foot length, arch length, heel-to-medial malleolus, heel-to-lateral malleolus, foot width, medial malleolus height, lateral malleolus height, ball girth, instep girth, and short heel girth) are available in the commercial software system. Those 10 measurements are henceforth referred to as the "common measurements" and the other 8 measurements (heel-to-fifth toe length, heel width, bimalleolar width, mid-foot width, height at 50% of the foot length, long heel girth, ankle girth, and waist girth) are referred to as "other measurements." The detailed definitions of foot dimensions are listed in the following.

Lengths

1. *Foot length*: The distance along the Brannock axis from pternion to the tip of the longest toe.
2. *Arch length*: The distance along the Brannock axis from pternion to the most medially prominent point on the first metatarsal head.

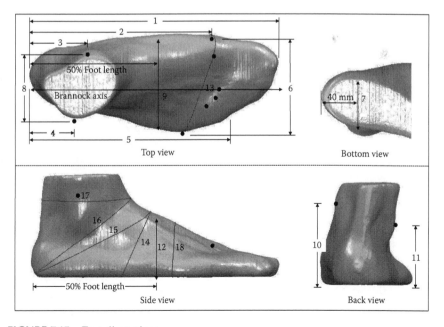

FIGURE 7.15 Foot dimensions.

3. *Heel-to-medial malleolus*: Length from pternion to the most medially protruding point of the medial malleolus measured along the Brannock axis.

4. *Heel-to-lateral malleolus*: Length from pternion to the most laterally protruding point of the lateral malleolus measured along the Brannock axis.

5. *Heel-to-fifth toe*: The distance along the Brannock axis from pternion to the
anterior fifth toe tip.

Widths

6. *Foot width*: Maximum horizontal breadth (*Y*-direction), across the foot perpendicular to the Brannock axis in the region in front of the most laterally
prominent point on the fifth metatarsal head.

7. *Heel width*: Breadth of the heel 40 mm forward of the pternion.

8. *Bimalleolar width*: Distance between the most medially protruding point
on the medial malleolus and the most laterally protruding point on the lateral malleolus measured along a line perpendicular to the Brannock axis.

9. *Mid-foot width*: Maximum horizontal breadth, across the foot perpendicular to the Brannock axis at 50% of foot length from the pternion.

Heights

10. *Medial malleolus height*: Vertical (*Z*-direction) distance from the floor to
the most prominent point on the medial malleolus.

11. *Lateral malleolus height*: Vertical (*Z*-direction) distance from the floor to
the most prominent point on the lateral malleolus.

12. *Height at 50% foot length*: Maximum height of the vertical cross section at
50% of foot length from the pternion.

Girths

13. *Ball girth*: Circumference of foot, measured with a tape touching the
medial margin of the head of the first metatarsal bone, top of the first
metatarsal bone, and the lateral margin of the head of the fifth metatarsal
bone.

14. *Instep girth*: Smallest girth over middle cuneiform prominence.

15. *Long heel girth*: The girth from instep point around back heel point.

16. *Short heel girth*: Minimum girth around back heel point and dorsal foot
surface.

17. *Ankle girth*: Horizontal girth at the foot and leg intersection.

18. *Waist girth*: Circumference at the approximate center of the metatarsal,
measured in a vertical plane, perpendicular to the Brannock axis.

7.6.2 Determination of Dimensions

For our convenience, 10 anatomical landmarks are identified and marked on the
foot surface as illustrated in Figure 7.16. Five of those landmarks are on the top of

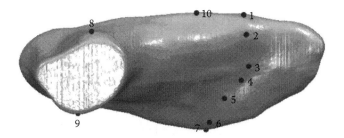

FIGURE 7.16 Landmarks of foot surface.

metatarsal-phalangeal joints (MPJs), one each at the side of the first and fifth MPJ, one each on the medial and lateral malleolus, and one at the arch point.

Lengths

1. Foot length is calculated as the distance between the minimum X value (mean of five minimum points) and the maximum X value (mean of five maximum points) along the X-axis.
2. Arch length is calculated as the distance between the point with the minimum X value and landmark-1 along the X-axis.
3. The heel-to-medial malleolus length is calculated as the distance between the point with the minimum X value and landmark-8 along the X-axis.
4. The heel-to-lateral malleolus length is calculated as the distance between the point with the minimum X value and landmark-9 along the X-axis.
5. The heel-to-fifth toe length is calculated as the distance between the point with minimum X value and the fifth toe along the X-axis.

Widths

1. Foot width is calculated as the distance between the point with the minimum Y value and the point with the maximum Y value along the Y-axis of the fore-foot (the area in which all points are ahead of landmark-7 [along the X-axis]). The mean of five points was used to determine both the minimum and maximum values in order to minimize the effects of "noise" in the scanned points.
2. Heel width is calculated as the distance between the point with the minimum Y value (mean of five minimum points) and the point with the maximum Y value (mean of five maximum points) along the Y-axis in the rear-foot region, that is, points within 40 mm from pternion along X-axis and no more than 20 mm above the platform.
3. Bimalleolar width is calculated as the distance between landmark-8 and landmark-9 along the Y-axis.
4. Mid-foot width is calculated as the distance along the Y-axis between the minimum Y value and the maximum Y value at the foot center: that is, points within a slab of half-thickness of 0.6 mm on either side of 50% of the foot length along the X-axis.

Heights

1. Medial malleolus height is calculated as the distance along the Z-axis from the platform (mean of five maximum Z points) to landmark-8 (the positive Z-axis is pointing downward, and hence, the scanner platform has the maximum Z value, see Figure 7.3).
2. Lateral malleolus height is calculated as the distance along the Z-axis between the platform and landmark-9.
3. The height at 50% of the foot length is calculated as the distance along the Z-axis from the platform to the minimum Z value at 50% of the foot length: that is, points within a slab of 70.6 mm distance from 50% of the foot length along X-axis.

Girths

1. Ball girth is calculated from the intersection points between the point cloud and the tape plane determined by landmark-1, landmark-2, and landmark-7.
2. Instep girth is calculated from the intersection points between the point cloud and the tape plane determined by three control points, P1, P2, and P3.
3. Long heel girth is calculated from the intersection points between the point cloud and the tape plane perpendicular to the XZ plane and located by two control points, P3 and P4.
4. Short heel girth is calculated from the intersection points between the point cloud and the tape plane perpendicular to the XZ plane and located by two control points, P5 and P6.
5. Ankle girth is calculated from the intersection points between the point cloud and the tape plane parallel to the XY plane and located by one control point, P6.
6. Waist girth is calculated from the intersection points between the point cloud and the tape plane parallel to the YZ plane and located by the mid-point of landmark-1 and one control point, P1.

Compared with the linear dimensions, girth measures are more difficult to determine as the digitized points are only discrete samples of the continuous surface. To overcome the discreteness, recently, Ding et al. (2010) presented the use of the radial basis function (RBF) surface modeling technique for measuring girths. Compared with the former methods (Xu et al. 2004, Witana et al. 2006, Zhao et al. 2008), it provides more stable and accurate measurements using relatively less time, proving its value in custom footwear manufacture. The algorithm is mainly performed in the following way:

Step 1: Extract the partial region from original foot point cloud, which is useful for girth measurement.
Step 2: Construct RBF surface of this partial region.

Step 3: Use tape plane to "cut" reconstructed RBF surface to get a series of intersection point.

Step 4: Sort those intersection points to form a polygon.

Step 5: Take the perimeter of this polygon as the ultimate girth value.

7.7 MODELS GEOMETRIC FITTING

Traditionally, foot sizing is performed using the Brannock Device Co. Inc. (2010) or similar devices for the measurement of length (and sometimes width), while a last (or shoe) is built using at least 30 dimensions. At the point of purchase, the matching between shoes and feet is performed using one (length) or sometimes two (length and width) dimensions. But a shoe has many other important features as well: arch position, arch height, arch width, flex (metatarsophalangeal) line, flex angle, shoe flare, and so on. However, the consumer is unable to properly match all of these characteristics when purchasing footwear. The existing research has shown the weaknesses of the existing grading and fitting techniques. Thus, theoretically, custom footwear allows a person to obtain a better fit between feet and footwear. The method for geometrical fitting among 3D models is always a difficult research problem. For a good 3D model fitting system, the matching method should be able to calculate the geometrical characteristics of all kinds of 3D models, and be robust to translation, rotation, scaling, noise, model degeneracy, etc.

Reported fitting methods (Chen et al. 2003, Ansary et al. 2004, Kazhdan et al. 2004, Bustos et al. 2005, Reuter et al. 2005, Daras et al. 2006, Xu et al. 2009) include outline-based matching using the distribution of vertices or polygons, visual-based matching using the visual projections, and topology-based matching using the topological structures of 3D models. There are still problems in the fitting of 3D objects. Definition and extraction of 3D features are difficult as good features should be quick to compute, invariant under transformations, insensitive to noise, robust to topological degeneracy, discriminating of similarities and differences, etc. Similarity measurement is also difficult, since it is not easy to build the correspondence between the obtained features automatically, and not easy to quantify the similarity.

For the application of footwear, here we introduce a ray-based method for model fitting by genetic algorithm. Efficiency of this method is tested not only on shoe-lasts, but also the other kinds of objects.

7.7.1 RAY-BASED DIFFERENCE COMPUTATION

For one triangular mesh model of last or foot with N triangles, its centroid is defined as

$$\text{Centroid} = \frac{\sum_{i=0}^{N-1} (\text{area}(i) * \text{center}(i))}{\sum_{i=0}^{N-1} \text{area}(i)} \tag{7.13}$$

where
   area(i) is the area of the ith triangle
   center(i) is the center of the ith triangle, that is, the average of its three vertices

After computation of the centroid, the 3D model is translated to move its centroid to the origin of the coordinate system.

Suppose that there are two foot models F_1 and F_2. We first need to translate them in the way described earlier so that they are centered with the same origin, and the fitting problem of these two 3D models becomes: rotating F_1 to search for its posture with the minimum difference from F_2. Rays are defined as the vectors from the centroid to the surfaces of the 3D models, and can be generated randomly.

To calculate the ray distance to surface (RDS), we must locate the intersection point between each ray and surface of the 3D model. Intersecting computation between 3D line and plane is a time-consuming operation; thus, trying on each triangle for one ray should be avoided. When the intersection for one ray is under consideration, the angle between the ray vector and the vector from centroid to each triangle is calculated first. The triangle with large angle means it has long distance from the ray vector. Therefore, only the triangles with angles less than one predefined threshold value are chosen for intersections' computing, and from them the real intersection point can be decided. RDS of one ray is defined as the 3D distance between the centroid and its intersection point. Obviously, RDS expresses not only shape but also size of the object.

7.7.2 ROTATION USING GENETIC ALGORITHM

F_1 can rotate about X, Y, and Z axes of the coordinate system. Given a set of angles (x, y, z) that are the rotational values around the three axes, respectively, and supposing $R(\ )$ is the abbreviation of rotational matrix, the posture of F_1 is determined by the following concatenation of matrices:

$$RM = RM(z) * RM(y) * RM(x) \qquad (7.14)$$

Here the fitting procedure is based on genetic algorithm, and application of genetic algorithm for 3D models' fitting is as follows:

Step 1: Randomly generate an initial population $G(0)$ of n individuals, each individual m is a set of randomly selected angles (float value within the rotation ranges) rotated around X, Y, and Z axes of the coordinate system.

Step 2: Evaluate and save fitness $f(m)$, which is the summed value of RDS differences from each individual m in the current population $G(t)$, then rearrange the individuals in $G(t)$ in the ascending order of $f(m)$ by bubble method.

Step 3: Calculate the selection probabilities $p(m)$ for each individual m in $G(t)$; here $p(m)$ is defined as

$$p(m) = \frac{\left(1/f(m)\right)}{\sum_{i=1}^{n}\left(1/f(i)\right)} \qquad (7.15)$$

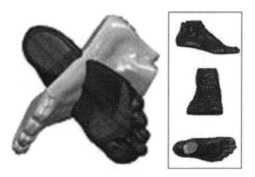

FIGURE 7.17   Result of model fitting: left—initial postures; right—fitted results.

Step 4: Probabilistically select individuals from $G(t)$ to produce a temporary population $N(t)$ with the same size as that of $G(t)$, those individuals with higher $p(m)$ values in $G(t)$ may appear many times in $N(t)$.

Step 5: Randomly select two individuals from $N(t)$ as the parent each time and then crossover them to produce their offspring until 80% of population $G(t + 1)$ is generated in this way; here the use of crossover operator cr is

```
cr = rand();

offspring1 = rc*parent1+(1-cr)*parent2;        (7.16)

offspring2 = (1-cr)*parent1+cr*parent2;
```

where *rand*() is a function to generate a random real number that is uniformly distributed in the interval of [0, 1]; thus, the generated offspring still fall in the same rotation ranges as their parents.

Step 6: Mutation operation is applied to generate the remaining 20% of population $G(t + 1)$ through randomly selecting new individuals, as convergence to the global optimum can be achieved quickly by using more global samplings.

Step 7: Repeat Steps 2 through 6 until an individual solution reaches a prespecified level of fitness or a predefined number of generations are exceeded. Figure 7.17 shows the result of model fitting with different postures.

7.8   SUMMARY

In this chapter, 3D point data processing techniques are introduced. For curves and surfaces, nonparametric and parametric representations are provided and compared. For reverse engineering, we orderly introduce data acquisition and preprocessing, model creation, and amendment. For processing of point cloud, point cloud denoising, transformation, registration including PCA and ICP are presented. For surface reconstruction, parametric method, implicit method, etc., are analyzed. For shape deformation, we take last and foot as examples to illustrate the deforming methods. For dimensional measurements, foot dimension definitions and determinations are provided. For geometric fitting, a practical method is given and analyzed.

QUESTIONS

7.1 Describe the generally used methods to represent curves and surfaces.

7.2 How can you eliminate noise from the noisy point cloud data?

7.3 Explain the functions of a translation matrix and a rotation matrix.

7.4 Explain the basic procedure to register two point clouds using coarse registration and subsequent fine registration.

7.5 How do you reconstruct a surface from a point cloud?

7.6 Outline a method to deform the shape of a last.

7.7 List the foot dimensions needed for footwear manufacture.

7.8 Is geometric-model fitting a more valuable method compared with dimension matching for footwear? Discuss.

REFERENCES

Amberg, B., Romdhani, S., and Vetter, T. 2007. Optimal step non-rigid ICP algorithms for surface registration. *Proceeding of IEEE Conference on Computer Vision and Pattern Recognition*, Minneapolis, MN, June 17–22, pp. 1–8.

Amenta, N., Bern, M., and Kamvysselis, M. 1998. A new Voronoi-based surface reconstruction algorithm. *SIGGRAPH 1998*, Orlando, FL, July 19–24, pp. 415–421.

Amenta, N., Choi, S., Dey, T. K. et al. 2002. A simple algorithm for homeomorphic surface reconstruction. *International Journal of Computational Geometry and Applications* 12: 125–141.

Amenta, N. and Kil, Y. J. 2004. Defining point-set surfaces. *ACM Transactions on Graphics* 23(3): 264–270.

Ansary, T., Vandeborre, J. P., Mahmoudi, S. et al. 2004. A Bayesian framework for 3D models retrieval based on characteristic views. *Proceedings of the Second International Symposium on 3D Data Processing, Visualization and Transmission*, Thessaloniki, Greece, September 6–9, pp. 139–146.

Bajaj, C. L., Bernardini, F., and Xu, G. 1995. Automatic reconstruction of surfaces and scalar fields form 3D scans. *SIGGRAPH 1995*, Los Angeles, CA, August 6–11, pp. 109–118.

Bendels, G. H., Degener, P., Wahl, R. et al. 2004. Image-based registration of 3D-range data using feature surface elements. *The Fifth International Symposium on Virtual reality, Archeology and Cultural Heritage*, Brussels and Ename Center, Oudenaarde, Belgium, December 7–10, pp. 115–124.

Besl, P. J. and McKay, N. D. 1992. A method for registration of 3-D shapes. *IEEE Transactions on Pattern Analysis and Machine Intelligence* 14(2): 239–256.

Boughorbel, F., Mercimek, M., Koschan, A. F. et al. 2010. A new method for the registration of three-dimensional point-sets: The Gaussian Fields framework. *Image and Vision Computing* 28: 124–137.

Brenner, C. and Dold, C. 2007. Automatic relative orientation of terrestrial laser scans using planar structures and angle constraints. *ISPRS Workshop on Laser Scanning and SilviLaser*, Espoo, Finland, September 12–14, pp. 84–89.

Bunch, R. P. 1988. Foot measurement strategies for fitting athletes. *Journal of Testing and Evaluation* 16(4): 407–411.

Bustos, B., Keim, D. A., Saupe, D. et al. 2005. Feature-based similarity search in 3D object databases. *ACM Computing Surveys* 37(4): 345–387.

Butdee, S. 2002. Hybrid feature modeling for sport shoe sole design. *Computers and Industrial Engineering* 42(2–4): 271–279.

Carr, J. C., Beatson, R. K., Cherrie, J. B. et al. 2001. Reconstruction and representation of 3D objects with radial basis functions. *SIGGRAPH 2001*, Los Angeles, CA, August 12–17, pp. 67–76.

Chang, Y. and Rockwood, A. P. 1994. A generalized de Casteljau approach to 3D free-form deformation. *SIGGRAPH 1994*, Orlando, FL, July 24–29, pp. 257–260.

Chen, Y. and Medioni, G. 1992. Object modeling by registration of multiple range images. *Image and Vision Computing* 10(3): 145–155.

Chen, D. Y., Tian, X. P., Shen, Y. T. et al. 2003. On visual similarity based 3D model retrieval. *Computer Graphics Forum* 22(3): 223–232.

Chua, C. and Jarvis, R. 1996. 3D free form surface registration and object recognition. *International Journal of Computer Vision* 17(1): 77–99.

Cohen, S. D. and Da, F. 2004. A greedy Delaunay-based surface reconstruction algorithm. *Visual Computer* 20: 4–16.

Daniels, II J., Ha, L. K., Ochotta, T. et al. 2007. Robust smooth feature extraction from point clouds. *Proceedings of Shape Modeling International*, Lyon, France, June 13–15, pp. 123–136.

Daras, P., Zarpalas, D., and Tzovaras, D. 2006. Efficient 3D model search and retrieval using generalized 3D Radon transforms. *IEEE Transactions on Multimedia* 8(1): 101–114.

De Boer, A., Van der Schoot, M. S., and Bijl, H. 2007. Mesh deformation based on radial basis function interpolation. *Computers and Structures* 85: 11–14.

Dey, T. K. and Goswami, S. 2006. Provable surface reconstruction from noisy samples. *Computational Geometry* 35: 124–141.

Dey, T. K., Gosmami, S., and Sun, J. 2004. Smoothing noisy point clouds with Delaunay pre-processing and MLS. Tech-report OSU-CISRC-3/04-TR17.

Ding, Y., Zhao, J., Ravindra, S. G. et al. 2010. An automatic method for measuring foot girths for custom footwear using local RBF implicit surfaces, *International Journal of Computer Integrated Manufacturing* 23(6): 574–583.

Du, S. Y., Zheng, N. N., Ying, S. H. et al. 2007. An extension of the ICP algorithm considering scale factor. *Proceedings of the14th IEEE International Conference on Image Processing*, Hyatt Regency San Antonio, TX, September 16–19, pp. 193–196.

Du, S. Y., Zheng, N. N., Ying, S. H. et al. 2010. Affine iterative closest point algorithm for point set registration. *Pattern Recognition Letters* 31: 791–799.

Farin, G. E., Hoschek, J., and Kim, M. S. 2002. *Handbook of Computer Aided Geometric Design*. North-Holland, Amsterdam, the Netherlands.

Feldmar, J. and Ayache, N. 1996. Rigid affine and locally affine registration of free-form surfaces. *International Journal of Computer Vision* 18(2): 99–119.

Fitzgibbon, A. W. 2003. Robust registration of 2D and 3D point sets. *Image and Vision Computing* 21(13–14): 1145–1153.

Fleishman, S., Cohen-Or, D., and Silva, C. T. 2005. Robust moving least-squares fitting with sharp features. *ACM Transactions on Graphics* 24(3): 544–552.

Fleishman, S., Drori, I., and Cohen-Or, D. 2003. Bilateral mesh denoising. *ACM Transactions on Graphics* 22(3): 950–953.

Flory, S. and Hofer, M. 2010. Surface fitting and registration of point clouds using approximations of the unsigned distance function. *Computer Aided Geometric Design* 27: 60–77.

Foley, J. D., VanDam, A., Feiner, S. K. et al. 1993. *Introduction to Computer Graphics*. Addison-Wesley Publishing Company, Reading, MA.

Gopi, M., Krishnan, S., and Silva, C. 2000. Surface reconstruction based on lower dimensional localized Delaunay triangulation. *The Proceeding of Eurographics*, Interlaken, Switzerland, August 21–25, Vol. 19, pp. 467–478.

Granger, S. and Pennec, X. 2002. Multi-scale EM-ICP: A fast and robust approach for surface registration. *Proceedings of European Conference on Computer Vision*, Copenhagen, Denmark, May 28–31, pp. 418–432.

Han, Y., Ma, L., He, S. P. et al. 2003. A study on data combination method in topography measurement of revolving objects. *Journal of University of Science and Technology of China* 33(6): 282–286.

Ho, J., Yang, M., Rangarajan, A. et al. 2007. A new affine registration algorithm for matching 2D point sets. *Proceeding of IEEE Workshop on Applications of Computer Vision*, Austin, TX, February 20–21, pp. 25–30.

Hoppe, H., DeRose, T., Duchamp, T. et al. 1992. Surface reconstruction from unorganized points. *SIGGRAPH 1992*, Chicago, IL, July 27–31, pp. 71–78.

Hoppe, H., DeRose, T. T., Duchamp, T. et al. 1994. Piecewise smooth surface reconstruction. *SIGGRAPH 1994*, New York, July 24–29, pp. 295–302.

Hu, H., Li, Z., Yan, J. et al. 2007. Anthropometric measurement of the Chinese elderly living in the Beijing area. *International Journal of Industrial Ergonomics* 37(4): 303–311.

Huang, J., Chen, L., Liu, X. H. et al. 2009a. Efficient mesh deformation using tetrahedron control mesh. *Computer Aided Geometric Design* 26(6): 617–626.

Huang, W. M., Li, Y. W., Wen, P. Z. et al. 2009b. Algorithm for 3D point cloud denoising. *Third International Conference on Genetic and Evolutionary Computing*, Guilin, China, October 14–17, pp. 574–578.

Jenke, P., Wand, M., Bokeloh, M. et al. 2006. Bayesian point cloud reconstruction. *Computer Graphics Forum* 25(3): 379–388.

Jian, B. and Vemuri, B. C. 2005. A robust algorithm for point set registration using mixture of Gaussians. *Proceeding of IEEE International Conference on Computer Vision*, Beijing, China, October 17–21, pp. 1246–1251.

Jiang, J., Cheng, J., and Chen, X. L. 2009. Registration for 3D point cloud using angular-invariant feature. *Neurocomputing* 72: 3839–3844.

Jost, T. and Heinz, H. 2003. A multi-resolution ICP with heuristic closest point search for fast and robust 3D registration of range images. *Proceedings of the Fourth International Conference on 3-D Digital Imaging and Modeling*, Alberta, Canada, October 6–10, pp. 427–433.

Kazhdan, M., Bolitho, M., and Hoppe, H. 2006. Poisson surface reconstruction. *Proceedings of SGP*, Cagliari, Sardinia, Italy, June 26–28, pp. 61–70.

Kazhdan, M., Funkhouser, T., and Rusinkiewicz, S. 2004. Shape matching and anisotropy. *ACM Transactions on Graphics* 23(3): 623–629.

Klassen, E., Srivastava, A., Mio, W. et al. 2004. Analysis of planar shapes using geodesic paths on shape spaces. *IEEE Transactions on Pattern Analysis and Machine Intelligence* 26(3): 372–383.

Lange, C. and Polthier, K. 2005. Anisotropic smoothing of point sets. *Computer Aided Geometric Design* 22(7): 680–692.

Lee, S., Wolberg, G., and Shin, S. Y. 1997. Scattered data interpolation with multilevel B-splines. *IEEE Transactions on Visualization and Computer Graphics* 3(3): 228–244.

Levoy, M., Pulli, K., Curless, B. et al. 2000. The digital Michelangelo project 3D scanning of large statues. *Proceedings of the 27th Annual Conference on Computer Graphics and Interactive Techniques*, New Orleans, LA, July 23–28, pp. 131–144.

Li, J. F. 2009. Feature-preserving denoising of point-sampled surfaces. *Proceedings of the Third WSEAS International Conference on Computer Engineering and Applications*, Stevens Point, WI, January 10–12, pp. 122–126.

Li, L., Schemenauer, N., Peng, X. et al. 2002. A reverse engineering system for rapid manufacturing of complex objects. *Robotics and Computer-Integrated Manufacturing* 18(1): 53–67.

Liu, L., Bajaj, C., Deasy, J. et al. 2008. Surface reconstruction from non-parallel curve networks. *Computer Graphics Forum* 27(2): 155–163.

Liu, W., Miller, J., Stefanyshyn, D., and Nigg, B. M. 1999. Accuracy and reliability of a technique for quantifying foot shape, dimensions and structural characteristics. *Ergonomics* 42(2): 346–358.

Long, X., Zhong, Y. X., Li, R. J. et al. 2002. 3D surface integration in structured light 3D scanning. *Journal of Tsinghua University* 42(4): 477–480.

Luximon, A. and Luxinmon, Y. 2009. Shoe-last design innovation for better shoe fitting. *Computer Industry* 60(8): 621–628.

Ma, W. and Kruth, J. P. 1995. Parameterization of randomly measured points for least square fitting of B-spline curves and surfaces. *Computer-Aided Design* 27(9): 663–675.

Mederos, B., Velho, L., and de Figueiredo, L. H. 2003. Robust smoothing of noisy point clouds. *SIAM Conference on Geometric Design and Computing*, Seattle, WA, November 9–13, pp. 1–13.

Miller, J. V., Breen, D. E., Lorensen, W. E. et al. 1991. Geometrically deformed models: A method for extracting closed geometric models from volume data. *SIGGRAPH 1991*, Las Vegas, NV, July 29–August 2, pp. 217–226.

Motavalli, S. 1998. Review of reverse engineering approaches. *The 23th International Conference on Computers and Industrial Engineering*, Chicago, IL, March 28–April 2, Vol. 35(1&2), pp. 25–28.

Ohtake, Y., Belyaev, A., Alexa, M. et al. 2003. Multi-level partition of unity implicits. *ACM Transactions on Graphics* 22(3): 463–470.

Ohtake, Y., Belyaev, A., and Seidel, H. P. 2004. 3D scattered data approximation with adaptive compactly supported radial basis functions. *Proceedings of the Shape Modeling International*, Genova, Italy, June 7–9, pp. 31–39.

Okabe, A., Boots, B., Sugihara, K. et al. 2000. *Spatial tessellations: Concepts and applications of Voronoi Diagrams*, 2nd edn., Wiley, New York.

Pauly, M. and Gross, M. 2001. Spectral processing of point-sampled geometry. *SIGGRAPH 2001*, Los Angeles, CA, August 12–17, pp. 379–386.

Pauly, M., Mitra, N. J., and Guibas, L. J. 2004. Uncertainty and variability in point cloud surface data. *Proceedings of Eurographics Symposium on Point-Based Graphics*, Zurich, Switzerland, June 2–4, pp. 77–84.

Perona, P. and Malik, J. 1990. Scale-space and edge detection using anisotropic diffusion. *IEEE Transactions on Pattern Analysis and Machine Intelligence* 12(7): 629–639.

Pottmann, H., Leopoldseder, S., and Hofer, M. 2004. Registration without ICP. *Computer Vision and Image Understanding* 95: 54–71.

Qiu, D. Y., May, S., and Nüchter, A. 2009. GPU-accelerated nearest neighbor search for 3D registration. *Proceedings of the Seventh International Conference on Computer Vision Systems*, Liege, Belgium, October 13–15, pp. 194–203.

Reuter, M., Wolter, F. E., and Peinecke, N. 2005. Laplace-spectra as fingerprints for shape matching. *Proceedings of Symposium on Solid and Physical Modeling*, Boston, MA, June 13–15, pp. 101–106.

Rusinkiewicz, S., Hall-Holt, O., and Levoy, M. 2002. Real-time 3D model acquisition. *Proceedings of the 29th Annual Conference on Computer Graphics and Interactive Techniques*, San Antonio, TX, July 23–26, pp. 438–446.

Schall, O., Belyaev, A., and Seidel, H. P. 2005. Robust filtering of noisy scattered point data. *Eurographics Symposium on Point-Based Graphics*, Stony Brook, NY, June 21–22, pp. 71–77.

Schall, O., Belyaev, A., and Seidel, H. P. 2008. Adaptive feature-preserving non-local denoising of static and time-varying range data. *Computer-Aided Design* 40: 701–707.

Schnabel, R., Degener, P., and Klein, R. 2009. Completion and reconstruction with primitive shapes. *Computer Graphics Forum* 28: 503–512.

Sharp, G. C., Lee, S. W., and Wehe, D. K. 2002. ICP registration using invariant features. *IEEE Transactions on Pattern Analysis and Machine Intelligence* 24(1): 90–102.

Stamos, I. and Leordeanu, M. 2003. Automated feature-based range registration of urban scenes of large scale. *IEEE Computer Society Conference on Computer Vision and Pattern Recognition*, Madison, WI, June 16–22, pp. 555–561.

Taubin, G. 1995. A signal processing approach to fair surface design. *SIGGRAPH 1995*, Los Angeles, CA, August 6–11, pp. 351–358.

Thapa, A., Pu, S., and Gerke, M. 2009. Semantic feature based registration of terrestrial point clouds. *Proceedings of Laser Scanning ISPRS* 38(3): 230–235.

The Brannock Device Co. Inc. 2010. http://www.brannock.com/

Várady, T., Martin, R. R., and Cox, J. 1997. Reverse engineering of geometric models—An introduction. *Computer-Aided Design* 29(4): 255–268.

Wang, Z. and Brenner, C. 2008. Point based registration of terrestrial laser data using intensity and geometry features. *International Archives of Photogrammetry, Remote Sensing and Spatial Information Sciences*, XXXVII, Part B5: 583–589.

Wang, J., Zhang, H., Lu, G. et al. 2011. Rapid parametric design methods for shoe-last customization. *International Journal of Advanced Manufacturing Technology* 54: 173–186.

Weyrich, T., Pauly, M., Keiser, R. et al. 2004. Post-processing of scanned 3D surface data. *Proceedings of Symposium on Point-Based Graphics*, Zurich, Switzerland, Jun 2–4, pp. 85–94.

Witana, C. P., Xiong, S. P., Zhao, J. H. et al. 2006. Foot measurements from 3-dimensional scans: A comparison and evaluation of different methods. *International Journal of Industrial Ergonomics* 36(9): 789–807.

Wyngaerd, J. V. and Gool, L. V. 1999. Invariant-based registration of surface patches. *Proceedings of IEEE International Conference on Computer Vision*, Kerkyra, Greece, September 20–25, pp. 301–306.

Wyngaerd, J. V. and Gool, L. V. 2002. Automatic crude patch registration: Toward automatic 3D model building. *Computer Vision and Image Understanding* 87(1–3): 8–26.

Xiao, C. X., Miao, Y. W., Liu, S. et al. 2006. A dynamic balanced flow for filtering point-sampled geometry. *Visual Computer* 22(3): 210–219.

Xie, Z. X., Xu, S., and Li, X. Y. 2010. A high-accuracy method for fine registration of overlapping point clouds. *Image and Vision Computing* 28: 563–570.

Xu, C., Liu, Y., Jiang, Y. et al. 2004. The design and implementation for personalized shoe last CAD system. *Journal of Computer-Aided Design and Computer Graphics* 16(10): 1437–1441.

Xu, C., Liu, J., and Tang, X. 2009. 2D shape matching by contour flexibility. *IEEE Transactions on Pattern Analysis and Machine Intelligence* 31(1): 180–186.

Yahara, H., Higuma, N., Fukui, Y. et al. 2005. Estimation of anatomical landmark position from model of 3-dimensional foot by the FFD method. *Systems and Computers in Japan* 36(6): 1–13.

Yaroslavsky, L. P. 1985. *Digital Picture Processing, An Introduction*. Springer Verlag, Berlin, Germany.

Yoon, M., Ivrissimtzis, I., and Lee, S. 2009. Variational Bayesian noise estimation of point sets. *Computers and Graphics* 33: 226–234.

Zhang, Z. 1994. Iterative point matching for registration of free-form curves and surfaces. *International Journal of Computer Vision* 13(2): 119–152.

Zhang, X., Xi, J., and Yan, J. 2006. A methodology for smoothing of point cloud data based on anisotropic heat conduction theory. *International Journal of Advanced Manufacturing Technology* 30(1–2): 70–75.

Zhao, H. K, Osher, S., Merriman, B. et al. 2000. Implicit and nonparametric shape reconstruction from unorganized points using variational level-set method. *Computer Vision and Image Understanding* 80: 295–319.

Zhao, J., Xiong, S., Bu, Y. et al. 2008. Computerized girth determination for custom footwear manufacture. *Computers and Industrial Engineering* 54(3): 359–373.

Part III

Footwear Design

8 Capturing Footwear Needs for Delighting Customers

Emily Y.L. Au and Ravindra S. Goonetilleke

CONTENTS

8.1 INTRODUCTION

After the Second World War began in 1939, there were significant restrictions on shoe designs (Pratt and Wolley, 1999): height of heels was limited to 2.5 cm in the United States and 5 cm in Britain. Leather was in short supply and as a result poor quality leather with bright hues and cheap materials such as cork, wood, and rubber were used for shoes. Shoes had canvas uppers, crepe soles, and plastic straps. With the removal of restrictions on footwear, high heels and the peep toes emerged in the early 1950s. Toward the late 1950s, the toe area became longer, heel heights reduced, and the court shoe became popular. With the price of leather increasing, the materials used for shoes changed in the 1960s. Plastics and other synthetic material became popular and were promoted by many fashion designers. The platform shoe started to re-emerge around 1967 after it had gone out of fashion in the 1940s. By the mid-1970s, platform shoes and boots, and bell-bottomed trousers were the most popular. Crepe rubber and leather-covered plastics or wood were common materials for the shoe heel and soles. In the 1980s, people were most concerned with designer labels and shoes made of materials that could breathe became popular. Platform shoes made another comeback in the early 1990s. All such changes over the last century have been driven by fashion.

Today, with hardly any material or manufacturing limitations, consumers want value for money. Hence, the footwear industry continues to grow and footwear trading is soaring around the world with footwear sold over a wide range of prices. Many footwear styles fly off the shelves in shoe stores to satisfy the diverse customer needs in varied market segments. Major retailers believe that the current trend of global footwear sales will continue to double every 20 years (Fry, 2010). With footwear customization becoming more popular (Boër et al., 2004), customers are seeking footwear configurators to look and feel unique. The low cost of manufacturing footwear makes manufacturers compete primarily on the value-added features (Solves et al., 2006) that meet the customers' intrinsic emotional needs and desires (Halliday and Setchi, 2009) rather than price. Piller and Müller (2004) indicated that fit is regarded as the most important attribute of footwear customization, followed by style and functionality. Hence, to gain market share in this fast growing business of customization, footwear designers and manufacturers have to clearly understand the customer needs and know how the design features influence the perception and customer satisfaction.

8.2 CONSUMER BUYING PROCESS

Aesthetics is an important criterion for buying a shoe (Clarks, 1976; Au and Goonetilleke, 2007). If a person does not like what they see, it is unlikely they will examine that product any further. Women tend to choose shoes based on fashion and appearance (Slater, 1985; Frey, 2000; Piller, 2002; Lutter, 2004), and people sacrifice physical comfort for psychological comfort, even when the risk of physical harm is obvious (Slater, 1985). Au and Goonetilleke (2007) found that both comfortable and uncomfortable shoes that people own have good aesthetics. In other words, the buying decision is heavily dependent on visual satisfaction as proven by a survey of 420 European consumers, where 63.1% of the females and 61.7% of the males indicated that they found it difficult to find a shoe with the right "design" (Piller, 2002). Thus, fit and form are as important as, if not more important than, the functional requirements when buying a product such as a shoe. Many of the advanced materials and technologies that manufacturers use for their performance footwear lines get incorporated to their casual collections over time (Leand, 2007) to improve customer satisfaction (Bloch, 1995; Tarasewich and Nair, 2001; Liu et al., 2007). For instance, Adidas torsion technology, first launched in 1989, was claimed to be a dynamic system that provided enhanced stability while reducing foot fatigue. In 2007, torsion was used in Rockport shoes for styles ranging from athletically inspired casual to modern, contemporary dress looks (Fibre2fashion.com, 2007). Hence, it is to the advantage of any manufacturer to attempt to elicit the physical and psychological needs of differing customer segments and to incorporate them into the various product attributes (MacMillan and McGrath, 1996).

8.3 COMFORT: ITS IMPORTANCE

Comfort is said to be (1) relief from distress, (2) a state of ease and quiet enjoyment, and (3) an aspect that makes life easy (Websters, 1990). The diversity of these phrases makes comfort to be a multifaceted entity (Goonetilleke and Luximon, 2001).

Comfort has varied definitions: for example, it has been defined as a lack of discomfort (Hertzberg, 1972). Based on Fredrick Hertzberg's dual factor theory, also called the motivation-hygiene theory (Herzberg et al., 1959), Zhang et al. (1996) associated comfort with feelings of relaxation and well-being and stated that poor biomechanics may turn comfort to discomfort even though good biomechanics is not a necessary and sufficient condition for comfort. Kano based his theory on the motivation-hygiene aspects to determine the necessary and attractive needs of a product. A sense of comfort is associated with a bean bag comprising macrospheres of rigifoam or the like. From an engineering perspective, it appears that such a cushion conforms to the buttocks, relieving high pressure and distributing force over a larger area of soft tissue thereby supporting the common sense approach of distributing the force rather than concentrating the loads. Is standing or walking on such macrospheres of rigifoam also comfortable even though they are able to distribute the forces? Or are bean bags really comfortable or is it that we use them in environments where the user is already in a relaxed state? Comfort is context dependent. We cannot evaluate the comfort of high-heeled shoes in an athletic setting or on a grass ground. Comfort has many components as well just like any other sense such as sound which has two subjective components of loudness and pitch. With respect to seating comfort, Kamijo et al. (1982) considered the five components of body pressure, posture, sensation of being cushioned, space, and the sensation of being held in place as the components of comfort. All such components may not be present in a scenario at any one time. Au and Goonetilleke (2007) did an extensive survey of the difference between comfortable and uncomfortable ladies' shoes and found that the differentiating factors were size of the shoes, unpleasant odors, texture, the feeling of the shoe, the sound the shoe emits, temperature and humidity inside the shoe, and amount of discomfort or pain when wearing the shoe. In other words, the differences were in the tactile (size, texture, feel, climate), auditory, and olfactory sensations. These perceptions were positive for the comfortable shoes and negative for the uncomfortable shoes supporting the notion that comfort is a positive sensation and discomfort a negative sensation (Goonetilleke, 2001). Au and Goonetilleke (2007) also found that styling, color, the looks, were not different between the comfortable and uncomfortable shoes. The lack of any difference in aesthetics does not imply that it is not important. Instead, it means that visual appearance is a key determinant in the buying decision (Bloch, 1995). The true differences between comfort and discomfort only emerge after some period of use (Kolarik, 1995). Visual and tactile perceptions are necessary at point of sale but these characteristics do not guarantee long-term comfort. Thus, the way we elicit comfort has to be based on psychophysical laws with a clear knowledge of the stimuli influencing it.

Perfumes and fragrances work when the odorant *molecules* bind to the receptor *membranes* and change the *mood* of a person. These events are known as the "3Ms" of perfumery (Dodd and Skinner, 1992) clearly indicating that the process is part subjective and part objective. Similarly, Slater (1985) proposed that comfort is part physical and part psychological with physical comfort related to tangible stimuli. For example, a shoe that is too short will have a physical component. Previous research has shown that physical comfort is influenced by foot anthropometry

(Hawes et al., 1994; Wunderlich and Cavanagh, 2001), shoe fit (Luximon et al., 2001; Witana et al., 2004; Xiong and Goonetilleke, 2006; Xiong et al., 2007, 2008; Au and Goonetilleke, 2009; Witana et al., 2009), skeletal alignment (Miller et al., 2000), inner shoe climate, material properties (Zhang et al., 1991; Goonetilleke, 1999; Mündermann et al., 2002; Lee and Hong, 2005), and so on. Some have focused on the psychological, affective, and emotional values (Kaiser et al., 1987; Benstock and Suzanne, 2001; McNeil and Riello, 2005). Psychological comfort, according to Slater (1985), relates to appearance, and is a quality that enhances the wearer's image (Slater, 1985). Fashion, style, personality, sexuality, class, and gender (McNeil and Riello, 2005) are all related to psychological comfort and play a critical role in the consumer buying decision.

8.3.1 Measurement of Sensation

Manufacturers and marketers of various products are always looking to enhance their product by adding on different attributes of a sensation or different sensation. These include food manufacturers looking for the crunchiness, snap, crackle and pop, and so on. All of these characteristics are based on sensation. Taste is unique but many of us have our preferences toward certain restaurants, brands, or even certain chefs. Food, like many other consumer products, can be made to suit an international audience or may be made to cater to a local audience. For example, the same meat can be prepared in quite different ways by Chinese, Sri Lankans, Indians, Americans, Japanese, and so on. That type of preparation gives the international flavor but the subtle differences within a country or region is accomplished through variations of the same or similar ingredients. In statistical terms, such a process can be compared with the mean and the standard deviation of a variable. Shifts in mean can correspond to the internationalization aspects while the standard deviations can represent the variations within a given locale. Even fields such as fluid mechanics and structural mechanics resort to using a similar approach to model systems. In those specialties, such modeling to represent variations is called perturbation theory, which is a study of small disturbances in a physical system. Similar approaches can be used to evaluate differences in subjective perceptions among different populations when the fundamental laws are clearly defined.

One's senses and internal state will determine one's preference for an item. Many items that we purchase in a grocery store go through a sniff-feel-and-sometimes-pinch test (Lindstrom, 2004). Sour, bitter, sweet, and salty have been identified as the basic categories of taste (Bartoshuk, 1988; Beauchamp et al., 1991). Sweetness and bitterness may be further subdivided into several subcategories using psychophysical procedures. If we rely on people to elicit these varied characteristics, should researchers have concerns about subjective opinions and ratings about footwear if the right sensations are captured?

The minimum amount of energy necessary to evoke a sensation is called the *absolute threshold* and the *difference threshold* is the minimum difference that is required to evoke a *just noticeable difference* (JND) in the sensation. Both these thresholds relate to changes in sensation, but what is important is to know the magnitude of the sensation. In the late 1950s, Stevens (1960) and colleagues proposed three ways to construct psychological scales: confusion scaling, partition scaling, and ratio scaling. Thereafter, Stevens established the now popular power law for many different stimuli:

$$S = aX^b$$

where
  S is the magnitude of the sensation
  X is the stimulus intensity
  a is a scaling constant
  b is an exponent that depends on the type of stimulus

Typical exponents of the power function are 1.0 for visual length, 1.1 for static force on palm, 1.45 for weights lifted by hand, 0.80 for squeezing rubber blocks, and so on. Most physical quantities have an absolute threshold (X_0) and the power law relationship is often modified (Stevens and Stevens, 1963) to reflect this threshold as

$$S = a(X - X_0)^b \text{ or alternatively, as } S = aX^b + X_0$$

Our senses adapt to whatever stimulus that exists whether it be smell, sound, taste, touch, and sight. The same can be said of discomfort and pain. In other words, with constant stimulation, the perceived intensity of the stimulus decreases over time. This adaptation can be viewed as a decrease in sensitivity or alternatively as an increase in threshold. The following formulation has been proposed to account for adaptation (Overbosch, 1986):

$$S = a(Xe^{-bt/X} - X_0)^b$$

Researchers have shown that the power law is only true up to about 60% of the maximum stimulus intensity and that the sensation plateaus with a maximum value rather than tend to infinity with increasing intensity (Laffort, 1966). All of these qualities have close equivalents in Kano's model described later. Furthermore, King (1986) found that the power exponent in Steven's law is not a constant over the range of the stimulus intensity but one that decrease with increasing stimulus intensity. These varied opinions led Mayer (1975) to suggest that the sensation–stimulus relationship follows a sigmoidal curve rather than a power curve. For saltiness, Easton (2005) has shown that the relationship follows a Gompertz sigmoidal function as follows:

$$S(X) = S_\infty e^{(-b/a)e^{(-aX)}}$$

where
  S_∞ is the maximum perceived intensity
  a is a retardation constant
  b represents the initial increase of rate constant

To account for these differences, Molski (2011) incorporated the Grompertz function into the original power function to extend Stevens power law to be as follows:

$$S(X) = k(X)X^{n(X)}$$

where the scaling factor k and the exponent n are both function of the stimulus intensity X.

The basis for using psychophysics is based on the principle of nomination proposed by Marks (1978), which states that the two identical stimuli that induce the same neural response in a person will give rise to the same sensory experience. Most physical measurements are made using ratio scales that have order, interval and origin, or absolute zero (Gescheider, 1985). Thus, measuring the sensation on a ratio scale is highly desirable. Psychologists started investigating ratio scales for sensation only after acoustic engineers became interested in the psychological variable of loudness. A common method to determine the psychological ratio scale is *magnitude estimation*, which can be performed in two ways. In the first, a standard stimulus is presented to the subject and the participant would be told that the sensation the standard stimulus would produce is 100 (or any positive number). Thereafter, whenever a stimulus is presented the subject has to assign a value to the sensation in terms of how many times greater or smaller the stimulus is relative to the standard stimulus (Witana et al., 2009). In the second method, a standard stimulus is not provided. Instead, the subject assigns a number to each stimulus that is presented as a proportion of their magnitudes. The instructions to the subjects would be as given by Stevens (1975) (p. 30):

"You will be presented with a series of stimuli in irregular order. Your task is to tell how intense they seem by assigning numbers to them. Call the first stimulus any number (that you think is appropriate). Then assign successive numbers in such a way that they reflect your subjective impression. There is no limit to the range of numbers that you may use. You may use whole numbers, decimals, or fractions. Try to make each number match the intensity as you perceive it."

After running the experiment with many subjects, the median or geometric mean of the data are used as the sensation for any given stimulus. The arithmetic mean is generally not used because it could be highly biased by outliers.

8.4   EMOTION

Digitization, measurement, scanning, mathematical modeling, and other techniques are used to evaluate the physical matching between the foot and the shoe (Luximon and Goonetilleke, 2004; Witana et al., 2004; Xiong and Goonetilleke, 2006; Xiong et al., 2007, 2008; Au and Goonetilleke, 2009; Witana et al., 2009). On the flip side, techniques such as Kansei engineering (Nagamachi, 1995, 2002, 2011; Ishihara et al., 1997; Nakada, 1997) and Kano's method (Au et al., 2006) have been used to identify footwear appeal to delight customers.

8.4.1 Kansei Engineering

Kansei engineering (KE) is used to develop new products or improve the existing products by linking the customer feelings (Kansei) with the physical design elements (Nagamachi et al., 2006; Nagamachi, 2011) thereby allowing consumer emotion to be incorporated into the design process (Pearce and Coleman, 2008). In KE, emotional words and phrases (called the semantic universe) associated with a product are collected from web pages, magazines, advertisements, promotion materials, and so on. In the next stage, consumers are asked to evaluate a sample of products using a questionnaire where the questions are formed using the emotive words. Typically, a five-point semantic differential scale is used. The collected data are then analyzed using multivariate techniques. Finally, design strategies are created to improve the design and the emotions of that product for a selected target segment.

Alcántara et al. (2005) investigated the semantic space of casual footwear using 74 adjectives without including color and tone. Sixty-seven participants (46 males and 21 females with age range 25–35 years) evaluated 36 shoes on a five-point scale as if they were looking at a shop window. A principal component analysis revealed 20 semantic axes. The first three axes were interpreted as comfort, usefulness, and practical; smart footwear; and modern and innovative. The identified semantic space claimed to provide the basis for assessing the user's perception of shoes. These results can be quite useful in improving the emotional aspects, but the axes do not provide sufficient information on the consumer needs.

Successful implementations of Kansei engineering (Nagamachi, 2011) include the Mazda Miata, Wacoal Good-Up Bra, Sharp Licked Crystal Viewcam, Matsusita sitting shower, Milbon Deesse's shampoo and treatment, and Mizuno IntageX3 golf club. Other applications include the design of ladies footwear (Ishihara et al., 1997), and men's everyday footwear and historic footwear (van Lottum et al., 2006).

However, there are two shortcomings of the method:

1. It is time consuming to extract the emotional response and a large number of respondents is required especially if small differences in opinion are to be detected (van Lottum et al., 2006; Pearce and Coleman, 2008).
2. Some characteristics such as "big–small" have a magnitude property, but others such as "beautiful–ugly" do not. Thus, the application of the statistical method to nonlinear characteristics is inappropriate (Nagamachi et al., 2006). Hence, other modeling techniques (e.g., rough set model) have to be used to solve the problem.

8.4.2 Kano's Method

The Kano's method (Kano et al., 1984) allows one to identify the attractive attributes of products using paired questions. It has been used to improve webpage designs (Tan and Shen, 2000; Zhang and Von Dran, 2002; Szmigin and Reppel, 2004), e-learning systems (Chen and Lin, 2007), online ticket booking systems (Witell and Fundin, 2005), cockpit weather information systems (Sireli et al., 2007), service quality (Yang, 2003), employee satisfaction (Matzler et al., 2004), consumer product packaging (Löfgren and Witell, 2005), ski equipment (Matzler and Hinterhuber, 1998), air-conditioners (Yang, 2005), and so on. However, Kano's method has not yet been extensively implemented in the design of footwear (Au et al., 2006).

In the Kano model, the *One-dimensional (O)*, *Attractive (A)*, and *Must-be (M)* qualities represent the different types of customer needs (Berger et al., 1993). These qualities are determined by having subjects respond to a series of paired questions comprising a functional and a dysfunctional question (Kano et al., 1984; Berger et al., 1993). An uncomfortable shoe does not have a poor fit everywhere (Au and Goonetilleke, 2007). Overall fit is determined by the degree of poor fit (pressure exceeding discomfort thresholds) in one or more regions (Witana et al., 2004; Rupérez et al., 2010). Thus, regional fit can be considered to be a must-be requirement for comfort.

A sample of 10 questions, for footwear evaluations, is given in the Appendix 8.A. The choices can be "I like it that way," "It must be that way," "I am neutral," "I can live with it that way," and "I dislike it that way." Depending on the paired question responses, some may be categorized as *Indifferent (I)*, *Reverse (R)*, or *Questionable (Q)* as well. Table 8.1 shows the initial classification of the differing qualities based on subject responses. This classification is based on three assumptions:

Assumption 1: The word "like" represents complete customer "delight." It is assumed to be a point at positive infinity and at the upper extreme of the y-axis (CS_1 in Figure 8.1).

Assumption 2: The word "dislike" represents extreme dissatisfaction. It is assumed to be a point at negative infinity and at the lower extreme of the y-axis (CS_5 in Figure 8.1).

Assumption 3: There is a relatively small difference in customer satisfaction between the terms "must-be" and "neutral." Similarly, there is a relatively small difference in customer satisfaction between "can live with" and "neutral." That is, $(CS_2 - CS_3)$ and $(CS_3 - CS_4)$ are relatively smaller than $(CS_1 - CS_2)$ and $(CS_4 - CS_5)$ (Figure 8.1).

In the original Kano model, nine combinations (cell (2,2), (2,3), (2,4), (3,2), (3,3), (3,4), (4,2), (4,3), and (4,4) in Table 8.1) are classified as I. Some researchers

TABLE 8.1
Classification Table

		Dysfunctional				
I like it that way (Like)		1. Like	2. Must-be	3. Neutral	4. Live with	5. Dislike
It must be that way (Must-be)						
I am neutral (Neutral)						
I can live with it that way (Live with)						
I dislike it that way (Dislike)						
Functional	1. Like	Q	A	A	A	O
	2. Must-be	R	I	I	I	M
	3. Neutral	R	I	I	I	M
	4. Live with	R	I	I	I	M
	5. Dislike	R	R	R	R	Q

A = attractive, M = must-be, O = one-dimensional, I = indifferent, R = reverse, Q = questionable.

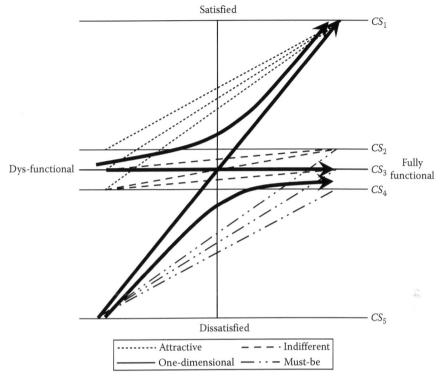

FIGURE 8.1   The four qualities: *A, O, M, I* in Kano's model. *Note:* "*CS$_i$*" represents the customer satisfaction level for five expressions, for $i = 1$–5 (1—"I like it that way"; 2—"It must be that way"; 3—"I am neutral"; 4—"I can live with it that way"; 5—"I dislike it that way). (Adapted from Berger, C. et al., *Center Qual. Manage. J.*, 2(4), 3, 1993.)

(Berger et al., 1993; Löfgren and Witell, 2005) believe that the five choices in the questionnaire may be context or product dependent and hence researchers have modified and used different wording (Berger et al., 1993; Terninko, 1997; Sireli et al., 2007). Berger et al. (1993) suggests, with a change of wording of the five choices, the two cells (2,2) and (4,4) would be more appropriately labeled as *Q* rather than *I* (Table 8.2).

To attain a reliable categorization for each product characteristic, the responses from a group of people are graded according to the frequency. The mode statistic is a measure of the overall grade but unfortunately falls apart if the surveyed population has more than one market segment. To overcome this problem, Berger et al. (1993) proposed a two-step method wherein the six categories are split into two groups. Group I consists of *A, M,* and *O* while Group II contains *R, Q,* and *I*. The grades are then determined based on the frequency in each category as follows:

If

$$(f(A) + f(M) + f(O)) > (f(R) + f(Q) + f(I)) \qquad (8.1)$$

TABLE 8.2
Categorization of Responses

I like it (CS_1)
I expect it (CS_2)
I am neutral (CS_3)
I can tolerate (CS_4)
I dislike it (CS_5)

		Dysfunctional				
		CS_1	CS_2	CS_3	CS_4	CS_5
Functional	CS_1	Q	A	A	A	O
	CS_2	R	Q	I	I	M
	CS_3	R	I	I	I	M
	CS_4	R	I	I	Q	M
	CS_5	R	R	R	R	Q

A = attractive, M = must-be, O = one-dimensional, I = indifferent, R = reverse, Q = questionable.

Then

$$\text{Grade} = \text{Category of } \max(f(A), f(M), f(O)) \qquad (8.2)$$

Else

$$\text{Grade} = \text{Category of } \max(f(R), f(Q), f(I))$$

To further validate the aforementioned measure, Löfgren and Witell (2005) employed an additional measure referred to as the classification agreement (Equation 8.3):

$$\text{Classification agreement } = \frac{f(\text{grade})}{N} 100\% \qquad (8.3)$$

where N is total number of responses.

In addition to the grade, Berger et al. (1993) proposed the use of a "better" and "worse" measure to indicate whether the customer satisfaction would be increased by providing that quality ("better" value) or customer satisfaction would be decreased if that aspect is not included ("worse" value):

$$\text{Better value} = \frac{f(A) + f(O)}{f(A) + f(O) + f(M) + f(I)} \qquad (8.4)$$

$$\text{Worse value} = -\frac{f(O) + f(M)}{f(A) + f(O) + f(M) + f(I)} \qquad (8.5)$$

8.4.3 Challenges of Applying Kano's Method

Although Kano's method has been applied to improve many products and services, the application of this method is not as simple as one would expect. Before analyzing the data from Equations 8.1 through 8.5, an evaluation table (Table 8.1) is used to classify each paired response. There are 25 possible combinations in the evaluation table (Table 8.1), but they are reduced to six categories ($O/A/M/I/R/Q$) after the classification process. The information reduction process can ease the data analysis, but this process decreases the resolution of the data and may also eliminate some useful information. For example, there are three possible combinations of the paired responses classified as A (i.e., cell (1,2), (1,3), and (1,4) in Table 8.1). Theoretically, grouping these three combinations into one category is possible only if the customer perceives the same decrease of satisfaction level when the particular quality changes from fully functional to not functional at all. In this case, the perceived customer satisfactory level on "must-be," "neutral," and "live with" should be equal to each other. In order to verify this assumption, the interpretations of five choices in terms of customer satisfaction levels are required.

8.4.3.1 Heterogeneity of the Population

It is common for the responses to spread out over several categories (A, O, M, and I) for various respondents (Berger et al., 1993). Matzler and Hinterhuber (1998) believed that this frequency spread can be explained by the fact that customers in different segments have different product expectations. According to an interview conducted by Piller (2002), the segments for footwear customization include comfort-oriented customers, fashion-oriented women, individualists, businessmen, and young people. Therefore, consumer heterogeneity should be considered when describing the importance of product attributes (Griffin and Hauser, 1993), since the preference of product attributes is different between market segments.

Some approaches, such as the quality improvement index (Matzler and Hinterhuber, 1998), importance-satisfaction model (Yang, 2003, 2005), and adaptive preference target (Rivière et al., 2006), can be combined with Kano's method to further understand customer segmentation (Löfgren and Witell, 2008). However, these proposed methodologies require respondents to answer additional questions (e.g., self-stated importance ranking) (Berger et al., 1993; Matzler and Hinterhuber, 1998; Löfgren and Witell, 2005) for each product attribute and hence the integration may be costly.

Others (Berger et al., 1993; Löfgren and Witell, 2005) have suggested collecting demographic data from respondents, so that the potential market segments can be identified, if they exist. However, Löfgren and Witell (2005) found that the demographic variable is insufficient for understanding segmentation. Behavioral variables should be considered, but they may not be easy to observe (Rivière et al., 2006). If the segmentation variables are known, then it would be easier to interpret the results of Kano's questionnaire in each market segment as they would have the same preference pattern (Rivière et al., 2006).

8.5 CONCLUSIONS

Subjective qualities are important in the design and development of any product. The buying decision for footwear is made on fit, form, and function. However, fitting for most people is secondary, and it is the looks that govern the attraction. Hence, the requirements that improve affect should be identified. Kansei Engineering and Kano's method are able to help identify the necessary qualities in footwear. Using any one method can facilitate the design process so that footwear will have a higher customer satisfaction.

QUESTIONS

8.1   What is comfort?

8.2   Explain the psychological component of footwear comfort.

8.3   How does footwear differ in short-term and long-term comfort?

8.4   Why is it that uncomfortable footwear have good aesthetics?

8.5   What methodologies can be used for investigating footwear comfort? What are the advantages and disadvantages of these methodologies?

8.6   What is Kano's model?

8.7   What are the assumptions of Kano's model?

8.8   How do you administer and analyze the data from Kano's questionnaire?

8.9   What are the challenges of applying Kano's method?

8.10 How does the Kano model differ from Kansei engineering?

APPENDIX 8.A: SAMPLE PAIRED QUESTIONS

Customer Requirement (CR)	(a) Functional Questions (b) Dysfunctional Questions
1. Black	(a) If leather shoes are black, how do you feel?
	(b) If leather shoes are not black, how do you feel?
2. Low heel (<3 cm)	(a) If leather shoes have a heel less than 3 cm (low heel), how do you feel?
	(b) If leather shoes do not have a heel less than 3 cm (low heel), how do you feel?
3. High heel (>5 cm)	(a) If leather shoes have a high heel of more than 5 cm, how do you feel?
	(b) If leather shoes do not have a high heel of more than 5 cm, how do you feel?
4. Round toe style	(a) If leather shoes have a round toe style, how do you feel?
	(b) If leather shoes do not have a round toe style, how do you feel?
5. Wide toe box	(a) If leather shoes have a wide toe box, how do you feel?
	(b) If leather shoes do not have a wide toe box, how do you feel?
6. Matches your clothes	(a) If leather shoes match your clothes, how do you feel?
	(b) If leather shoes do not match your clothes, how do you feel?
7. Shiny upper materials	(a) If leather shoes are made of shiny upper materials, how do you feel?
	(b) If leather shoes are not made of shiny upper materials, how do you feel?
8. Looks durable	(a) If leather shoes look durable, how do you feel?
	(b) If leather shoes do not look durable, how do you feel?
9. Looks light in weight	(a) If leather shoes look light in weight, how do you feel?
	(b) If leather shoes do not look light in weight, how do you feel?
10. Adjustable strap	(a) If leather shoes have an adjustable strap, how do you feel?
	(b) If leather shoes do not make you look casual, how do you feel?

REFERENCES

Alcántara, E., Artacho, M. A., González, J. C., and García, A. C. 2005. Application of product semantics to footwear design: Part I—Identification of footwear semantic space applying differential semantics. *International Journal of Industrial Ergonomics* 35: 713–725.

Au, E. Y. L. and Goonetilleke, R. S. 2007. A qualitative study on the comfort and fit of ladies' dress shoes. *Applied Ergonomics* 38(6): 687–696.

Au, E. Y. L. and Goonetilleke, R. S. 2009. A psychophysical model for predicting footwear fit. *Proceedings of IEEE International Conference on Virtual Environments, Human-Computer Interfaces, and Measurement Systems (VECIMS 2009)*, Hong Kong, May 11–13.

Au, E. Y. L., Li, W. Y., and Goonetilleke, R. S. 2006. The challenges of applying Kano's method to footwear design. *BME2006 Biomedical Engineering Conference*, Hong Kong, September 21–23.

Bartoshuk, L. M. 1988. Taste. In *Steven's Handbook of Experimental Psychology, Vol. 1. Perception and Motivation*, 2nd edn. Eds. R. S. Atkinson, R. J. Hernstein, G. Lindzey, and R. D. Luce Wiley, New York, pp. 461–499.

Beauchamp, G. K., Cowart, B. J., and Schmidt, H. J. 1991. Development of chemosensory sensitivity and preference. In *Smell and Taste in Health and Disease*. Eds. T. V. Getchell, R. L. Doty, L. M. Bartoshuk, and J. B. Snow. Raven, New York, pp. 405–416.

Benstock, S. and F. Suzanne. 2001. *Footnotes: On Shoes*. Rutgers University Press, New Brunswick, NJ.

Berger, C., Blauth, R., Boger, D. et al. 1993. Kano's methods for understanding customer-defined quality. *Center for Quality Management Journal* 2(4): 3–35.

Bloch, P. H. 1995. Seeking the ideal form: Product design and consumer response. *Journal of Marketing* 59(3): 16–29.

Boër, C. R., Dulio, S., and Jovane, F. 2004. Editorial: Shoe design and manufacturing. *International Journal of Computer Integrated Manufacturing* 17(7): 577–582.

Chen, L. H. and Lin, H. C. 2007. Integrating Kano's model into E-learning satisfaction. *Proceedings of the 2007 IEEE International Conference on Industrial Engineering and Engineering Management*, Singapore, December 2–5.

Clarks. 1976. *Manual of Shoemaking*, Training Department, Clarks, UK.

Dodd, G. and Skinner, M. 1992. From moods to molecules: The psychopharmacology of perfumery and aromatherapy. In *The Psychology and Biology of Perfume*. Eds. S. Van Toller and G. H. Dodd. Elsevier, London, U.K., pp. 113–142.

Easton, D. M. 2005. Gompertzian growth and decay: A powerful descriptive tool for neuroscience. *Physiology and Behavior* 86: 407–414.

Fibre2fashion.com 2007. USA: Rockport stylish collection features adidas TORSION Technology. Available URL: http://www.fibre2fashion.com/news/company-news/adidas-salomon/newsdetails.aspx?news_id=40928

Frey, C. 2000. Foot health and shoewear for women. *Clinical Orthopaedics and Related Research* 372: 32–44.

Fry, C. 2010. Kicking the landfill habit. *Engineering and Technology* 23: 54–56.

Gescheider, G. A. 1985. *Psychophysics: Method, Theory, and Application*. Lawrence Erlbaum Associates, Hillsdale, NJ.

Goonetilleke, R. S. 1999. Footwear cushioning: Relating objective and subjective measurements. *Human Factors* 41(2): 241–256.

Goonetilleke, R. S. 2001. The comfort-discomfort phase change. In *International Encyclopedia of Ergonomics and Human Factors*. Ed. W. Karwowski. Taylor and Francis Group, Boca Raton, FL, pp. 399–402.

Goonetilleke, R. S. and Luximon, A. 2001. Designing for comfort: A footwear application. *Proceedings of the Computer-Aided Ergonomics and Safety Conference 2001 (Plenary Session, CD-ROM)*, Maui, HI, July 28–August 2.

Griffin, A. and Hauser, J. R. 1993. The voice of the customer. *Marketing Science* 12(1): 1–27.

Halliday, D. and Setchi, R. 2009. A comparative study of using traditional user-centered and Kansei Engineering approaches to extract user's requirements. *I*PROMS'09*, Cardiff, U.K., July 6–17.

Hawes, M. R., Sovak, D., Miyashita, M., Kang, S. J., Yoshihuku, Y., and Tanaka, S. 1994. Ethnic differences in forefoot shape and the determination of shoe comfort. *Ergonomics* 37(1): 187–196.

Hertzberg, H. T. E. 1972. The human buttock in sitting: Pressures, patterns, and palliatives. *American Automobile Transactions* 72: 39–47.

Herzberg, F., Mausner, B., and Snyderman, B. B. 1959. *The Motivation to Work*. John Wiley, New York.

Ishihara, S., Ishihara, K., Nagamachi, M., and Matsubara, Y. 1997. An analysis of Kansei structure on shoes using self-organizing neural networks. *International Journal of Industrial Ergonomics* 19: 93–104.

Kaiser, S. B., Schutz, H. G., and Chandler, J. 1987. Cultural codes and sex-role ideology: A study of shoes. *American Journal of Semiotics* 5(1): 13–34.

Kamijo, K., Tsujimura, H., Obara, H., and Katsumata, M. 1982. Evaluation of seating comfort. SAE Technical Paper Series, 820761.

Kano, N., Seraku, N., Takanashi, F., and Tsuji, S. 1984. Attractive quality and must-be quality (English Translation). *Journal of the Japanese Society for Quality Control* 14(2): 39–48.

King, B. M. 1986. Odor intensity measured by an audio method. *Journal of Food Science* 51: 1340–1344.

Kolarik, W. J. 1995. *Creating Quality: Concepts, Systems, Strategies, and Tools*. McGraw-Hill, New York.

Laffort, P. 1966. Recherche d'une to de I'intensite odorante supraliminaire, conforme aux diverses donnees experimental. *Journal of Physiology* 58: 551.

Leand, J. 2007. Easy does it. *SGB* 40(7): 26–33.

Lee, Y. H. and Hong, W. H. 2005. Effects of shoe inserts and heel height on foot pressure, impact force, and perceived comfort during walking. *Applied Ergonomics* 36(3): 355–362.

Lindstrom, M. 2004. *Brand Sense: Build Powerful Brands through Touch, Taste, Smell, Sight, and Sound*. Free Press, New York.

Liu, X., Zhang, W. J., Tu, Y. L., and Jiang, R. 2007. An analytical approach to customer requirement satisfaction in design specification development. *IEEE Transactions on Engineering Management* 55(1): 94–102.

Löfgren, M. and Witell, L. 2005. Kano's theory of attractive quality and packaging. *Quality Management Journal* 12(3): 7–20.

Löfgren, M. and Witell, L. 2008. Two decades of using Kano's theory of attractive quality: A literature review. *Quality Management Journal* 15(1): 59–76.

van Lottum, C., Pearce, K., and Coleman, S. 2006. Features of Kansei engineering characterizing its use in two studies: Men's everyday footwear and historic footwear. *Quality and Reliability Engineering International* 22(6): 629–650.

Lutter, L. D. 2004. Sexy shoes or sorry feet. *Foot and Ankle International* 25(1): 1–2.

Luximon, A. and Goonetilleke, R. S. 2004. Foot shape modeling. *Human Factors* 46(2): 304–315.

Luximon, A., Goonetilleke, R. S., and Tsui, K. L. 2001. A fit metric for footwear customization. *Proceedings of the 2001 World Congress on Mass Customization and Personalization*, Hong Kong, October 1–2.

MacMillan, I. C. and McGrath, R. G. 1996. Discover your product's secret potential. *Harvard Business Review* 74(3): 58–73.

Marks, L. E. 1978. *The Unity of the Senses: Interrealtions Among the Modalities*. Academic Press, New York.

Matzler, K., Fuchs, M., and Schubert, A. K. 2004. Employee satisfaction: Does Kano's model apply? *Total Quality Management* 15(9/10): 1179–1198.

Matzler, K. and Hinterhuber, H. H. 1998. How to make product development projects more successful by integrating Kano's model of customer satisfaction into quality function deployment. *Technovation* 18(1): 25–38.

Mayer, D. 1975. On the approximation of psychophysical functions by power functions. *Acta Instituti Psychologici Universitatis Zagrabiensis*, 78: 39–45.

McNeil, P. and Riello, G. 2005. The art of science and walking: Gender, space, and the fashionable body in the long eighteenth century. *Fashion Theory: The Journal of Dress, Body and Culture* 9(2): 175–204.

Miller, J. E., Nigg, B. M., Liu, W., Stefanyshyn, D. J., and Nurse, M. A. 2000. Influence of foot, leg and shoe characteristics on subjective comfort. *Foot Ankle International* 21(9): 759–767.

Molski, M. 2011. Extended Stevens' power law. *Physiology and Behavior* 104(5): 1031–1036.

Mündermann, A., Nigg, B. M., Stefanyshyn, D. J., and Humble, R. N. 2002. Development of a reliable method to assess footwear comfort during running. *Gait & Posture* 16(1): 38–45.

Nagamachi, M. 1995. Kansei engineering: A new ergonomic consumer-oriented technology for product development. *International Journal of Industrial Ergonomics* 15: 3–11.

Nagamachi, M. 2002. Kansei engineering as a powerful consumer-oriented technology for product development. *Applied Ergonomics* 33: 289–294.

Nagamachi, M. 2011. *Kansei/Affective Engineering*. CRC Press, Boca Raton, FL.

Nagamachi, M., Okazaki, Y., and Ishikawa, M. 2006. Kansei engineering and application of the rough sets model. *Proceedings of the Institution of Mechanical Engineers, Part I: Journal of Systems and Control Engineering* 220: 763–768.

Nakada, K. 1997. Kansei engineering research on the design of construction machinery. *International Journal of Industrial Ergonomics* 19: 129–146.

Overbosch, P. 1986. A theoretical model for perceived intensity in human taste and smell as a function of time. *Chemical Senses*, 11: 315–329.

Pearce, K. F. and Coleman, S. Y. 2008. Modern-day perception of historic footwear and its links to preference. *Journal of Applied Statistics* 35(2): 161–178.

Piller, F. 2002. *EuroShoe Consortium: The Market for Customized Footwear in Europe—Market Demand and Consumers' Preferences*. Technische Universität München, Munich, Germany.

Piller, F. T. and Müller, M. 2004. A new marketing approach to mass customization. *International Journal of Computer Integrated Manufacturing* 17(7): 583–593.

Pratt, L. and Wolley, L., 1999, *Shoes*. V & A Publications, London, U.K.

Rivière, P., Monrozier, R., Rogeaux, M., Pagès, J., and Saporta, G. 2006. Adaptive preference target: Contribution of Kano's model of satisfaction for an optimized preference analysis using a sequential consumer test. *Food Quality and Preference* 17: 572–581.

Rupérez, M. J., Monserrat, C., Alemany, S., Juan, M. C., and Alcañíz, M. 2010. Contact model, fit process and, foot animation for the virtual simulator of the footwear comfort. *Computer-Aided Design* 42: 425–431.

Sireli, Y. Kauffmann, P., and Ozan, E. 2007. Integration of Kano's model into QFD for multiple product design. *IEEE Transactions on Engineering Management* 54(2): 380–390.

Slater, K. 1985. *Human Comfort*. Thomas Publisher, Springfield, IL.

Solves, C., Such, M. J., and Gonzalez, J. C. et al. 2006. Validation study of Kansei engineering methodology in footwear design. In *Contemporary Ergonomics*. Ed. P. D. Bust. Taylor & Francis Group, London, U.K., pp. 164–168.

Stevens, S. S. 1960. Ratio scales, partition scales, and confusion scales. In *Psychological Scaling: Theory and Applications*. Eds. H. Gulliksen, and S. Messick. Wiley, New York.

Stevens, S. S. 1975. *Psychophysics: Introduction to Perceptual, Neural and Social Prospects.* Wiley, New York.

Stevens, J. C. and Stevens, S. S. 1963. Brightness function: Effects of adaptation. *Journal of the Optical Society of America* 53: 375–385.

Szmigin, I. and Reppel, A. E. 2004. Internet community bonding: The case of macnews.de. *European Journal of Marketing* 38(5/6): 626–640.

Tan, K. C. and Shen, X. X. 2000. Integrating Kano's model in the planning matrix of quality function deployment. *Total Quality Management* 11(8): 1141–1151.

Tarasewich, P. and Nair, S. K. 2001. Designer-moderated product design. *IEEE Transactions on Engineering Management* 48(2): 175–188.

Terninko, J. 1997. *Step-by-Step QFD: Customer-Driven Product Design*, 2nd edn. St. Lucie Press, Boca Raton, FL.

Websters New World Dictionary. 1990. *Pocket Books*. New York.

Witana, C. P., Goonetilleke, R. S., Au, E. Y. L., Xiong, S., and Lu, X. 2009. Footbed shapes for enhanced footwear comfort. *Ergonomics* 52(5): 617–628.

Witana, C. P., Goonetilleke, R. S., and J. Feng. 2004. Dimensional differences for evaluating the quality of footwear fit. *Ergonomics* 47(12): 1301–1317.

Witell, L. N. and Fundin, A. 2005. Dynamics of service attributes: A test of Kano's theory of attractive quality. *International Journal of Service Industry Management* 16(2): 152–168.

Wunderlich, R. E. and Cavanagh, P. R. 2001. Gender differences in adult foot shape: Implications for shoe design. *Medicine and Science in Sports and Exercise* 33(4): 605–611.

Xiong, S. and Goonetilleke, R. S. 2006. Midfoot shape for the design of ladies' shoes. *BME2006 Biomedical Engineering Conference*, Hong Kong, September 21–23.

Xiong, S., Goonetilleke, R. S., Witana, C. P., and Au, E. Y. L. 2008. Modelling foot height and foot shape-related dimensions. *Ergonomics* 51(8): 1272–1289.

Xiong, S., Witana, C. P., and Goonetilleke, R. S. 2007. The use of a foot dorsal height model for the design of wellington boots. *Proceedings of Agriculture Ergonomics Development Conference*, Kuala Lumpur, Malaysia, November 26–29.

Yang, C. C. 2003. Establishment and applications of the integrated model of service quality measurement. *Managing Service Quality* 13(4): 310–324.

Yang, C. C. 2005. The refined Kano's model and its application. *Total Quality Management* 16(10): 1127–1137.

Zhang, L., Drury, C. G., and Wooley, S. M. 1991. Constrained standing: Evaluating the foot/floor interface. *Ergonomics* 34(2): 175–192.

Zhang, L., Helander, M. G., and Drury, C. G. 1996. Identifying factors of comfort and discomfort in sitting. *Human Factors* 38(3): 377–389.

Zhang, P. and Von Dran, G. M. 2002. User expectations and rankings of quality factors in different web site domains. *International Journal of Electronic Commence* 6(2): 9–33.

9 Shoe-Last Design and Development

Ameersing Luximon and Yan Luximon

CONTENTS

9.1 INTRODUCTION

The mechanization of the footwear industry started during the seventeenth century after the invention of sewing machines and equipment to roll sole leather (Quimby, 1944), and prior to that the shoe production was completely manual. In today's society, footwear has become increasingly specialized for a variety of tasks and functions. Shoes have increased in complexity with the incorporation of functionality to improve user performance and comforts. Hence, it is not surprising to see hundreds of new varieties of sport shoes and fashion shoes every season.

It is apparent that the proper fit can be achieved if the shoe is shaped like the foot; however, due to the traditional method of shoemaking, the shoe shape and style are dependent on a shoe-last (Cheskin, 1987). The shoe-last development depends entirely on the skill and artistry of the model maker who shapes and sands a shoe-last until it looks right (Adrian, 1991; National Footwear Research Institute, 1998). The model shoe-last maker is usually guided by girth and length measurements (Adrian, 1991). It is not surprising that the way the shoe-last has been sanded will affect the shape, style, and shoe fits. The model shoe-last is then graded in order to make shoe-lasts of different sizes (ISO 9407, 1991). The sized shoe-lasts are used to make shoes to fit the entire selected population. In fact, the shoe design should start with an understanding of the foot (Goonetilleke et al., 1997; Luximon and Zhang, 2006; Luximon et al., 2003) and then the shoe-last should be designed (Luximon and Luximon, 2009).

9.2 HISTORY

The shoe-last (Figure 9.1) comes from the old Norse (a northern Germanic language) word "laest," meaning a footprint, or from the Anglo-Saxon language, meaning footstep. Shoemaking is an age-old craft rich in tradition, heritage, and prestige (Pivecka and Laure, 1995). Due to the increasing use of footwear and the increased injuries and illnesses associated with it, shoe-last has been recognized as an important fashion and health factor (Quimby, 1944). The shoe-last is the heart and the single most important element of the shoe (Rossi, 1983). It is the most scientific and complex part of the whole shoemaking process and it is the foundation upon which much of the shoe-related foot health depends (Rossi and Tennant, 1984). It is responsible for the size, fit, shape, feel, wear, style, tread, and even the making of the shoe (Clarks, 1989).

The shoe-last is a reproduction of the approximate shape of the human foot. A shoe when properly constructed (SATRA, 1993) using a well-designed shoe-last furnishes support and protection without undue pressure, binding, or constriction at any point (Quimby, 1944). The first American shoe-last maker on record who was granted a patent for a minor shoe-last improvement in the year 1807 was William Young, Philadelphia, Pennsylvania. In 1815, Thomas Blanchard from Sutton, Massachusetts, invented a lathe for turning of irregular shaped objects. It was first used for turning out gun-stock but was later adopted for shoe-lasts. The Gilman lathe now used throughout the shoe-last industry was an improvement to the Blanchard lathe. During the nineteenth century, until about the time of the civil war, the same wooden shoe-last was used in the construction of both right and left shoes. The shoe-lasts were straight in shape with no allowance for the foot contour. The shoes can be swapped from left to right. Before 1880, shoes were made in full sizes and were

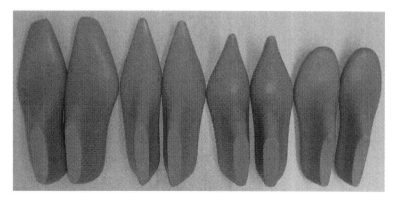

FIGURE 9.1 Shoe-lasts.

seldom more than three widths. After 1880, half sizes were introduced. In the twentieth century, quarter size was introduced but was unsuccessful due to the additional inventory costs. In 1940, shoe-lasts were slightly improved. By adding a pad over the front of the shoe they created two sizes and called them slim and fat. Now, shoes are made on shoe-lasts graded from AAA to EE widths. Because of the fact that no feet are exactly alike, one can readily appreciate the necessity of making shoes on scientifically designed shoe-lasts (Quimby, 1944).

9.3 NOMENCLATURE

This section explains the basic nomenclature of the shoe-lasts. The dimensions provided are relative to the size 36 shoe-last. The *cone* (Adrian, 1991) is the curved upper surface of the back part of the shoe-last (Figure 9.2). The *cone top profile* is the profile of the cone top when viewed from the side. The *cone top surface* is the surface of the cone when viewed from top. The *cone top surface outline* is the outline of the cone top surface. The *cone length* or *heel top length* is the length of the cone. If C_1 and C_2 are the front and rear extreme points of the cone of the shoe-last respectively, then the length from C_1 to C_2 is the cone length. It is approximately equal to 90 mm. The *cone width* is the maximum width of the cone top surface outline, and is approximately equal to 20 mm.

Feather line or *Feather edge* is the edge of the shoe-last (Figure 9.3). It is essential to have a sharp edge in a shoe-last to enable easy shoemaking (Adrian, 1991).

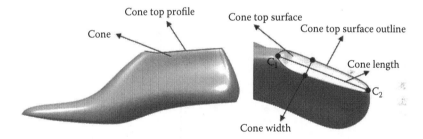

FIGURE 9.2 Cone of the shoe-last.

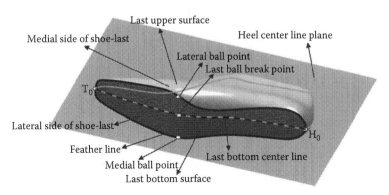

FIGURE 9.3 Shoe-last center line and related nomenclatures.

Also, it gives a beautiful sharp edge between the upper and the outsole. The *last bottom surface* is the surface of the shoe-last that matches with the outsole. The *last upper surface* is the last surface that is used for making the shoe upper. The feather line separates the last bottom surface to the last upper surface. The *last center line plane* is the plane points C_1 and C_2 and H_0. H_0, the *heel point*, is the rear extreme point at the heel lying on the feather line. *The last center line plane* separates the last into the *lateral side of shoe-last* and the *medial side of shoe-last*. The medial side of shoe-last is the side that will have the big toe, while the lateral side of shoe-last will have the small toe. *Bottom curve* or *last bottom center line curve* is the intersection of the bottom surface of the shoe-last with the last center plane. The length of the bottom curve is dependent on the toe length and is greater than 256 mm (246 + 10 mm [or more] for toe allowance).

Stick length is the overall length of a shoe-last as measured by a Last stick or Ritz stick (Figure 9.4). For women, size 36 it is about 260 mm. It should be noted that the stick length is dependent upon toe spring and heel height. Pointed toe shoes will have longer stick length. For the same toe length, the stick length decreases with increasing heel height. *Last ball break point* separates the front part of the shoe-last with the back part of the shoe-last (Figure 9.5). The last break point is around 150 mm from heel point (H_0) along the last bottom center line curve. The *shank curve* (Pivecka and Laure, 1995) is the curve on the back part of the shoe-last along the last bottom center line. The shank curve will influence the pressure of the foot

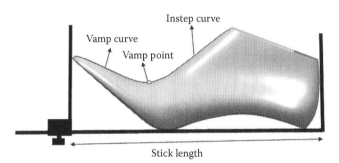

FIGURE 9.4   Stick length.

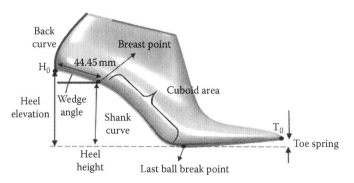

FIGURE 9.5   Shoe-last side view.

and hence to some extent the comfort qualities of the shoe. The *cuboid* is an important factor in the shoe-last design (Adrian, 1991). It is on the outer rim of the shoe-last. It serves as a firm stabilizer for the foot on weight bearing. It also influences the fit of top line of the finished shoe.

Toe point, T_0, is the toe extreme point on the last bottom center line curve. *Toe spring* is the height of the toe point from the horizontal. Toe spring is around 5–25 mm, but in most cases it is about 10 mm. The amount of toe spring is related to the sole material and the heel height. The harder the sole, the greater the toe spring. The higher the heel height, less toe spring is required. The *heel elevation*, based on American system (Adrian, 1991) and defined as heel height in Chinese system (Chen, 2005; National Footwear Research Institute, 1998), is the height of the heel point. It varies based on the shoe design and ranges from 0 to 152 mm (6 in.). The heel elevation for most shoes is around 50 mm. For sports shoes, it is around 25 mm. The *breast point* defines the approximate location of the heel seat and is equal to 44.45 mm (1¾ in.) from the heel point along the last bottom center line curve. The *breast line* defines the arbitrary boundary of the heel seat. It also defines the approximate location of the center of pressure at the rear part of the foot. The *heel height* (as defined by the American system) is the height at the breast point. *Wedge angle* is the angle of the line from breast point to heel point to the horizontal. The heel elevation (H_E), heel height (H_H), and wedge angle (θ) are related by this simple equation, $H_E = H_H + 44.45 \times \sin(\theta)$. Also, the wedge angle is related to the heel height. The higher the heel height, the larger is the wedge angle. For heel height below 12.5 mm, the wedge angle is 0°. For heel height greater than 12.5 mm, wedge angle is given by the following equation: wedge angle (θ) = 0.4054 H_H − 4.7774. For illustration, different shank curves for different heel heights are shown in Figure 9.6. The *back curve* is the curve of the back part of the shoe-last. The *back height* is the height of the back part of the shoe-last, and is equal to around 70 mm. The back curves for different heel heights are shown in Figure 9.7.

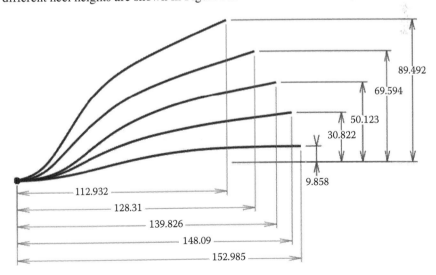

FIGURE 9.6 Shoe-last shank curves (length of all the curves are 155 mm).

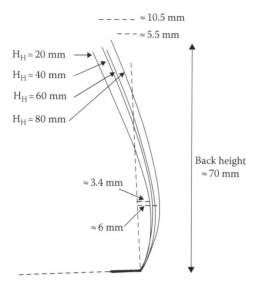

FIGURE 9.7   Shoe-last back curves.

The *lateral ball point* is the point on the feather edge such that the distance extended along the last bottom center line curve is equal to 143 mm (Figure 9.3). The *medial ball point* is the point on the feather edge such that the distance extended along the last bottom center line curve is equal to 162.5 mm. The *ball curve* or *joint curve* is the curve between the lateral ball point and the medial ball point making a geodesic (shortest) curve on the top and bottom surface of the shoe-last (Figure 9.8). The *ball girth* is the length of the joint curve. The *vamp point* is the point on the upper part of the geodesic curve intersection with the last center line plane. In most cases, the vamp point is the minimum point on the combined vamp and instep curves (Figure 9.4). The *instep curve* is the curve on the front side of the shoe-last on the last center line plane.

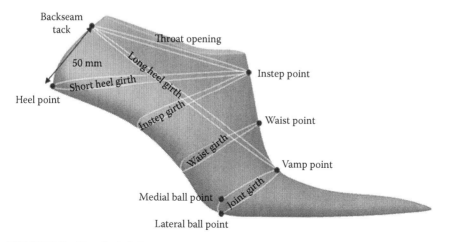

FIGURE 9.8   Shoe-last girths.

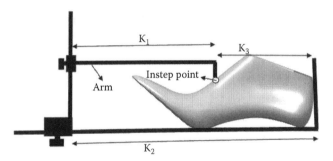

FIGURE 9.9 Krippendorf finder.

Vamp length is the length from the vamp point to the toe point. The vamp length will be based on toe design (pointed toe, square toe, round toe, bump, etc.).

The *instep point* is located using a Krippendorf finder (Figure 9.9). The arm of the Krippendorf finder is fixed, thus length K_1 is fixed and is equal to 190.5 mm. The length K_2, called the Krippendorf finder settings for instep, is based on a normal Ritz stick length. It is different for different heel height (shown in Table 9.1). The *instep curve* is the geodesic curve at the instep point. The *instep girth* is the length of the instep curve. The *waist point* is the midpoint between the instep point and the vamp point. The *waist curve* is the geodesic curve at the waist point. The *waist girth* is the length of the waist curve. The ball girth, waist girth, and instep girth for size 36 shoe is 211, 209, and 222 mm, respectively.

The *short heel curve* is the geodesic curve generated between the heel point and the instep point. For easy computation, the short heel curve is computed by the intersection of the shoe-last with a plane formed by the heel point and the instep point, perpendicular to the heel center plane. The *short heel girth* is the length of the short heel curve. The *long heel curve* is the geodesic curve generated between the back seam tack point (2 1/16 in. = 52.4 mm from heel point) and the vamp point (Figure 9.8). For easy computation, the long heel curve is computed by the intersection of the shoe-last with a plane formed by the back seam tack point and the vamp point, perpendicular to the heel center plane. The *long heel girth* is the length of the long heel curve. The *throat opening* is the length of the geodesic curve between

TABLE 9.1
Krippendorf Finder Settings for Instep

Heel Height (in.)	Heel Height (mm)	Finder Setting (K_2 in Stick Length)	K_2 (mm)	K_1 (mm)	K_3 (mm)
1/8–11/8	3.2–34.9	13	319.6	190.5	129.1
12/8–15/8	38.1–47.6	12 3/4	317.5	191.5	126
16/8–19/8	50.8–60.3	12 1/2	315.4	192.5	122.9
20/8–22/8	63.5–69.9	12	311.2	193.5	117.7
23/8–26/8	73.0–88.9	11 1/2	306.9	194.5	112.4

the back seam tack point and the instep point. The short heel girth, the long heel girth, and the throat opening of a 36 shoe-last with 72 mm heel elevation is about 306, 356, and 261.3 mm, respectively.

9.4   SHOE-LAST MATERIAL

Shoe-lasts, used extensively in the making of footwear, are forms that are made to varying degrees in the shape of the human foot, depending on their specific purpose. They come in many styles and sizes, depending on the exact job they are designed for. They range from simple one-size shoe-lasts used for repairing soles and heels, to hard-wearing shoe-lasts used in modern mass production, and to custom-made shoe-lasts used in the making of customized footwear. The shoe-last must represent the anatomical information of the foot (Luximon, 2001) at the same time giving the finished shoe a pleasing and fashionable appearance.

Shoe-lasts are made from a number of materials. These materials should have shape retaining characteristics and the shoe-last shape must not change with heat, humidity, and other environmental factors. The material used to make modern shoe-lasts must be strong enough to withstand the forces of mass production machinery, such as that applied by the pullover machines when bottoming the shoe, and must also be able to hold tacks (a type of nail), which are used to hold shoe parts together temporarily before the sole is added. The most common used material for making shoe-lasts is wood, high density polyethylene (HDPE), and aluminum.

Wooden shoe-lasts are mostly used for customized shoemaking. These shoe-lasts are expensive to make and also have low durability. These shoe-lasts may swell or shrink with temperature and humidity. But HDPE is used extensively for all kinds of shoe-lasts. The shoe-lasts made of HDPE are less expensive and can be recycled for new shoe-lasts. The shoe-lasts made of aluminum are used for solid or scoop shoe-lasts but are very expensive. These shoe-last shapes are also dependent on temperature. Aluminum shoe-lasts are used for making rubber boots.

9.5   TYPES OF SHOE-LASTS

There are many kinds of shoe-lasts (Pivecka and Laure, 1995). These are used in different shoe industries. The most used types of shoe-lasts are described in Table 9.2.

9.6   COMPARISON OF FOOT AND SHOE-LAST

It is very essential to understand the difference between a shoe-last and a human foot (Figure 9.10). The shoe-last is for shoemaking while the foot is used for weight bearing and locomotion (Luximon, 2001). The outline of a shoe-last is regular and continuous with a sharp feather edge around the seat and forepart to assist shoe-lasting and gives a clear defined edge to the finished shoe. The foot has no feather edge. The shoe-last increases gradually in height from the feather line, but not in foot. The shoe-last surface is smooth to enhance the appearance of the shoe and to enable the upper to be mounded more easily to shape. The surfaces of the feet are irregular and vary with individuals.

TABLE 9.2
Types of Shoe-Lasts

Types of Shoe-Lasts

Solid last: These kinds of shoe-lasts are the simplest and are used for low-heel shoes and sandals.

Scoop black last: These lasts are used for the manual shoe production. The shoe-lasts have a wedge on the top and can be detached from the main body. The lasts can be easily taken out of the lasted shape by removing the wedge.

Hinge last: These lasts are used for all kinds of shoe production. The lasts have a fore part and a back part and are connected by a spring. When slipping, the last is bent to shorten at the V-cut hinge. Then the last is removed from the shoe without damaging or deforming the back part of the shoe.

Telescopic last: These lasts are similar as the Hinge lasts but without a V-cut. The last will slid and reduce in length when slipping.

Three-piece lasts: These lasts are mainly used for the forced lasting boot or reversed slipper in the past. These kinds of shoe-lasts consist of three pieces. The center and the back pieces will be removed when slipping. After the removal of the center and back parts, it becomes easier to remove the front part.

The shoe-last is hard and firm, while the foot is softer and more flexible. The foot has separate toes, while the toe end of a shoe-last is solid. The back curve is greater on shoe-last to help the shoe grip the foot. The heel height is present in shoe-last but not on the foot. The front part of the shoe-last is thinner to help the shoe to grip the foot around the quarters (front part of shoe). The shoe-last length is greater than the foot to prevent pressure on the foot. Toe spring is not present on foot but is included on shoe-lasts. Girth and size intervals are regular on shoe-lasts but irregular on feet. The dimensions are identical on a pair of shoe-lasts but rarely identical on a pair of feet.

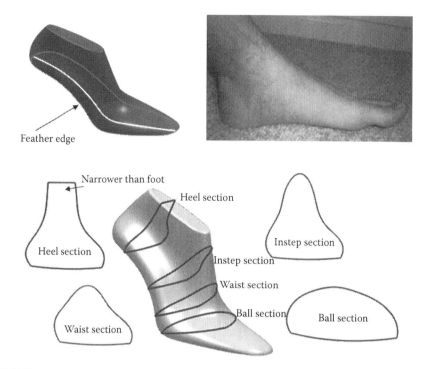

FIGURE 9.10   The different sections of a shoe-last.

9.7   SHOE-LAST DESIGN

There are many dimensions used for the shoe-last making. The most used terms are defined by the American Footwear Manufacturers Association (AFMA). Although AFMA has defined 61 terms for the explanation of shoe-last, there is no direct relation between the foot and the shoe-last. The shoe-last is a complex 3D shape having no straight lines but the manufacturers use four reference points—vamp, instep, ball break, and heel point (Backseam tack)—for the initial shoe marking. Then the shoemakers usually use six measures of which five are circumference measures such as ball girth, waist girth, instep girth, long heel measure, and short heel around the shoe-last and the last length. Sometimes throat opening is also used. The model maker shapes, sands, and files until the six measurements match and the model "looks right" (Cavanagh, 1980).

The way the model has been sand will affect the fit. The shoe-last making is evolved on a women's American size 6B shoe-last. This is equivalent to continental size 36, but sometimes size 37 is also used. The size 6B will accommodate a foot of 235 mm, and due to toe and heel back allowance, the last bottom length is around 246 mm. The ball girth is 8 5/16 in. = 211 mm, waist girth is 8 1/4 in. = 209.5 mm, and the instep girth is 8 3/4 in. = 222 mm. The shoe-last making is evolved on a men's American size 7½C shoe-last. This is equivalent to continental size 40, but sometimes size 41 is also used. The size 7½C will accommodate a foot of 254 mm and due to toe allowance the last bottom length is around 270 mm.

The ball girth is 9 1/8 in. = 231.8 mm, waist girth is 9 1/16 in. = 230 mm, and the instep girth is 9 7/16 in. = 239.7 mm.

The degree of the toe spring in a shoe-last depends on several factors, namely, the heel height, the shoe style, the upper material, and the general flexibility of the shoes. Furthermore, toe spring depends on the purpose of the footwear. Normally, the walking shoes need more toe spring than the dress shoes, while a ballet shoe does not have toe-spring.

Shoe manufacturers in Germany developed AKA64 shoe-last from an extensive foot measurement study of children for the improvement of the shoe fit. This shoe-last system is now widely used in most of the shoe factories. The developed shoe-last has three available widths of W (weit = wide), M (Mittel = Medium), and S (Schmal = Narrow). So this shoe-last is also referred as WMS system. The shoe system is quite elaborate on the foot dimensions. The detailed information about the foot dimension is explained for the ladies shoe size 6B and the dimensions are marked in Figure 9.11 and the dimensions measured in millimeters. CK = 15% ball girth and CM = 23% ball girth. MJ = 20% AD. AB = 1/6 AD. OB = BN = 1/3 (KM) +1 mm.

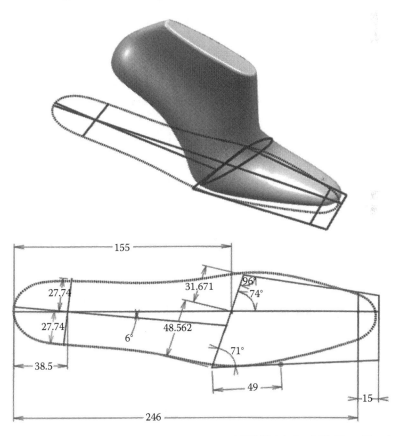

FIGURE 9.11   Shoe-last dimensions used for AKA64/WMS (girth = 211). *Note*: The size 6B in Adrian (1991) has ball girth = 204.8, but current shoe-last factories refer to size 6B which has a ball girth = 211. This is equivalent to size 6C in Adrian (1991).

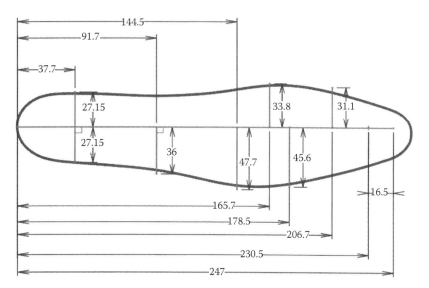

FIGURE 9.12   Shoe-last dimensions used for Chinese system (Chinese size 23.5 width 1.5; ball girth = 220 mm).

The other system is the Chinese shoe-last system (Chen, 2005; National Footwear Research Institute, 1988). This system is more detailed and provides information about the templates for heel curves and instep curves for differing heel heights and differing shoe constructions. The dimensions in Chinese shoe-last system for size 23.5 of width grade 1.5 having a heel height of 20 is explained in Figure 9.12. The marked dimensions in Figure 9.12 are in millimeters. In spite of being relatively detailed, the Chinese system is not widely used due to its complexity and the lack of software to implement the shoe-last system.

9.7.1   Shoe-Last Design Using Traditional Method

Traditional method for designing shoe-last can be done in many different ways. Since traditional method is based on skill and experience, many different techniques have been used to modify shoe-last based on existing shoe-last; copy new toe shape from existing shoes; modify shoe-last to adapt for different heel type; and modify toe spring. The steps for shoe-last design based on existing shoe-last and shoe-last based on existing shoe are discussed in the following.

Traditional shoe-last design based on existing shoe-last
The steps are as follows:

1. Select a standard shoe-last with a somewhat similar toe design (Shoe-last A). Most shoe-last factories will have a large collection of existing shoe-lasts.
2. Select a shoe-last with the desired heel height (Shoe-last B).
3. Cut shoe-last A at the ball break point and use the toe part (Toe part A).
4. Cut shoe-last B at the ball break point and use the back part (Back part B).

5. The toe part from shoe-last A and the back part from shoe-last B are joined together by adding melted HDPE material.

6. Many times the cross section shape of different shoe-lasts is not same. Thus, the last maker will sand and file until the shape looks good at the joint. The measurements such as ball girth, instep girth, waist girth, and shoe-last length are also measured to match with standards. The shoe-last designer will also use the AKA64 template to see the orientation of the bottom of the shoe-last. This may involve several iterations and the skill of the shoe-last maker will determine the quality of the shoe-last.

7. A shoe-last with the desired heel height but somewhat similar toe-design is obtained.

8. Melted HDPE material is added on top of the shoe-last toe surface.

9. The shoe-last is filed and sanded.

10. Continue Steps 8 and 9 until the desired toe shape is obtained.

11. A shoe-last with the desired heel height and the toe shape is obtained.

12. The shoe-last is digitized, machined, and given to the client for sample shoemaking.

13. The sample shoe is produced and the footwear fit is tested.

14. Based on the footwear fit testing, the shoe-last may be sent to the shoe-last maker for modification (to Step 8 or to Step 6). Step 6 is repeated if the shoe does not fit properly at the rear part of the shoe.

Traditional shoe-last design by coping toe shape from existing shoes.
The steps are as follows:

1. Select a shoe-last with the desired heel height (Shoe-last A).

2. Cut shoe-last A at the ball break point and use the back part (Back part A).

3. Use putty to acquire the 3D shape of the desired toe shape from a shoe (Toe design B).

4. The Toe design B and the back part A are joined together by adding melted HDPE material.

5. The last maker will sand and file until the shape looks good and the measurements such as ball girth, instep girth, waist girth, and shoe-last length match. This may involve several iterations and the skill of the shoe-last maker will determine the quality of the shoe-last. Also, the extracted toe from existing shoe does not have sharp feather line. This is created to by the shoe-last maker.

6. A shoe-last with the desired heel height and copied toe shape (from existing shoe) is obtained.

7. A shoe-last with the desired heel height and toe shape is obtained.

8. The shoe-last is digitized, machined, and given to the client for sample shoemaking.

9. The sample shoe is produced and the footwear fit is tested.

10. Based on the footwear fit testing, the shoe-last may be send to the shoe-last maker for modification (to Step 5).

9.7.2 SHOE-LAST DESIGN USING DIGITAL METHOD

There are many commercial shoe-last design software such as EasyShoe-last3D™, LastElf™, and Shoemaster™. These software enable sizing, grading, and changes to the master shoe-last. Luximon and Luximon (2009) have also proposed new method for shoe-last design based on spline curves and surfaces. Digital shoe-lasts can be modified by providing options such as toe length, toe type, vamp curve, and toe section curves. For example, the toe can be designed by lateral spline curve to match with the feather line at the lateral side of the shoe-last, the medial spline curve to match the feather line at the medial side of the shoe-last, vamp curve, and several toe cross section curves. All these curves are defined by points and tangents, and different designs can be easily created and stored. Also, automated process can be developed for design exploration. On the other hand, since all the dimensions can be related to shoe-last length and girth, modification of length and girth can be easily carried out to create different sizes. The advantages of computer aided design (CAD) technique for shoe-last design include high accuracy, less time, more design options, and storage space for digital shoe-lasts.

9.8 SIZING AND GRADING

The existing shoe sizing systems are primarily based on foot length and girth. It has evolved in various parts of the world with a view to assist consumers to select suitable footwear. The commonly used sizing systems in the world are the English sizing, American sizing, Continental sizing, Chinese, and Mondopoint system. Continental sizing is commonly used in Western Europe, and widely used as a second reference sizing in size labels for shoes using English or American sizing. The fundamentals of the various sizing systems are explained in Table 9.3. The English footwear sizing system was the first foot sizing system. It is based on the English measurement units of "foot" and "inch." After the year 1880, half sizes were introduced (1/2 sizes = 4.23 mm) to improve fit and, in the twentieth century, quarter sizes were introduced but not adopted due to the added production and inventory costs. The United States adopted the English sizing system, but instead of starting at 4 in., it starts at 3 and 11/12 in., hence the size designations of any given length are different between the English and the American systems. A woman's shoe in the American system would be 1½ sizes larger than the English sizing system while a men's shoe

TABLE 9.3
Various Dimensions Used for Different Sizing Systems

Sizing System	Starting Unit	Length Increment	Girth Increment
English	4 in. = 101.6 mm	1/3 in. = 8.46 mm	1/4 in. = 6.35 mm
American	3 11/12 in. = 99.48 mm	1/3 in. = 8.46 mm	1/4 in. = 6.35 mm
Continental	100 mm	2/3 cm = 6.66 mm	5 mm
Chinese	90 mm	5 mm	3.5 mm
Mondopoint	—	10 mm	5 mm

would differ by one full size. The American system has widths designated as AAAA, AAA, AA, A, B, B, D, E, EE, and EEE, where AAAA is the narrowest and EEE is the widest. Most European countries use the French system also known as the Continental system or the Paris points. The ratio of length increase to girth increase is 1.33 and is same as in the American and English systems. However, the increment is not same. The Chinese system even though rigorous is not so common. In Chinese system the width grade ranges are denoted as 1, 1.5, 2, 2.5, 3, 3.5, and 4. The Mondopoint system, proposed by the International Standards Organization (ISO 9407, 1991), is based on foot length and foot width. In contrast to the other systems, the length and width of the foot that will fit the shoe is used. The shoe length is equivalent to the foot length plus 1 cm. The shoe size given as 24/95 represents a foot that fits the shoe having a length of 24 cm and a width equivalent to 95% of 24 cm. Sometimes, the shoe size is given as 240/95 representing a foot length of 240 mm and a width of 95 mm. The Mondopoint shoe sizing system is mainly used for military boots sizing. In order to generate the different sizes from the model size, grading rule is used. There are different grading rules, namely, proportional, arithmetic, and geometric. Usually, proportional grading is used. There are many techniques used for shoe-last modification with relation to sizing and grading in order to save production costs (Adrian, 1991). It should be noted that shoe size is not fixed. The difference between shoe size and foot length depends on the shape of the shoe-lasts, which is influenced by fashion and design. Since the current sizing is not accurate and not optimal, researches are still ongoing in an attempt to find better sizing and grading methods (Cheng and Perng, 1999; Goonetilleke et al., 1997; Luximon and Goonetilleke, 2003).

9.9   CUSTOM SHOE-LAST DESIGN

A considerable amount of automation has been achieved by the implementation of the CAD system in the footwear industry (Mitchell et al., 1995), where CAD tools are increasingly demanded to assist the whole footwear design process (Kim and Park, 2007; Lee et al., 2002; Lévy et al., 2002).

9.9.1   CUSTOM SHOE-LAST DESIGN USING TRADITIONAL METHOD

This method of shoe-last making is very tedious. Even now, the method for making model shoe-last is highly labor intensive. The technique for making custom shoe-last manually is similar to the technique for making model shoe-lasts. The experienced worker grinds, files, and sands the shoe-last to create new model, each time checking the girth measurements (ball girth, waist girth, instep girth, short heel girth, long heel girth, and throat opening) together with length measurement. Some examples of custom-made shoes include Esatto (www.esatto.biz), Battle Ground, Washington. Esatto provides custom-made shoes and custom orthotics; however, the varieties are limited. In order to make custom shoes, Esatto provides a fit Kit to consumers. The fit Kit provides methods for creating a foot print, foot outline on partial weight bearing and full weight bearing, ball circumference, waist circumference, instep circumference, and short heel girth. On the other hand, Surefit (www.surefitlab.com) from Coral Springs, Florida, provides solutions for shoes and inserts. In the *Surefit*

Instruction Guide, people (Therapeutic footwear program user's guide) should use brannock device to measure the heel to toe length (foot length), heel to ball length (arch length), foot width, and semi-circumference of the ball (only the top portion of the ball). They also need foot print, foot outline, and details of any foot problems to make shoes and inserts. The British Standard (BS 5943, 1980) provides the measurements needed for orthopedic footwear design.

Manual custom-made shoes require custom shoe-lasts. The shoe-last production is tedious, time consuming, and expensive; most consumers do not buy custom shoes because of the lack of variety and high price. In most cases, the shoe-last is not 100% custom-made. The shoe-lasts are graded and sized using normal sizing to create shoe-lasts that nearly match the foot shape in terms of foot length and foot girth. Since the upper shoe-last is custom designed, most consumers are not aware of the shoe-last used. Some examples of custom-made shoes include NikeID (NikeID.nike.com) and Addidas Mi (www.miadidas.com). In these cases, only the shoe is customized, but normal sizing is used to make the shoe; hence, there is no fit customization. In order to reduce production cost, recent studies are making use of information technology to make custom shoe-last.

9.9.2 DIGITAL CUSTOM SHOE-LAST DESIGN

Consumers are always willing to buy customized products; however, they demand lower prices than manual custom-made shoes. Thus, Hwang et al. (2005) have developed template shoe-lasts for efficient fabrication of custom ordered shoe-lasts. In addition, consumers not only want to have custom shoes but they also demand more varieties of shoes that manual custom shoemaking does not provide. Consumers wish to personalize the style and fit of their shoes (Zhang et al., 2010). This unique characteristic of the consumer can be achieved by using the information and CAD technology for shoe design and development (Denkena and Scherger, 2001; Zhang et al., 2010). Recent studies indicate the possibility of involving consumers in the design and selection process of shoe-last and shoes, hereby reducing the production of unwanted design (Leng and Du, 2006; Zhang et al., 2010). In addition to technology, the concept of mass customization (Bae and May-Plumlee, 2005; Lee and Chen, 2000) enables design and production of near custom-made product at near mass production costs. Information technology is enabling the production of the right product at the right time, hence reducing wastage, inventory cost, improving consumer satisfaction, and eventually improving market share (Denkena and Scherger, 2001). Many researchers (Lee et al., 2002; Luximon and Luximon, 2008; Sheffer and Sturler, 2001) have discussed the framework for custom manufacturing and mass customization. New research is now focusing on the details of shoe customization (Zhang et al., 2010).

Bao et al. (1994) have proposed an integrated approach to design and develop shoe-last for orthopedic use. This technique can still be used for custom shoe-last production for normal population. Recently, researchers have proposed an innovative customized shoe-last design cycle by using CAD system and their framework of the customized shoe-last design cycle (Zhang et al., 2010). In their system, consumers can choose both fit and style customization before machining the required shoe-last. The shoe-last is developed using spline curves and surfaces (Luximon and

Luximon, 2009) and the designed digital shoe-last can be manufactured using special shoe-last CNC machine (Figure 9.13). In the proposed framework (Zhang et al., 2010), the consumer can scan the feet and store the feet in the form a digitized foot. The consumer is also free to choose the style of the footwear from the database. Once the selection procedure is finished, the shoe manufacturer can create the desired footwear of the particular consumer according to the shape and size of the consumer feet according to the scanned feet data. The CNC machine (Figure 9.13) can be used for the making the desired shoe-last. It is important to note that design in this form links the consumer and the manufacturer in one system. Manufacturers create specifications of shoe-last styles. Selected shoe-last is presented to consumers.

(a)

(b)

FIGURE 9.13 Shoe-last CNC machine. (a) shoe last CNC machine with 2 pairs of cutter and (b) rough cutting.

The consumer then can modify the design slightly according to their needs. After style customization, the consumer can decide on normal sizes or fit customization based on the cost and fit requirements. This results not only in production of desired shoe-last, but also, more importantly, saves time and effort. Further researches are still being done to make custom shoes acceptable, widely available and affordable.

9.10   CONCLUSION

Past research has shown that the shoes have a variety of effects on human movement and are responsible for a number of sports injuries. So, shoes have increased in complexity with the incorporation of much functionality to improve user performance and comforts. It is apparent that the proper fit can be achieved if the shoe is shaped like the foot. Since the shoe shape is dependent on the shoe-last, the last can take the shape of the foot with some modifications. The last is then graded in order to make different sizes for the whole population. In fact, the interfacing of the foot and the last should be made with the knowledge of the foot anatomy and shapes. The mass produced shoes do not fit properly to the population; hence, customers need customization. Custom-made footwear if manually made is very expensive, hence information technology is being considered to reduce cost. Recent development in footwear research has enabled design and customization of shoe-last quickly such that better fitting shoes can be made. In addition to custom-made (customization), mass customization (near custom made at near mass production cost) in footwear is also being considered to reduce cost while improving fit.

QUESTIONS

9.1 What are the similarities and differences between a shoe-last and a foot?

9.2 A shoe-last is used to make shoes, then why is it that the inside shape of a shoe is not exactly the same shape as the shoe-last?

9.3 List the different types of shoe-lasts. Discuss the advantages and disadvantages of the different types of shoe-lasts?

9.4 What is the difference between sizing and grading? List and discuss three different sizing and grading systems?

9.5 What are the materials used for making shoe-last? Discuss the advantages and disadvantages of the different materials used for making shoe-last?

9.6 What is the difference between custom shoe-last and mass-customized shoe-last?

9.7 Discuss the shoe-last production using traditional techniques?

9.8 Make an AKA64/WMS template for size women 39D?

REFERENCES

Adrian, K. C. 1991. *American Last Making: Procedures, Scale Comparisons, Sizing and Grading Information, Basic Shell Layout, Bottom Equipment Standards.* Arlington, MA: Shoe trades publishing.

Bae, J. H. and May-Plumlee, T. 2005. Consumer focused textile and apparel manufacturing systems: Toward an effective e-commerce model. *Journal of Textile and Apparel Technology and Management*, 4: 1–19.

Bao, H. P., Soundar, P., and Yang, T. 1994. Integrated approach to design and manufacture of shoe lasts for orthopaedic use. *Computers and Industrial Engineering*, 26 (2): 411–421.

BS 5943 1980. *Methods for Measurement and Recording for Orthopaedic Footwear*. London, U.K.: British Standards Institution.

Cavanagh, P. R. 1980. *The Running Shoe Book*. Mountain View, CA: Anderson World.

Chen, G. X. 2005. *The Last Design*. Beijing, China: China Light Industry Press (in Chinese).

Cheng, F. T. and Perng, D. B. 1999. A systematic approach for developing a foot size information system for shoe last design. *International Journal of Industrial Ergonomics*, 25 (2): 171–185.

Cheskin, M. P. 1987. *The Complete Handbook of Athletic Footwear*. New York: Fairchild Publications.

Clarks, Ltd. 1989. *Training Dept. Manual of Shoe Making*, Training Department Clarks.

Denkena, B. and Scherger, S. 2001. A concept for shoe last manufacturing in mass customization. *CIRP Annals Manufacturing Technology*, 54 (1): 341–344.

Goonetilleke, R. S., Ho, C. F., and So, R. H. Y. 1997. Foot sizing beyond the 2-D Brannock method. *Annual Journal of IIE (HK)*, December: 28–31.

Hwang, T. J., Lee, K., Oh, H. Y., and Jeong, J. H. 2005. Derivation of template shoe-lasts for efficient fabrication of custom-ordered shoe-lasts. *Computer-Aided Design*, 37 (12): 1241–1250.

ISO 9407 1991. Shoe sizes—Mondopoint system for sizing and marking. International Organization for Standardization.

Kim, S. and Park, C. K. 2007. Basic garment pattern generation using geometric modeling method. *International Journal of Clothing Science and Technology*, 19: 7–17.

Lee, S. E. and Chen, J. C. 2000. Mass-customization methodology for an apparel industry with a future. *Journal of Industrial Technology*, 16: 1–8.

Lee, S. E., Kunz, G. I., Fiore, A. M., and Campbell, J. R. 2002 Acceptance of mass customization of apparel: Merchandising issues associated with preference for product, process, and place. *Clothing and Textiles Research Journal*, 20 (3): 138–146.

Leng, J. and Du, R. 2006. A CAD approach for designing customized shoe last. *Computer-Aided Design and Application*, 3 (1–4): 377–384.

Lévy, B., Petitjean, S., Ray, N., and Maillot, J. 2002. Least squares conformal maps for automatic texture atlas generation. *ACM Transactions on Graphics*, 21: 372–379.

Luximon, A. 2001. Foot shape evaluation for footwear fitting. PhD thesis. Hong Kong, China: University of Science and Technology.

Luximon, A. and Goonetilleke, R. S. 2003. Critical dimensions for footwear fitting. *IEA 2003 Conference*, Seoul, Korea (CD-ROM).

Luximon, A., Goonetilleke, R. S., and Tsui, K. L. 2003. Foot landmarking for footwear customization. *Ergonomics*, 46: 364–383.

Luximon, A. and Luximon, Y. 2008. Creation of design variations using CAD. *Proceedings of the 86th Textile Institute World Conference*. Hong Kong, China, pp. 1465–1471.

Luximon, A. and Luximon, Y. 2009. Shoe last design innovation for better shoe fitting. *Computers in Industry*, 60 (8): 621–628.

Luximon, A. and Zhang, M. 2006. Foot biomechanics. In *2nd Edition of the International Encyclopedia of Ergonomics and Human Factors*, W. Karwowski (ed.). London, U.K.: Taylor & Francis Group, pp. 333–337.

Mitchell, S. R., Jones, R., and Newman, S. T. 1995. A structured approach to the design of shoe lasts. *Journal of Engineering Design*, 6 (2): 149–166.

National Footwear Research Institute. 1998. *Chinese Shoe Size and Last Design*. Beijing, China: China Light Industry Press, 1984 (in Chinese).

Pivecka, J. and Laure, S. 1995. *The Shoe Last: Practical Handbook for shoe Designers*. Slavicin, Czech Republik: Pivecka Jan Foundation.

Quimby, H. R. 1944. *The Story of Lasts*. New York: National Shoe Manufacturers Association.

Rossi, W. A. 1983. The high incidence of mismatched feet in the population. *Foot Ankle*, 4: 105–112.

Rossi, W. A. and Tennant, R. 1984. *Professional Shoe Fitting*. New York: National Shoe Retailer Association.

SATRA. 1993. *How Shoes are Made*. England, U.K.: Shoe and Allied Trades Research Association Footwear Technology Centre.

Sheffer, A. and de Sturler, E. 2001. Parameterization of faceted surfaces for meshing using angle based flattening. *Engineering with Computers*, 17: 326–337.

Zhang, Y. F., Luximon, A., Pattanayak, A. K., and Leung, K. L. 2010. The research and exploration for customized shoe last design. *The Textile Institute Centenary Conference*, Manchester U.K., November 3–4, 2010.

10 Computer-Aided Design of Footwear

Ajay Joneja and Fan Sai Kit

CONTENTS

The use of computer-aided design (CAD) for footwear is almost as old as CAD technology itself, although it is only in recent decades that there have been complete and integrated solutions integrating various aspects of the design and manufacture of footwear. In this chapter, our goal is to introduce the technology, and, to some extent, the geometric ideas that form the basis of such integrated systems. Our presentation here primarily targets readers who may be interested in using a standard CAD system to design footwear; we do not restrict ourselves to proprietary systems that focus purely on the CAD of footwear, although the techniques we describe are generic enough. These materials should be easily comprehensible to a reader who is familiar with any standard CAD system. At some stages, we also take a slightly deeper look at the underlying geometric complexity of various operations; these advanced materials require some background in computational geometry or differential geometry, and are separated into boxes within the chapter body.

10.1 BACKGROUND

10.1.1 TYPICAL FUNCTIONALITY OF FOOTWEAR CAD SYSTEMS

The functionality of footwear CAD systems parallels the more traditional methods of design. There are several reasons for this. Design methods are driven partially by aesthetics, but largely by the manufacturing technology and available materials and their properties. It follows therefore that the CAD systems are driven by the same concerns as traditional methods. So we begin with a brief introduction to the traditional footwear design process. At the same time, we shall introduce some of the standard terminologies that we shall use throughout the chapter.

As with many mechanical products, footwear design often begins with conceptual design. At this stage, the designer often uses sketches to lay out the important design ideas, aesthetics, and possibly color schemes. Styling and aesthetics strongly influence this stage. The outcome of conceptual design may be set of sketches, or sometimes even photorealistic artwork created using image processing software tools (e.g., Adobe™ Photoshop). Occasionally, this step is followed up by construction of physical samples, called mock-ups. Such prototypes may be constructed manually, with relatively little automation—standard shoe lasts, possibly modified to approximate the desired shape, are covered in cloth tape on which the designer sketches the upper patterns. The tape is then cut, peeled off the last, and flattened to obtain the necessary pattern shapes. The uppers can then be constructed by manual cutting and stitching of the designed materials. Heels, for example, those used in women's shoes, may be constructed out of plastic by manual machining or occasionally by the use of 3D printing techniques such as fused deposition modeling. The most complex component is often the outsole, especially for athletic footwear. Fabricating the outsole requires careful and expensive machining of metal molds. Therefore, if time and cost are constrained, the mock-up may often use surrogate outsoles. The next stage in the design, and the one with which this chapter is mainly concerned, is the construction of the detailed design of the shoe. While this was traditionally done by making engineering drawings, we shall focus more on the process and technology of specifying detailed design using modern 3D CAD systems. Loosely, the process of constructing the 3D models includes the following steps, usually, although not necessarily, in the order mentioned. The design begins with the selection of an appropriate shoe last of standard size. The geometry of the last model is modified until it matches exactly the design requirements and various constraints, for example, those related to comfort. Using this shape as a basis, the upper is designed, along with the patterns and layers of materials as required. At this stage, two related sets of geometric objects are often available: the geometric shape of the upper in the 3D model, wrapped over the shoe last, and its equivalent flattened shape on a 2D plane, called its surface development. The insole and possibly inserts are designed by using the geometry on the last bottom surface. These provide the necessary geometric information to design the midsole, which in turn is used to construct the model of the outsole. This is followed by the design of the heel, if required. During the detailed design, any ornaments, logos, and similar details may be added as necessary. Finally, the geometric model may be enhanced by adding rendering artifacts, for example, material-related textures, stitches, etc., to facilitate generation of photorealistic renderings useful for communication, promotions, marketing, etc.

At each stage of the CAD process, the designer must be keenly aware of the manufacturing process and materials that will eventually be used during the production—in many modern CAD systems, this is done via providing various design rule checks, a mechanism that simulates the now-common practice of concurrent engineering to achieve design for manufacturability. In the following sections, we shall occasionally introduce some examples of these considerations.

10.1.2 CAD Modeling Background

Since shoes are 3D objects, we need to be concerned with how to specify 3D shapes, and furthermore, how to modify such shapes in order to conveniently generate precise models of the designs that we wish to produce. At the same time, the software that we use to perform the design must also provide adequate support for transforming the CAD model data into production or fabrication information.

Our emphasis in this chapter is to understand the important stages of the process of footwear design. These operations will be discussed in terms of the design and manufacturing requirements along with some examples. These examples are not taken from the use of any particular commercial CAD system. However, much of the functionality of popular commercial systems covers each of the operations we shall discuss, so a generic discussion adequately captures the core operation sets of almost all of such systems.

10.2 TWO TYPES OF CAD SYSTEMS

As we discuss the various functions of CAD modeling of shoes, we shall encounter the different problems of geometry, often using analytical, computational, and differential geometry. It is beyond the scope of this chapter to give ample mathematical insight on such details, and so the reader will be referred to other sources for further reading. Since CAD systems work with geometry of objects, their underlying data structures influence the capability and behavior of the shape modification tools (or operators) that they provide. We will refer to two popular types of data structures in the context of footwear CAD: surface modelers that store all geometric information as tessellated or mesh models and solid modelers that store boundary representation-based data structures and have explicit analytical equations for boundary elements such as curves, surfaces, etc. (Figure 10.1). Most solid modelers use BSplines (Piegl 1997) for curves and surfaces. Furthermore, solid modelers are parametric in the sense that the 3D models are constructed by utilizing a sequence of operations, with each operation defining a shape feature and the shape of each feature being defined by the specification of a small set of parameter values. The geometric

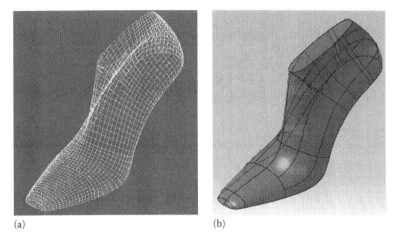

(a)   (b)

FIGURE 10.1   (a) A mesh-based surface model of a shoe last. (b) A boundary representation-based analytical model of a last.

relationships and constraints between the features are specified by the designer, and the integrity of these constraints is managed during any shape manipulation done on the model at any stage of the design.

Apart from the internal data structure differences, the type of shape manipulation that is provided by these two types of systems may also differ (we show an example in the following). In some mesh-based modelers, for example, a very local change to the geometry can easily provide very localized shape modification options to the user (see the example in Figure 10.2, where a sharp local change made to some cross-sectional curves result in the creation of large ridges and regions of low smoothness); this feature can be good or bad—sometimes local control is useful, while at other times, it is more natural for a change made in one location to lead to a smoothly dissipating change across a broader region of the model.

However, as we shall soon see, most footwear CAD design operations mimic operations and practices followed by the traditional shoe designer who worked with physical (not computer) models. As a consequence, the look and feel, that is, the graphical user interface (GUI), of both types of systems may be similar to the user.

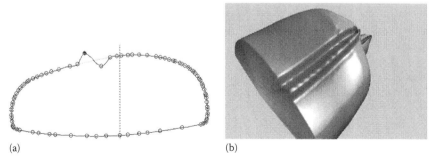

(a)   (b)

FIGURE 10.2   (a) A cross-sectional curve on a last model is modified to introduce a localized variation. (b) Effect of the shape modification on the model is highly unsmooth.

It is only the underlying geometric engines that are totally different in the types of algorithms that they use to implement any given shape modification operator.

10.3 FOOTWEAR CAD

In this section, we discuss various stages of the construction of CAD models of footwear.

10.3.1 CONCEPTUAL DESIGN

At this stage, often the designer will sketch out the key design features for the shoe. Some sketches illustrate on the styling. For example, it is customary for many brands to release a new line of shoes once every few months often according to seasons. Thus, a summer fashion line may have a more open upper, while a winter line may be dominated by boots. In other cases, there may only be some aspects of the shoe styling that is changed from earlier styles. For instance, an athletic shoe designer may choose to change only the upper style and materials, keeping the sole relatively unchanged (Figure 10.3).

Conceptual designs traditionally are sketched physically on paper. Some modern companies use computer tools for generating sketches. The tools for this are 2D image processing software systems, for instance Photoshop™ from Adobe Systems.

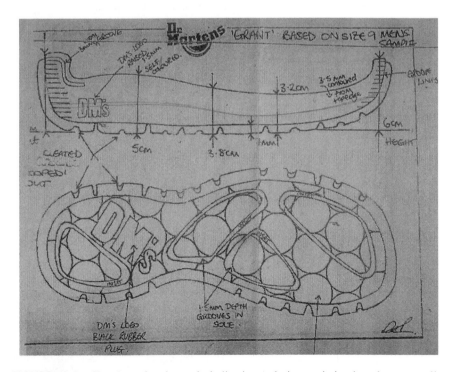

FIGURE 10.3 Sketches of a shoe sole indicating a design variation based on an earlier model.

FIGURE 10.4   Sketch of a new style made on an image editor.

Figures 10.3 and 10.4 give two examples of conceptual designs: The first is a physical sketch highlighting some key features in the sole of a walking shoe; and the second shows a potential design for a ladies shoe sketched in an image editor.

It is important to note that image editors do not create 3D models; the outcome is merely an artist's impression (or, more correctly, the industrial designer's impression) of the potential product. However, they offer several advantages, including easy archiving and retrieval, ease of modification, powerful graphics filters including colors, texture maps, and even lighting effects to add photorealism.

To create the actual 3D CAD model, the designer can only use the sketch as a reference, while the entire geometry has to be recreated using the CAD system.

10.3.2   Last Design

10.3.2.1   Background

The design process begins with the specification of the shoe last. The geometry of the last is dictated partly by concerns of fit, and partly by the aesthetics and styling. The initial design is made for a single size, and the prototype shoe is constructed and tested using this last. For mass production, it is only economical to produce a few discrete sizes of shoes. These variations of sizes are based on a nonlinear scaling of the prototype, generated by a process called grading. An understanding of the related issues will be useful for the reader, and we refer the reader to Chapter 9 of this book (Luximon 2012). In the following, we will consider how the geometry of a (series of) shoe lasts can be made in a CAD system. Obviously, it is fairly difficult to describe the required geometry from scratch. It is customary to use an existing shoe last model and modify its shape to create a new one as required. But how do we start this process? We must have some models, or templates, to bootstrap our library. A popular way is to use existing physical models as a basis, as we illustrate in the following.

10.3.2.2   Constructing CAD Models from Physical Models

The first step is to measure the geometry of the existing physical last. Currently, a fast and inexpensive way to do so is by the use of laser scanning. There are also methods for contact-scanning of the last. For example, most last fabrication machines come equipped with the tools and sensors required to generate the coordinates of a dense set of points on the surface of a mounted last; this functionality is useful in creating multiple copies of a hand-crafted last to be used for batch production of the shoes. This set of coordinates, hereafter referred to as a point cloud, can be used to generate a surface representation of the shoe last by almost any modern CAD system. There are several specialized software systems that can automatically fit surfaces onto a given point cloud, including user-controlled noise suppression; this is an important stage in the popular field of reverse engineering. Naturally, this function is provided by almost all commercial shoe or last CAD software systems (including ShoeMaster, Delcam-Crispin Systems, EasyLast™ from Newlast Inc., etc.). Some commercial systems that provide robust surface-fitting to point clouds include Geomagic™ and Visi reverse from Vero™ software. Also, all general purpose CAD systems provide this functionality either directly (see the following example) or in the form of imported modules (e.g., an external module called RhinoResurf™ can be plugged into Rhinoceros®).

Here, we illustrate the steps of construction of a CAD model from an imported point cloud data of a shoe last, using CATIA™ from Dassault Systems Inc.

Step 1. The point cloud is just a collection of point coordinates in 3D. The underlying model is first given a surface by creating interpolating triangles between carefully selected triplets of neighboring points. This results in a well-defined, oriented tessellated model. The model is well defined in the sense that the resulting mesh model is regular, no pair of triangles intersect in their interior, and there are no gaps between neighboring triangles (see Box 10.1) (Figure 10.5).

Step 2. The tessellated surface model is divided, or cut up, into sets of regions. The boundaries of these regions are curves, which can then be faired, or smoothed. This step usually helps in suppressing random variations in shape, such as tiny bumps, kinks, etc., that may arise due to noise in the scanning data (Figure 10.6).

Step 3. Selected subsets of the smoothed curves are interpolated to create algebraic surfaces. This interpolation is done repeatedly to cover the entire surface of the last with a set of surface patches. During this surface interpolation, if required, different levels of continuity are specified, for instance C^0 (e.g., along a ridge, such as the one along the last bottom surface and the last upper surface), G^1 (which forces the surface tangents at the common boundary of neighboring patches to be continuous), or perhaps G^2 (which constraints the curvature continuity) (Figure 10.7).

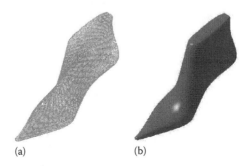

(a)   (b)

FIGURE 10.5   (a) Point cloud data of a last. (b) Tessellated model of the point cloud.

FIGURE 10.6 Tessellated model is subdivided into patches by forming intersection curves. The intersection curves are subsequently smoothed.

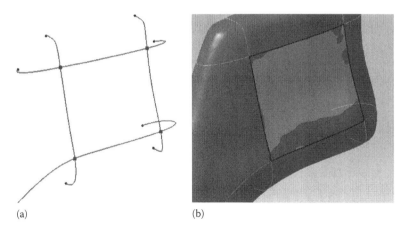

(a)   (b)

FIGURE 10.7 (a) Bounding curves of a subset of the last surface. As a consequence of the smoothing process in Step 2, pairs of curves may not intersect at all, so the smoothing operators must be constrained to guarantee that indeed the curves in the two directions intersect at the points shown in red dots. (b) A smooth surface is fitted to interpolate the bounding curves as well as the relevant scanned points in the region.

An interesting little problem arises after the last model is built using the aforementioned approach: We need to precisely locate the reference coordinate frame attached to the last, with respect to which we can perform the various measurements that are required. An engineering solution to this problem is to attach special markers or locators on the body of the shoe last before scanning it; these locators can then be identified on the point cloud, and their coordinates can be used to create the reference coordinate frame. An alternate approach is to attempt and recognize some features on the surface of the last itself, and use them to localize the model. Examples of such features include the (sharp edged)

BOX 10.1   SURFACE RECONSTRUCTION

Several different approaches are known for producing a surface that interpolates a cloud of points. The problem is nontrivial due to two factors: (a) When main points are close to each other in the same vicinity, it is possible to interpolate them by using several distinct surfaces and (b) the measurement device that generates the point cloud, for example, a laser scanner, is not totally accurate, that is, noisy. There are a few robust algorithms for surface reconstruction that create a triangulated mesh (see Dey 2007) for an excellent exposition. Let us look briefly at one algorithm, called ball-pivoting (Bernardini 1999), to get an appreciation of the problem.

The main idea of the ball-pivoting algorithm is to start with a ball of appropriate diameter, d, and to locate it at some initial point such that (a) it touches exactly three of the points in the cloud and (b) no other point of the cloud is inside the ball in this position. The three points that the ball rests upon are connected to form three edges of the seed triangle. Next, the ball is pivoted around an axis formed by any of the three edges, until the first time that it just touches another point in the cloud. This point and the vertices defining the pivot edge are then candidates for another triangle. The algorithm advances the triangulated surface by propagating outward along its boundary, until all points have been consumed.

It is easy to see that the selection of the diameter of the ball used in this process is of considerable importance. If the ball is too small, there are times when it will not add any point of the cloud even if it pivots over an entire revolution; this prematurely stops the advancing front. On the other hand, if the diameter is too large, then the ball may leave some points of the cloud untouched: Consider, for example, the case where the pivoting ball is advancing the front along the points that are on a flat terrain, with a narrow but deep hole in the middle; the ball will roll right over the hole, without being able to touch any point defining the surface of the hole. Figure 10.8 illustrates the basic idea of the ball-pivoting algorithm.

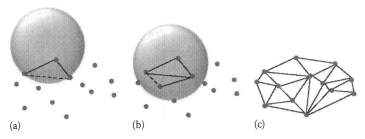

(a)   (b)   (c)

FIGURE 10.8   The ball-pivoting algorithm: the input is a point cloud (red dots) and the ball radius. (a) The ball rests on a seed triangle; the pivot edge is shown as a dashed line; (b) the ball pivots to add a triangle; (c) a possible outcome of the algorithm, with a triangulated surface.

(*continued*)

BOX 10.1 (continued)   SURFACE RECONSTRUCTION

A simplified version of the algorithm is presented in the following. As soon as a triangle is formed by a pivot operation, its edges are classified either as active (if it is eligible to be used for pivoting) or boundary (if pivoting around it will not yield a new triangle). A seed triangle is formed by finding an unused vertex, and testing amongst pairs of its nearby points if that trio of vertices can support the ball without intersecting any other points.

Algorithm *Surface Reconstruction by Ball Pivoting*

```
while (true) do
(u, v): = select the first active edge
if (u, v) ! = null
      w: = find-pivot-vertex (u, v)
      if w ! = null and (w is unused or it is on the
      front)
          add triangle (u, v, w) to the Surface;
          add edges (u, w) and (v, w) to the set of
          active edges
      else mark edge (u, v) as a boundary edge
      (u, v, w): = form-seed-triangle(unused_vertices,
      ball_radius);
else
      (u, v, w): = find a new seed triangle;
      if (u, v, w) ! = null
        add triangle (u, v, w) to the Surface;
        add edges (u, v), (v, w), (w, u) to the active
          edges
      else exit;
end while
```

For a more detailed coverage of the algorithm, the interested user is referred to Bernardini (1999).

---

profile of the last bottom, the flat surface above the ankle part of the last, etc. It has to be noted that there is some inaccuracy that can result from either of these localization approaches. Reasons for this include noise in the scanned data, as well as the fact that scanning machines only report a finite number of points on the surface. On the other hand, once we have a reference CAD model, it can easily be fabricated with great accuracy on a last cutting lathe. This allows us the opportunity to add locating features on the CAD model, which will then be machined with good accuracy in the test model. This modified model can then be measured on a last measurement machine (this is basically a jig that provides for accurate localization of physical last samples). Using measurements on this jig allows us to adjust the coordinate frame on the CAD model to within very small tolerance.

Once we have a set of last models in our CAD library, constructing a new model requires us to select one that is close enough to the desired model, and modify its shape to get the desired result. This process of modification of the shape is not unconstrained; it is imperative that the basic measurements related to the fit, and therefore comfort, are not violated. In the following sections, we shall discuss a series of useful shape operators and also how we can use the CAD system to measure the design for conformance to any particular standard for fit.

10.3.2.3 Shape Modification

Traditional design of shoe lasts was (and in many instances, still is) done with physical prototypes. The physical sample shape is manipulated in the workshop, by operations including cutting, plastic deformation (achieved by heating the plastic to just under the melting point), adding extra plastic in some regions, etc. Such modifications are driven by aesthetic concerns as well as ergonomic ones, and the design intent is communicated and recorded in traditional manner, for example, using sketches of engineering drawings. In discussing these operations, it will be useful if the reader is familiar with the basic shape descriptors and standard nomenclature referring to various geometric elements of last. These have been covered in the first few sections of Chapter 9 in this book (Luximon 2012; Pivecka 1995). Figure 10.9 illustrates the

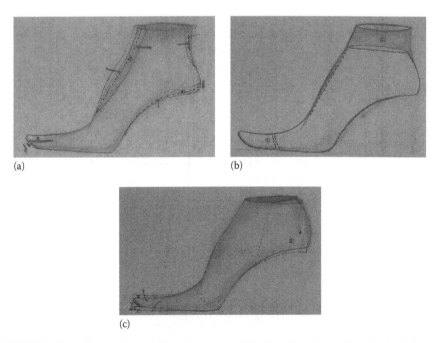

(a)   (b)

(c)

FIGURE 10.9 Examples of last shape modification functions for physical lasts. (a) (1) Widening or tightening the toe part, (2) modification on instep and waist, (3) modification of the back curve, and (4) modification of the shank profile, which is a part of the last bottom profile. (b) (1) Shortening or elongation of front part and (2) transforming between shoe and boot lasts. (c) (1) Changing the toe spring and (2) changing the heel height.

traditional modification instructions on physical lasts; such modifications are often made by designers to create new shoe designs by using older ones. Unsurprisingly, similar operations are offered by several commercial CAD systems dealing with last design, including EasyLast3D™ from Newlast Group Ltd., or the e-Last™ module of the ShoeMaster suite from CSM3D Int. Ltd. (which branched out of the Clarks shoe company in the United Kingdom).

The operations shown in the earlier illustrations are indeed commonly provided by most commercial CAD systems. The exact mechanism by which a user of these systems can realize them can vary, partly because of the particular user interface in the system, and partly due to the underlying data structures used by the system. But the basic interface provided by various systems use a few common ideas, including the following:

1. Volumetric shape modification by changing one or more cross-sectional curves
2. Volumetric shape change by changing some boundary curves
3. Replacing one volume by another

Each of these operations requires the user to only input a few parameter values, which are used by some underlying algorithms in the CAD system to make global changes in the surface of the last. Such algorithms, called free form shape deformation in computational geometry, (Coquillart 1990) usually follow some high-level guidelines, for example, minimization of the overall change in the volume or the shape of the last while effecting the required change. This is still an active area of research within the CAD community. In the following examples, we shall illustrate examples of how a typical CAD system may be used to make the type of modifications illustrated by Figure 10.9, by using techniques (1)–(3) (Figure 10.10).

Figure 10.10 illustrates a shoe last model, with a series of section curves. Conversely, one may think of the surface of this shoe last as a shape that is constrained to interpolate, that is, pass through each of these planar curves (notice that each curve is shown along with a small square indicating the plane containing it [Figure 10.11]).

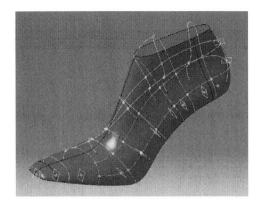

FIGURE 10.10 Shoe last model interpolating some skeleton curves.

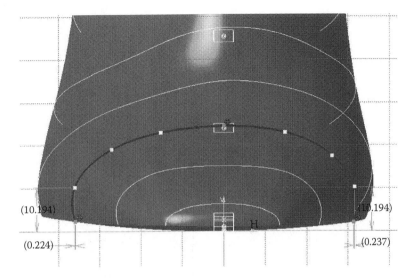

FIGURE 10.11 Modifications on interpolating section curves can be made by moving curve points (shown as white squares in the figure).

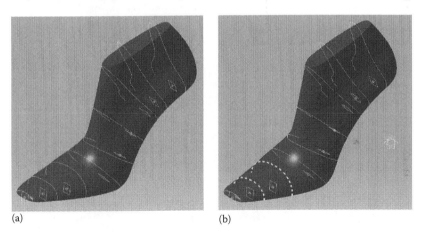

(a)   (b)

FIGURE 10.12 (a) Shoe last model showing interpolating section curves. (b) The two curves shown in dashed line were modified, resulting in a slimmer toe as shown.

Thus, if we modify the first two curves on the left side, for example, by either bulging them slightly upward or by squeezing them downward, the changed shape of the modified last surface would have a slightly more spacious (respectively, less spacious) toe region. A similar shape change in the waist or instep region of the last can be realized by modification of the middle three of the skeleton curves (Figure 10.12).

Figure 10.13 shows an example where the shape of the back curve of a last is modified to generate a different last back shape. In this example, we also illustrate a different style of constraint on such modifications. Ergonomics data lead to a guideline on the angle that the back curve makes with the last bottom profile curve (more precisely, this is the angle between the tangents of these two curves at the

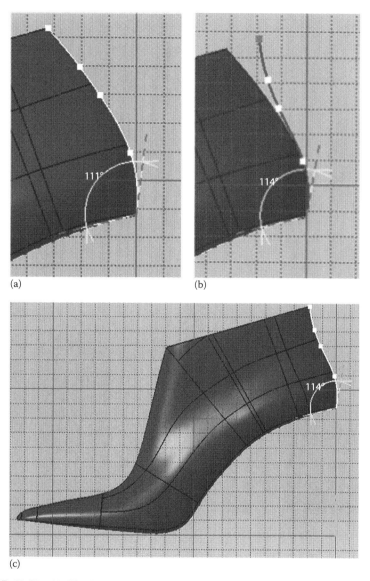

FIGURE 10.13 (a) The back curve profile of a last. (b) Modification of the back curve profile; the shape of the curve, as well as the angle it makes with the last bottom is changed. (c) The modified 3D last model, highlighting the changed back curve.

point where they meet, the back point). The value of this angle is largely dictated by the heel height. Any modification to the back curve should normally respect this constraint.

Figure 10.14 shows another example of shape modification by changing some skeleton curves; in this case, the object is to modify some of the back section curves, in order to thicken (or shrink) the back portion of the last. In this example, a different shape constraint is illustrated. The end points of the back section curves are

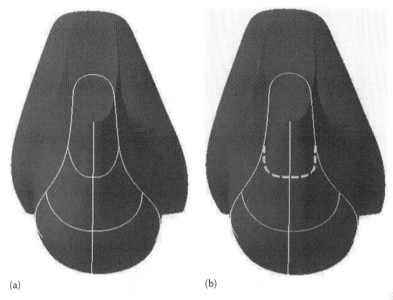

FIGURE 10.14 (a) The two orange curves indicate interpolating back section curves. (b) The upper back section curve geometry is modified to yield a more square-shaped back on the last; here, the modified curve's end points are constrained to stay on the vertical sections curve shown in white, and its mid-point is constrained to pass through the back curve. Continuity constraints on the interpolating surface guarantee the smoothness of the last surface.

restricted to coincide with the vertical section curve that is used in the formation of the surface ahead of the back section. Thus, any modification to the back sections may also modify this skeleton curve, and the designer may have to edit this curve later in order to control the shape of the back region as desired.

The reader may have noticed from Figure 10.14 that the last models used for illustrations have three interconnected, yet separate surfaces: (a) the top profile of the last (Figure 10.15), (b) the main surface forming the last, and (c) the last bottom surface (Figure 10.16). All the aforementioned examples have dealt with modifications of the last shape by a direct action on the main surface of the last. Let us look at some examples of modifications on the other two.

The preceding example is interesting in its effect. Notice that the shape of the last is generated by skinning the skeletal curves as indicated in Figure 10.10. We may imagine the last bottom profile as a spine, holding the skeletal curves via a set of last bottom section curves, as shown in Figure 10.17. Thus, modification of the last bottom curve (e.g., to slightly raise or lower the arch) will cause the entire set of section curves to move along with this spine curve. In order to make this operation better defined, a CAD model may further specify the relationship of the section curves with the spine (e.g., by fixing the angle between the plane of the section curve and the tangent of the spine curve at the point where it intersects this plane) (Box 10.2).

Our next example of last geometry manipulation is a volume-replacement operation, which follows a practice quite common in shoe design. The idea here is to create a new last shape that uses the back part of one last, and the toe part of a different one.

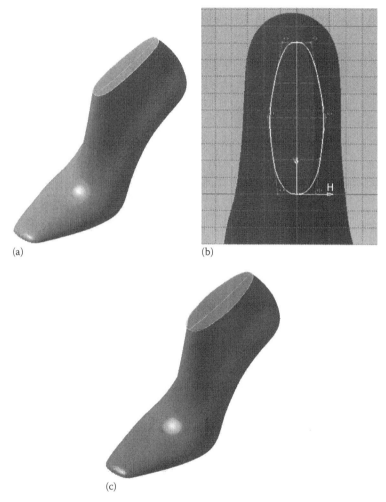

(a)   (b)

(c)

FIGURE 10.15   (a) Last with top surface profile highlighted. (b) Shape modification of the top profile. (c) Resulting model of the last.

This operation follows a commonly used practice in shoe design, since it allows designers to create a large number of design variations by matching together pairs of previously designed lasts. Clearly, this operation is nontrivial: It is quite unlikely that one cannot join the required toe and back parts directly: the geometry is bound to be different, and therefore, the resulting surface is discontinuous. In practice, the technique used is to eliminate a slice of material from both pieces, and fill this space with a slice of plastic material of equal thickness. This slice can subsequently be sculpted to generate a smooth blend between the new toe and the back parts. In CAD systems, a common way to achieve this function is to initially scale the geometry of the replacement toe part, followed by the generation of a controlled blending surface that smoothly morphs between the shape of the back part and the toe part; see for example (Guo 2004). An example of this operation is shown in Figure 10.18.

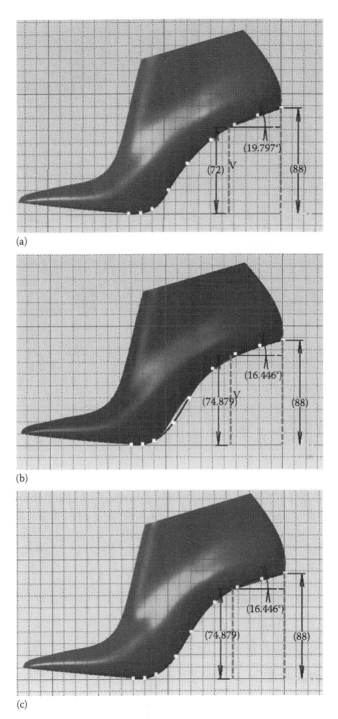

(a)

(b)

(c)

FIGURE 10.16 (a) Model of a last with the last bottom profile curve highlighted. (b) Modification of the last bottom profile curve. (c) The modified last shape.

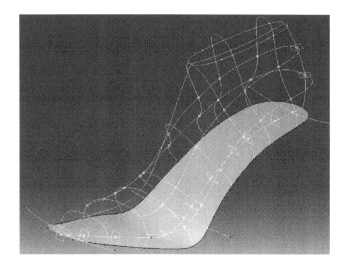

FIGURE 10.17 One way to construct a CAD model for a shoe last. The algebraic surfaces are constructed by skinning a set of skeleton curves that are held by one or more spine curves. The last bottom profile, the bottom section curves, and the last section curves have spatial relationships; the dots in the figure indicate coincidence of intersecting curves, and surfaces are constrained to pass through the curves. The surface patches making up the last body (respectively, last top and last bottom) have $G1$ continuity along their shared boundaries.

BOX 10.2   CONSTRAINED SHAPE MODIFICATION

In our discussion, we have assumed that the surface is represented by an analytical surface, whose geometric shape is stored in the CAD system. In this text, we shall further restrict our focus to BSpline surfaces, and get some insight into the mathematical reasoning behind the shape modifications described earlier. A BSpline surface (Piegl 1997) is a piecewise parametric tensor product representation usually written in vector form as

$$S(u,v) = \sum_{i=0}^{n} \sum_{j=0}^{m} N_{i,p}(u)\, N_{j,q}(v)\, P_{i,j}$$

where the surface is defined in piecewise patches corresponding to rectangular regions in the (u, v) parameter space defined by the knot vectors U and V, specified as

$$U = \{0,\dots,0,u_{p+1},\dots,u_{r-p-1},1,\dots,1\}, \quad V = \{0,\dots,0,u_{q+1},\dots,u_{s-q-1},1,\dots,1\}$$

The equation of the surface in any parameter-space patch, for example, $u_i \le u < u_{i+1}$, $v_j \le v < v_{j+1}$, interpolates the set of control points $P_{i,j}$ weighted by the

BOX 10.2 (continued)   CONSTRAINED SHAPE MODIFICATION

corresponding basis functions (defined in the following) over that range. The $N_{i,j}$ are the Bernstein basis functions defined as

$$N_{i,0}(u) = \begin{cases} 1 & \text{if } u_i \leq u \leq u_{i+1} \\ 0 & \text{otherwise} \end{cases}$$

$$N_{i,p}(u) = \frac{u - u_i}{u_{i+p} - u_i} N_{i,p-1}(u) + \frac{u_{i+p+1} - u}{u_{i+p+1} - u_{i+1}} N_{i+1,p-1}(u)$$

The v-basis functions are defined analogously. Thus, the equation is a spline, and this allows us to define complicated shapes by using multiple knot spans of lower degree. In the preceding example, the surface is defined by u-polynomials of degree p and v-polynomials of degree q, and the exact geometry of the surface is defined by specifying the $(n + 1) \times (m + 1)$ 3D coordinates of the control points.

To modify the shape of such a surface, an intuitive method is to move the control points, since BSplines have the nice property that the surface smoothly follows the shape of mesh formed by the control points. There are various techniques that facilitate direct manipulation of BSpline surfaces, and an excellent exposition can be found in Piegl (1997). However, even a typical shoe last shape requires several hundred control points to define its shape accurately. To modify its shape interactively by direct manipulation of control points is cumbersome and proves very crude control over the other constraints we wish for the surface to obey (e.g., the length of the various girth measurements of the last). An excellent mechanism to overcome this problem has its roots in a method called free form deformation (FFD), first proposed by Sederberg and Parry (1986). The main idea if FFD is fairly simple: the object we wish to deform is embedded in a simple 3D polyhedral shape, for example, a parallelepiped. This embedding shape is defined by a simple lattice of points, and can be deformed by moving any point of the lattice. A trivariate Bezier function is used to define a continuous and invertible mapping between the lattice points and the control points of the shape we wish to change. Using this mapping, any (gross, easily defined) change in the shape of the embedding shape is used to compute the image of the original control points of the shape to their corresponding new coordinates. The deformed shape is therefore obtained by the BSpline formed by using the mapped control points. This approach is simple, extremely versatile, and can even be easily adapted to use other enveloping shapes that more closely approximate the original shape. However, it does not provide a mechanism for constraining the deformation. Consider any of the last shape modification operations described earlier, say, the replacement of

(continued)

BOX 10.2 (continued)   CONSTRAINED SHAPE MODIFICATION

the last bottom profile curve with an alternate one, for example, to convert a last designed for oriental population to one for a western population. We can try to move some of the lattice points of an embedding parallelepiped around the last, but may require several trials before the modified last bottom profile shape matches the desired target shape. An elegant solution to alleviate this problem was proposed by Hu (2001). We now describe a simplified version of their approach. The embedding shape is a rectangular parallelepiped with $(l + 1) \times (m + 1) \times (n + 1)$ uniformly distributed lattice points; the embedding volume is defined by a BSpline volume $Q(e, f, g)$ as

$$Q(e,f,g) = \sum_{i,j,k=0}^{l,m,n} N_{i,p}(e)N_{j,q}(f)N_{k,r}(g)P_{i,j,k}, \quad 0 \leq e,f,g \leq 1$$

Suppose that we wish to deform the surface $S(u, v)$ by moving a point $S_o(u_o, v_o)$ lying on it; since this point also lies inside the trivariate BSpline solid, let its coordinates in the parameter space of this solid be given by $S_o(e_o, f_o, g_o)$, and the deformation itself moves each lattice point by some displacement, $\delta_{i,j,k}$. Then the transformed image of S_i under the deformation is given by

$$T_o = \sum_{i,j,k=0}^{l,m,n} (P_{i,j,k} + \delta_{i,j,k})N_{i,p}(e_o)N_{j,q}(f_o)N_{k,r}(g_o)$$

$$= S + \delta_{i,j,k}N_{i,p}(e_o)N_{j,q}(f_o)N_{k,r}(g_o)$$

Since T_o is given, we wish to find the values of $\delta_{i,j,k}$'s that will satisfy this equation. Clearly, this is an underconstrained problem, and Hu et al. suggest to get a unique solution by finding the values that minimize the change in the $\delta_{i,j,k}$'s in the least squares sense. This leads to a Lagrangian formulation: λ

$$\min L = \sum_{i,j,k=0}^{l,m,n} \left\| \delta_{i,j,k} \right\|^2 + \lambda \left(T_o - S_o - \delta_{i,j,k}N_{i,p}(e_o)N_{j,q}(f_o)N_{k,r}(g_o) \right)$$

where $\lambda = [\lambda_1, \lambda_2, \lambda_3]^T$ is the vector Lagrangian multiplier and the $\| * \|$ is the usual L^2 norm. This minimization can be solved explicitly, yielding the solution

$$\delta_{i,j,k} = \frac{N_{i,p}(e_o)N_{j,q}(f_o)N_{k,r}(w_s)}{\displaystyle\sum_{i,j,k=0}^{l,m,n} \left(N_{i,p}(e_o)N_{j,q}(f_o)N_{k,r}(g_o) \right)^2}(T_o - S_o)$$

BOX 10.2 (continued)   CONSTRAINED SHAPE MODIFICATION

Furthermore, Hu et al. show that the solution of the analogous problem where a set of k points $S_1, ..., S_k$ must be mapped to respective target positions $T_1, ..., T_k$ is equivalent to solving the problem one point at a time. In other words, the composition of a set of single-point constrained deformations is commutative in this formulation. Using the previous results, we can implement an algorithm for constrained shape modifications that can be used for many of the last shape modification functions described earlier.

Algorithm *Constrained Shape Deformation*

```
Inputs:
Surface S(u, v)
List of surface points to be moved I = {(u₁, v₁), …,
(uₖ, vₖ)}
Target locations for the k points F = {(x₁, y₁, z₁), …
(xₖ, yₖ, zₖ)}
```

1. For each point (u_r, v_r) in I

$$(e_r, f_r, g_r) = \text{uv\_to\_efg}(u_r, v_r);$$

2. For each point (x_s, y_s, z_s) in F

$$(e_s, f_s, g_s) = \text{xyz\_to\_efg}(x_s, y_s, z_s);$$

3. For each point (e_r, f_r, g_r) to be deformed
   For each control point of the S

$$P_{i,j,k} := P_{i,j,k} + \delta_{i,j,k};$$

4. Output S based the mapped control points $P_{i,j,k}$

In the preceding algorithm, the function uv _ to _ efg computes the coordinates of a surface point, $S_i(u_i, v_i)$, in the parameter space of the embedding rectangular solid, while function xyz _ to _ efg computes the coordinates of any point in the global coordinate frame to the parameter space of the embedding parallelepiped. If we use one corner of the parallelepiped as the origin, and the three edges of the lattice incident upon this corner to form an orthonormal coordinate frame, then the construction of these functions is trivial (Sederberg 1986). In Step 4, the computation of $\delta_{i,j,k}$ uses the derivation given earlier.

We mention in passing that this is a simplified version to illustrate the main ideas behind our shape deformation problem. Please see Question 10.5 in the Exercises at the end of this chapter.

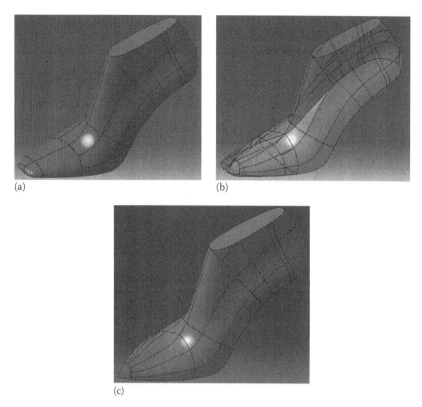

(a)

(b)

(c)

FIGURE 10.18 (a) A shoe last. (b) The toe of a different last is merged smoothly onto the back part of the last in (a). (c) The new last, which retains the back part of the original merged with the toe shape of the second one.

We now look at one of the more complex shape modification examples in the context of last design. As we saw earlier, the last bottom and the main last surfaces are maintained within the CAD system as two distinct entities. One big reason for this separation is that the last bottom is the basis for the generation of other components of the shoe last, in particular the shoe insole. More precisely, the shoe insole profile is obtained by flattening the last bottom surface onto a plane. This flattening operation, called a surface development, is in fact one of the most studied operations in mechanical design, since it has numerous designs for manufacturing applications ranging from sheet metal designs to ship building. It is well known that a general curved surface cannot be developed into an isometric flat surface (i.e., the length of any line on the 3D surface is exactly equal to the length of its image on the developed surface). Intuitively, flattening the last bottom surface on to a plane can only be achieved by stretching it in some regions, and squeezing it in others. There are several surface development algorithms, each one designed to minimize some measure of the distortion of the curved surface. Some attempt to minimize the area distortion, while others look at the approximate amount of shear, or skewing (see Box 10.3). The outcome of

---

BOX 10.3   SURFACE DEVELOPMENT VIA
MESH PARAMETERIZATION

Consider a (piecewise) smooth surface, S, in 3D; it can be represented by a function of two parameters, thus $S(u, v) = (x(u, v), y(u, v), z(u, v)) \subset R^3$.

Ideally, a surface development is an isometric mapping $f: R^3 \to R^2$ such that a surface is developable if there exists such a map. If a surface has anywhere nonzero Gaussian curvature, K, then no such map exists.

Gaussian curvature $K = (LM - N^2)/(EG - F^2)$, where $E = S_u \cdot S_u$, $F = S_u S_v$, $G = S_v \cdot S_v$, $L = S_{uu} \cdot \mathbf{n}$, $M = S_{uv} \cdot \mathbf{n}$, and $N = S_{vv} \cdot \mathbf{n}$, where the subscripts denote the surface partial derivative with respect to the corresponding parameter(s) and $\mathbf{n} = S_u \times S_v /\| S_u \times S_v \|$ is the unit normal. A surface is developable if its Gaussian curvature is zero at all points. Unfortunately, this is not the case for almost any surface on the shoe. The practical implication is that to generate the equivalent flattened geometry of any 3D surface on the CAD model of the shoe involves some amount of geometric distortion. Practically, this means that any flat pattern used for shoemaking must undergo some stretching, and possibly wrinkle slightly in some regions in order to form the shape of the shoe. Since it is natural for physical objects to settle down to a least energy (i.e., stable equilibrium) state when distorted in this fashion, therefore, most surface development algorithms are designed to minimize some measure of distortion. Suppose we develop a mapping $f: S \to R^2$. We say that the mapping f is *isometric*, if the length of any curve, c, on S is equal to the length of its image $f(c)$ in R^2. If instead, the mapping f guarantees that the angle between any pair of intersecting curves, c_1 and c_2 on S is equal the angle of their respective images $f(c_1)$ and $f(c_2)$ in R^2, then we say that f is *conformal*. Similarly, f is said to be equiareal if the area of any closed region on S is equal to the area of its map in R^2. It is well known that a mapping that is equiareal as well as conformal is isometric (and in this sense, isometry is the most stringent requirement we may put on our mapping). To compute a mapping that will yield a good development of a 3D surface, we typically need to solve systems of first-order or second-order partial differential equations. If we are working with BSpline surfaces, such systems do not in general have closed form solutions, and therefore we need to employ numerical methods. The most robust and computationally efficient approaches solve for a discrete approximation of the problem by using piecewise linear approximation of the input surface, typically a triangulation. In the following, we describe a fast and robust algorithm that yields a mapping to minimize the discrete conformal distortion, the least squares conformal mapping (LSCM) algorithm, developed by Levy (2002).

Consider a mapping from a domain in 2D parameter space to the surface S, $X: (u, v) \to R^3$; if X is conformal, it must obey the Cauchy–Riemann conditions, namely, $\mathbf{n} \times X_u = X_v$. It is convenient to use the complex plane to represent the-components of X, that is, $X = x + iy$. Using this notation and applying the Cauchy–Riemann criterion to the inverse mapping of X, say $\mathcal{U}$, we get $\mathcal{U}_u + i\mathcal{U}_v = 0$.

(continued)

BOX 10.3 (continued)   SURFACE DEVELOPMENT VIA MESH PARAMETERIZATION

The key notion in Levy et al.'s approach is to determine the mapping $\mathcal{U}$ that will minimize the deviation from conformality, measured as the sum of the weighted conformal deviation $\|\mathcal{U}_u + i\mathcal{U}_v\|^2 A_i$ over all triangles forming the approximation of the 3D surface. The weighting factor A_i is the area of the triangle T_i in the triangulation $\mathcal{T}$. After some algebraic manipulations, this results in a surprisingly elegant unconstrained minimization problem expressed as

$$\min \| \mathcal{A}x - \boldsymbol{b} \|^2 \tag{10.1}$$

where
  $\| * \|$ denotes the norm
  x is the vector of the unknown (u, v) parameter values corresponding to the vertices of the triangulation

The conformal mapping is invariant with respect to translations in the plane, and the aforementioned minimization is under-constrained; in order to bring the solution close to isometric, we can specify the parametric coordinates of two or more vertices (these vertices are then said to be pinned); in our case, we select two vertices that are far apart, set the coordinates of the first one at $(0, 0)$, and of the other one at $(L, 0)$, where L is the geodesic distance between the vertices in R^3. Thus, x is a vector, $x = \left(U_{1,f}^T, U_{2,f}^T\right)^T$, where subscripts 1, 2, and f denote parameter u, parameter v, and free vertices, respectively. Since there are totally n vertices, and we pinned two, U has $2(n - 2)$ entries. Further,

$$A = \begin{bmatrix} M_{1,f} & -M_{2,f} \\ M_{2,f} & M_{1,f} \end{bmatrix} \quad \text{and} \quad \boldsymbol{b} = \begin{bmatrix} M_{1,p} & -M_{2,p} \\ M_{2,p} & M_{1,p} \end{bmatrix}$$

where
  the subscript p refers to the two pinned vertices
  the entries m_{ij} of the sparse matrices $M_{i,j}$ refer to weights corresponding to the conformal error at the corresponding vertex in the surface approximation, computed as follows:

$$m_{ij} = \begin{cases} \dfrac{w_{ji}}{\sqrt{2}A_i} & \text{if vertex } j\{1,2,3\} \text{ belongs to triangle } i \\ 0 & \text{otherwise} \end{cases}$$

and the weights W_{ji} for triangle j are computed from

$$W_{j1} = (u_{j3} - u_{j2}) + i(v_{j3} - v_{j2})$$

$$W_{j2} = (u_{j1} - u_{j3}) + i(v_{j1} - v_{j3})$$

$$W_{j3} = (u_{j2} - u_{j1}) + i(v_{j1} - v_{j1})$$

**BOX 10.3 (continued) SURFACE DEVELOPMENT
VIA MESH PARAMETERIZATION**

Using this, the LSCM algorithm can be summarized by the following pseudocode:

Algorithm *Least Squares Conformal Mapping*

```
Inputs: Surface S, tolerance: τ
T = triangulate(S, τ); //approximate S by triangles,
within the specified tolerance
V₁, V₂: = the two diametric (i.e., farthest) vertices on
boundary of T;
L = geodesic_distance(V₁, V₂, T);
(u₁, v₁): = (0, 0); (u₂, v₂): = (L, 0); //pin the vertices
corresponding to V₁, V₂.
For each vertex in T {compute mᵢⱼ;}
parameterization: = Solve linear system corresponding to
the minimization problem (1)
Output the parameterization;
  //(uji, vji) are the coordinates of vertex i of triangle j
  //in the development of the 3D surface S
```

---

this process is a 2D profile of the last bottom. All surface development algorithms are essentially iterative methods that converge to some (and often local) minimum. Consequently, the surface development function does not have an analytical inverse. The 3D curve that defines the boundary of the last bottom surface is called a last bottom featherline; its image under surface development on the 2D plane is called the last bottom profile (Figure 10.19).

FIGURE 10.19 A last model showing the last bottom surface, the last bottom featherline (in orange), the last bottom centerline, and the surface development of the last bottom onto the bottom plane (in red).

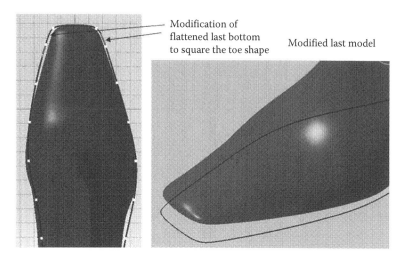

Modification of
flattened last bottom
to square the toe shape

Modified last model

FIGURE 10.20 Modification of developed profile of the last bottom. Mapping this modification back to modify the last model requires complex computations, possibly involving solving on an optimization problem.

It is again customary to control the last bottom profile shape for reasons of fit; also, designers often modify its shape for aesthetic reasons. For example, Figure 10.22 shows the AKA64/WMS* guideline followed by many last designers for flattened last bottom; its details are described in Adrian (1991) and Luximon (2009). Thus, after performing an initial design of the shoe last, the designer may compute the last bottom profile, and find that it violates the guidelines of AKA64. If the designer then decides to modify the bottom profile in order to conform to the guidelines, such changes must be mapped back to equivalent changes on the last bottom featherline (and consequently to the shape of the last). This mapping is nontrivial, due to the reasons mentioned earlier, and again some iterative algorithms are used to implement this. Mathematically, such an algorithm may be set up to move a set of prescribed points on the featherline by the minimum amount, so that the resulting last bottom surface would flatten to best fit the new last bottom profile.

The preceding examples show many, although not all, shape operations useful for last design. We now switch our focus to the determination of whether a designed last conforms to the required size constraints.

10.3.2.4 Last Measurements

Size and shape controls on shoe lasts are imposed to ensure a certain amount of comfort to any person who uses the shoe of the best fitting size made by this last. There are several different standards that prescribe how a standard last should

---

* The AKA64 system was introduced in Germany in the 1960s as a standard for lasting children's shoes out of ergonomics concerns; later, the system was extended for adult last measurements, into what is now referred to as WMS, or AKA64/WMS system.

be measured. Prominent among them are the Shoe and Allied Trade Research Association (SATRA) measurement system popular in the United Kingdom, the U.S. standard (Adrian 1991), etc. In a modern globalized context, if a CAD system provides for the design of footwear, it is likely that it provides the tools to apply any of this set of standard measurement systems. For our purpose, we shall focus on one particular system, the U.S. standard measurement system. We shall also see how to adopt the AKA64/WMS system into this standard.

We now return to the theme touched briefly at the outset: The measurements made on any physical object depend on first locating it with respect to the measurement device. In particular, measurements on a shoe last depend on accurate identification of some landmark points, lines, and planes on the last. We first look at how some of these landmarks are identified, and next at how these are used to generate measures on the last. Finally, we look at how the measurements are used.

10.3.2.4.1   Last Measurement Landmarks

Geometrically, there is some amount of ambiguity about precise definitions of the important measurement landmarks of a shoe last; this is not necessarily a reason for concern, since all mechanical objects have some degree of inaccuracy in establishing the locating data on their surface. Let us first give some definitions of the commonly used terms in last measurements. It will be useful to assume that the last bottom featherline, which is the curve where the body of the last surface joins the last bottom surface, is distinguishable. This is reasonable for most lasts, as there is a clearly visible sharp edge except in many cases in the inner arch region (see Figure 10.21a). Although the featherline is a 3D curve, it is planar in the heel region; so it is possible to accurately fit an imaginary plane, the *heel plane*, to it in this region. We shall further assume that a plane, perpendicular to the heel plane, can be

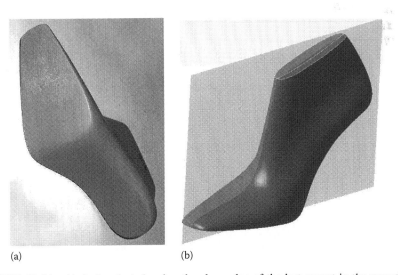

(a)   (b)

FIGURE 10.21   (a) A shoe last showing the clear edge of the last except in the smoother inner arch region. (b) The last center plane.

discovered to define a center plane for the back part of the shoe, and indeed, we shall call this the *last center plane* (see Figure 10.21b). A simple process for identification of this plane is as follows:

Step 1: Identify the back part of the last, where its geometry is symmetric.

Step 2: Intersect a series of planes, parallel to the heel plane, with the surface obtained in Step 1.

Step 3: Compute an axis for each intersection curve in Step 2, for example, by using a medial axis algorithm.

Step 4: Sample enough points on each axis, and fit a plane through these points that is perpendicular to the heel plane. We shall call the plane obtained by this process as the *last center plane*.

Finally, it will be useful to transform the last into a position that we shall call the *normal position*. In normal position, the last center plane is vertical, and the heel is raised from the plane of repose, or the *x*–*y* plane, so that the heel point (defined in the following) is at a height equal to the designated heel height of the last. In normal position, the last touches the *x*–*y* plane at exactly one point on the last bottom surface. We need a few more definitions.

The back curve: The intersection curve of the back part of the last body surface and the last center plane is called the back curve. Given a shoe size, style, and heel height, there are standard shapes for the back curves that are prescribed for best fitting different populations. Ordinarily, manufacturers use cardboard cutout templates to check that a fabricated last conforms closely to the required back curve profile.

The back point: The end point of the back curve is the point of intersection of the back curve with the last bottom featherline. This is the back point, also sometimes called the *heel point*.

Last bottom centerline: The intersection curve of the last center plane with the last bottom surface is the last bottom centerline. Typically, this curve is further divided into three regions—the region toward the front of the last is the toe region, the region at the back end is the heel region, and the bridging region is called the shank region.

The last bottom length: The length of the last bottom centerline curve.

The toe point: The intersection point of the last center-plane with the foremost part of the last featherline is the toe point. It may be noted that under the AK64/WMS system, the toe point is in fact a point on the last bottom centerline, at a distance equal to the ideal foot length measured along this curve from the back point. Since the last length is obviously longer than the foot, so the toe point is in the interior of the front region of the bottom surface.

The ball points: The ball points identify the points on the last that correspond to the outer metatarsal-phalangeal joints. In a sense, these are the points where the foot is the widest. A simple method to identify these points on the shoe last is to find the tangential points on planes parallel to the last center plane.

The tread line: The tread line approximates the line of flexion on the sole during normal walking. On the last, it is identified as a line joining the two points where the shoe last bottom is at its widest (with respect to the center plane). Consider a pair of planes, parallel to the last center plane and tangent to the last bottom featherline (one on each side). The points of tangency will define the end points of the tread line. The intersection of a vertical plane passing through these two points and the last bottom surface gives the tread line.

The tread point: When the last is held in normal position, the point on the tread line that just touches the x–y plane is the tread point. Ideally, the tread point should lie on the last bottom centerline.

10.3.2.4.2 Computing Characteristic Last Measurements

Figure 10.22 shows a last model with several representative measurements marked on it. In production settings, there are several tools, jigs, and templates that are used to perform the measurements, many of which are described in Adrian (1991). Therefore, if a CAD system following the current practice should provide functions that simulate the measurements made by these devices, an alternate approach would be to create a new set of measurements that have, say, a better ergonomic and economic justification. But in this chapter, we focus on the essence of such measurements, and comment briefly on the algorithmic tools that are used to implement these; and we do so by looking at a few examples.

Last length: This is the length of the last bottom centerline curve. This is measured from the back point to the toe point. The length of an arbitrary curve is computed fairly easily in any scenario. In the case that the last bottom centerline is actually sketched out during the design phase, this curve will be represented and stored internally by the CAD system in terms of its selected data structure. For conventional

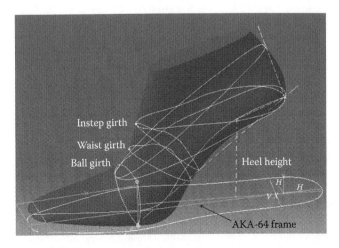

FIGURE 10.22   U.S. last measurements illustration.

solid modelers, this is usually a piecewise polynomial parametric format called a BSpline curve (Piegl 1997). Suppose we represent the curve in vector polynomial form as $r(t)$, where $a \leq t \leq b$; $r(a)$ is the back point, and $r(b)$ is the toe point. Then the last length is given by

$$L = \int_a^b \| r'(t) \| \, dt$$

which can be computed numerically.

In most cases, however, the last bottom centerline must be computed as the intersection of the last bottom surface and the last center plane. Since the surface is stored as a biparametric vector BSpline function, $S(u, v)$, its intersection with the plane must be computed numerically (Patrikalakis 2002). This yields a set of points on the curve, which are then used to fit an interpolating BSpline curve; see Piegl (1997). Thus, the computation of the arc length requires three sets of numerical computations: a surface–plane intersection, an approximating BSpline fitting, and finally, a curve length computation. However, typically the error at each stage can be controlled under 1e-6, so accuracy is not an issue. Finally, we note that some last CAD systems may just use tessellated surface models as their underlying data structure. In such cases, computing curve lengths is reduced to a large number of very efficient plane–triangle intersections, followed by a summation.

Back height: This is the height of the back point of the last; once the last is located in the normal position, we can directly lookup this value from the z-coordinate of the back point.

Heel height: This is the vertical distance (i.e., the z-coordinate in normal position) of the *heel point*, which is located on the last bottom centerline at a fixed Euclidean distance from the back point (e.g., on U.S. ladies size 6B shoe, the heel point is 1.75 in. from the back point). The weight of the person wearing the shoe acts directly down along a line through the heel point. Thus, finding the heel point requires us to solve for x in the equation: $\int_a^x \| r'(t) \| \, dt = 1.75$, where $r(t)$ is the equation of the last bottom centerline and $r(a)$ is the heel point. Once the heel point is known, computing the heel height is trivial.

Girth: There are several girth measurements that are made on physical lasts. These are made by using a flat, 0.25 in. wide flexible tape that is pressed along the surface of the last in such a way that it passes through the required landmarks. For example, to measure the *ball girth*, this tape must be wrapped tightly around the shoe last such that it holds both the *ball points* along the same edge. Mathematically, this is equivalent to computing the shortest path (this is called a geodesic) along the surface between the two ball points.

Let $C(s)$ be a curve constrained to remain on a given surface $S(u, v)$. If $C(s)$ is a geodesic, its curvature on the tangent plane of $S(u, v)$ should be zero everywhere; this property can be used to generate the governing equations of a geodesic curve as a set of two second-order differential equations:

$$u'' + 1/2E\left(E_u u'^2 + 2F_u u'v' + G_u v'^2\right) = 0$$

and

$$v'' + 1/2G\left(-E_v u'^2 + 2F_v u'v' - G_v v'^2\right) = 0$$

where E, F, and G are the inner products, $E = \langle S_u, S_u \rangle$, $F = \langle S_u, S_v \rangle$, and $G = \langle S_v, S_v \rangle$, and the subscript denotes the partial derivative with respect to the indicated parameter. Given the required boundary conditions (e.g., coordinates of the two end points), this system can be solved numerically to yield solution points along the geodesic curve joining these two points; these points can then be used to fit a BSpline that approximates the required geodesic curve. Thus, it is not difficult to compute the ball girth curve, which is the geodesic curve on the surface of the last joining the two ball points. Once the curve is computed, the same methods as discussed earlier can give us the ball girth length.

We note in passing that in almost all practitioners' publications about lasts, the illustrations indicate that the profile view of these girth curves are straight lines; this may be a mere simplification, since geodesic curves on general surface are not necessarily planar. Or it is possible that using planar section to get an approximate girth curve may yield a girth value within acceptable error. Under this interpretation, finding the girth curves requires solving an optimization problem. Consider the case of ball girth. We wish to find the plane passing through the two ball points whose intersection with the last surface is the shortest length curve among all such possible planes.

Finally, we close this discussion by considering how the girths can be computed when using CAD systems that use tessellated models. Fortunately, even in this case there are several efficient algorithms for computing the shortest path curves between given pairs of points (Kanai 2001; Kimmell 1994).

AKA64-WMS measures: The WMS system gives guidelines on the lengths and angles between a set of geometric objects drawn on top of the development of the last bottom surface. In Figure 10.22, this structure is shown in green lines. This structure can be generated fairly easily once we have (a) the surface development of the last bottom profile and (b) the mapping of a few measurement points onto the 2D profile, including the back point, the heel point, the toe point, and the last bottom centerline. This mapping is obtained trivially, irrespective of which numerical approach we use to perform the development. For details of several approaches, see Hinds (1991), Azariadis (2000), and Tam (2007). We note in passing that the flattening of the last bottom surface is a constrained problem, since we have the additional restriction that the last bottom centerline must map onto a straight line after the development procedure (this is indicated by the WMS figures).

10.3.2.5 Last Grading

The prototypes for a new shoe design are based usually on a standard size (e.g., U.S. ladies size 6B). Once the design has been tested and approved, all the sizes required for the target market must be designed and produced. For off-the-shelf shoes, the size ranges are discrete step sizes along two variables: the length and the width. Grading is the process by which the models for all these sizes are generated. The step sizes for length and width vary depending on the measurement standard adopted (see Chapter 9 of this book (Luximon 2012) for more details). To simplify our discussion of how such grading can be achieved by using a CAD system, we will limit our discussion to length grades. The methods to handle width grading at any given size can be handled analogously. Suppose we wish to impose the following size grading requirements:

1. The last length changes by 8.47 mm for each full size step (e.g., from size 6 to 7).
2. The ball girth, waist girth, and instep girth change by 6.35 mm for each full size step.
3. The ball-to-heel girth changes by 12.7 mm for each full size step.
4. The ball tread changes by 2.12 mm for each full size step.
5. The wedge angle, heel height, and toe spring remain unchanged across grades.

Obviously, the simplest way to retain the shape and therefore the styling of the last while resizing it is to scale it uniformly along the coordinate axes. But it is evident that the grading operation is not merely a matter of scaling the CAD model along the axes of a Cartesian coordinate frame. For instance, any uniform scaling along the z-coordinate in the normal position would immediately change the toe spring and possibly the heel height. Recall also that while we can construct a fully parametric CAD model of the last (as we have in the example shown in Figure 10.17), the independent variables of the grading process indicated by the size grading requirements (1)–(5) cannot easily be used as parameters controlling the geometry of our CAD model. For example, a numerical method is used to compute the last length for a given model; since such functions are not directly invertible, the length of a curve cannot be conveniently used as a parameter controlling the curve sketch. Therefore, in practical CAD systems, grading can only be achieved by applying a series of shape modification operations to different parts of the last. If we are working within a commercial CAD system, it is possible to automate this using *design tables* or simple application programs that automatically apply a prescribed sequence of operations to the input model.

One can think of a design table as an array of numeric values; each row in the array specifies the values of a set of parameters that instantiate a particular shape/size of the CAD model of a part. By using different rows of data from this table, the same parametric CAD model can be used to generate a family of models of the same shape and different sizes.

On the other hand, it may be difficult to merely use a set of parametric values to achieve grading with acceptable accuracy. Constraints on girth measurements and template curves are not parametric. For example, if the geometry of waist region is changed by moving control points of some curve, then this results in a new value for the waist girth. This girth can be computed accurately once the geometry is fixed. But the reverse is not true: suppose that we prescribe a new value of the waist girth; we now want to know which control points describing the last surface must be moved, and by how much, such that the resulting last shape will have the target waist girth measurement. Obviously, this problem is under-constrained, and, therefore, it is not clear which solution is the best one. The case of template curves (an example is when the grading system prescribes a series of last bottom centerline curves, one for each graded size). Describing a robust algorithm to achieve accurate last grading is beyond the scope of this chapter; we shall restrict ourselves to a relatively brief outline of a possible implementation within a parametric CAD system.

The inputs are the reference last model, a list of standard curves, and a list of target parameter values.

1. Using the ratios of the required parameter values to the corresponding values in the reference model, compute the x-, y-, and z-scaling factors.
2. Apply the nonisotropic scaling factors to the reference model to create an approximate target model.
3. Apply the target parametric values to the appropriate elements of the model geometry and update the model shape, respecting any geometric constraints among the elements.
4. Insert the input standard curves into the appropriate places in the geometric description (i.e., the feature tree), replacing the corresponding elements in the standard model; update the model shape, again respecting all geometric constraints between various elements.
5. Evaluate the modified model by computing the target feature sizes and comparing the actual feature curve shapes with the target shapes. If the measurements conform to the required specifications, report the graded model shape and exit; else using the modified model obtained in Step 5 as the input (i.e., reference model), return to Step 1.

The aforementioned approach is an iterative one; we cannot say with any certainty whether it will converge to an acceptable solution (in which case we may wish to terminate the algorithm, e.g., after some fixed number of iterations). In Step 1, several heuristics can be used to make good initial choices for the scaling; the heel height and the last bottom length requirements can usually give a fairly good estimate on the x-direction scale factor; similarly, the required change in the ball girth can be used to get a good estimate of the y-scale factor. In Step 4, we can use the constrained shape deformation algorithm described earlier, since it aims to make the least change to the last shape while modifying it to conform to the required template curves. With this we conclude our discussion of the last design process, and move on to the other parts of the shoe.

10.3.3   SHOE DESIGN

10.3.3.1   Insole

The insole is usually a sheet of uniform thickness made of fiberboard, synthetic foam, or even leather. Geometrically, since the insole is attached flush to the last bottom surface, generating its model is quite straightforward—it merely requires the use of a surface offset function. A simple example is shown in Figure 10.23. From a CAD operator point of view, computing offsets of surfaces is a fairly well-studied problem. Given an analytical surface $S(u, v)$, a surface that is offset by a distance d from it is given by

$$S^\circ(u,v) = S(u,v) + dS_u \times S_v / \| S_u \times S_v \|$$

The norm $\| S_u \times S_v \|$ in the denominator of the second term is not a polynomial, and therefore the offset surface of a BSpline cannot be represented exactly by a BSpline. Internally, CAD systems compute offset curves by approximating the offset function by a BSpline that remains within a very small tolerance of the analytical value. A difficult issue in computing offsets is that of detecting and removing self-intersections in the offset surface (Rossignac 1986; Jung 2004). However, most modern CAD systems have geometric kernels with robust offset operations, which can handle the relatively fair (with very few highly curved regions) last surfaces without any problem.

10.3.3.2   Midsole, Supports, and Inserts

The midsole plays an important role in many different types of shoes. In women's dress shoes, the midsole provides significant mechanical strength to the shank and the heel regions. Most high-heel shoes get structural strength from a strip of steel, approximately 1 cm wide, which is fixed on one side to the heel, and follows the curve of the last bottom centerline along the shank. The midsole houses this element also. Such midsoles usually are thicker in the region near the heel, and become progressively thinner as they run down the bottom centerline toward the tread line.

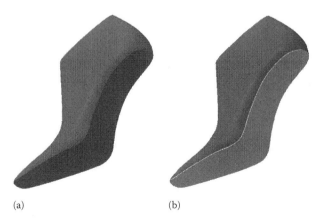

(a)   (b)

FIGURE 10.23   (a) A shoe last model. (b) Surface offset operator on the last bottom surface used to create a 1 mm uniform thickness insole.

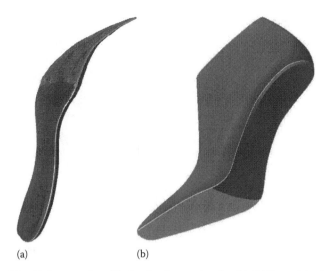

(a) (b)

FIGURE 10.24 (a) Image of a midsole glued to an insole. (b) CAD model of a last model along with an insole and a graduated thickness midsole.

Figure 10.24a shows an image of one such midsole and insole combination. Figure 10.24b shows a CAD model of a similar midsole. In most modern CAD systems, this is not a difficult geometry to construct. One possible technique that may be used is to create a graduated offset of the bottom surface of the insole model. A graduation of the offset is defined as monotonically increasing thickness along the centerline: It has the minimum thickness near the tread line and the maximum thickness near the heel back point. The edge of the midsole is usually chamfered, as can be seen in the image, as well as in the model.

The midsole also features prominently in athletic shoes. In particular, soft, cushioning ethylene vinyl acetate (EVA) midsoles are often seen in running, basketball, or tennis shoes. For such midsoles, essentially the same CAD operation (namely, graduated offset) may be used to define the main shape. Subsequently, a variety of shape deformations or sculpting operations can be applied to create the required geometry, especially along the side walls.

10.3.3.3 Outsole

Outsoles form the outermost layer of the sole assembly on shoes. Outsoles are usually constructed from some type of rubber, for example, thermoplastic rubber (TPR), leather (e.g., in men's dress shoes), or even plastic. The design of the outsole is the outcome of various concerns: Aesthetics are important, as we see more and more aggressive shape patterns on shoe soles; functionality also plays a role, for example, the herringbone patterns on outsoles of the tennis shoes that provide additional traction on clay courts; some outsole designs are also claimed to provide better shock absorption during heel strike; wear resistance and expected life of the shoe dictate the minimum thickness that the outsole must have; and finally, since rubber is relatively dense, the weight of the sole must be controlled by possibly eliminating excess rubber from regions that will not be subjected to high wear rates.

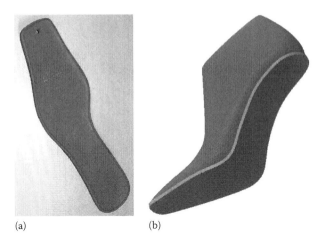

(a)   (b)

FIGURE 10.25   (a) Image of simple outsole. (b) CAD model showing a last and an outsole covering the insole and midsoles.

The geometric variations of the shape elements found on outsole designs are so large that practically every function or operator provided by any CAD system would at some stage be required. In this section, we shall briefly look at a few examples to get a feel for the process of designing outsoles in a parametric CAD system.

We begin with a simple example useful for the design of women's shoes, where the aesthetics are controlled by the upper and the heel, while the outsole is a relatively simple and flat rubber cover for the midsole and insoles. Figure 10.25a shows an image of such an outsole, and Figure 10.25b shows a model of a similar outsole for our ladies pump example. Notice in the model that the outsole actually extends slightly beyond the profile of the last bottom. Therefore, we cannot directly use the bottom surface of the midsole and offset it to generate the outsole. In our example, a possible method is to first extrapolate the midsole surface outward along its entire boundary by the required amount (typically a few millimeters). This operation, called, surface extension, is simple in an analytical CAD system, since we already have a parametric equation for the surface that we wish to extrapolate. Generating the surface extension can be a matter as simple as merely changing the domain of the surface parameters, u and v. The major task is to compute the range and combination of the values of these parameters that will yield the region of the extension. Once the extrapolated surface is computed, the sole can be created by using simple offset operations, followed by some filleting along the edges.

Next, we look at a very common style of men's dress shoe outsole. Figure 10.26 shows two views of our target model.

As in the previous example, the profile of the outsole can be generated by a surface extension of the midsole bottom surface. Since the side walls of our model are devoid of any features, we can generate a solid block by extruding this profile downward, and slightly upward. The basic bottom geometry only requires a few extrusion cut operations to yield a block as shown in right side on Figure 10.27.

FIGURE 10.26   Two views of an outsole for a men's dress shoe.

FIGURE 10.27   Initial steps in generation of the CAD model of the outsole.

Notice that the inside surface of the outsole is hollowed out—this helps in reducing the weight, at the same time giving better cushioning as well as materials savings. Generation of the cutout shapes is fairly straightforward—most CAD systems allow the user to define a simple shape and then repeat it in regular (following arithmetic or geometric steps) patterns. Notice that the inner surface of the outsole must be glued eventually to the midsole and, therefore, should follow its geometric shape; in other words, it is a curved surface. To generate this geometry, the curved surface of the midsole should be intersected with the block-shaped outsole, eventually using this intersection result to perform a Boolean subtraction. Such surface–surface intersections (Patrikalakis 2002) are also computed numerically in modern CAD systems and are quite robust.

The shape features on the outer surface of the outsole are also, in this case, fairly regular and can be generated by using repeated patterns of some simple geometric sketches.

Our final example is a simple sporting shoe outsole, as shown in Figure 10.28. Essentially, the steps in generation of such outsole models are similar to the ones in the previous example, with two significant differences. The first difference is that the side profile of the model and the side walls are complex curved shapes. The second is that the shape features on the outsole bottom are complex and not repeated patterns. For the first issue, depending on the complexity of the sidewall shapes, some powerful surface modeling operations may be required. In our example, however, we can generate this seemingly complex curved sidewall by the use of a variable radius fillet—an operator that is readily available in most modern CAD systems. The computational geometry for creating variable radius fillets (or, in general, variable radius blends) has been studied quite extensively, and the curious reader is referred to Vida (1994) and Lukacs (1998) (Figure 10.29).

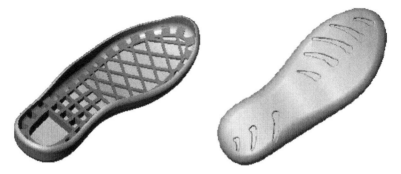

FIGURE 10.28   Two views of an athletic shoe outsole.

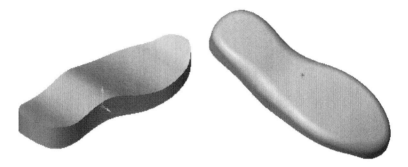

FIGURE 10.29   A variable radius blend operator along the bottom face of the (partially completed) outsole model.

The discussion of details of CAD operators used in and useful for complex outsoles commonly seen in modern shoes is unfortunately beyond the current scope; however, we hope that the earlier discussion has given a flavor of the modeling techniques useful in doing this step of footwear design.

10.3.3.4   Heel Design

In most women's shoes, the heel is not a part of the outsole—rather it is a highly functional component of the design. While there are some extreme creative designs of heels, in most practical cases all heels are merely parametric variations of a few basic styles: Cuban, Hooked, Spanish, Louis, and Dutch (see Figure 10.30).

Due to the relatively simple shapes, heels are perfectly suited for semiautomated design in any modern CAD system. The top surface of the heel should essentially be identical to the bottom surface of the midsole (or outsole, depending on where the heel is attached). A standard shape profile can be stored for each style in a template (in most CAD systems, this is called a design catalog and is useful for designing standard components such as nuts, bolts, etc.). The particular instance of the catalog is created by merely setting the parameter values that control the size of the geometric elements of the selected template. Figure 10.31 shows our familiar ladies pump design, with a Spanish style heel attached to the outsole.

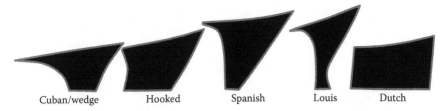

Cuban/wedge   Hooked   Spanish   Louis   Dutch

FIGURE 10.30 Profiles of common women's shoe heel styles. (From HKPC, *Information Handbook for Merchandisers of the HK Footwear Industry*, Hong Kong Productivity Council, Kowloon Tong, Hong Kong, ISBN 962-8040-48-0, 2001.)

FIGURE 10.31 A CAD model with a Spanish style heel.

10.3.3.5 Upper Design

Upper design is one of the most skillful tasks in footwear design; it is also perhaps the most creative. Apart from last design, this is also the function that traditionally has been very strongly integrated with computer-aided manufacturing (CAM) operations. In this section, we shall look at a few typical upper design operations and also briefly introduce the main issues of CAD.CAM integration. To maintain coherence with the main example in our coverage throughout this chapter, the discussion will be limited to women's dress shoes.

The design of uppers is closely tied to the task of pattern making (Anzelc 1994). Here, the term "upper design" shall be used to refer to the CAD operations relating to the 3D model, while pattern design will refer to the task of designing the 2D shapes of leather, cloth, etc., that must be stitched together to form the upper. Figure 10.32 shows two views of a typical stitched upper. The outer shell of upper is made up of a few pieces of leather (and also some other accessories, such as zippers, buckles, etc.). Often an upper may be constructed out of two to four layers of materials, including the outer shell, one or two layers of inner shells or lining, usually made of a softer fabric, and possibly a sheet of thermosetting resin sandwiched in between these shells. An example of this last item is the toe puff, which is inserted in the front, or the vamp, region of the upper. After the upper is pulled tightly over the last during manufacturing, a hot roller is used to set the toe puff into taking the shape of the last; this helps the shoe retain its designed shape over its life.

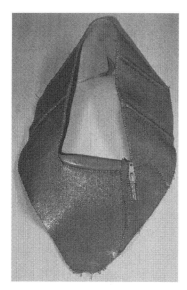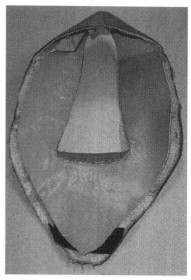

FIGURE 10.32   Views of outer and inner sides of a stitched shoe upper.

We shall restrict our coverage to three stages of the CAD of uppers. The first stage is the style and layout design, where we specify the overall shape of the upper on top of the last. The second stage is the surface subdivision; this specifies the stitch lines and therefore the number and shape of the pieces of leather that will be required to make the upper. The third stage is the pattern design, which determines the shape of the 2D patterns that will be required to make the stitched upper.

The style and layout design is basically required for the designer to sketch on the surface of the last the lines that will define the shoe. This process depends on the style of the shoe, as well as the aesthetic and functional concerns. Figure 10.33 shows the basic styling of the most common styles of ladies shoes. The most important CAD function in this stage is that of drawing curves on top of the 3D surface of the last. Mathematically, as well as from the user-interface point of view, this is a difficult operator to provide in CAD systems. The user typically is looking at a computer monitor, which shows a 2D projection of the 3D model. Suppose that we wish to add a point on the surface, through which the curve will pass. An intuitive method to add a point is by providing the input via a mouse click. Now the mouse click provides two pieces of data: the x and y coordinates of the click point on the projection plane of the current view of the last. The CAD system can project this point along the normal of the projection plane onto the last surface, or it can map the input point to the nearest point on the last surface (this is called a normal projection). In either case, the CAD system must solve a set of algebraic equations numerically to compute the result. Often, the mapped image of the input point is not quite what the designer expects, and some more coordinate manipulation is required to get the point where the designer wants. For a good discussion on drawing curves on surfaces, the reader is referred to Nikolski (2009) and Belchich (2010). Another issue

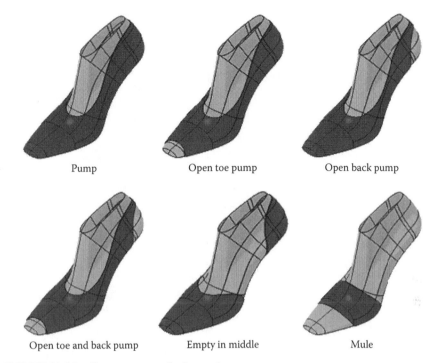

Pump   Open toe pump   Open back pump

Open toe and back pump   Empty in middle   Mule

FIGURE 10.33   Common women's shoe styles.

with sketching curves on 3D nonsymmetric surfaces such as the shoe last is that it is difficult to manage symmetry constraints (e.g., between the left and right sides of the back portion in a pump). This problem is usually solved by a different technique: The user draws a profile of the curve on a 2D sketch, and then extrudes this profile along a direction perpendicular to the sketch plane. Intersecting this surface with the shoe last surface yields the desired sketched curve. An example of this approach is shown in Figure 10.34. The computational geometry of this type of operation is fairly well studied: The equation of the extrusion surface is trivially derived from the equation of the sketched curve. The intersection curve requires a surface–surface intersection, which is solved numerically. Finally, the surface curve is generated by interpolating a dense enough sample of the intersection points.

It is not difficult, in any CAD system, to set up a large number of templates for the standard shoe styles, and perhaps also for a set of standard shapes such as straps. The designer can then directly apply these standard templates onto the last surface, and modify the boundaries or parameters of the templates to create the desired design effect.

The final step in this process is to add thickness equivalent to the leather's thickness, which is easily done by applying the surface offset operators that we have discussed earlier.

Recall our earlier discussion on surface development. If we attempt to develop, that is, flatten, the entire upper surface as designed earlier, obviously the resulting mapping will have very large distortions. Such distortions can be avoided in

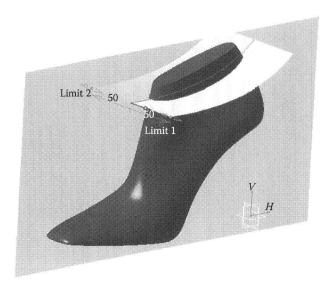

FIGURE 10.34 Using an extruded surface to generate design curves on the last. The red curve on the last center plane is extruded to form a surface, which is intersected with the last to yield the required design curve.

several different ways. The first is to subdivide the upper surface into patches, such that each patch can individually be developed with little distortion. It is also possible to use larger single patches, but with some strategically introduced cuts, such that stitching along the cut allows the surface to assume a curved 3D shape. Such cuts are called darts, and textile/shoe designers often use standard locations for darts in many designs. Another technique is to introduce a cutout in some high distortion region of the upper and use a more stretchable material in this hole, or gusset, for constructing the upper. Geometrically, the operations used for the stage of the upper design are essentially the same as the ones used in the previous stage.

The patterns that we design on the 3D last surface can be used to derive the actual 2D shapes of leather that will be needed to construct the shoe. The main operation at this stage is to map each designed piece onto its planar development. Surface development is a traditionally important problem with many engineering applications, see, for example, Hinds (1991), Azariadis (2000), Tam (2007), Floater (2005), and Levy (2002); Box 10.3 sketches an algorithm for one approach. Following the flattening, we also need to add additional materials along the entire borders of the pieces to account for the following: (a) along the boundary curves where two pieces will be stitched together, there must be some extra material that will be hemmed inside the stitch; (b) along the edges that will remain unstitched but will be pulled over the last and fixed to the last insole on the bottom of the shoes, we must provide ample allowance to create a strong bond; (c) the material of the upper, for example, leather, also tends to stretch when the upper is pulled over the last. Usually, at the stage of surface development, it is difficult to apply boundary conditions relating to the pulling forces, and therefore, we cannot account for this stretching. This allowance is usually used to adjust the allowances along the boundaries of the patterns.

Once the 2D pattern shapes are determined, several CAM functions can be implemented. For example, it is possible to directly cut profiles of all patterns into leather or cardboard by the use of a 3D plotter that uses a knife edge (or laser) instead of a pen. The cardboard cutouts are useful as templates for manually cutting patterns out of leather or cloth raw stock. Alternatively, the pattern shapes can be used to fabricate 2D steel-rule cutting dies that can be used in stamping machines to mass produce pattern pieces. There are also laser-operated pattern-cutting machines that are integrated with vision systems. These machines have planning software that can process the image of a raw hide, determine the shape and size of the hide, and, using software for nesting, determine the orientation and position of the largest (or near largest, since no optimal nesting algorithms exist) number of patterns that can fit on that hide.

10.3.3.6 Accessories

Once the main design of a shoe is completed, the designer may also choose to add several accessory components to the model. These include laces, fasteners, and ornaments. The geometric variation on these components is almost limitless, and therefore almost any CAD operation provided by any system may be required to design these models. Any footwear design company will typically maintain large libraries of such components, to allow designers to quickly browse through earlier, similar designs for reuse. Most CAD systems provide for users to create libraries, or catalogs, of such components or assemblies. Figure 10.35 shows a few examples of metal accessories. A component or subassembly model of these accessories can easily be incorporated at the appropriate location on the shoe model. The steps for this process are to load the models as components of an assembly model, and to specify the positioning constraints between the shoe model and the accessory model (e.g., coincidence of pairs

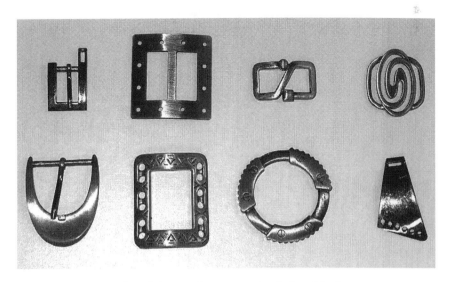

FIGURE 10.35   Just a few of the countless accessories designed for footwear.

FIGURE 10.36 A sandal style shoe with a buckle assembled to the ankle strap.

of planes, points, or axes selected from the respective models). Figure 10.36 shows a simple example of a shoe design with a strap carrying a simple buckle.

10.3.3.7 Industrial Design

The final topic that we shall briefly touch upon in this chapter relates to the rendering of the images of the 3D design models. It is often desirable to get as realistic a view of the 3D model as possible, to the extent that the computer model may be used for replacing the requirement of building a physical mock-up prototype altogether, during the early stages of product design. Apart from the improved product design communication, photorealistic rendered images of the model can also be useful in generating marketing and other publicity materials. Many CAD systems provide very sophisticated methods to generate such rendered views. We will briefly mention a few of these techniques. The simplest technique used in controlling the images of the model is the control on the color and texture of the surfaces. A large library of texture maps corresponding to various material types are provided in CAD systems. Almost all of these libraries can be easily extended by the users, who can directly add images of any material they wish to the library of textures. When applying a texture, the system generates a tiling of the texture image on the surface, with special functions that create seamless boundaries of the repeating tiles.

Another useful tool provided by rendering engines is an accurate and effective control of the lighting. The user may specify diffuse, specular, and directed lighting, associating different intensities with each, to create very realistic renderings of the model and its shadows under various conditions. Figure 10.37a shows a 3D CAD model of a shoe; Figure 10.37b shows a rendering of the same model with a different material and lighting effects.

It is useful to have some background about the two common prevalent techniques of specifying the geometry of complex shapes in 3D space: *tessellated surface models* and *solid models* using algebraic boundary representations. Most footwear CAD systems use one or the other, and this has strong implications on how the shape manipulation for various CAD functions is achieved and to some extent, this also reflects upon the system capabilities.

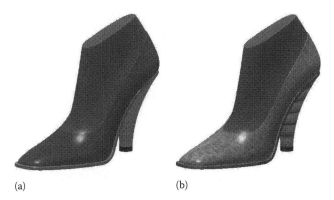

(a)   (b)

FIGURE 10.37   (a) CAD model of a shoe. (b) The same CAD model rendered with a leather material applied to create a different texture for realistic rendering.

10.4   DISCUSSIONS

In this chapter, we described the basic issues in CAD modeling of footwear. All steps of the design process were described, and examples were shown using several different CAD systems. At various stages, the algorithms that are used in the current state-of-the-art are not necessarily optimal. Therefore, apart from giving a general reference to readers about the process of footwear CAD, we hope that the chapter also opens the door to future research on these problems, leading to more powerful functions.

A topic that we did not cover in much depth in this chapter is the connectivity of CAD and CAM. Design for manufacturability is very important in modern companies, and therefore, much CAD functionality is linked or constrained by manufacturing concerns. In this area also, there are several interesting and open problems that the interested reader can pursue.

EXERCISES AND PROJECTS

Note to the reader: The following exercises are for advanced readers. They are formulated in a fashion suitable to be used as projects for a postgraduate level CAD applications course.

10.1 In many mechanical CAD applications, it is desirable to maintain a G^2 level continuity between neighboring surface patches. Do you think this level of continuity is required for the design of shoe lasts?

In constructing your answer, consider the following aspects:

(a) The degree of smoothness (i.e., surface continuity) of a last has perceivable effect on the comfort level of the corresponding shoe. (*Hint*: how would you design an experiment to test this?)

(b) Assume that the boundaries of the surface patches on the last are not aligned with the boundaries of the patterns on the surface of the shoe outer. What do you think is the effect of the level of continuity of the last on the surface of the shoe outer?

10.2 Suppose that a CAD system using a triangulated surface mesh is being used to design and maintain the model of a shoe last. Derive the discrete form of the equations for G^0 and G^1 continuity at the vertices and edges where multiple mesh elements meet.

10.3 [Mini-project] Several excellent open source software systems provide support for shape and surface manipulation functions, for example, MeshLab and CGAL. Use any standard CAD software to create 3D surface patterns for a typical shoe. Convert the surface into its triangulated approximation (e.g., you can simply output the shape as an STL file).

(a) Write a simple program incorporating the CGAL parameterization libraries to create surface development of the pattern. Compare the 3D surface and its 2D development in terms of (i) the total area and (ii) the maximum area distortion of a triangle.

10.4 [Mini-project] The CAD model of a shoe last requires several girth measurements to conform to the corresponding ergonomic standards. Let us consider the example of the ball girth. Suppose that we are given a CAD model of a shoe last, with fixed (given) ball points. Develop a simple mechanism to measure the ball girth based on this input.

(a) [Approach 1] Develop an algorithm to measure the ball girth based on the minimum length *planar* cross section of the last passing through the given ball points.

(b) [Approach 2] Develop an algorithm to measure the ball girth based on the geodesic curve passing through the given ball points.

(c) Consider the traditional (manual) method of using a standard flexible tape to measure girth values on physical models of lasts during design. Do you think the algorithm in part (a) or part (b) better approximates this traditional method? Give reason for your belief. Describe in detail how you can set up an experiment to test the validity of your answer.

10.5 Revisit our algorithm for constrained shape deformation. It is based on the approach developed in Hu (2001). We compute the change in position, $\delta_{i,j,k}$, for each control point $P_{i,j,k}$, of our input surface, using a least squares formulation that minimizes the sum of squares of the displacements, subject to the constraint that the input point moves to the prescribed target position. This constrained optimization problem is converted to an unconstrained one by using a Lagrangian formulation. Suppose that our input surface, for example, the surface of the shoe last, is composed of a set of surface patches, which must also obey other constraints across their shared boundaries. For example, the tangential angle between the last bottom centerline and the back curve may be restricted to some fixed value even when we change, say, some back section of the last.

(a) Will such constraints be satisfied by the resulting last geometry after it has undergone some shape deformation according to our algorithm? Give reason.

(b) Will a surface continuity constraint (e.g., G^1 continuity across the common boundary of two patches defining some portion of the last shape) be respected by our shape deformation operation? Give reason.

(c) Suppose that we use our algorithm to modify the shape of some region of the last near the waist; under what conditions will such a change affect the ball girth? Suppose now that we wish to modify the shape, but keep the ball girth constant, suggest a modified version of our algorithm that can achieve this additional constraint.

10.6 Develop a method to generate models of shoe insoles, given the last bottom surface as the input.

(a) [Approach 1] In this part, we shall assume that you are using a commercial CAD system. Further assume that you are given the following inputs: (i) a CAD model of the last surface; (ii) a trimmed surface that is based on the last model, clearly identifying a bounded, possibly composite surface (sometimes called a joined-surface) as the last bottom; and (iii) the last bottom centerline.

First, develop the sequence of CAD operations (e.g., surface selection and surface offset) provided by your CAD system that you can use to create the geometric model of an insole based on the inputs. Note that the thickness of the insole can vary as we move along the centerline. For simplicity, assume that this thickness is measured along the outward normal of the last bottom surface. Further, assume that the thickness of the insole remains constant in any cross section of the insole generated by the plane formed by a point on the featherline and the outward normal of last bottom surface at that point.

(b) [Approach 2] In this part, you will develop the algorithm to generate the insole model given the last bottom surface as a triangulated model, as follows. Your method should be based on the following approach. First, generate a large sampling of (randomly distributed) points on the last bottom surface; next, develop a function that maps each sample point to the appropriate offset point; then use a ball-pivoting algorithm to generate the offset surface across all the offset points of the point cloud created in the first step.

REFERENCES

Adrian, K. C., 1991, *American Last Making*, Shoe Traders Publishing, Arlington, MA, ISBN: 999408836X.

Anzelc, D., 1994, *Practical Pattern Making*, Shoe Traders Publishing, Arlington, MA, ASIN: B0006QLT4Y.

Azariadis, P. N., Aspragathos, N. A., 2000, Geodesic curvature preservation in surface flattening through constrained global optimization, *Computer-Aided Design*, 33, 581–591.

Belchich, M., 2010, sketch based design of 2D and 3D freeform geometry, MSc thesis, Department of Computer Science, Technion, Israel.

Bernardini, F., Mittleman, J., Rushmeier, H., Silva, C., Taubin, G., 1999, The ball pivoting algorithm for surface reconstruction, *IEEE Transactions on Visualization and Computer Graphics*, 5, 349–359.

CGAL. Computational Geometry Algorithms Library, http://www.cgal.org

Coquillart, S., 1990, Extended free-form deformation, a sculpting tool for 3D geometric modeling, *Proceedings of SIGGRAPH'90*, in *Computer Graphics*, 24 (4), 187–196.

Dey, T. K., 2007, *Curve and Surface Reconstruction: Algorithms with Mathematical Analysis*, Cambridge University Press, Cambridge, U.K.

Floater, M. S., Hormann, K., 2005, Surface parameterization: A tutorial and survey, in *Advances in Multiresolution for Geometric Modelling*, N. A. Dodgson, M. S. Floater, and M. A. Sabin (eds.), Springer-Verlag, Heidelberg, Germany, pp. 157–186.

Guo, Li, Ajay Joneja, A., 2004, Morphing-based surface blending operator for footwear CAD, *Symposium on Computational Geometry, Design and Manufacturing, 2004 ASME IMECE Conference*, Anaheim, CA, November 13–19.

Hinds, B. K., McCartney, J., Woods, G., 1991, Pattern development for 3D surfaces, *Computer-Aided Design*, 23 (8), 583–592.

HKPC, 2001, *Information Handbook for Merchandisers of the HK Footwear Industry*, Hong Kong Productivity Council, Kowloon Tong, Hong Kong, ISBN 962-8040-48-0.

Hu, S. M., Zhang, H., Tai, C. L., Sun, J. G., 2001, Direct manipulation of FFD: Efficient explicit solutions and decomposible multiple point constraints, *The Visual Computer*, 17, 370–379.

Jung, W., Shin, H., Choi, B. K., 2004, Self-intersection removal in triangular mesh offsetting, *Computer-Aided Design and Applications*, 1 (1–4), 477–484.

Kanai, T., Suzuki, H., 2001, Approximate shortest path on a polyhedral surface and its applications, *Computer-Aided Design*, 33, 801–811.

Kimmel, R., Amir, A., Bruckstein, A. M., 1994, Finding shortest paths on surfaces, in *Curves and Surfaces in Geometric Design*, Laurent, Le. Méhauté and Schumaker (eds.), AK Peters/CRC Press, pp. 259–268.

Levy, B., Petitjean, S., Ray, N., Maillot, J., 2002, Least squares conformal maps for automatic texture atlas generation, *ACM Transactions on Graphics, Proceedings of the ACM SIGGRAPH 02*, San Antonio, TX, Vol. 21 (3), pp. 657–664, July 2002.

Lukasc, G., 1998, Differential geometry of G^1 variable radius rolling ball blend surfaces, *Computer Aided Geometric Design*, 15 (6), 585–613.

Luximon, A., Luximon, Y., 2009, Shoe-last design innovation for better shoe fitting, *Computers in Industry*, 60 (2009), 621–628.

Luximon, A., Luximon, Y., 2012, Shoe-last design and development, Chapter 9 in *The Science of Footwear*, R. Goonetilleke (ed.), Taylor & Francis Group, London, U.K.

MeshLab. meshlab.sourceforge.net, 3D-COFORM project.

Nikolski, M., Elber, G., 2009, Sketch based design of 2D and 3D freeform geometry, *Computer-Aided Design and Applications*, Reno, NV.

Patrikalakis, M. P., Maekawa, T., 2002, *Shape Interrogation for Computer Aided Design and Manufacturing*, Springer Verlag, Heidelberg, Germany.

Piegl, L., Tiller, W., *The NURBS Book*, 2nd edn., Springer, New York, 1997.

Pivecka, J., Laure, S., 1995, *The Shoe Last: Practical Handbook for shoe Designers*, Pivecka Jan Foundation, Slavicin, Czech Republik.

Rossignac, J. R., Requicha, A. A. G., 1986, Offsetting operations in solid modeling, *Computer Aided Geometric Design*, 3 (2), 129–148.

Sederberg, T. W., Parry, S. R., 1986. Free-form deformation of solid geometric models, *Proceedings of SIGGRAPH'86*, in *Computer Graphics*, 20 (4), 151–160.

Tam, A., Joneja, A., Tang, K., Yao, Z., 2007. A surface development method with application in footwear CAD/CAM, *Computer-Aided Design and Applications*, 4 (1–4), 67–77.

Vida, J., Martin, R. R., Várady, T., 1994. A survey of blending methods that use parametric surfaces, *Computer-Aided Design*, 26, 341–365.

11 High-Heeled Shoes

Makiko Kouchi

CONTENTS

11.1 INTRODUCTION

According to Linder and Saltzman (1998), wearing high heels spreads from the upper classes to working-class women after the latter part of the nineteenth century. Medical scientists have warned about the health hazards of high-heeled shoes for 250 years. It has been said that wearing high-heeled shoes has the following effects:

1. Puts weight on the toes
2. Changes the posture—the wearer's body is thrown forward to maintain balance
3. Increases lumber lordosis and anterior pelvic tilt
4. Increases the activities of leg muscles to maintain balance
5. Increases the activity of muscles of the lower spine
6. Changes the gait

Therefore, high-heeled shoes induce adverse effects such as the following:

7. Muscle fatigue and pain
8. Low back pain
9. Corns and bunions
10. Deformation of the toes
11. Increased risk of ankle sprain

However, high-heeled shoes also have the following supposed benefits:

12. The calves look slimmer
13. The feet look smaller

These effects are divided into those that happen only while wearing high-heeled shoes and those that continue after taking off high-heeled shoes. Based on common sense, 1–6, 12, 13 are short-term effects, and 7–11 are long-term effects.

Many studies on the physical measurements of human subjects in relation to heel height and problems caused by wearing high-heeled shoes have answered questions if the aforementioned effects are true or not. Studies related to high-heeled shoes can be divided into three categories. The first category is related to the interaction between high-heeled shoes and humans. Studies of this category are based on biomechanics, including morphology and electromyography. The second category includes epidemiological surveys on the usage of shoes and/or problems and pathology of the foot. The third category includes studies on the relation between perception and physical measurements. In the third category, the goal of studies is to improve the comfort of high-heeled shoes. In this chapter, findings from many studies are reviewed, and possible topics of further studies related to the high-heeled shoes are presented.

11.2 EFFECTS OF HIGH-HEELED SHOES

The results from studies that compared physical measurements taken under high-heeled (≥ 6 cm) and low-heeled (≤ 2 cm) or barefoot conditions are not always identical. However, the results from many studies are summarized as follows.

11.2.1 POSTURE AND MORPHOLOGY

11.2.1.1 Weight Distribution

When the foot is plantarflexed, the distance from the heel to the ankle joint (AJ) projected on the floor increases, while the distance from the metatarso-phalangeal joint (MPJ) to the AJ projected on the floor decreases (Broch et al., 2004), as shown in Figure 11.1. In a standing position, the moment around the pivot point is balanced. Therefore, $F_1 \times L_1 = F_2 \times L_2$, as shown on the left side of Figure 11.1. When the heel is passively elevated, as on the right side of Figure 11.1, the lever arm for MPJ becomes shorter ($L_1' < L_1$), and that for the heel becomes longer ($L_2' > L_2$). Therefore, the force at MPJ becomes larger ($F_1' > F_1$), and the force at the heel becomes smaller ($F_2' < F_2$).

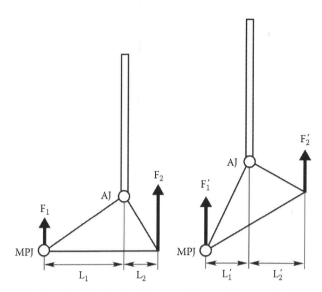

FIGURE 11.1 Load under the forefoot becomes larger when wearing high-heeled shoes (left: barefoot, right: high heels). AJ: ankle joint, MPJ: metatarso-phalangeal joint.

Snow and Williams (1994) measured the weight under the forefoot of the subjects who wore shoes of different heel heights by placing the forefoot part on a force plate. The heights of the heels were 1.91, 3.81, and 7.62 cm, respectively. The weights under the forefoot under the three conditions were 39%, 48%, and 57% of the total body weight, respectively.

In-shoe foot pressure measurement while walking in high-heeled shoes shows that the peak pressure increases at the medial forefoot and toes, and decreases at the rearfoot, midfoot, and lateral forefoot in proportion to heel height (Eisenhardt et al., 1996; Hong et al., 2005; Nyska et al., 1996; Soames and Clark, 1985). Peak in-shoe foot pressures under the first and second metatarsal heads are larger than those under the third, fourth, and fifth metatarsal heads irrespective of the height of the heel, and the difference increases with heel height (Soames and Clark, 1985). According to Nyska et al. (1996), who calculated the contact area as the sum of the area of sensors with a non-zero output, the contact area decreased at the midfoot and the lateral forefoot in high-heeled shoes. The maximal plantar force increased in the medial forefoot, lateral toes, and hallux, whereas the force decreased in the rearfoot, the midfoot, and the lateral forefoot. The pressure–time integral increased at the forefoot and at the toes in high-heeled shoes.

11.2.1.2 Lumbar Lordosis, Anterior Pelvic Tilt, and Posture

Traditionally, it was assumed that wearing high-heeled shoes increased lumbar lordosis and anterior pelvic tilt (Linder and Saltzman, 1998). Experimental results to quantify lumbar lordosis during standing, with landmarks set on the posterior median line, confirmed that this is not true. No significant difference was found in the degree of lumbar lordosis between the high-heel condition and the low-heel or barefoot condition (de Lateur et al., 1991; Snow and Williams, 1994). In fact, in one study, lumbar lordosis

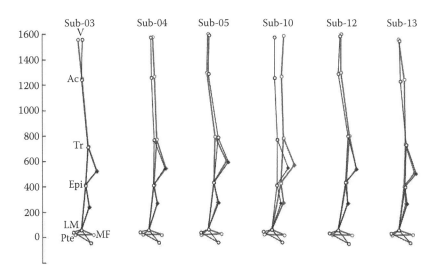

FIGURE 11.2 Side view of the locations of landmarks during standing. Gray line: barefoot, black line: high heels (7 cm). V: vertex, Ac: acromion, Tr: center of greater trochanter, Epi: most lateral point of lateral epicondyle, LM: center of lateral malleolus, MF: metatarsale fibulare, Pte: pternion.

was even smaller in the high-heel condition (Opila et al., 1988). In a photograph of the lateral side of the body, anterior pelvic tilt is quantified as the angle between the horizontal line and the line connecting the anterior superior iliac spine and the posterior superior iliac spine (or L5). Either no difference in this angle was reported for different heel height conditions (de Lateur et al., 1991; Snow and Williams, 1994) or else the angle was reported to be less inclined in the high-heel condition (Opila et al., 1988).

Lumbar lordosis during walking is not different for varying heel height conditions (female subjects of de Lateur et al., 1991), or is smaller in the high-heel condition (male subjects of de Lateur et al., 1991; Lee et al., 2001), so no trend exists according to heel height (Snow and Williams, 1994).

The relative position between body segments changes by wearing high-heeled shoes. Ankle plantarflexion increases (de Lateur et al., 1991; Gollnick et al., 1964, Opila et al., 1988; Snow and Williams, 1994) and the knee is slightly hyperextended (Gollnick et al., 1964; Snow and Williams, 1994) in a standing position when wearing high-heeled shoes.

Figure 11.2 shows the locations of several landmarks of the same subject standing in barefoot and high-heel conditions superimposed at the lateral malleolus. The relative positions of body segments by wearing high-heeled shoes are not consistent from person to person. Figure 11.2 suggests that there is more than one strategy to keep one's balance when wearing high-heeled shoes.

11.2.1.3 Foot Morphology

Kouchi and Tsutsumi (2000) compared the shapes of the right foot for 39 females when standing barefoot and when standing on platforms simulating the footbed curves of high-heeled shoes (4, 8 cm). Figure 11.3 shows examples of the measured

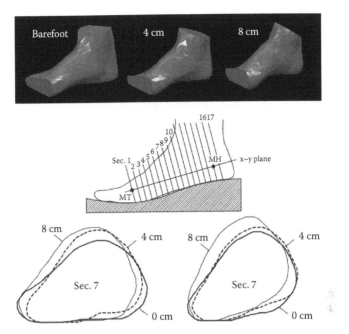

FIGURE 11.3 Differences in the three-dimensional foot shape according to heel height. Top: Three-dimensional foot form of a subject under three different heel conditions. Middle: Definitions of cross sections of the foot. Bottom: Changes in Section 7 (front view) under three different conditions for two subjects. (Modified from Kouchi, M. and Tsutsumi, E., *Anthropol. Sci.*, 108(4), 331, 2000. With permission.)

foot shapes. The instep length becomes shorter as the heel height increases. The shapes of cross sections of the midfoot become laterally rotated (supinated) and thus narrower, higher, and less flat with the increasing heel height (Figure 11.3, bottom). The midline of cross sections at the rearfoot (S16 and S17 in Figure 11.3, middle) is inclined more laterally by 5.4° in an 8 cm heel than in the barefoot condition.

Measurement of the foot pronation/supination angle by using an electric goniometer (Adrian and Karpovich, 1966) or a motion capture system (Snow and Williams, 1994) shows that the foot is more supinated in the high-heel condition than in the low-heel or barefoot condition. In the front view, the AJ is higher on the medial side and lower on the lateral side, and this geometry causes lateral rotation of the tibia when standing with the heel raised, as in wearing high-heeled shoes, and this in turn causes supination of the foot (Sangeorzan, 1991).

A shorter instep length or medial arch length when wearing high-heeled shoes is also observed by x-ray measurements (Schwartz and Heath, 1959). The medial arch length is shorter when wearing high-heeled shoes than when only the heel is elevated by a block. Schwartz and Heath considered that the medial arch length is shorter when the tarsal bones anterior to the calcaneus are supported by the shank curve. The main cause of reduction of the medial arch length may be the windlass effect of plantar aponeurosis (Hicks, 1954), though muscle activities may play a role.

Ohta (1987) measured circumferences and shapes of the lower leg by using a tape measure and moire topography for 16 females under barefoot and 4 other heel conditions (heel heights: 2, 4, 6, and 8 cm). When the heel is raised passively by placing the heel on a block, the maximum calf circumference becomes larger, the minimum leg length becomes smaller, and the location of the most posterior position of the calf becomes higher with the increasing heel height. Ohta considered that the maximum calf circumference increases due to muscle constriction, and the minimum leg circumference decreases due to the fact that the location of attachment of the Achilles tendon on the calcaneus becomes closer to the center of the calf.

11.2.2 Gait

11.2.2.1 Time Factors and Distance Factors

According to some studies, in overground walking at a self-selected speed, the walking speed is slower (Esenyel et al., 2003; Opila-Correia, 1990) and the stride length is shorter (Esenyel et al., 2003; Merrifield, 1971; Opila-Correia, 1990) in the high-heel condition than in the barefoot or lower-heel conditions. However, de Lateur et al. (1991) reported no significant difference in walking speed and stride length between high-heeled, flat-heeled, and middle-heeled gaits. No significant difference is observed in cadence (de Lateur et al., 1991; Esenyel et al., 2003; Opila-Correia, 1990). In treadmill walking, the stride length is smaller in the high-heeled gait than in the flat-heeled gait (Gehlsen et al., 1986; Snow and Williams, 1994). No difference is observed in the step width between the high-heeled gait and the flat-heeled gait (Merrifield, 1971) or between the barefoot and the high-heeled gait (Adrian and Karpovich, 1966).

11.2.2.2 Joint Angles

The most conspicuous difference due to heel height is observed in the AJ angle. Plantarflexion becomes larger as the heel becomes higher (de Lateur et al., 1991; Ebbeling et al., 1994; Esenyel et al., 2003; Gollnick et al., 1964; Snow and Williams, 1994; Stefanyshyn et al., 2000). In overground walking at a self-selected speed, the knee is more flexed at heel contact (HC) and in the early half of the stance phase in the high-heeled gait compared to the barefoot or flat-heeled gait (Esenyel et al., 2003; Opila-Correia, 1990; Stefanyshyn et al., 2000), although the knee is less flexed in the swing phase in the high-heeled gait than in the barefoot or flat-heeled gait (Esenyel et al., 2003; Opila-Correia, 1990). In treadmill walking, the knee is less flexed in the swing phase in the higher-heeled gait (Gollnick et al., 1964; Snow and Williams, 1994), but no differences are found at the HC between the barefoot and the high-heel conditions (Gehlsen et al., 1986; Gollnick et al., 1964; Snow and Williams, 1994) or the knee is more flexed in the high-heeled gait (Ebbeling et al., 1994). For the hip joint angle, many studies report no significant differences between heel height conditions (de Lateur et al., 1991; Ebbeling et al., 1994; Snow and Williams, 1994) both in overground walking and in treadmill walking. Opila-Correia (1990) reported that the hip joint is more extended in the swing phase, whereas Esenyel et al. (2003) reported that the hip joint is more flexed in the stance phase.

In the movement on a horizontal plane, the tibia is more internally rotated in the stance phase in the high-heeled gait than in the flat-heeled gait (Opila-Correia, 1990), or no significant difference exists (Gehlsen et al., 1986; Stefanyshyn et al., 2000). The foot abduction is smaller (out-toeing is smaller) in the high-heeled gait (Adrian and Karpovich, 1966; Snow and Williams, 1994; Stefanyshyn et al., 2000), or no significant difference exists between the flat-heeled and the high-heeled gait (Merrifield, 1971).

In the movement in the frontal plane, foot eversion (pronation) is smaller in the higher-heeled gait (Adrian and Karpovich, 1966; Snow and Williams, 1994; Stefanyshyn et al., 2000). A decrease in foot eversion and a decrease in medial rotation of the tibia may be caused by the coupling effect of the foot and the lower leg (Stefanyshyn et al., 2000).

11.2.2.3 Ground Reaction Force

The high-heeled gait is less fluent because both the maximum brake and the maximum propulsion are larger than the barefoot or low-heeled gait (Ebbeling et al., 1994; Hong et al., 2005; Snow and Williams, 1994; Stefanyshyn et al., 2000). These studies used a treadmill, or the walking speed or the cadence was controlled. In overground walking at a self-selected speed, no significant difference was observed in the antero-posterior component of the GRF (Esenyel et al., 2003). No difference between heel height conditions was observed in the lateral component of the GRF (Ebbeling et al., 1994; Hong et al., 2005; Snow and Williams, 1994).

In high-heeled walking, both active peaks of the vertical component of GRF are larger (Hong et al., 2005; Snow and Williams, 1994; Stefanysyn et al., 2000). The impact force at the HC is reported to be large (Hong et al., 2005; Lee and Hong, 2005), although Stefanyshyn et al. (2000) reported that the impact force is smaller in high-heeled shoes.

11.2.2.4 Muscle Activities and Energy Cost

Ono (1969) measured the activities of 13 muscles during standing and walking in 5 shoe conditions for 14 females (barefoot and heel heights of 1, 3, 5, and 7 cm). Between-condition differences in muscle activities during standing were observed mainly in the foot and the leg muscles (abductor hallucis, abductor digiti minimi, gastrocnemius medialis, peroneus longus, and tibialis anterior), and small differences were observed in the thigh muscles (rectus femoris, inner hamstring muscles, and adductor longus), gluteal muscles (gluteus maximus and gluteus medius), and trunk muscles (sacrospinalis and rectus abdominis). Five muscles that showed differences due to heel height were the abductor hallucis, abductor digiti minimi, gastrocnemius medialis, peroneus longus, and inner hamstring. The activities of these muscles in lower-heel conditions (1, 3 cm) were either not much different or lower than those in the barefoot condition, but their activities increased in the higher-heel condition (5, 7 cm). Joseph and Nightingale (1956) reported that activity of the soleus was higher in the high-heel condition than in the low-heel condition.

While walking, the heel height affects the activities of the foot and the leg muscles, whereas the thigh and the trunk muscles are only slightly affected (Ono, 1969; see Figure 11.4). In seven muscles (abductor pollicis, flexor digitorum brevis, abductor digiti minimi, tibialis anterior, peroneus longus, gastrocnemius medialis, and rectus femoris), the activity level and/or the activity duration increased in high-heeled shoes.

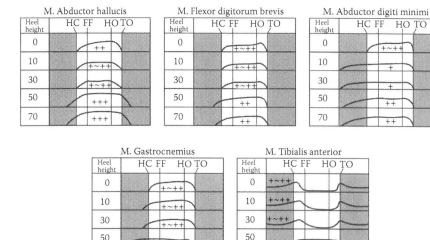

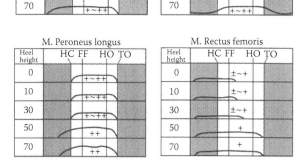

FIGURE 11.4 Activities of muscles during walking. HC: heel contact, FF: foot flat, HO: heel off, TO: toe off. +++: >60% of maximum contraction, ++: 30%–60% of maximum contraction, +: ≤30% of maximum contraction, -: no action potential. (Modified from Ono, H., *J. Jpn. Orthop. Assoc.*, 43, 527, 1969. With permission.)

The activity level increased for the following: abductor hallucis, flexor digitorum brevis, abductor digiti minimi, peroneus longus, and rectus femoris. However, activities of the plantar muscles in heel heights of 1 and 3 cm were lower than or not much different from those in the barefoot condition, and increased only in heel heights of 5 and 7 cm. Activities of the tibialis anterior and the rectus femoris in the stance phase were observed only in heel heights of 5 and 7 cm.

The activity of the soleus increases in high-heeled walking (Joseph, 1968; Stefanyshyn et al., 2000), but the activity of the gastrocnemius does not increase (Ono, 1969; Stefanyshyn et al., 2000) or decrease (Lee et al., 1990). The activity of the peroneus longus increases (Ono, 1969; Stefanyshyn et al., 2000), but not significantly (Stefanyshyn et al., 2000). The activity of the tibialis anterior does not change (Joseph, 1968; Lee et al., 2001; Ono, 1969; Stefanyshyn et al., 2000) or decrease (Lee et al., 1990).

The activity of the rectus femoris increases in high-heeled walking (Ono, 1969; Stefanyshyn et al., 2000). No between-heel height differences were observed in activities of the following: vastus lateralis (Lee et al., 2001), biceps femoris (Joseph, 1968; Ono, 1969; Stefanyshyn et al., 2000), abductor longus, gluteus medius, sacrospinalis, rectus abdominis (Ono, 1969), gluteus maximus (Joseph, 1968; Ono, 1969), and semitendinosus (Stefanyshyn et al., 2000). Joseph (1968) reported no difference in the activity of the erector spinae, but Lee et al. (2001) reported an increase in the activity with heel height.

Ebbeling et al. (1994) measured motion, GRF, heart rate, and oxygen consumption of 15 females during treadmill walking (4.2 km/h) under four different heel height conditions (1.25, 3.81, 5.08, and 7.62 cm). No significant difference was observed between the three lower heel conditions, but in the 7.62 cm heel condition, the heart rate and the oxygen consumption were significantly higher. Higher oxygen consumption was also reported by Mathews and Wooten (1963) for treadmill walking (3.219 km/h) wearing high-heeled shoes (7.62 cm) compared to low-heeled or barefoot walking.

11.2.2.5   Motion and Force

Studies by Esenyel et al. (2003) and Stefanyshyn et al. (2000) are summarized as follows. While walking in high-heeled shoes, the AJ is always more plantarflexed than wearing low-heeled shoes. To compensate for the forward movement of the center of mass due to the larger plantarflexion of the AJ, flexion of knee and hip joints increases. The plantarflexion moment of the ankle is smaller in the early half of the stance phase. The activity of the tibialis anterior increases in high-heeled shoes (Ono, 1969) but the negative power (concentric plantarflexor power) is smaller than that in the lower-heeled condition, and thus shock absorption by the AJ is limited. The increased activity of the rectus femoris and the larger extension moment of the knee control and stabilize the knee flexion.

During the latter half of the stance phase, the maximum plantarflexion moment of the ankle is smaller in higher-heeled shoes though the activity of soleus muscle increases. This is because the plantarflexed position of the ankle makes the lever arm of the Achilles tendon shorter (Stefanyshyn et al., 2000). In contrast, the flexion moment of the hip joint increases and concentric hip flexor power increases, which helps kick off and propulsion (Esenyel et al., 2003).

In the frontal plane, the maximum inversion moment of the knee during the early stance phase is larger in high-heeled shoes. This is because increased adduction of the hip joint makes the position of the foot more medial to the hip joint. The abduction moment of the hip increases in order to control the hip joint (Esenyel et al., 2003). Increased activity of the peroneus longus may control the increased supination of the ankle caused by high-heeled shoes or may stabilize the AJ.

11.2.2.6   Stability

Lord and Bashford (1996) examined static balance (body sway at waist level), maximal antero-posterior balance range, and coordinated stability for 30 females aged 60–89 years under barefoot, low-heel, and high-heel conditions. In all tests, the performance was worse when wearing high-heeled shoes than when barefoot or wearing low-heeled shoes.

The relative position of the heel base and the heel counter affects the stability. Phillips et al. (1991) measured the angle between the lower leg and the counter for five subjects during walking in five different high-heel conditions (height 6.3 cm, width of the base 8 mm). In these conditions, the center of the heel was offset from the center of the heel counter by 0, ±2, and ±4 mm in the medio-lateral direction. They found that the eversion (acute angle between the centerline of the lower leg and the center line of the counter) was the smallest and the stability feeling was the largest when the heel was moved 2 or 4 mm medially from the original position. These results suggest that the lateral tilt (inversion) and the extreme medial tilt (eversion) of the heel cause the perception of instability.

Gefen et al. (2002) found that habitual high-heeled shoe wearers demonstrated an imbalance of gastrocnemius lateralis versus gastrocnemius medialis activity in fatigue conditions. The gastrocnemius lateralis and peroneus longus muscles were more vulnerable to fatigue in habitual high-heel wearers, and this correlates with the abnormal lateral shift in the foot-ground or shoe-ground center of pressure (COP). The lateral shift of the COP could result in an inverting moment that acts to incline the foot laterally and thus may increase the risk of ankle sprains.

11.3   EPIDEMIOLOGICAL SURVEYS: OUTCOME OF WEARING HIGH-HEELED SHOES

The American Orthopaedic Foot and Ankle Society conducted a survey of 356 females aged 20–60 years (Frey et al., 1993). Among 72% of the women who answered that the shoes they wore during work time were comfortable, 27% wore high-heeled shoes. Among 28% of the women who answered that the shoes they wore during work time were not comfortable, 62% wore high-heeled shoes. The results indicate that the percentage of wearers who think the shoes are uncomfortable is 47% of the high-heeled shoe wearers, 20% of the flat-heeled shoe wearers, and 7% of the sneaker wearers. Discomfort is a more serious problem for high-heeled shoe wearers than for flat-heeled shoe or sneaker wearers. Eighty percent of the high-heeled shoe wearers had pain in their feet, and 76% had one or more deformations in the forefoot. The most frequent deformations were hallux valgus (71%), hammer toe (50%), and bunionettes (18%). Eighty-eight percent wore shoes narrower than their foot width (average difference: 12 mm). The difference between the foot width and the shoe width was 5.6 mm for women without foot pain, and 6 mm for women without foot deformation. The authors conclude that women need to be instructed in proper shoe fit.

De Castro et al. (2010) conducted a survey on foot dimensions, foot pain, and custom of wearing high-heeled shoes in São Paulo, Brazil for 172 males and 227 females aged 60 and over. A larger percentage of women had foot pain when wearing shoes than men (females: 60.8%, males: 29.6%). The custom of wearing high-heeled shoes was not related to the foot pain. The foot dimensions were not significantly different between male groups with or without foot pain. However, the female group with foot pain had significantly larger values of foot circumference, instep circumference, and height of the first toe relative to the foot length than the female group without foot pain. The results suggest that women have more shoe problems regardless of the heel height, and that one or more of these three dimensions are related to these problems.

Dawson et al. (2002) conducted a questionnaire survey, a foot examination, and an interview for 96 females aged 50–70 years. More than 80% of the examinees had foot problems; blisters (62%), bunions (38%), and hammer toes (37%) were the most common problems. However, no relationship existed between these problems and the custom of wearing high-heeled shoes.

The shoe appears to be an essential extrinsic factor contributing to the development of foot deformities in women (Frey et al., 1993). However, doctors have not been able to persuade consumers to stop wearing high-heeled shoes because they have not been able to prove the cause and effect relationship between the custom of wearing high-heeled shoes and foot problems (Linder and Saltzman, 1998). Since many individuals wear high-fashion shoes without developing foot pain or deformities, other factors must contribute to the foot pain and deformity (Frey et al., 1993).

11.4   COMFORT

The possible outcome of high-heeled shoes may be divided into those resulting from all high-heeled shoes and those resulting from high-heeled shoes that are not proper for the wearers. Since high-heeled shoes have been accepted as a part of culture and females choose to wear high-heeled shoes, it is important to reduce discomfort, which is a short-term effect of wearing high-heeled shoes.

The perception of comfort is evaluated by the sensory evaluation method. Then, a semantic differential such as a visual analog scale (Hong et al., 2005; Mündermann et al., 2002) or a Borg scale (Miller et al., 2000) is used to quantify the perceived intensity of stimulation. With the increase of heel height, the comfort rating decreases (Hong et al., 2005; Lee and Hong, 2005). However, since the evaluation of comfort is subjective, what exactly each individual evaluates may be different. To improve comfort, it is essential to reduce "comfort" into more objective factors and represent each factor by physical measurements. Kouchi and Mochimaru (2011) conducted interviews using the evaluation grid method (Sanui, 2000), which is based on the personal construct theory (Kelly, 1955). Seven females walked in five high-heeled shoe conditions that differed in footbed shape, friction of the sock liner, or heel shape. Each subject ranked the five conditions according to individual preference. Then the interviewer asked for detailed reasons for this ranking. In this way, causal associations in the subject's mind between shoe characteristics, perceptual benefit, and psychological benefit were extracted. The psychological benefits of the preferred shoe were stability and a lower feeling of fatigue. Perceptual benefits that led to the psychological benefits were that no unpleasant sensation was felt on the sole of the foot, the weight was distributed on the entire sole of the foot, extra muscle power was not required, and the heel and ankle did not wobble. These perceptual benefits were related to the foot–shoe relationships, which in turn were related to shoe characteristics such as footbed shape, width of the heel, friction of the sock liner, and the design and structure of the heel. These relationships between shoe characteristics and perception were validated by experiment (Kouchi and Mochimaru, 2011).

Witana et al. (2009) evaluated the relationship between the characteristics of the footbed shape (wedge angle and heel sheet length) and perception of comfort on the assumption that shoe comfort is related to footbed shape. The comfort rating was higher when the heel sheet length was decided based on the individual's foot dimension rather than when the heel sheet length was a constant value (45 mm). Since the tarsal bones anterior to the talus and calcaneus were on the heel sheet in the individual heel sheet, the average individual heel sheet length was longer than the constant value by 22–23 mm, whereas only the calcaneus was on the heel sheet in the constant length heel sheet. In the most preferred footbed shape, the higher the heel, the larger the wedge angle. The optimum value in the wedge angle was found when the sheet length was individually determined, but no optimum value was found when the constant heel sheet length was used. By a multiple regression function, 80% of the variance of comfort ratings was explained by the contact area of the sole (number of cells with nonzero output × area of a cell), percentage of force on the forefoot (pressure × area), and peak plantar pressure.

Hong et al. (2005) evaluated the effects of a total contact insert (TCI) on comfort ratings. Comfort ratings were estimated from peak foot pressure in the medial forefoot, impact force, and first peak of the vertical GRF ($R^2 = 0.756$). The comfort rating is higher when all three values are smaller. Lee and Hong (2005) found that the heel cup and the TCI reduce the peak pressure in the rearfoot and the impact force, and the arch pad and the TCI reduce the peak foot pressure in the medial forefoot. Further investigation on the mechanism of a shoe insert to change these measurements is necessary.

Au and Goonetilleke (2007) compared the ratings of the fit for the most comfortable and the least comfortable fashion shoes of 20 females. A significant difference was observed in the toes, the MP joint, the plantar arch, and the top line. The fit of the most comfortable shoes was perfect at the plantar arch, perfect or loose at the toes and the MP joint, and loose at the top line.

According to the surveys discussed in the previous section, foot deformation and problems are most frequent in the forefoot. According to Dohi et al. (2001), who conducted a questionnaire investigation of 50 Japanese females aged 60 and over, 32 females (64%) had complaints about their shoes. The most frequent complaint was about the first toe (41%), followed by the fifth toe (34%) and the fourth toe (22%). Among the complaints during walking (N = 24), the most frequent complaint was that the heel slipped out (50%), and the second most frequent complaint was that the top line of the shoe rubbed the lateral malleolus (42%).

Studies on footbed shape may lead to a solution of foot and shoe problems. However, it may take some time to find a compromise between fashion design and improvement of comfort in the forefoot. Some guidelines need to be developed to incorporate comfort into the design factors of high-heeled shoes. The fact that foot pain is more prevalent in females than males (de Castro et al., 2009) suggests that the design of women's shoes, including lower-heeled shoes, needs improvement.

11.5 FUTURE STUDIES

Figure 11.5 summarizes the results of the interaction between high-heeled shoes and wearers. The rearfoot is passively elevated by high-heeled shoes. This in turn changes the posture (the relative positions between body segments). Therefore, the forces acting on body segments change and the gait changes. Prolonged wearing of high-heeled shoes may cause long-lasting body changes. Causal associations

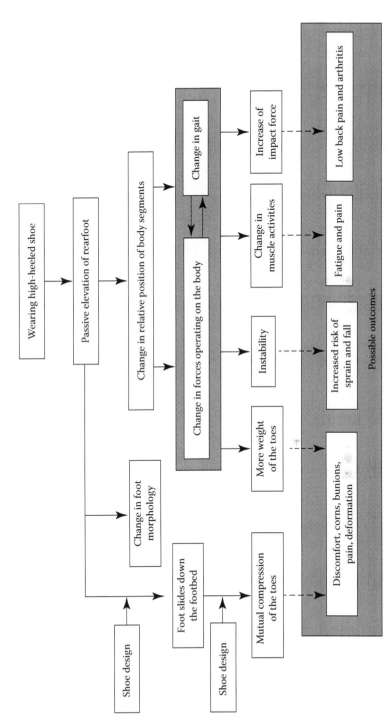

FIGURE 11.5 Effects of high-heeled shoes on wearers. Note that the increases of lumber lordosis and anterior pelvic tilt were rejected.

of possible outcomes are self-evident for some (short-term discomfort, foot pain, increased risk of sprain, and muscle fatigue), but not for others (long-term foot deformation, low back pain, and arthritis).

Studies on walking in high-heeled shoes revealed that high-heeled shoes cause changes in joint angles, foot morphology, gait, muscle activities, GRF, in-shoe foot pressure, and perception of comfort. Among the confirmed short-term effects are changes in weight distribution, increases in the activity of leg muscles, and changes in gait. However, increases of lumber lordosis and anterior pelvic tilt were not confirmed. The supposition that high-heeled shoes make the calf look slimmer was also disproved. Among the adverse effects of high-heeled shoes, muscle fatigue and pain and the increased risk of ankle sprain are supported on the basis of physical measurements. As for low back pain, corns and bunions, and deformations of the toes, causal associations have not yet been proved. Though it is a challenge to predict the long-term outcomes of a specific foot in a specific shoe, recent studies using finite element analysis may provide a solution (Cheung et al., 2009).

Previous studies revealed morphological, biomechanical, and physiological reactions of humans when wearing high-heeled shoes. On the other hand, we still do not know much about how we can improve the comfort of high-heeled shoes. Shoes are mass produced products that need to fit feet with large morphological as well as kinematic variations. A shoe last needs to be modified to improve the shoe comfort, because the shoe last determines the shape of a shoe. When modifying a shoe last, we try to maximize the perceived comfort resulted from the interaction between a foot and a shoe, which is made using a shoe last. A problem here is that "comfort" is subjective, and different persons may evaluate different things using the same word. It is important to reduce ambiguities when evaluating the "comfort." Also, effects of the height of heel have been the main target of research, and effects of other design factors on the perception of comfort have not been studied. Effects of material properties on comfort may also need further studies (Voloshin and Loy, 1994). The following topics deserve further studies:

1. Surveys on relationships between the foot, the shoe, and the foot–shoe interaction. Efforts to reveal what foot has what kind of trouble when it wears what shoe made on what shoe last may help to derive insights into cause–effect relationship between the foot trouble and the shoe design.
2. Studies to separate different aspects of "comfort" into more specific evaluation factors.
3. To clarify what physical phenomenon is represented by each of the separated evaluation factors. These studies are essential to better understand shoe comfort.
4. Studies on effects of design factors of the heel other than the height on perception of comfort or biomechanics of gait. If heel shapes, such as the size of the top lift (base of the heel) or heel pitch (directional slant of the heel), have significant effects on the comfort, such information is useful for designing comfortable high-heeled shoes.
5. Studies on effects of material properties of the heels, soles, or inserts on perception of comfort or biomechanics of gait.

QUESTIONS

11.1 Is the posture while standing in high-heeled shoes systematically different from the posture while standing barefoot or in low-heeled shoes? If different, how?

11.2 Does the weight supported by forefoot actually increase with the heel height? If so, why?

11.3 Is the foot shape while wearing high-heeled shoes systematically different from the foot shape while wearing low-heeled shoes or in barefoot? If different, how?

11.4 Is the gait in high-heeled shoes systematically different from that in low-heeled shoes or in barefoot? If different, how?

11.5 Are muscle activities while standing/walking in high-heeled shoes systematically different from those while standing/walking barefoot or in low-heeled shoes? If different, how? Also, does the energy cost increase with the heel height?

11.6 Are the forces acting on the joints of lower limbs while walking in high-heeled shoes systematically different from those while walking barefoot or in low-heeled shoes?

11.7 Does the stability while standing decrease with the heel height?

11.8 What is the prevalence of the foot deformities or foot troubles? Is it actually related to the habit of wearing high-heeled shoes?

11.9 What design factors and physical measurements are related to the comfort of high-heeled shoes, and exactly what high-heel shoe wearers evaluate for the comfort?

REFERENCES

Adrian M and Karpovich P, 1966: Foot instability during walking in shoes with high heels. *Research Quarterly*, 37: 168–175.

Au EYL and Goonetilleke RS, 2007: A qualitative study on the comfort and fit of ladies' dress shoe. *Applied Ergonomics*, 38: 687–696.

Broch NL, Wyller T, and Steen H, 2004: Effects on heel height and shoe shape on the compressive load between foot and base. *Journal of the American Podiatric Medical Association*, 94: 461–469.

Cheung JT-M, Yu J, Wong DW-C, and Zhang M, 2009: Current methods in computer-aided engineering for footwear design. *Footwear Science*, 1: 31–46.

Dawson J, Thorogood M, Marks SA et al., 2002: The prevalence of foot problems in older women: A cause for concern. *Journal of Public Health Medicine*, 24: 77–84.

de Castro AP, Rebelatto JR, and Aurichio TR, 2010: The relationship between foot pain, anthropometric variables and footwear among older people. *Applied Ergonomics*, 41: 93–97.

de Lateur BJ, Giaconi RM, Questad K, Ko M, and Lehmann JF, 1991: Footwear and posture: Compensatory strategies for heel height. *American Journal of Physical Medicine and Rehabilitation*, 70(5): 246–254.

Dohi, M, Mochimaru M, and Kouchi M, 2001: Foot shape and shoe fitting comfort for elderly Japanese women. *The Japanese Journal of Ergonomics*, 37: 228–237. (In Japanese with English abstract)

Ebbeling CJ, Hamill J, and Crussemeyer JA, 1994: Lower extremity mechanics and energy cost of walking in high-heeled shoes. *Journal of Orthopaedic and Sports Physical Therapy*, 19(4): 190–196.

Eisenhardt JR, Cook D, Predler I, and Foehl HC, 1996: Change in temporal gait characteristics and pressure distribution for bare foot versus heel heights. *Gait and Posture*, 4: 280–286.

Esenyel M, Walsh K, Walden JG, and Gitter A, 2003: Kinetics of high-heeled gait. *Journal of American Podiatric Medical Association*, 93: 27–32.

Frey C, Thompson F, Smith J, Sanders M, and Horstman H, 1993: American orthopaedic foot and ankle society women's shoe survey. *Foot Ankle*, 14: 78–81.

Gefen A, Megido-Ravid M, Itzchak Y, and Arcan M, 2002: Analysis of muscular fatigue and foot stability during high heeled gait. *Gait and Posture*, 15: 56–63.

Gehlsen G, Braatz JS, and Assmann N, 1986: Effects of heel height on knee rotation and gait. *Human Movement Science*, 5: 149–155.

Gollnick PD, Tipton CM, and Karpovich PV, 1964: Electrogoniometric study of walking in high heels. *Research Quarterly for Exercise and Sports*, 35: 370–378.

Hicks, JH, 1954: The mechanics of the foot. II. The plantar aponeurosis and the arch. *Journal of Anatomy, London*, 88, 25–30.

Hong WH, Lee YH, Chen HC, Pei YC, and Wu CY, 2005: Influence of heel height and shoe insert on comfort perception and biomechanical performance of young female adults during walking. *Foot and Ankle International*, 26: 1042–1048.

Joseph J, 1968: Patterns of activity of some muscles in women walking on high heels. *Annals of Physical Medicine*, 9: 295–299.

Joseph J and Nightingale A, 1956: Electromyography of muscles of posture: Leg and thigh muscles in women including the effects of high heels. *Journal of Physiology*, 132: 465–468.

Kelly G, 1955: *The Psychology of Personal Constructs*. W. W. Newton, New York.

Kouchi M and Mochimaru, M, 2011: Factors for evaluating high-heeled shoe comfort. *Footwear Science*, 3(S1): S90–S92.

Kouchi M and Tsutsumi E, 2000: 3D foot shape and shoe heel height. *Anthropological Science*, 108(4): 331–343.

Lee YH and Hong WH, 2005: Effects of shoe inserts and heel height on foot pressure, impact force and perceived comfort during walking. *Applied Ergonomics*, 36: 355–362.

Lee CM, Jeong EH, and Freivalds A, 2001: Biomechanical effects of wearing high-heeled shoes. *International Journal of Industrial Ergonomics*, 2: 321–326.

Lee KH, Shieh JC, Matteliano A, and Smiehorowski T, 1990: Electromyographic changes of leg muscles with heel lifts in women: Therapeutic implications. *Archives of Physical Medicine and Rehabilitation*, 71: 31–33.

Linder M and Saltzman CL, 1998: A history of medical scientists on high heels. *International Journal of Health Service*, 28(2): 201–225.

Lord S and Bashford G, 1996: Shoe characteristics and balance in older women. *Journal of American Geriatrics Society*, 44: 429–433.

Mathews DK and Wooten EP, 1963: Analysis of oxygen consumption of women while walking in different styles of shoes. *Archives of Physical Medicine and Rehabilitation*, 99: 567–571.

Merrifield HH, 1971: Female gait patterns in shoes with different heel heights. *Ergonomics*, 14: 411–417.

Miller JE, Nigg BM, Liu W, Stefanyshyn DJ, and Nurse MA, 2000: Influence of foot, leg and shoe characteristics on subjective comfort. *Foot and Ankle International*, 21(9): 759–767.

Mündermann A, Nigg BM, Stefanyshyn DH, and Humble RN, 2002: Development of a reliable method to assess footwear comfort during running. *Gait and Posture*, 16(1): 38–45.

Nyska M, McCabe C, Linge K, and Klenerman L, 1996: Plantar foot pressure during treadmill walking with high-heel and low-heel shoes. *Foot and Ankle International*, 17: 662–666.

Ohta S, 1987: A study on the change of the shape of the calf of females by contraction of tri-
    ceps surae muscle. *Bulletin of the Faculty of Education Hirosaki University*, 58: 67–74
    (In Japanese).

Ono H, 1969: Heel height and muscle activity. *Journal of the Japanese Orthopedic Association*
    43: 527–547 (In Japanese).

Opila KA, Wagner SS, Schiowitz S, and Chen J, 1988: Postural alignment in barefoot and
    high-heeled stance. *Spine*, 13: 542–547.

Opila-Correia KA, 1990: Kinematics of high-heeled gait. *Archives of Physical Medicine and
    Rehabilitation*, 71: 304–309.

Phillips RD, Reczek DM, Fountain D et al., 1991: Modification of high-heeled shoes to
    decrease pronation during gait. *Journal of the American Podiatric Medial Association*,
    81: 215–219.

Sangeorzan BJ, 1991: Biomechanics of the subtalar joint. In Stiehl JB ed., *Inman's Joints of
    the Ankle*, 2nd edn. Williams & Wilkins, Baltimore, MD, pp. 65–73.

Sanui J, 2000: Visualization of user needs: Introduction of the evaluation grid method.
    Japanese Journal of Ergonomics, 36(suppl.): 30–31 (In Japanese).

Schwartz BP and Heath AL, 1959: Preliminary findings from a roentgenographic study of the
    influence of heel height and empirical shank curvature on osteo-articular relationships
    of the normal female foot. *Journal of Bone and Joint Surgery*, 46-A: 1065–1076.

Snow RE and Williams KR, 1994: High heeled shoes: Their effect on center of mass posi-
    tion, posture, three-dimensional kinematics, rearfoot motion and ground reaction forces.
    Archives of Physical Medicine and Rehabilitation, 75: 568–576.

Soames RW and Clark C, 1985: Heel height induced changes in metatarsal loading patterns
    during gait. In Winter D et al., eds., *Biomechanics IX-A*, Human Kinetics Publishers
    Inc., Champaign, IL, pp. 446–450.

Stefanyshyn DJ, Nigg BM, Fisher V, O'Flynn B, and Liu W, 2000: The influence of high
    heeled shoes on kinematics, kinetics and muscle EMG of normal female gait. *Journal of
    Applied Biomechanics*, 16: 309–310.

Voloshin AS and Loy DJ, 1994: Biomechanical evaluation and management of the shock
    waves resulting from the high-heel gait: I—Temporal domain study. *Gait and Posture*,
    2: 117–122.

Witana CP, Goonetilleke RS, Au EYL, Xiong S, and Lu X, 2009: Footbed shapes for enhanced
    footwear comfort. *Ergonomics*, 52(5): 617–628.

12 Footbed Design

Ravindra S. Goonetilleke and
Thilina W. Weerasinghe

CONTENTS

12.1 INTRODUCTION

Drawings more than 15,000 years old have shown that people wore some kind of foot-wear even in very ancient times (Kurup et al., 2012). The first custom shoes, on the other hand, can be traced back about 5500 years (Pinhasi et al., 2010). In those days, footwear were used to protect people's feet from external hazards (DeMello, 2009). Today, they have become an integral part of a person's appearance more than function.

In 2010, the U.S. civilian population spent 17 min each day on average on weekdays and 21 min each day on weekends on sports, exercise, and recreation (Bureau of Labor Statistics, 2012). Data for other developed countries are quite similar: 21 min/day in Australia (Australian Bureau of Statistics, 2012). Footwear are used for standing or walking during a large part of the day. Thus, it is not surprising that the customer needs of footwear have somewhat changed from ancient times. That is, footwear form has to delight customers, the fit between foot and footwear has to make the person comfortable, and the footwear functions have to be suitable for the chosen activity. A survey carried out by Piller et al. (2002) with 420 European subjects revealed that 58.8% of females and 51.5% of males have difficulty of selecting shoes due to comfort issues primarily due to a poor fit between foot and footwear. A similar survey carried out in Hong Kong showed that style and com-fort were ranked higher than quality, price, function, or the brand (Chong and Chan, 1992).

12.2 PERFORMANCE AND FIT

Appearance and comfort have dominated the footwear needs, primarily, because many of the performance enhancements have not been scientifically proven (Stonebrook, 2010). However, the Load 'N launch technology by Athletic Propulsion

Labs (APL, 2012) claims that their technology allows a vertical leap increase of 3.5 in. Similarly, many other patents exist in relation to performance enhancing shoes even though the specific claims have not been investigated thoroughly. On the flip side, there is no doubt that some of the current footwear eliminate and minimize the risk factors associated with injury. This aspect has been well researched even though the design parameter manipulations to achieve the biomechanical parameters have been fuzzy in relation to the fit between foot and footwear. In traditional mechanical engineering, the fit between parts can be a loose fit, an interference fit, or a transition fit depending on the functional requirements (Norton, 2000). Unfortunately, with the structural complexity of the foot, the foot–footwear fit is one of multiple parts that require differing kinds of fit in differing locations. Alemany et al. (2012) have outlined these requirements for the dorsal side of the foot. One important part that has been neglected or not that well researched in the past has been the footbed. A poor fit between foot and footwear results in undue stress on the feet: normal stress expressed as pressure or shear stress expressed as friction. Our feet are such that when we stand bare-footed, the whole sole area of the foot contacts the ground unless the person has a very high arch (Xiong et al., 2012). With a heel lift, the plantar surface of the foot would change and fitting that shape requires some thought. As the heel height increases, one would expect that the foot shape changes and would even affect the dorsal sides. Hence, it is important that the primary fitting be formed at the plantar or foundation level and then on the rest of the foot (Au et al., 2011; Luximon and Goonetilleke, 2003). The contours on the foot and the shoe will determine the exact positioning of the foot inside a shoe (Goonetilleke and Luximon, 2001). Any misfit resulting in a "clearance" between foot and footbed would unduly increase the pressures in regions where the foot contacts the footbed. Thus, distributing the loads optimally on the plantar surface, with the right fit, is very important for proper functioning. Low peak pressures can be achieved with conforming footbeds or through varying levels of cushioning (Goske et al., 2006; Mientjesa and Shorten, 2011). Many researchers have investigated different types of materials and shapes to relieve the high pressures linked to foot pain and discomfort (Godfrey et al., 1967; Silvino et al., 1980; Tsung et al., 2004). Leber and Evanski (1986) performed a study to compare the effectiveness of reducing plantar pressure using seven different materials. With changes in material technology, such comparisons among different materials are not viable.

12.3   CUSHIONING

Low stiffness materials can give adequate conformance to eliminate such clearances between foot and footbed. In the biomechanics literature, injury has been linked to shock absorption while stiffness and hardness have been linked to discomfort in the ergonomics literature (Figure 12.1). Discomfort precedes pain or injury. Does discomfort lead to pain or injury or are these two effects independent of each other? Any existence of such a relationship can be found through an understanding of the material properties that contribute to perception and/or injury. Shock absorption is generally related to damping phenomenon in the engineering literature. The literature on injury seems to imply that shock absorption is related to a reduction in impact force magnitude. Nigg et al. (1988) has shown that there is no reduction in vertical

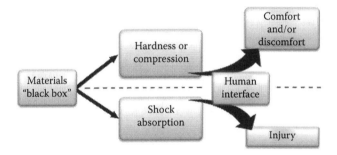

FIGURE 12.1   Foot supporting material implications.

impact forces during running when viscoelastic insoles were compared with conventional running shoe insoles, probably due to bottoming-out of the insoles. However, others such as Aguinaldo and Mahar (2003) have shown that impact force and loading rates depend on midsole construction. Impact loads generated at the feet during activity have been related to a range of injury problems such as cartilage breakdown (Simon et al., 1972), osteoarthritis (Radin et al., 1972), knee injuries (Newell and Bramwell, 1984), stress fractures (Milgrom et al., 1985), Type 1 shin splints (Detmer, 1986), and low back pain (Voloshin and Wosk, 1982). Alternatively, shock "absorption" is meant to be a gradual increase in the load with slow deceleration (as opposed to an impact load which is a rapid increase in load). A cushioning system functions by increasing the duration of an impact, thereby reducing the shock load transmitted to the musculoskeletal system. Cushioning has been found to reduce the local pressures and axial shock in the lower extremity (NIKE Sport Research Review, 1988) even though some claim that the expensive cushioning systems fail to attenuate shock (Shorten, 2002). The various physical properties that contribute to the measure of cushioning have been investigated using commercially available compression and impact testers (NIKE Sport Research Review, 1990) from companies such as Instron, Exeter Research, MTS, and SATRA. The ASTM F1976-06 outlines measuring the impact attenuation properties of athletic shoes using an impact tester. The ASTM F1614-99 (2006) gives the "Standard Test Method for Shock Attenuating Properties of Materials Systems for Athletic Footwear." This test gives the force–displacement curves for materials under uniaxial compression conditions that replicate heel strike conditions of forces up to 2 kN in a time frame of 10–20 ms. Such physical tests quantify the mechanical properties of the footwear materials in controlled conditions. However, when a shoe is worn, the kinetics between foot and footwear may not be the same as the test conditions and the foot–footwear fit will determine the exact interaction.

The human body has the ability to sense damaging motions as discomfort or pain. Poor cushioning ratings are a good predictor of discomfort and pain in feet (Grier et al., 2011). Xiong et al. (2012) have shown that the pressure pain threshold (PPT) is related to rate of loading, amount of compression, stiffness, and time of loading. Thus, human tissue and materials are quite similar. In other words, people should be able to perceive most of the material qualities that can be mechanically tested as long as they exceed the minimum thresholds or the just noticeable differences.

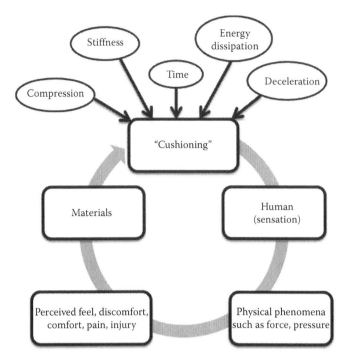

FIGURE 12.2   Material property effects in footwear.

Goonetilleke (1999) showed that time to peak deceleration and/or stiffness (or compression) on an impact tester can predict the perceived level of cushioning (PLC) in the heel during standing and running, and the magnitude of the peak deceleration on the impact tester is a predictor of PLC during walking (Figure 12.2). In the same study, cushioning was described by the subjects as softness, mushiness, comfort, shock absorption, reducing jarring motion, support, the sinking feeling, stability, protection, deceleration, and so on. These descriptors have a related physical metric and hence it is not surprising that there is a relationship between the objective measures and the subjective measures. Interestingly, Goonetilleke (1999) found no relationship between the percent energy loss in the force–displacement curve in the heel area and PLC. Similarly, Xiong (2008) showed that percent energy loss had no effect on PPT. Percent energy loss and rebound resilience are somewhat similar. Rebound resilience, a dimensionless quantity, is defined as the ratio of the energy returned to the energy applied to a test piece as a result of an impact whereas percent energy loss, also another dimensionless quantity, quantifies the proportion of energy lost compared to the amount of energy applied. Many "mechanical" systems of differing characteristics are present at the foot–ground interface. All these systems tend to act serially (Figure 12.3). One important system is the floor surface. A second is the foot–ground interface, that is, the shoe. Depending on the number of systems present, the overall characteristics of a composite surface can be significantly different when compared to any one surface. Even though the suitability of new materials can be tested on machines to test their appropriateness for

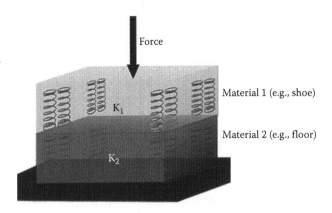

FIGURE 12.3 The "composite" structure between foot and floor.

footwear, the actual interaction should be tested with psychophysical techniques and/or equipment at the foot–shoe interface.

12.4 PLANTAR PRESSURE

The shape of a shoe footbed is generally based on previously successful models of shoes or shapes that have been modified through trial and error methods over many years. Some attempts have been made to reverse engineer good fitting shoes to determine the last shape. As little as 2 mm in height can be sensed on the plantar surface by people. A 2 mm difference in shape can result in large differences in pressure distribution. Anatomically shaped footbeds are common in orthotics and some insoles. These are means to distribute the plantar surfaces somewhat uniformly. However, with an anatomically shaped footbed, areas such as the arch are unable to function as they should. Hence, a footbed that is not as closely shaped should work better to support the foot. Most designers assume that the foot can deform to any shape as it has so many bones and joints. However, the lengths of the various bones and the way they are joined together constrain the movement and the deformation and knowledge of these bones and joints are important to design the optimal footbed.

The heel lift tends to move the center of pressure (COP) forward. In general, the higher the heel height, the larger the force on the forefoot and the larger the anterior shift of COP (Gefen et al., 2002; Han et al., 1999; McBride et al., 1991; Shimizu and Andrew, 1999; Snow and Williams, 1994). Corrigan et al. (1993) found that a 1¾ in. heel had 50% more pressure under the ball of the foot when compared to wearing a ¾ in. heel. Holtom (1995) demonstrated plantar foot pressure increases of 22%, 57%, and 76% with heel heights of 2, 5, and 8.25 cm, respectively. Nonconforming or ill-fitting footbeds may not have significant effects at low heel heights. But, with high-heeled shoes, the footbed plays an important role to modulate the force distribution throughout the plantar surface. Unfortunately, most studies on high-heeled shoes have been performed using commercially available shoes or shoes specially designed for testing (Chiu and Wang, 2007; Hansen and Childress, 2004; Hong et al., 2005;

Lin et al., 2007; Mandato and Nester, 1999; Nyska et al., 1996; Snow and Williams, 1994). As a result, the effects of footbed shape are somewhat unknown.

Bauman et al. (1963) and Silvino et al. (1980) have shown that clinical pain, in the plantar areas, tends to occur when pressures exceed 255 kPa. The pressure discomfort threshold is around 0.4–0.5 of the pressure pain threshold (Xiong et al., 2011). Johansson et al. (1999) found that the discomfort to pain thresholds for the finger and the palm are 0.38 and 0.40, respectively, suggesting that the discomfort threshold to pain threshold ratio is somewhat consistent across structures. In other words, if the pressure in the plantar regions can be controlled to be around discomfort threshold, a reasonably comfortable pressure distribution may be achieved.

12.5   FOOTBED SIMULATOR

The footbed has many different parameters that give the footbed its shape. These are seat length, wedge angle, shank shape, and so on (Figure 12.4). Unfortunately, these parameters have not been systematically explored until recently, primarily due to the lack of an instrument to simulate the footbed (Figure 12.5). The effect of the wedge angle on plantar pressure is shown in Figure 12.6. When the wedge angle is low as in Figure 12.6a, most of the load is borne by the heel and as a result the high pressures are at the heel. As the wedge angle increases, a balance between heel and forefoot is obtained (Figure 12.6b). Further increases of wedge angle will tip the COP more to the anterior resulting in higher forces and pressures in the forefoot (Figure 12.6c). Any further increase will not allow the midfoot to contact the footbed (Figure 12.6d), and this void results in higher forces on both forefoot and heel. Witana et al. (2009a) have shown that the midfoot can change shape even when the heel and forefoot have not changed. This shows the importance of having the right shank shape in the shoe. Having an insole with very soft material or an insert (Burgess et al., 1997) will allow the foot to contact the footbed, but it may not be ideal from the point of view of performance and function. Thus, the optimum level of contact is critical to support the foot and the body weight. The footbed simulator developed by Goonetilleke and Witana (2010) allows a person to "ride" on any shape until the subjective as well as the objective measures are satisfactory. The objective measure, pressure, can be monitored as the subject controls the shape of the footbed. Numerous experiments have been conducted using the simulator. In all previous experiments, it has been clear that higher comfort is related to a more distributed

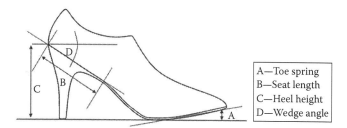

FIGURE 12.4   Footbed parameters.

(a)   (b)

FIGURE 12.5 The footbed simulator (a) user operating the simulator with the remote controller; (b) close-up view of an adjusted footbed and the display panel/controls. In this view, the tested foot is the left one so that a comparison can be made against the right foot when wearing a shoe.

force that is achieved by manipulating the various footbed parameters for any one individual or group of individuals. Witana et al. (2009a) showed the existence of optimal wedge angles for various heel heights. These were 4°–5° at a heel height of 25 mm, 10°–11° at 50 mm, and 16°–18° at 75 mm. We have also proposed the necessary criteria for a comfortable footbed:

1. A contact area greater than that when standing on a flat surface.
2. A peak pressure lower than approximately 100 kPa when standing, which is incidentally 39.2% of the PPT reported by Bauman et al. (1963) and Silvino et al. (1980). Thus, this value may be viewed as representing the discomfort threshold.
3. Forefoot force of approximately 15%–22.5% of that on one foot.

In a different study, Witana et al. (2009b) showed that stiffness and percentage of energy loss in the region of 35%–45% of foot length have an influence on the subjective ratings. Even though Goonetilleke (1999) showed that energy loss in the heel had no influence on perceived ratings, the energy loss in the midfoot area does have an influence on perceived feel. The heel material has to have the right amount of rebound to lessen the impact and aid propulsion. However, the midfoot is in the preparation stage of take-off and will be aided by a low percent energy loss. Thus, a poorly fitting midfoot area can affect the ride quality of a shoe, and may strain the plantar fascia resulting in plantar fasciitis (Cole et al., 2005; Riddle et al., 2003; Singh et al., 1997).

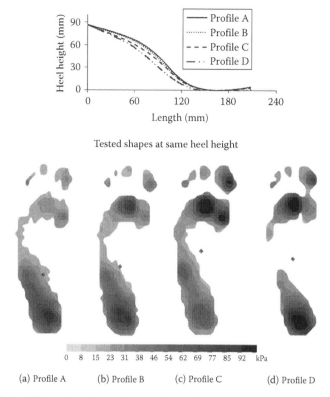

FIGURE 12.6   Effect of varying wedge angle on plantar pressure. (a) Higher heel pressure, (b) balanced forces between heel and forefoot, (c) higher forefoot pressure, and (d) lack of midfoot contact resulting in higher forefoot and heel pressures.

Weerasinghe and Goonetilleke (2011) have shown that comfort is inversely proportion to COP having the following form of a relationship:

$$Comfort \ = \ 87.2 - 0.798 \times COP$$

Hence, controlling the location of the COP can change the comfort of a shoe. Contrary to all studies that have shown an anterior shift of COP with increase of heel height, we have shown that COP can be shifted close to a barefoot stance with the ideal footbed geometry (Weerasinghe and Goonetilleke, 2011). A suboptimal wedge angle and footbed curvature will tilt the body away from a neutral posture or make the foot slide forward causing discomfort due to looseness in the shoe. Then, the foot is squeezed in the toe area resulting in high pressures that increase the tissue and joint deformations and hinder movement compromising the foot's performance. Such effects can result in temporary or permanent impairments some of which can be detrimental to the functioning of the feet. Common problems such as callouses and corns are due to undue pressure and relative movement between footbed and foot due to a poor fit of shape and a mismatch of material properties. Hallux valgus is a long-term effect of unwanted pressure in the MPJ area.

The effects of poorly fitting footbeds can propagate up to the spine and beyond and reflect as body tilt or awkward posture, as people respond like an inverted pendulum. A common belief is that high heels make the body tilt so that the buttocks and breasts are emphasized (Danesi, 1999). However, some researchers have found opposite effects (Hansen and Childress, 2004). For example, Franklin et al. (1995) used a wooden board 5.1 cm high under the heels to study the standing posture and found that lumbar lordosis actually decreases as a result of a posterior tilt of the pelvis. Decreased lumbar lordosis is one of the common observations in the high-heeled shoe wearers (Franklin et al., 1995; Lee et al., 2001; Opila et al., 1988). The inconsistencies among high-heeled posture–related studies are possibly due to the use of heel blocks or shoes of a certain height with no control on fit, which is affected by the footbed parameters such as surface geometry (Franklin et al., 1995; Lee et al., 2001). More research is needed to investigate the postural effects when wearing shoes with differing footbed shapes.

12.6   CONCLUSION

The footbed is an important element in footwear. A rigid footbed shape is critical for the right fit between the plantar surface of the foot and the surface of the shoe. A poor fit between foot and footwear can result in instability, discomfort, and injury in the long term. A footbed simulator is an ideal tool for investigating the effects of the various footbed parameters and related biomechanics. The ideal footbed can improve the pressure distributions, postures, and comfort of an individual. More emphasis should be placed to design the bottom shape of a last and then cushioning ought to provide the finesse to improve the sensations when wearing a shoe.

ACKNOWLEDGMENTS

The authors would like thank the Research Grants Council for supporting the numerous studies under GRF grants 612711, 613205, and 613607. The help of Gemma Maria Sanz Signes and Qu Yan are also appreciated.

QUESTIONS

12.1   What is a footbed and what are its primary characteristics?

12.2   Describe cushioning. What are possible subjective or objective measures that describe cushioning?

12.3   What are the various standards related to cushioning? What apparatus are used for quantifying cushioning?

12.4   Plot a representative force–deflection curve of a midsole material such as polyurethane. How would the shape of this curve change if you use another material?

12.5   How could you characterize the various cushioning variables using a force–displacement curve?

12.6   Determine the length of a shoe corresponding to a given wedge angle, heel height, and varying midfoot shape. Assume the necessary parameters of the foot of the person wearing the shoe to be known. Plot a curve to determine the variation in shoe length in the differing parts with varying shape and wedge angle.

12.7 Why is it necessary to consider wedge angle, seat length, and midfoot shape in high-heeled shoes?

12.8 Explain why comfort decreases when the center of pressure shifts anteriorly.

12.9 What are suitable wedge angles for heel heights of 25, 50, and 75 mm?

12.10 What is the implication of discomfort threshold on footbed design?

REFERENCES

Aguinaldo, A. and Mahar, A. (2003). Impact loading in running shoes with cushioning column systems. *Journal of Applied Biomechanics*, 19(4), 353–360.

Alemany, S., Olaso, J., Puigcerver, S., and González, J. C. (2012). Virtual shoe test bed. In: *Science of Footwear*. Ed. R. S. Goonetilleke, CRC Press, Boca Raton, FL.

APL. (2012). Stop dreaming, jump higher. Accessed March 10, 2012. http://www.athleticpropulsionlabs.com/load-n-launch-technology.html

Au, E. Y. L., Goonetilleke, R. S., Witana, C. P., and Xiong, S. (2011). A methodology for determining the allowances for fitting footwear. *International Journal of Human Factors Modelling and Simulation*, 2(4), 341–366.

Australian Bureau of Statistics. (2012). Average time spent on sport and recreation. Accessed March 12, 2012. http://www.abs.gov.au/ausstats/abs@.nsf/Products/7CCDD38A5FC83 E0FCA25796B0015180B?opendocument

Bauman, J. H., Girling, J. P., and Brand, P. W. (1963). Plantar pressures and trophic ulceration: An evaluation of footwear. *Journal of Bone Joint Surgery*, 45B, 652–673.

Bureau of Labor Statistics (2012). Economic news release. Table 11. Time spent in leisure and sports activities for the civilian population by selected characteristics, 2010 annual averages. Accessed March 12, 2012. http://www.bls.gov/news.release/atus.t11.htm

Burgess, S., Jordan, C., and Bartlett, R. (1997). The influence of a small insert, in the footbed of a shoe, upon plantar pressure distribution. *Clinical Biomechanics* (Bristol, Avon), 12(3), S5–S6.

Chiu, M. C. and Wang, M. J. (2007). Professional footwear evaluation for clinical nurses. *Applied Ergonomics*, 38, 133–141.

Chong, W. K. F. and Chan, P. P. C. (1992). Consumer buying behavior in sports footwear industry. Business Research Center, Hong Kong Baptist College, Hong Kong, China.

Cole, C., Seto, C., and Gazewood, J. (2005). Plantar fasciitis: Evidence-based review of diagnosis and therapy. *American Family Physician*, 72(11), 2237–2242.

Corrigan, J. P., Moore, D. P., and Stephens, M. M. (1993). Effect of heel height on forefoot loading. *Foot and Ankle*, 14, 148–152.

Danesi, M. (1999). *Of Cigarettes, High Heels, and Other Interesting Things*. St. Martin's Press, New York.

Demello, M. (2009). *Feet and Footwear: A Cultural Encyclopedia*. ABC-Clio LLC, Santa Barbara, CA.

Detmer, D. E. (1986). Chronic shin splints. Classification and management of medial tibial stress syndrome. *Sports Medicine*, 3, 436–446.

Franklin, M. E., Chenier, T. C., Brauninger, L., Cook, H., and Harris, S. (1995). Effect of positive heel inclination on posture. *Journal of Orthopedic Sports Physical Therapy*, 21(2), 94–99.

Gefen, A., Megido-Ravid, M., Itzchak, Y., and Arcan, M. (2002). Analysis of muscular fatigue and foot stability during high-heeled gait. *Gait and Posture,* 15, 56–63.

Godfrey, C. M., Lawson, G. A., and Stewart, W. A. (1967). A method for determination of pedal pressure changes during weight-bearing: Preliminary observations in normal and arthritic feet. *Arthritis and Rheumatism*, 10(2), 135–140.

Goonetilleke, R. S. (1999). Footwear cushioning: Relating objective and subjective measurements. *Human Factors*, 41(2), 241–256.

Goonetilleke, R. S. and Luximon, A. (2001). Designing for comfort: A footwear application. *Proceedings of the Computer-Aided Ergonomics and Safety Conference 2001* (July 28–August 2, 2001), Maui, HI.

Goonetilleke, R. S. and Witana, C. P. (March 30, 2010). Method and apparatus for determining comfortable footbed shapes. U.S. Patent No. 7,685,728 B2.

Goske, S., Erdemirb, A., Petreb, M., Budhabhattib, S., and Cavanagh, P. R. (2006). Reduction of plantar heel pressures: Insole design using finite element analysis. *Journal of Biomechanics*, 39(13), 2363–2370.

Grier, T. L., Knapik, J. J., Swedler, D., and Jones, B. H. (2011). Footwear in the United States Army Band: Injury incidence and risk factors associated with foot pain. *The Foot*, 21(2), 60–65.

Han, T. R., Paik, N. J., and Im, M. S. (1999). Quantification of the path of centre of pressure using an F-scan in shoe transducer. *Gait and Posture*, 10, 248–254.

Hansen, A. H. and Childress, D. S. (2004). Effects of shoe heel height on biologic rollover characteristics during walking. *Journal of Rehabilitation Research and Development*, 41, 547–553.

Holtom, P. D. (1995). Necrotizing soft tissue infections. *Western Journal of Medicine*, 163(6), 568–569.

Hong, W. H., Lee, Y. H., Chen, H. C., Pei, Y. C., and Wu, C. Y. (2005). Influence of heel height and shoe insert on comfort perception and biomechanical performance of young female adults during walking. *Foot and Ankle International*, 26, 1042–1048.

Johansson, L., Kjellberg, A., Kilbom, A., and Hagg, G. A. (1999). Perception of surface pressure applied to the hand. *Ergonomics*, 42(10), 1274–1282.

Kurup, H. V., Clark, C. I. M., and Dega, R. K. (2012). Footwear and orthopedics. *Foot and Ankle surgery*, 18(2), 79–83.

Leber, C. and Evanski, P. M. (1986). A Comparison of shoe insole materials in plantar pressure relief. *Prosthetics and Orthotics International*, 10, 135–138.

Lee, C. M., Jeong, E. H., and Freivalds, A. (2001). Biomechanical effects of wearing high-heeled shoes. *International Journal of Industrial Ergonomics*, 28, 321–326.

Lin, C. L., Wang, M. J. J., and Drury, C. G. (2007). Biomechanical, physiological and psychophysical evaluations of clean room boots. *Ergonomics*, 50, 481–496.

Luximon, A. and Goonetilleke, R. S. (2003). Critical dimensions for footwear fitting. *Proceedings of the IEA 2003 XVth Triennial Congress*, August 24–29, 2003, Seoul, South Korea.

Mandato, M. G. and Nester, E. (1999). The effects of increasing heel height on forefoot peak pressure. *Journal of the American Podiatric Medical Association*, 89, 75–80.

McBride, I. D., Wyss, U. P., Cooke, T. D., Murphy, L., Phillips, J., and Olney, S. J. (1991). First metatarsophalangeal joint reaction forces during high-heel gait. *Foot and Ankle*, 11, 282–288.

Mientjesa, M. I. V. and Shorten, M. (2011). Contoured cushioning: Effects of surface compressibility and curvature on heel pressure distribution. *Footwear Science*, 3(1), 23–32.

Milgrom, C., Giladi, M., and Stein, M. (1985). Stress fractures in military recruits. A prospective study showing an unusually high incidence. *Journal of Bone Joint Surgery*, 67B, 732–735.

Newell, S. G. and Bramwell, S. T. (1984). Overuse injuries to the knee in runners. *Physician SportsMed*, 12(3), 81–92.

Nigg, B. M., Herzog, W., and Read, L. J. (1988). Effect of viscoelastic shoe insoles on vertical impact forces in heel-toe running. *American Journal of Sports Medicine*, 16, 70–76.

NIKE Sport Research Review. (September/October 1988). Athletic shoe cushioning, Beaverton, Oregon, NIKE.

NIKE Sport Research Review. (January/February 1990). Physical tests, Beaverton, Oregon, NIKE.

Norton, R. L. (2000). *Machine Design*. Prentice Hall, Upper Saddle River, NJ.

Nyska, M., McCabe, C., Linge, K., and Klenerman, L. (1996). Plantar foot pressure during treadmill walking with high-heel and low-heel shoes. *Foot and Ankle International*, 17, 662–666.

Opila, K. A., Wagner, S. S., Schiowitz, S., and Chen, J. (1988). Postural alignment in barefoot and high-heeled stance. *Spine*, 13(5), 542–547.

Pillar, F. (2012). Euroshoe consortium report: The market for customized footwear in Europe market demand and consumer's preferences, Technishe University Munchen.

Pinhasi, R., Gasparian, B., Areshian, G., Zardaryan, D., and Smith, A. (2010). First direct evidence of chalcolithic footwear from the near eastern highlands. *PLoS One*, 5(6), e10984. doi:10.1371/journal.pone.0010984.

Radin, E. L., Paul, I. L., and Rose, R. M. (1972). Role of mechanical factors in pathogenesis of primary osteoarthritis. *Lancet*, 7749, 519–522.

Riddle, D. L., Pulisic, M., Pidcoe, P., and Johnson, R. E. (2003). Risk factors for plantar fasciitis: A matched case-control study. *Journal of Bone and Joint Surgery*, 85(5), 872–877.

Shimizu, M. and Andrew, P. D. (1999). Effect of heel height on the foot in unilateral standing. *Journal of Physical Therapy Science*, 11(2), 95–100.

Shorten, M. R. (2002). The myth of running shoe cushioning. *IV International Conference on the Engineering of Sport*; September 2002, Kyoto, Japan.

Silvino, N., Evanski, P. M., and Waugh, T. R. (1980). The Harris and Beath footprinting mat: Diagnostic validity and clinical use. *Clinical Orthopaedics*, 151, 265–269.

Simon, S. R., Radin, E. L., Paul, I. L., and Rose, R. M. (1972). The response of joints to impact loading—II. In vivo behavior of subchondral bone. *Journal of Biomechanics*, 5, 267–272.

Singh, D., Angel, J., Bently, G., and Trevino, S. G. (1997). Plantar fasciitis. *British Medical Journal*, 315, 172–177.

Snow, R. E. and Williams, K. R. (1994). High heeled shoes: Their effect on center of mass position, posture, three dimensional kinematics, rearfoot motion, and ground reaction forces. *Archives of Physical Medicine and Rehabilitation*, 75, 568–576.

Stonebrook, I. (2010). Jump offs: A history of performance enhancing basketball shoes. Accessed March 14, 2012. http://www.nicekicks.com/2010/10/jump-offs-a-history-of-performance-enhancing-basketball-shoes/

Tsung, B. Y., Zhang, M., Mak, A. F., and Wong, M. W. (2004). Effectiveness of insoles on plantar pressure redistribution. *Journal of Rehabilitation Research and Development*, 41(6A), 767–774.

Voloshin, A. and Wosk, J. (1982). An in vivo study of low back pain and shock absorption in the human locomotor system. *Journal of Biomechanics*, 15, 21–27.

Weerasinghe, T. W. and Goonetilleke, R. S. (2011). Getting to the bottom of footwear customization. *Journal of Systems Science and Systems Engineering*, 20(3), 310–322.

Witana, C. P., Goonetilleke, R. S., Au, E. Y. L., Xiong, S., and Lu, X. (2009a). Footbed shapes for enhanced footwear comfort. *Ergonomics*, 52(5), 617–628.

Witana, C. P., Goonetilleke, R. S., Xiong, S., and Au, E. Y. L. (2009b). Effects of surface characteristics on the plantar shape of feet and subjects' perceived sensations. *Applied Ergonomics*, 40(2), 267–279.

Xiong, S. (2008). Pressure perception on the foot and the mechanical properties of foot tissue during constrained standing among Chinese. PhD thesis of Industrial Engineering and Logistic Management, Hong Kong University of Science and Technology, Kowloon, Hong Kong.

Xiong, S., Goonetilleke, R. S., and Jiang, Z. (2011). Pressure thresholds of the human foot: Measurement reliability and effects of stimulus characteristics. *Ergonomics*, 54(3), 282–293.

Xiong, S., Rodrigo, A. S., and Goonetilleke, R. S. (2012). Foot characteristics. In: *Science of Footwear*. Ed. R. S. Goonetilleke. CRC Press, Baca Raton, FL.

13 Design of Insoles

Tammy M. Owings and Georgeanne Botek

CONTENTS

13.1 WHAT ARE INSOLES AND WHY DO WE NEED THEM?

Insoles, also referred to as inserts or inlays, are the interior liner placed in a shoe that is in contact with the plantar surface of the foot. Many shoes are manufactured with a removable insole that can be replaced with an insole designed for a specific purpose such as odor control, moisture control, arch support, cushioning, reducing foot pain, or foot realignment. Depending on the required purpose, insoles can be classified as prefabricated, customized, or custom-made.

Prefabricated, or off-the-shelf, insoles are non-customized and are manufactured for some of the basic needs mentioned earlier—odor and moisture control, arch support, and extra-cushioning (Figure 13.1). These insoles are sold over-the-counter at most drug stores, shoe stores, and even grocery and retail stores. They are mass produced and come in a variety of shapes and sizes that can be trimmed to fit a specific shoe size.

Customized, or modified, insoles begin as a prefabricated insole, but have added extra features based on the specific needs of the individual (Figure 13.2). Examples include an additional layer of cushioning along the length of the insole for individuals who spend long periods of time walking or standing on hard surfaces such as concrete, or adding a heel lift to the rear portion of the insole for those with a limb length discrepancy. Minor biomechanical changes can be made to the foot by adding additional pads or wedges to the insole (Table 13.1). A frequent modification is the metatarsal pad that is added to the forefoot region of the insole to redistribute forefoot pressures. Another example of a modifiable feature is the addition of a heel lift added to the posterior region of the insole to correct for an anatomical leg length discrepancy or to treat Achilles tendonitis or ankle equinus.

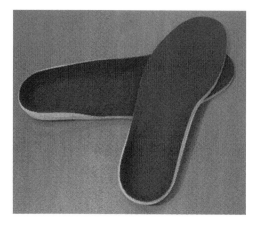

FIGURE 13.1 Prefabricated insoles manufactured to provide extra cushioning. These insoles can be trimmed to match the contour of the shoe.

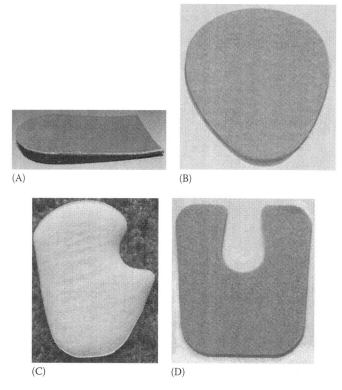

FIGURE 13.2 Examples of modification that can be added to prefabricated insoles. (A)—heel lift for limb length discrepancies and for other therapeutic needs, (B)—metatarsal pad to off-load the metatarsal heads, (C)—dancer pad for off-loading the first metatarsal head, (D)—callus pad for off-loading corns or calluses (traditionally applied directly to the foot, but can also be applied to the foot orthotic).

TABLE 13.1
Common Accommodations for Foot Orthotics

Posts	Used to correct the positioning of the foot.
	Extrinsic posts add material directly to the bottom of the orthotic, whereas intrinsic posts are made to the positive cast or by grinding material away from the shell of the orthotic.
	Forefoot posts: used to invert or evert the forefoot.
	Rearfoot posts: material added to the medial region of the rear of the foot orthotic used to correct overpronated feet or to the lateral region for supinated feet.
Pads	*Metatarsal pad* (Figure 13.2B): A pad placed proximal to the metatarsal heads. Functions to displace pressure from the metatarsal heads to the metatarsal shafts. Used to treat problems at the forefoot or at the transverse arch.
	Dancer pad (Figure 13.2C): A variation of a metatarsal pad that is used to off-load the hallux joint. Used for conditions such as hallux limitus or sesamoiditis.
	Neuroma pad: A smaller, narrow metatarsal pad placed in the innerspace between the plantar metatarsal heads. Functions to increase the space between the metatarsals to relieve symptoms caused by an interdigital neuroma or intermetatarsal bursitis.
	Forefoot pad: Additional padding placed at the distal portion of orthotic to cushion metatarsal heads.
	Arch pad: A pad along the medial aspect of the foot orthotic to provide support and cushioning to the medial arch.
	Cuboid pad: A pad beneath the cuboid bone that raises the medial border of the bone.
	Heel pad: A pad covering the heel cup to cushion the heel. Use to increase shock absorption at the rearfoot and to treat plantar fasciitis.
	Horseshoe pad: A U-shaped pad that surrounds the outer rim of the heel cup to create a depression in the central area to off-load the calcaneus. Used to treat heel spurs or calcaneal bursitis.
Depressions	Indentations in the shell material to off-load specific areas. Examples include a plantar fascial groove, navicular accommodations, and sub-metatarsal accommodations.
Flange	*Medial flange*: Extension of material along the medial border of the foot orthotic to control the motion of the foot. Used for foot pronation, pes planus, and stabilization of the midtarsal and subtalar joints.
	Lateral flange: Extension of material along the lateral border of the foot orthotic to control the motion of the foot. Used for foot supination, pes cavus, and protection from inversion ankle sprains.
Cut-outs	*First metatarsophalangeal cut out*: A notch cut from the orthotic shell below the first metatarsophalangeal joint which allows pronation of the forefoot and for slight plantarflexion of the first metatarsal. Used to treat functional hallux limitus and forefoot supination.
	Heel spur cut-out: A notch cut from the orthotic shell under the center of the heel. Used to treat heel spurs.

(continued)

TABLE 13.1 (continued)
Common Accommodations for Foot Orthotics

Fillers	*Toe filler* (Figure 13.4): Material added to the top of the foot orthosis to fill the region vacated following a toe amputation.
	Arch filler: Material added to the bottom of the foot orthotic under the arch to reinforce the region. Used to prevent arch collapse for heavier patients.
Extensions	*Morton's extension*: Material placed under the first metatarsophalangeal joint to limit the range of motion of the hallux. Used for treating turf toe and hallux limitus or rigidus.
	Reverse Morton's extension: Material placed under the second to fifth metatarsophalangeal joints to increase the range of motion of the hallux. Used to treat hallux limitus, sesamoiditis, and valgus deformity of the forefoot.
	Varus extension: Material placed in the forefoot region that is thicker on the medial side and thinner on the lateral side. Used to treat a varus forefoot.
	Valgus extension: Material placed in the forefoot region that is thicker on the lateral side and thinner on the medial side. Used to treat a valgus forefoot.
Heel lift (Figure 13.2A)	Material added to the posterior region of the orthotic heel as a rearfoot post which functions to raise the calcaneus. A heel lift can also be a separate wedge not attached to the foot orthotic. Used to correct a congenital or acquired limb length discrepancy, or used to treat Achilles tendonitis or ankle equinus.
Deep heel cup	Increased depth of the heel cup to provide increased stability for an overpronated foot.
Metatarsal bar	A bar added across the foot orthotic proximal to all metatarsal heads. Functions to displace pressure from the metatarsal heads to the metatarsal shafts.

Custom-made insoles, or custom foot orthotics, are fabricated from a partial or non-weight-bearing three-dimensional impression of the individual's foot and require a prescription from a physician. Custom-made insoles are individualized for each patient and designed to change the function and biomechanics of the foot secondary to deformities in the shape of the foot, improper positioning of the foot during gait, or other discomforts caused by injury, overuse, or disease (Figures 13.3 and 13.4; Table 13.1).

The choice between prefabricated/customized insoles and custom foot orthotics is at the discretion of the health care provider (i.e., podiatric physician, chiropractor, and physical therapist) and is dependent on a number of variables (Pedorthic Association of Canada 2008). For convenience and cost effectiveness, a prefabricated or customized insole would be chosen for minor abnormalities or to determine if a patient could tolerate the type of support provided by a foot orthotic. Prefabricated insoles can also be used as an immediate treatment option while a permanent custom foot orthotic is being fabricated. Once a foot orthotic is dispensed, a prefabricated insole could also be used within a secondary pair of shoes that are only worn for a limited period of time. Prefabricated insoles may also be indicated for self-limited foot pain caused by a temporary condition such as pregnancy or an acute injury. Custom-made foot orthotics often are recommended for a chronic condition that requires more accurate

FIGURE 13.3   Custom-made insoles designed with a metatarsal pad on the left and a depression at the base of the fifth metatarsal on the right.

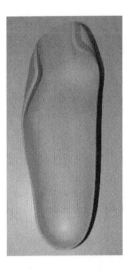

FIGURE 13.4   Custom insole with toe fillers to accommodate multiple toe amputations.

and long-standing intervention. Custom foot orthotics should have a longer lifespan than prefabricated insoles and should be able to provide more options for modifications. The design of custom-made insoles is initiated with an identification of the treatment goals and will be the focus of the remainder of this chapter.

13.2   CUSTOM FOOT ORTHOTICS

Depending on the underlying etiology, custom-made insoles (henceforth referred to as custom foot orthotics) are classified as being either functional or accommodative. *Functional* foot orthotics are designed to correct and maintain the positioning and motion of the foot during gait, primarily by controlling the subtalar and midtarsal joints.

To control the motion of the foot, a rigid material is needed to maintain stability without allowing deformity of the material. This material is used as the shell of the orthotic, which cradles the foot from the back of the heel to the proximal forefoot, thus supporting the heel and midfoot arches. Excessive pronation or supination of the foot is an example of an abnormal foot function that can be controlled with functional foot orthotics. A rearfoot or forefoot post can be utilized to tilt the foot into proper alignment.

Accommodative foot orthotics are designed to protect and off-load problematic regions of the foot by redistributing high plantar pressures to other regions of the foot. People with diabetes, especially those with peripheral neuropathy, begin to develop excessive plantar pressures at the metatarsal heads and toes due to changes in the foot structure, such as improper alignment of the bones caused by muscle atrophy (Myerson and Shereff 1989) and displacement of the submetatarsal fat pads (Bus et al. 2004). Accommodative foot orthotics are often prescribed for these individuals to prevent calluses from developing into foot ulcers. To reduce pain and redistribute high plantar pressures, a softer material that will absorb shock and cushion the foot is needed.

13.3 TRADITIONAL METHODS FOR DESIGNING AND FABRICATING INSOLES

The success of foot orthotics is initiated with the written prescription from a physician who is experienced in treating foot problems. A thorough biomechanical examination of the foot as well as an observation of gait is essential for determining the underlying etiology. A few of the assessments that the physician will make include the mobility of the first metatarsophalangeal and ankle joint, arch height (i.e., pes planus, pes cavus, or neutral), hindfoot valgus or varus, and any bony prominences. Once the etiology is identified, the physician must decide on treatment goals such as reducing pain, redistribution of excessive plantar pressures, improving the functional ability of the patient, and/or slowing or preventing further development of the disorder. These goals provide guidelines as to the modifications that should be made to the foot orthotic (Table 13.1), as well as the materials to be used (The American College of Foot & Ankle Orthopedics & Medicine 2006).

The next step is to capture the morphology or shape of the patient's foot. This step is equally as important as the written prescription for determining the quality of the foot orthotic. Typically, the impressions are obtained using plaster casts, fiberglass stockings, foam blocks, or wax. Proper positioning of the foot, with the subtalar joint in a neutral position, is necessary to produce an accurate cast. Any deviations for this position will influence the quality and effectiveness of the foot orthotic by not providing a true impression of the foot shape. Historically, a plaster slipper cast is one of the most common casting methods (Marzano 1998). With the patient seated comfortably with the leg extended, the casting splint is dipped into water, applied around the outside of the foot, and then folded over to cover the entire plantar surface. While the plaster is still wet, the fourth and fifth metatarsal heads are gently pushed until the subtalar joint is in a neutral position. This position must be maintained until the plaster has hardened. The slipper cast is then removed

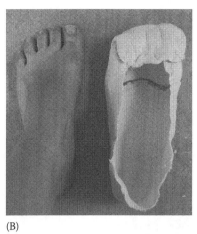

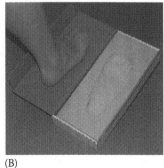

FIGURE 13.5 (A)—Plaster slipper cast for determining foot shape. (B)—Resulting negative foot impression.

resulting in a negative impression of the plantar surface of the foot (Figure 13.5). Any areas that will be modified can be marked directly on the cast.

An alternative casting method that is cleaner and less time-consuming is impression foam casting. The patient should be seated comfortably with their thigh parallel to the floor, their knee bent to a 90° angle, and the lower shank perpendicular to the floor. With the patient's foot placed gently onto the center of the foam, the clinician stabilizes the patient's ankle while slowly pushing downward on the patient's knee forcing the foot into the foam. Additional force is applied to the top of the foot and the toes, so that the heel, forefoot, and toes have been depressed to the same depth. The foot is then carefully removed from the foam revealing the negative impression (Figure 13.6). Problematic areas to be off-loaded can be marked directly on the foam.

Regardless of the casting method employed, the next step is to create a positive plaster cast of the foot by filling the negative foot impression with plaster. Once the plaster has hardened, the positive cast is sanded to a smooth finish and alterations

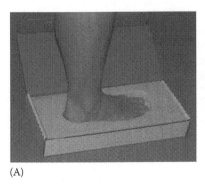

FIGURE 13.6 (A)—Foam box cast for determining foot shape. (B)—Resulting negative foot impression.

can be applied based on the physician's prescription for modifications. This may include the addition or removal of plaster to form a raised or depressed region. Next, the base material for the foot orthotic is heated and molded around the positive cast using a vacuum former. Depending on the prescription, additional layers of other materials are added to obtain the desired function. The excess material is removed by cutting and grinding the orthotic and is shaped to fit the intended footwear (Figure 13.7). Slight modifications may be required once the patient dons the footwear and begins to ambulate.

In most cases, a patient's initial set of foot orthotics will be fabricated to fit within a pair of running or walking shoes. These initial orthotics will be designed in a way to provide maximal control and protection of the foot and run along the entire

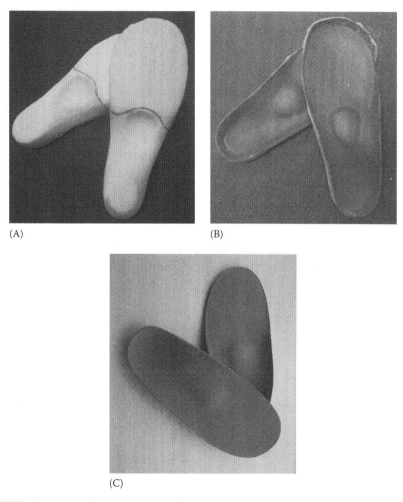

(A)   (B)

(C)

FIGURE 13.7   (A)—Plaster mold from foam box impressions. Plaster has been removed to allow for placement of a metatarsal pad to off-load the metatarsal heads. (B)—Untrimmed fabricated orthotic from the plaster mold with the resulting metatarsal pad. (C)—Finished orthotic to be distributed to patient.

length of the foot. Depending on the pathology, additional or future orthotics can be made to fit into other shoe types such as dress shoes or sport-specific shoes. In these cases, alterations in materials and design can be used to produce foot orthotics that are shorter, thinner, or narrower while maintaining sufficient control and protection.

13.4   MATERIALS USED FOR MAKING INSOLES

There is a broad spectrum of materials from which to choose, and multiple materials (or combination of materials) may satisfy these objectives (Table 13.2). The mechanical properties of the materials, such as compression, density, durability, flexibility, friction, hardness, resilience, response to temperature, and stiffness, dictate their role in the design of an insole (Nicolopoulos et al. 2000; Paton et al. 2007). Using a combination of various materials can enhance the effectiveness of the orthotic. The desired characteristics of one material can be utilized while the undesired effects can be minimized by introducing additional materials with differing properties. Physicians and orthotists must remain up to date and knowledgeable on available material choices. The final selection of materials for a specific orthotic may be based on preference of the physician or physician experience, cost, and the available resources of local orthotic laboratories.

Based on the material properties, foot orthotics can be classified as rigid, semi-rigid, or soft. Functional foot orthotics are generally rigid to control the motion of the foot and thus should be made of materials with minimal flexibility, such as plastics and carbon fiber composites. Accommodative foot orthotics should be soft to provide shock absorption and cushion the foot, or to disperse high pressure from problem areas of the foot. Materials include polyethylene foams, ethylene vinyl acetate (EVA), leather, and cork. A combination of materials can be utilized to fabricate a semi-rigid foot orthotics that provides the control of the rigid orthotic, but additional cushioning for comfort. In general, when a multilayered orthotic is designed, the top layer is primarily for durability and to offer a comfortable interface between the foot and the orthotic. The middle layer, or layers, is for cushioning and shock absorption, while the bottom layer (shell) provides a rigid support to the orthotic and protects it from wear. Depending on the prescription, the length of the foot orthotic or of a specific layer can be specified according to the distal length. *To Toes* refers to a foot orthotic or layer that extends the entire length of the foot, *To Sulcus* extends to the distal end of the metatarsal heads, and *To Metatarsals* extends to the proximal end of the metatarsal heads.

Additional factors that contribute to the selection of materials employed during fabrication include the weight of the patient, the activity level of the patient, and the type/style of shoe the patient will be wearing. For example, an accommodative foot orthotic made of Plastazote® may be sufficient for an 80 kg patient, but would be too porous (i.e., containing small pores or holes) to be used for a 115 kg patient. In this case, a layer of a denser material such as EVA would be added to prevent the orthotic from compressing to an ineffective depth (i.e., bottoming out). Similarly, the orthotic of a young, active patient would require a more durable material than that of an older, more sedentary patient. As for shoe style, a majority of foot orthotics are made to fit within an athletic or walking shoe, and would be interchangeable within

TABLE 13.2

Common Materials Used to Fabricate Foot Orthotics

Thermoplastics	Description: Plastic that turns soft when heated and hard when cooled
	Polypropylene
	Usage: shell material
	Advantages: lightweight, high stiffness
	Disadvantages: prone to cracking if notches or grooves are introduced
	Trade name: none
	Polyethylene
	Usage: shell material
	Advantages: durable, lightweight, flexible, malleable, and resilience to compression
	Disadvantages: limited shock absorption
	Trade names: Ortholen and Subortholen
Acrylic	Description: methyl methacrylate polymer
	Usage: shell material
	Advantages: tough and remoldable
	Disadvantages: prone to cracking
	Trade names: Polydur and Plexidur
Carbon fiber composite	Description: combination of acrylic plastic with carbon fibers
	Usage: shell
	Advantages: thin, ultralight weight, rigid, extremely durable, and retains shape
	Disadvantages: difficult to cut and inability to remold
	Trade names: as Carboplast, Graphite, TL-series
Foams	Description: cellular material classified as closed cell (isolated air pockets) or open cell (porous air pockets) with varying densities, such as soft or firm
	EVA (ethyl vinyl acetate) foam
	Description: closed-cell material
	Usage: topcover (soft), accommodations (firm), shell (firm)
	Advantages: shock absorption (soft), durable, resists tearing, rigid (firm), heat moldable
	Disadvantages: compressibility (soft), limited cushioning (firm)
	Trade name: P-cell®
	Polyurethane foam
	Description: open-cell material
	Usage: top layer, cushioning
	Advantages: high shock absorption, resists compression
	Disadvantages: durability as a top cover and not heat moldable
	Trade names: Poron and Sorbothane
	Polyethylene foam
	Description: closed-cell material
	Usage: top cover (soft), shell (firm)
	Advantages: tough, durable, malleable, lightweight, flexible, heat moldable
	Disadvantages: moderate shock attenuation, easily compressed
	Trade names: Plastazote®, Pelite, Aliplast, Dermaplast

TABLE 13.2 (continued)
Common Materials Used to Fabricate Foot Orthotics

Leather	Description: cowhide
	Usage: top cover
	Advantages: low friction, durable, thin, tear resistant, absorbs moisture
	Disadvantages: no cushioning, remains damp, and stains
Cork	Description: natural cellulose
	Usage: posting
	Advantages: lightweight
	Disadvantages: durability
	Trade names: Korex, Thermocork
Spenco	Description: closed-cell expanded rubber
	Usage: top cover, extensions
	Advantages: soft cushioning system to help absorb shock and reduce friction
	Disadvantages: durability
Vinyl	Description: Synthetic leather-type material
	Usage: top cover
	Advantages: tough, long-lasting, extremely thin, resists tearing, does not absorb moisture, minimal friction
	Disadvantages: no cushioning

Source: Nicolopoulo, C.S. et al., *Foot*, 10, 1, 2000.

similar shoe styles. In certain situations, a slimmer orthotic would be required to accommodate a patient who primarily wears a dressier style of shoe. For shoes having less available toe room, a three-quarter length orthotic may be required. This type of orthotic provides support from the heel to the proximal forefoot and thus is not appropriate for pathologies involving the metatarsal heads or the toes.

13.5   SPECIALIZED INSOLE DESIGNS FOR SPECIFIC PATHOLOGIES

Foot orthotics can be used to manage numerous pathologies, where altered mechanics or redistribution of plantar forces is needed. While patients may have similar pathologies, the etiology may be different and require slightly different prescriptions to optimize the performance of their foot orthotics. As mentioned earlier, the success of custom foot orthotics relies on a prescription from a physician who is experienced in treating foot problems. The following are a few examples of pathologies that are often treated with foot orthotics.

Plantar fasciitis is caused by inflammation of the connective tissue (plantar fascia) located on the plantar surface of the foot that extends from the calcaneus to the toes of the foot. This inflammation can cause sharp heel pain that is post-static dyskinetic or more prominent in the morning or after periods of inactivity but generally dulls as the day progresses. Causes of plantar fasciitis include excessive tension on the plantar fascia due to pronation, collapsing of the arch, tightness in the heel cord, or overuse. Equinus, excessive weight, and walking on hard or concrete surfaces have been associated with the onset. Conservative treatment options may include rest, stretching, night splints, strengthening, anti-inflammatory medications, ice, and orthotic devices (Dyck and Boyajian-O'Neill 2004; McPoil et al. 2008). A common design for a custom foot orthotic used to treat plantar fasciitis is a three-quarter to full-length, semi-rigid device with a longitudinal arch support (Gross et al. 2002). These types of devices have been shown to be effective in controlling some of the biomechanical factors that may lead to plantar fasciitis, such as overpronation, pes planus, motion of the first metatarsal head, and valgus heel alignment (Dyck and Boyajian-O'Neill 2004). Evidence-based guidelines from The Orthopaedic Section of the American Physical Therapy Association states that custom foot orthoses, as well as prefabricated insoles, can provide short-term (3 months) pain reduction and functional improvement, although no evidence is currently available to support the use of these devices for long-term (1 year) use (McPoil et al. 2008).

Rheumatoid arthritis is a chronic inflammatory disorder affecting the lining of the peripheral joints resulting in painful swelling, bone erosion, and joint deformity. Within the foot, these conditions can lead to hallux valgus, subluxation of the metatarsophalangeal joint, hammer toes, altered angle motion, and increased forefoot plantar pressures (Grondal et al. 2008). The problem to be addressed determines the appropriate design of the foot orthotic for patients with rheumatoid arthritis. A rigid foot orthotic would be suitable for controlling pronation, reducing pressure at the forefoot, and reducing the risk of hallux valgus deformity, whereas soft orthotics would be used for reducing generalized foot pain (Riskowski et al. 2011). Limited research has shown that foot orthotics can reduce pain, improve physical function, and lead to ankle joint stability in patients with rheumatoid arthritis (Clark et al. 2006; Hawke et al. 2008; Oldfield and Felson 2008).

Osteoarthritis is a degenerative joint disease causing pain, stiffness, and reduced mobility. It is generally accepted that medial compartmental knee osteoarthritis is associated with increased knee adduction moments during walking. Foot orthotics with a lateral wedge are commonly prescribed to reduce that adduction moment at the knee. While biomechanical studies have shown that the use of lateral heel wedges will decrease knee adduction moment (Butler et al. 2007; Hinman et al. 2008), a systematic review of the literature presents conflicting recommendations for the use of lateral wedges (Gelis et al. 2008; Zhang et al. 2008; Richmond et al. 2009). It has been suggested that lateral wedge foot orthotics can be beneficial if the wedge is placed along the full length of the orthotic and not only at the heel, the wedge is worn in shoes with flat heels and without a medial arch support, and the patient is younger, less obese, and has less severe symptoms (Hinman and Bennell 2009).

Posterior tibial tendon dysfunction is an overuse injury resulting in inflammation, overstretching, or partial/complete tearing of the posterior tibial tendon in the foot. Symptoms include pain, swelling, flattening of the arch, and ankle pronation. Conservative treatment for early stage posterior tibial tendon dysfunction may include taping, anti-inflammatory medications, physical therapy, and orthotic devices (The American College of Foot & Ankle Orthopedics & Medicine 2006). A full-length, functional foot orthotic with medial posting to control rearfoot pronation and support the medial arch is most commonly recommended (Neville and Houck 2008).

Turf toe is an injury affecting the ligament of the metatarsophalangeal joint generally caused by excessive hyperextension of the hallux. This type of injury most commonly occurs in athletes playing on artificial surfaces. A treatment option for an athlete who has sustained this type of injury may include a foot orthotic to limit toe dorsiflexion. This may include posting under the first metatarsal head or a rigid (carbon fiber composite) insert. In either case, a rigid soled shoe is also recommended.

Hallux limitus, where motion of the metatarsophalangeal joint is limited, and hallux ridigus, where motion of the metatarsophalangeal joint is restricted, are other conditions that affect the mobility of the hallux. Similar to turf toe, a treatment option would be a rigid (carbon fiber composite) insert or Morton's extension that is designed to limit the range of motion of the hallux joint.

Diabetic peripheral neuropathy is associated with a loss of sensation in the extremities. Patients with this disorder should examine their feet daily for detection of redness, blisters, calluses, or ulceration. To reduce the risk of foot trauma, it is essential for these patients to avoid walking barefoot and to be in proper footwear. Accommodative foot orthotics are generally prescribed to off-load areas of high pressure to reduce callus formation and the risk of ulceration. Some of the most commonly prescribed modifications include metatarsal bars, metatarsal pads, or depressions to off-load the problem areas of the forefoot. Patients with diabetes and who have peripheral neuropathy can progress to an inflammatory condition known as Charcot arthropathy, which results in deterioration of the bone and soft tissue of the foot and ankle leading to dislocation, fracture, and deformities. Once these patients move from the acute to a chronic stage of this condition, management typically advances to braces or custom orthotics (Cavanagh et al. 2010; Lavery and Brawner 2010). Care must be taken to accommodate the resulting foot shape and to provide padding and protection to the midfoot prominence. A combination of materials will generally be used to provide both cushioning and support of the foot. In some instances, a more severely deformed Charcot foot cannot be managed with a custom foot orthotic and would require an ankle foot orthotic or a patellar tendon bearing brace to provide additional support.

Regardless of the pathology, custom foot orthotics can only be effective if the patient is compliant about wearing the device. The patient needs to be educated about their foot problems, how the foot orthotic is being designed to treat the problem, and the expected outcomes. It is also important for the patient to return for follow-up visits to verify that treatment goals are being attained with the foot orthotics or if minor modifications are required. Long-term follow-up visits will allow the physician to monitor the effectiveness of the foot orthotics in managing the foot problems and determining when or if new foot orthotics are needed.

13.6   NEWER/FUTURE TRENDS

Over the last decade, newer technologies have emerged for determining the morphology of the foot (Figure 13.8); the two most prominent methods being contact digitizing and laser scanning. With contact digitizing, the patient's foot is placed upon an array of small pins that conform to the shape of the foot. As with plaster casting and foam impressions, the foot is placed with the subtalar joint in a neutral position. A computer then records the positioning of the pins and a 3D image of the foot is rendered. While contact digitizing requires the computer

FIGURE 13.8   Laser scanner for creating a 3D image of the foot morphology.

to interpolate the foot surface occurring between adjacent pins, laser scanning employs multiple lasers to create a "continuous" 3D image of the foot. For both methods, 3D graphical images can be modified using computer-aided design software, thus eliminating the messiness of the plaster cast. Following modifications to the 3D image, a computer-aided milling machine can carve the foot orthotic. Research has shown that these 3D scanning systems can provide accurate representation of the patient's foot for use within the design of custom foot orthotics (Telfer and Woodburn 2010).

Incorporating plantar pressure distribution into the foot orthotic design is a newer method for off-loading the metatarsal heads. Historically, placement of metatarsal pads and bars within the foot orthotic is subjective and reliant on the experience of the orthotist or pedorthist. Research investigating plantar pressures has shown that the efficacies of metatarsal pads are highly sensitive to small changes in placement (Hsi et al. 2005; Hastings et al. 2007). With recent developments, the pressure distribution at the plantar surface of the feet as a patient walks across an instrumented platform can be input into computer algorithms, which optimize the metatarsal placement for off-loading (Cavanagh and Owings 2006). This provides a quantitative measurement that has shown enhanced off-loading capabilities (Owings et al. 2008).

As technology continues to advance, orthotic designs that incorporate sensors for determining step count, temperature, humidity, pressure, or shear will become available. Feedback from these sensors could alert the patient when they reach potentially harmful levels. The future of foot orthotic design and manufacturing is slowly progressing into utilizing digital technology (Foster 2011). While the use of digital technology to design and manufacture foot orthotics may be less time-consuming and more cost-effective, ultimately, the effectiveness and quality of these foot orthotics will need to be investigated.

13.7 SUMMARY

The design of foot orthotics is based on a number of factors. After the etiology of the foot problem is identified and treatment goals are determined, the decision between prefabricated, customized, and custom-made insoles is made. If a custom-made insole is prescribed, the physician then chooses between a functional and an accommodative orthotic, identifies suitable modifications, and selects the appropriate materials. Finally, the effectiveness of custom foot orthotics begins with patient compliance; the best foot orthotics are worthless if they are never used. Overall, the success of foot orthotics is a result of a good design, proper fabrication, patient education, and long-term follow-up with the health care provider.

QUESTIONS

13.1 What are some possible advantages in using a customized foot insole versus a custom-made foot orthotic?

13.2 Why would a health care provider be more likely to prescribe accommodative foot orthotics instead of functional foot orthotics for a patient with diabetic neuropathy?

13.3 What are the different techniques used for casting custom-made foot orthotics? Why is capturing the shape of the foot an important step in the design of foot orthotics?

13.4 Does the weight of the patient have any effect on the type of insoles prescribed? Are there specific material properties that might be preferred?

13.5 Are there specific conditions that should be treated with custom-made foot orthotics?

13.6 When determining the material to be used for a functional, custom-made, foot orthotic, what advantages do carbon fiber composites offer over thermoplastics?

13.7 How has advances in technology changed the foot orthotic fabrication process?

13.8 When designing a foot orthotic that will function to control foot motion such as supination or pronation, what are some potential accommodations to consider?

REFERENCES

Bus, S. A., Maas, M., Cavanagh, P. R., Michels, R. P., Levi, M. 2004. Plantar fat-pad displacement in neuropathic diabetic patients with toe deformity: A magnetic resonance imaging study. *Diabetes Care* 27:2376–2381.

Butler, R. J., Marchesi, S., Royer, T., Davis, I. S. 2007. The effect of a subject-specific amount of lateral wedge on knee mechanics in patients with medial knee osteoarthritis. *J Orthop Res* 25:1121–1127.

Cavanagh, P. R., Botek, G., Owings, T. M. 2010. Biomechanical factors in Charcot's neuroarthropathy. In: *The Diabetic Foot: Principles and Management*, ed. R. G. Frykberg, pp. 131–142. Brooklandville, MD: Data Trace Publishing Company.

Cavanagh, P. R., Owings, T. M. 2006. Nonsurgical strategies for healing and preventing recurrence of diabetic foot ulcers. *Foot Ankle Clin* 11:735–743.

Clark, H., Rome, K., Plant, M., O'Hare, K., Gray, J. 2006. A critical review of foot orthoses in the rheumatoid arthritic foot. *Rheumatology (Oxford)* 45:139–145.

Dyck, D. D. Jr., Boyajian-O'Neill, L. A. 2004. Plantar fasciitis. *Clin J Sport Med* 14:305–309.

Foster, J. B. 2011. Custom orthotic insoles technology forum. *Lower Extremity Rev* 6:25–28.

Gelis, A., Coudeyre, E., Hudry, C., Pelissier, J., Revel, M., Rannou, F. 2008. Is there an evidence-based efficacy for the use of foot orthotics in knee and hip osteoarthritis? Elaboration of French clinical practice guidelines. *Joint Bone Spine* 75:714–720.

Grondal, L., Tengstrand, B., Nordmark, B., Wretenberg, P., Stark, A. 2008. The foot: Still the most important reason for walking incapacity in rheumatoid arthritis: Distribution of symptomatic joints in 1,000 RA patients. *Acta Orthop* 79:257–261.

Gross, M. T., Byers, J. M., Krafft, J. L., Lackey, E. J., Melton, K. M. 2002. The impact of custom semirigid foot orthotics on pain and disability for individuals with plantar fasciitis. *J Orthop Sports Phys Ther* 32:149–157.

Hastings, M. K., Mueller, M. J., Pilgram, T. K., Lott, D. J., Commean, P. K., Johnson, J. E. 2007. Effect of metatarsal pad placement on plantar pressure in people with diabetes mellitus and peripheral neuropathy. *Foot Ankle Int* 28:84–88.

Hawke, F., Burns, J., Radford, J. A., du Toit, V. 2008. Custom-made foot orthoses for the treatment of foot pain. *Cochrane Database Syst Rev* CD006801.

Hinman, R. S., Bennell, K. L. 2009. Advances in insoles and shoes for knee osteoarthritis. *Curr Opin Rheumatol* 21:164–170.

Hinman, R. S., Payne, C., Metcalf, B. R., Wrigley, T. V., Bennell, K. L. 2008. Lateral wedges in knee osteoarthritis: What are their immediate clinical and biomechanical effects and can these predict a three-month clinical outcome? *Arthritis Rheum* 59:408–415.

His, W. L., Kang, J. H., Lee, X. X. 2005. Optimum position of metatarsal pad in metatarsalgia for pressure relief. *Am J Phys Med Rehabil* 84:514–520.

Lavery, L. A., Brawner, M. 2010. Footwear in Charcot arthropathy. In: *The Diabetic Foot: Principles and Management*, ed. R. G. Frykberg, pp. 157–164. Brooklandville, MD: Data Trace Publishing Company.

Marzano, R. 1998. Fabricating shoe modifications and foot orthoses. In: *Introduction to Pedorthics*, ed. D. Janisse, pp. 147–152. Columbia, MD: Pedorthic Footwear Association.

McPoil, T. G., Martin, R. L., Cornwall, M. W., Wukich, D. K., Irrgang, J. J., Godges, J. J. 2008. Heel pain—Plantar fasciitis: Clinical practice guidelines linked to the international classification of function, disability, and health from the orthopaedic section of the American Physical Therapy Association. *J Orthop Sports Phys Ther* 38:A1–A18.

Myerson, M. S., Shereff, M. J. 1989. The pathological anatomy of claw and hammer toes. *J Bone Joint Surg Am* 71:45–49.

Neville, C. G., Houck, J. R. 2008. Science behind the use of orthotic devices to manage posterior tibial tendon dysfunction. *Tech Foot Ankle Surg* 7:125–133.

Nicolopoulo, C. S., Black, J., Anderson, E. G. 2000. Foot orthoses materials. *Foot* 10:1–3.

Oldfield, V., Felson, D. T. 2008. Exercise therapy and orthotic devices in rheumatoid arthritis: Evidence-based review. *Curr Opin Rheumatol* 20:353–359.

Owings, T. M., Woerner, J. L., Frampton, J. D., Cavanagh, P. R., Botek, G. 2008. Custom therapeutic insoles based on both foot shape and plantar pressure measurement provide enhanced pressure relief. *Diabetes Care* 31:839–844.

Paton, J., Jones, R. B., Stenhouse, E., Bruce, G. 2007. The physical characteristics of materials used in the manufacture of orthoses for patients with diabetes. *Foot Ankle Int* 28:1057–1063.

Pedorthic Association of Canada. 2008. Position statement on casting techniques for custom foot orthoses. www.pedorthic.ca

Richmond, J., Hunter, D., Irrgang, J., Jones, M. H., Levy, B., Marx, R., Snyder-Mackler, L. et al. 2009. Treatment of osteoarthritis of the knee (nonarthroplasty). *J Am Acad Orthop Surg* 17:591–600.

Riskowski, J., Dufour, A. B., Hannan, M. T. 2011. Arthritis, foot pain and shoe wear: Current musculoskeletal research on feet. *Curr Opin Rheumatol* 23:148–155.

Telfer, S., Woodburn, J. 2010. The use of 3D surface scanning for the measurement and assessment of the human foot. *J Foot Ankle Res* 3:19.

The American College of Foot & Ankle Orthopedics & Medicine. 2006. Prescription custom foot orthoses practice guidelines. www.acfaom.org

Zhang, W., Moskowitz, R. W., Nuki, G., Abramson, S., Altman, R. D., Arden, N., Bierma-Zeinstra, S. et al. 2008. OARSI recommendations for the management of hip and knee osteoarthritis, Part II: OARSI evidence-based, expert consensus guidelines. *Osteoarthritis Cartilage* 16:137–162.

14 Design of Custom Shoe Lasts for Challenging Feet

Carl G. Saunders, Claudia Kieserling, and Johan Steenwyk

CONTENTS

14.1 INTRODUCTION

This chapter discusses the automated design of custom shoe lasts for challenging feet. The term "challenging" implies a difficulty in being able to achieve both comfortable fit and attractive styling.

Two distinct case studies are presented. The first is a "retail" situation in which a customer wants to wear fashionable shoes, but is limited by an abnormality on the big toe of one foot. The second involves a diabetic patient with neurological pathology who desperately needs functional shoes in order to ambulate. Style will be seen to influence the clients' acceptance of their finished footwear. However, in one case the shape of a proven shoe last drives the custom shoe last design process, while in the other, the shape of the customer's foot is the overriding element. It should be noted that the conventional footwear industry offers no solutions for these two individuals and many others like them.

14.2 CASE STUDY OF A "DIFFICULT-TO-FIT" FOOT IN A RETAIL SETTING

14.2.1 PROBLEM FOOT

In women's retail footwear, a typical "difficult-to-fit" scenario is a foot with an enlargement of the bone and tissue around the joint at the base of the great toe, also known as a bunion (hallux valgus). This mostly genetic condition is usually accompanied with an inward deviation of the big toe toward the second toe. In terms of foot morphology, the net effect is typically a width increase at the ball of the foot, while the rest of the foot remains of normal or narrow width. This will often preclude a client from being able to wear fashionable or elegant shoes. If she purchases for increased width to accommodate her bunion, then the rest of the shoe will simply be too wide. Feminine styles, like pumps without laces or other fastenings, become impossible to fit.

In this case study, a woman presents with a moderate bunion on her left foot (see Figure 14.1).

To date, she has compromised on materials (softer and more open uppers) whenever she has chosen stylish shoes with pointed toe character (see Figure 14.2).

No longer willing to compromise fashion for function, she has visited a specialty retail shoe company to help realize her desire to wear attractive black patent leather pumps that fit her feet well and do not cause her pain. In this example, style is the primary objective; function is secondary, albeit a very high second priority.

14.2.2 PROPOSED SOLUTION

The bunion on the left foot, combined with the client's desire for a fashion shoe, necessitated the design of a custom left shoe last. It was proposed that three-dimensional (3D) shape scans be taken of both feet. The right foot would be fitted using stock sizes of classic pumps. Once size and style were determined for the

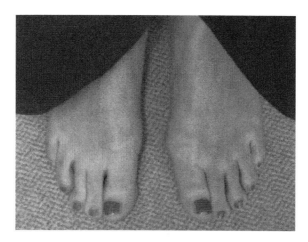

FIGURE 14.1 Typical, moderate bunion on the left foot.

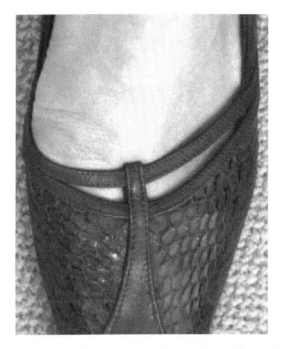

FIGURE 14.2 Soft upper material as a compromise to achieve fit in standard shoe.

right foot, the 3D data corresponding to the shape of the left foot could be compared to the mirror image of the 3D shoe last for the right foot. If deemed aesthetically feasible, the in-store shoe designer could then digitally modify the mirrored right shoe last to accommodate the client's bunion. If not, an alternative style could be recommended and tried.

14.2.3 PROCESS IMPLEMENTED

At the specialty retail shoe store, the first step was to take weight-bearing foot scans of both feet using the Yeti™ Foot Scanner. Measurements were extracted and reviewed to determine potential sample shoes and sizes from a collection of standard pumps of varying toe shapes and heel heights. Physical samples were then tried on her right foot, until a mostly comfortable fit was attained. The client was specifically encouraged to point out any areas of discomfort, so that these could be addressed during the manufacturing of the shoe.

Once the best-fitting right shoe was determined, the corresponding left shoe was tried on, and as expected, it did not fit. The left shoe was too narrow at the ball of the foot (where the bunion was). The in-house shoe last designer then used the Canfit™ FootWare™ Design software to review and compare the 3D data corresponding to the client's feet and the shoe lasts for the selected pump style. Since the right foot had normal anatomy and shape, no changes were needed to the stock right shoe last. For the left foot, the software was used to create a mirrored version of the right shoe last and then expand it medially in the area of the bunion (see Figure 14.3). In such a case,

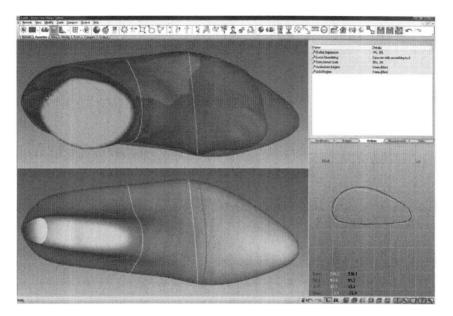

FIGURE 14.3 CAD design screen showing custom expansion of shoe last in the area of the bunion.

width will be adjusted to be the same (±0.5 mm) and bounds set to blend, in order to create an attractive shape. Care was taken to ensure that the aesthetic appeal of the last shape was maintained.

Next, the cosmetic and construction aspects of the custom shoes were specified. These included color, stitching details, outsole and heel option, as well as inner lining materials. This information, the selected shoe style and size for the right foot and the 3D custom last design data for the left foot, comprised the client's detailed order. This was forwarded electronically to the store's custom retail shoe supplier.

When the order was received by the fabrication facility, the right shoe last was located in inventory, whereas the custom-shaped left last was carved out on a numerically controlled machine. Once both lasts were ready, the shoes were manufactured to the order specifications and shipped to the specialty retailer. The client returned to the store to try on the custom black pumps. After testing the shoes in the store, she confirmed that both shoes fit well and she was pleased with their appearance.

14.2.4 Final Result

The following photograph (see Figure 14.4) shows the client wearing the custom black patent leather pump on her left foot. Despite the stiff leather upper, the shoe was comfortable. The custom last provided expansion around the bunion and maintained the narrowness and elegance of the classic pump design. The client experienced no pressure or pain and can wear an attractive shoe.

This case study shows clearly that where style is the highest priority in the custom shoe last design process, the best approach is to start with a proven retail last and make as few adjustments as possible to accommodate the shape of the client's feet.

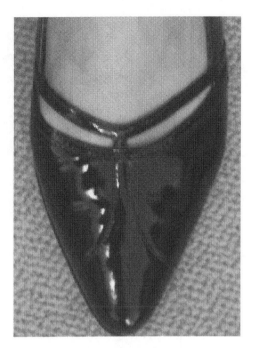

FIGURE 14.4   Custom black leather pump for client's left foot.

14.3   CASE STUDY OF A MEDICAL PATIENT REQUIRING CUSTOM LASTS AND SHOES

14.3.1   PROBLEM FEET

In this example, a diabetic patient is presented with diabetic peripheral neuropathy (see Figure 14.5). Additional complications include Charcot joint disease (a condition which causes the progressive degeneration, fracture, and deformation of bones in the feet) and, for the left foot, a missing hallux (big toe). The latter was surgically removed 12 years ago.

Due to lack of sensation, diabetic patients are unable to detect pressure sores, blisters, or minor injuries. Because diabetes also affects blood circulation, lesions in the diabetic foot often do not heal properly, putting the patient at risk for infection, gangrene, and potential amputation. A diabetic patient with neuropathy may also manifest complications such as bunions (hallux valgus) or hammer toes. These situations often lead to abnormal pressure distributions which in turn increase the risk of skin breakdown on bony prominences and the plantar aspect of the foot. Careful and thorough foot examinations must be performed regularly to monitor for such problems.

14.3.2   PROPOSED SOLUTION

Due to the deformities associated with this patient's Charcot feet, custom shoes were required in order to provide (1) enough volume to accommodate the large feet and ankles and (2) sufficient anchorage to prevent the feet from sliding forward into the toe box. As a

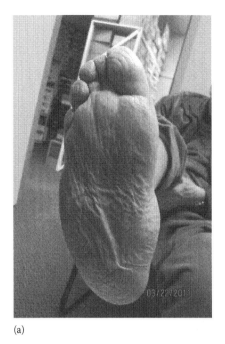 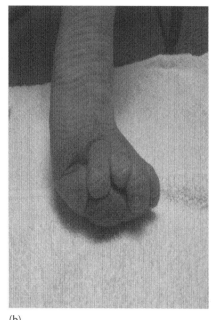

(a)   (b)

FIGURE 14.5   (a) Right foot and (b) left foot.

result, a short ankle boot design was proposed. To compensate for the fused ankle joints, rocker bottom soles were recommended; these would help propel the patient forward during walking. Finally, to optimize plantar surface pressure distribution and minimize friction, custom shaped, variable density, and ultrasoft foam insoles were planned.

14.3.3 Process Implemented

The first step was to take weight-bearing imprints of both feet. These provided valuable information about which surfaces on the plantar aspect of the foot could be used by the custom shoe last/insole designer to support the patient's weight.

Next, casting socks (tubular socks wherein the fabric is impregnated with a water-activated resin) were used to take impressions of the patient's feet. During this process, downward pressure was exerted over the top of each foot to effect soft tissue displacement (the natural splaying of the foot when the person is in a weight-bearing position). The external surface of the resulting molds was then scanned with the Yeti 3D Foot Scanner to produce 3D data corresponding to the shape of each foot (see Figure 14.6).

Some may wonder why the patient's feet were casted and not scanned directly in the Yeti Foot Scanner. There are two main reasons for this. First, diabetic patients with Charcot feet are highly susceptible to skin breakdown; walking without shoes, even a short distance to a scanner, is risky. Second, a skilled practitioner can manipulate the shape of the casting sock during his molding process to obtain idealized plantar surface contours for the final shoe (e.g., simulated weight-bearing under the

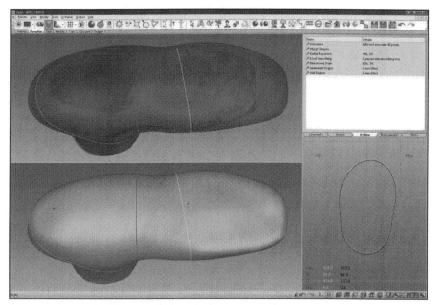

(a)

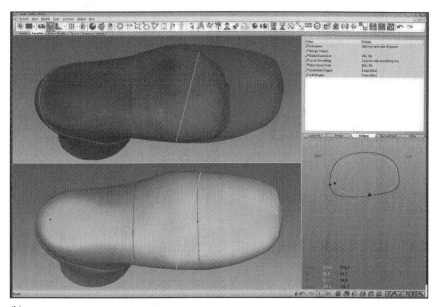

(b)

FIGURE 14.6 On-screen view of scanned 3D data of right foot with superimposed weight-bearing imprint and toe box extension. Left foot: scanned 3D data with superimposed weight-bearing imprint and toe box. (a) Right foot and last, plantar aspect and (b) left foot and last, plantar aspect.

forefoot combined with non-weight-bearing under the rest of the foot). This reduces the time needed in the subsequent shoe last and insole design stages.

The next step was to input the 3D scanned data into the Canfit Design software. Unlike the retail example earlier, much more of the actual foot shape had to be incorporated into the design of appropriate lasts for this patient. To ensure patient compliance, the shoemaker strove to provide symmetric custom shoes, but this goal was secondary to ensuring proper clinical function. In the subsequent digital design process, the 3D scan data were smoothed and repaired to remove wrinkles from the casting process. Pre-stored 3D toe boxes were then merged onto the front of this data and viewed until a suitable configuration was obtained. Next, interactive modification tools were used to blend the selected toe box into the rest of the foot, while taking care not to introduce any pressure zones. Finally, specific local reliefs* were implemented to offload potential pressure areas due to the Charcot deformity. Because the software allowed the clinician to align and view the foot and shoe last on the same screen, the clinician was able to validate the fit of each last digitally (see Figures 14.7). Once he was satisfied with the fit, he would then send his data to a numerically controlled carver so that the physical lasts could be produced.

With the shoe last design phase completed, it was then necessary to design the custom insoles for this patient. Using the weight-bearing imprints and the 3D shape data corresponding to the bottoms of the feet, modifications were made to accommodate the Charcot deformities, as well as distribute weight over pressure tolerant areas. These insoles were then manufactured and, in combination with the aforementioned custom shoe lasts, used to fabricate custom ankle boots.

The next step involved the patient trying on the new boots. However, the patient's neuropathy posed a significant challenge in assessing their suitability. Having no sensory perception in his feet, he was unable to provide feedback on the presence of any high pressure areas, or on overall comfort or discomfort. Without this crucial information, potential wear/friction problems could have been missed, resulting in sores and lesions at a later date.

To compensate for this lack of feedback on the fit, visual and tactile examinations were performed by the shoemaker. First, the overall fit was assessed. With the patient standing and wearing the shoes, the fit of the toe box was checked by feeling the top and sides of the forefoot to ensure adequate volume. Next, the shoemaker confirmed that the ankles were properly anchored so that the feet would remain fixed in place throughout the gait cycle.

The patient was then asked to wear the shoes for 10 min, during which time he was encouraged to walk in them, as much as was practical. He then removed his shoes and socks and the plantar aspects of both feet were examined for any obvious pressure areas. If present, these zones were accommodated by manually reshaping the custom soft foot bed.

Once the overall fit and plantar surface pressure distributions were deemed satisfactory by the shoemaker, the patient was allowed to leave the clinic with his shoes. Note that he was instructed to wear the new boots no longer than one hour at a time

---

* Typical relief is 3–5 mm from patient surface.

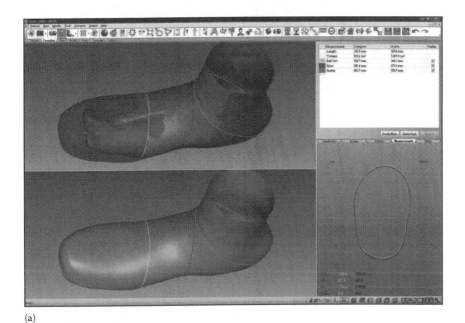

(a)

(b)

FIGURE 14.7 On-screen fit assessment of custom shoe lasts for right and left feet. (a) Right foot and last, dorsal aspect and (b) left foot and last, dorsal aspect.

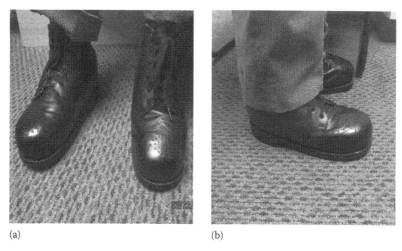

(a)   (b)

FIGURE 14.8 Custom ankle boots: (a) front view and (b) side view.

and immediately afterward his feet were monitored for any signs of redness or other discoloration. When it was confirmed that there were no problematic pressure areas, the patient was advised that he could gradually increase wearing time.

14.3.4 FINAL RESULT

The final shoes for this patient are shown in the following (see Figure 14.8). Despite the significant differences in the morphology of the two feet, the shoemaker achieved shoes which were remarkably symmetric in appearance. The specific clinical needs of the patient were successfully addressed. As a result of proper footwear management, the foot pathology has remained stable, and the patient has required no further medical intervention.

For complex medical conditions, where function is the primary goal of the footwear, the most effective and efficient shoe last design methodology is to start with the 3D foot shape and modify it into a suitable last.

14.4 SUMMARY

Custom shoe lasts are needed by clients who present with abnormally shaped feet and/or ankles. A fundamental truism of custom footwear is that no matter how complicated the foot morphology, the styling of the finished footwear solution is always important to the client's final acceptance of the shoes.

A practical implementation of an automated process for fitting a difficult retail client has been described. For this person, appearance of her footwear was first priority, and comfort was a close second. It was shown that the most effective and efficient approach for designing a suitable custom shoe last was to start with a proven industry/retail last and make as few modifications (to accommodate the foot) as possible.

Also presented was a complex medical footwear case, wherein specially shaped shoe lasts were required. Proper function was first priority. Aesthetic appeal was a secondary goal, but still a very important objective. It was shown that the most efficient design approach involved starting with the patient's 3D foot shape and modifying it to incorporate suitable areas of support and relief.

Hence, the best methodology for the automated design of custom shoe lasts varies according to the complexity of the fitting problem. That said, both examples demonstrate the significant benefit of being able to see foot and shoe last simultaneously on a computer screen. Interactive 3D graphics tools allow a shoe last to be modified in a forgiving manner; the designer/clinician can visualize the effect on the foot, without needing to create a test shoe, and if a specific modification is not quite right, it can simply be redone.

Experienced shoe last designers refine their craft over time. Modern computers, 3D scanners, and software provide the opportunity to collect a digital record of the modification processes used to create well-fitting shoe lasts. The documentation of these processes is a key step in developing our understanding of what is required to design well-fitting shoes for challenging feet, and will ultimately lead to further advancements in the science of footwear.

ACKNOWLEDGMENTS

The author greatly appreciates the contributions of Claudia Kieserling, Johan Steenwyk, and Jeffrey Chang to this chapter.

QUESTIONS

14.1 Explain what causes a bunion to form and how this affects the overall shape of the foot. How does this impact foot comfort when wearing shoes?

14.2 When purchasing shoes, what will people with bunions typically do to accommodate their foot morphology?

14.3 What was the proposed solution for creating fashionable footwear for the customer with a bunion?

14.4 How did the shoemaker create fashionable footwear that was also comfortable to wear?

14.5 Where style is the highest priority, what is the best approach to creating a custom shoe last?

14.6 How does diabetes affect foot health?

14.7 Explain what accommodations were built into the orthopedic shoe in order to address the diabetic foot concerns. How was this achieved?

14.8 Why is it important to apply downward pressure on the foot during the casting process?

14.9 How is fit assessed when a diabetic patient is trying on new footwear?

14.10 What challenges does the patient's neuropathy present to the fitting process? How was this managed?

14.11 Where function is the primary goal of the footwear, what is the most effective and efficient shoe last design methodology?

15 Finite Element Analysis Methods in Footwear Design

Philip Azariadis

CONTENTS

15.1 INTRODUCTION

Footwear manufacturing is evolving from a labor-intensive activity to a knowledge-based manufacturing process. This is evident from the amount of new and breakthrough technologies that have been developed during the last decade covering a wide spectrum of applications in engineering, informatics, materials, information and communication technologies, etc. (CEC, 2009; Chituc et al., 2008; Pedrazzoli, 2009; Wong et al., 2006). Several efforts are being devoted to making shoe industry human-centered by developing new concepts for customizing or personalizing the final products (Azariadis and Papagiannis, 2010; Lee, 2006; Leng and Du, 2006; Luximon et al., 2003).

From the design perspective, several technologies are emerging which are mainly focused on the creative or the industrial design of the shoe. Computer-aided design (CAD) systems play a dominant role in these two phases allowing the designer to exploit better/easier his/her creativity since detailed product designs can now be derived with much less effort. On the other hand, design support with respect to the final product properties is still limited. Existing methods and associated technologies can be divided into two groups: "Methods and technologies for supporting shoe design in terms of general functional properties" and "Methods and technologies for supporting shoe design in terms of certain, usually human-oriented, biomechanics properties." Indicative examples of the first group include the "Virtual She Test Bed" (Azariadis et al., 2007, 2010) and the "T-smart" (Mao et al., 2008) systems. Although the latter is mainly focused on apparel design, in principle, it is possible to consider shoe data too.

On the other hand, during the last decade, a significant number of works have been published for the study of footwear biomechanics and its relation with the human-foot kinematics or gait cycle. Modern approaches use computer-aided engineering (CAE) with detailed 3D foot models in order to derive results that allow the assessment and optimization of the shoe design. These tools are based on numerical methods, and in particular on the finite element method (FEM), for computing realistic simulations of the foot–footwear interface. With these tools, one is able to compute biomechanical information such as internal stress and strain distributions of modeled structures. This chapter provides a review on FEMs in footwear design and concludes with indicative examples of applications in the footwear industry. The presented review is mainly focused on healthy foot cases and casual shoes.

15.2   3D FOOT MODELS FOR FINITE ELEMENT ANALYSIS

Today, the FEM is a very popular tool for computing several parameters affecting foot biomechanics, such as internal stresses and strain states of the foot–ankle complex, and the load distribution between foot and insole. With finite elements analysis (FEA), one is able to make adjustments to insole shape and to the materials selection without needing to pass through a time-consuming trial-and-error cycle, which involves physical proto- types. On the other hand, the FEM results rely heavily on the accuracy of the underlying geometric model used to represent the actual human foot, the selection of the materi- als, and the determination of the various loads and boundary conditions. The higher the accuracy of the FE model, the closer to the reality the derived results will be.

Several researchers have proposed 3D foot models for FEA in order to derive accu- rate information related to foot biomechanics. Although early works of Nakamura et al. (1981), Chu et al. (1995), Lemmon et al. (1997), Shiang (1997), Jacob and Patil (1999), Kitagawa et al. (2000), Gefen et al. (2000), and later of Chen et al. (2001, 2003), Gefen (2002, 2003), Cheung and Zanhg (2005), Lewis (2003), and Verdejo and Mills (2004) include several assumptions and simplifications in the foot struc- tures and the corresponding geometrical representations, these works verified that FEA can be used with success in predicting the foot biomechanics parameters and, therefore, paved the way for producing more accurate foot models such as those proposed in Camacho et al. (2002), Cheung et al. (2005), Cheung and Zhang (2005, 2006), Yu et al. (2007), Hsu et al. (2008), and Antunes et al. (2008).

The basic workflow/methodology for developing the model of a human foot is depicted in Figure 15.1. Four main steps are identified, which are briefly analyzed in the sequel.

Step 1—Foot CT or MR scan images: The initial step involves the computed tomog- raphy (CT) or the coronal magnetic resonance (MR) scan of the lower part of the foot. This is realized using CT or MR equipment, respectively, with the foot in neutral position in order to minimize any potential tension or pressure on tendons, muscles, or bones. In principle, MR scans are preferred since it is easier to capture more anatomical details, such as tendons and ligaments, compared to CT. In both cases, the density and the resolution of the scanned images play a crucial role in the accuracy of the reconstructed anatomical elements.

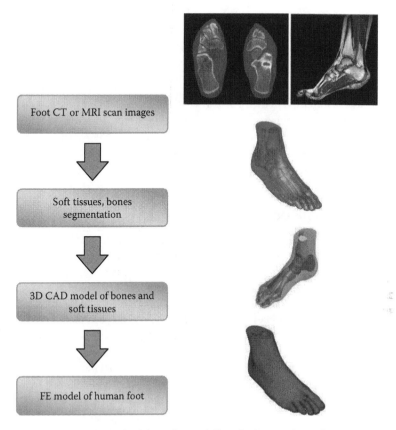

FIGURE 15.1   The basic methodology for modeling the human foot. (From Antunes, P.J., Non-linear finite element modeling of anatomically detailed 3D foot model, Materialise (www.materialise.com), report paper, 2008.)

Step 2—Soft tissues and bones segmentation: In this task, 3D models of anatomical elements such as bones and soft tissues are generated. This is achieved using specialized software capable of performing density segmentation. All gray-scaled images are processed by the software in order to distinguish anatomical elements such as bones, cartilages, tendons, ligaments, and soft tissues. At this stage, researchers usually perform additional simplifications in order to fuse the articular cartilages of the bones with their corresponding bone surfaces (Cheung and Zhang, 2006), or to merge tarsal bones with rigid structures (Chen et al., 2001), or to model plantar fascia and other major ligament structures as tension-only elements (Chen et al., 2003; Yu et al., 2008). Usually this stage requires the users' intervention in order to specify the density masks between different bone structures and the various thresholds required for the successful segmentation of soft tissues.

Step 3—3D CAD model of bones and soft tissues: All generated models of the anatomical elements are further processed with CAD software in order to produce smooth surface boundaries for bones and to reconstruct (when required) foot parts, such as phalanges and cartilages, which may not be segmented in the previous step.

All elements are carefully aligned together to avoid undesirable collisions between structural elements such as bones or bones and cartilages. The rest of the soft tissue is modeled by performing volumetric Boolean operations by subtracting the generated elements from the foot volume.

Step 4—FE model of human foot: In the final step, all 3D models of the foot anatomical elements are imported in the FEA software where several tasks are performed that include (a) 3D models discretization according to the selected finite elements, (b) geometric definition of potentially missing elements such as Achilles tendon or cartilages, (c) selection of loads and boundary conditions, and (d) definition of material properties. These tasks are subject to user's selection and depend on the purpose of the FEA.

15.2.1 GEOMETRY PROCESSING

Chen et al. (2001) developed a preliminary 3D FE model of a foot in order to estimate the stress distribution in the foot during midstance to push-off phase during barefoot gait. In their model, major bones and soft tissues were identified using CT sectional images. Also, the bones in the five phalanges were modeled by five integrated parts, while the rest of the metatarsal and tarsal bones were modeled with two rigid columns (medial and lateral) without segmenting each of the individual bone in order to reduce the complexity of the produced model. The joint spaces between each of the five phalanges and its connective metatarsal bones were modeled with cartilage elements to allow deformation and to simulate the actual metatarsophalangeal joints. Additionally, the joint space between the medial and the lateral columns representing the metatarsal and the tarsal bones was also modeled with cartilage elements. The 3D CAD models have been created using a public-domain program—NUAGES (INRIA, Sophia Antipolis, France)—and then imported to a commercial finite element pre- and post-processing program—Mentat (MSC/MARK, Los Angeles, CA)—to generate 4-node tetrahedral elements for each surface model. Figure 15.2 depicts the produced simplified surface models for each of the bone and cartilage structures.

Similar simplifications can be found in the foot model proposed by Gefen (2002, 2003). The model consists of five planar longitudinal cross sections throughout the foot, which together yield a convenient representation of its complex half-dome-shaped structure. The geometric data of the foot's skeletal cross sections were detected using MR imaging and transferred to ANSYS (ANSYS, Inc., Pennsylvania, PA) FEA software package for the construction of planar models for the five rays of the foot. Cartilage layers and ligaments were introduced in the joints based on anatomic data. Bony elements and cartilages were assigned as having the properties of linear, elastic, and isotropic materials, while ligaments, fascia, and soft tissue fat pad were considered as being nonlinear materials (Figure 15.3).

A detailed model of the human foot and ankle, incorporating realistic geometrical properties of both bony and soft tissue components, has been proposed by Camacho et al. (2002). In total, 286 CT images have been acquired from a cadaveric human foot and imported to image-processing software produced by the National Institute of Health in order to perform density segmentation and to derive the boundary description of each bony element. Then a custom software program (Polylines developed

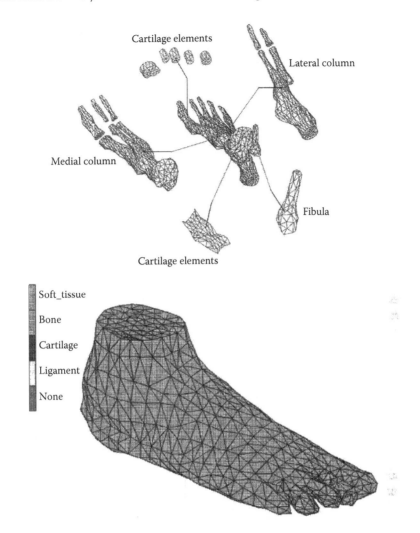

FIGURE 15.2 The bones and cartilages models and the final FE model of the human foot produced by Chen et al. (2001). (*Clin. Biomech.*, 16, Chen, W.P., Tang, F.T., Ju, C.W., Stress distribution of the foot during midstance to push-off in barefoot gait: A 3-D finite element analysis, 614–620, Copyright 2001, with permission from Elsevier.)

by Randal P. Ching) was used to pile together the sectional images for each bone, and finally the form-Z (AutoDesSys, Inc.) program was employed to produce the corresponding 3D CAD models. All 3D CAD models of the bones were imported to TrueGrid software (XYZ Scientific Applications, Inc.) in order to generate the necessary mesh consisting of 4-noded shell elements. With the same procedure (but with different segmentation thresholds), the mesh of the plantar soft tissue has been generated using 8-noded hexahedral elements. The 3D cartilage elements were generated from the 3D surface models of the bone, since the cartilage borders were not readily viewable in the CT scans. Once the bone models were exported into

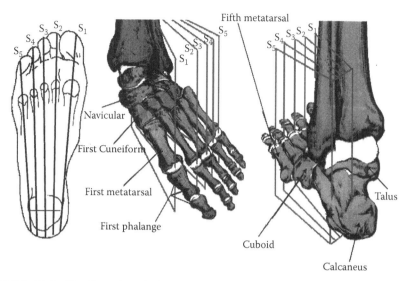

FIGURE 15.3   The five selected planar cross sections (marked as S_1–S_5) of the foot FE model proposed by Gefen (2002). (*J. Biomech.*, 35, Gefen, A., Stress analysis of the standing foot following surgical plantar fascia release, 629–637, Copyright 2002 with permission from Elsevier.)

form-Z, representative cartilage bodies have been generated by creating a solid volume around the joint of interest. The mesh of the cartilage objects has been generated using 8-noded hexahedral elements, too. Figure 15.4 shows the 3D mesh models of the bones structures and the final FE model of the foot including bones, plantar soft tissue, and cartilage layers.

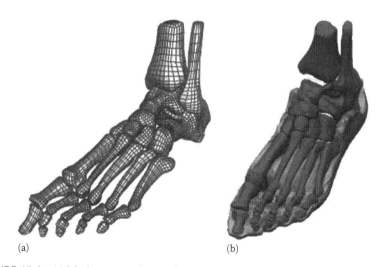

(a)   (b)

FIGURE 15.4   (a) Mesh representing surface geometry of all bones of foot. (b) Final model of the foot, containing bones, plantar soft tissue, and cartilage layers. (From Camacho, D.L.A. et al., *J. Rehabil. Res. Dev.*, 39, 401, 2002.)

Another detailed model of a real human foot has been proposed by Cheung and Zhang (2005, 2006) and Cheung et al. (2005). The geometry of the FE model was obtained from the 3D reconstruction of MR images from the right foot of a normal male. The images were segmented using MIMICS (Materialise, Leuven, Belgium) to obtain the boundaries of skeleton and skin surface. The boundary surfaces of the skeletal and skin components were processed using SolidWorks (SolidWorks Corporation, Massachusetts) to form solid models for each bone and the whole foot surface. The solid model was then imported and assembled in the FE package ABAQUS (Hibbitt, Karlsson & Sorensen, Inc., Pawtucket, RI). The final FE model consists of 28 bony segments, including the distal segments of the tibia and fibula and 26 foot bones: talus, calcaneus, cuboid, navicular, 3 cuneiforms, 5 metatarsals, and 14 components of the phalanges. The phalanges are fused together with 2 mm thick solid elements to simulate the connection of the cartilage and other connective tissues. The interaction among the metatarsals, cuneiforms, cuboid, navicular, talus, calcaneus, tibia, and fibula is defined by contact surfaces, which allow relative articulating movement. To simulate the frictionless contact between the joint surfaces, ABAQUS-automated surface-to-surface contact option was used. Compressive stiffness resembling the cartilage structure (Athanasiou et al., 1998) is prescribed between each pair of joint contact surfaces to simulate the covering layers of articular cartilage. Apart from the collateral ligaments of the phalanges and other connective tissues, a total number of 72 ligaments and the plantar fascia were included and defined by connecting the corresponding attachment points on the bones. All the bony and ligamentous structures were embedded in a volume of soft tissues. The bony and soft tissue structures were meshed with 4-noded tetrahedral elements and the ligaments were defined with tension-only truss elements. Figure 15.5 depicts the

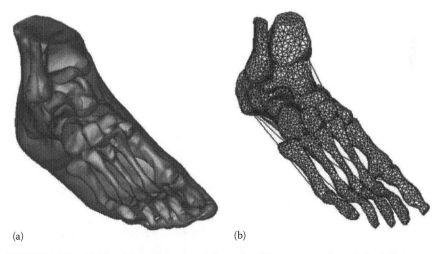

(a)   (b)

FIGURE 15.5 (a) The 3D model created from the 3D reconstruction of the MR images. (b) The FE model of bony and ligamentous structure. (*J. Biomech.*, 38, Cheung, T.M., Zhang, M., Leung, K.L. and Fan, Y. Bo., Three-dimensional finite element analysis of the foot during standing—A material sensitivity study, 1045–1054, Copyright 2005 with permission from Elsevier.)

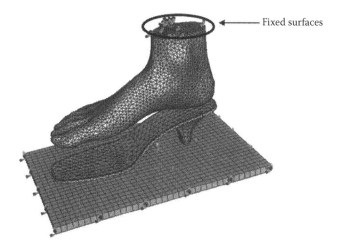

Fixed surfaces

FIGURE 15.6 The FE mesh model of the soft tissue, foot bones, ligamentous structures, and support showing boundary conditions for simulating balanced standing with a high-heeled shoe. (*Clin. Biomech.*, 23, Yu, J., Cheung, J.T.M., Fan, Y., Zhang, Y., Leung, A.K.L., Zhang M., Development of a finite element model of female foot for high-heeled shoe design, S31–S38, Copyright 2007 with permission from Elsevier.)

final FE foot model. A similar FE model has been developed by Yu et al. (2008) for a female foot in order to study the effects of wearing high-heel footwear on the musculoskeletal system of the foot–ankle complex (see Figure 15.6).

Hsu et al. (2008) developed another detailed FE model of a real human foot in order to examine the role of shoe insole shape in lowering plantar fascia pressures. The basic foot geometry has been captured from CT images of a young male's foot in unloaded neutral position. The 3D FE model consisting of plantar fascia, bones, ligaments, and skin was reconstructed using the FE software ANSYS (ANSYS, Inc., Pennsylvania, PA). All phalanges, cartilages, bones, and skin were simulated by SOLID 45 elements. The SOLID 45 element used for the 3D modeling of solid structure is defined by 8 nodes having 3 degrees of freedom at each node. The plantar fascia and major ligamentous structures such as the deltoid ligament, the lateral collateral ligament, the short plantar ligament, the long plantar ligament, and the spring ligament were created using tension-only LINK 10 elements that have 3 degrees of freedom at each node. The final foot FE model comprised of 34,251 nodes and 38,908 elements is shown in Figure 15.7.

15.2.2 Assignment of Materials

In principle, five distinct types of foot parts are modeled for FEA: bony structures, ligaments, cartilage, plantar fascia, and soft tissue. Most of the researchers consider bony elements and cartilage as linear, elastic, and isotropic materials, while ligaments, plantar fascia, and soft tissue are considered as being nonlinear materials. Depending on the specific application and the types of elements used for the FE meshing there are some differences in the choices of material properties among researchers.

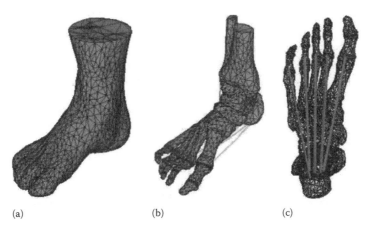

(a)   (b)   (c)

FIGURE 15.7 (a) The final FE model of foot. (b) The FE model of bony structures and cartilages. (c) The FE model of the plantar fascia. (With kind permission from Springer Science + Business Media: *Ann. Biomed. Eng.*, Using an optimization approach to design an insole for lowering plantar fascia stress—A finite element study, 36, 2008, 1345–1352, Hsu, Y.C., Gung, Y.W., Shih, S.L., Feng, C.K., Wei, S.H., Yu, C.H., Chen, C.S.)

Gefen (2002, 2003) adopt for material properties the Young's modulus and Poisson's ratio for (a) bone structures according to Nakamura et al. (1981) and (b) cartilage according to Patil et al. (1996). The nonlinear properties of the plantar fascia are determined based on the results of Kitaoka et al. (1997). For the Poisson ratios of the ligaments and plantar fascia, the author adopts the works of Chu et al. (1995) and Nakamura et al. (1981), respectively. For the nonlinear stress–deformation relation of the ligaments, the work of Race and Amis (1994) is adapted who used an Instron system to test normal lower-limb ligaments under tension. Finally, the mechanical behavior of the soft tissue fat pad is taken from Nakamura et al. (1981) who obtained the nonlinear stress–deformation curve of a specimen taken from the heel of a fresh cadaver. The earlier-given experimental data are fitted to the polynomial expressions in the form of

$$\sigma = a_1\lambda^5 + a_2\lambda^4 + a_3\lambda^3 + a_4\lambda^2 + a_5\lambda + a_6 \tag{15.1}$$

where
  σ is the stress in MPa
  λ is the resultant stretch ratio
  the constraints a_i, $i = 1,\ldots, 5$ are specified according to Table 15.1

Finally, Table 15.2 summarizes the values of the material properties adopted by Gefen (2002, 2003) for building the corresponding foot models.
  Yu et al. (2008), Cheung and Zhang (2005, 2006), Cheung et al. (2005), and Antunes et al. (2008) treat all elements except soft tissues as homogeneous, isotropic, and linearly elastic materials. The Young's modulus and Poisson's ratio for the bony structures is assigned according to the model developed by Nakamura et al. (1981).

TABLE 15.1

Coefficients a_i (in MPa) Used in Equation 15.1 for the Determination of the Stress–Deformation Relations of Ligaments, Plantar Fascia, and Soft Tissue Fat Pad

Tissue	a_1	a_2	a_3	a_4	a_5	a_6
Ligaments	−412640.5	2235967.7	−4841544.8	5236972.7	−2829945.7	611190.6
Plantar fascia	−488737.9	2648898.5	5736967.6	6206986.7	−3354935.1	724755.5
Soft tissue fat pad	0	59.2	275.5	480.4	−371.9	107.7

Source: Gefen, A., *J. Biomech.*, 35(5), 629, 2002.

TABLE 15.2

Material Properties Adopted by Gefen (2002, 2003)

Component	Element Type	Young's Modulus (MPa)	Poisson's Ratio
Bone	Flat 8-node	7300	0.3
Cartilage	Flat 8-node	10	0.4
Soft tissue	Flat 8-node	—	0.49
Ligaments	Flat 8-node	—	0.4
Fascia	Flat 8-node	—	0.4

TABLE 15.3

Material Properties Adopted in Yu et al. (2008), Cheung and Zhang (2005, 2006), Cheung et al. (2005), Antunes et al. (2008), and Hsu et al. (2008)

Component	Element Type	Young's Modulus (MPa)	Poisson's Ratio
Bone	3D-Tetrahedra/8-node*	7300	0.3
Soft tissue	3D-Tetrahedra/8-node*	Hyperelastic	—
Soft tissue*	8-node	0.15	0.45
Cartilage	3D-Tetrahedra/8-node*	1	0.4
Ligaments	Tension-only Truss/2-node*	260	—
Fascia	Tension-only Truss/2-node*	350	—

Sources: Yu et al., *Clinical Biomechanics*, 23, S31–S38, 2008; Cheung, J.T.M. and Zhang, M., *Arch. Phys. Med. Rehabil.*, 86, 353, 2005; Cheung, T.M. and Zhang, M., Finite element modeling of the human foot and footwear, *2006 ABAQUS Users' Conference*, Boston, MA, pp. 145–149, 2006; Cheung, T.M. et al., *J. Biomech.*, 38(5), 1045, 2005; Antunes, P.J. et al., Non-linear finite element modeling of anatomically detailed 3D foot model, Materialise (www.materialise.com), report paper, 2008; Hsu, Y.C. et al., *Ann. Biomed. Eng.*, 36(8), 1345, 2008.
The values marked with asterisk (*) are adopted in Hsu et al. (2008).

TABLE 15.4

Material Properties Adopted in Chen et al. (2001)

Component	Element Type	Young's Modulus	Poisson's Ratio
Bone	3D-Tetrahedra	10 GPa	0.34
Soft tissue	3D-Tetrahedra	1.15 MPa	0.49
Cartilage	3D-Tetrahedra	10 MPa	0.4
Ligaments	2-node	11.5 MPa	—

Source: Chen, W.P. et al., *Clin. Biomech.*, 16, 614, 2001.

The Young's modulus of the cartilage is selected according to Athanasiou et al. (1998), the ligaments according to Siegler et al. (1988), and the plantar fascia according to Wright and Rennels (1964). The ligaments and the plantar fascia are assumed to be incompressible, while the cartilage is assigned with a Poisson's ratio of 0.4 for its nearly incompressible nature. Similar material properties can be found in Hsu et al. (2008); the authors therein adopt a different selection of soft tissue properties and element types used in FEA. All material properties adopted in Yu et al. (2008), Cheung and Zhang (2005, 2006), Cheung et al. (2005), Hsu et al. (2008) are listed in Table 15.3.

Finally, Chen et al. (2001) use simplified material properties by considering all components as homogeneous, linear, and elastic solids. They make the corresponding values selection according to Chu et al. (1995), Jacob et al. (1996), and Lemmon et al. (1997); see Table 15.4.

15.3   FINITE ELEMENT ANALYSIS FOR SHOE DESIGN

A typical sole structure for a casual shoe can be found in Figure 15.8 and consists an insole (or insert), midsole (or mounting insole), and an outer sole layer. The insole is the interior bottom of a shoe, which sits directly beneath the foot under the footbed. The outsole is the layer in direct contact with the ground and may comprise a single piece, or may be an assembly of separate pieces of different materials. Finally, midsole is the layer in between the outsole and the insole and is typically there for shock absorption. Note that special types of shoes, such as athletic, often have modifications in the design of these components to comply with different specifications.

Many experimental techniques have been developed for the quantification of foot biomechanics including gait analysis (Balmaseda et al., 1988; Whittle, 2003), pressure sensing platforms (Gefen, 2003; Lundeen et al., 1994; Orlin and McPoil, 2000; Rosenbaum et al., 1996; Tanimoto et al. 1998; Urry, 1999), in-shoe pressure transducers and films (Alexander et al., 1990; Brown et al., 1996; Cavanagh et al., 1992; Cobb and Claremont, 1995; Nevill et al., 1995; Soames, 1985; Wiegerinck et al., 2009; Yung-Hui and Wei-Hsien, 2005), and in vivo force measurements (Jacob, 2001; Jacob and Zollinger, 1992; Nagel et al., 2008). In many cases, as it is mentioned in Chen et al. (2001), it is very difficult to quantify the in vivo bone and soft tissue stresses with experimental techniques, while with in vitro studies it is difficult to define proper loading conditions representing the

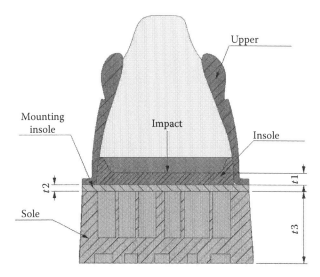

FIGURE 15.8   A typical structure of the sole of a casual shoe consists of an insole, midsole (or mounting insole), and an outer sole layer.

actual physiological loading situation of a real foot. The need to develop an analytical method for measuring foot biomechanics led many researchers to the use of FEA as a primary tool for studying and quantifying the aforementioned parameters during the last decade. These studies play an important role in shoe design: They allow the redesign or improvement of insole, midsole, or outsole parts without the necessity of producing physical prototypes in a time-consuming trial-and-error cycle. The major volume of published works deals with the analysis of plantar stress distribution with subjects wearing casual or therapeutic footwear. Some indicative examples are presented in the sequel.

A preliminary computational model for estimating the distribution of stress in a normal foot during gait is presented in Chen et al. (2001), where a 3D foot model has been developed using CT images. The authors focused on the midstance to push-off gait phases and applied loading and boundary conditions accordingly. A rigid plane simulating the floor was created and set to be moving relatively with respect to the foot model and an angular velocity calculated from kinematic data obtained by studying the gait cycle of the male subject. The results displaying normal stresses at the plantar surface and the von Mises equivalent stresses at the skeletal parts of the foot at the 0.12 s analysis instance are given in Figure 15.9. The authors claim that more accurate results could be derived with more accurate loading and boundary conditions and with a better geometric model of the foot.

A more detailed foot model incorporating realistic geometric and material properties of bony and soft tissue components has been incorporated by Cheung and Zhang (2005) and Cheung et al. (2005) in order to investigate the effect of insole material stiffness on plantar pressures and stress distribution in the bony and ligamentous structures during balanced standing. In Cheung and Zhang (2005), authors studied two kinds of insole designs, flat and custom-mold (Figure 15.10a), and concluded that the shape of the insole is more important in reducing peak plantar pressures than the stiffness of the used material (Figure 15.10b). Later, Cheung and Zhang (2008)

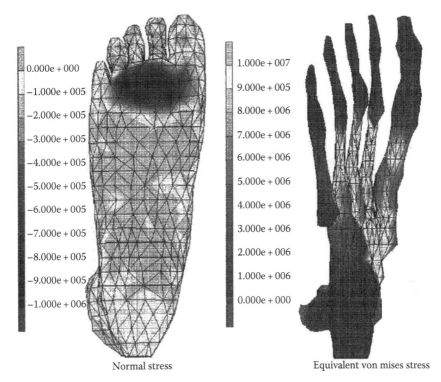

Normal stress	Equivalent von mises stress
0.000e + 000	1.000e + 007
−1.000e + 005	9.000e + 005
−2.000e + 005	8.000e + 006
−3.000e + 005	7.000e + 006
−4.000e + 005	6.000e + 006
−5.000e + 005	5.000e + 006
−6.000e + 005	4.000e + 006
−7.000e + 005	3.000e + 006
−8.000e + 005	2.000e + 006
−9.000e + 005	1.000e + 006
−1.000e + 006	0.000e + 000

FIGURE 15.9 The normal stress distribution on the plantar surface (left), and the von Mises stress distribution of the bone at time instance 0.12 s (right). (*Clin. Biomech.*, 16, Chen, W.P., Tang, F.T., Ju, C.W., Stress distribution of the foot during midstance to push-off in barefoot gait: A 3-D finite element analysis, 614–620, Copyright 2001, with permission from Elsevier.)

studied the pressure-relieving capabilities of different material and structural configurations of foot orthosis using the 3D FE model of a human foot (Figure 15.11a). They used five different design factors (arch type, insole material, insole thickness, midsole material, and midsole thickness), and concluded that the use of an arch conforming foot orthosis and a softer insole material was found to be effective in the reduction of peak plantar pressure with the former providing a larger pressure reduction. Insole thickness, midsole stiffness, and midsole thickness were found to play less important roles in peak pressure reduction. They fabricated flat and custom-mold foot orthoses using a computerized numerical control (CNC) milling machine with the aid of an orthorist to fit the size of the shoes of the male subject (Figure 15.11b).

High-heeled shoes have been studied in Yu et al. (2008) by incorporating a detailed FE foot model for evaluating the biomechanical effects of heel elevation on the foot–ankle complex. Goske et al. (2006) studied the relation of pressure relief in the heel area with (a) insole conformity, (b) insole thickness, and (c) insole material. Heel geometry was obtained from coronal plane MR images of the right heel of a healthy adult male. The heel pad was represented by a lumped soft tissue model with a hyperelastic strain energy function. A barefoot simulation was conducted during which the heel was loaded vertically to simulate maximal loading of the heel at first

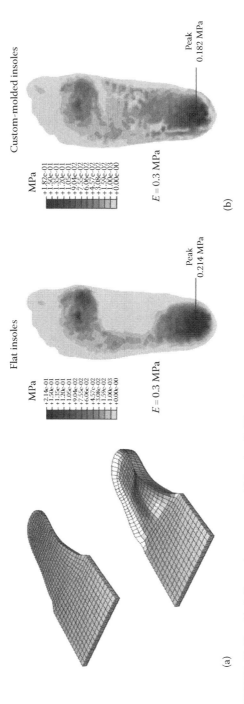

FIGURE 15.10 (a) A flat and a custom-mold insole. (b) The plantar pressure distributions supported by the flat and the custom-mold insole. (*Arch. Phys. Med. Rehabil.,* 86, Cheung, J.T.M and Zhang, M., A 3-Dimensional finite element model of the human foot and ankle for insole design, 353–358, Copyright 2005 with permission from Elsevier.)

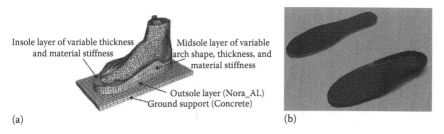

Insole layer of variable thickness and material stiffness

Midsole layer of variable arch shape, thickness, and material stiffness

Outsole layer (Nora_AL)
Ground support (Concrete)

(a)

(b)

FIGURE 15.11 (a) The FE meshes of the ankle–foot structures, foot orthosis, and ground support. (b) The fabricated flat and custom-mold foot orthosis. (*Med. Eng. Phys.*, 30, Cheung, J.T.M., Zhang, M., Parametric design of pressure-relieving foot orthosis using statistics-based finite element method, 269–277, Copyright 2008 with permission from Elsevier.)

step of walking. This simulation provided the baseline peak pressure that needed to be relieved by insole intervention. For simulations regarding insole design, footwear was included into the model (Figure 15.12). A 20 mm thick midsole was modeled as Firm Crepe (compressible, hyperfoam) and the shoe sidewalls were modeled as stiff leather (linearly elastic). Three insole design variables were investigated: conformity (flat, half, and full); thickness (6.3, 9.5, and 12.7 mm); and compressible hyperfoam material (hard, medium, and soft). In all simulations, the percentage reduction in peak heel pressure was calculated compared to model prediction of peak barefoot pressure. In accordance with Cheung and Zhang (2005, 2008), it was concluded that the material had a limited effect on pressure reduction since pressure relief slightly improved with softer insole. On the other hand, conformity of insole had the most profound effect in pressure reduction, possibly caused by the larger contact area of surfaces. The authors report that with a careful selection of insole properties, it was possible to reduce pressure peaks by 44% compared to barefoot values.

Similarly, Hsu et al. (2008) achieved a reduction of peek pressure up to 38.9% by developing an insole with "optimal" design. In order to perform design optimization, they used 15 design variables and an objective function expressing the von Mises stress between the plantar fascia and the calcaneus. Stress analysis for the design of therapeutic footwear has been studied in Lemmon et al. (1997), Gefen (2002, 2003), Erdemir et al. (2005), and Owings et al. (2008). The goal of the work reported in Gefen (2002) was the development of a computational model for analyzing the static structural behavior of the foot in the standing posture following surgical plantar fascia release. The effects of the plantar fascia degree of release on the distributions of stresses and deformations in the foot during standing were characterized, thus providing a biomechanical tool that can be applied clinically in a presurgical evaluation of the partial/total release results. Furthermore, stress distribution in plantar soft tissue under the medial metatarsal heads of simulated diabetic versus normal feet has been studied in Gefen (2003). Design guidelines for therapeutic footwear have been proposed in Erdemir et al. (2005) in order to relieve focal loading under the prominent metatarsal heads. The authors studied 36 plug designs, a combination of 3 materials, 6 geometries, and 2 placements using a 2D FE model developed for the purpose of this research.

The determination of plantar stresses using computational footwear models that include temperature effects has been studied in Shariatmadari et al. (2010) in order

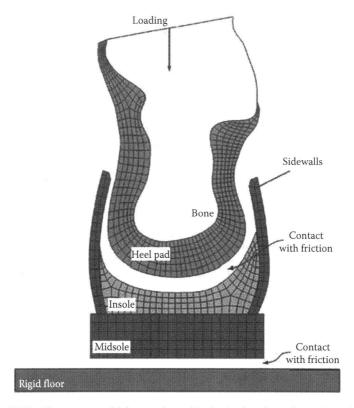

FIGURE 15.12 Footwear model interacting with the heel to investigate the influence of insole parameters on heel pressure relief. A full-conforming insole with 9.5 mm thickness is illustrated. (*J. Biomech.*, 39, Goske, S., Erdemir, A., Petre, M., Budhabhatti, S., Cavanagh, P.R., Reduction of plantar heel pressures: Insole design using finite element analysis, 2363–2370, Copyright 2006 with permission from Elsevier.)

to investigate the effect of varying footwear temperature on plantar stresses and, thus, to provide design guidelines for optimizing shoe design. Using a FE model and simulation, they concluded that temperature variation can affect the cushioning properties of insole and midsole layers made from various foam materials. The loss of footwear cushioning effect can have important clinical implications for those individuals with a history of lower limb overuse injuries or diabetes.

The effect of inversion foot angles on the deformation and stresses within the metatarsal bones and surrounding tissues during landing was investigated in Gu et al. (2010). A detailed FE model based on subject-specific CT images was developed. The deformations of metatarsals at different inversion angles (normal landing, 10° inversion and 20° inversion angles) were comparatively studied. The results showed that in the lateral metatarsals stress increased while in the medial metatarsals stress decreased with increasing inversion angles. The peak stress point was found to be near the proximal part of the fifth metatarsal, and this is in reasonable agreement with reported clinical observations on metatarsal injuries.

15.4 CONCLUSIONS

Clearly, the FEMs play a prominent role in studying the foot biomechanics in the last decade. The current advances have a benefit for both the consumers and the manufacturers. From the consumers' point of view, with existing methods and technologies one is able to produce a custom-mold insole fitted to his or her foot geometric data. This is particularly important for people with diseases such as diabetes where the effect of peak pressures on the plantar fascia can have a serious impact on the person's health. On the other hand, a shoe manufacturer is able to evaluate his product during the design phase avoiding time- and money-consuming trial-and-error tests with real prototypes.

Finally, there are several subjects for future work. Clearly, more advanced foot models including further geometric details and more realistic loads and boundary conditions will be developed in the near future. Parametric foot models could provide a solution to the variety of differences that exist in the shape and internal structure/geometry of bones, cartilages, and ligaments from person to person, which have a significant effect to the FEA results. In addition, another research direction concerns the evaluation of the shape of the outsole with respect to the required level of friction with the floor or soil (Sun et al., 2005).

QUESTIONS

15.1 What is the role of CAD in the footwear industry? Give indicative examples with real production processes.

15.2 Describe the difference between a CAD and a CAE system in the footwear industry.

15.3 What is a "human-centered" CAE system?

15.4 What biomechanics data can be measured with traditional and/or analog measuring devices? What are the pros and cons of these devices?

15.5 What biomechanics data can be measured with FEA? What are the pros and cons of these methods?

15.6 Describe the main steps required to develop a FEM of a real human foot from CT or MRI sectional images.

15.7 What are the basic data required for performing FE analysis on a 3D foot model?

15.8 Make a sketch of an algorithm for converting CT or MRI sectional images to 3D geometry.

15.9 Describe some common diseases or anatomical anomalies related to human foot and explain the role of footwear and CAE in these cases.

REFERENCES

Alexander I.J., Chao E.Y.S., Johnson K.A., 1990. The assessment of dynamic foot-to-ground contact forces and plantar pressure distribution: A review of evolution of current techniques and clinical applications. *Foot and Ankle International*, 11: 152–167.

Antunes P.J., Dias G.R., Coelho A.T., Rebelo F., Pereira T., 2008. Non-linear finite element modelling of anatomically detailed 3D foot model. Materialise (www.materialise.com), report paper.

Athanasiou K.A., Liu G.T., Lavery L.A., Lanctot D.R., Schenck R.C., 1998. Biomechanical topography of human articular cartilage in the first metatarsophalangeal joint. *Clinical Orthopaedics*, 348: 269–281.

Azariadis P., Moulianitis V., Alemany S., Olaso J., de Jong P., van der Zande M., Brands D., 2007. Virtual shoe test bed: A computer-aided engineering tool for supporting shoe design. *Computer-Aided Design and Applications*, 4(6): 741–750.

Azariadis P., Olaso J., Moulianitis V., Alemany S., González J-C., de Jong P., Dunias P., van der Zande M., Brands D., 2010. An innovative virtual-engineering system for supporting integrated footwear design. *International Journal of Intelligent Engineering Informatics (IJIEI)*, 1(1): 53–74.

Azariadis P., Papagiannis P., 2010. A new business model for integrating textile/clothing and footwear production, *The 3rd International Conference on Advanced Materials and Systems -ICAMS 2010*, Bucharest, Romania, September 16–18, pp. 235–240.

Balmaseda M.T., Koozekanani S.H., Fatechi M.T., Gordon C., Dreyfuss P.H., Tanbonliong E.C., 1988. Ground reaction forces, center of pressure, and duration of stance with and without an ankle-foot orthosis. *Archives of Physical Medicine and Rehabilitation*, 69: 1009–1012.

Brown M., Rudicel S., Esquenazi A., 1996. Measurement of dynamic pressures at the shoe-foot interface during normal walking with various foot orthosis using the FSCAN system. *Foot and Ankle International*, 17: 152–156.

Camacho D.L.A., Ledoux W.R., Rohr E.S., Sangeorzan B.J., Ching R.P., 2002. A three-dimensional, anatomically detailed foot model: A foundation for a finite element simulation and means of quantifying foot-bone position. *Journal of Rehabilitation Research and Development*, 39(3): 401–410.

Cavanagh P.R., Hewitt F.G., Perry J.E., 1992. In-shoe plantar pressure measurement: A review. *The Foot*, 2: 185–194.

CEC, 2009. CEC-made-shoe: Custom, Environment and Comfort made shoe. 2004–2008, Contact No 507378 - 2, FP6 IST-NMP (Manufacturing, Products and Service Engineering 2010), http://www.cec-made-shoe.com

Chen W.P., Ju C.W., Tang F.T., 2003. Effects of total contact insoles on the plantar stress redistribution: A finite element analysis. *Clinical Biomechanics*, 18(6): S17–S24.

Chen W.P., Tang F.T., Ju C.W., 2001. Stress distribution of the foot during mid-stance to push-off in barefoot gait: A 3-D finite element analysis. *Clinical Biomechanics*, 16: 614–620.

Cheung J.T.M., Zhang M., 2005. A 3-dimensional finite element model of the human foot and ankle for insole design. *Archives of Physical Medicine and Rehabilitation*, 86: 353–358.

Cheung T.M., Zhang M., 2006. Finite element modeling of the human foot and footwear. *2006 ABAQUS Users' Conference*, Boston, MA, pp. 145–159.

Cheung J.T.M., Zhang M., 2008. Parametric design of pressure-relieving foot orthosis using statistics-based finite element method. *Medical Engineering and Physics*, 30: 269–277.

Cheung T.M., Zhang M., Leung K.L., Fan Y.Bo., 2005. Three-dimensional finite element analysis of the foot during standing—A material sensitivity study. *Journal of Biomechanics*, 38(5): 1045–1054.

Chituc C.M., Toscano C., Azevedo A., 2008. Interoperability in collaborative networks: Independent and industry-specific initiatives—The case of the footwear industry, *Computers in Industry*, 59(7): 741–757.

Chu T.M., Reddy N.P., Padovan J., 1995. Three-dimensional finite element stress analysis of the polypropylene, ankle-foot orthosis: Static analysis. *Medical Engineering and Physics*, 5: 372–379.

Cobb J., Claremont D.J., 1995. Transducers for foot pressure measurement: Survey of recent developments. *Medical and Biological Engineering and Computing*, 33(4): 525–532.

Erdemir A., Saucerman J.J., Lemmon D., Loppnow B., Turso B., Ulbrecht J.S., Cavanagh P.R., 2005. Local plantar pressure relief in therapeutic footwear: Design guidelines from finite element models. *Journal of Biomechanics*, 38: 1798–1806.

Gefen A., 2002. Stress analysis of the standing foot following surgical plantar fascia release. *Journal of Biomechanics*, 35(5): 629–637.

Gefen A., 2003. Plantar soft tissue loading under the medial metatarsals in the standing diabetic foot. *Medical Engineering and Physics*, 25(6): 491–499.

Gefen A., Megido-Ravid M., Itzchak Y., Arcan M., 2000. Biomechanical analysis of the three-dimensional foot structure during gait: A basic tool for clinical applications. *Journal of Biomechanical Engineering*, 122: 630–639.

Goske S., Erdemir A., Petre M., Budhabhatti S., Cavanagh P.R., 2006. Reduction of plantar heel pressures: Insole design using finite element analysis. *Journal of Biomechanics*, 39: 2363–2370.

Gu Y.D., Ren X.J., Li J.S., Lake M.J., Zhang Q.Y., Zeng Y.J., 2010. Computer simulation of stress distribution in the metatarsals at different inversion landing angles using the finite element method. *International Orthopaedics (SICOT)*, 34: 669–676.

Hsu Y.C., Gung Y.W., Shih S.L., Feng C.K., Wei S.H., Yu C.H., Chen C.S., 2008. Using an optimization approach to design an insole for lowering plantar fascia stress—A finite element study. *Annals of Biomedical Engineering*, 36(8): 1345–1352.

Jacob H.A.C., 2001. Forces acting in the forefoot during normal gait—An estimate, *Clinical Biomechanics*, 16(9): 783–792.

Jacob S., Patil M.K., 1999. Stress analysis in three-dimensional foot models of normal and diabetic neuropathy. *Frontiers in Medical and Biological Engineering*, 9: 211–227.

Jacob S., Patil M.K., Braak L.H., Huson A., 1996. Stresses in a 3D two arch model of a human foot. *Mechanics Research Communication*, 23: 387–393.

Jacob H.A.C., Zollinger H., 1992. Biomechanics of the foot—Forces in the forefoot during walking and their clinical relevance. *Orthopade*, 21: 75–80.

Kitagawa, Y., Ichikawa, H., King, A.I., Begeman, P.C., 2000. Development of a human ankle/foot model. In: Kajzer J., Tanaka E., Yamada H. (Eds.), *Human Biomechanics and Injury Prevention*. Springer, Tokyo, Japan, pp. 117–122.

Kitaoka H.B., Luo Z.P., An K.N., 1997. Mechanical behavior of the foot and ankle after plantar fascia release in the unstable foot. *Foot and Ankle International*, 18: 8–15.

Lee K., 2006. CAD system for human-centered design. *Computer-Aided Design and Applications*, 3(5): 615–628.

Lemmon D., Shiang T.Y., Hashmi A., Ulbrecht J.S., Cavanagh P.R., 1997. The effect of insoles in therapeutic footwear-a finite element approach. *Journal of Biomechanics*, 30(6): 615–620.

Leng J., Du R., 2006. A CAD approach for designing customized shoe last. *Computer-Aided Design and Applications*, 3(1–4): 377–384.

Lewis G., 2003. Finite element analysis of a model of a therapeutic shoe: Effect of material selection for the outsole. *Biomedical Materials and Engineering*, 13(1): 75–81.

Lundeen S., Lundquist K., Cornwall M.W., 1994. Plantar pressure during level walking compared with other ambulatory activities. *Foot and Ankle International*, 15: 324–328.

Luximon A., Goonetilleke R.S., Tsui K.L., 2003. Footwear Fit Categorization. In: Tseng, M.M. and Piller, F.T. (Eds.), *The Customer Centric Enterprise: Advances in Mass Customization and Personalization*. Springer, Berlin, Germany, pp. 491–500.

Mao A., Li Y., Luo X., Wang R., Wang S., 2008. A CAD system for multi-style thermal functional design of clothing. *Computer-Aided Design*, 40(9): 916–930.

Nagel A., Fernholz F., Kibele C., Rosenbaum D., 2008. Long distance running increases plantar pressures beneath the metatarsal heads: A barefoot walking investigation of 200 marathon runners. *Gait and Posture*, 27(1): 152–155.

Nakamura S., Crowninshield R.D., Cooper R.R., 1981. An analysis of soft tissue loading in the foot—A preliminary report. *Bulletin of Prosthetics Research*, 18: 27–34.

Nevill A.J., Pepper M.G., Whiting M., 1995. In-shoe foot pressure measurement system utilising piezoelectric film transducers. *Medical and Biological Engineering and Computing*, 33(1): 76–81.

Orlin M.N., Mc Poil T.G., 2000. Plantar pressure assessment. *Physical Therapy*, 80(4): 399–409.

Owings T.M., Woerner J.L., Frampton J.D., Cavanagh P.R., Botek G., 2008.Custom therapeutic insoles based on both foot shape and plantar pressure measurement provide enhanced pressure relief. *Diabetes Care*, 31(5): 839–844.

Patil K.M., Braak L.H., Huson A., 1996. Analysis of stresses in two-dimensional models of normal and neuropathic feet. *Medical and Biological Engineering and Computing*, 34: 280–284.

Pedrazzoli P., 2009. Design Of customeR dRiven shoes and multisiTe factory—DOROTHY. In: Thoben K.-D., Pawar K.S., and Goncalves R. (Eds.), *15th International Conference on Concurrent Enterprising*, Leiden, Netherlands, June 22–24, 2009.

Race A., Amis A., 1994. The mechanical properties of the two bundles of the human posterior cruciate ligament. *Journal of Biomechanics*, 27: 13–24.

Rosenbaum D., Bauer G., Augat P., Claes L., 1996. Calcaneal fractures cause a lateral load shift in chopart joint contact stress and plantar pressure pattern in vitro. *Journal of Biomechanics*, 29: 1435–1443.

Shariatmadari M.R., English R., Rothwell G., 2010. Finite element study into the effect of footwear temperature on the forces transmitted to the foot during quasi-static compression loading. *IOP Conference Series: Materials Science and Engineering*, Bristol, U.K., 10;012126.

Shiang T.Y., 1997. The nonlinear finite element analysis and plantar pressure measurement for various shoe soles in heel region. *Proceedings of National Science Council Republic of China Part B*, 21(4): 168–174.

Siegler S., Block J., Schneck C.D., 1988. The mechanical characteristics of the collateral ligaments of the human ankle joint. *Foot and Ankle*, 8: 234–242.

Soames R.W., 1985. Foot pressure patterns during gait. *Journal of Biomedical Engineering*, 7: 120–126.

Sun Z., Howard D., Moatamedi M., 2005. Finite-element analysis of footwear and ground interaction. *Strain*, 41: 113–115.

Tanimoto Y., Takechi H., Nagahata H., Yamamoto H., 1998. The study of pressure distribution in sitting position on cushions for patient with SCI (Spinal Cord Injury). *IEEE Transactions on Rehabilitation Engineering*, 47(5): 1239–1243.

Urry S., 1999. Plantar pressure-measurement sensors. *Measurement Science and Technology*, 10(1): R16–R32.

Verdejo R., Mills N.J., 2004. Heel-shoe interactions and the durability of EVA foam running-shoe midsoles. *Journal of Biomechanics*, 37(9): 1379–1386.

Whittle M., 2003. *Gait Analysis: An Introduction*. Elsevier Ltd, Edinburgh, U.K.

Wiegerinck J.I., Boyd J., Yoder J.C., Abbey A.N., Nunley J.A., Queen R.M., 2009. Differences in plantar loading between training shoes and racing flats at a self-selected running speed. *Gait and Posture*, 29: 514–519.

Wong K.H.M., Hui P.C.L., Chan A.C.K., 2006. Cryptography and authentication on RFID passive tags for apparel products. *Computers in Industry*, 57(4): 342–349.

Wright D., Rennels D., 1964. A study of the elastic properties of plantar fascia. *The Journal of Bone and Joint Surgery. American Volume*, 46: 482–492.

Yu J., Cheung J.T.M., Fan Y., Zhang Y., Leung A.K.L., Zhang M., 2008. Development of a finite element model of female foot for high-heeled shoe design. *Clinical Biomechanics*, 23(2008): S31–S38.

Yung-Hui L., Wei-Hsien H., 2005. Effects of shoe inserts and heel height on foot pressure, impact force, and perceived comfort during walking. *Applied Ergonomics*, 36: 355–362.

16 Footwear—The Forgotten Treatment—Clinical Role of Footwear

Helen Branthwaite, Nachiappan Chockalingam, and Aoife Healy

CONTENTS

16.1 INTRODUCTION

Footwear in general plays an important role in protecting the foot from the environment by reducing the risks associated with trauma and enables users to have pain-free locomotion over a range of walking surfaces (McPoil 1998). However, the prolonged use of unsuitable footwear can also result in detrimental changes to occur that alter the protective nature of the shoe into a barrier between the contact surface and the natural behavior of the foot. These changes can result in altered foot morphology, reduced or impaired postural stability, neurophysiological alterations, muscle imbalance, and the development of a sensitive foot (Menz and Lord 1999, Robbins and Waked 1997). Alterations in footwear design, structure, and manufacture have been attributed to these mechanical and physiological observations.

As footwear evolves into specific subgroups, the basic structure of the shoe changes to reflect the use of the wearer. Fashion and style play a more important role as the principle use of footwear in protecting the foot changes into a color-orientated accessory (Branthwaite and Chockalingam 2009). By following this fashion pathway, the essential components required to provide stability and support are lost in the design process (McPoil 1998). These changes in behavior over the last 500 years in shoe-wearing populations have been attributed to reported footwear pathologies (Paiva de Castro et al. 2010, Rebelatto and Aurichio 2010).

The last decade, however, has seen more attention being focused on using footwear as a means of improving function, in particular running, and athletic shoes that have been developed to enhance performance. These shoes that have been designed to alter behavior can also be used as a therapeutic device. This chapter will focus on the design aspects incorporated into footwear in an attempt to improve performance or treat/prevent injuries, examine how functional shoes can change gait and pathology, and discuss some of the new concepts in footwear design.

16.2 FOOTWEAR DESIGN AND TECHNOLOGIES

Developments made along the road of technological improvements have included many functional and style modifications to alter fit, comfort, and performance within running and athletic shoes. Brand specifications have become a marketing point with enhancements including air, gel, fluid, and changes in midsole material. Companies have employed scientific researchers to test and adapt the designs to improve motion control. Certain design parameters are agreed upon as being beneficial including the heel geometry, stiffness, and heel collar (Figures 16.1 and 16.2).

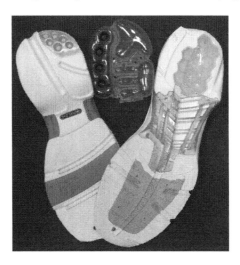

FIGURE 16.1 Development of brand specific technologies to enhance motion control and impact during use.

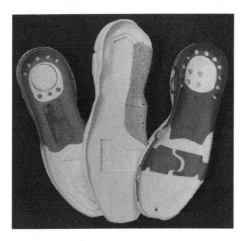

FIGURE 16.2   Midsole material designs altered to effect function.

16.2.1   HEEL COUNTER AND COLLAR

A stiffer well-fitted heel counter has been found to reduce rearfoot motion and improve comfort (Anthony 1987, Van Gheluwe et al. 1999). Injuries associated with increased rearfoot motion include tibialis posterior dysfunction, Achilles tendonitis, and plantar fasciitis, all of which can improve when a stiff well-fitted heel counter are incorporated into footwear design. Heel collar height has been suggested as a desirable feature to prevent ankle sprains and improve postural stability. This design feature is thought to provide mechanical support around the ankle region and improve the proprioceptive feedback of the ankle position. The presence of cushioned material around the ankle region is thought to provide mechanical lateral ankle stability, such that rapid excursions of the foot into eversion or inversion are restricted by the shoe (Menz and Lord 1999, Stacoff et al. 1996). It is also thought that the presence of the high heel collar may provide additional tactile cues, thereby improving proprioceptive input to the central nervous system (CNS). Furthermore, there have been suggestions that the extrasensory feedback provided by a high collar, similar to ankle taping, ankle bracing, and knee bracing, may facilitate joint position sense and improve medio-lateral balance (Menz and Lord 1999).

16.2.2   SHOE WEIGHT

Increases in shoe weight have been associated with oxygen consumption and performance. A lighter shoe can improve oxygen consumption by 1% for every 100 g lost in weight (Morgan et al. 1989, Williams and Cavanagh 1987). For an individual who chooses running for general exercise, this difference may not be significant to their performance; however, to the elite athlete such changes could be critical to their career. Choosing the correct footwear that supports the foot and allows for normal function as well as not weighing too much can be difficult as a reduction in weight often changes the composition of materials which correlates to a reduction in structure and support.

16.2.3   SOLE AND MIDSOLE

16.2.3.1   Thickness and Density

Sole thickness and density and the type of materials used in manufacture and design also vary between brands. Dual density sole structures have been adopted in running shoes to add stability and give cushioning to the wearer in the appropriate areas of the foot. Changes in density and thickness have been shown to affect the stability and balance of the user. Shoes that are manufactured with a thicker and softer sole have been associated with increases in dynamic instability and a reduction is postural balance (Robbins et al. 1992, Sekizawa et al. 2001). Conversely, shoes that have a thinner stiffer sole improve both stability and balance parameters (Menant et al. 2008, Perry et al. 2007, Robbins and Waked 1997, Robbins et al. 1994). These design features affect afferent feedback to the CNS regarding foot position. This method of altering sole density and thickness can be utilized when there is diminished or impaired feedback; for example, in an elderly population who are at risk of falling due to altered stability and balance, a thinner-stiffened sole can improve feedback reducing falls (Robbins et al. 1992). Similarly, a softer thicker sole may help in the rehabilitation and strength conditioning of ankle instability where the shoe is used as a training device to improve feedback.

The concept of a dual density midsole has been adopted to alter the stiffness and cushioning properties of the shoe. The medial border of the sole is manufactured from a stiffer material than the midsole attempting to reduce pronation. Stacoff et al. (2000) used a dual density shoe to evaluate motion during running and found minimal changes. These results were also reported by Morio et al. (2009) who found that midsole hardness and stiffness did not affect the foot pronation during walking or running. However, Heidenfelder et al. (2006) reported benefits of using a dual density sole. Their study measured impact shock rather than assessing motion and demonstrated that the cushioning properties of the shoe are more beneficial than the motion control element. The application of such research will benefit clinicians in advising appropriate footwear for particular conditions. The prescriptive nature of footwear advice will assist in patient-specific management plans, giving individual recommendations dependent on presenting features rather than generalizing specific injuries for a specific shoe.

16.2.3.2   Sole Design

The angle and position of flexion at the forefoot of the shoe is known as the sole flex. It is important in design and manufacture because flexion at the metatarsophalangeal joints is a normal requirement for gait (Menz and Sherrington 2000). If the shoe does not flex with the natural angle of the metatarsals during the propulsive phase of gait, the shoe may either flex against the foot or direct motion and propulsion laterally causing instability. Matching footwear to the patient morphology and foot type will assist in preventing injury and discomfort from the shoe; forefoot flexion at propulsion is important to maintain stability and forward progression of the body's centre of mass (Menz and Lord 1999) (Figures 16.3 and 16.4).

Cleat design and slip resistance of the outer sole is another aspect of footwear design that is important in motion control. Football, rugby, and hockey where the

FIGURE 16.3   Sole flex angles differ between sole design, altering forefoot propulsion.

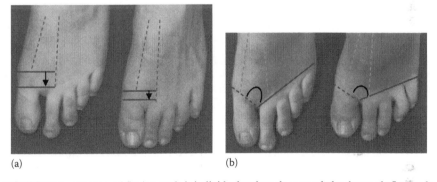

(a)   (b)

FIGURE 16.4   Metatarsal flexion angle is individual and needs to match the shoes sole flex angle.

cleat orientation and size are critical to provide grip have been shown to increase rotational forces around the knee joint, induce cruciate ligament injuries, and stress (Villwock et al. 2009). However, changes in the position and orientation of the cleats have not been shown to reduce the incidence of such injuries and it is thought that the cutting manoeuvres involved in such sports are more influential than the footwear. Cleat design in running shoes has had little attention although the type of activity and terrain have been identified as important considerations when looking at cleats and outsole construction (McPoil 2000).

16.2.3.3   Insert Adaptations

Other technologies incorporated into footwear design to improve or alter motion include modifications that allow for the windlass mechanism to function efficiently. This type of modification within the shoe would be particularly useful in the management of plantar fascial pain and first metatarsal phalangeal joint pathology associated with a dysfunctional windlass mechanism. Payne et al. (2005) examined whether the windless mechanism was enhanced in all shoes marketed as having windlass technology and found positive changes, but the results differed with each brand of shoe. Other trends have come and disappeared depending on the success of that shoe model: heel flare, rearfoot and forefoot valgus adaptations, cobra padding, and an arch cookie are to name but a few. The generalization of such features available to the mass market reduces specificity as the incorrect match and advice regarding shoe and foot type causing ill-fitting uncomfortable shoes. It is becoming more common for specific running and athletic shoes to be provided by retailers with advanced knowledge into the characteristic and design features available from each brand and model. Various motion analysis systems available within retailer's premises provide further detail on the gait pattern of an individual, allowing for informed judgment on the specific models, brand features, and composition of the footwear matching the recorded data.

16.2.3.4   Shock Absorption

The human body's ability to absorb shock involves several complex mechanisms, whilst running increases the load and shock that has been associated with joint disease, particularly knee pathology (Lafortune et al. 1996). Footwear modifications to enhance shock absorbency are common place and it has been suggested that these design features in running footwear must be able to provide adequate and durable cushioning systems to enable safe participation in sports (Frederick 1995). Changes in material properties and behavior provide shock absorption characteristics for the wearer in an attempt to prevent overuse injuries resulting from repetitive impact loading. It can be theorized that a thicker sole material would deform more than a thinner sole material, and thus, impact forces will be attenuated more in the thicker sole giving that shoe better shock absorption (Barnes and Smith 1994). However, studies looking at the effect of running shoes on impact forces have not been able to show a systematic difference (Aguinaldo and Mahar 2003, Cole et al. 1995).

The cushioned shoe has been developed to attenuate shock. However, research that has investigated other parameters that are altered with the introduction of shock absorbency shoes have found that impact force may not be the significant variable that is altered. Plantar pressure can be altered in cushioned shoes, yet the type of cushioning material used in the design does not affect the degree of pressure changes (Clinghan et al. 2008). Loading rates during gait and timing for each of the observed phases have also been investigated and shown to alter with the use of cushioned shock absorbency shoes (Hennig and Milani 1995, O'Leary et al. 2008).

Oxygen consumption and muscle activity have both been investigated when wearing different shock absorbency shoes. Whilst previous literature indicate that, softer, more compliant and resilient shoes required less oxygen to run in than a stiffer shoe, thus improving the rate of fatigue (Frederick 1986), more recent research contradicts this finding. Nigg et al. (2003) demonstrated that a visco-elastic heel and an elastic heel did not affect oxygen consumption or alter muscle activity. However, some of the study participants showed a systematic and consistent difference in oxygen consumption with varying heel material properties. Nigg (2001) also suggested that impact forces during normal physical activity are important because rather than being a direct cause of potential injuries, they affect muscular work; therefore, attenuating those impacts could result in a change in muscular function.

When the human body senses it is being exposed to larger impact shocks due to large high frequency ground reaction force components acting through the plantar region of the foot, it has been shown that the human body will adapt its running style (De Wit et al. 2000, Lafortune et al. 1996). Furthermore, various studies have identified that by purposefully adapting the running style of participants, the loading characteristics can be significantly changed (Laughton et al. 2003, Lieberman et al. 2010, Oakley and Pratt 1988).

16.2.3.5 Material Properties

The midsoles of running shoes are typically made from ethylene vinyl acetate (EVA) and/or polyurethane of varying density. While research has found that midsole hardness had little or no effect on lower extremity muscle activity, impact peak, and ankle joint forces, it has been shown to effect rearfoot movements during touchdown and muscle forces on the medial side of the subtalar joint (Nigg and Gérin-Lajoie 2011, Stacoff et al. 1988). One important aspect of the materials used in running shoes is their durability with research suggesting that running shoe age can contribute to injuries (Taunton et al. 2003). Verdejo and Mills (2004a) examined the durability of EVA foam running shoe midsoles with their results showing a 100% increase in peak plantar pressure after 500 km run and structural damage of the foam after 750 km run. They related the decrease in the cushioning properties to a decrease in the initial compressive collapse stress and not to a change in the air content (Verdejo and Mills 2004b). Furthermore, research carried out by Cook et al. (1985) found a reduction in shock absorption of running shoes with increased mileage. Compression forces applied to the outer sole material whilst running can affect the cushioning properties by up to 30% (Kenoshita and Bates 1996). Similarly, it has been shown that EVA can take up to 24 h to return to its original state, and therefore, people who use running shoes more than once in a day should think about the changes in the materials and how that may impact on cushioning and potentially increase the risk of injury occurrence (McPoil 2000). This evidence supports the thought that a running shoe with an EVA sole should only be used to run with for a set distance and after that the material properties change and could enhance injuries related to shock absorbency.

16.2.3.6 Other Design Aspects

Innovation and concept shoes have given inspiration to changing the mould for manufacture of the running shoe. The designs and theories have focused on being prophylactic in nature by providing a better performance for the wearer and a reduction in the incidence of injury. For example, Heeless technology™—this shoe development began from the experience of a marathon runner who was plagued by Achilles tendon injuries. The design attempts to alter foot strike pattern and the loading pattern on the Achilles tendon. Similar innovations include the Hoka™ One One shoe which is based on a rocker shoe concept but aims to increase the contact surface area of the shoe with the floor. The aim of this concept appears to help improve stability and users gain the benefit of the additional volume of the shoe.

The volume of the forefoot of a shoe has been indicated to be significant in the development of foot pathology (Paiva de Castro et al. 2010, Rebelatto and Aurichio 2010). Increasing the volume of the forefoot allows for the metatarsals to function without medial and lateral compression and therefore reduces the instability and impingement pathologies commonly seen in the forefoot. The use of shoes with increased volume and stability could be used in sufferers of forefoot pathology. As well as users of shoes that alter loading patterns of the Achilles can alleviate strain around potential injury sites. It is essential to be advised as to which shoe would suit the wearer's foot type and mechanics as adverse effects may be encountered from the use of concept shoes.

16.2.4 COMFORT

Oxygen consumption and therefore fatigue have been strongly correlated with comfort. Research by Luo et al. (2009) found that the participants who were most comfortable had a significant decrease in oxygen consumption. If a shoe is making the user consume more oxygen and the muscle activity is higher, then it is understandable that the user will feel uncomfortable in that shoe. Comfort has been used as a predictive factor for muscle activity and intensity in the lower limb (Mundermann et al. 2003). Improving comfort of a shoe can result in a reduction in stress-related injuries (Mundermann et al. 2001). If a shoe is comfortable and fits well, it is less likely to be compressing the foot in any way. Forefoot width and depth have been shown to be positively correlated to foot pain and pathology. The narrower and shallower the shoe, the higher the incidence of forefoot injury (Paiva de Castro et al. 2010, Rebelatto and Aurichio 2010). The breadth of the forefoot can influence the perception of comfort and any compressive forces around the forefoot altering the structure increase localized stress on the metatarsals (Morio et al. 2009). However, Miller et al. (2000) found that the fit of the shoe is not solely sufficient for comfort. They found that skeletal alignment, shoe torsional stiffness, and cushioning seem to be mechanical variables which may be important for comfort. Comfort has been correlated with a number of measures and differences in comfort have been associated with changes in kinetics, kinematics, muscle activity, foot shape, fit between foot and footwear, skeletal alignment, foot sensitivity, weight of the shoe, temperature, and joint range of motion (Mundermann et al. 2003, Wegener et al. 2008).

Most people are able to identify what is a comfortable and noncomfortable shoe. This measure therefore can be adopted in the assessment of footwear fitting and give confidence to the person giving advice about which shoe to wear that if the shoe is perceived as comfortable then there is a reduced chance of developing injury.

Adaptation and individual variability can be used to describe why studies investigating the outcomes of footwear modifications do not generate systematic differences. As individuals, the human body responds in many different ways to intervention. This has been shown in several studies around shock absorption and impact forces where the conclusions derived from the work have been that outcomes vary according to the individual and therefore cannot be generalized (Frederick 1986, Kersting et al. 2006, Nigg 2001, Nigg and Wakeling 2001). Translating this information into an injured population will possibly give the same results. One patient group may respond well to a shock absorption shoe when they have developed plantar fasciitis, whereas others may not get any relief from wearing a cushioned shoe.

16.2.5 Neutral Shoe

The original running shoe design did not have the characteristics of support along the medial side of the shoe and the dual density midsole; these motion control features were added later in the development based on the research of the time. The basic model of a running shoe is still manufactured today and provides the consumer with a shoe that is known as being neutral. This shoe may be preferred by orthoses users as the last used within the manufacturing process and the resulting structure provides a base to add a specific orthotic prescription to (Baker et al. 2007). Neutral shoes have been shown to reduce the muscle activity in tibialis anterior and soleus compared to motion control or cushioning shoes (O'Conner and Hamill 2004) providing positive enhancement for a neutral shoe to be included in a management plan for overuse disorders of these muscle groups. However, Cheung and Ng (2007) compared the effects of motion control and neutral shoes on muscle fatigue and found that the motion control shoe was superior in minimizing muscle fatigue. Fatigue has been identified as a strong link in the development of overuse injuries with increases in metatarsal loading and impact accelerations being observed in exhausted subjects (Mizrahi et al. 2000, Nagel et al. 2008). Footwear that can improve the fatigue rate of users is desirable to use as the impact on injury occurrence may be beneficial.

16.3 OVERUSE INJURIES

While previous physical activity would seem to predispose individuals to develop a musculoskeletal system that was less likely to suffer an overuse injury due to conditioning, research has suggested this may not be the case. No correlation was found between army recruits who participated in sporting activities prior to training and those who did not with the occurrence of stress fractures (Swissa et al. 1989). Studies such as this have to be approached with caution as a participant who is physically fitter than another may have increased their body's ability to withstand larger forces leading to stress fractures. However, due to this, the participant may expose their body to more intense exercise over a longer duration. In many studies, this may be a problem that affects the findings. If an athlete is more physically fit and more protected from injury due to previous training, this may lead them to expose their body to more intense exercise and more volatile movement strategies.

It would appear that getting the balance between enough exposures to lower extremity impacts to promote health in the musculoskeletal system without reaching the point of injury may be the correct way to avoid stress fractures. However, controlling an athlete's movement characteristics strictly is not necessarily possible and thus dictates the stresses to which their body is subjected in a competitive environment (Ekenman et al. 2001). Using army recruits who undergo controlled physical activity provides sports scientists with large amounts of comparable data. Using army recruits ($n = 1357$), research has found that a week's rest in the middle of an 8 week training course did not significantly reduce the incidence of overuse lower extremity injuries (Popovich et al. 2000). The results of this relatively large study of army recruits found that the lowest injury group was the one who ran the most miles with intermittent rests, recommending running and marching with single day of rest in between as having a positive effect on lower extremity overuse injuries such as stress fractures.

16.4 NEW CONCEPTS

The trend and focus on the development of sports footwear has been to provide a shoe that makes the body exercise muscles and provide a workout within the normal working day. Other more specialized running shoes have focused on increasing or improving performance.

16.4.1 INSTABILITY SHOE

The concept of introducing a rocker sole to footwear has been used within orthopedic and rehabilitation medicine for centuries with a specific aim to off-load compression forces from joints within the lower limb (Bauman et al. 1963). More recently, the therapeutic intervention of improving stability by exercising the systems that provide locomotive stability has been adopted into the footwear industry. The MBT® shoe was the first commercial instability shoe to affect the mass market.

MBT training shoes have been associated with variable changes between subjects due to different strategies for compensation being employed (Nigg et al. 2006, Stewart et al. 2007). The marketed aim of the MBT, unstable shoe, has been to provide an exercise workout whilst walking in the shoe. However, it is not clear from research to date as to the specific effects of the shoe on lower limb function. Landry et al. (2010) used an EMG circumferential linear array with MRI scanning to evaluate the activity of the extrinsic foot muscles whilst standing in the MBT shoe. As with previous studies, the anterior and posterior sway increased in the MBT shoe (Nigg et al. 2006, 2010). However, the activity of the soleus muscle remained unchanged. There was increase in activity in the anterior leg compartment and peroneal muscle group, suggesting that these muscles are recruited as a method to improve stability whilst wearing the shoe.

The APOS shoe has a similar strategy of creating instability for the user, unlike the MBT—the instability comes from a medial and lateral direction and can be altered to balance loads. The use of this shoe has been shown to be effective in the management of knee osteoarthritis where subjects showed a 70% improvement in pain over a 8 week period which were sustained after 1 year (Bar-Ziv et al. 2007). Elbaz et al. (2010) support the conclusion and the use of the shoe is advocated in the management of knee osteoarthritis. Patient selection and correct management with the specific shoe will be critical to recovery with reports of limitations in the use of the footwear and patient compliance. However, Reeboks launch of the Tone ups® provide a commercially available product that mimics the more prescriptive APOS shoe. Independent research evaluating the effect of the Tone ups has yet to be completed.

16.4.2 GENDER

Although there is a clear need for distinguishing footwear between genders as previous research has reported that females have higher-arched feet and a shorter outside foot length (Wunderlich and Cavanagh 2001), yet in general running shoes are manufactured for unisex populations. During standing results, a significant difference ($P < 0.05$) in plantar contact areas was reported, when comparing genders suggesting different requirements from a sports shoe (Gravante et al. 2003). Changes in the manufacture and design of female-specific running shoes will help with fit and comfort for the female runner (Krauss et al. 2010). The changes in fit are not the only differences between male and female. Ferber et al. (2003) highlighted that lower extremity movement patterns are more diverse in women than men. Female changes and requirements and studies comparing mixed gender samples should consider this in any evaluation of results.

It has also been reported that footwear size selection in female populations tend toward choosing footwear that is designed for smaller feet which has been shown to be linked to injury occurrence (Frey et al. 1995). Shoe manufacturers construct slender shoes for females compared to males to allow for the more slender anatomy of the female foot. However, research has found that although the female-specific shoes were reported to have provided a better fit by the female participants, the shoes did not improve the cushioning or rearfoot control characteristics during running (Hennig 2001). Overall, gender-specific footwear seems sensible considering the differences reported in plantar regions and foot size. One of the major footwear brands has gone on to develop some gender-specific features within their footwear range based on published research (Bryant et al. 2008). These include varied shock absorption capability within midsoles of running shoes and special support mechanisms which are altered to provide support for changes in arch height over a 28 day ovulation cycle. Furthermore, as research indicates that Achilles tendonitis is more prevalent amongst female runners, some of the footwear design has a higher heel gradient on running shoes for women, which relieves much of the loading that causes Achilles tension.

16.5   FUTURE DEVELOPMENT

Over the years, preventing overuse injuries has been the focus for footwear scientists and designers. This is an area which still needs a variety of sport-specific or event-specific prospective studies. This should include how previous training, duration, and intensity of activities can be adapted along with specific footwear to reduce the risk of overuse injuries.

QUESTIONS

16.1 How have technology enhanced strategies that influence design impacted on lower limb injuries?

16.2 What design features have been incorporated into footwear design to improve function and reduce onset of pathology?

16.3 How do shock absorbency and comfort effect fatigue and overuse injury?

16.4 Instability shoes have been shown to alter postural stability, how could these shoes be used to improve stability in those who are unstable?

16.5 How can footwear design improve the female runner's experience?

REFERENCES

Aguinaldo, A. and Mahar, A. 2003. Impact loading in running shoes with Cushioning column systems. *Journal of Applied Biomechanics* 19(4): 353–360.

Anthony, R.S. 1987. The functional anatomy of the running training shoe. *The Chiropodist* December: 451–59.

Baker, K., Goggins, J., Xie, H., Szumowski, K., LaValley, M., Hunter, D.J., and Felson, D.T. 2007. A randomized crossover trial of a wedged insole for treatment of knee osteoarthritis. *Arthritis and Rheumatism* 56(4): 1198–1203.

Barnes, R.A. and Smith, P.D. 1994. The role of footwear in minimizing lower limb injury. *Journal of Sports Sciences* 12(4): 341–353.

Bar-Ziv, Y., Beer, Y., Ran, Y., Benedict, S., and Halperin, N. 2007. Dynamic wedging: A novel method for treating knee osteoarthritis. A 1-year follow-up study. *Osteoarthritis and Cartilage* 15(3 Suppl): C83–C84.

Bauman, J.H., Girling, J.P., and Brand, P.W. 1963. Plantar pressures and trophic ulceration: An evaluation of footwear. *The Journal of Bone and Joint Surgery* 45: 652–673.

Branthwaite, H.R. and Chockalingam, N. 2009. What influences someone when purchasing new trainers? *Footwear Science* 1: 71–72.

Bryant, A.L., Clark, R.A., Bartold, S., Murphy, A., Bennell, K.L., Hohmann, E., Marshall-Gradisnik, S., Payne, C., and Crossley, C.M. 2008. Effects of estrogen on the mechanical behavior of the human Achilles tendon in vivo. *Journal of Applied Physiology* 105(4): 1035–1043.

Cheung, R.T.H. and Ng, G.Y.F. 2007. Efficacy of motion control shoes for reducing excessive rearfoot motion in fatigued runners. *Physical Therapy in Sport* 8: 75–81.

Clinghan, R., Arnold, G.P., Drew, T.S., Cochrane, L.A., and Abboud, R.J. 2008. British Journal of Sports Medicinebjsportmed.com Do you get value for money when you buy an expensive pair of running shoes? *British Journal of Sports Medicine* 42: 189–193.

Cole, G.K., Nigg, B.M., and Fick, G.H. 1995. Internal loading of the foot and ankle during impact in running. *Journal of Applied Biomechanics* 11: 25–46.

Cook, S.D., Kester, M.A., and Brunet, M.E. 1985. Shock absorption characteristics of running shoes. *American Journal of Sports Medicine* 1: 248–253.

De Wit, B., De Clercq, D., and Aerts, P. 2000. Biomechanical analysis of the stance phase during barefoot and shod running. *Journal of Biomechanics* 33(3): 269–278.

Ekenman, I., Hassmén, P., Koivula, N., Rolf, C., and Felländer-Tsai, L. 2001. Stress fractures of the tibia: Can personality traits help us detect the injury-prone athlete? *Scandinavian Journal of Medicine and Science in Sports* 11(2): 87–95.

Elbaz, A., Mor, A., Segal, G., Debbi, E., Haim, A., Halperin, N., and Debi, R. 2010. APOS therapy improves clinical measurements and gait in patients with knee osteoarthritis. *Clinical Biomechanics* 25(9): 920–925.

Ferber, R., Davis, I.M., and Williams, D.S. 2003. Gender differences in lower extremity mechanics during running. *Clinical Biomechanics* 18: 350–357.

Frederick, E.C. 1986. Kinematically mediated effects of sports shoe design. *Journal Sports Science* 4: 169–184.

Frederick, E.C. 1995. Biomechanical requirements of basketball shoes. In: Shorten, M., Knicker, A., and Brüggemann, G-P. (Eds.), *Proceedings of the Second Symposium of the ISB Working Group on Functional Footwear*, International Society of Biomechanics, Cologne, Germany, pp. 18–19.

Frey, C., Thompson, F., and Smith, J. 1995. Update on women's footwear. *Foot and Ankle International* 16(6): 328–331.

Gravante, G., Russo, G., Pomara, F., and Ridola, C. 2003. Comparison of ground reaction forces between obese and control young adults during quiet standing on a baropodometric platform. *Clinical Biomechanics* 18(8): 780–782.

Heidenfelder, J., Odenwald, S., and Milani, T. 2006. Mechanical properties of different midsole materials in running shoes. *Journal of Biomechanics* 39: S550.

Hennig, E.W. 2001. Gender differences for running in athletic footwear *Proceedings of the Fifth Symposium on Footwear Biomechanics,* Zuerich, Switzerland, pp. 44–45.

Hennig, E.M. and Milani, T.L. 1995. In-shoe pressure distribution for running in various types of footwear. *Journal of Applied Biomechanics* 11(3): 299–310.

Kenoshita, H. and Bates, B.T. 1996. The effect of environmental temperatures on the properties of running shoes. *Journal of Applied Biomechanics* 12: 258–264.

Kersting, U.G., Kriwet, A., and Brüggemann, G.P. 2006. The role of footwear-independent variations in rearfoot movement on impact attenuation in heel–toe running. *Research in Sports Medicine* 14(2): 117–134.

Krauss, I., Valiant, G., Horstmann, G., and Grau, S. 2010. Comparison of female foot morphology and last design in athletic footwear—Are men's lasts appropriate for women? *Research in Sports Medicine* 18(2): 140–156.

Lafortune, M., Lake, M., and Hennig, E.M. 1996. Differential shock transmission response of the human body to impact severity and lower limb posture. *Journal of Biomechanics* 29(12): 1531–1537.

Landry, S.C., Nigg, B.M., and Tecante, K.E. 2010. Standing in an unstable shoe increases postural sway and muscle activity of selected smaller extrinsic foot muscles. *Gait and Posture* 32(2): 215–219.

Laughton, C.A., Davis, I.M., and Hamill, J. 2003. Effect of strike pattern and orthotic intervention on tibial shock during running. *Journal of Applied Biomechanics* 19: 153–168.

Lieberman, D.E., Venkadesan, M., Werbel, W.A., Daoud, A.I., D'Andrea, S., Davis, I.S., Mang'Eni, R.O., and Pitsiladis, Y. 2010. Foot strike patterns and collision forces in habitually barefoot versus shod runners. *Nature* 463: 531–535.

Luo, G., Sterigou, P., Worobets, J., Nigg, B.M., and Stefanyshyn, D. 2009. Improved footwear comfort reduces oxygen consumption during running. *Footwear Science* 1(1): 25–29.

McPoil, T.G. 1998. Footwear. *Physical Therapy* 68(12): 1857–1865.

McPoil, T.G. 2000. Athletic footwear: Design, performance, and selection issues. *Journal of Science and Medicine in Sports* 3(3): 260–267.

Menant, J.C., Steele, J.R., Menz, H.B., Munro, B.J., and Lord, S.R. 2008. Effects of footwear features on balance and stepping in older people. *Gerontology* 54: 18–23.

Menz, H.B. and Lord, S.R. 1999. Footwear and postural stability in older people. *Journal of the American Podiatric Medical Association* 98(7): 346–357.

Menz, H.B. and Sherrington, C. 2000. The footwear assessment form: A reliable clinical tool to assess footwear characteristics of relevance to postural stability in older adults. *Clinical Rehabilitation* 14(6): 657–664.

Miller, J.E., Nigg, B.M., Liu, N., and Stefanyshyn, D. 2000. Influences of foot, leg and shoe characteristics on subjective comfort. *Foot and Ankle International* 21(9): 759–767.

Mizrahi, J., Verbitsky, O., Isakov, E., and Daily, D. 2000. Effect of fatigue on leg kinematics and impact acceleration in long distance running. *Human Movement Science* 19: 139–151.

Morgan, D.W., Martin, P.E., and Krahenbuhl, G.S. 1989. Factors affecting running economy *Sports Medicine* 7: 310–330.

Morio, C., Lake, M.J., Gueguen, N., Rao, G., and Baly, L. 2009. The influence of footwear on the foot motion during walking and running. *Journal of Biomechanics* 42(13): 2081–2088.

Mundermann, A., Nigg B.M., Humble, R.N., and Stefanyshyn, D.J. 2003. Foot orthotics affect lower extremity kinematics and kinetics during running. *Clinical Biomechanics* 18: 254–262.

Mundermann, A., Stefanyshyn, D.J., and Nigg, B.M. 2001. Relationship between footwear comfort of shoe inserts and anthropometric and sensory factor. *Medicine and Science in Sports and Exercise* 33(11): 1939–1945.

Nagel, A., Fernholz, F., Kibele, C., and Rosenbaum, D. 2008. Long distance running increases plantar pressures beneath the metatarsal heads: A barefoot walking investigation of 200 marathon runners. *Gait and Posture* 27(1): 152–155.

Nigg, B.M. 2001. The role of impact forces and foot pronation—A new paradigm. *Clinical Journal of Sports Medicine* 11: 2–9.

Nigg, B.M., Emery, C., and Hiemstra, L.A. 2006. Unstable shoe construction and reduction of pain in osteoarthritis patients. *Medicine and Science in Sports and Exercise* 38: 1701–1708.

Nigg, B.M. and Gérin-Lajoie, M. 2011. Gender, age and midsole hardness effects on lower extremity muscle activity during running. *Footwear Science* 3(1): 3–12.

Nigg, B.M., Stefanyshyn, D., Cole, G., Stergiou, P., and Miller, J. 2003. The effect of material characteristics of shoe soles on muscle activation and energy aspects during running. *Journal of Biomechanics* 36(4): 569–575.

Nigg, B.M., Tecante, K.E., Federolf, P., and Landry, S.C. 2010. Gender differences in lower extremity gait biomechanics during walking using an unstable shoe. *Clinical Biomechanics* 25(10): 1047–1052.

Nigg, B.M. and Wakeling, J. M. 2001. Impact forces and muscle tuning—A new paradigm. *Exercise and Sport Sciences Review* 29(1): 37–41.

Oakley, T. and Pratt, D.J. 1988. Skeletal transients during heel and toe strike running and the effectiveness of some materials in their attenuation. *Clinical Biomechanics* 3: 159–165.

O'Conner, K.M. and Hamill, J. 2004. The role of selected extrinsic foot muscles during running. *Clinical Biomechanics* 19: 71–77.

O'Leary, K., Vorpahl, K.A., and Heiderscheit, B. 2008. Effect of cushioned insoles on impact forces during running. *Journal of the American Podiatric Medical Association* 98(1): 36–41.

Paiva de Castro, A., Rebelatto, J.R., and Aurichio, T.R. 2010. The relationship between foot pain, anthropometric variables and footwear among older people. *Applied Ergonomics* 41: 93–97.

Payne, C., Zammitt, G., and Patience, D. 2005. Predictors of a response to windlass mechanism enhancing running shoes. *ISB Seventh Symposium on Footwear Biomechanics*, Cleveland, OH, pp. 56–57.

Perry, S.D., Radtke, A., and Goodwin, C.R. 2007. Influence of footwear midsole material hardness on dynamic balance control during unexpected gait termination. *Gait and Posture* 25(1): 94–98.

Popovich, R.M., Gardner, J.W., Potter, R.P., Knapik, J.J., and Jones, B.H. 2000. Effect of rest from running on overuse injuries in army basic training. *American Journal of Preventive Medicine* 18(3 Suppl): 147–155.

Robbins, S., Gouw, G., and McClaren, J. 1992. Shoe sole thickness and hardness influence balance in older men. *Journal of the American Geriatrics Society* 40(11): 1089–1094.

Robbins, S. and Waked, E. 1997. Hazards of deceptive advertising of athletic footwear. *British Journal of Sports Medicine* 31: 299–303.

Robbins, S., Waked, E., Gouw, G.J., and McClaran, J. 1994. Athletic footwear affects balance in men. *British Journal Sports Medicine* 28: 117–122.

Sekizawa, K., Sandrey, M.A., Ingersoll, C.D., and Cordova, M.L. 2001. Effects of shoe sole thickness on joint position sense. *Gait and Posture* 12: 221–228.

Stacoff, A., Denoth, J., Kaelin, X., Stuessi, E. 1988. Running injuries and shoe construction: Some possible relationships. *International Journal Sport Biomechanics* 5: 342–357.

Stacoff, A., Reinschmidt, C., Nigg, B.M., van den Bogert, A.J., Lundberg, A., and Denoth, J. 2000. Effects of foot orthoses on skeletal motion during running. *Clinical Biomechanics* 15(1): 54–64.

Stacoff, A., Steger, J., Stüssi, E., and Reinschmidt, C. 1996. Lateral stability in sideward cutting movements. *Medicine and Science in Sports and Exercise* 28(3): 350–358.

Stewart, L., Gibson, J.N., and Thomson, C.E. 2007. In-shoe pressure distribution in "unstable" (MBT) shoes and flat-bottomed training shoes: A comparative study. *Gait and Posture* 25(4): 648–651.

Swissa, A., Milgrom, C., Giladi, M., Kashtan, H., Stein, M., Margulies, J., Chisin, R., and Aharonson Z. 1989. The effect of pretraining sports activity on the incidence of stress fractures among military recruits. A prospective study. *Clinical Orthopaedic and Related Research* 245: 256–260.

Taunton, J.E., Ryan, M.B., Clement, D.B., McKenzie, D.C., Lloyd-Smith, D.R., and Zumbo, B.D. 2003. A prospective study of running injuries: The Vancouver sun run "in training" clinics. *British Journal of Sports Medicine* 37: 239–244.

Van Gheluwe, B., Kerwin, D., Roosen, P., and Tielemans, R. 1999. The influence of heel fit on rearfoot motion in running shoes. *Journal of Applied Biomechanics* 15: 361–372.

Verdejo, R. and Mills, N.J. 2004a. Heel–shoe interactions and the durability of EVA foam running-shoe midsoles. *Journal of Biomechanics* 37(9): 1379–1386.

Verdejo, R. and Mills, N.J. 2004b. Simulating the effects of long distance running on shoe midsole foam. *Polymer Testing* 23(5): 567–574.

Villwock, M.R., Meyer, E.C., Powell, J.W., Fouty, A.J., and Haut, R.C. 2009. Football playing surface and shoe design affect rotational traction. *The American Journal of Sports Medicine* 37(3): 518–525.

Wegener, C., Burns, J., and Penkala, S. 2008. Effect of neutral-cushioned running shoes on plantar pressure loading and comfort in athletes with cavus feet: A crossover randomized controlled trial. *American Journal Sports Medicine* 36: 2139–2146.

Williams, K.R. and Cavanagh, P.R. 1987. Relationship between distance running mechanics, running economy, and performance. *Journal of Applied Physiology* 63(3): 1236–1245.

Wunderlich, R. and Cavanagh, P. 2001. Gender differences in adult foot shape: Implications for shoe design. *Medicine and Science in Sports and Exercise* 33(4): 605–611.

Part IV

Testing

17 Foot Pressure Measurements

Ewald M. Hennig

CONTENTS

17.1 INTRODUCTION

Cinematographic techniques are not well suited to determine the forces and accelerations experienced during locomotion and sports activities. Transducers are necessary to provide accurate data for these mechanical quantities. Ground reaction forces and the path of the centre of gravity can be determined with force platforms. These data are used to estimate internal and external loads on the body, its bones, muscles, and joints during locomotion and sports activities. However, the knowledge of the ground reaction forces provides no information about the stress, acting on various anatomical foot structures. The understanding of the etiology of foot stress fractures, for example, requires a more detailed analysis of foot loading. Many separate force measuring sensors that cover the area of contact between the foot and the ground provide this information. For many years, researchers have explored methods to measure pressure distribution. Nowadays, different commercial pressure measuring devices are available, and the determination of static and dynamic pressure distribution patterns under the foot has become a standard evaluation tool for clinical applications and footwear design. The knowledge of individual foot loading characteristics is important for clinical applications and the prescription of orthotic devices and therapeutic footwear. Excessive pressures under the foot may lead to ulcers in neuropathic feet, as they can be present at the later stages of diabetes. To avoid these high plantar loads, in-shoe pressure distribution measurements are widely used for the prescription of footwear and orthotics for diabetic patients. Plantar pressures are closely related to shoe comfort. Therefore, many athletic footwear companies use pressure distribution measuring devices in the process of designing new products. For example, the "Nike Free" footwear concept is based on the plantar pressure pattern from barefoot running on natural grass. Several stages of sole modifications were built until the same in-shoe pressure patterns were measured as found during barefoot running on grass. After a short historical review, different transducer technologies, plantar pressure variables, and the reliability of pressure distribution measurements will be discussed in the following pages. The use of pressure distribution in the design and testing of casual, athletic, and therapeutic footwear will conclude this chapter.

17.2   HISTORICAL REVIEW OF PRESSURE DISTRIBUTION TECHNOLOGIES

There has been interest in knowing the distribution of loads under the human foot for more than 150 years. An early recording of foot-to-ground interaction was performed by letting subjects stand on and walk across plaster of Paris and clay (Beely, 1882). These foot imprints identified more foot shape rather than the plantar pressure distribution. Abramson (1927) used small steel balls upon which a thin lead plate was positioned. Weight bearing on this plate left impressions of the steel balls in the lead. The depth of the impressions was an estimate of the pressures under the foot. Morton (1930) used an inked rubber mat with pyramidal projections which was placed on top of paper. After walking across this rubber mat, the size of the pyramid imprints on the paper allowed an estimate of the magnitude of the local forces under the foot. A similar approach was used by Elftman (1934). He also used a mat with rubber pyramids, which was placed on a glass plate with the pyramids making contact with the glass. The contact area of the pyramids on the glass plate increased as a function of the applied load to the mat. Elftman was the first researcher who recorded dynamic pressure patterns by filming these changes in pyramid contact areas through the glass plate (72 frames/s) for subjects walking across the rubber mat. Several researchers improved Elftman's principle. They gained a better graphic representation by color densitometry (Miura et al., 1974) and higher resolution (Beierlein, 1977). An optical interference sandwich was introduced by Arcan and Brull (1976). These researchers used monochromatic light which created an interference patterns upon compression of the sandwich. The contact between the foot and the sandwich was made via aluminum half spheres. Circular interference rings resulted from this device. The interference ring diameter increased with higher loads onto the half spheres. Using this measuring device, Cavanagh and Ae (1980) performed cinematographic recording (100 Hz) of the interference patterns of a single subject during walking barefoot and jumping. Each circular ring of the interference pattern from all loaded half spheres under the foot was digitized manually for the analysis. In their publication, the authors mention that this measuring technique would be too laborious for data collection in clinical applications. Force sensors, providing an electric signal, are necessary for more effective recording and evaluation techniques.

Stokes et al. (1975) used twelve parallel strain gage instrumented beams for recording the load distribution across these beams. To generate a matrix pressure distribution pattern, the subjects had to walk twice across this linear beam arrangement in anterior–posterior direction and perpendicular to it. Only in recent years, the availability of inexpensive force transducers and modern data acquisition systems has made the construction of pressure distribution measuring systems possible that offer good spatial and temporal resolutions. In 1976, a measurement method for matrix-based capacitance pressure distribution devices was introduced and patented by Nicol and Hennig. This technology is still used and marketed by NOVEL® Inc. In their first pressure measuring mat, the inventors built an array of 256 capacitors using the overlap in area from two sets of 16 conducting strips in orthogonal directions on both sides of an elastomer layer. This layer, a foam rubber material, served as dielectric for the capacitive sensors. The mechanical characteristics of the dielectric elastomer are essential for the quality of the transducers. A multiplex technique was used to scan each single transducer up to 100 times per second. In 1982, Hennig et al. reported the use of inexpensive piezoceramic materials for plantar pressure measurements. A piezoelectric insole and a pressure distribution platform with 1000 elements were constructed and used in a variety of applications (Cavanagh and Hennig, 1983; Cavanagh et al., 1985; Hennig and Cavanagh, 1987). The use of force sensing resistors (FSR) as an inexpensive transducer technology for pressure distribution devices has been described by Maalej et al. (1987). FSR can be very thin and have conductive paint between the electrodes. The recording of shear stress distributions has been of interest for many years and attempts have been made to develop and apply a suitable transducer technology (Heywood et al., 2004). However, until today no commercial shear distribution measurement device is available. Therefore, Yavuz et al. (2009) tried to predict shear distribution from vertical load parameters with a modeling approach. However, the authors came to the conclusion that there is no direct relationship between pressure and shear distribution magnitudes. The difficulty in measuring or predicting shear stress distributions originates from the differences in friction characteristics between the foot and the ground. Barefoot recordings against measurements with socks will differ substantially, because the friction conditions are different. Especially, for in-shoe shear stress measurements, the sensor construction itself modifies the friction characteristics between the foot (sock) and the shoe—thus influencing the measurement results considerably.

17.3 TRANSDUCER TECHNOLOGIES

Pressure is calculated as force divided by the contact area on which this force acts. It is measured in units of kPa ($100\,kPa = 10\,N/cm^2$). Force measuring elements are needed to determine pressures and pressure distributions. All force measurement technologies are based on the registration of strain of materials. In force transducers, a change in electrical properties is produced when subjected to mechanical loads. Depending on the type of transducer, forces can create electrical charges, cause a change in capacitance, modify the electrical resistance, or influence inductance. Because inductive transducers are based on relatively large displacements, they are rarely used in measuring instruments for biomechanical applications. Desirable transducer characteristics for biomechanical purposes can differ from the quality criteria of sensors for engineering applications. Pressure measurements during sitting, lying on a bed, and in a running shoe require a soft and pliable transducer matrix that will adapt to the shape of the human anatomy. However, such transducers can only have limited quality in technical specifications such as linearity, hysteresis, and frequency response. Based on different technologies, several pressure distribution systems are sold today. Most of them are used for the analysis of the foot-to-ground interaction as pressure platforms or as insoles for in-shoe measurements.

17.3.1 RESISTIVE SENSORS

In resistive transducers, the electrical resistance changes under tension or compression of the transducer. Volume conduction has been used as a method for measuring forces. Silicone rubber sensors, filled with silver or other electrical conducting particles, were produced in the past. With increasing pressure, the conducting particles are pressed closer together, increasing the surface contact between the conducting particles and thus lowering the electrical resistance. Resistive contact sensors work on a similar principle to the volume conduction transducers (Tekscan Inc., Boston, MA, RScan, Zebris). Typically, two thin and flexible polyester sheets with electrically conductive electrodes are separated by a semiconductive ink layer between the electrical contacts (rows and columns). Exerting pressure on two intersecting strips causes an increased and more intense contact between the conductive surfaces, thus causing a reduction in electrical resistance. Reduced accuracy, as it is frequently seen in resistive transducers, is mainly caused by a lack of material elasticity of the resistive layer between the electrodes. Because of insufficient elastic recoil in a contact sensor, they often demonstrate large hysteresis.

17.3.2 CAPACITIVE SENSORS

Electrical capacitors typically have two conducting plates or foils in parallel to each other with an insulating dielectric material in between. In a capacitive force transducer, the capacitance changes as a function of distance between the two conducting surfaces and a change in dielectric material properties. Applying a force on such a transducer causes a change in the distance between the two conducting capacitor surfaces as well as a change in the dielectric constant of the compressed material between the electrodes. This principle allows simple construction and low material costs for manufacturing inexpensive pressure distribution mats with up to several thousand discrete capacitive transducers (Hennig and Nicol, 1978; Nicol and Hennig, 1976). Figure 17.1 shows the first capacitive pressure distribution mat, based on this measurement principle which was patented in Germany and

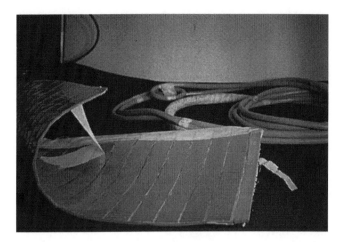

FIGURE 17.1 First matrix type capacitive pressure distribution mat (256 sensors), built in 1975.

the United States. These patents were licensed to NOVEL Inc. (Munich, Germany) which is the leading manufacturer of pressure distribution measurement systems for biomechanical applications.

As compared to rigid piezoelectric transducers, relatively large deformations are necessary for capacitive transducers. A better compressibility and the visco-elastic nature of the dielectric interface material limit measurement accuracy. The elastic behavior of the dielectric material influences the magnitude of the hysteresis effects in capacitive transducers. Silicone rubber mats with good elastic recoil properties are used for the production of commercially available pressure distribution measuring platforms and insoles (NOVEL Inc., München, Germany). Based on this technology, capacitive pressure mats show good accuracy for dynamic loads under the foot for walking and running. Capacitive transducers can be built flexible to accommodate body curvatures and contours. Due to their soft and pliable nature, capacitive transducers can be employed for the measurements of seating pressures and the registration of body pressures on mattresses (Nicol and Rusteberg, 1993).

17.3.3 PIEZOELECTRIC SENSORS

High precision force transducers often use quartz as sensor material. However, the electrical charge that is generated on the quartz surfaces is very low (2.30 pC/N), and costly high performance charge amplifiers have to be used for electronic processing. Piezoceramic materials also generate charges when they are compressed, which are approximately 100 times higher as compared to the electrical output of quartz. This high charge generation allows the use of inexpensive charge amplifiers and thus permits the construction of inexpensive pressure distribution devices. For dynamic loads, piezoceramic transducers show good linearity and low hysteresis characteristics. Whereas temperature has only a minor influence on the piezoelectric properties of quartz, piezoceramics also show pyroelectric properties.

FIGURE 17.2 Piezoelectric 1000 element pressure distribution platform with charge amplifier electronic.

Therefore, thermal insulation or temperature equilibrium, as it is normally present inside of shoes, is necessary for reliable force measurements. High precision and inexpensive piezoceramic materials have been used for pressure distribution measurements (Hennig et al., 1982). Figure 17.2 shows a pressure distribution measuring platform with 1000 single piezoceramic force transducers, each one connected to its own charge amplifier. Piezoelectric polymeric films (polyvinylidene fluoride [PVDF]) and piezoelectric rubbers were also tested and have been proposed for the measurements of forces. However, due to large hysteresis effects and unreliable reproducibility, these materials have not been successful for pressure distribution measurement devices.

17.4   PRESSURE VARIABLES AND DATA VISUALIZATION

For pressure distribution measurements with many sensors and high frame rates, data acquisition, storage, and the evaluation are challenging. 100,000 measuring values will be stored after collecting plantar pressures during a single walking trial, if data are sampled for 1 s at a frame frequency of 100 Hz on a pressure plate with 1000 elements. Data reduction is important for a meaningful presentation of the results. Peak pressure graphical representations are often used to illustrate the foot contact with the ground (Figure 17.3). The wire frame image in Figure 17.3 was created by using the highest pressure values under the foot, as they have occurred at any time during the ground contact. To include time information, pressure–time integral images can also be presented by displaying the impulse values of each sensor in a similar graph. However, this kind of presentation is less commonly used. For numerical and statistical analyses, a division of the plantar contact of the foot into meaningful anatomical areas (masks) is usually performed. Depending on the research question, the foot is normally divided into up to 10 anatomical regions.

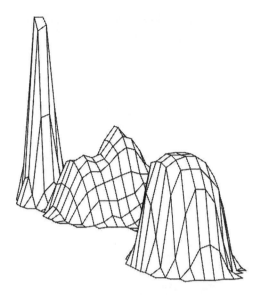

FIGURE 17.3   Peak pressure wireframe image of barefoot walking.

Division of the foot anatomy by masks can either be done by an interactive process on the computer or it is based on a predefined geometric algorithm.

A standardized anatomical division is, for example, implemented in the current software by NOVEL (PRC mask). This mask was created from footprints during half body weight standing of 107 randomly selected individuals (Cavanagh et al., 1987). The authors suggested dividing the footprint into 10 anatomical regions: lateral and medial heel and midfoot areas; first, second, and lateral metatarsal head areas; and first, second, and lateral toe areas. For these anatomical areas, different pressure variables can be calculated: regional peak pressures (kPa), maximum pressure rates (kPa/s), regional impulses (Ns), and relative loads (%). The regional peak pressure during foot contact reveals information about the highest pressures in the anatomical region. Regional impulses are calculated by determining the local force–time integral under the specific anatomical region. This is done by summing up the force–time integral from each sensor in this area. These impulse values can then be used to calculate a relative impulse distribution under the foot. The following equation serves to determine the relative load RL_i in a foot region i.

$$RL_i(\%) = \frac{\int F_i(t)dt}{\sum_{j=1-n}\int F_j(t)dt} \times 100$$

The relative load analysis method allows a comparison of the load distribution patterns between individuals that is less dependent on the subjects' weight and anthropometric foot dimensions.

17.5 RELIABILITY OF PRESSURE DISTRIBUTION MEASUREMENTS IN BIOMECHANICAL APPLICATIONS

Hughes et al. (1991) investigated the reliability of pressure distribution measurements during gait with a capacitive pressure distribution platform (EMED, Novel). These authors reported that a determination of peak pressures can be achieved with a reliability coefficient of $R = 0.94$, if an average of five trial repetitions is used. McPoil et al. (1999) also investigated the number of repetitive trials needed to obtain a reliable representation of plantar pressure patterns. They reported that three to five walking trials are sufficient to obtain reliable peak pressure and pressure–time integral values. Maiwald et al. (2008) investigated the reproducibility of plantar pressure distribution parameter for 95 subjects during barefoot running across a capacitive pressure distribution platform. They reported intraclass correlation coefficients between 0.58 and 0.99, depending on the variable and plantar foot location. The authors also found that force–time integral variables had the best within-subject repeatability. Similarly, Gurney et al. (2008) concluded from their study on between-day reliability of repeated plantar pressure distribution measurements good repeatability results allowing comparative clinical evaluations. Kernozek et al. (1996) studied the reliability of in-shoe foot pressure measurements (PEDAR insole; NOVEL Inc.) during walking at three speeds on a treadmill. Depending on the pressure variable, a maximum of eight steps was needed to achieve an excellent reliability (>0.90). As compared to timing variables, peak pressures and pressure–time integrals needed fewer trials for a good representation of foot loading. Using the same instrumentation, Kernozek and Zimmer (2000) investigated test retest reliabilities for slow treadmill running (2.24–3.13 m/s). Depending on foot region and the measurement variable, the intraclass correlation coefficients (ICC's) ranged from 0.84 to 0.99.

17.6 APPLICATION TO FOOTWEAR

Pressure distribution measurements have widely been used for basic research to evaluate the mechanical behavior of the foot in static and dynamic loading situations. Plantar pressures have been recorded to explore gender differences in the foot mechanics of adults (Hennig and Milani, 1993) and to compare the adult foot with the feet of children in different age groups (Hennig and Rosenbaum, 1991; Hennig et al., 1994). Furthermore, the effect of body weight and obesity on foot function and pain was studied with pressure distribution instrumentation (Hills et al., 2001). This basic research was performed for standing on and walking barefoot across pressure distribution platforms. In-shoe plantar pressure measurements are necessary to evaluate the effect of shoe construction on foot loading. Most of the footwear-related pressure distribution research has a clinical background with an emphasis on the reduction of high pressures in diabetic and rheumatoid feet. For the prescription and adaptation of orthotic devices, pressure distribution measuring insoles are increasingly used by orthopedists and orthopedic technicians. Athletic footwear is also a field where the plantar pressure measurement technology has been used for footwear design. For casual and fashion footwear, little research has been published until today.

17.6.1 ATHLETIC FOOTWEAR

Pressure distribution analyses have been used to understand the foot loading behavior in various sport disciplines. Early applications looked at the in-shoe pressures during downhill skiing (Schaff and Hauser, 1987). Milani and Hennig (1994, 1996) identified the in-shoe pressure patterns in track and field activities (running, high jump, broad jump, triple jump). The highest pressures were found for the step in the triple jump. The effect of bicycle shoes on the transfer of momentum to the pedal was demonstrated by Hennig and Sanderson (1995) (Figure 17.4). For the design and testing of running and soccer shoes, pressure distribution technologies are used frequently to evaluate foot comfort and shoe function. Jordan et al. (1997) studied the relationship between perceived comfort and pressure distribution data in 10 different commercially available shoe models. Plantar and dorsal in-shoe pressures were measured with 20 subjects walking in each shoe condition. Peak pressures and maximum forces under the heel and forefoot as well as the dorsum of the foot were found to be lower for the more comfortable shoes. The authors concluded that the measurement of in-shoe pressure distribution is a good method for footwear manufacturers to improve shoe comfort of their products.

17.6.1.1 Running Shoes

As mentioned before, shoe comfort is related to plantar and dorsal foot pressures. This is important for long distance running. Plantar pressure patterns show great differences between individuals and are dependent on foot architecture and foot function during loading (Molloy et al., 2009). In-shoe pressures are also largely influenced by the midsole material and stability features of running shoes (Hennig and Milani, 1995a). For the testing of running shoes, pressure distribution measurements were proposed as one of several important evaluation tools (Hennig and Milani, 2000). Even the shoe upper and different shoe lacing patterns have an influence on plantar (Hagen and Hennig, 2009) and dorsal (Hagen et al., 2010) pressures. Peak pressures and peak rates of loading were used by Dixon (2008) to study the influence of six running shoe models on the plantar foot loading of runners. Large differences in both variables were found between the shoe models. Dixon (2006) also compared the amount of pronation during running to centre of pressure (CoP) data from plantar pressure measurements. She found only small correlations between the CoP measurements and rearfoot range of motion ($r = 0.46$). Therefore, plantar pressure in-shoe measurements are not well suited to predict stability features of running shoes. A deterioration of footwear material occurs when shoes are exposed to many repetitive impacts with high mechanical loads. The protective functions of footwear such as the restriction of excessive rearfoot motion and shock attenuation changes with the number of miles run with the shoe. A study by Hennig and Milani (1995b) identified the changes of footwear properties after wearing 19 different running shoes for a distance of 220 km. The changes in maximum pronation increased on average by 21.5% and peak tibial acceleration by 10.1%. Plantar peak pressures under the heel, midfoot, and forefoot were on average less than 10% in the used shoes. However, one of the shoe models showed a substantial increase in peak pressures under the heel and the first metatarsal head—a clear indication of material deterioration after only 220 km of use (Figure 17.5).

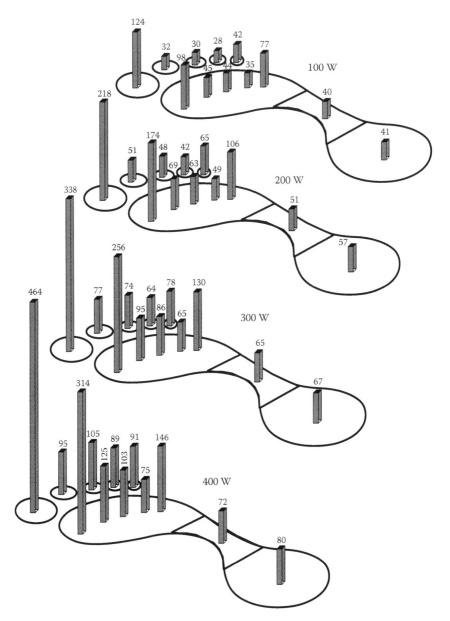

FIGURE 17.4 Peak pressures (kPa) during bicycling at different powers at a pedaling rate of 80 rpm.

17.6.1.2 Soccer Shoes

Fit and comfort are the most important properties that soccer players expect from their shoes (Hennig and Sterzing, 2010). Because pressures under the foot are closely related to the perception of comfort, pressure distribution measurements are a good method to evaluate the comfort in soccer shoes. However, pressure analyses may

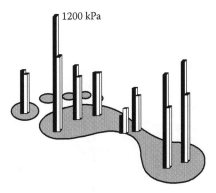

FIGURE 17.5 Peak pressures for running in a used (black bars) against a new shoe model (white bars).

also help to identify overuse injury risks for soccer players. High impact loads on the foot during soccer are suspected to cause foot problems such as metatarsal stress fractures, metatarsalgia, and interdigital neuroma (Hockenbury, 1999).

Ford et al. (2006) investigated the effect of playing football on natural grass and synthetic turf on foot loading. Based on the results of 17 football players, the authors reported increased peak pressures in the central forefoot region on the synthetic turf surface. Eils et al. (2004) measured the in-shoe foot pressures for four soccer-specific movements: running, sprinting, cutting, and shots on goal. The authors reported substantial differences in the loading patterns between the different movements but no differences between a grass and cinder playing surface. They concluded from their results that the medial side of the plantar surface may be more prone to injuries in soccer. Increased plantar pressures under the medial foot during four soccer-related movements were also reported by Wong et al. (2007).

Our laboratory tested the plantar pressure distribution patterns during soccer-specific movements under field testing conditions. The pressures were measured during running, cutting movements, and kicking. In Figure 17.6, the peak pressure averages from 18 subjects are shown for a cutting movement during a sudden change of direction. For this movement, high medial forefoot pressures are apparent which may lead to overuse injuries. Nihal et al. (2009) identified a high incidence of first ray disorders for soccer players. High traction during a cutting movement allows a rapid change in movement direction for the player. Therefore, it is important for being successful in soccer. The knowledge of the foot loading pattern can be used for designing soccer shoes with better traction properties. High medial forefoot forces cause a large penetration of the cleat under this part of the foot into the ground. Thus, the cleats under the medial forefoot are the most important outsole structures in providing the necessary traction for better performance during cutting movements.

17.6.2 CLINICAL APPLICATIONS

The majority of clinical pressure distribution studies have been published in the diagnosis and treatment of neuropathic feet. Depending on the severity of the condition, diabetes mellitus can cause neuropathic and vascular changes at a later stage of

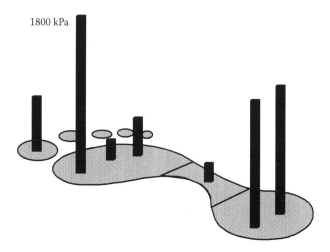

1800 kPa

FIGURE 17.6 Peak plantar pressures for a cutting movement in a soccer shoe.

the disease. The sensory neuropathy will result in a progressive, distal to proximal, loss of sensation in the lower extremities, often resulting in ulceration at locations of high plantar pressures. Ulceration may result in partial or total amputation of the foot. Many studies have proven the usefulness of pressure distribution measurements for the prescription of therapeutic footwear. The importance and limitations of plantar pressure measurements for the diagnosis and therapy of the diabetic foot have been summarized by Cavanagh et al. in 1993 and in 2000 and more recently by Leung (2007).

In 2006, Cavanagh and Owings emphasize the importance of off-loading as the main factor for success in healing diabetic foot ulcers and preventing their recurrence. The authors suggest the use of computerized design for custom-made insoles which allow a reduction of foot pressures by taking foot shape into account. Cavanagh and Bus (2011) point out that conventional or standard therapeutic footwear is not effective in ulcer healing. They propose that total-contact casts and other nonremovable devices in the shoe should be used for the patients. The authors complain that many clinics still use methods that are known to be ineffective in the treatment of ulcer healing. Lott et al. (2008) measured peak plantar pressures, pressure–time integrals, and peak pressure gradients under the forefoot of diabetic patients and calculated maximal shear stress. The authors concluded that peak pressure gradients and peak maximal shear stresses provide better information than peak plantar pressures alone for recognizing individuals who are at risk for developing foot ulcers.

Woodburn and Helliwell (1996) studied the influence of heel alignment on plantar pressures under the forefoot and on skin callosities in rheumatoid arthritis (RA) patients. The plantar pressures from 102 rheumatoid patients with normal and valgus heel alignment and 42 healthy, age-matched adults were measured. For the normal heel alignment, patient feet and the feet of the healthy adults peak pressures were greater under the forefoot except for the first metatarsal head. Highest peak pressures were found for all groups under the central metatarsal heads. Only for the rheumatoid feet callosities were found at these sites.

However, no clear relationship was found between plantar peak pressures and callus formation. For the rheumatoid valgus heel subjects, the lateral metatarsal heads showed reduced pressures, thus shifting the load to the medial forefoot. Novak et al. (2009) investigated the effectiveness of functional foot orthoses against an unshaped orthotic material insole for reducing forefoot pain. From 40 rheumatoid arthritis patients, 20 subjects received functional orthoses and the other 20 subjects the unshaped insoles. Plantar pressures and foot pain were measured for all subjects at the beginning of the study, after 1 week and following the 6 months intervention period. In general, peak plantar pressures were higher under painful against the nonpainful foot areas. A reduction of foot pain and an improvement in walking ability was observed for both groups after the 6 months intervention period. However, no clear advantage of the functional orthoses against the nonshaped insole was found. Different orthotics were studied by Brown et al. (1996). These researchers found orthotics to be effective in relieving high pressures in shoes. The authors point out that this stress relief can only occur at the cost of increasing pressure under other areas under the foot.

Hallux valgus is a common forefoot deformity at the metatarsophalangeal joint. It is more frequently seen in women. Congenital factors—poorly fitting footwear and high-heeled shoes—are mentioned by many authors as etiologic factors in many cases of hallux valgus. Using pressure distribution instrumentation, lateral forefoot load shifts in hallux valgus feet have been described by Blomgren et al. (1991). Nyska et al. (1996) studies the effect of heel height with 10 women walking in shoes with low and high heels. In the high-heeled shoes, load was lower under the heel and increased under the forefoot. The forefoot load shifted medially with higher force and pressure–time integrals. The authors concluded that the higher medial forefoot loads may aggravate the symptoms of hallux valgus patients.

Foot and back pain is a common complaint for women during the later stages of pregnancy. The influence of pregnancy on foot pain and plantar pressures was studied by Karadag-Saygi et al. (2010). For 35 last trimester pregnant women and 35 age and body mass index matched nonpregnant women, plantar pressures and postural balance were determined. The pregnant women showed higher forefoot pressures during standing and walking. They also had an increased postural sway in the anterior–posterior direction. The authors concluded that exercise and appropriate footwear may help to reduce foot pain for women in their later stages of pregnancy.

In a prospective geriatric study, Mickle et al. (2010) studied the effect of foot pain and plantar pressures on the likelihood of falling in the elderly. 158 men and 154 women with an age over 60 years participated in this study. Based on the falls occurrence over a 12 months period, the participants were classified as fallers ($n = 107$) and nonfallers ($n = 196$). The fallers showed significantly higher peak pressures and pressure–time integrals under their feet than the nonfallers. Subjects with foot pain had higher plantar peak pressures and pressure–time integrals than subjects with no foot pain. The authors concluded that high plantar pressures may contribute to foot pain as well as risk of falls. Providing therapeutic interventions to reduce foot pain and high plantar pressures may reduce the falls risk of older people.

Using the CoP path and other pressure variables, plantar pressure measurements have also been used for the analysis of postural control (Hohne et al., 2009; Nurse and Nigg, 2001) and the analysis of neurological disorders such as Parkinsonian gait (Kimmeskamp and Hennig, 2001).

17.7 CONCLUSIONS

For almost 150 years, the foot loading behavior of the human foot was already subject of interest for anatomists, orthopedists, and biomechanists. Originally, the research aimed to understand the static and dynamic function of the human foot during standing, walking, running, and other athletic activities. Furthermore, gender differences in foot function, the foot development during childhood, and foot problems in different patient populations were studied. Especially, the use of pressure distribution for the recognition and treatment of diabetic foot problems was studied by many research groups. Resistive, capacitive, and piezoelectric sensor technologies are commonly used to measure pressure distributions. The availability of inexpensive transducers, rapid data acquisition, and efficient data processing has resulted in easy-to-use, commercially available pressure distribution instrumentation. Therefore, pressure distribution measurements have become a standard clinical procedure for the diagnosis of foot problems and therapeutic interventions. For casual and athletic footwear, the primary use of in-shoe pressure measurements is the evaluation of foot comfort. However, pressure analyses are also used in athletic footwear research to study the possibilities of shoe design features to prevent injuries and improve performance.

QUESTIONS

17.1 What are the advantages and disadvantages of capacitive, resistive, and piezoelectric transducers for pressure distribution devices?

17.2 How are relative loads calculated from pressure distribution data?

17.3 Why are heel pressures present in a bicycle shoe, although the force to the pedal is only applied by the forefoot?

17.4 How are the pressures distributed under the foot during a sudden cutting movement of soccer players?

17.5 What are the differences between wireframe diagrams against isobarographic representations of pressure patterns?

17.6 How are peak pressure images created from pressure distribution data?

REFERENCES

Abramson, E. (1927). Zur Kenntnis der Mechanik des Mittelfusses. *Scand Arch Physiol*, 51, 175–234.

Arcan, M. and Brull, M. A. (1976). A fundamental characteristic of the human body and foot, the foot ground pressure pattern. *J Biomech*, 9, 453–457.

Beely, F. (1882). Zur Mechanik des Stehens. *Arch Klin Chir*, 27, 457–471.

Beierlein, H. R. (1977). Geraet zur zeitsynchronen Messung von Druckverteilung und Komponenten der resultierenden Kraft unter der menschlichen Fussohle. *Zeitschrift f Orthopaedie*, 115, 778–782.

Blomgren, M., Turan, I., and Agadir, M. (1991). Gait analysis in hallux valgus. *J Foot Surg*, 30(1), 70–71.

Brown, M., Rudicel, S., and Esquenazi, A. (1996). Measurement of dynamic pressures at the shoe-foot interface during normal walking with various foot orthoses using the FSCAN system. *Foot Ankle Int*, 17(3), 152–156.

Cavanagh, P. R. and Ae, M. (1980). A technique for the display of pressure distributions beneath the foot. *J Biomech*, 13, 69–75.

Cavanagh, P. R. and Bus, S. A. (2011). Off-loading the diabetic foot for ulcer prevention and healing. *Plast Reconstr Surg*, 127(1), 248–256.

Cavanagh, P. R. and Hennig, E. M. (1983). Pressure distribution measurement—A review and some observations on the effect of shoe foam materials during running. Paper presented at the *Biomechanical Aspects of Sport Shoes and Playing Surfaces*, Calgary, Alberta, Canada.

Cavanagh, P. R., Hennig, E. M., Rodgers, M. M., and Sanderson, D. J. (1985). The measurement of pressure distribution on the plantar surface of diabetic feet. In M. Whittle and D. Harris (Eds.), *Biomechanical Measurement in Orthopaedic Practice* (Vol. 5, pp. 159–166). Oxford, England: Clarendon Press.

Cavanagh, P. R. and Owings, T. M. (2006). Nonsurgical strategies for healing and preventing recurrence of diabetic foot ulcers. *Foot Ankle Clin*, 11(4), 735–743.

Cavanagh, P. R., Rodgers, M. M., and Iiboshi, A. (1987). Pressure distribution under symptom-free feet during barefoot standing. *Foot Ankle*, 7(5), 262–276.

Cavanagh, P. R., Simoneau, G. G., and Ulbrecht, J. S. (1993). Ulceration, unsteadiness, and uncertainty: The biomechanical consequences of diabetes mellitus. *J Biomech*, 26(1), 23–40.

Cavanagh, P. R., Ulbrecht, J. S., and Caputo, G. M. (2000). New developments in the biomechanics of the diabetic foot. *Diabetes Metab Res Rev*, 16(1), S6–S10.

Dixon, S. J. (2006). Application of center-of-pressure data to indicate rearfoot inversion-eversion in shod running. *J Am Podiatr Med Assoc*, 96(4), 305.

Dixon, S. J. (2008). Use of pressure insoles to compare in-shoe loading for modern running shoes. *Ergonomics*, 51(10), 1503–1514.

Eils, E., Streyl, M., Linnenbecker, S., Thorwesten, L., Volker, K., and Rosenbaum, D. (2004). Characteristic plantar pressure distribution patterns during soccer-specific movements. *Am J Sports Med*, 32(1), 140–145.

Elftman, H. (1934). A cinematographic study of the distribution of pressure in the human foot. *Anat Rec*, 59(4), 481–487.

Ford, K. R., Manson, N. A., Evans, B. J., Myer, G. D., Gwin, R. C., Heidt, R. S., Jr. et al. (2006). Comparison of in-shoe foot loading patterns on natural grass and synthetic turf. *J Sci Med Sport*, 9(6), 433–440.

Gurney, J. K., Kersting, U. G., and Rosenbaum, D. (2008). Between-day reliability of repeated plantar pressure distribution measurements in a normal population. *Gait Posture*, 27(4), 706–709.

Hagen, M. and Hennig, E. M. (2009). Effects of different shoe-lacing patterns on the biomechanics of running shoes. *J Sports Sci*, 27(3), 267–275.

Hagen, M., Hömme, A.-K., Umlauf, T., and Hennig, E. (2010). Effects of different shoe-lacing patterns on dorsal pressure distribution during running and perceived comfort. *Res Sports Med*, 18(3), 176–187.

Hennig, E. M. and Cavanagh, P. R. (1987). Pressure distribution under the impacting human foot. In B. Jonsson (Ed.), *Biomechanics X-A*, 375–380.

Hennig, E. M., Cavanagh, P. R., Albert, H., and Macmillan, N. H. (1982). A piezoelectric method of measuring the vertical contact stress beneath the human foot. *J Biomed Eng*, 4(July), 213–222.

Hennig, E. M. and Milani, T. L. (1993). Die Dreipunkunterstützung des Fußes—Eine Druckverteilungsanalyse bei statischer und dynamischer Belastung (The tripod support of the foot. An analysis of pressure distribution under static and dynamic loading). *Z Orthop Ihre Grenzgeb*, 131(3), 279–284.

Hennig, E. M. and Milani, T. L. (1995a). In-shoe pressure distribution for running in various types of footwear. *J Appl Biomech*, 11(3), 299–310.

Hennig, E. M. and Milani, T. L. (1995b). Biomechanical profiles of new against used running shoes. In K. R. Williams (Ed.), *Nineteenth Annual meeting of the American Society of Biomechanics*, Stanford University: American Society of Biomechanics, Palo Alto, CA, pp. 43–44.

Hennig, E. M. and Milani, T. L. (2000). Pressure distribution measurements for evaluation of running shoe properties. *Sportverletz Sportschaden*, 14(3), 90–97.

Hennig, E. M. and Nicol, K. (1978). Registration methods for time-dependent pressure distribution measurements with mats working as capacitors. In E. Asmussen and K. Joergensen (Eds.), *Biomechanics VI-A* (Vol. 2A, pp. 361–367). Baltimore, MD: University Park Press.

Hennig, E. M. and Rosenbaum, D. (1991). Pressure distribution patterns under the feet of children in comparison with adults. *Foot Ankle*, 11(5), 306–311.

Hennig, E. M. and Sanderson, D. J. (1995). In-shoe pressure distributions for cycling with two types of footwear at different mechanical loads. *J Appl Biomech*, 11(1), 68–80.

Hennig, E. M., Staats, A., and Rosenbaum, D. (1994). Plantar pressure distribution patterns of young school children in comparison to adults. *Foot Ankle Int*, 15(1), 35–40.

Hennig, E. M. and Sterzing, T. (2010). Review Article: The influence of soccer shoe design on playing performance: A series of biomechanical studies. *Footwear Sci*, 2(1), 3–11.

Heywood, E., Jeutter, D., and Harris, G. (2004). Tri-axial plantar pressure sensor: Design, calibration and characterization. *Conf Proc IEEE Eng Med Biol Soc*, 3, 2010–2013.

Hills, A. P., Hennig, E. M., McDonald, M., and Bar-Or, O. (2001). Plantar pressure differences between obese and non-obese adults: A biomechanical analysis. *Int J Obes Relat Metab Disord*, 25(11), 1674–1679.

Hockenbury, R. T. (1999). Forefoot problems in athletes. *Med Sci Sports Exerc*, 31(7), S448–S458.

Hohne, A., Stark, C., and Brueggemann, G. P. (2009). Plantar pressure distribution in gait is not affected by targeted reduced plantar cutaneous sensation. *Clin Biomech*, 24(3), 308–313.

Hughes, J., Pratt, L., Linge, K., Clark, P., and Klenerman, L. (1991). Reliability of pressure measurements: The EMED F system. *Clin Biomech*, 6, 14–18.

Jordan, C., Payton, C., and Bartlett, R. (1997). Perceived comfort and pressure distribution in casual footwear. *Clin Biomech*, 12(3), S5.

Karadag-Saygi, E., Unlu-Ozkan, F., and Basgul, A. (2010). Plantar pressure and foot pain in the last trimester of pregnancy. *Foot Ankle Int*, 31(2), 153–157.

Kernozek, T. W., LaMott, E. E., and Dancisak, M. J. (1996). Reliability of an in-shoe pressure measurement system during treadmill walking. *Foot Ankle Int*, 17(4), 204–209.

Kernozek, T. W. and Zimmer, K. A. (2000). Reliability and running speed effects of in-shoe loading measurements during slow treadmill running. *Foot Ankle Int*, 21(9), 749–752.

Kimmeskamp, S. and Hennig, E. M. (2001). Heel to toe motion characteristics in Parkinson patients during free walking. *Clin Biomech*, 16(9), 806–812.

Leung, P. C. (2007). Diabetic foot ulcers—A comprehensive review. *Surgeon*, 5(4), 219–231.

Lott, D. J., Zou, D., and Mueller, M. J. (2008). Pressure gradient and subsurface shear stress on the neuropathic forefoot. *Clin Biomech*, 23(3), 342–348.

Maalej, N., Zhu, H. S., Webster, J. G., Tompkins, W. J., Wertsch, J. J., and Bach-y-Rita, P. (1987). Pressure monitoring under insensate feet. Paper presented at the *IEEE/9th Conference of the Engineering in Medicine and Biology Society*, Boston, MA.

Maiwald, C., Grau, S., Krauss, I., Mauch, M., Axmann, D., and Horstmann, T. (2008). Reproducibility of plantar pressure distribution data in barefoot running. *J Appl Biomech*, 24(1), 14–23.

McPoil, T. G., Cornwall, M. W., Dupuis, L., and Cornwell, M. (1999). Variability of plantar pressure data. A comparison of the two-step and midgait methods. *J Am Podiatr Med Assoc*, 89(10), 495–501.

Mickle, K. J., Munro, B. J., Lord, S. R., Menz, H. B., and Steele, J. R. (2010). Foot pain, plantar pressures, and falls in older people: A prospective study. *J Am Geriatr Soc*, 58(10), 1936–1940.

Milani, T. L. and Hennig, E. M. (1994). Druckverteilungsanalysen im Sportschuh beim Weitsprung unterschiedlicher Leistungsklassen. *Dt Z für Sportmedizin*, 45(1), 4–8.

Milani, T. L. and Hennig, E. M. (1996). Druckverteilungsmuster unter dem Fuß beim Absprung zum Fosbury-Flop Hochsprung. *Dt Z für Sportmedizin*, 47(6), 371–376.

Miura, M., Miyashita, M., Matsui, H., and Sodeyama, H. (1974). *Photographic Method of analyzing the pressure distribution of the foot against the ground (Vol. IV)*. Baltimore, MD: University Park Press.

Molloy, J. M., Christie, D. S., Teyhen, D. S., Yeykal, N. S., Tragord, B. S., Neal, M. S. et al. (2009). Effect of running shoe type on the distribution and magnitude of plantar pressures in individuals with low- or high-arched feet. *J Am Podiatr Med Assoc*, 99(4), 330–338.

Morton, D. J. (1930). Structural factors in static disorders of the foot. *Am J Surg*, 19, 315–326.

Nicol, K. and Hennig, E. M. (1976). Time-dependent method for measuring force distribution using a flexible mat as a capacitor. In P. V. Komi (Ed.), *Biomechanics V-B* (Vol. 1B, pp. 433–440). Baltimore, MD: University Park Press.

Nicol, K. and Rusteberg, D. (1993). Pressure distribution on mattresses. In *Biomechanics XIV*, International Society of Biomechanics, Paris, France, pp. 942–943.

Nihal, A., Trepman, E., and Nag, D. (2009). First ray disorders in athletes. *Sports Med Arthrosc*, 17(3), 160–166.

Novak, P., Burger, H., Tomsic, M., Marincek, C., and Vidmar, G. (2009). Influence of foot orthoses on plantar pressures, foot pain and walking ability of rheumatoid arthritis patients—A randomised controlled study. *Disabil Rehabil*, 31(8), 638–645.

Nurse, M. A., and Nigg, B. M. (2001). The effect of changes in foot sensation on plantar pressure and muscle activity. *Clin Biomech*, 16(9), 719–727.

Nyska, M., McCabe, C., Linge, K., and Klenerman, L. (1996). Plantar foot pressures during treadmill walking with high-heel and low-heel shoes. *Foot Ankle Int*, 17(11), 662–666.

Schaff, P. and Hauser, W. (1987). Druckverteilungsmessungen am menschlichen Unterschenkel in Skischuhen [Measuring pressure distribution on the human tibia in ski boots]. *Sportverletz Sportsch*, 1(3), 118–129.

Stokes, I. A. F., Faris, I. B., and Hutton, W. C. (1975). The neuropathic ulcer and loads under the foot in diabetic patients. *Acta Orthop Scand*, 46, 839–847.

Wong, P. L., Chamari, K., Mao de, W., Wisloff, U., and Hong, Y. (2007). Higher plantar pressure on the medial side in four soccer-related movements. *Br J Sports Med*, 41(2), 93–100.

Woodburn, J. and Helliwell, P. S. (1996). Relation between heel position and the distribution of forefoot plantar pressures and skin callosities in rheumatoid arthritis. *Ann Rheum Dis*, 55(11), 806–810.

Yavuz, M., Ocak, H., Hetherington, V. J., and Davis, B. L. (2009). Prediction of plantar shear stress distribution by artificial intelligence methods. *J Biomech Eng*, 131(9), 091007.

18 Plantar Pressure Analysis

Noël Keijsers

CONTENTS

Plantar pressure measurements, that is, pedobarography, have played an essential role in the research of foot and ankle and footwear science in the last three decades. Especially in the diabetic and rheumatic foot, many studies have been performed using plantar pressure measurements (Boulton et al., 1983; Veves et al., 1992; Armstrong et al., 1998; Mueller et al., 2003; Otter et al., 2004; Tuna et al., 2005; Schmeigel et al., 2008; Novak et al., 2009). In addition, plantar pressure measurements have revealed increased plantar pressure in certain areas in subjects with foot deformities such as claw toes or metatarsophalangeal joint subluxations or dislocations (Armstrong and Lavery, 1998, Waldecker, 2002; Mueller et al., 2003; Bus et al., 2005; Burns et al., 2005). The advantage of plantar pressure measurements is that it can easily be performed without any substantial effort, in contrast to 3D gait analysis. Plantar pressure can be measured without marker placement, which is required in 3D gait analysis. Another advantage of plantar pressure measurements compared to 3D gait analysis is that the equipment is relatively inexpensive.

Although plantar pressure measurements are easy to perform, the analysis and interpretation of these measurements are rather difficult. In most biomechanical analysis, the data is restricted or can easily be reduced to the value of interest (e.g., ground reaction force, joint angles, and joint moments) over time, for example, 2D data. Plantar pressure data, however, can be seen as 4D data, pressure on x and y location over time. Therefore, the analysis of plantar pressure is complicated and several reducing steps have to be made to perform statistical analysis of plantar pressure measurements between groups. Figure 18.1 shows an example of a plantar pressure measurement of two subjects. Clear differences can be seen between both feet, such as foot length and number of sensors activated in the midfoot.

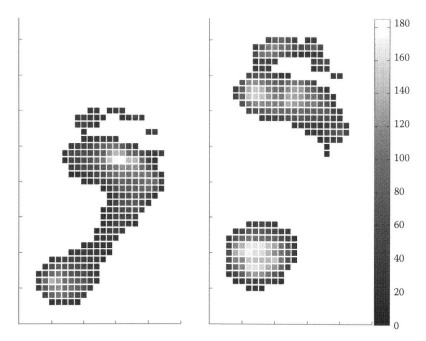

FIGURE 18.1   Two examples of the mean pressure of a plantar pressure measurements during walking. Gray scale indicates the pressure in kPa.

One of the first steps in the analysis of plantar pressure data is to indicate areas of interest or divide the pressure data into a fixed number of anatomical regions, which will be referred to as the masking method. In general, 6–11 of such regions are considered and an example of masking the foot into 10 regions is shown in Figure 18.2. From these regions, various parameters can be extracted, which will be described in the next paragraph. Although the masking method extracts important information from the plantar pressure data and describes the overall foot loading, subtle differences within a region cannot be seen. Recently, new methods have been developed that spatially normalize plantar pressure images for foot size and foot progression angle (Keijsers et al., 2009; Pataky et al., 2008a). In other words, the foot is divided in a larger number of regions. The main advantages of these methods is that plantar pressure can be studied at a detailed sensor level, more sophisticated analysis techniques can be used, and pressure patterns between groups can be studied. As a consequence of this technique, the plantar pressure pattern of the whole foot can be studied in contrast to the pressure under a limited amount of anatomical regions in the masking method. The purpose of this chapter is to gain more understanding of the assessment and analysis of plantar pressure data. Firstly, the assessment of plantar pressure will be described. Subsequently, analysis of the plantar pressure data will be described. Then, the relation between important subject characteristics and plantar pressure will be discussed. Finally, differences between groups will be described and cluster analysis will be used to group plantar pressure measurements of a large number of subjects.

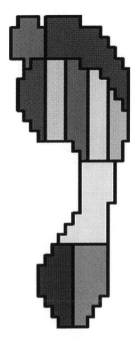

FIGURE 18.2   Example of dividing a plantar pressure pattern in 10 regions.

18.1   ASSESSMENT OF PLANTAR PRESSURE

The most commonly used pressure plates have the size of about 50 cm in length and 40 cm in width just like the regular force plate sizes. However, pressure plates can differ in size and can even be smaller than 50 cm or be as big as 2 m nowadays. The advantage of larger plates is that they can measure two consecutive steps. With two consecutive steps, additional information such as step length, step frequency, and walking velocity can be assessed. Another advantage is that subjects will be less focused on aiming to place their foot on the plate, which is often the case in smaller pressure plates. The midgait protocol in which plantar pressure is measured during steady-state walking is most favorable. Because steady-state walking requires a relatively long walkway and is time-consuming (McPoil et al., 1999), a three-step protocol in which the third step is on the pressure plate is often used as a good alternative. During the first two steps, subjects are mainly accelerating, which will have its effect on the plantar pressure pattern. The third step is usually a good indicator of normal walking. The three-step protocol appeared to have a high reproducibility not only during several measurements at one time but also between different days (Bus and de Lange, 2005; Gurney et al., 2008). It can be avoided that the subject is focusing on placing the foot on the plate by starting from a marked position, which automatically leads to foot on the plate with a regular walking pattern. However, subjects with diseases that affect gait seldom have a regular walking pattern and the three-step protocol can then be replaced by the two-step protocol. The walking speed of patients is often reduced and the acceleration of the second step will be less compared to healthy subjects, and therefore an appropriate alternative. Furthermore, the two-step protocol reduces the number of steps, which is beneficial for patients who often have difficulty walking and are quickly exhausted or who suffer from pain during walking. Many studies have been performed to compare different step protocols or comparing one-step or two-step protocols with a midgait method (Harrison and Folland, 1997; Bryant et al., 1999; McPoil et al., 1999; Peters et al., 2002). In general, no significant differences between protocols in regional peak pressure and/or pressure–time integrals were found. In diabetic patients, the two-step protocol required the least amount of repeated trials for obtaining reliable pressure and may be recommended for assessment of these patients (Bus and de Lange, 2005). Hence, the three-step protocol might be the most favorable protocol, even though the one-step protocol has to be chosen due to research group of interest. To ensure that subjects walk normally over the plate during the measurement, it is recommended to watch them walk freely, since many subjects will alter their walking pattern when entering a laboratory setting.

Plantar pressure plates should have at least 1 sensor/cm^2 but more are preferable. Pressure plates with a density of 4 sensors/cm^2 will result in around 300 activated sensors for a subject with a foot size of 42 (U.S. 9), whereas a sensor density of 1 sensor/cm^2 will result in only 75 activated sensors. The plantar pressure measurements shown in Figure 18.1 are of a subject with foot size 34 (U.S. 1, left panel) and foot size 44 (U.S. 10, right panel). The data were obtained on an RSscan pressure plate with 2 sensors/cm in width and 1.3 sensors/cm in length. Finally, the sample frequency or sample rate, which is the number of samples per time (usually in seconds), is of importance. A sample frequency of at least 100 Hz is preferable for walking but the minimal required sample frequency will depend on the parameter of interest.

There are many different plantar pressure plates on the market and they all have their individual advantages. There are no optimal specifications of a pressure plate, whereas the specific requirements will depend on the research question. For example, a higher sample frequency will be needed for running compared to walking, and sample frequency of 100 Hz might be not sufficient to assess rapid changes in, for example, the center of pressure (COP) during walking in a patient with foot problems. In addition, if you are interested in a particular region of the foot, a higher sensor density will be preferable. Most plantar pressure plates meet the minimal requirements for studying walking, but for specific research-related questions one should be alert that the pressure plate is able to measure it accurately.

In the following paragraph, information about plantar pressure during walking is provided using data of 412 subjects. The subject population was heterogeneous for age, sex, length, weight, insoles, and walking velocity. The data were collected with a 0.5 m pressure plate of RSscan (RSscan, Olen, Belgium). The pressure plate was mounted on top of a force plate (Kistler, Winterthur, Switzerland). Both systems were synchronized with the RSscan 3D-box. Plantar pressure data were collected at 500 Hz. The pressure plate had 2 sensors/cm in width and 1.3 sensors/cm in length. Foot contact on the plate was calculated using RSscan software with a threshold level of 5 N. Subjects walked barefoot at their preferred walking speed using the three-step protocol. Before the actual measurement started, the subject walked a couple of times to get familiar with the task. Subsequently, subjects walked 10 times, starting alternately with the left or right foot.

18.2 MASKING AND NORMALIZATION

One of the first and major problems in analyzing plantar pressure data is the differences in foot size between subjects. In order to solve this problem, the pressure data are divided into a fixed number of anatomical regions or areas of interest are indicated. Dividing the foot into a fixed number of anatomical regions is also referred to as masking of the foot. In general, the following 11 foot regions are mostly described in the literature: Bigger toe/Hallux, lesser toes, five metatarsal regions, medial and lateral midfoot, and medial and lateral rearfoot. Division can be based on geometrical (longitudinal bisection or horizontal trisection) or on anatomical factors. Most masking methods are based on a general anatomical foot model. For example, the forefoot is divided into five regions corresponding to the five metatarsals, and the width of the regions corresponds to the width of the general foot model. Division of the forefoot in a smaller number of regions can therefore be seen as a more geometrical division. In the literature, three forefoot regions (medial, center, and lateral) or four forefoot regions (medial, distal and proximal center, and lateral) have been used (Postema et al., 1998; Zammit et al., 2010). Another way of masking is to indicate the metatarsals based on the plantar pressure image. The metatarsal heads can be individually indicated on the pressure pattern for some subjects. However, in many subjects this is often not the case and certainly not all five metatarsals can be clearly separated. For example, in Figure 18.1 the metatarsal I of the subject in the right panel can be clearly indicated whereas metatarsals II and III cannot be distinguished from each other. Deschamps et al. (2009) used an RSscan plate and studied the inter- and intraobserver reliability of placing manual masks at the forefoot. They found a good to high intraobserver reliability but a lower interobserver reliability. They concluded that placing small-sized masking in order to define pressure characteristics in the forefoot should be done with care. In contrast to the metatarsals, the toes are mostly analyzed as one region or as two separate regions: the bigger toe and the four lesser toes. The midfoot and rearfoot are mostly divided in a medial and lateral part or just studied as a rearfoot and midfoot. For the division in a medial and lateral part, a longitudinal bisection is used.

The masking procedures are often not precisely described and depend on the software programs of the different plantar pressure plate companies. Consequently, the shape and the number of masking regions differ between companies and therefore also between studies. For the longitudinal division of the foot, the forefoot, midfoot, and hindfoot will each count for approximately 1/3. The widely used Novel 10 region mask has been described precisely and has set the boundary between the hindfoot and midfoot at 73% and between the midfoot and forefoot at 45% of the length from the toes to the heel. For the medial and lateral division of the foot, a longitudinal foot axis has to be determined. In general, the longitudinal foot axis is defined as the line through the base of the heel and metatarsal II. For the division of the forefoot in five metatarsals regions, different proportions for each of the metatarsals are used. The metatarsal regions of the automask of Novel are defined as 30% for the first; 17% for the second, third, and fourth; and 19% for the fifth of the long plantar angle. The division lines are based on the longitudinal foot axis. The medial and lateral hindfoot and midfoot are also based on the longitudinal foot axis. Ellis et al. (2011) studied the accuracy of the standard Novel 10 region mask and found a high accuracy especially for dynamic measurements. Zammit et al. (2010) studied the reliability of the masking method of the TekScan MatScan® system, which divides the foot into seven regions (bigger toe, lesser toes, three forefoot regions, midfoot, and hindfoot). The intraclass correlation coefficient values ranging from 0.96 to 1.00 demonstrated that the mask application has good reliability. In conclusion, the reliability of masking methods is high. However, different masking methods are difficult to compare because they differ in the number of masks and the shape of the masks.

Combining plantar pressure measurements with a motion analysis system could be a solution for a more uniform masking method. Based on markers, which are placed on relevant anatomical landmarks, the regions are defined. The position of each of the markers is projected onto the pressure plate at a point corresponding to midstance. Giacomozzi et al. (2000) proposed this method and achieved encouraging results. The identification of anatomical landmarks is most likely more accurate than the arbitrary division of the foot. Stebbins et al. (2005) used anatomical landmarks to divide the plantar pressure images (i.e., foot print) in five areas in children. Peak force and peak pressure appeared to be reliable for each subarea in healthy children. However, although the method can be used whenever a motion system is available, it is too time-consuming in applying it routinely.

Although the masking process extracts important information out of the plantar pressure and describes the overall foot loading, subtle differences within a region cannot be seen due to the masking process. Therefore, plantar pressure pattern scaled to a standard size and progression angle would be helpful to study the plantar pressure in more detail. Moreover, scaling the plantar pressure to a standard foot allows more sophisticated analysis techniques such as pattern recognition and machine learning. In addition, clinicians and foot experts will rather use pressure images (plantar pressure pattern) than the masking method in their daily (clinical) routine. Shorten et al. (1989) presented a normalization method at the conference of the *International Society of Biomechanics* (ISB). Shorten developed a shape transformation algorithm, which enabled pressure distributions from different subjects to be normalized to a standard foot shape, size, and orientation. Although the method of Shorten has potential, it has not been used in further studies in the literature. Pataky et al. (2009) described a normalization method based on statistical parametrical mapping. Shortly after the publication of Pataky et al. (2008a) and Pataky and Goulermas (2008), the normalization method of Keijsers et al. (2009) was published. The normalization methods of Shorten et al. (1989), Patakay et al. (2008), Pataky and Goulermas (2008), and Keijsers et al. (2009) differ slightly from each other, but have a common goal in mind: rescale the plantar pressure measurement to a standard foot. Recently, the normalization method has been used more often to study plantar pressure measurements (Keijsers et al., 2010; Stolwijk et al., 2010;

Pataky et al., 2011a,b; Koenraadt et al., 2012). These studies show some new insights and demonstrate some drawbacks of the masking method. Pataky et al. (2008) concluded that masking methods obscure and may even reverse statistical trends and stated that studies using plantar pressure measurement should consider pixel-level data where possible. However, it is important to note that some information might be lost by using a normalization method due to the rescaling of the plantar pressure data. Marsden et al. and Keijsers et al. use linear methods that minimally affect the data, but the nonlinear method of Pataky et al. could compromise morphological information that may otherwise be clinically relevant. A bony protuberance, for example, may be indicative of pathology, and nonlinear warping could distort or destroy this information.

The methods of Shorten and Keijsers are most similar to each other and use the foot size and orientation (i.e., foot progression angle) to normalize the plantar pressure data. In contrast, the method of Pataky et al. is based on statistical parametrical mapping (SPM) used in neuroscience on fMRI data and is referred to as pedobarographic SPM (pSPM). pSPM consists of three main steps: (1) image generation, (2) registration, and (3) statistical test. The image generation and registration can be best compared with the Shorten and Keijsers method and has as goal to optimally overlap the plantar pressure of the various subjects to a reference template (mean image). Subsequently, statistical tests will be used to indicate differences between images. The pSPM technique is more pixel based than the two other methods because each pixel is scaled individually. The pSPM is a nonlinear technique and nonlinear methods have the ability to warp any foot shape into another shape. As a result, morphological information that might be clinically relevant could be compromised. A bony protuberance, for example, may be indicative of pathology, but could be distorted or destroyed. Because the method of Keijsers et al. (2009) is used in the current chapter, it will be described in more detail in the following paragraph.

The normalization method of Keijsers et al. (2009) uses mean pressure per sensors over the stance phase (i.e., from heel strike to toe-off). Before the normalization procedure, the data were interpolated to a square grid with 2 sensors/cm in width and length (4 sensors/cm²), due to the 2 sensors/cm in width and 1.3 sensors/cm in length of the RSscan plate. Subsequently, the mean plantar pressure pattern was interpolated to 16 sensors/cm² using spline interpolation. As a consequence of this interpolation, the term "pixels" will be used instead of sensors since the pressure is calculated at locations between sensors. Figure 18.3 shows the mean pressure after interpolation of the sensors. The contour line of 10 kPa of the mean plantar pressure pattern was used to calculate the foot progression angle and foot size. The foot progression angle was defined as the average angle of the tangent lines to the medial and lateral side of the foot. Foot length was defined as the distance between the back of the heel and the forefoot line. Foot width was defined as the mediolateral distance between the most medial and most lateral point of the 10 kPa contour line of the forefoot. To calculate the normalized foot, the plantar pressure pattern was rotated over the foot progression angle and normalized for foot size. Figure 18.4 shows an example of the contour lines of six feet before (left panel) and after (right panel) the normalization procedure.

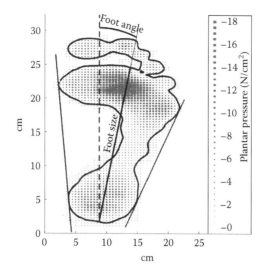

FIGURE 18.3 Plantar pressure distribution pattern of a subject with the 10 kPa contour line. The middle black line indicates foot length and the curved line the foot progression angle. Black lines on the outside indicate tangent lines.

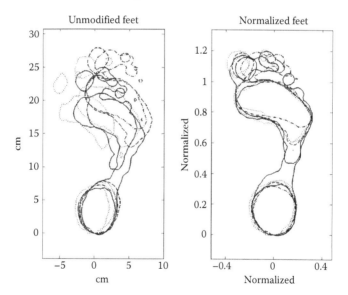

FIGURE 18.4 Contour lines of 10 kPa of six randomly chosen feet before (left panel) and after normalization (right panel).

18.3 PLANTAR PRESSURE PARAMETERS

For the masking method as well as the normalization method, various parameters can be calculated during the stance phase. The stance phase is defined as the phase between heel strike and toe-off. In most gait analysis, the swing phase is also often studied, but will be of no interest for plantar pressure measurement. Various other

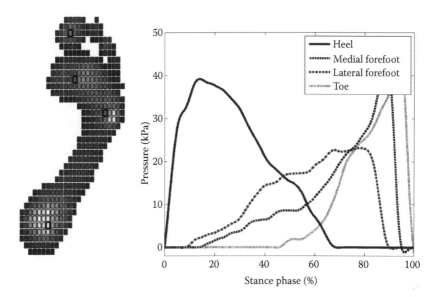

FIGURE 18.5 Left panel shows a typical example of the mean pressure during walking. Gray scale indicates the pressure. The right panel shows the relation between pressure and time for four sensors: heel, medial forefoot, lateral forefoot, and toe.

moments and phases during stance phase are defined and can be determined. Foot flat and heel-off are often used in addition to heel strike and toe-off and can be assessed by plantar pressure measurements. But also more general terms as early, late, and midstance are used. Midstance is not only used as a phase but also as 50% of the stance phase. In the gait analysis of Perry and Burnfield (2010), the gait cycle is well described.

Figure 18.5 shows the pressure in relation to time of sensors located in the heel, lateral forefoot, medial forefoot, and bigger toe during the stance phase. The heel sensor is activated from heel strike to heel-off, which is in the example of Figure 18.5 at approximately 68% of the stance phase. The forefoot sensors are activated at foot flat, which is between 15% and 20% of the stance phase. Finally, the toe sensor is activated at 45% of the stance phase. The pressure under the toe and forefoot sensors rises gradually, whereas the plantar pressure at the heel sensor increases rapidly at heel strike.

Based on these pressure–time relationships, various parameters can be calculated. Table 18.1 presents the plantar pressure parameters and their definitions, which will be described in this chapter. The most frequently used parameters are peak pressure, mean pressure, and pressure–time integral (also referred to as pressure impulse). In Figure 18.6, the peak pressure, mean pressure, and pressure–time integral are schematically indicated for a sensor at the distal/lateral midfoot. The mean values and the standard deviations for the peak, mean, and pressure–time integral for our large group of 412 subjects after normalization are shown in Figure 18.7. The metatarsal II–III and heel area showed the largest values for all three parameters. The average peak pressure image shown in the upper center panel (Figure 18.7) is very

TABLE 18.1

Plantar Pressure Parameters and Their Definitions

Parameter	Definition
Peak pressure	Maximum plantar pressure of a pixel or within a region
Mean pressure	Mean plantar pressure of a pixel or all sensors within a region
Pressure–time integral	Mean pressure–time integral of a pixel or all sensors within a region
Mean loading rate	Mean loading rate from pixel-on to maximal loading rate
Peak loading rate	Maximal loading rate
Pixel-on	Difference between first contact of a pixel and initial heel contact
Pixel-off	Difference between last contact of a pixel and initial heel contact
Pixel-time	Difference between last and first contact of a pixel
Center of pressure (COP)	Point at which all pressures are centered
Single Plantar Pressure Parameters	
Contact time	Time between heel strike and toe-off
Foot progression angle	Angle between the longitudinal axis of the foot and a straight line representing the progression of the body in walking
Foot length	Length of the foot
Foot width	Width of the foot
Total contact area	Surface of plantar area that has contact with the floor
Arch index	Ratio of the area of the middle third to the entire footprint area (excluding the toes)

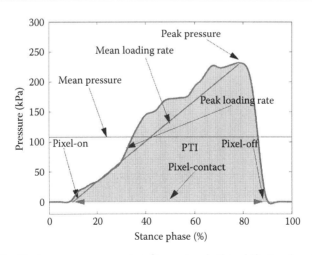

FIGURE 18.6   Plantar pressure parameters for a sensor in the midfoot region of Figure 18.5.

identical to the unbiased pedobarographic template computed with pSPM method by Pataky using 104 healthy subjects (Pataky et al., 2011a). The plantar pressure images of the three parameters are very similar. Moreover, the values of these parameters appeared to be highly correlated, especially the mean pressure and pressure–time integral when analyzed at a pixel level (Keijsers et al., 2010). In addition to the mean

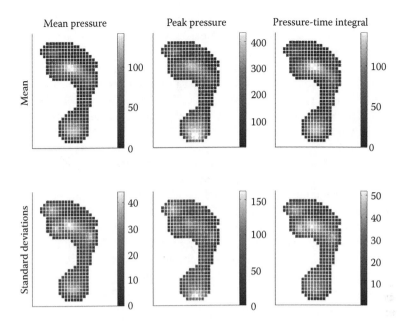

FIGURE 18.7   Mean (upper panels) and standard deviation (lower panels) of mean pressure, peak pressure, and pressure–time integral for 412 subjects. Gray scale indicates the pressure in kPa.

values of the pressure parameters for each pixel, it is also important to know what the variation in pressure parameters is between subjects, which is indicated by the standard deviation. The standard deviation of the mean, peak, and pressure–time integral for each pixel is shown in the lower panels of Figure 18.7. The standard deviation is high under the heel and metatarsal II and III, at which the mean pressure is also high. In addition, large variations between subjects are found under the metatarsal I and V and the bigger toe, which is in contrast to the mean values for the parameters in these regions. The high standard deviations under metatarsal I and V and the bigger toe are mainly due to the large difference in roll-off between subjects. Subjects with a more medial roll-off have larger pressure values under metatarsal I, whereas subjects with a more lateral roll-off have large values under metatarsal V.

For the assessment of mean pressure, peak pressure, and pressure–time integral for the regions, all sensors of a region are used. For the mean pressure and the pressure–time integral of a region, the mean of the mean pressure and the mean of the pressure–time integral of all sensors of a region are most often used. The peak pressure of a region is defined as the value of the sensor with the highest peak. The correlation coefficient between peak pressure, mean pressure, and pressure–time integral per region is lower than per-pixel.

In addition to pressure values of a region, the force of a region can be calculated. The force of a region can be calculated by summing the pressure of all sensors in a region and dividing by the sensor density of the pressure plate. Subsequently, the mean force, peak force, and force–time integral can be calculated by evaluating the force over time. Even more than the pressure values, the values of these force-related parameters will be influenced by the shape and number of masking regions. Therefore, one should be cautious in comparing studies that use different software.

Another interesting plantar pressure parameter is the loading rate. Loading rate is defined as the differential in pressure and can be presented in peak or average and is expressed in MPa/s. In other words, loading rate describes how fast the pressure increases. The upper panel of the left side of Figure 18.8 shows the mean loading rate and the peak loading rate of the 412 subjects. The mean and peak loading rates are especially high for pixels in the heel region. The high loading rate in the heel is due to heel strike. At heel strike, the heel area is loaded rapidly because the weight of the subjects is transferred from one leg to the other until foot flat. Subsequently, a subject will roll off and the pressure will smoothly shift to the toes. Therefore, the loading rate of the midfoot, forefoot, and toes is much less than the heel. The right side of Figure 18.8 shows only the loading rate for the midfoot and forefoot to indicate the relative differences in these areas. The loading rate is relatively the same for all metatarsal heads, especially for the peak loading rate. The peak loading rate is higher than the mean loading rate but the pattern is very similar.

Plantar pressure parameters related to time are sporadically reported in literature. Time-related pressure parameters involve the time that the sensors are activated or deactivated and can be expressed in relation to heel strike or toe-off or to another defined instant in the stance phase. The timing parameters provide information about when certain areas are loaded and unloaded and, maybe even more important,

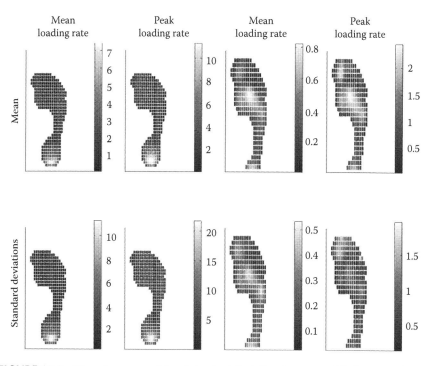

FIGURE 18.8   Mean (upper panels) and standard deviation (lower panels) of mean and peak loading rate for 412 subjects. Gray scale indicates loading rate in MPa/s.

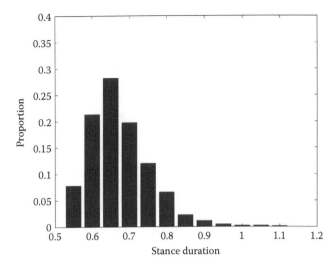

FIGURE 18.9   Histogram of stance duration for the 412 subjects.

the total time these areas are loaded. Table 18.1 shows the definitions of the timing parameters. Important in describing the timing parameters is the stance time also referred to as contact time, which is the time that the foot has contact with the floor, for example, time between heel strike and toe-off. Figure 18.9 shows an histogram of the stance duration for the 412 subjects. Subsequently, the timing parameters can be related to heel strike or toe-off and normalized to the stance time. In this chapter, the difference in time between first contact for the particular pixel and initial heel contact (referred to as pixel-on) and the difference between last contact and initial heel contact (referred to as pixel-off) will be described. In addition, the difference between pixel-off and pixel-on indicates the contact time for each pixel (referred to as pixel-contact) and will be the third timing parameter. Figure 18.10 shows the mean (upper panels) and standard deviation (lower panels) of the 412 subjects for the pixel-on, pixel-off, and pixel-contact parameter in milliseconds and are not normalized for stance duration. The pixel-contact shows that the pixels in the forefoot area are activated longest. The largest variation between subjects is in the midfoot and toes region, which is caused by the large difference in activated sensors in these areas. Sensors that are not activated will have a pixel-contact value of zero. All subjects do activate the sensors under the heel and the metatarsals but not in the midfoot or toes region. Pixel-off and pixel-contact have a relatively large standard deviation for the pixels because the data were not normalized for stance duration. Describing timing parameters per region using a masking method is rather difficult, but you can choose the mean values for all sensors in an area. However, it is important to indicate which sensors will be taken into account because the sensors at the outer part of the image are only active for a small period of time as can be seen in Figure 18.10.

The COP is another parameter that can be derived from plantar pressure measurements, and it provides information about a subject's roll-off. The COP is the point at which all pressures are centered. However, for comparing the COP between subjects, ith has to be related to a certain footline. Often the longitudinal foot axis, which is the

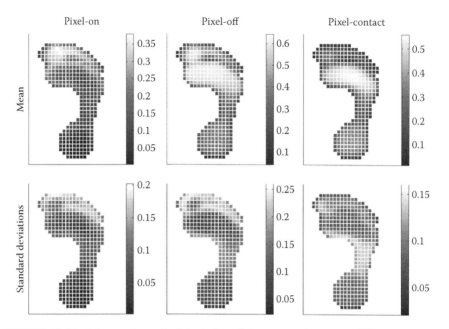

FIGURE 18.10 Mean and standard deviation of the center of pressure (COP). Left panel shows the COP in relation to the foot area. Upper right panel (lower right panel) shows the medial/lateral COP displacement (longitudinal) in relation to stance duration.

line through the base of the heel and metatarsal II, is used. However, indicating the exact middle of metatarsal II can be troublesome and induces inter- and intravariability when defining the foot axis line. Therefore, to compute the footline based on the outer parts of the heel and forefoot, as is done in the normalization procedure used in this chapter, is a better solution. After defining the footline, the distance of the COP in relation to the footline can be used in providing information about the medial and lateral roll-off of a subject. The upper right panel of Figure 18.11 shows the mean and the standard deviation of the medial/lateral displacement of the COP of the 412 subjects in relation to the footline. For the longitudinal displacement of the COP also, the length of the foot has to be taken into account. For the normalization method, the foot length is defined as the length between the heel and the forefoot. The lower right panel shows the mean and standard deviation of the longitudinal displacement of the COP in relation to foot length. The left panel of Figure 18.11 shows the mean and standard deviation of the COP and the average 10 kPa contour (see Figure 18.3) of 412 subjects. The standard deviation is relatively small, indicating that the COP does not differ largely between subjects. The largest variation in COP is seen in the push-off phase.

Finally, several single plantar pressure parameters can be derived from plantar pressure measurements. The mean and standard values for the single plantar pressure parameters for the left and right foot of the 412 subjects are presented in Table 18.2. In addition, a paired t-test was used to test for significant differences between the left and right foot. The first parameters is the aforementioned contact time or stance phase, which is the difference in time between heel strike and toe-off. Contact time is moderately correlated with walking velocity and is therefore often used as an

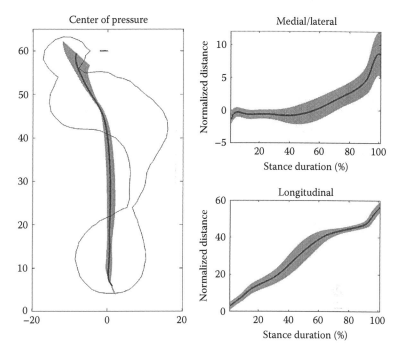

FIGURE 18.11 Mean (upper panels) and standard deviation (lower panels) of pixel-on, pixel-off, and pixel-contact (see Table 18.1 for definitions) for 412 subjects.

TABLE 18.2

Mean and Standard Deviation for the Left and Right Foot for the Single Plantar Pressure Parameters

Parameter	Left	Right	Difference
Contact time (s)	0.68 (0.09)	0.68 (0.09)	$p = 0.01$
Foot progression angle (°)	11.1 (6.7)	13.7 (6.5)	$p < 0.01$
Foot length (cm)	21.0 (1.25)	20.9 (1.24)	$p = 0.06$
Foot width (cm)	9.23 (0.66)	9.27 (0.68)	$p = 0.03$
Total contact area (cm²)	100.2	100.7	$p = 0.12$

Paired *t*-test was used to indicate significant differences between left and right foot.

indicator of walking velocity. Mean contact time was 0.67 s and ranged between 0.53 and 1.11 s. A histogram of contact time can be seen in Figure 18.11. The difference in contact time between left and right foot was small (0.002 s, right foot larger than left foot) but appeared to be significantly different ($p = 0.01$, see Table 18.2).

Secondly, foot progression angle can be assessed and is defined as the angle between the longitudinal axis of the foot and a straight line representing the progression of the body in walking. The assessment of the base of the heel and especially metatarsal II for assessing the longitudinal axis is difficult. In the current chapter, the progression angle was defined as the average angle of the tangent lines to the medial and lateral side of the foot (see Figure 18.3). The mean foot progression angle for both feet was 12.8° (SD 6.3) outward and ranged between 3.4° inward to 31.1° outward. The foot progression angle of the right foot appeared to be significantly larger for the right foot compared to the left foot (see Table 18.2).

Parameters related to foot size (foot length and foot width) can be derived from the plantar pressure measurement. To assess the correct foot width, the outer parts of the plantar area of the forefoot have to be visual. To assess the correct foot length, the bigger and to a lesser extent the second toe needs to be visual. During walking, the heel is almost always visual, except for subjects with a pes equinus or large foot deformities in, for example, hereditary motor sensory neuropathy (HMSN) patients. As an alternative for foot length measurement, the distance between the heel and the forefoot as used in the normalization method is described (see Figure 18.3). The mean foot length was 21.0 cm and mean foot width was 9.25 cm. The left foot seems to be a little bit larger than the right foot (see Table 18.2). Related to foot width and foot length is the total contact area or the number of pixels that are activated. In the current analysis, a pixel was defined as activated when the mean pressure during stance was at least above 5 kPa. Mean total contact area was 100.4 cm² (SD 17.0) and ranged between 63.5 and 168.6 cm². The total contact area mainly depends on foot size and arch height. Subjects with a high arch will have a relatively small total contact area (see, e.g., Figure 18.1). An often-reported parameter to describe the arch height is the arch index, which can be assessed using plantar pressure data (Cavanagh and Rodgers, 1987). The correlation coefficient between arch height (navicular bone height) and arch index is approximately 0.7 (McCrory et al., 1997). Therefore, subjects with a low arch index will have a flat foot (i.e., pes planus), whereas subjects with a high arch index will have a hollow foot (i.e., pes cavus).

18.4 RELATION BETWEEN PLANTAR PRESSURE AND SUBJECT CHARACTERISTICS

For a better understanding of plantar pressure data, it is important to know which characteristics are responsible for the plantar pressure pattern of a certain person. Especially, the relation between a certain plantar pressure parameter and a subject's specific characters such as body weight, walking speed, and foot progression angle will be described. Certain subject characteristics, such as body weight or walking velocity, have been studied extensively in the literature using the masking method. In this chapter, the correlation coefficient between a pressure parameter (in most cases mean pressure) of a pixel and a certain characteristic is calculated. The subject characteristics are body weight, body length, age, contact time, and foot progression angle. A negative (positive) correlation means that the pressure parameter will decrease (increase) in relation to an increase (increase) of the subject characteristic.

The most important subject characteristic that affects the plantar pressure is body weight. For a better understanding of the relation between body weight and plantar pressure, we will start with the vertical ground reaction force. The sum of the total plantar

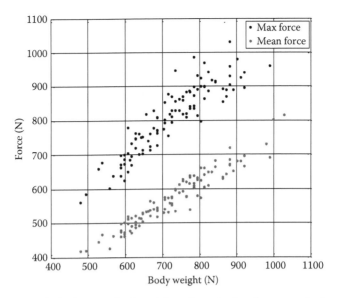

FIGURE 18.12   Relation between body weight and maximal and mean vertical ground reaction force.

pressure under the foot divided by the contact area of the foot equals the total vertical ground reaction force. The vertical ground reaction force during walking mainly depends on the weight of a subject. Figure 18.12 shows the average relation between body weight and mean vertical ground reaction force during stance for 114 subjects ($r = 0.96$). The linear fit can be described as GRF = 0.67 × BW + 93 (BW in Newton). The lower mean vertical ground reaction force during stance compared to body weight is mainly caused by the double support phase during walking. The maximal ground reaction force and body weight are highly correlated ($r = 0.90$) with a linear fit described by GRF = 0.80 * BW + 243 (BW in Newton). Hence, the sum of the total plantar pressure under the foot is mainly caused by the subject's body weight. To investigate if other subject characteristics affect the total pressure after normalization, a step-forward regression analysis was performed. In the step-forward regression analysis, the dependent variable was the sum of the total pressure after normalization and as independent variables were used: body weight, body length, stance duration, age, and foot progression angle. The regression analysis resulted in body weight ($r = 0.646$), contact time (increase to $r = 0.682$), and body length (increase to $r = 0.705$) as important factors. Contact time and body length were added to body weight, because they are important factors in relation to walking velocity. The factors age and foot progression angle did not have any effect on the total normalized plantar pressure.

More important than the total plantar pressure is the plantar pressure value for each sensor in relation to the plantar pressure value of the other sensors. Therefore, the combination of total plantar pressure and contact area of the foot plays an important role; larger contact areas will result in lower pressure values for subjects with the same weight. Moreover, the way the plantar pressure is distributed over the contact area, which can also be referred to as plantar pressure pattern, is the most important issue in plantar pressure measurements. However, the normalization method influences the contact area; for example, large feet will be more decreased in size than small ones. As a consequence, the correlation between the sum of the total pressure after normalization and body weight is reduced, but is still considerable ($r = 0.646$).

To determine the relation between plantar pressure pattern and subject characteristics, the correlation of the subject characteristics body weight, stance duration, body length, and foot progression angle with mean plantar pressure for each pixel was calculated. Figure 18.13 shows the correlation coefficients of the subject characteristics and mean plantar pressure for each pixel. In general, correlation coefficients were low. Due to the number of subjects, a correlation coefficient above 0.10 is significant ($p = 0.05$). However, a correlation coefficient of 0.10 means that only 1% of the variance is explained by the variable. Therefore, only correlations of at least 0.22 (explained variance of 5%) are considered to be relevant and are indicated with a black border. About 49.4% of the pixels showed a positive correlation coefficient larger than 0.22 between mean pressure and body weight (upper left panel of Figure 18.13). In addition, many pixels (40.7%) had a significant correlation between stance duration and mean pressure. The pixels with a correlation coefficient of 0.22 and larger were mainly in the midfoot, distal heel, and proximal forefoot. Body weight showed a correlation coefficient of 0.33 with stance duration. For the foot progression angle, seven pixels in the distal midfoot showed significantly positive correlation coefficients with mean pressure. Body length appeared to have three small areas (four pixels) with a correlation coefficient larger than 0.22.

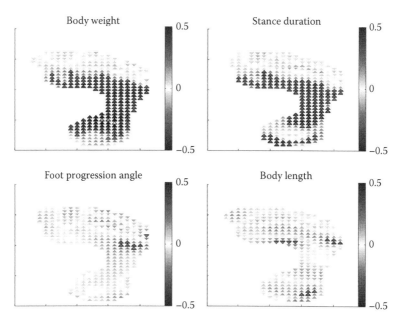

FIGURE 18.13 Correlation coefficient between mean pressure for each pixel and body weight (upper left panel), stance duration (upper right panel), foot progression angle (lower left panel), and body length (lower right panel). Pixels with a black border have a correlation coefficient larger than 0.22 (5% explained variance).

The positive correlation coefficients shown in Figure 18.13 are mostly related to body weight. To reduce the influence of body weight, the mean pressure can be normalized for total pressure. Normalization for total pressure means that each pixel is divided by the sum of the pressure of all pixels. Subsequently, it is multiplied by the mean of the sum of the pressures of all pixels for all subjects to represent average pressure values. After normalization for total pressure, the relation of the plantar pressure distribution with the subject characteristic is determined. Figure 18.14 shows the correlation coefficients of the subject characteristics and mean plantar pressure for each pixel normalized for total pressure. The correlation coefficient values and number of pixels with a correlation larger than 0.22 are lower when normalized for total pressure compared with not normalized for total pressure. For body weight and stance duration, pixels in the heel and forefoot showed a negative correlation with mean pressure. Only 11% of the pixels showed a correlation coefficient larger than 0.22 between stance duration and normalized mean pressure. Foot progression angle and body length had a minimal number of pixels with correlation coefficients larger than 0.22. Stance duration is moderately correlated with contact time and is often used as a measure for walking velocity. Rosenbaum et al. (1994) showed that an increase in walking speed results in an increase in peak pressure under the heel and the medial part of the forefoot. Significant decreases in peak pressure were found under the midfoot and lateral forefoot by other studies (Burnfield et al., 2004; Pataky et al., 2008). The aforementioned studies are all based on within-subject comparison; subjects were instructed to walk at different walking velocities. The main advantage

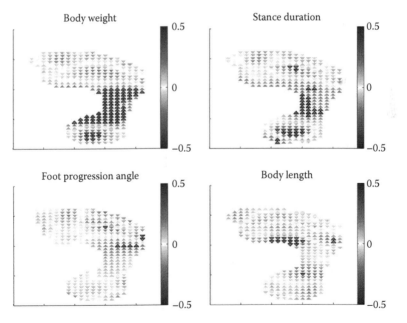

FIGURE 18.14 Correlation coefficients between mean pressure *normalized for total pressure* for each pixel and body weight (upper left panel), stance duration (upper right panel), foot progression angle (lower left panel), and body length (lower right panel). Pixels with a black border have a correlation coefficient larger than 0.22 (5% explained variance).

of within-subject comparison is that confounding factors such as age, weight, and foot morphology can be excluded because the pressure data at different speed conditions are obtained from the same subject. In the current chapter, the between subject effect on plantar pressure is described. Therefore, the minimal effect of contact time on plantar pressure pattern suggests that walking velocity minimally influences the individual differences between subjects but will be effected when a subject differs in his/her walking velocity. The effect of body weight has been studied by comparing obese and nonobese subjects (e.g., between subjects; Hills et al., 2001; Birtane and Tuna, 2004) or by adding extra weight to a subject (e.g., within subjects; Vela et al., 1998; Arnold et al., 2010). Both types of studies found that the plantar pressure significantly increased under the heel, midfoot, and metatarsal heads. Hills et al. (2001) supported the finding that the greatest effect of body weight on pressure was found under the longitudinal arch.

In conclusion, stance duration, body length, and foot progression have minimal effect on the plantar pressure pattern. In contrast, body weight highly affects plantar pressure pattern. The negative correlations in the heel and forefoot with the positive correlations in the midfoot for body weight indicate that the midfoot is relatively more loaded than the heel and the forefoot when the body weight of a subject increases. For example, a foot becomes more flat when the weight of the subject increases.

In addition to mean pressure, the correlation coefficient of the subject characteristics—body weight, stance duration, body length, and foot progression angle—with contact time for each pixel was calculated. Figure 18.15 shows the

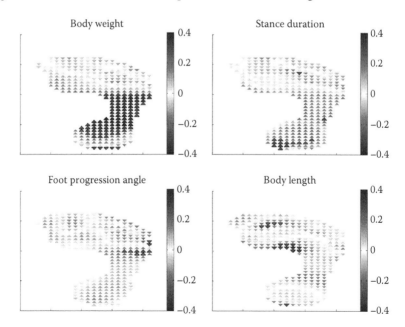

FIGURE 18.15 Correlation coefficients between contact time *normalized for stance duration* for each pixel and body weight (upper left panel), stance duration (upper right panel), foot progression angle (lower left panel), and body length (lower right panel). Pixels with a black border have a correlation coefficient larger than 0.22 (5% explained variance).

correlation coefficients of the subject characteristics and pixel-contact for each pixel. Pixel-contact was normalized for stance duration by dividing pixel-contact by stance duration for each subject. Only body weight had a potential number of pixels with a correlation larger than 0.22 (30.6%), which were located in the heel and midfoot (upper left panel of Figure 18.15). For stance duration, foot progression angle, and body length, a minimal number of pixels showed a correlation larger than 0.22 (4.6%, 2.8%, and 4.9%, respectively). Therefore, body weight seems to be the most important factor that influences the loading of the foot.

18.5 DIFFERENCES IN PLANTAR PRESSURE BETWEEN GROUPS

The normalization method provides the ability to compare plantar pressure patterns between groups but also between the left and the right foot. The plantar pressure pattern between the left and right foot is slightly different, while the contour lines of the foot are similar (see Figure 18.16). The left foot has more pressure under the lateral side of the forefoot compared to the right foot. Individual comparison between left and right foot can be interesting in indicating symmetry between left and right foot. Symmetry is interesting from a clinical point of view since many patients show asymmetry when one side is affected. Patients with knee or hip osteoarthritis, for example, show asymmetrical leg loading (Boonstra et al., 2010). Asymmetry in plantar pressure pattern has also been used in patients after an ACL rupture (Huang et al., 2012). Asymmetry in plantar pressure pattern can be defined as the mean absolute difference between left and right foot for each pixel (or sound and affected side). Figure 18.17 shows a histogram of the asymmetry of the 412 subjects. The mean asymmetry score was 9.0 kPa (SD = 3.2 kPa) for the 412 subjects. In patients after an ACL rupture, copers showed an increase in asymmetry in plantar pressure pattern around the MTP 1 joint compared with noncopers, which might indicate that copers stiffen their hallux more effectively to compensate for torsional instability of the knee with an ACL rupture (Huang et al., 2012).

When comparing normalized plantar pressure pattern between groups such as men and women, body weight can easily affect the result. Therefore, when interested

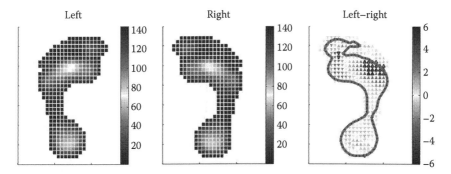

FIGURE 18.16 Mean of the mean pressure for the left (left panel) and right foot (middle panel). The right panel shows the difference in mean pressure between left and right foot and the 10 kPa contour lines of the right (thick line) and left (dotted line) foot.

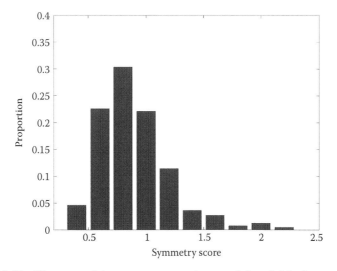

FIGURE 18.17   Histogram of the symmetry score between left and right foot.

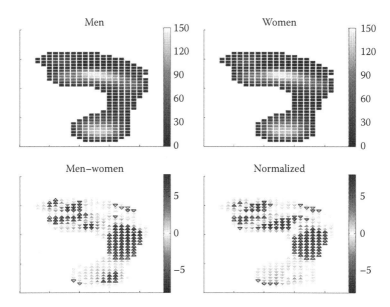

FIGURE 18.18   Mean pressure image for the men and women (upper panels). The right panel shows the difference in mean pressure between men and women, not normalized (lower left panel) and normalized (lower right panel) for total pressure. Pixels with a black border have a significant difference between men and women.

in the plantar pressure pattern, normalized plantar pressure should be normalized for total pressure. Figure 18.18 shows the mean pressure image for men and women and the difference in mean pressure between men and women. When not normalized for total pressure, 64.5% of the pixels of men have a larger value than women, whereas normalized for total pressure, the mean difference is zero and only 45.7% of the pixels of men have a larger value than women. Therefore, the lower right panel of Figure 18.18 shows the difference in plantar pressure between men and women best. Men appeared to have relatively larger pressure values under metatarsal I and III–V and the midfoot and lower pressure values under the metatarsal II. The heel loading was relatively the same, but might be a little bit more laterally positioned for men.

Foot progression angle showed almost no correlation with mean pressure (see Figures 18.13 and 18.14). Another way to investigate the effect of foot progression angle is to divide the group in two subgroups: subjects with a more inward-rotated foot progression angle and more outward-rotated foot progression angle. For the division, we used a split-half method (median foot angle). The mean pressure image for the inward and outward group is shown in Figure 18.19. When not normalized for total pressure, 73.8% of the pixels of subjects with an outward foot progression angle were larger than subjects with a more inward foot progression angle. The total pressure was significantly different between the two groups ($p < 0.01$), which was most likely a consequence of the significant difference in body weight between the two groups ($p < 0.001$, outward: 73 kg ± 12 inward: 78 kg ± 15). The correlation coefficient of 0.22 between progression angle and body weight suggests that heavier subjects walk with a more outward-rotated foot progression angle. After normalization for total pressure, the number of pixels was reduced to 48.2%. Subjects with a more outward foot progression angle show reduced mean pressure values under the midfoot and lateral forefoot.

Finally, the subjects were divided into a young age and old age group using the split-half method (median age was 51.5 years). Body weight was not significantly

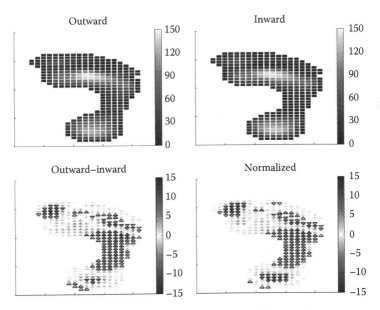

FIGURE 18.19 Mean pressure image for the outward and inward foot progression angle groups (upper panels). The right panel shows the difference in mean pressure between outward and inward, not normalized (lower left panel) and normalized (lower right panel) for total pressure. Pixels with a black border have a significant difference between men and women.

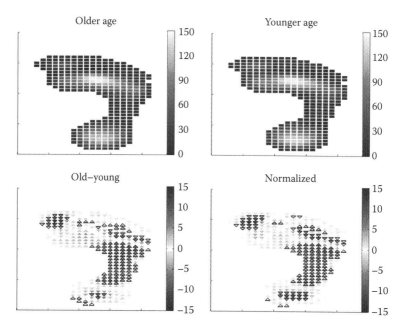

FIGURE 18.20 Mean pressure image for older age and younger age group (upper panels). The right panel shows the difference in mean pressure between the two age groups, not normalized (lower left panel) and normalized (lower right panel) for total pressure. Pixels with black border indicate a significant difference between older age and younger age group.

different between the age groups. Body length (170 cm versus 173 cm) and stance duration (0.70 s versus 0.66 s) were significantly different between the older age group and the younger age group. Figure 18.20 shows the mean pressure image for the age groups and the difference between the groups. Old age group had a significantly larger mean pressure under the midfoot and proximal part of the lateral forefoot. In addition, the heel and toe loading is smaller for the old age group compared with the young age group. The relative increase in midfoot loading and decrease in heel and toe loading suggests that a foot might be flatter for older age subjects.

18.6 PRINCIPAL COMPONENT AND CLUSTER ANALYSIS ON DATA-SET

An advantage of the normalized method is that sophisticated analysis techniques such as pattern recognition techniques can be used. In the final section of this chapter, principal component and cluster analysis will be used to analyze the mean pressure image of the 824 feet. The main advantage of principal component analysis is that it reduces the number of important variables. Principal component analysis transforms the original variables (324 pixels) into new, uncorrelated variables called principal components. Each principal component is a linear combination of the original variables. The principal component's variance expresses the amount of information contained in that principal component. The principal components are derived in decreasing order of variance. Thus, the first component

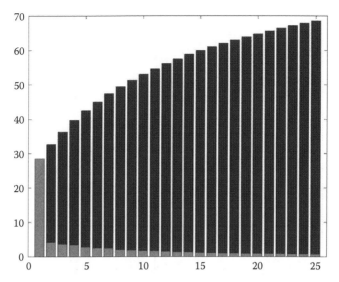

FIGURE 18.21   Explained variance of the first 25 principal components.

contains the most information, the last the least. Figure 18.21 shows the explained variance of the first 25 principal components. The first component explained almost 30% of the total variance. In contrast, the most informative pixel only explained approximately 1.5% of the total variance. The high percentage of the first principal component indicates that the explained variance reduces fast for each subsequent principal component. The second principal component added 4.2%, whereas the 16th principal component was the last principal component that added more than 1% to the total explained variance. The first 25 principal components explained in total 68.5% of the variance and the first 100 principal components explained in total 90.1% of the variance. The minimal increase of 10% in the explained variance by the remaining 234 principal components indicates that many of the sensors are highly correlated with each other. Therefore, principal component analysis can reduce the number of important variables and still describe almost the complete plantar pressure pattern.

More interesting than the explained variance are the eigenvectors used in the principal component analysis. The larger (or more negative) the eigenvector, the larger the contribution of the corresponding pixel to the calculation of the principal component score. Pixels with an eigenvector close to zero will minimally contribute to the principal component value. Figure 18.22 illustrates the eigenvectors of the plantar pressure image for the first nine principal components. For the first principal component, the pixels in the metatarsal II–III region and the heel region are most dominant. In contrast, the value of principal component can also be determined by a small region of the foot, which is the case for principal component 7. The value of principal component 7 will be determined by the pressure values under the hallux. The eigenvector images can also be seen as the difference in mean pressure image between subjects with a large value versus subjects with a small value for the principal component.

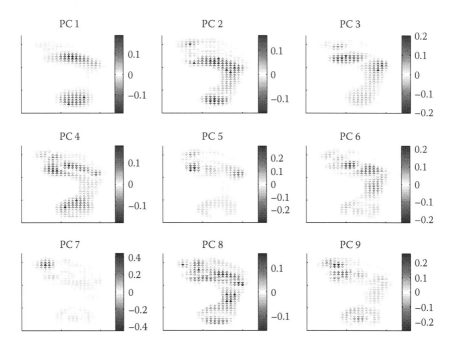

FIGURE 18.22   Weights of the first nine principal components.

To indicate the role of each principal component, the correlation coefficient between the principal components and subject characteristics can be calculated or the eigenvector images can be described. The first two and the fourth principal component scores showed correlation coefficient values of 0.44, 0.35, and 0.30 with body weight, respectively. These high correlation coefficient values indicate the importance of body weight on plantar pressure images. Principal component 1, 2, and 4 also had a correlation coefficient above 0.22 with contact time (0.24, 0.34, and 0.26, respectively). The subject characteristics age, body length, and foot progression angle had correlation coefficient below 0.22 with all nine principal components. Based on the eigenvector image, it can be observed that the first two principal components and especially the second principal component will also discriminate between a flat and hollow foot. The second principal component score will be larger for subjects with a flatter foot. The third and fourth principal components discriminate between subjects with lateral and medial loading of the forefoot. Principal components 5, 6, 7, and 9 focus on small differences in forefoot loading. The value for principal component 5, for example, will be high for subjects with a relatively large loading of metatarsal II.

Another way to indicate groups in the data is by cluster analysis. Cluster analysis or clustering is the task of assigning a set of objects into groups (clusters) so that the objects in the same cluster are more similar to each other than to those in other clusters. The number of clusters has to been assigned in advance. In the current chapter, K-means clustering method was used to assign the 824 feet into nine clusters. The smallest cluster consisted of 30 feet whereas the largest cluster consisted of 164 feet.

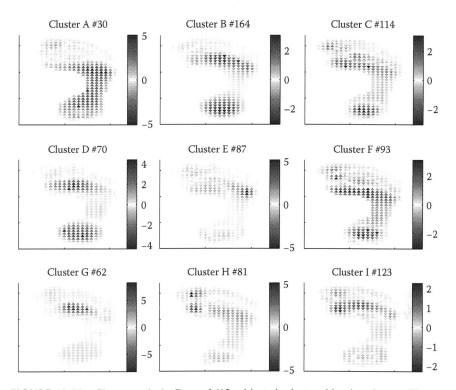

FIGURE 18.23 Cluster analysis. Data of 412 subjects is clustered in nine clusters. Figures show the difference between the mean of the subjects in a cluster and the mean of all subjects. Title indicates the number of feet in the particular cluster.

Figure 18.23 shows the difference between the mean pressure image of a cluster and the mean pressure image of the total group as shown in the left upper panel of Figure 18.7. For 61% of the subjects, the left and right feet fall into the same cluster. Cluster A was the cluster with the largest body weight (99 kg ± 16) and the longest stance duration (0.79 s ± 0.08). The other clusters with relatively large body weight were clusters D (86 kg ± 15) and F (86 kg ± 12). Cluster B compromised the largest number of feet and these feet were from subjects with the smallest body weight (66 kg ± 3) compared to the other clusters. Body weight appeared to be the subject characteristic that differed most between the clusters. The three clusters with the relatively largest body weight A, D, and F, differ from each other in the way the pressure is distributed. Cluster D contains hollow feet, whereas clusters A and F contain more flat feet. Body length was smallest for cluster B (168 cm ± 8) and largest for cluster E (178 cm ± 8). The cluster with the oldest feet was cluster A (57 years ± 13) and the one with the youngest feet was cluster H (44 years ± 13). In cluster C, the smallest contact time was found (0.63 s ± 0.07).

In contrast to principal component analysis, clustering can and will often result in two opposite clusters. In our case, clusters B and D and clusters C and F are each opposites. The pressure distribution under the metatarsals resulted in the three clusters E, G, H, and I. In clusters E, G, and H, the pressure was largest under metatarsal III–V, metatarsal II, and metatarsal I and the bigger toe, respectively, whereas the metatarsals of the feet in cluster I were relatively less loaded. The clustering related to the metatarsals might be due to the large number of subjects with metatarsalgia in the used group.

18.7 CONCLUSION AND PLANTAR PRESSURE IN FOOTWEAR

Although plantar pressure measurements are easy to collect, the analysis of plantar pressure measurements is complicated. Many parameters can be derived from plantar pressure measurements. In the scientific literature, mostly pressure values of a region are reported whereas in (clinical) practice mainly the total image of the pressure will be used for application of the foot treatment and/or optimal footwear. The recently developed normalization methods provide the opportunity to analyze plantar pressure measurements in a manner that is closer to plantar pressure use in daily routine. The most important factor that influences the plantar pressure is body weight. Body weight will increase the pressure over the whole foot, but relatively more under the midfoot. Therefore, when less interested in absolute plantar pressure values and more interested in plantar pressure pattern or distribution, plantar pressure should be corrected for body weight or total plantar pressure. Factors such as side, foot progression angle, age, and sex have only a minimal effect on plantar pressure and pressure distribution when corrected for body weight. Therefore, the differences in plantar pressure pattern between subjects will mostly be a result of differences in foot shape and loading (roll-off) of the foot.

Footwear can change the plantar pressure pattern by changing the way the foot is loaded. Walking with footwear over a pressure plate and comparing it to walking barefoot over a pressure plate can be used to indicate the effect of the footwear on the roll-off by looking at the COP. However, a better way to measure the effect of insoles and footwear on plantar pressure pattern is to use a pressure insole system. The commonly used pressure insole systems are the Novel pedar insole system and the F-scan system. Pressure insole systems can be put in a shoe and measures the pressure between the insole and the foot. Pressure insoles have to be thin and flexible in order to minimize the effect of the system on the effect of footwear and insoles. As a result, the pressure insole systems are often of less accuracy than the pressure plates, which are not restricted by the thin and flexible requirements.

When studying the effect of footwear and insoles, it is important to keep in mind that footwear cannot change the total pressure because body weight will not be affected by footwear. Footwear can only alter the plantar pressure pattern by reducing the pressure at certain areas and increasing the pressure at other areas. Only the high impact force at heel strike due to the high deceleration of the lower leg might be reduced by footwear. In this case, the footwear reduces the high deceleration of the lower leg. However, the high impact occurs during running and is less or hardly seen in walking. In the literature, the focus is mainly on the reduction of peak plantar pressure by insoles or footwear. In the prevention of ulcers in patients with diabetics, reduction of plantar pressure under certain areas is indeed the most important issue (Perry et al., 1995; Mueller et al., 1997; Praet and Louwerens, 2003;

Hennessy et al., 2007; Bus et al., 2008). The importance of reduction of pressures is supported by the successful relief of foot pain by reducing the peak plantar pressure under the painful areas of the foot by properly fitting insoles (Postema et al., 1998; Hodge et al., 1999; Tsung et al., 2004; Burns et al., 2006). As a result, many studies report the effectiveness of reduction in peak plantar pressure by insoles but only a few describe the increase in plantar pressure in other areas (Postema et al., 1998; Stolwijk et al., 2010). Postema et al. (1998) showed a proximal shift of pressure from the metatarsal heads II and III toward the metatarsal bones. Stolwijk et al. (2010) described the changes in plantar pressure distribution by insoles in a large group of subjects with minor foot complaints for a large number of sensors using a normalization method. Plantar pressure was reduced under all metatarsals and heel and increased under the whole midfoot especially under the metatarsal bones. By analyzing a large number of sensors (e.g., normalization method), the redistribution of the pressure can be determined. For the masking method, however, the redistribution of the pressure cannot be indicated. A mask of the metatarsals will consist of the metatarsal heads as well as the metatarsal bones. The peak pressure of metatarsal regions will be found at the metatarsal heads and will be much larger than the pressure under the metatarsal bones (see Figure 18.7). Therefore, the increase of the pressure under the metatarsal bones cannot be determined by using masks, which cover the whole metatarsal. Although decrease in peak plantar pressure might be most important, the redistribution of plantar pressure will reveal additional information of the effect of footwear and insoles and could influence the person's satisfaction with footwear. In addition to redistribution of plantar pressure, characteristics of the change in the COP could be used as an easier alternative. Although normalization of the COP to a standard foot size is beneficial, describing the mediolateral displacement relative to the midline of the foot could also be used. Nigg et al. (2003) found significant changes in the path of the COP for a full lateral insert. However, the effect of other inserts was not consistent over subjects, indicating that it is difficult to generalize over subjects. They concluded that prescription of inserts and/or orthotics is a difficult task and that methods must be developed to test and assess these effects. So far, there are no methods available that can predict the effect of footwear and insoles on the plantar pressure of an individual. Normalization of plantar pressure will be the first step in this process because plantar pressure patterns of individuals can be compared.

QUESTIONS

18.1 What is the preferable number of steps before the plantar pressure plate?

18.2 What are the optimal specifications of a plantar pressure plate?

18.3 Describe the differences between masking methods and normalization methods.

18.4 Mention parameters that can be derived from plantar pressure measurements.

18.5 What is the most important factor that influences the plantar pressure?

18.6 Describe the average COP line.

18.7 What regions have the largest pressure for an average foot?

18.8 What is the effect of age on plantar pressure pattern?

18.9 What are the largest differences in plantar pressure in a large group of subjects?

18.10 What is the easiest way to indicate the effect of footwear using a plantar pressure plate?

REFERENCES

Armstrong DG, Lavery LA. Elevated peak plantar pressures in patients who have Charcot arthropathy. *J. Bone Joint Surg. Am*. 1998, 80(3): 365–369.

Armstrong DG, Peters EJG, Athanasiou KA, Lavery LA. Is there a critical level of plantar foot pressure to identify patients at risk for neuropathic foot ulceration? *J. Foot Ankle Surg*. Jul 1998, 37(4): 303–307.

Arnold JB, Causby R, Jones S. The impact of increasing body mass on peak and mean plantar pressure in asymptomatic adult subjects during walking. *Diabet. Foot Ankle* 2010, 1: 5518, DOI: 10.3402/dfa.v1i0.5518.

Birtane M, Tuna H. The evaluation of plantar pressure distribution in obese and non-obese adults. *Clin. Biomech*. 2004, 19: 1055–1059 [Crossref].

Boonstra MC, Schwering PJ, De Waal Malefijt MC, Verdonschot N. Sit-to-stand movement as a performance-based measure for patients with total knee arthroplasty. *Phys. Ther*. Feb 2010, 90(2): 149–156. Epub Dec 10, 2009.

Boulton AJ, Hardisty CA, Betts RP, Franks CI, Worth RC, Ward JD, Duckworth T. Dynamic foot pressure and other studies as diagnostic and management aids in diabetic neuropathy. *Diabetes Care* 1983, 6: 26–33.

Bryant A, Singer K, Tinley P. Comparison of the reliability of plantar pressure measurements using the two-step and midgait methods of data collection. *Foot Ankle Int*. 1999, 20: 646–650.

Burnfield JM, Few CD, Mohamed OS, Perry J. The influence of walking speed and footwear on plantar pressures in older adults. *Clin. Biomech. (Bristol, Avon.)* 2004, 19: 78–84.

Burns J, Crosbie J, Hunt A, Ouvrier R. The effect of pes cavus on foot pain and plantar pressure. *Clin. Biomech*. Nov 2005, 20(9): 877–882.

Burns J, Crosbie J, Ouvrier R, Hunt A. Effective orthotic therapy for the painful cavus foot: A randomized controlled trial. *J. Am. Podiatr. Med. Assoc*. 2006, 96: 205–211.

Bus SA, de Lange A. A comparison of the 1-step, 2-step, and 3-step protocols for obtaining barefoot plantar pressure data in the diabetic neuropathic foot. *Clin. Biomech*. 2005, 20(9): 892–899.

Bus SA, Maas M, de Lange A, Michels RP, Levi M. Elevated plantar pressures in neuropathic diabetic patients with claw/hammer toe deformity. *J. Biomech*. 2005, 38: 1918–1925.

Bus SA, Ulbrecht JS, Cavanagh PR. Pressure relief and load redistribution by custom-made insoles in diabetic patients with neuropathy and foot deformity. *Clin. Biomech. (Bristol, Avon)* 2004, 19: 629–638. http://dx.doi.org/10.1016/j.clinbiomech.2004.02.010.

Bus SA, Valk GD, van Deursen RW, Armstrong DG, Caravaggi C, Hlavacek P, Bakker K, Cavanagh PR. The effectiveness of footwear and offloading interventions to prevent and heal foot ulcers and reduce plantar pressure in diabetes: A systematic review. *Diabetes Metab. Res. Rev*. May–Jun 2008, 24(Suppl 1): S162–S180.

Cavanagh PR, Rodgers MM. The arch index: A useful measure from footprints. *J. Biomech*. 1987, 20: 547–551.

Deschamps K, Birch I, Innes JM, Desloovere K, Matricalia GA. Interand intra-observer reliability of masking in plantar pressure measurement analysis. *Gait Posture* 2009, 30(3): 379–382.

Ellis SJ, Stoecklein H, Yu JC, Syrkin G, Hillstrom H, Deland JI. The accuracy of an automask-ing algorithm in plantar pressure measurement. *HSS J.* 2011, 7(1): 57–63 [7 page(s) (article)].

Giacomozzi C, Macellari V, Leardini A, Benedetti MG. Integrated pressure-force-kinematics measuring system for the characterisation of plantar foot loading during locomotion. *Med. Biol. Eng. Comput.* 2000, 38: 156–163.

Gurney JK, Kersting UG, Rosenbaum D. Between-day reliability of repeated plantar pressure distribution measurements in a normal population. *Gait Posture* 2008, 27: 706–709.

Harrison AJ, Folland JP. Investigation of gait protocols for plantar pressure measurement of non-pathological subjects using a dynamic pedobarograph. *Gait Posture* 1997, 6: 50–55.

Hennessy K, Burns J, Penkala S. Reducing plantar pressure in rheumatoid arthritis: A com-parison of running versus off-the-shelf orthopaedic footwear. *Clin. Biomech.* Oct 2007, 22(8): 917–923. Epub Jun 19, 2007.

Hills AP, Hennig EM, McDonald M, Bar-Or O. Plantar pressure differences between obese and nonobese adults: A biomechanical analysis. *Int. J. Obes.* 2001, 25: 1674–1679 [Crossref].

Hodge MC, Bach TM, Carter GM. Novel Award First Prize Paper. Orthotic management of plantar pressure and pain in rheumatoid arthritis. *Clin. Biomech. (Bristol, Avon)* 1999, 14: 567–575.

Huang H, Horemans H, Keijsers N, Jansen R, Stam H, Praet S. Dynamic pedobarography to evaluate gait symmetry after anterior cruciate ligament reconstruction. 2012. Submitted.

Keijsers NLW, Stolwijk NM, Nienhuis B, Duysens J. A new method to normalize plantar pres-sure measurements for foot size and foot progression angle. *J. Biomech.* Jan 5, 2009, 42(1): 87–90. Epub Dec 3, 2008.

Keijsers NLW, Stolwijk NM, Pataky TC. Linear dependence of peak, mean, and pressure-time integral values in plantar pressure images. *Gait Posture* Jan 2010, 31(1): 140–142.

Koenraadt KLM, Stolwijk NM, van den Wildenberg D, Duysens J, Keijsers NLW. Effect of a metatarsal pad on the forefoot during gait. *J. Am. Podiatr. Med. Assoc.* 2012, 102(1): 18–24.

McCrory JL, Young MH, Boulton AJM, Cavanagh R. Arch index as a predictor of arch height. *The Foot* 1997, 7: 79–81.

McPoil TG, Cornwall MW, Dupuis L, Cornwell M. Variability of plantar pressure data. A compar-ison of the two-step and midgait methods. *J. Am. Podiatr. Med. Assoc.* 1999, 89: 495–501.

Mueller MJ, Hastings M, Commean PK, Smith KE, Pilgram TK, Robertson D et al. Forefoot structural predictors of plantar pressure during walking in people with diabetes and peripheral neuropathy. *J. Biomech.* 2003, 36: 1009–1017.

Mueller MJ, Stube MJ, Allen BT. Therapeutic footwear can reduce plantar pressures in patients with diabetes and transmetatarsal amputation. *Diabetes Care* Apr 1997, 20(4): 637–641.

Nigg BM, Stergiou P, Cole G, Stefanyshyn D, Münderman A, Humble N. Effect of shoe inserts on kinematics, center of pressure, and leg joint moments during running. *Med. Sci. Sports Exerc.* 2003, 35(2): 314–319.

Novak P, Burger H, Tomsic M, Marincek CRT, Vidmar GAJ. Influence of foot orthoses on plantar pressures, foot pain and walking ability of rheumatoid arthritis patients—A randomised controlled study. *Disabil. Rehabil.* 2009, 31(8): 638–645.

Otter SJ, Bowen CJ, Young AK. Forefoot plantar pressures in rheumatoid arthritis. *J. Am. Podiatr. Med. Assoc.* 2004, 94(3): 255–260.

Pataky TC, Bosch K, Mu T, Keijsers NLW, Segers V, Rosenbaum D, Goulermas JY. An ana-tomically unbiased foot template for inter-subject plantar pressure evaluation. *Gait Posture* Mar 2011a, 33(3): 418–242. Epub Jan 11, 2011.

Pataky TC, Caravaggi P, Savage R, Parker D, Goulermas JY, Sellers WI, Crompton RH. New insights into the plantar pressure correlates of walking speed using pedobarographic statistical parametric mapping (pSPM). *J. Biomech.* 2008a, 41(9): 1987–1994. Epub May 22, 2008.

Pataky TC, Goulermas JY. Pedobarographic statistical parametric mapping (pSPM): A pixel-level approach to foot pressure image analysis. *J. Biomech.* Jul 19, 2008b, 41(10): 2136–2143. Epub Jun 13, 2008.

Pataky TC, Keijsers NLW, Goulermas JY, Crumptton RH. Nonlinear spatial warping for between-subjects pedobarographic image registration. *Gait Posture* Apr 2009, 29(3): 477–482. Epub Dec 27, 2008.

Pataky TC, Mu T, Bosch K, Rosenbaum D, Goulermas JY. Gait recognition: Highly unique dynamic plantar pressure patterns among 104 individuals. *J. R. Soc. Interface* April 2011, 9(69): 790–800.

Perry J, Burnfield JM. *Gait Analysis, Normal and Pathological Function*, 2nd edn. Thorofare, NJ: Charles B. Slack, 2010.

Perry JE, Ulbrecht JS, Derr JA, Cavanagh PR. The use of running shoes to reduce plantar pressures in patients who have diabetes. *J. Bone Joint Surg. Am.* Dec 1995, 77(12): 1819–1828.

Peters EJ, Urukalo A, Fleischli JG, Lavery LA. Reproducibility of gait analysis variables: One-step versus three-step method of data acquisition. *J. Foot Ankle Surg.* 2002, 41: 206–212.

Postema K, Burm PE, Zande ME, Limbeek J. Primary metatarsalgia: The influence of a custom moulded insole and a rockerbar on plantar pressure. *Prosthet. Orthot. Int.* Apr 1998, 22(1): 35–44.

Praet SF, Louwerens JW. The influence of shoe design on plantar pressures in neuropathic feet. *Diabetes Care* Feb 2003, 26(2): 441–445.

Rosenbaum D, Hautmann S, Gold M, Claes L. Effects of walking speed on plantar pressure patterns and hindfoot angular motion. *Gait Posture* September 1994, 2(3): 191–197.

Schmeigel A, Vieth V, Gaubitz M, Rosenbaum D. Pedography and radiographic imaging for the detection o foot deformities in rheumatoid arthritis. *Clin. Biomech.* 2008, 23: 648–652.

Shorten MR, Beekman-Eden K, Himmelsbach JA. Plantar pressure during barefoot walking. #121 in *Proc. XII International Congress of Biomechanics* (Eds. R.J. Gregor, R.F. Zernicke, and W.C. Whiting). Los Angeles, CA: University of California, 1989.

Stebbins JA, Harrington ME, Giacomozzi C, Thompson N, Zavatsky A, Theologis TN. Assessment of subdivision of plantar pressure measurement in children. *Gait Posture* 2005, 22: 372–376.

Stolwijk NM, Duysens J, Louwerens JWL, Keijsers NLW. Plantar pressure changes after long-distance walking. *Med. Sci. Sports Exerc.* Dec 2010, 42(12): 2264–2272.

Taylor AJ, Menz HB, Keenan AM. The influence of walking speeds on plantar pressure measurements using the two-step gait initiation protocol. *The Foot* 2004, 14: 49–55.

Tsung BY, Zhang M, Mak AF, Wong MW. Effectiveness of insoles on plantar pressure redistribution. *J. Rehabil. Res. Dev.* 2004, 41: 767–774. http://dx.doi.org/10.1682/JRRD.2003.09.0139.

Tuna H, Birtane M, Tastekin N, Kokino S. Pedobarography and its relation to radiologic erosion scores in rheumatoid arthritis. *Rheumatol. Int.* 2005, 26: 42–47.

Vela SA, Lavery LA, Armstrong DG, Anaim AA. The effect of increased weight on peak pressures: Implications for obesity and diabetic foot pathology. *J. Foot Ankle Surg.* 1998, 37: 416–420 [Crossref].

Veves A, Murray HJ, Young MJ, Boulton AJ. The risk of foot ulceration in diabetic patients with high foot pressure: A prospective study. *Diabetologia* 1992, 35: 660–663.

Waldecker, U. Metatarsalgia in hallux valgus deformity: A pedographic analysis. *J. Foot Ankle Surg.* 2002, 41: 300–308.

Zammit GV, Menz HB, Munteanu SE. Reliability of the TekScan MatScan(R) system for the measurement of plantar forces and pressures during barefoot level walking in healthy adults. *J. Foot Ankle Res.* Jun 18, 2010, 3: 11.

19 Virtual Shoe Test Bed

Sandra Alemany, José Olaso, Sergio Puigcerver, and Juan Carlos González

CONTENTS

19.1 INTRODUCTION

Functional and comfort properties are more and more a decisive value for customers and many footwear companies perform test batteries to analyze aspects like fitting, shock absorption, cushioning, flexibility, torsion, friction, perspiration, thermal isolation, or global comfort. Despite the lack of standards in this area, research organizations have developed their own machine and user tests to assess these footwear functional aspects.

However, the speed in the development of new product collections and the wide variety of models make very expensive and almost unfeasible for companies the evaluation of functional and comfort attributes of footwear. The development of new tools for the virtual simulation of shoe performance would eliminate the need to manufacture and test physical prototypes, ensuring footwear comfort and functionality while reducing the time and costs to create new collections.

Although in the last years computer-aided engineering (CAE) software in high technological industrial sectors has been developed very fast, there are no commercial applications addressed to the footwear industry to predict functional performance of shoes, and the integration of CAE tools in the design department of footwear companies is very low. In this context, many investigations in the research and scientific fields have been done to generate knowledge supporting this approach.

In this chapter, current and emerging technologies developed for the virtual evaluation of footwear comfort and performance are presented, followed by a review of the main advances in the scientific field of foot and footwear modeling, generating the knowledge basis for the development of such technologies.

19.2 SIMULATION TECHNOLOGIES IN FOOTWEAR DESIGN

The development of footwear design tools has centered its efforts during the last decades on design and manufacturing applications. Those developments have taken place under the computer-assisted design (CAD) approach, and different types of CAD commercial systems depending on the footwear component focused (e.g., sole, upper) exist in the market (Table 19.1).

Benefits provided by these CAD tools are wide and can be characterized by the type of support they give to the designer: three-dimensional (3D) design, two-dimensional (2D) patterning from 3D design, scaling to different sizes, and others.

Some of these CAD tools for footwear design are able to provide a more or less realistic visual simulation of the designed footwear. This approach to the final aspect of the product is worthy from the aesthetical point of view, for example, to build virtual catalogues. However, a lot of work is necessary to achieve a realistically plausible virtual prototype, so its benefit is not too much in terms of time costs.

Concerning virtual testing of footwear, some commercial systems provide the ability to predict costs based on material consumption, through machining simulations in CAD environment.

Based on the CAE technology, finite elements analysis (FEA) has become very popular in the last years due to its versatility and accuracy. With this technology,

TABLE 19.1
Some CAD Commercial Systems Based on Footwear Design

CAD Tool	Company	Link
LastElf; ImagineElf	Digital Evolution System	www.desystem.com.tw
RhinoShoe	TDM Solutions	www.rhino3d-design.com/rhinoshoe.htm
RomansCAD	Lectra software	www.lectra.com
Shoemaker	Delcam software	www.footwear-cadcam.com
Shoemaster	CSM 3D	www.shoemaster.co.uk
Forma 3D, Sipeco, etc.	INESCOP	www.inescop.es

both 2D and 3D models of foot, sole, insole, and upper have been developed in the scientific and research fields (see Section 19.3.2) and has been also applied in the industry (Tak-Man et al. 2009) to reduce the time and cost of the footwear development process. However, FEA has mainly been used in a scientific environment, entering only in footwear industry in certain companies, prosperous enough to support its costs. The fact is that FEA is a complex and expensive technology, not affordable by the great majority of companies.

In the last years, a new approach merging the CAE approach with a new technological frame (CAT: computer-aided test) has been presented in footwear industry, developed under the scope of a European project CECMadeShoes* (Azariadis et al. 2007, Olaso 2010). The objective of this approach was to develop a proof of concept of a new virtual shoe test bed aiming to achieve a trade-off between a technically and economically affordable (by the majority of companies) technology, and a reliable and precise tool providing worthy information to achieve the targeted quality in footwear design.

We have developed a virtual shoe test bed considering different levels of user-footwear functional interaction (fitting, plantar pressures, shock absorption, flexibility, friction, and thermoregulatory aspects) and perceptual interaction (comfort) (Figure 19.1).

Under that approach, functional performance of each footwear component (sole, insole, upper, and last) is characterized through machine tests and conforming a database of footwear components ready for their selection during the design process. These test results feed a model of the footwear components interaction, which is built to simulate complete footwear functional performance. Finally, the result of the virtual analysis for the complete footwear functional performance is linked to users' comfort perception though a regression model. In that sense, the designer receives worthy info of the footwear design at two levels: first, a functional performance characterization of different aspects of the user–footwear biomechanical interaction (fitting, pressure distribution, shock absorption) and, second, a prediction of the users' perception comfort when wearing such products.

19.3   BIOMECHANICAL MODELING

Biomechanical simulation of the foot–footwear interaction is necessary when aiming to develop a virtual test bed. In fact, this is a field where researchers have been focused on since long time ago. Ground reaction forces, plantar pressure distribution pattern, foot movement, and internal stress and strain of different foot and footwear elements are some of the variables that can be predicted by simulation methods. First studies were based on very simplified rheological or multibody models, but after years of research the complexity of the models increased, since the biomechanical interaction of the human foot–footwear was better understood and the computational efficiency was highly increased. FEA methods have also been used for biomechanical simulation, being currently the most common ones.

---

* CEC-Made-Shoe Integrated Project funded by the European Commission—6 FP Priority IST—NMP (Manufacturing, Products and Service Engineering 2010) Contract No. 507378.

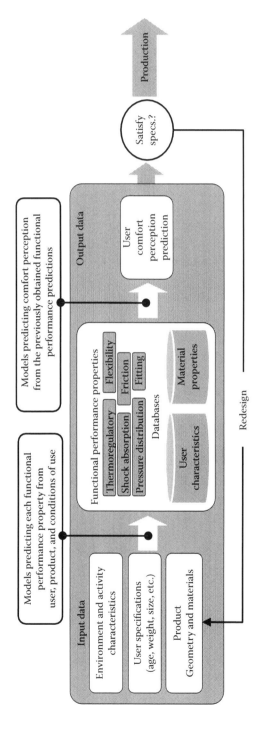

FIGURE 19.1 Conceptual layout of the virtual shoe test bed tool for footwear designers developed in the CECMadeShoes European project.

Biomechanical modeling can be applied with several purposes, such as evaluating the functionality and biomechanical behavior of different foot elements during gait or running, predicting foot lesions on sports, or optimizing the geometry and materials of any footwear component. A review of different biomechanical simulations between foot–footwear–ground using rheological models and FEA methods are included in this section.

19.3.1 Rheological Models

Rheological models have been used for years by many authors for biomechanical modeling obtaining good results, especially for simulations in which relatively big simplifications can be assumed. Rheological models are based on a network of springs and dampers, connected among them either in parallel or in series, being computationally fast and easy to pre- and postprocess (Azariadis 2007). They have been widely used for material modeling, which is one of the key points for obtaining proper results when doing CAE simulations.

Many components and elements of very diverse characteristics have an influence on the interaction between foot, footwear, and ground. The foot is mainly composed of rigid bones, soft tissues, ligaments, tendons, and muscles; footwear is normally made of materials like rubber, ethylene-vinyl acetate (EVA), polyurethane (PU), or latex; and the ground can be made up of many different materials such as asphalt, soil, ceramic tiles, grass, rubber, stones, sand, etc. Most of all these materials show a viscoelastic or hyperelastic nonlinear behavior, needing a proper constitutive model in order to simulate the foot–footwear–ground interaction with a proper accuracy.

Research works in the literature including rheological models present differences on number of springs and dampers, the connexion types (in parallel or in series), whether the models are linear or nonlinear, and the values of the model parameters. Flügge (1967) summarized the differential equations of several basic linear rheological models (Table 19.2).

These models have been applied for several topics such as heel pad modeling, ground reaction forces estimation, and the material modeling of shoes and grounds. Following sections summarize some of the most relevant works.

19.3.1.1 Heel Pad Modeling

Heel pad is a fat tissue placed under the calcaneus, which acts as a cushion for the heel. It is especially relevant for attenuating impacts produced during walking or running. Gefen et al. (2001) used a modified Kelvin–Voight model in order to describe the in vivo stress–strain behavior of the human heel pad during gait. The model contained a linearly elastic spring in parallel to a nonlinear damper. Models that are much more complex are available in the literature, such as the one published by Scott and Winter (1993), who modeled the heel pad by means of a nonlinear spring and a nonlinear damper in parallel, with the damper being dependent upon both strain and strain rate. Damping coefficients were different for the compression and decompression phases. They also used the same model for simulating the soft tissue placed under the five heads of the metatarsal bones as well as the one under the hallux.

TABLE 19.2
Differential Equations of Some Basic Linear Rheological Models

Model	Name	Differential Equation
	Elastic solid	$\sigma = q_0 \cdot \varepsilon$
	Maxwell fluid	$\sigma + p_1 \cdot \dot{\sigma} = q_1 \cdot \dot{\varepsilon}$
	Kelvin solid (or Kelvin–Voight)	$\sigma = q_0 \cdot \varepsilon + q_1 \cdot \dot{\varepsilon}$
	3-Parameter solid	$\sigma + p_1 \cdot \dot{\sigma} = q_0 \cdot \varepsilon + q_1 \cdot \dot{\varepsilon}$ $(q_1 > p_1 \cdot q_0)$
	4-Parameter solid	$\sigma + p_1 \cdot \dot{\sigma} = q_0 \cdot \varepsilon + q_1 \cdot \dot{\varepsilon} + q_2 \cdot \ddot{\varepsilon}$ $(q_1 > p_1 \cdot q_0, \quad q_1^2 > 4 \cdot q_0 \cdot q_2,$ $q_1 \cdot p_1 > q_0 \cdot p_1^2 + q_2)$

Source: Flügge, W., *Viscoelasticity*, Blaisdell Publishing Company, Waltham, MA, pp. 16–17, 1967.

σ and $\dot{\sigma}$ are the stress and stress ratios, respectively; ε, $\dot{\varepsilon}$, and $\ddot{\varepsilon}$ are the strain ratios; p_1 define the stress ratio coefficient; and q_0, q_1, and q_2 define the strain and strain ratios.

The following equation was used by Scott and Winter (1993):

$$F_p = j \cdot \varepsilon^k + l \cdot \varepsilon^m \cdot |\dot{\varepsilon}|^n \tag{19.1}$$

where
  F_p is the force transferred by the tissue
  ε is the strain
  $\dot{\varepsilon}$ is the strain rate
  j and k define the spring characteristics
  l, m, and n define the damper ones

D'Andrea et al. (1997) also generated a complex rheological model for predicting the behavior of the heel pad. The model was based on 6 spherical balls, connected among them by 15 arms (8 of them were modeled as simple springs and the resting ones as springs and dampers in parallel). According to D'Andrea et al., this model could be used to describe the effects of some diseases, which have a detrimental effect on the soft tissue of the heel.

19.3.1.2   Ground Reaction Force Modeling

One of the simplest analyses was done by Alexander (1988) who used some spring–damper–mass systems in order to replicate ground reaction forces in human locomotion. His model was based on a mass and a spring, which was systematically improved by adding new rheological elements: an additional smaller mass on the bottom, a second spring, and a dashpot in parallel to the second spring. Similar models were defined by McMahon and Cheng (1990), who used a mass–spring model for simulating a leg during hopping, running, or walking. Walker and Blair (2002) modified the McMahon and Cheng model incorporating a damping effect during running, using a spring and a dashpot in parallel. Kim et al. (1994) simulated the heel strike transients during running using a more complex model based on three masses, three springs, and three dashpots.

Although most of these models used for better understanding the dynamics during running, they were based on rigid bodies without taking into account the wobbling masses, such as muscles, ligaments, tendons, and other soft tissues. Nigg and Liu (1999) proposed a more complex model for simulating the ground reaction forces during human locomotion, especially during running, taking into account both rigid and wobbling masses. The model included five springs, three dashpots, and four masses for representing the human body. This system is supposed to reconstruct almost any vertical ground reaction force that can be measured during normal heel–toe running. This model was later improved by Zadpoor et al. (2007).

Recently, Nikooyan and Zadpoor (2011) have modified this model taking into account the muscle activity. The stiffness and damping properties of the lower body wobbling mass is controlled by a mechanism that mimics the functionality of the central nervous system.

19.3.1.3   Shoe Materials and Ground Surface Modeling

Since shoe materials, especially insole and midsole, and ground play a very important role for shock absorption and overpressures cushioning, many authors have developed models to simulate their behavior during walking and running activities.

Bretz (2000) modeled the ground as a rigid body connected by a spring and dashpot in parallel. This model simulated the effects of deck surface on the initial impact impulse forces in human gait. For this purpose, Ly et al. (2010) extended the model proposed by Nigg and Liu (1999) to explicitly define both ground and shoe-soles material properties. They replaced the original nonlinear element (referred to both shoe and ground surface) with a bilayered spring–damper–mass model: the first layer represented the shoe midsole (assuming a viscoelastic material) and the second one the ground (using a specific elastic model).

A more sophisticated model was proposed by Olaso (2010) for predicting the nonlinear viscoelastic behavior of insole and sole materials available in the current market. He developed a set of algorithms for fitting the model parameters to a previously tested material. After the optimal material parameters were determined, the model was used for predicting the material behavior depending on the applied load.

The following equation was used by Olaso (2010):

$$\sigma = a + b \cdot \varepsilon^c + d \cdot \varepsilon^e \cdot \dot{\varepsilon}^f \cdot \left| \dot{\varepsilon} \right|^g - h \cdot \dot{\sigma} \tag{19.2}$$

where
   σ and $\dot{\sigma}$ are the stress and stress ratio
   ε and $\dot{\varepsilon}$ are the strain and strain ratio
   a, b, c, d, e, f, g, and h define the model characteristics

19.3.2 FINITE ELEMENT MODELS

Among different CAE simulation techniques, the finite element (FE) method is becoming more and more popular because of its versatility and accuracy in modeling irregular geometrical structures, complex material properties, and complicated loading and boundary conditions in both static and dynamic simulations (Cheung et al. 2009). Although several aspects related to biomechanical modeling could still be highly improved, a lot of work and advances have already been done during the last years. There are several aspects that vary from one work to another, like objectives and simulation strategies, generating very different results. Aspects such as foot parts and footwear components to simulate, number, type and size of elements, material properties (linear or nonlinear, elastic, hyperelastic, or viscoelastic behavior), type of analysis (2D or 3D, static or dynamic), and load and boundary conditions must be defined taking into account the objectives of the simulation, the computational cost, and the desired accuracy. An overview of different approaches that are possible when simulating the biomechanical interaction between foot–footwear–ground using FEA methods are presented in the following.

19.3.2.1 Foot Structural Modelization

The foot is one of the most complex mechanical structures of the body, made up of bones, joints, muscles, ligaments, as well as other tissues such as the plantar fascia, fat pads, and skin. All these elements are connected among them generating a complex movement, load support, and contact pressure distribution. Several FE models have been performed using different foot geometries, components, simplifications, and approximations.

Early models were based on isolated parts of the foot, such as Lemmon et al. (1997) and Nakamura et al. (1981). With the improvement of computational performance, more realistic models were performed considering detailed geometries, better material properties definition, inclusion of more foot elements, and more realistic load and boundary conditions (Actis et al. 2006, Chen et al. 2001, Gefen 2002a,b, 2003). Currently, works describe the human foot in a quite realistic way, such as Cheung and Zhang (2005, 2008), who used an ankle–foot model considering 28 bony structures, 72 ligaments, and the plantar fascia embedded in a volume of encapsulated soft tissue.

19.3.2.2 Biomechanics of Foot Elements

FEA simulations have been done aiming to research the functionality of different foot elements during specific activities and conditions. One of the aspects most widely studied is the plantar pressures over the foot plant, for example, Cheung et al. (2004) studied the plantar foot pressures, the contact area, and the biomechanical interaction among the bony and ligamentous structures for different stages of diabetic neuropathy by varying the plantar fascia stiffness. Actis et al. (2006) also focused his work on determining the effect on the forefoot plantar pressures during push off when varying the cartilage's properties and the bone, fascia, and flexor tendon Young's modulus. Stress distribution in the foot has also been predicted during barefoot gait (Chen et al. 2001) or for studying the role of the skin to conform the fat pad. More recently, Spears et al. (2007) compared the compression stress distribution of fat pad, rear foot plantar pressures, and skin tension on two heel models (unified bulk soft tissue and fat pad and skin as two different elements).

More specific biomechanics effects have been analyzed by means of FEA models, such as the influence of heel elevation on female foot, comparing a flat versus a high-heeled support during static standing (Yu et al. 2008) or the effect of muscular fatigue induced by intensive military or athletic marching on the structural stability of the foot and on its internal stress state during heel-strike and push off (Gefen 2002b).

19.3.2.3 Interaction between Foot and Footwear

The first approaches to simulate interaction between foot and footwear applied the resultant load on the upper surface of the footwear, avoiding modeling both footwear and foot geometries. Loads can be applied by means of vertical forces (Lewis 2003) or by means of normal pressure distribution pattern on the upper surface of the insole (Figure 19.2) (Alemany et al. 2003, Barani et al. 2005).

Next works in the field included foot geometry in the simulations between foot and footwear as two independent deformable interacting bodies. In fact, most of the

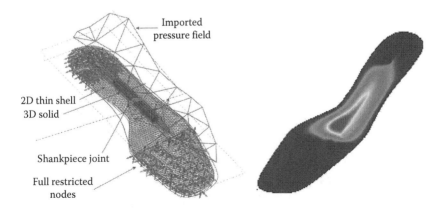

FIGURE 19.2 FEA model of the shankpiece place on a female heeled footwear and deformation pattern. (From Alemany, S. et al., Integration of plantar pressure measurements in a finite element model for the optimization of shankpiece design for high-heeled shoes, in *Proceedings of the 6th Symposium on Footwear Biomechanics*, Queenstown, New Zealand, 2003.)

works aim to study the effect of varying material properties of insoles or midsoles, as well as upper surface insole's geometry on the biomechanical behavior of the foot (Budhabhatti et al. 2007, Erdemir et al. 2005, Even-Tzur et al. 2006, Goske et al. 2006, Hsu et al. 2008).

Other footwear components different than the sole or insole have also been modeled in order to either study the biomechanical effect on the foot or optimize its geometrical or material design. Heel counters have been analyzed in some works (Ankrah and Mills 20004, Spears et al. 2007), as well as upper material (Cheung et al. 2007), shankpiece (Alemany et al. 2003), and outsole slip-resistant tread geometries (Sun et al. 2005), which have also been simulated using FEA techniques.

Despite FEA models of the foot and footwear components have been developed for different purposes, a complete FEA model of both foot and footwear is not available for virtual testing, and neither there is an industrial application based on this method.

Biomechanical modeling is a key point for obtaining a reliable virtual shoe test bed. Different techniques can be applied for predicting the interaction between foot and footwear, such as rheological models or FEA. Although big scientific progresses have been made on this topic during the last years, there are still several aspects that can be improved in order to generate more realistic models and to increase accuracy on the simulation of the biomechanical foot–footwear interaction.

19.4   MODELING FOOTWEAR FITTING

Fitting is one of the main factors for footwear comfort. It is a basic quality requirement taken for granted when fulfilled but resulting in dissatisfaction when it is not adequate.

Shoe fitting is modified actuating on the design of the shoe last. This process is still done through a trial and error process, applying modifications in the shoe-last shape, building shoe prototypes, and performing fitting trials. Being able to define and control footwear fitting in a virtual environment would enable companies to improve fitting reducing costs. However, despite the recent evolvement of CAD tools for shoe-last design, nowadays there are no commercial systems integrating fitting assessment functionalities (i.e., virtual testing capabilities).

With this open challenge, several investigations are focused on the development of new models to evaluate and predict fitting.

19.4.1   MODELING FOOTWEAR FITTING THROUGH A GEOMETRIC APPROACH

First studies to understand shoe fitting were focused on the optimization of the sizing system. In such studies, length and width at the flex line were the key anthropometric parameters of the foot used to classify feet into sizes using bivariate normal distributions (Cheng and Perng, 1999, Haber and Haber 1997).

In the last decade, the fast development of commercial shoe scanners has promoted new business models based on footwear selection and footwear customization. This trend boosted the development of foot anthropometrical studies and their implication in the geometry of the last. In this line, a fitting model for footwear customization was proposed by Alemany et al. (2003), considering as key sections of the last

TABLE 19.3

Optimal Ratio (Dimension of the Last/Dimension of the Foot) and Tolerances to Obtain a Comfortable Customized Last

Dimensión	Optimal Ratio for Woman Footwear	Optimal Ratio for Man Footwear
Total length	1.025 ± 0.0021	1.037 ± 0.0042
Ball girth	0.986 ± 0.0104	1.008 ± 0.0104
Ball width	0.917 ± 0.0125	0.942 ± 0.0104
Toe width	0.697 ± 0.0166	0.610 ± 0.0166
Toe height	2.450 ± 0.0707	1.100 ± 0.0354
Heel width	0.983 ± 0.0166	1.011 ± 0.0146

Source: Obtained from Alemany, S., et al., A novel approach to define customized functional design solution from user information, 3rd Interdisciplinary World Congress on Mass Customization and Personalization, Hong Kong, 2005.

Shoe-last dimension = coefficient · foot dimension.

related to footwear fitting the heel, ball girth, instep, and toes. This model relates anthropometric dimensions of the foot with dimensions of the shoe last providing a ratio coefficient. The optimal point and tolerance (Table 19.3) were fixed performing fitting trials with footwear manufactured with controlled lasts.

Based on this geometric approach, a virtual shoe test bed was developed by Azariadis et al. (2007) to assess the fitting of the last in a CAD environment. A set of shoe-last measurements are compared with corresponding measures of the foot obtained from an anthropometric database representing the target population (Figure 19.3). This is the first trial enabling the fitting assessment with CAD tools. The customized design of shoe lasts using CAD tools was proposed by Luximon (2009) using a method to build customized sections based on the German system AKA64-WMS. This new design software enables shoe-last designers to create shoe-last of different toe styles, heel heights and heel bottom styles quickly and accurately.

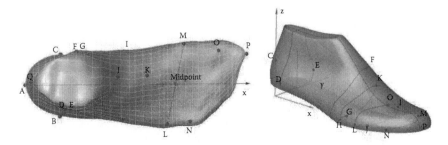

FIGURE 19.3 Key sections and anatomical points in the foot and their corresponding measures in the last. (From Azariadis, P. et al., *Comput.-Aided Des. Appl.*, 4(6), 741, 2007.)

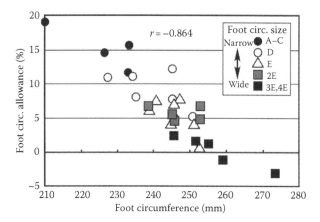

FIGURE 19.4 Correlation between the foot circumference allowance in the metatarsal area of the best fitting shoe size selected by the user (circumference of last–circumference of foot)/ circumference of foot [%]) and the foot circumference of the user in the same section for different footwear widths (A–C, D, E, 2E, 3E, 4E). (From Kouchi, M. et al., Morphological fit of running shoes: Perception and physical measurements, in *Proceedings of the 7th Symposium on Footwear Biomechanics*, Cleveland, OH, pp. 38–39, 2005.)

The process to select appropriate size for a specific foot was studied by Kouchi et al. (2005) for running footwear. The result of this study (Figure 19.4) shows a high correlation between the foot circumference allowance in the metatarsal area of the best fitting shoe size selected by the user (circumference of last– circumference of foot)/circumference of foot [%]) and the foot circumference of the user in the same section. The fact that runners with wider feet prefer tighter shoes strongly suggests that the perception of fit or comfort is significantly affected by the fitting past experience: since the availability of the widest sizes F and C is limited, runners with wider feet used to wear tighter shoes, and they are accustomed to wearing tight shoes.

A different approach was proposed by Wang (2010) whose main purpose was to design a process to analyze and evaluate the most suitable shoe last in a database for a specific 3D feet considering the three main girths of shoe last (instep, waist, and ball). Fuzzy theory was used to build a membership function to relate foot and last girths considering an optimal relationship with a linear progression from this point to model tight fitting with a positive slope and loose fitting with a negative. As the implication of each section of the last in the global footwear fitting is not the same, an analytical hierarchy process (AHP) was applied to find the weighting factors for each girth to determine the fitness. Despite these studies using 3D foot models, fitting assessment is based on linear dimensions (girths and widths).

A 3D approach for fit modeling was proposed by Luximon et al. (2001) who developed a process to obtain the 3D differences between foot and last as a way to quantify fit (Figure 19.5). The resulting 3D map of geometric differences between a foot and their proper last integrates several factors such as the requirements for a proper dynamic adaptation, the upper effect, and the user perception. The method can be further enhanced through perception studies to generate guidelines for the clearances between feet and shoes.

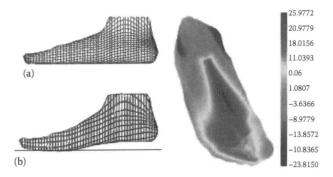

FIGURE 19.5 3D differences between foot and last. (a) 3D scanned foot and (b) foot with heel height and toe spring adjustment. (From Luximon, A. et al., A fit metric for footwear customization, in *Proceedings of the World Congress on Mass Customization and Personalization* (CD-ROM), Hong Kong, October 1–2, 2001. With permission.)

Mochimaru (Mochimaru et al., 2000 and Mochimaru and Kouchi, 2003) presented a novel methodology using the free form deformation (FFD) technique. The FFD method is a way to deform the shapes of an object smoothly by moving control lattice points set around the object. The reference body form is automatically deformed to coincide with the other body forms. The dissimilarity is defined by summation of control point movements, which is a type of morphological distance. Two applications of FFD on footwear fitting have been published by this author, one related to the improvement of the shoe-last grading methods and the other addresses footwear customization (Figure 19.6).

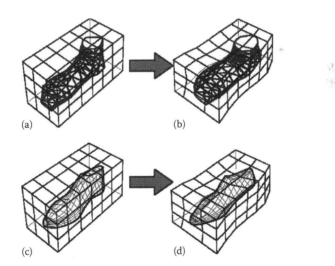

FIGURE 19.6 Application of FFD method to customize a shoe last for a specific foot: (a) normal foot form, (b) different foot form, (c) last shape for normal foot, and (d) last shape for different foot. (From Mochimaru, M. and Kouchi, M., Last customization from an individual foot form and design dimensions, in *6th ISB Footwear Biomechanics*, Queenstown, New Zealand, 2003, pp. 62–63. With permission.)

The use of the 3D foot data to assess the fitting of a shoe last presents some difficulties, which have been the subject of the earliest research and hence require further investigations:

- The geometric criteria to assess fitting interaction dramatically depend on the alignment of foot and last shapes.
- The foot is scanned on a flat position while the shoe last usually has a certain heel height; therefore, a geometric comparison of both volumes requires a shape treatment which influences fitting analysis.
- The specific geometric parameter to choose to represent foot–shoe interaction will influence fitting characterization.

19.4.2 FITTING ANALYSIS THROUGH QUANTIFICATION OF DYNAMIC PRESSURES OVER THE FOOT DORSUM

Footwear fitting is a result of a dynamic mechanical interaction and apart from the geometry, the mechanical properties of the upper material should be considered.

Current investigations considering the upper material in shoe fitting use pressure sensors to quantify fitting in walking conditions. First studies performed using this technology (Jordan and Barlett 1995, Jordan et al. 1997) found significant differences on dorsal pressure between footwear types grouped into comfortable and uncomfortable.

The effect of the shoe-last geometry was studied by Olaso et al. (2007) using a similar foot dorsal pressure system. Three boots changing only the width of the ball section of the last was tested in five different walking conditions. Significant differences were obtained among the boots for the anatomical points situated on the metatarsophalangeal joint sides (ap1, ap2, ap4) (Figures 19.7 and 18.8). In general, the maximum peak pressure was significantly greater ($p < 0.05$) with the narrower boot (boot 1) compared to the widest boot (boot 3).

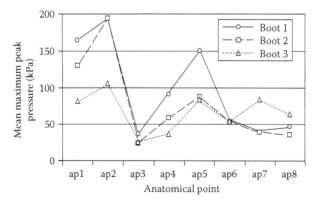

FIGURE 19.7 Mean maximum peak pressures obtained in each anatomical point for the three boots. (From Olaso, J. et al., Study of the influence of fitting and walking condition in foot dorsal pressure, in E.C. Frederick and S.W. Yang (Eds.), *Proceedings of Eighth Footwear Biomechanics Symposium*, Taipei, Taiwan, 2007, pp. 41–42.)

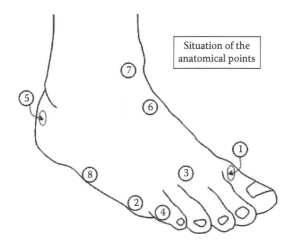

FIGURE 19.8 Situation of the anatomical points. 1, external face of the first metatarsal head; 2, external face of the fifth metatarsal head; 3, third metatarsal head; 4, external face of the proximal phalange head; 5, most prominent point of the heel near the Achilles insertion; 6, most prominent point of the instep; 7, the distal point of the talotibial articulation; 8, apophasis of the fifth metatarsus. (From Olaso, J. et al., Study of the influence of fitting and walking condition in foot dorsal pressure, in E.C. Frederick and S.W. Yang (Eds.), *Proceedings of Eighth Footwear Biomechanics Symposium*, Taipei, Taiwan, 2007, pp. 41–42.)

The link between pressures on the foot dorsum and subjective fitting perception was analyzed by Cheng and Hong (2010) in a recent study of running shoes. Stepwise multiple regressions were used to find out a model ($R^2 = 0.594$) relating the pressure of the metatarsal fibulare (MF), pressure of the medial calcaneous (Mcal), and 3D differences—heel breath, ball girth, and foot length (see Table 19.4).

Although these studies confirm the contribution of the upper material to the shoe fitting, modeling footwear fitting from a mechanical point of view involve several difficulties. Mechanical properties of the upper leather and foot tissues are very complex, since both present nonlinear with a viscoelastic component behavior. Furthermore, the 3D shape of the foot varies during walking, and foot–shoe interaction is dynamic with a contact model interface. A study of the viability of obtaining a generic animation of the foot while walking for the virtual testing of footwear using dorsal pressures was carried out by Rupérez et al. (2009, 2010). The aim of this work is to develop a 3D virtual model,

---

TABLE 19.4

Best Fitting Model Provided by Cheng and Hong

Subjective overall fit = 22.811 + 0.433 (PMF) + 2.62 (DDHb) + 0.680 (DDBg – 2.962 (PMcal) – 1.286 (DFl)

Source: Cheng, Y.L. and Hong, Y., *Footwear Sci.*, 2(3), 149, September 2010.

PMF, Metatarsal fibulare; PMcal, Pressure of the medial calcaneous; DDHb, Dimensional differences in the heel breath; DDBg, Dimensional differences in the ball girth; DFl, Dimensional differences in the foot length.

which simulates how the upper material is deformed when contacting foot surface adapting a FE model for the shoe upper and a set of foot meshes to build a complete step.

Despite these advances, there are still open challenges in order to obtain a complete understanding of the shoe fitting enabling the commercial integration of a virtual fitting test in a CAD environment:

- Characterization of fitting properties of the upper materials
- Interrelationship of geometric and nongeometric variables:

Shoe shape + last shape + upper material properties = pressure distribution

- Deformation of the foot during walking (4D anthropometry)
- Consideration of the dynamic nature of foot–shoe interaction in the analysis of fitting perception

19.5 FOOTWEAR THERMOPHYSIOLOGICAL MODELING

Thermal comfort is a feature increasingly demanded for footwear users. However, the design or selection of the optimal materials for a shoe addressed to a specific activity and climate is an unresolved problem. This usually leads to comfort problems due to poor isolating characteristics, poor ventilation and, most importantly, sweating accumulation on footwear and socks. The advances and application of virtual footwear thermal testing could lead to the improvement of the thermal performance of footwear designed according to the guidelines obtained.

A huge number of studies have been developed for modeling the foot–footwear biomechanics; however, only a few specific research works have addressed the thermal modeling of foot–footwear–environment interaction. Moreover, typically, thermal models involving the foot/footwear come from whole body/clothing models.

19.5.1 THERMOREGULATORY MODELS OF THE FOOT

There have been many attempts to model human thermoregulation systems to understand the physiological behavior under cold or warm conditions in different fields such as workplace, climatic control, or clothing evaluation and design. Human thermoregulatory models are usually divided into two systems (Fiala et al. 2001, Lotens 1993, Parsons 2003, Stolwijk and Hardy 1977): (1) A passive system incorporating thermal and geometrical properties of body tissues (taking the form of basic shapes such as cylinders and spheres) and simulating the heat transfer phenomena within and in the body surface; recently, more realistic simulations have been carried out with FEA of the whole body in 3D (McGuffin et al. 2002) and (2) an active control system simulated with cybernetic models that predict the thermoregulatory response (vasoconstriction/vasodilatation, sweating, and shivering) from the thermal state of the whole body (core temperature, skin temperature, heat flow, etc.). These models have been combined with clothing models (Li et al. 2004, Lotens 1993).

Whole body thermal models typically define the foot as a body segment (Fiala et al. 2001, Stolwijk and Hardy 1977). However, their definition is insufficient and they are not validated for footwear analysis. Lotens (1989) and Lotens et al. (1989)

used a similar approach for the development of a 1D model of the foot only capable to predict the average foot skin temperature for cold conditions. The foot is divided into two compartments (nodes): the skin with a nonlinear control of the thermoregulatory blood flow and the core, thermally passive except by a small nutritional blood flow. The model considers the efficiency reduction in the blood heat transport due to the countercurrent heat exchange. This model presents a good correlation between predicted and measured results, but it does not include the evaporative heat loss and the sweating transmission, resulting inappropriate for high temperatures where most of the heat is lost due to evaporation.

Kuklane et al. (2000) carried out an exhaustive validation of the Lotens model (Lotens 1989, Lotens et al. 1989), finding differences between the skin temperature in the instep and the model prediction which depend on the environmental temperature and the activity level. These differences could be due to the metabolic heat produced when walking which is not considered. On the other hand, model prediction was better when compared with the instep temperature than with the mean skin temperature, suggesting that this could be due to the fact that the model does not consider the cooling of local areas such as toes.

Xu et al. (2005) developed a model for the toes based on the Lotens model (Lotens 1989, Lotens et al. 1989). However, this model only included the nutritional blood flow, not considering the thermoregulatory blood flow.

More recently, Covill et al. (2008) developed a 2D FE model of foot, sock, and shoe to study the heat transfer in footwear (Figure 19.9). This model considers a more real geometrical approach and heat transfer by conduction, convection, and radiation through each material. However, it uses a simplified estimation of the metabolic heat generated in the foot and does not consider the evaporative heat exchange. Further improvements described are to include the countercurrent heat exchange as a control for heat input to the foot and to include a dynamic heat loading to simulate activity.

Finally, only a specific foot model makes an attempt to predict the production and evaporation of foot moisture (Gonzalez 2007). The numeric thermal model of the foot has been based on the thermoregulation of the foot and in the heat and mass transfer equations consisting of two modules: (1) A model of heat transfer, developed from the work carried out by Lotens (1989). That incorporates the heat loss due to the evaporation of the sweating. (2) A model of mass transfer, developed from the equations of Fiala (2001) for the production of sweating and the equations of mass balance between the foot and the environment.

Although foot model results are promising for cold and also for warm environments, further work is necessary in order to understand the variability in the thermophysiologic response of subjects and the complexity of the mass transfer mechanisms.

19.5.2 THERMAL MODELS OF FOOTWEAR

Clothing thermal models with diverse degrees of complexity have been widely used considering from the simplest, which only consider thermal resistance and water vapor resistance (ISO 9920 1995, ISO 11079 1999), to the most complex models, which also include ventilation due to movement, clothing fitting, water absorption/desorption in textiles, condensation, and radiation (Havenith et al. 2000,

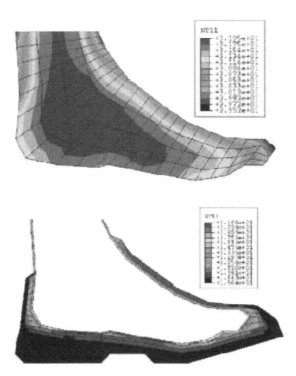

FIGURE 19.9 Finite element model simulation of (left) foot temperatures (°C) and (right) sock/ shoe temperatures (°C) in 15°C. (From Covill, D. et al., *Eng. Sport 7*, 2, 247, 2008. With permission.)

Lotens 1993). These models are based on the thermophysical properties of textiles and in the heat and mass transfer through clothing, allowing their combination with the mathematical models of human thermoregulation.

Thermal footwear models have been less explored, and existing ones are mainly based on basic clothing models. Lotens foot model (Lotens 1989) and toes model developed by Xu et al. (2005) only include a shoe component with a global value of thermal resistance. Physiologic simulation machines were used in order to obtain the values of the thermal properties of footwear. These tests are based on foot models based on electrically heated processes that allow the measurement of the heat exchanged with the environment (Bergquist and Holmer 1997, Kuklane et al. 2000, Uedelhoven et al. 2002).

Covill et al. (2008) developed a 2D FEM shoe model differentiating between the materials used for the upper and the sole. A sock layer between the shoe and foot was also included. Material properties for the model were based on values obtained from the literature.

The most completed footwear model presented in the literature (Gonzalez 2007) is based on Lotens' clothing model (Lotens 1993) and allows the prediction of thermal properties of footwear (thermal and water vapor resistances) from the properties of its components. The footwear was simplified to a cylinder with heat and humidity loss only considered in radial direction (Lotens 1993), over which homogenous surfaces (S1, counter; S2, toecap; S3, sole; S4, upper; S5, lacing; and S6, seams) were distributed (Figures 19.10 and 19.11).

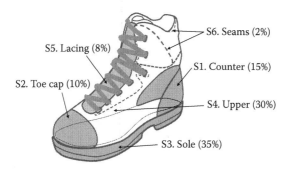

FIGURE 19.10 Footwear simplification in homogenous surfaces (in brackets: average surface percentage over the whole shoe area).

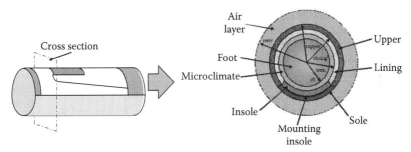

FIGURE 19.11 Left: Lateral view of the footwear cylindrical model. Right: Cross-section of the footwear model.

To calculate the global heat and the water vapor resistances of the footwear defined areas, the properties of the different materials were added, correcting by the area factor (Lotens 1993).

The model was validated comparing the data collected in the test of the complete boot, with the calculated ones from the properties of the footwear components.

The development of the model and the comparison between the properties of the components and the complete footwear allowed obtaining relevant conclusions for footwear design:

- The footwear heat resistance is influenced to a great extent by the air trapped between materials and by their contact resistance.
- The use of non-breathable components (sole, insole, counter, foams, etc.) in footwear manufacturing reduces the effective area for evaporation. In average, evaporation is only possible through approximately the 30% of the total surface of the footwear.

Further work is necessary to improve footwear thermal models, mainly in relation with sweat management. This process is very complex, in addition to the direct evaporation of the sweat on the skin; different phenomena take place as the water vapor diffusion in the empty spaces, the humidity absorption of fibers, condensation/evaporation inside the materials, and the liquid transport by capillarity.

19.6   FOOTWEAR COMFORT PERCEPTION MODELING

Classical approaches to understand comfort are based on the supposition that comfort is the result of the psychological integration of the physical sensations that the user is receiving when using a specific footwear model. In that sense, several studies have focused on extracting the relations (Figure 19.12) between the perception of comfort and the physical parameters characterizing footwear (i.e., design parameters like shoe-last shapes, insert shapes, and/or materials, etc.), foot (i.e., anthropometry, age, etc.), or the interaction between them (i.e., pressures, forces, geometrical interaction, etc.).

Among them, only some of these studies propose a mathematical model relating the footwear design aspects with the users' comfort perception, able to be implemented in a virtual test bed (Gonzalez 2007, Olaso, 2010, Witana et al. 2004, 2009a,b).

The functional properties focused in such studies are fitting, analyzed through foot and shoe-last geometrical interaction, and cushioning, analyzed considering insole geometric and material properties and thermoregulation, mainly centered on upper, sole, and insole material thermal properties.

Referring to fitting perception, there is a study with men's dress shoes in which Witana et al. (2004) evaluated the quality of footwear fit using 2D foot outlines. In this study, fitting perception depended not only on the foot region considered but also on the shoe-last and foot geometrical differences. They provided two regression models (Figure 19.13) to quantify forefoot and midfoot fitting from the shoe-last and foot shape dimensional differences.

These results allowed them to propose a theoretical mathematical expression to quantify a discomfort threshold related to poor fit in different foot regions depending on individual sensitivity:

$$\text{Discomfort } J_\text{D} = \sum_{i=1}^{N} a_i F_i \quad \text{for all } i = 1 \ldots N$$

where
   F_i is dimensional difference between shoe-last and foot shapes in each foot region i
   a_i are weightings corresponding to each foot region i, depending on each person's sensitivity

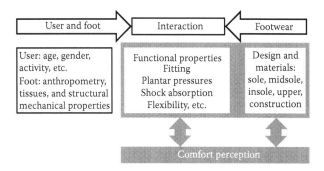

FIGURE 19.12   Methodological approach of the comfort existing studies.

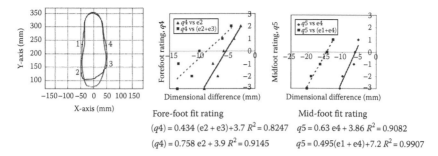

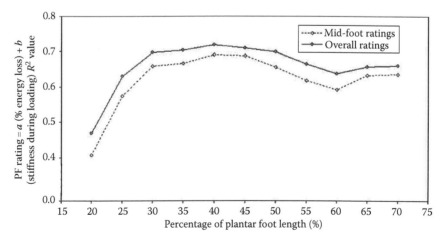

Fore-foot fit rating   Mid-foot fit rating

$(q4) = 0.434\,(e2 + e3) + 3.7\ R^2 = 0.8247$ $q5 = 0.63\,e4 + 3.86\ R^2 = 0.9082$

$(q4) = 0.758\,e2 + 3.9\ R^2 = 0.9145$ $q5 = 0.495(e1 + e4) + 7.2\ R^2 = 0.9907$

FIGURE 19.13 Regression models predicting forefoot and midfoot fitting from the shoe-last and foot shape dimensional differences. (Extracted from Witana, C.P. et al., *Ergonomics*, 47(12), 1301, 2004.)

They theorize about the existence of a discomfort threshold represented by a certain value of the J_D, under which the user tends to maximize the contact area with the footwear. However, although being an interesting proposal, it has not been validated and so its reliability remains unclear.

Focusing on the influence of the footwear insole design, Witana et al. (2009a) developed a study analyzing the effects of surface geometrical and cushioning properties on the plantar shape of the midfoot region and perceived sensations. They found that both surface shape and material properties (tensile strength, Young's modulus, flexural rigidity, rebound resilience, and percentage of energy loss) influenced ratings of perceived feelings in different foot regions. They developed a regression analysis (Figure 19.14) linking material properties with perceived feeling ratings, although they do not provide the equations but the relationship strength (R^2, p).

FIGURE 19.14 Regression models predicting perceived feeling ratings from footwear material properties (energy loss and stiffness during loading). (Extracted from Witana, C.P. et al., *Ergonomics*, 52(5), 617, 2009a.)

Yung-Hui and Wei-Hsien (2005) analyzed the impact of heel height and viscoelastic inserts on perceived discomfort while walking. They concluded that the higher is the heel, the higher is the perceived discomfort. Further, focusing on heeled ladies' shoes, Witana et al. (2009b) analyzed the effect of heel geometric characteristics (height, seat length, and inclination) by measuring plantar pressure patterns and perceived feeling. They found a positive relation between seat inclination and the heel height. Interestingly, they propose a regression model explaining approximately the 80% of the variation of perceived feeling with the foot–footwear contact area, peak plantar pressure, and percentage of force acting on the forefoot region. Although it is not a model directly relating footwear design properties but footwear–user interaction issues with users' comfort, it provides worthy information to understand the effect of footwear design in users' perception.

Thermal comfort perception and their relation with thermal properties of shoe was analyzed in a study conducted by Gonzalez (2007) with mountaineering footwear. The author performed experiments under different real weather conditions (cold, dry-hot, and wet-hot) while wearing footwear with different thermal insulation (resistance conductivity) and perspiration (water vapor resistance) properties. The study provides statistically significant regression models relating users' discomfort perception and footwear thermal insulation under cold ($TC = -0.44\ Rt$; $R^2 = 0.2$) and dry-hot ($TC = 0.32\ Rt$; $R^2 = 0.09$, where TC is the thermal comfort, Rt is the thermal resistance) environments.

Finally, some studies exist that analyze the influence of different footwear functional properties together on users' comfort perception (Miller et al. 2000, Olaso 2010). One of these studies (Olaso 2010) proposes a comfort prediction model considering more than one functional property. The author sets out the hypothesis that footwear comfort perception can be explained as the combined effect of the functional properties of footwear–user interaction (fitting, shock absorption, cushioning, thermal comfort, etc.) and other psychological factors, including aesthetics preferences and social acceptance. An experimentation to test such hypothesis was conducted with 25 users and 35 footwear models, measuring users' foot shape, footwear functional properties with machines tests (shoe-last shape, cushioning, shock absorption, flexibility, thermal properties), and their interaction in wear trials under specific weather and physical conditions simulating the shopping experience (flat and hard surface, $T = 25°C$, dry, between 5 and 10 min test). As a result (Table 19.5), the study provides three regression models to predict comfort perception under such specific conditions from fitting and cushioning under high pressures received on the heel ($R^2 = 0.706$), fitting and cushioning under low pressures received on the metatarsals ($R^2 = 0.671$), and fitting and shock absorption ($R^2 = 0.743$). These results support the fact that fitting and cushioning have been traditionally accepted as the most related to comfort aspects in concordance with the results of Goonetilleke (2003).

A lot of studies have focused the analysis of comfort perception. However, only some of them have dealt with the difficulty of developing a prediction model useful to virtually test a footwear model. Aspects like the role of specific functional properties on global comfort perception or the level and type of influence of aesthetics and social acceptance have on comfort perception are still a challenge from the research point of view.

TABLE 19.5
Comfort Prediction Models from Different Functional Properties

Models	R^2	Where
Cg = 22.206 – 12.061 * Lhp – 6.924 * Mhp + 1.119 * Dl	0.706	**Cg**, global comfort perception; **Lhp**, length shoe last/foot ratio; **Mhp**, flex line perimeter shoe last/foot ratio;
Cg = 16.672 – 8.820 * Lhp – 4.676 * Mhp + 0.414 * Dh	0.671	**Dl**, maximum deformation (mm) under low pressures applied on the metatarsals; **Dh**, maximum deformation
Cg = 17.138 – 8.233 * Lhp – 4.761 * Mhp – 0.001 * SA	0.743	(mm) under high pressures applied on the heel; **SA**, shock absorption (dynamical rigidity)

Source: Extracted from Olaso, J., Predicción del confort a partir de prototipos virtuales de calzado y modelos de ingeniería asistida por ordenador (Eng. Tr. Comfort prediction of virtual prototypes of footwear using computer aided engineering models), Ph. D. Thesis, Universitat Politècnica de València (UPV), Valencia, Spain, 2010.

QUESTIONS

19.1 Which virtual tests for footwear are currently available with commercial applications?

19.2 What kind of methods could be used to model footwear biomechanics and how to use it depending on the application?

19.3 What are the main difficulties to predict footwear fitting?

19.4 What are the footwear components (geometric design and materials) influencing footwear fitting?

19.5 How the modelization of the foot could help in the virtual footwear assessment?

19.6 What are the main aspects of the footwear influencing comfort?

19.7 What is the role of the functional aspects of foot–shoe interaction in the comfort perceived by the users?

REFERENCES

Actis, R.L., Ventura, L.B., Smith, K.E., Commean, P.K., Lott, D.J., Pilgram, T.K., Mueller, M.J. (2006). Numerical simulation of the plantar pressure distribution in the diabetic foot during the push-off stance. *Medical and Biological Engineering and Computing*, 44, 653–663.

Alemany, S., Garcia, I., Alcantara, E., Gonzalez, J.C., Castillo, L. (2003). Integration of plantar pressure measurements in a finite element model for the optimization of shank-piece design for high-heeled shoes. In: *Proceedings of the 6th Symposium on Footwear Biomechanics*, Queenstown, New Zealand, pp. 5–6.

Alemany, S., González, J.C., Garcia, A.C., Olaso, J., Montero, J., Chirivella, C., Prat, J., and Sánchez, J. (2005). A novel approach to define customized functional design solution from user information, 3rd Interdisciplinary World Congress on Mass Customization and Personalization, Hong Kong.

Alexander, R.M. (1988). *Elastic Mechanisms in Animal Movement*. Cambridge University Press, Cambridge, U.K.

Ankrah, S., Mills, N.J. (2004). Analysis of ankle protection in association football. *Sports Engineering*, 7, 41–52.

Asai, T., Carré, M.J., Akatsuka, T., Haake, S.J. (2002). The curve kick of a football I: Impact with the foot. *Sports Engineering*, 5, 183–192.

Asai, T., Nunome, H., Maeda, A., Matsubara, S., Lake, M. (2005). Computer simulation of ball kicking using the finite element skeletal foot model. *Science and Football V*, 12, 77–82.

Azariadis, P., Moulianitis, V., Alemany, S., González, J.C., Jong, P., Zande, M., Brands, D. (2007). Virtual shoe test bed: A computer-aided engineering tool for supporting shoe design. *Computer-Aided Design and Applications*, 4(6), 741–750.

Barani, Z., Haghpanahi, M., Katoozian, H. (2005). Three dimensional stress analysis of diabetic insole: A finite element approach. *Technology and Health Care*, 13, 185–192.

Baroud, G., Goerke, U., Guenther, H., Stefanyshyn, D.J. (1999). A non-linear hyperelastic finite element model of energy return enhancement in sport surfaces and shoes. In: *Proceedings of the 4th Symposium on Footwear Biomechanics*, Canmore, Alberta, Canada, pp. 18–19.

Bergquist, K., Holmer, I. (1997). A method for dynamic measurement of the resistance to dry heat exchange by footwear. *Applied Ergonomics*, 5(516), 383–388.

Bretz, D.A. (2000). A computer simulation approach to the study of effects of deck surface compliance on initial impact impulse forces in human gait. PhD thesis, Naval Postgraduate School, Monterey, CA.

Budhabhatti, S.P., Erdemir, A., Petre, M., Sferra, J., Donley, B., Cavanagh, P.R. (2007). Finite element modelling of the first ray of the foot: A tool for the design of interventions. *Journal of Biomechanical Engineering*, 129(5), 750–756.

Cheng, F.-T., Perng, D.-B. (1999). A systematic approach for developing a foot size information system for shoe last design. *International Journal of Industrial Ergonomics*, 25(2), 171–185.

Chen, W.P., Tang, F.T., Ju, C.W. (2001). Stress distribution of the foot during mid-stance to push-off in barefoot gait: A 3D finite element analysis. *Clinical Biomechanics*, 16, 614–620.

Cheng, Y.L., Hong, Y. (September 2010). Using size and pressure measurement to quantify fit of running shoes. *Footwear Science*, 2(3), 149–158.

Cheung, J.T.M., Bouchet, B., Zhang, M., Nigg, B.M. (2007). A 3D finite element simulation of foot-shoe interface. In: *Proceedings of the 8th Symposium on Footwear Biomechanics*, Taipei, Taiwan, pp. 45–46.

Cheung, J.T.M., Yu, J., Wong, D.W., Zhang, M. (2009). Current methods in computer-aided engineering for footwear design. *Footwear Science*, 1(1), 31–46.

Cheung, J.T.M., Zhang, M. (2005). A 3-dimensional finite element model of the human foot and ankle for insole design. *Archives of Physical Medicine and Rehabilitation*, 86, 353–358.

Cheung, J.T.M., Zhang, M. (2008). Parametric design of pressure-relieving foot orthosis using statistics-based finite element method. *Medical Engineering and Physics*, 30, 269–277.

Cheung, J.T.M., Zhang, M., An, K.N. (2004). Effects of plantar fascia stiffness on the biomechanical responses of the ankle-foot complex. *Clinical Biomechanics*, 19, 839–846.

Cheung, J.T., Yu, J., Zhang, M. (2009). Computational simulation of high heeled shoe fitting and walking. *Footwear Science Special Issue: Proceedings of the Ninth Footwear Biomechanics Symposium*, Vol. 1, Supplement 1, Stellenbosch, South Africa, pp. 53–55.

Covill, D., Zhongwei, G., Bailey, M., Pope, D. (2008). Finite element analysis of the heat transfer in footwear. *Engineering of Sport 7*, 2, 247–254.

D'Andrea, S.E., Lord, D.R., Davis, B.L. (1997). A rheological model of the human heel pad. In: *XXI Annual Meeting of the American Society of Biomechanics*, Clement, SC.

Erdemir, A., Saucerman, J.J., Lemmon, D., Loppnow, B., Turso, B., Ulbrecht, J.S., Cavanagh, P.R. (2005). Local plantar pressure relief in therapeutic footwear: Design guidelines from finite element models. *Journal of Biomechanics*, 38, 1798–1806.

Even-Tzur, N., Weisz, E., Hirsch-Falk, Y., Gefen, A. (2006). Role of EVA viscoelastic properties in the protective performance of a sport shoe: Computational studies. *Bio-Medical Materials and Engineering*, 16, 289–299.

Fiala, D., Lomas, K.J., Stohrer, M. (2001). Computer prediction of human thermoregulatory and temperature responses to a wide range of environmental conditions. *International Journal of Biometeorol*, 45(3), 143–159.

Flügge, W. (1967). *Viscoelasticity*. Blaisdell Publishing Company, Waltham, MA, pp. 16–17.

Gefen, A. (2002a). Stress analysis of the standing foot following surgical plantar fascia release. *Journal of Biomechanics*, 35, 629–637.

Gefen, A. (2002b). Biomechanical analysis of fatigue-related foot injury mechanisms in athletes and recruits during intensive marching. *Medical and Biological Engineering and Computing*, 40, 302–310.

Gefen, A. (2003). Plantar soft tissue loading under the medial metatarsals in the standing diabetic foot. *Medical Engineering and Physics*, 25, 491–499.

Gefen, A., Megido-Ravid, M., Itzchak, Y. (2001). In vivo biomechanical behavior of the human heel pad during the stance phase of gait. *Journal of Biomechanics*, 34, 1661–1665.

Gonzalez, J.C. (2007). Modelization of the influence of the footwear and the environmental conditions in the foot thermophysiological response and the thermal comfort. Doctoral dissertation. Universidad Politécnica de Valencia, Valencia, Spain.

Goonetilleke, R. (2003). Designing footwear: Back to basics in an effort to design for people. In: Khalid, H.M., Lim, T.Y., and Lee, N.K. (Eds.), *Proceedings of SEAMEC 2003*, Kuching, Sarawak, Malaysia, pp. 21–25.

Goske, S., Erdemir, A., Petre, M., Budhabhatti, S., Cavanagh, P.R. (2006). Reduction of plantar heel pressures: Insole design using finite element analysis. *Journal of Biomechanics*, 39, 2363–2370.

Haber, R.N., Haber, L. (January 1997). One size fits all? *Ergonomics in Design: The Quarterly of Human Factors Applications*, 5(1), 10–17(8).

Havenith, G., Heus, R., Lotens, W.A. (2000). Clothing ventilation, vapour resistance, and permeability index: Changes due to posture, movement and wind. *Ergonomics*, 33, 989–1005.

Hsu, Y.C., Gung, Y.W., Shih, S.L., Feng, C.K., Wei, S.H., Yu, C.H., Chen, C.S. (2008). Using an optimization approach to design an insole for lowering plantar fascia stress—A finite element study. *Annals of Biomedical Engineering*, 36, 1345–1352.

ISO 9920. (1995). Ergonomics of the thermal environment—Estimation of the thermal insulation and evaporative resistance of a clothing esemble. International Standards Organization, Geneva, Switzerland.

ISO 11079. (1999). Evaluation of cold environments—Determination of required clothing insulation (IREQ). International Standards Organization, Geneva, Switzerland.

Jordan, C., Barlett, R. (1995). Pressure distribution and perceived comfort in casual footwear. *Gait and Posture*, 3(4), 215–220.

Jordan, C., Payton, C., Bartlett, R. (1997). Perceived comfort and pressure distribution in casual footwear. *Clinical Biomechanics*, 12(3), S5.

Kim, W., Voloshin, A.S., Johnson, S.F. (1994). Modelling of heel strike transients during running. *Human Movement Science*, 13, 221–244.

Kouchi, M., Mochimaru, M., Nogawa, H., Ujihashi, S. (2005). Morphological fit of running shoes: Perception and physical measurements. In: *Proceedings of the 7th Symposium on Footwear Biomechanics*, Cleveland, OH, pp. 38–39.

Kuklane, K., Holmer, I., Havenith, G. (2000). Validation of a model for prediction of skin temperatures in footwear. *Applied Human Science*, 19(1), 29–34.

Lemmon, D., Shiang, T.Y., Hashmi, A., Ulbrecht, J.S., Cavanagh, P.R. (1997). The effect of insoles in therapeutic footwear—A finite element approach. *Journal of Biomechanics*, 30, 615–620.

Lewis, G. (2003). Finite element analysis of a model of a therapeutic shoe: Effect of material selection for the outsole. *Bio-Medical Materials and Engineering*, 13, 75–81.

Li, Y., Li, F., Liu, Y., Luo, Z. (2004). An integrated model for simulating interactive thermal processes in human-clothing system. *Journal of Thermal Biology*, 29, 567–575.

Lotens, W.A. (1989). A simple model for foot temperature simulation. IZF 1989-8, TNO Institute for Perception, Soesterberg, Utrecht, the Netherlands.

Lotens, W.A. (1993). Heat transfer from humans wearing clothing. Doctoral dissertation, Delft University of Technology, Delft, the Netherlands.

Lotens, W.A., Heus, R., Van de Linde, F.J.G. (1989), A 2-node thermoregulatory model for the foot. In: J.B. Mercer (ed.) *Thermal Physiology*. Elsevier Science Publishers B.V., Amsterdam, the Netherlands, pp. 769–775.

Luximon, A., Goonetilleke, R.S., Tsui, K.L. (2001). A fit metric for footwear customization. In: *Proceedings of the 2001 World Congress on Mass Customization and Personalization*, October 1–2, 2001, Hong Kong (CD-ROM).

Luximon, A., Luximon, Y. (2009). Shoe-last design innovation for better shoe fitting. *Computers in Industry*, 60, 621–628.

Ly, Q.H., Alaoui, A., Erlicher, S., Baly, L. (2010). Towards a footwear design tool: Influence of shoe midsole properties and ground stiffness on the impact force during running. *Journal of Biomechanics*, 43, 310–317.

McGuffin, R., Burke, R., Huizenga, C., Hui, Z., Viahinos, A., Fu, G. (June 4, 2002). Human thermal comfort model and manikin. In: *2002 SAE Future Car Congress* (SAE paper # 2002-01-1955), Arlington, VA.

McMahon, T.A., Cheng, G.C. (1990). The mechanics of running: How does stiffness couple with speed? *Journal of Biomechanics*, 23, 65–78.

Miller, J.E., Nigg, B.M., Liu, W., Stefanyshyn, D.J., Nurse, M.A. (2000). Influence of foot, leg and shoe characteristics on subjective comfort. *Foot and Ankle International*, 21(9), 759–767.

Mochimaru, M., Dohi, M., Kouchi, M. (2000). Analysis of 3D human foot forms using the FFD method and its application in grading shoe last. *Ergonomics*, 43(9), 1301–1313.

Mochimaru, M., Kouchi, M. (2003). Last customization from an individual foot form and design dimensions. In: *6th ISB Footwear Biomechanics*, Queenstown, New Zealand, pp. 62–63.

Nakamura, S., Crowninshield, R.D., Cooper, R.R. (1981). An analysis of soft tissue loading in the foot—A preliminary report. *Bulletin of Prosthetics Research*, 18, 27–34.

Nigg, B.M., Liu, W. (1999). The effect of muscle stiffness and damping on simulated impact force peaks during running. *Journal of Biomechanics*, 32(8), 849–856.

Nikooyan, A.A., Zadpoor, A.A. (2011). An improved cost function for modelling of muscle activity during running. *Journal of Biomechanics*, 44, 984–987.

Olaso, J. (2010). Predicción del confort a partir de prototipos virtuales de calzado y modelos de ingeniería asistida por ordenador (Eng. Tr. Comfort prediction of virtual prototypes of footwear using computer aided engineering models). PhD thesis, Universitat Politècnica de València (UPV), Valencia, Spain.

Olaso, J., González, J.C., Alemany, S., Medina, E., Ló pez, A., Martín, C., Prat, J., Soler, C. (2007). Study of the influence of fitting and walking condition in foot dorsal pressure. In: E.C. Frederick and S.W. Yang (Eds.), *Proceedings of Eighth Footwear Biomechanics Symposium*, Taipei, Taiwan, pp. 41–42.

Parsons, K. (2003). *Human Thermal Environments: The Effects of Moderate to Cold Environments on Human Health, Comfort and Performance*. London, U.K.: Taylor & Francis.

Rupérez, M.J., Alemany, S., Monserrat, C., Olaso, J., Alcañíz, M., González, J.C. (2009). A study of the viability of obtaining a generic animation of the foot while walking for the virtual testing of footwear using dorsal pressures. *Journal of Biomechanics*, 42, 2040–2046.

Rupérez, M.J., Monserrat, C., Alemany, S., Juan, M.C., Alcañíz, M. (2010). Contact model, fit process and, foot animation for the virtual simulator of the footwear comfort. *Computer-Aided Design*, 42, 425–431.

Scott, S.H., Winter, D.A. (1993). Biomechanical model of the human foot: Kinematics and kinetics during the stance phase of walking. *Journal of Biomechanics*, 26(9), 1091–1104.

Spears, I.R., Miller-Young, J.E., Sharma, J., Ker, R.F., Smith, F.W. (2007). The potential influence of the heel counter on internal stress during static standing: A combined finite element and positional MRI investigation. *Journal of Biomechanics*, 40, 2774–2780.

Stolwijk, J.A.J., Hardy, J.D. (1977). Control of body temperature. In D.H.K. Lee (Ed): *Handbook of Physiology. Reactions to Environmental Agents*, American Physiological Society, Bethesda, MD, Section 9, Chapter 4, pp. 45–67.

Sun, Z., Howard, D., Moatamedi, M. (2005). Finite element analysis of footwear and ground interaction. *Strain*, 41, 113–117.

Uedelhoven, W.H., Kurz, B., Rösch, M. (2002). Wearing comfort of footwear in hot environments. In: *Blowing Hot and Cold: Protecting Against Climatic Extremes*. Papers presented at the *RTO Human Factors and Medicine Panel (HFM) Symposium* held in Dresden, Germany, October 8–10, 2001 [CD-ROM]. Neuilly-Sur-Seine, France: Research and Technology Organisation (RTO), North Atlantic Treaty Organisation (NATO).

Waker, C.A., Blair, R. (2002). Leg stiffness and damping factors as a function of running speed. *Sports Engineering*, 5(3), 129–139.

Wang, C.S. (2010). An analysis and evaluation of fitness for shoe lasts and human feet. *Computers in Industry*, 61, 532–540.

Witana, C.P., Feng, J., Goonetilleke, R.S. (2004). Dimensional differences for evaluating the quality of footwear fit. *Ergonomics*, 47(12), 1301–1317.

Witana, C.P., Goonetilleke, R.S., Au, E.Y.L., Xiong, S., Lu, X. (2009a). Footbed shapes for enhanced footwear comfort. *Ergonomics*, 52(5), 617–628.

Witana, C.P., Goonetilleke, R.S., Xiong, S., Au, E.Y.L. (2009b). Effects of surface characteristics on the plantar shape of feet and subjects' perceived sensations. *Applied Ergonomics*, 40, 267–279.

Xu, X., Endrusick, T.L., Santee, B., Kolka, M.A. (June 2005). Simulation of toe thermal responses to cold exposure while wearing protective footwear. In: *Digital Human Modelling for Design and Engineering Symposium* (SAE paper # 2005-01-2676), Iowa City, IA.

Yu, J., Cheung, J.T.M., Fan, Y., Zhang, Y., Leung, A.K.L., Zhang, M. (2008). Development of a finite element model of female foot for high-heeled shoe design. *Clinical Biomechanics*, 23(S1), S31–S38.

Yung-Hui, L., Wei-Hsien, H. (2005). Effects of shoe inserts and heel height on foot pressure, impact force, and perceived comfort during walking. *Applied Ergonomics*, 36, 355–362.

Zadpoor, A.A., Nikooyan, A.A., Reza, A.A. (2007). A model-based parametric study of impact force during running. *Journal of Biomechanics*, 40, 2012–2021.

20 Measuring the Motion Control Properties of Footwear

Assessment of Footwear and Foot Function Should Be Harmonized

Stephen Urry, Lloyd Reed, and William Gordon

CONTENTS

20.1 BACKGROUND

Ask an individual who has worn shoes if they perceive that their footwear has any impact on their feet or for that matter on other parts of their body such as their knees or back, and they will often quickly affirm the notion. Similarly, ask a clinician who deals with foot problems if footwear is an important factor in assessment and treatment and they can, almost without exception, supply anecdotal evidence to support the contention. Furthermore, ask the researcher who attempts to glean insight into foot function whether the effect of footwear needs to be controlled in their experiments and they will usually agree that findings from studies where footwear has been controlled are generally more robust than where it has not. However, scrutiny of the scientific literature on footwear, foot function, comfort, pain, and dysfunction reveals a glaring paradox; very few scientific investigations incorporate quantitative data regarding both the footwear and the feet! In the majority of cases by far, investigations focus on one aspect and omit details of the other—or at best, give a relatively superficial qualitative description. Why should such an incongruous situation exist? Is it a research convenience? Are the omissions considered minor and simply accepted? Is the effort required to rectify such omissions considered too great? Should some attempt be made to improve the status quo? And, if anything can be done, what options might furnish the most fruitful improvement at an acceptable cost?

The omission of pertinent data regarding footwear in clinical studies of the foot has not gone completely unaddressed. Menz and Sherrington (2000) chose to develop and validate a footwear assessment tool in an attempt to broaden and strengthen research findings. The intention was to develop an instrument that would provide valid information regarding a number of footwear characteristics, especially those thought to influence postural stability, and which would subsequently enable clinicians and researchers to differentiate between appropriate and inappropriate shoes. Barton et al. (2009) later developed this approach with substantial focus on what were referred to as the motion control properties of the shoe, stating that health professionals have a responsibility to consider footwear characteristics in the etiology and treatment of patient problems. While neither Menz and Sherrington (2000) nor Barton et al. (2009) cite strong evidence that shoes can control motion at the internal joints of the foot, both advocate that aspects of footwear which might provide such a role are worthy of research attention. Menz and Sherrington (2000) stated that stiff heel counters are generally regarded as being a beneficial feature, to assist in retaining the shape of the upper and to control a certain amount of heel movement within the shoe during gait, while Barton et al. (2009) defended the selected motion control property items on the basis of general consensus within the literature. Between them, these two studies identified the following four footwear characteristics that, it was conjectured, are widely considered to influence motion at the internal joints of the foot:

1. Heel counter stiffness and stability
2. Midfoot sole sagittal flexibility, referred to as longitudinal stability by Menz and Sherrington (2000)
3. Sole flexion point, referred to as the forefoot flexion point by Barton et al. (2009)
4. Midfoot sole frontal or torsional stability

Having settled on these characteristics, Menz and Sherrington (2000) and Barton et al. (2009) assessed the ability of clinicians to rate them in a variety of footwear styles and thus determined the intra- and inter-tester reliability. Menz and Sherrington (2000) determined that each of the variables could be reliably documented both between two examiners and within each examiner over time despite the inherent subjectivity of the task, and concluded the Footwear Assessment Form to be a reliable clinical tool for the assessment of heel counter stiffness and longitudinal sole rigidity. Barton et al. (2009) reported inferior intra-rater reliability but similar inter-rater reliability for midfoot sagittal stability, while intra-rater and inter-rater reliability for heel counter stiffness was found to be superior to that of Menz and Sherrington (2000). Barton et al. (2009) concluded that the reliability in both studies for each item was high, strengthening the claim that these items possess adequate reliability for future use.

In a similar assessment of intra- and inter-tester reliability using the exact method described by Menz and Sherrington (2000), Reed (2007) reported a broader range of kappa (k) and percentage agreement statistics than either Menz and Sherrington (2000) or Barton et al. (2009) (Tables 20.1 and 20.2). While it is possible that these differences may be attributable to factors such as the tester's experience (Menz and Sherrington (2000) and Barton et al. (2009) recruited only testers who were qualified clinicians, while Reed (2007) used final year podiatry students as well as experienced clinicians), the findings indicate that reliability may, at times, be moderate to poor rather than generally good as reported by Menz and Sherrington (2000) and Barton et al. (2009). If the level of reliability is dependent on the experience and skill of the assessor, then the use of the tools must be restricted to only experienced or highly trained individuals. Alternatively, if experience and skill are not the reasons for low reliability, then the tools should be used with caution until the cause is understood fully.

Even if reliability is not a significant issue and it can be concluded that the respective tools may be used with confidence in the clinical research setting, there is still

TABLE 20.1
Results for Inter-Tester Reliability

		Inter Tester		
Author		Menz and Sherrington (2000)	Barton et al. (2009)	Reed (2007) (Wt k)
Heel counter stiffness	Min k	0.64	0.81	0.39 (0.23)
	Max k	0.75	0.86	1.0 (1.0)
	Min % agreement	75	93	60
	Max % agreement	83	96	100
Longitudinal sole rigidity	Min k	−0.04	0.69	0.2 (0.2)
	Max k	0.47	0.87	1.0 (1.0)
	Min % agreement	92	88	60
	Max % agreement	92	95	100
Sole flexion point	Min k	0.75	0.82	0.19 (na)
	Max k	1.0	0.83	0.22 (na)
	Min % agreement	92	97	50
	Max % agreement	100	97	70

k, kappa statistic; Wt k, weighted kappa.

TABLE 20.2
Results for Intra-Tester Reliability

		Intra Tester		
Author		Menz and Sherrington (2000)	Barton et al. (2009)	Reed (2007) (Wt k)
Heel counter	Min k	0.77	0.86	0.39 (0.5)
stiffness	Max k	0.86	0.87	1.0 (1.0)
	Min % agreement	83	96	60
	Max % agreement	92	96	100
Longitudinal	Min k	−0.04	0.71	0.5 (0.57)
sole rigidity	Max k	1.0	0.78	1.0 (1.0)
	Min % agreement	92	88	70
	Max % agreement	100	92	100
Sole flexion	Min k	0.4	0.91	0.00 (na)
point	Max k	0.62	0.92	0.62 (na)
	Min % agreement	83	98	40
	Max % agreement	92	98	90

k, kappa statistic; Wt k; weighted kappa.

the question of validity. Both Menz and Sherrington (2000) and Barton et al. (2009) expressed some reservation with regard to clinical validity. They acknowledged the paucity of established clinical guidelines regarding shoe factors, and conceded that the selected scales were based on a conceived notion of general consensus leading to rather arbitrary divisions for heel counter stiffness, midfoot sole sagittal stability, and midfoot sole torsional stability (i.e., 0°–10°, 10°–45°, >45°). Furthermore, with respect to the motion control properties scale, Barton et al. (2009) admitted that the clinical significance of each items inclusion and the weightings (score) for each category requires further evaluation. Overall then, while simple clinical tests may provide some distinction between a "stiff" shoe and a "flexible" shoe and thus allow shoes that are grossly deficient to be set apart from shoes that appear clinically acceptable, no evidence is provided to indicate that this approach will be adequate in a broader range of clinical research settings. Put frankly, subjective assessment is simply too imprecise to differentiate between shoes that have a similar feel but are quantitatively different—could a group of shoes which all seem to have fairly stiff heel counters be reliably ranked by different testers and, could a level in the ranking hierarchy subsequently be identified as providing the appropriate level of stability for a particular foot or individual? It is clear then that subjective assessment is unlikely to allow optimal footwear performance characteristics to be determined. Therefore, objective measurement of the structural characteristics of shoes should be the option of choice.

Currently, how are shoes quantitatively evaluated, and are these methods adequate to meet the needs of health care professionals or, for that matter, manufacturers and retail suppliers who wish to provide consumers with suitable and appropriate footwear or footwear advice? Historically, the quantitative testing of footwear seems to give attention to two major aspects; material testing and performance testing. Material testing has a long history and has been well integrated with the manufacturing industry. Organizations such as SATRA, which originated in 1919 as the British Boot, Shoe and Allied Trades Research Association, direct and develop appropriate testing techniques, but much of this work is inevitably related to aspects of manufacturing quality and material endurance—issues of construction and wear and tear. While construction techniques and material qualities will impact on the function of any shoe, on their own they do not address the functional interaction of the foot and the shoe. Performance testing to assess the functional characteristics of shoes is undertaken but, in general, is restricted to research environments where sophisticated testing equipment is the norm. Moreover, performance-related studies tend to focus on sport shoes rather than school shoes or shoes for everyday wear. Relevant examples from the contemporary literature include investigations of torsional rigidity and sole stiffness, and bending stiffness at the metatarsophalangeal joint region (Tanaka et al. 2003, Hillstrom et al. 2005, Kleindienst et al. 2005, Oleson et al. 2005, Nishiwaki 2008). These investigations demonstrate clearly the modern trend toward the use of material testing machines, bespoke micro-controlled actuators, or motion analysis systems incorporating force plates and 3D motion-capture systems. While investigations of this caliber generate undeniably useful findings and expand our overall knowledge, usually they are narrow in their focus and more pertinent to the specialist investigator or sport shoe designer than the clinician or consumer. Furthermore, because of the equipment costs and the time required to follow complex test procedures, these types of investigation are unlikely to ever find a role in routine, repetitive, mass testing of footwear.

There is, therefore, something of a gap between simple but imprecise subjective testing and high-end research that is narrow, expensive and time consuming to perform, and it seems that a compromise between the two approaches is needed. Published studies in this middle ground are not common in the literature; however, when found they reveal the potency of such a judicious approach. For example, Stacoff et al. (1991) constructed a simple testing machine to evaluate the torsional stiffness of running shoes and spikes. The machine was based on earlier works that investigated the torsional stiffness of the human foot (Rasmussen and Andersen 1982, Kjaersgaard-Andersen et al. 1987). In a similar manner, Van Gheluwe et al. (1995) demonstrated a simple method for measuring heel counter rigidity. Stacoff's findings (Stacoff et al. 1991) placed the characteristics of the tested footwear on a spectrum which incorporated an equivalent measure of participant's feet, while Van Gheluwe's findings (Van Gheluwe et al. 1995) allowed a range of shoes to be effectively ranked and related to the subsequent shod performance of participant's feet. These studies illustrate how biomechanical findings can be massively enhanced by the fusion of quantitative data from both footwear and feet—the interplay between the allied factors can be revealed so clearly! Can a way be found to allow contemporary investigators to refine and fortify their investigations in a similar manner without undue burden? Perhaps it could if simple, efficient, and inexpensive methods for shoe testing were widely available.

To explore this possibility and search for generic solutions, a series of fundamental footwear tests were developed and evaluated, using very simple equipment. Overall, the goal was to determine whether inexpensive equipment, controlled by inexperienced or unskilled operators, could return reliable, quantitative measures for the motion control footwear factors highlighted by Menz and Sherrington (2000) and Barton et al. (2009). While a variety of potential options were identified for each of the tests, the final or preferred methods were selected, in part, because they reflected the approach adopted commonly by clinicians. By taking this approach it was hoped that the quantitative findings might be readily understood and applied not only by researchers but also by others more directly involved with the provision or assessment of footwear in clinical or commercial settings.

20.2   COMMON METHODS FOR TESTING FOOTWEAR BY HAND

Precisely when subjective testing of shoes became a routine part of clinical assessment is impossible to determine, but there is no doubt that it is historically well established. Menz and Sherrington (2000) took this imprecise but somewhat intuitive process and refined it in the hope of establishing a more methodical and consistent approach.

To assess heel counter stiffness, Menz and Sherrington (2000) instructed the assessor to exert firm pressure halfway up the posterior aspect of the heel counter, then visually estimate the degree of buckling relative to vertical and categorize it as minimal, less than 45°, or greater than 45°. Barton et al. (2009) offer more-or-less the same instructions but stated that the force should be applied approximately 20 mm from the base of the heel counter. This maneuver is demonstrated in Figure 20.1.

To assess sole rigidity in the sagittal plane (flex), Menz and Sherrington (2000) and Barton et al. (2009) instructed the assessor to exert firm pressure to the front of the shoe while the rear part is stabilized, then visually estimate the degree of sole

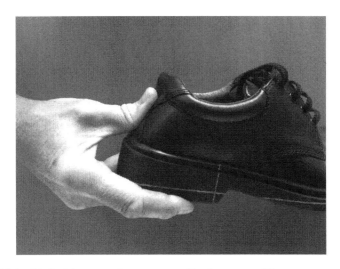

FIGURE 20.1   Method for manual assessment of heel counter stiffness.

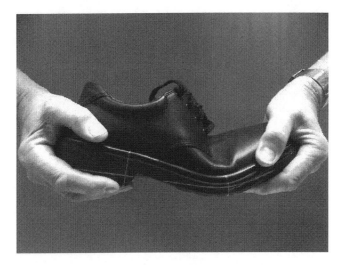

FIGURE 20.2 Method for manual assessment of flexion stiffness.

flexion relative to the horizontal plane and categorize it as minimal, less than 45°, or greater than 45°. This maneuver is demonstrated in Figure 20.2.

Barton et al. (2009) added the assessment of sole rigidity in the frontal plane (torsional stiffness) and instructed the assessor to grasp both the rearfoot and forefoot components of the shoe and attempt to bend it at the midfoot in the frontal plane. As can be seen in Figures 20.3 and 20.4, this maneuver involves a twist rather than a flex type of action.

Therefore, the objective of simple, machine-based, quantitative footwear assessment is to retain the essence of the equivalent manual test and put a figure on the

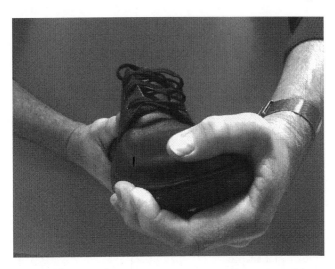

FIGURE 20.3 Method for manual assessment of torsional stiffness showing inversion of forefoot.

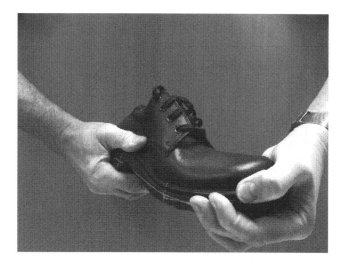

FIGURE 20.4 Method for manual assessment of torsional stiffness showing twist through length of the shoe.

outcome rather than categorize the finding according to an arbitrary scale. This then was the design philosophy adopted when constructing the test rigs used in his investigation.

20.3   DESIGN, CONSTRUCTION, AND OPERATION OF THE TEST RIGS

The heel counter stiffness test rig, Figure 20.5, was constructed essentially using two clamps and a force gauge, all of which were commercially available and relatively inexpensive. Both clamps had a quick release action. The first clamp was used to fasten the shoe to the test bench and prevent any displacement during testing. The force from the restraining clamp was distributed through the midsection and heel via a steel plate placed inside the shoe. The second clamp had a trigger-grip compression

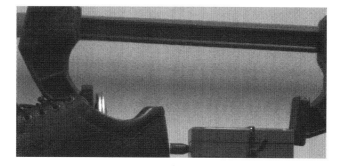

FIGURE 20.5 Setup for heel counter stiffness test rig showing first clamp in throat of shoe to maintain stability and second clamp with trigger grip providing compression through force gauge.

FIGURE 20.6 Alignment of soft-faced indenter at posterior aspect of heel counter.

mechanism that provided a remarkably consistent displacement of the jaws each time the grip was squeezed fully. The trigger-grip clamp was used to exert force in incremental steps, via a force gauge, to the heel counter. The force gauge was supplied with a series of interchangeable hooks and indenters and, for this rig, a circular, soft faced, indenter with a contact area of $1\,cm^2$ was selected, it being the closest approximation to the fat pad of a thumb, Figure 20.6. The height of the indenter relative to the heel counter could be adjusted between 5 and 40 mm by using shims beneath the force gauge, which was secured above a low friction slide fabricated from a strip of high-density polyethylene. Once the shoe was clamped in position, the counter could be step-loaded by repeatedly squeezing the trigger, and the linear displacement and force could be recorded for each increment. The number of increments could be varied depending on the robustness of the footwear being tested, but five increments, each of approximately 3.8 mm (19 mm total), generally provided a satisfactory test.

A simple hinged platform formed the basis of the forefoot flex test rig, Figure 20.7. In addition to the hinged platform, a single clamp, an appropriately sized shoe last, a force gauge, and an inclinometer were needed to secure the shoe and derive the measurements. The force from the restraining clamp was distributed through the heel and midsection via the truncated last, which had been shortened to 60% of its

FIGURE 20.7 Flexion stiffness test rig setup showing restraining clamp acting through truncated last and hinged platform with inclinometer.

original length by the removal of the toe region. This method of restraint differed from the steel plate used to hold the shoe in the heel counter test rig. The shortened last was chosen because it prevented the heel counter and vamp from collapse or excessive distortion when the shoe was flexed during testing, thereby simulating the effect of the foot in the shoe and probably producing more realistic results. The shoe was located in the rig so that the hinge was aligned with the metatarsal flex region.

To determine the position of the metatarsal flex region on the shoe, groups of footwear of different styles were measured prior to any testing in the rig. The metatarsal flex region appeared to be located in marginally different positions depending on the style and design of the footwear. For example, in school shoes the metatarsal flex region appeared to be located, on average, at 66% of the length of the shoe, while for safety boots the flex region was closer to 69% of the length of the boot (apparently due to the provision of extra length allowance in the safety toe-cap region). In general, the flex region would seem to be located in a zone that is between 65% and 70% of the length of the footwear.

With the shoe correctly located in the rig, the hinged platform was raised until it made gentle contact with the front section of the sole, and the inclinometer was zeroed. A torque, centered at the axis of the hinge mechanism, could then be applied to the shoe via the platform. The handheld force gauge was used to generate the torque, having replaced the indenter by a hook and sling. The sling allowed tensile load to be applied to pull the platform up, and this was physically easier to control than trying to push the platform up from below. The magnitude of the applied torque was determined as *force × distance*, where the force was measured by the load cell and the distance was the distance from the hinge to the point of attachment of the sling along the platform. Raw data were corrected to ensure that the component of force perpendicular to the platform was used rather than the vertical force acting through the sling and gauge, and to make an allowance for the torque required to displace the unloaded platform (i.e., the platform without a shoe). With a shoe in place, the effective torque was recorded at 5°, 10°, and 15° displacements from the zeroed start position, which reflected the inherent toe spring formed at the time of manufacture.

The torsional stiffness test rig is comprised of two platforms: a fixed rear platform on which the heel was clamped and a rotatable front platform on which the forefoot of the shoe could be constrained by means of an adjustable strap, Figures 20.8 and 20.9. As in the other tests, the heel was clamped using a quick release clamp and the load distributed to the appropriate internal areas of the shoe. To distribute the clamp load, a flat wooden block, approximately 15 mm thick, was cut to the outline shape of the heel cup and inserted in the shoe. The inferior surface of the block was mildly contoured to ensure that it sat snugly in the shoe. The block size was such that its straight leading edge aligned with the leading edge of the heel at the boundary between the heel and the midsection of the shoe.

To determine the position of the leading edge of the heel, groups of footwear of different styles were measured prior to any testing on the rig. On average, the leading edge of the heel was found to be at 30% the length of the shoe. The 30% length mark was subsequently used to locate the heel section on the rear platform, while the forefoot flex mark (65%–70% of length depending of footwear type) was similarly

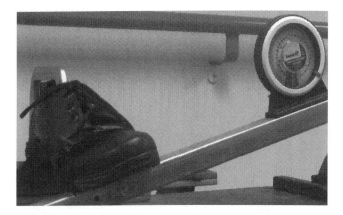

FIGURE 20.8 Torsional stiffness test rig setup showing front platform with central axis of rotation.

FIGURE 20.9 Torsional stiffness test rig setup showing clamp to grip heel and restraining strap for front platform. Note that the shoe is located according to 30% and 66% marks.

used to align the forefoot on the front platform, Figure 20.9. To minimize collapse of the toe box, an appropriately shaped, stiff foam plug was inserted into the shoe prior to tightening the restraining strap around the forefoot. During the setup procedure, care was taken to ensure that the axis of rotation of the front platform was aligned with the long axis of the shoe.

On completion of the setup process, the shoe was deemed to be correctly located in the rig if the length marks were aligned as described, the front platform was horizontal, and the inclinometer indicated zero. A torque could then be applied to the shoe via the platform in the same way that it was when testing flex—the force gauge and sling was used to generate the torque by pulling the platform up. Force vector corrections and compensations to allow for the inherent resistive torque of the unloaded platform were made as they were for the flex tests. The effective torque was then derived at 5°, 10°, and 15° displacements from the horizontal start position. The shoe was subjected only to torques that inverted the forefoot.

20.4 FINDINGS

Examples of test results for heel counter stiffness, forefoot flex, and torsional stiffness are given in Figures 20.10 through 20.12, respectively. The data are from a group of five school shoes.

20.5 RELIABILITY

Since the goal of this work was to determine whether inexpensive equipment, controlled by inexperienced or unskilled operators, could return reliable, quantitative measures of motion control characteristics, an assessment of reliability was undertaken. Ten articles of footwear were selected for testing by two operators.

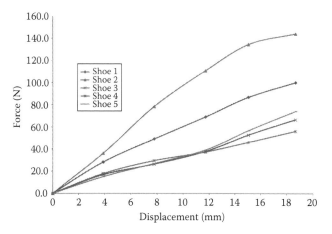

FIGURE 20.10 Example of heel counter stiffness curves obtained for a group of five school shoes.

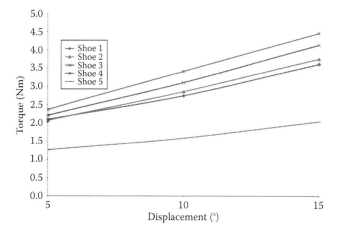

FIGURE 20.11 Example of flexion stiffness curves obtained for a group of five school shoes.

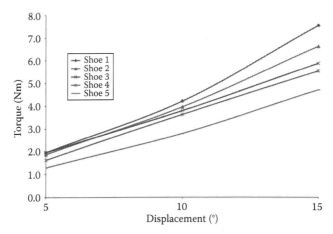

FIGURE 20.12 Example of torsional stiffness curves obtained for a group of five school shoes.

The footwear comprised of five school shoes from three different manufactures, and five safety boots from three different manufacturers. The two operators were both experienced health care professionals (podiatrists), one had used prototype versions of the test equipment over a 12 month period while the other had not used the equipment at all. For each test rig, the novice operator was given a single demonstration and then supervised and given feedback during a practice trial using a maximum of two items of footwear (not from the test groups). Each operator tested the 10 items of footwear on two separate occasions between 1 week and 2 months apart. Intra- and inter-tester reliability was assessed.

20.6 HEEL COUNTER STIFFNESS: SUMMARY

The coefficient of variation for inter-tester reliability was determined to be within the range 0.78%–3.25% when the heel counter was displaced between 0 and 20 mm. The coefficient of variation for intra-tester reliability was determined to be within the range 9.7%–14.2% under the same conditions.

20.7 FOREFOOT FLEX: SUMMARY

The coefficient of variation for inter-tester reliability was determined to be within the range 4.2%–7.3% when the forefoot was flexed from 5° to 15°. The coefficient of variation for intra-tester reliability was determined to be within the range 4.0%–6.8% under the same conditions.

20.8 TORSIONAL STIFFNESS: SUMMARY

The coefficient of variation for intra-tester reliability was determined to be within the range 5.4%–11.4% when the shoe was twisted from 5° to 15°. The coefficient of variation for inter-tester reliability was determined to be within the range 11.1%–14.8% under the same conditions.

20.9 INTERPRETATION OF TEST RIG RELIABILITY RESULTS

These early findings indicate that inexpensive rigs may be used to reliably derive quantitative measures of the mechanical or motion control characteristics of a diverse range of footwear, even when the rigs are used by operators with minimal training. It should be remembered that these rigs were prototypes, and were constructed with minimum expense using off-the-shelf components and therefore represent the worst case scenario. For one operator, the tests were repeated after a fairly long period (>2 months) during which time the rigs had been disassembled and stored. To complete the second round of tests, the rigs had to be reassembled, and this is likely to have increased the variance. This series of tests tended to produce intra-tester coefficients of variation with values greater than 10%. In contrast, when the rigs were left in situ and the tests were repeated after a short period (next week), the coefficient of variation was found to be in the range 1%–5%. Therefore, highly reliable results, with coefficients of variation not exceeding 5%, could be anticipated if the rigs are not disassembled, even if unskilled operators conduct the tests.

20.10 FUTURE WORK

The findings indicated that optimal results might best be achieved by building refined versions of these rigs, especially if they are integrated in a stand-alone design that is not intended to be disassembled. An engineering concept project was therefore initiated. An initial design for a single stand-alone test module was completed, Figures 20.13 through 20.15. While such a design could incorporate all of the tests outlined here, it is anticipated that the cost of construction would be considerably less than the cost of a material test machine. The characteristics for a suitable design might be an appropriate topic for discussion among interested researchers and potential users.

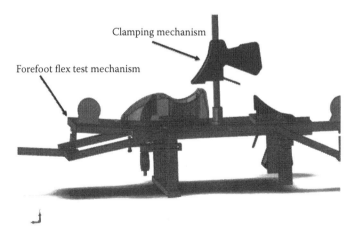

FIGURE 20.13 Design concept for an integrated, stand-alone test rig showing main clamping mechanism and flexion test mechanism.

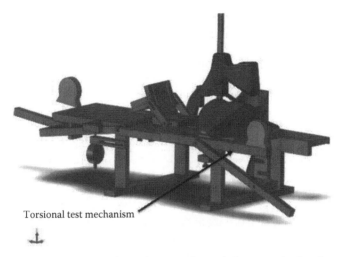

Torsional test mechanism

FIGURE 20.14 Design concept for an integrated, stand-alone test rig showing torsion test mechanism.

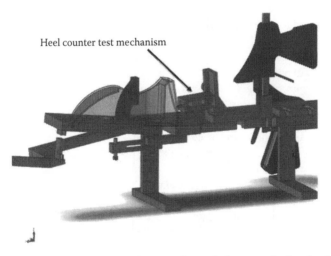

Heel counter test mechanism

FIGURE 20.15 Design concept for an integrated, stand-alone test rig showing heel counter test mechanism.

20.11 DISCUSSION AND CONCLUSION

It seems that test equipment to provide reliable quantitative assessments of specific shoe characteristics could be constructed at relatively low cost. However, would quantitative information on the motion control properties of footwear really be useful in clinical investigations and studies of foot biomechanics?

In this respect, the investigation into heel motion by Van Gheluwe et al. (1995) was exemplary because it linked the measurement of heel counter stiffness with the functional action of the heel. Further research subsequently revealed that greater heel counter rigidity can lead to reductions of heel motion, provided a sufficiently tight heel fit is achieved (Van Gheluwe et al. 1999). Heel dynamics can therefore be influenced by footwear but the effects are dependent on fit. If the fit is loose then movement between the shoe and foot arises, possibly during the airborne phase when the shoe is unloaded and able to slip (Stacoff et al. 1992). Other evidence indicates that a stiff, snug heel counter may even reduce lower limb muscle load (Jørgensen and Ekstrand 1988, Jørgensen and Bojsen-Moller 1989, Jørgensen 1990). Overall, the evidence demonstrates that findings regarding rearfoot function may be best appreciated when set in a footwear context, and indicates a pressing need for quantitative data on heel counter stiffness to be included in similar studies whenever possible.

With regard to torsion, the investigations of Stacoff et al. (1991) and Eslami et al. (2007) clearly revealed that rearfoot–forefoot coupling is affected by footwear. While Stacoff et al. (1991) focused on frontal plane interactions, Eslami et al. (2007) considered transverse plane motion to be more significant. This disparity possibly reflects the mechanical complexity of the midfoot and probably forewarns that individual responses to particular footwear will be difficult to predict with confidence, at least until our knowledge and understanding improves. In addition to coupling with the forefoot, the rearfoot links with the leg, but Eslami et al. (2007) did not uncover any tibial rotation effects despite the fact that rearfoot motion was altered by the motion control footwear in that study. In a different study, however, motion control shoes were found to affect tibial rotations, but only for low arched feet (Butler et al. 2007). Thus, while links between torsional motion (coupling) and torsional stiffness (footwear) have been found, our knowledge is far from complete and, at this time, cannot be used to predict footwear effects with certainty. For shoe effects to be more predictable, our knowledge of midtarsal mechanics must obviously be improved, but not in isolation; some future studies must investigate the torsion characteristics of feet and footwear simultaneously, rather than address them as separate issues.

While these fragments of evidence hint tantalizingly toward the notion that motion control footwear might affect foot function in a systematic manner, they simultaneously demonstrate how frustratingly difficult it is to predict effects with confidence. Reflecting on this quandary, Butler et al. (2006) speculated that it may be more important to match footwear to specific foot-mechanics rather than foot characteristics such as arch type, and suggested that individual assessment of lower limb biomechanics may be a better way to make progress. However, for this to be an optimally effective strategy, the proposal needs to be balanced and the motion control characteristics of shoes should be individually assessed at the same time. To simply designate a shoe as controlling or cushioning, stiff or flexible provides insufficient information; the complex problem of foot–shoe interaction is unlikely to be resolved without quantitative footwear data. The motion control characteristics of footwear should therefore be quantitatively defined and incorporated with studies of foot function or dysfunction whenever possible.

QUESTIONS

20.1 In general, are detailed footwear characteristics incorporated in a significant proportion of published studies regarding foot function or dysfunction?

20.2 How many key motion control characteristics of footwear were identified by Menz and Sherrington (2000) and Barton et al. (2009) and what were they?

20.3 Why may a subjective approach to the assessment of motion control character-istics of footwear be considered as less than optimal?

20.4 From a broad perspective, contemporary quantitative investigations of the functional effects of footwear have at least two disadvantages, what are they?

20.5 What are the main benefits in harmonizing research factors regarding footwear and foot characteristics when undertaking an investigation of foot and ankle biomechanics?

20.6 Considering the design and use of a rig for testing and measuring the motion control characteristics of footwear, what factor might have a significant effect on achieving consistent results?

REFERENCES

Barton CJ, Bonanno D, and Menz HB. 2009. Development and evaluation of a tool for the assessment of footwear characteristics. *Journal of Foot and Ankle Research* 2: 10 (www.footankleres.com/content/2/1/10).

Butler RJ, Davis IS, and Hamill J. 2006. Interaction of arch type and footwear on running mechanics. *American Journal of Sports and Medicine* 34: 1998–2005.

Butler RJ, Hamill J, and Davis I. 2007. Effect of footwear on high and low arched runners' mechanics during a prolonged run. *Gait and Posture* 26: 219–225.

Eslami M, Begon M, Farahpour N, and Allard P. 2007. Forefoot-rearfoot coupling patterns and tibial internal rotation during stance phase of barefoot versus shod running. *Clinical Biomechanics* 22: 74–80.

Hillstrom H, Song J, Heilman B, and Richards C. 2005. A method for testing shoe torsional and toe break flexibilities. *7th International Symposium on Footwear Biomechanics*, Cleveland, OH, July 27–29.

Jørgensen U. 1990. Body load in heel-strike running: The effect of a firm heel counter. *American Journal of Sports and Medicine* 18(2): 177–181.

Jørgensen U and Bojsen-Moller F. 1989. Shock absorbency factors in the shoe/heel interaction—With special focus on role of the heel pad. *Foot and Ankle* 9(11): 294–299.

Jørgensen U and Ekstrand J. 1988. Significance of heel pad confinement for the shock absorp-tion at heel strike. *International Journal of Sports and Medicine* 9: 468–473.

Kjaersgaard-Andersen P, Wethelund JO, and Nielsen S. 1987. Lateral calcaneal instability fol-lowing section of the calcaneofibular ligament. *Foot and Ankle* 7: 355–361.

Kleindienst FI, Michel KJ, and Krabbe B. 2005. Influence of midsole bending stiffness on the metatarsophalangeal joint based on kinematic and kinetic data during running. *7th International Symposium on Footwear Biomechanics*, Cleveland, Ohio, July 27–29.

Menz HB and Sherrington C. 2000. The footwear assessment form: A reliable clinical tool to assess footwear characteristics of relevance to postural stability in older adults. *Clinical Rehabilitation* 14: 657–664.

Nishiwaki T. 2008. Running shoe sole stiffness evaluation method based on eigen vibration analysis. *Sports Technology* 1(1): 76–82.

Oleson M, Adler D, and Goldsmith P. 2005. A comparison of forefoot stiffness in running and running shoe bending stiffness. *Journal of Biomechanics* 38: 1886–1894.

Rasmussen O and Andersen K. 1982. Ligament function and joint stability elucidated by a new technique. *Engineering in Medicine* 11: 77–81.

Reed L. 2007. An investigation of foot and ankle problems experienced by nurses. PhD thesis, Queensland University of Technology, Brisbane, Queensland, Australia.

Stacoff A, Kalin X, and Stussi E. 1991. The effects of shoes on the torsion and rearfoot motion in running. *Medicine and Science in Sports and Exercise* 23(4): 482–490.

Stacoff A, Reinschmidt C, and Stussi E. 1992. The movement of the heel within a running shoe. *Medicine and Science in Sports and Exercise* 24(6): 695–701.

Tanaka K, Uwai H, and Ujihashi S. 2003. A method of measurement and evaluation of the mechanical properties on the stability of running shoes. *6th International Symposium on Footwear Biomechanics*, Queenstown, New Zealand.

Van Gheluwe B, Kerwin D, Roosen P, and Tielemans R. 1999. The influence of heel fit on rearfoot motion in running shoes. *Journal of Applied Biomechanics* 15: 361–372.

Van Gheluwe B, Tielemans R, and Roosen P. 1995. The influence of heel counter rigidity on rearfoot motion during running. *Journal of Applied Biomechanics*, 11: 47–67.

Part V

Footwear Effects

21 Footwear Effects on Running Kinematics

Joseph Hamill, Allison H. Gruber,
and Ross H. Miller

CONTENTS

21.1 INTRODUCTION

In the 1970s, there was a significant increase in the popularity of running as a form of exercise. It was acknowledged that footwear-induced changes in movement could alleviate some of the stress and strain imposed on runners as their mileage increased (Frederick 1986). It was also suggested that running footwear could be a risk factor for cumulative microtrauma or overuse injury (Taunton et al. 2002). At this time, biomechanists sought to understand the mechanics of running and to relate the mechanics to injury prevention. Researchers also attempted to use the findings of their research to help design footwear that would accomplish the goal of reducing overuse injury. In order to respond to this goal, many biomechanists used a kinematic approach to solve the problem. Kinematics is the description of movement (i.e., position, velocity, or acceleration) without regard to the forces that cause the movement.

The starting place in the kinematic analysis of footwear on the human body is the analysis of the movements of the foot and ankle. The foot and ankle are mechanically linked such that foot movements during the ground contact phase influence knee and hip motion. The particular emphasis, however, was on the kinematics of the foot and ankle. Very little research actually took into account the kinematics of the knee and hip although it is now thought that what occurs at the hip, in particular, can influence the onset of overuse injuries (Pollard et al. 2004).

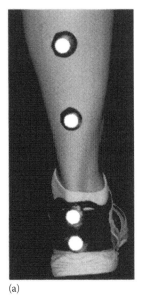 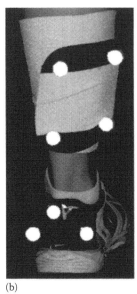

(a)   (b)

FIGURE 21.1 Marker placement for (a) 2-D rearfoot analysis and (b) 3-D RF analysis. In a 2-D marker arrangement only two markers per segment are required whereas at least three noncollinear markers are required for a 3-D set-up.

Initial research on the kinematics of footwear was accomplished using two-dimensional (2-D) methods (see Figure 21.1a). In a 2-D method, two markers, representing the long axis of the segment, are placed on lower extremity segments. That is, many studies focused on planar kinematics in either the sagittal or frontal planes. This led to many problems in interpretation because much of human movement, especially at the foot and ankle, is not planar. It should be clear that a 2-D analysis does not take into account the "cross-talk" between planes. More recently, three-dimensional (3-D) analyses have been used resulting in more complete description of the motions of the lower extremity (see Figure 21.1b). In 3-D, at least three noncollinear markers are placed on each segment.

Regardless, the parameters that have been primarily investigated are the same in 2-D and 3-D. These parameters describe angles, velocities, or accelerations at discrete points in time. These points in time generally occurred during the ground contact or support phase of the running stride. It has only been recently that researchers have investigated and compared the complete time course of the kinematic profile to determine the cause of injury and how footwear can affect this cause (Hamill et al. 1999).

The majority of runners (approximately 75%) use a heel–toe footfall pattern as opposed to a midfoot (MF) or forefoot (FF) footfall pattern (Hasegawa et al. 2007). Each of these patterns is determined by the initial contact of a portion of the foot on the ground. The majority of research on the kinematics of running and of running footwear has emphasized the heel–toe runner and thus have modeled the foot as a rigid structure and emphasized. In particular, the research has emphasized the motions of the rearfoot (RF) (i.e., pronation/supination) during running. RF motion is defined as the motion of the tibia relative to the calcaneus in the frontal plane.

21.2   KINEMATICS OF THE FOOT AND ANKLE

In 3-D kinematics, the frontal plane ankle angle (inversion/eversion) is defined as the motion of the foot relative to the leg. During running (i.e., in runners with no injury), the foot contacts the ground in a supinated or calcaneal inverted position as measured in the frontal plane (Figure 21.2a). Associated with the inverted position of the ankle are the motions of plantar flexion (in the sagittal plane) and adduction (in the transverse plane). The supinated position at touchdown ranges from 0 to approximately 10°. Frontal plane pronation or calcaneal eversion then takes place as the foot rotates to become flat on the running surface and the center of mass is directly over the base of support (Figure 21.2b). Two other actions of the ankle joint also take place in conjunction with eversion are dorsiflexion (sagittal plane) and abduction (transverse plane). Pronation can range up to 18° and may exceed 20° in some individuals (McClay and Manal 1997). From a maximally pronated position, the foot–ankle complex resupinates until the foot leaves the ground.

Footwear have been generally designed around the principle of "overpronation" as a cause or at least a contributor to overuse injury. Manufacturers have used this principle to design devices that limit or control the amount of pronation in running footwear. The major problem with this concept is that there is no clinical definition for overpronation. Clarke et al. (1983) suggested that 8°–13° of pronation was a normal range of maximum pronation. However, McClay and Manal (1999) reported magnitudes much greater than this. Therefore, without a concrete definition, overpronation cannot and probably should not be used as an indicator of injury. Suffice to say that overpronation for one individual may not be overpronation for another. Research on the relationship of pronation to injury indicates that the mechanism of overuse injuries is not well understood at this time.

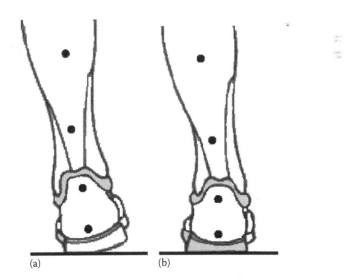

(a)   (b)

FIGURE 21.2   RF position for a left foot contact at (a) touchdown (the RF is slightly supinated) and (b) maximum RF eversion (the RF is maximally everted). The black dots refer to the positions of markers in a 2-D marker placement.

Runners often classify themselves as "pronators" or "supinators" and so they attempt to purchase footwear that will control "pronation" or "supination." This self-categorization is often carried out without the aid of any analysis and, in many instances, without knowledge of what constitutes an individual who has a problem of pronating too much or too little. In a very interesting study, Stefanyshyn et al. (2003) asked 81 runners to categorize themselves into "pronators" or "normal." The self-categorized "pronator" group had 41 subjects. The subjects all performed running trials barefoot and with a neutral shoe. The results indicated that 25% of the runners were actually categorized as "pronators."

21.3   FOOTWEAR EFFECTS OF THE FOOT AND ANKLE MOTION

Running footwear are considered to accomplish two primary tasks: cushioning and mediolateral control. Cushioning refers to the attenuation of impact shock as the foot collides with the ground. Mediolateral control concerns the motion of the subtalar joint and refers to the degree of stability in the pronatory action of the RF during the first portion (i.e., to approximately 45%) of the support phase. Both of these tasks are attributed to the function of the midsole of the running shoe. As a result, the great majority of the research on the kinematics of running footwear has used differences in the midsole characteristics as the independent variable. Many of these studies have been undertaken to relate the kinematics of the foot and ankle to overuse-type injuries. In particular, the effectiveness of footwear in controlling or limiting RF motion is well-documented (Ryan et al. 2010).

The two main foams that are used in the midsoles of modern running footwear are ethyl vinyl acetate (EVA) and polyurethane (PU). The kinematic influences of different midsole characteristics that have been studied extensively involve the hardness (density) of the midsole foam, height of the midsole, and width of the midsole. The main parameters that have been investigated concerning these influences are (1) pronation angle at touchdown, (2) maximum pronation angle, (3) time to maximum pronation angle, and (4) angle range of motion (Figure 21.3).

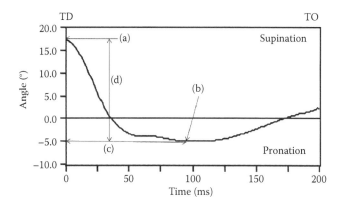

FIGURE 21.3   An exemplar RF angle profile illustrating the parameters normally used in footwear evaluation: (a) RF angle at touchdown, (b) maximum RF angle, (c) time to maximum RF angle, and (d) total range of motion.

The dual function of midsoles (i.e., cushioning and mediolateral control) has led to a number of studies on the hardness of midsoles. It was a logical thought that making midsoles softer (i.e., less firm or less dense) would result in greater cushioning. However, softer midsoles, while possibly accomplishing this cushioning task, had the effect of increasing RF motion or the pronation motion (Stacoff et al. 1988). Hamill et al. (1992) showed that softer midsole shoes resulted in greater RF motion than firmer midsole shoes and that the peak eversion occurred earlier in the support period. It has been suggested that the increase in RF motion or calcaneal eversion may contribute to overuse injuries although prospective studies have shown that the occurrence of injury may not be related to the degree of motion (Nigg 2001).

A further examination of the effect of increasing RF motion, similar to increasing midsole softness, used footwear that were modified to accentuate RF motion (valgus) or limit RF motion (varus). In these studies, there was a significant difference in the amount of pronation between the valgus and the neutral and varus shoes and no difference between the varus and neutral shoes (van Woensel and Cavanagh 1992; Kersting and Newman 2001; Hamill et al. 2009).

The implication of the midsole hardness dichotomy is that a soft midsole would increase pronation angles, which was thought to be a negative in footwear design, while a firm midsole would decrease cushioning (Bates et al. 1983; Stacoff et al. 1988). In order to counteract the effect of a soft or hard midsole was the development of a dual-density midsole with the softer midsole material placed in the lateral heel area where the foot–ground collision occurs and a firmer material on the medial border of the heel to control the degree of pronation (see Figure 21.4). A number of footwear manufacturers instituted the dual-density midsole deign in their product line. In fact, several companies used a tri-density midsole that was graded much like the dual-density midsole but with a moderate level of hardness between the soft lateral and firm medial densities. Shoe manufacturers also conceived of uniform durometer midsoles with pronation control devices attached to the medial border of the shoe to essentially fulfill the same function as the dual-density midsole.

 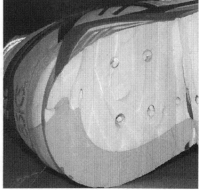

FIGURE 21.4 The dual-density midsole of the ASICS Epirus, a running shoe developed in the 1980s. In this example of a right shoe, the softer (white) material is on the lateral border while the firmer (gray) material is on the medial border.

Another factor that has been studied in terms of the effect on pronation is the width or flare of the midsole. It has been suggested that some midsole flare is beneficial and that a large heel flare can increase both maximum pronation and the total inversion/eversion range of motion (Clarke et al. 1983). Nigg and Morlock (1987) investigated this problem using identical shoes with different degrees of heel flare: (1) a conventional shoe with a 16° flare, (2) a shoe with no heel flare, and (3) a shoe with a rounded or negative heel flare. In this study, increasing heel width increased the magnitude of pronation at initial foot contact. However, there was no difference in the magnitude of total pronation. Interestingly, the shoes with the negative heel width reduced the initial pronation. It was suggested that this effect may have the benefit of preventing overuse injuries such as anterior medial compartment syndrome.

A final parameter, midsole height, has been studied again to determine the appropriate height for stability of the RF during ground contact. While other factors such as midsole durometer and midsole width have an effect on RF kinematics, it has been reported that midsole height has little or no effect on pronation (Clarke et al. 1983). However, other studies have contradicted this assertion. The RF football patern is characterized by a dorsiflexed and slightly supinated foot position at touchdown (Bates et al. 1978a) (Figure 21.10a). In the former study, it was shown that the height to which the heel is raised relative to the FF reduced both maximum pronation and the period of pronation. Using a different definition of heel height (i.e., the distance between the RF plantar surface and the ground), Stacoff and Kaelin (1988) increased heel height from 1.8 to 4.3 cm. They reported that pronation was reduced in heel heights from 2.3 to 3.3 cm but increased above and below that range. On the other hand, Clarke et al. (1983) found no difference in RF kinematics across heel heights ranging from 1.0 to 3.0 cm.

The results of many of the studies previously mentioned must be cautiously interpreted because of methodological limitations of the studies. In addition to the aforementioned limitations of 2-D versus 3-D data capture, another

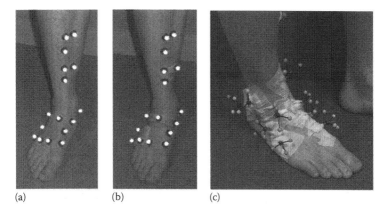

(a)   (b)   (c)

FIGURE 21.5 Marker arrangement using (a) skin mounted markers, (b) markers mounted on plates, and (c) markers on bone pins inserted into nine bones. (Reprinted from Nester, C. et al., *J. Biomech.*, 40, 3412, 2007. With permission.)

significant limitation was that the markers that defined RF motion were placed on the shoe and not on the calcaneus itself (see Figure 21.1). Therefore, it is difficult to ascertain whether it was foot movement that was in fact measured or whether it was the movement of the shoe that was determined. Reinschmidt et al. (1997) performed a study in which they compared foot and shoe movement. In this study, intra-cortical pins with triads of retro-reflective markers were inserted into the calcaneus and tibia of healthy runners. In this study, the inversion/eversion angles were similar across subjects and the shape of the profiles was the same between the shoe markers and the bone pins (see Figure 21.5). However, the shoe markers generally overestimated the magnitude of the angles with the shoe marker having an average magnitude of 16.0° and the bone pin markers 8.8°. The authors concluded that this difference was caused by the relative movement between the shoe and the calcaneus. This study was repeated by O'Connor and Hamill (2002) using external markers placed on the calcaneus and markers placed on the shoe. The results of this study were consistent with those of Reinschmidt and colleagues.

21.4   FOREFOOT KINEMATICS

FF flexion/extension refers to the motion of the metatarsophalangeal (MTP) joint during the latter phase of support. The range of motion of the MTP joint is approximately 30° (i.e., the included angle between the foot and the phalanges with the MTP joint as the axis of rotation). When the heel lifts off the ground at midstance, the MTP joint begins to flex and continue to flex until the foot leaves the ground. This is usually measured kinematically in the sagittal plane although it should be pointed out that the MTP joint does not form a line that is perpendicular to the sagittal plane. Some studies have corrected this by calculating the angle in 3-D using a line across the ball of the foot (i.e., from the first metatarsal to the fifth metatarsal) relative to the longitudinal axis of the foot.

The advantages/disadvantages of flexibility in the FF region have been a source of wide speculation. It has generally been suggested by clinicians that reduced FF flexion may cause or even exacerbate foot injuries (Brody 1980). There are few studies that determine whether reduced or increased flexion is either a positive or negative quality in running footwear.

Recently, Roy and Stefanyshyn (2006) investigated the role of the MTP joint in running and sprinting with a particular emphasis on footwear design. They have suggested that bending of the MTP joint causes a loss of energy during running (Stefanyshyn and Nigg 1997). These researchers suggested that a stiff plate could be placed across the metatarsal heads in order to return energy. However, they reported no significant differences in MTP joint behavior with the stiff FF intervention (see Figure 21.6).

21.5   TORSION ALONG THE LONGITUDINAL AXIS OF THE FOOT

The longitudinal axis of the foot is defined from the midpoint of the calcaneus to the center of the second toe. Until the 1980s, footwear were constructed that generally assumed that the whole foot acted as a rigid body with the whole foot rotating

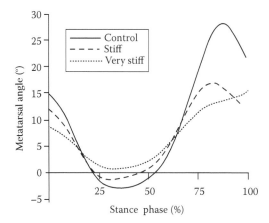

FIGURE 21.6 Metatarsal angle during the stance phase for a typical running subject while wearing shoes of different stiffness. (Reprinted from Stefanyshyn, D.J. and Nigg, B.M., *Med. Sci. Sports Exerc.*, 32(2), 471, 2000. With permission.)

about the longitudinal axis as a single unit. However, by the 1990s the concept of foot torsion about the longitudinal axis evolved. Stacoff et al. (1991) demonstrated that, if you consider the RF and the FF to be two rigid bodies, there is actually an out-of-phase angular relationship between the two segments. That is, the foot does not move as a single rigid body. These authors suggested that foot eversion can be affected by the torsional stiffness of a shoe. A running shoe that was manufactured by Adidas referred to as the torsion shoe was manufactured to take into consideration the torsional characteristics of the foot.

21.6 FOOTWEAR EFFECTS ON COORDINATION

The majority of studies on footwear and lower extremity motion have reported on the action of the joints individually rather that investigating the interaction between these joints (McClay and Manal 1997). It is known that the lower extremity joints and segments are coordinated such that there is a specific pattern to their actions. It is well documented that calcaneal eversion is related to internal tibial rotation and knee flexion. In many instances, the relationships between joints have been assessed using angle–angle diagrams of calcaneal eversion and tibial rotation (Figure 21.7). Van Woensel and Cavanagh (1992) studied the influence of varus and valgus shoes on the timing between calcaneal eversion and knee flexion. They reported that the timing was increased in both of the varus and valgus shoes compared to a neutral shoe condition. While most studies reported the mean time across a number of subjects thus possibly masking individual differences, in this study, the authors reported individual relative times.

The coupling or interaction between eversion, tibial internal rotation, and knee flexion or the lack of timing between these actions has been cited as a possible mechanism for knee pain in runners (Bates et al. 1978a). Bates et al. (1978a) suggested that peak RF eversion should occur at the same time as maximal internal tibial rotation and maximum knee flexion. Using a timing analysis, it was demonstrated that

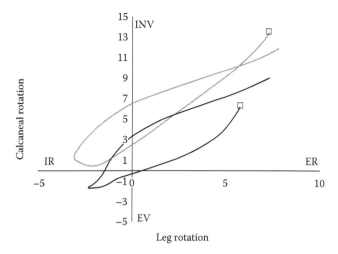

FIGURE 21.7 Angle–angle plot of frontal plan calcaneal inversion (INV) and eversion (ER) and transverse plane tibial internal rotation (IR) and external rotation (ER) during RF (black line) and FF (gray line) running footfall patterns. Data represent the mean of a sample of runners participating in a study at the University of Massachusetts Amherst.

the durometer of the midsole could disrupt this timing (Hamill et al. 1992) (see Figure 21.8). However, it was recently reported that there was much greater variability in this timing that resulted in less tight interactions between these joints (Hamill et al. 1999). In fact, Hamill and associates suggested that greater variability in the coupling coordination of joints might be the preferred state of the system while lower variability may be the less desirable or possibly pathological state.

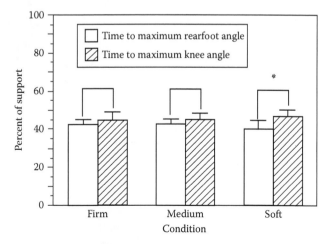

FIGURE 21.8 Timing relationship between the maximum RF angle (maximum eversion) and maximum knee flexion in shoes of different midsole firmness. There was a significant difference in the timing only in the softest midsole. (Reprinted from Hamill, J. et al., *Med. Sci. Sports Exerc.*, 24(7), 807, 1992. With permission.)

It should be clear that footwear can influence the coupling relationships between the lower extremity joints. Most footwear designs attempt to control the motions of the RF and thus control the motions of the other joints. However, as mentioned earlier in this chapter, shoe motion may not truly reflect foot motion inside the shoe. Therefore, future footwear research should investigate the coordination of the lower extremity joints in situations in which markers are placed on the foot and not on the shoe. This relationship may give insight into lower extremity injuries.

21.7 FOOTWEAR AND FOOTFALL PATTERNS

Seventy-five percent of all runners make initial contact with the ground on the heel of the foot (Hasegawa et al. 2007), which is known as the RF footfall pattern. All other runners exhibit a MF pattern or a FF pattern. The MF pattern is characterized by an initial contact on the forward portion, or ball, of the foot before heel contact is made. The FF pattern is characterized by initial contact with the forward portion of the foot while the heel is prevented from touching the ground (Figure 21.9). Until recently, footwear manufacturers created shoes that were built on the characteristic of an RF runner because of the predominant use of this footfall pattern. However, some manufacturers are now recognizing that specific running footwear for MF and FF runners is necessary.

There are significant functional differences between these three footfall patterns that suggest a footwear design for each pattern. The RF footfall pattern is characterized by a dorsiflexed and slightly supinated foot position at touchdown (Bates et al. 1978) (Figure 21.10a). The point of initial contact is made on the lateral boarder of the heel and is within 33% of the foot length relative to the heel (Williams and

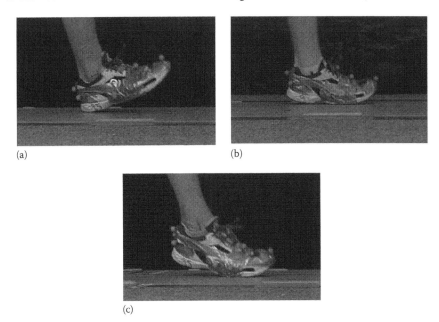

(a)   (b)

(c)

FIGURE 21.9 Footfall patterns exhibited in human running are categorized by the point of initial contact made between the foot and the surface: (a) RF, (b) MF, and (c) FF.

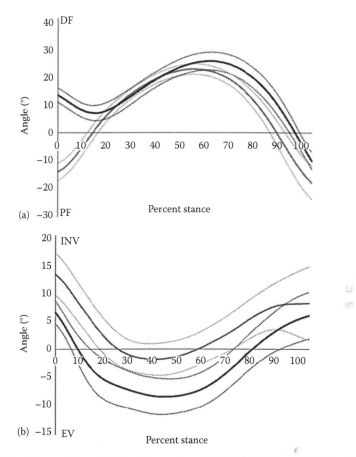

FIGURE 21.10 Mean ± SD ankle joint kinematics of the RF (black line) and FF (gray line) footfall patterns during the stance phase of running for the (a) sagittal and (b) frontal planes. Data represent the mean of a sample of runners participating in a study at the University of Massachusetts Amherst.

Cavanagh 1987). Additionally, an increased heel height has been suggested to lead to greater RF instability (Pratt 1989), whereas others have found a raised heel limits pronation excursion in excessive pronators (Bates et al. 1978b). Maximum eversion is typically coupled with maximum knee flexion. With the FF pattern, the foot lands in a plantar flexed and supinated position, then dorsiflexes and everts as the heel is lowered to the ground (Pratt 1989). The MF pattern will have slightly less plantar flexion at initial contact and heel contact will be made. It should be noted that the MF pattern is an intermediate between the RF and FF patterns; therefore, certain kinematic characteristics of the MF pattern will be similar to the MF pattern in some cases and the FF pattern in other cases.

RF eversion excursion and peak inversion is greater in FF running and remains in inversion for all of stance (Laughton et al. 2003; Gruber and Hamill 2010) (Figure 21.10b). Additionally, FF runners tend to land at a greater degree of inversion which may contribute to greater eversion excursions and eversion velocities.

Greater calcaneal eversion and eversion velocity has been implicated as a significant factor relating to greater incidence of running injuries (McClay and Manal 1997, 1998). Therefore, those who land with the FF may benefit from a motion control shoe. However, a previous study determined orthotic intervention did not greatly affect the RF motion of those that utilize the FF pattern (Stackhouse et al. 2004). Since an orthotic intervention does not greatly affect the FF, orthotic intervention plus a shoe that controls RF motion may not provide a beneficial effect on the kinematics of a FF runner.

FF runners, however, may benefit from a shoe with a stiffer midsole in order to reduce eversion and allow foot and knee joint actions to occur with appropriate timing (Hamill et al. 1992). Controlling the timing between foot eversion and tibial internal rotation may help reduce knee joint stress in both RF and FF runners (Hamill et al. 1992; McClay and Manal 1998). A trend of greater tibial internal rotation has been found to occur with FF running in a previous study (Gruber and Hamill 2010). Since FF running has been shown to have greater eversion excursion and a trend for increased leg internal rotation, FF running may require an appropriate MF shoe construction in order to possibly avoid knee injury.

Some researchers have suggested that the raised heel of modern footwear forces runners to execute the RF pattern (Pratt 1989; Lieberman et al. 2010). These authors suggest an increased heel height will cause the foot to increase the angle between the sole of the shoe and the ground. Additionally, an increased heel height has been suggested to lead to greater RF instability (Pratt 1989), whereas others have found a raised heel limits pronation excursion in excessive pronators (Bates et al. 1978). In a recent study, shoes with a range of midsole heights did not have a significant effect on running footfall pattern until comparing shod and barefoot conditions (Hamill et al. 2011b). All subjects in this study altered footfall pattern to an MF pattern during the barefoot condition. The authors of the Hamill et al. (2011b) study as well as others (DeWit et al. 2000; Squadrone and Gallozzi 2009) suggested the alteration in footfall pattern when running barefoot may occur in order to limit heel pressure reduce the pain that would otherwise occur by hitting an unprotected heel on a hard surface. However, not all habitually shod, RF runners will switch to a FF pattern when barefoot, even on a soft surface (Hamill et al. 2011a). Alternatively, others have found that those who have switched to habitual barefoot running and those that grew up not wearing shoes tend to run with a FF pattern during both shod and barefoot running (Lieberman et al. 2010). These authors believe there is an evolutionary explanation for the occurrence of FF pattern and modern humans were not meant to wear substantial footwear during running.

At present, the benefits of barefoot running are unknown but many anecdotal claims have been made. If barefoot running provides a benefit above that of shod running, it is also currently unknown if the benefits are due to the alteration in footfall pattern or from being unshod. Only a few studies have been published which investigated the kinematic differences between shod and barefoot running. As previously discussed, barefoot running with a FF footfall pattern will produce a flatter, more plantar flexed foot position at initial contact than shod, RF running (DeWit et al. 2000). Barefoot running is also characterized by a smaller calcaneal eversion angle at initial contact and less maximum knee flexion and knee flexion range of motion during stance (Bergman et al. 1995; DeWit et al. 2000; Morio et al. 2009). Barefoot running also results in little change in tibial internal rotation compared to shod running (Stacoff et al. 2000; Eslami et al. 2007). Additionally, barefoot running results in a shorter stride length and increased stride frequency compared to shod, RF running which may be due to foot placement at impact (DeWit et al. 2000).

21.8 FOOTWEAR FOR SPECIFIC TYPES OF RUNNING

Running shoes come in a variety of designs that are distinguished by factors such as weight, material properties, traction, and RF motion. The optimal shoe properties for a particular athlete will depend on the characteristics of their training and competition. Since the late 1970s, footwear manufacturers have developed what are referred to as "anti-pronation" or "motion control" shoes and are categorized by manufacturers as stability (some attempt to limit foot motion) or control (significant pronation control) shoes. These shoes are an example of a shoe used for a specific type of running, in this case distance running. Motion control shoes typically have a firm midsole and a deep heel counter, and are intended to reduce the pronation excursion and slow the pronation velocity. Motion control shoes are intended to protect the knee joint from stresses induced by extraneous subtalar joint motion. Shoes with harder midsoles typically allow less maximum ankle eversion during the stance phase (Clarke et al. 1983). However, these effects are generally small, not systematic, and subject-specific (Nigg 2001). There are currently no definitive links between specific knee injuries and the ability of motion control running shoes to prevent injury (Cheung et al. 2006; Fields et al. 2010).

The other common type of distance running shoe is a cushioned shoe, which has a softer midsole and is intended to attenuate the impact forces between the foot and the ground with little or no attempt to control the pronation actions of the foot. These shoes are intended for runners who do not pronate a great deal and are thus not designed to explicitly reduce or control foot and ankle motion. Running in harder or soft shoes has no significant effect on the magnitude of vertical impact force peaks (Nigg et al. 1987) although this effect is also subject-specific (Bates et al. 1983). Midsole hardness has more consistent effects on estimates of lower extremity internal loading such as muscle and joint contact forces (Wright et al. 1998; Miller and Hamill 2009). There is currently no strong evidence that impact forces cause injuries, or that cushioned running shoes are effective at preventing injuries (Nigg 2001).

Competitive distance runners often train in cushioned or motion control shoes and then compete in a much lighter shoe such as a racing flat or a spike. Running in a racing flat increases the peak RF eversion and the total RF excursion compared to running in training shoes (Hamill et al. 1988). The rate of oxygen consumption needed to run at a particular speed increases as mass is added to the foot (Burkett et al. 1985), suggesting that runners may improve their economy by racing in flats rather than heavier training shoes. The runner sacrifices the benefits of the training shoe for better oxygen consumption. The biomechanics of distance running in racing spikes have not been well studied as they are not well suited to use within a laboratory, although they presumably have similar effects to racing flats as their minimalist designs are similar. The spikes are essentially a lightweight running shoe with spikes in the FF for increased traction. Spikes for sprint running are similar, but have longer spikes and are even more minimal, typically having no heel cushioning at all.

The newest trend in running footwear is the "minimalist" running shoe. This type of footwear is designed with a thin, flexible outsole, a light, very basic upper and little or no heel counter. Minimalist footwear appears to be a response of the footwear industry to the interest in barefoot running. Very little research has been reported on this type of footwear but it is clear that it is a new and potentially large category of footwear for all manufacturers.

21.9   SUMMARY

Footwear design has resulted from the study of the kinematics of the lower extremity with the express purpose of reducing running injuries. In particular, the kinematics of the foot–leg relationship, commonly referred to as RF motion, has been investigated because of the proposed relationship between RF eversion and overuse injury. Researchers have investigated many aspects of footwear construction that appear to influence foot–leg kinematics. While some of these constructions do affect the kinematics of the lower extremity, there is no direct relationship between altered kinematics and injury. Running injuries are multifactorial in nature and not one specific factor will necessarily cause an injury. Footwear in itself may be a risk factor for injury but cannot be the sole factor in causing injury. Nevertheless, running footwear have been subsequently categorized by the industry as cushioning footwear, stability footwear, and control footwear based on the amount of control of RF motion.

Recently, however, this paradigm has been challenged. Several researchers have challenged the notion that the kinematics of the lower extremity result in injury by referring to the fact that the response to changing footwear is very small. That is, the relationship between lower extremity kinematics (i.e., pronation) and injury is very tenuous (Nigg 2001). The relationship, however, is well ingrained in the conscious of footwear manufacturers.

Recent trends in running have influenced footwear manufacturers. The movement toward "barefoot" running has produced the minimal shoe concept. Footwear that have little or no midsole and thus minimal protection for the plantar surface of the foot are being advertised as "barefoot" shoes. The resultant effect of these footwear and in running barefoot is that runners oftentimes change their footfall pattern—from a RF pattern to a FF pattern. More research in the long-term effects of minimal footwear and in altering one's footfall pattern is needed.

QUESTIONS

21.1   Describe each of the three footfall patterns that can be used by runners in terms of the initial foot contact.

21.2   Describe the sequence of events that take place in ankle inversion/eversion during the support phase of running.

21.3   What is the generally accepted role of footwear and how do manufacturers' take this into consideration in footwear design?

21.4   What is the purpose of the dual-density midsole design in running footwear?

21.5   What factors in the design of running footwear can affect foot function during running?

21.6   How are the lower extremity joints coordinated during the support phase of running?

21.7   How are footwear designed for specific types of running?

21.8   What are the industry classifications of footwear and what structures differentiate these classifications?

21.9   What possible alterations in the kinematics of the lower extremity may be altered by barefoot running?

21.10 What are the characteristics of a "minimalist" running shoe?

REFERENCES

Bates, B.T., S.L. James, Osternig, L.R. 1978a. Foot function during the support phase of running. *Running* 3:24–31.

Bates, B.T., L.R. Osternig, B.R. Mason et al. 1978b. Lower extremity function during the support phase of running. In: *Biomechanics VI-B* (edited by E. Asmussen and K. Jorgensen), pp. 30–39. Baltimore, MD: University Park Press.

Bates, B.T., L.R. Osternig, J.A. Sawhill et al. 1983. An assessment of subject variability, subject interaction and the evaluation of running shoes using ground reaction force data. *Journal of Biomechanics* 16:181–191.

Bergman, G., H. Kniggendorf, F. Graichen et al. 1995. Influence of shoes and heel strike on loading of the hip joint. *Journal of Biomechanics* 28(7):817–827.

Brody, D.M. 1980. Running injuries. *CIBA Clinical Symposia* 32:1–36.

Burkett, L.N., W.M. Kohrt, R. Buchbinder. 1985. Effects of shoes and foot orthotics on VO2 and selected frontal plane knee kinematics. *Medicine and Science in Sports and Exercise* 17(1):159–163.

Cheung, R.T., G.Y. Ng, B.F. Chen. 2006. Association of footwear with patellofemoral pain syndrome in runners. *Sports Medicine* 36(3):199–205.

Clarke, T.E., E.C. Frederick, C.L. Hamill. 1983. The effects of shoe design parameters on rearfoot control in running. *Medicine and Science in Sports and Exercise* 15:376–381.

DeWit, B.D., D. DeClercq, P. Aerts. 2000. Biomechanical analysis of the stance phase during barefoot and shod running. *Journal of Biomechanics* 33(3):269–278.

Eslami, M., M. Begon, N. Farapour et al. 2007. Forefoot–rearfoot coupling patterns and tibial internal rotation during the stance phase of barefoot versus shod running. *Clinical Biomechanics* 22(1):74–80.

Fields, K.B., J.C. Sykes, K.M. Walker, J.C. Jackson. 2010. Prevention of running injuries. *Current Sports Medicine Reports* 9(3):176–182.

Frederick, E.C. 1986. Kinematically mediated effects of sport shoe design: A review. *Journal of Sports Sciences* 4:169–184.

Gruber, A.H., J. Hamill. 2010. Segment coordination responses to alterations in footfall pattern. *Brazilian Journal of Biomechanics* 11:1–10.

Hamill, J., B.T. Bates, K.G. Holt. 1992. Timing of lower extremity joint actions during treadmill running. *Medicine and Science in Sports and Exercise* 24(7):807–813.

Hamill, J., P.S. Freedson, W. Boda, F. Reichsman. 1988. Effects of shoe type on cardiorespiratory responses and rearfoot motion during treadmill running. *Medicine and Science in Sports and Exercise* 20(5):515–521.

Hamill, J., A.H. Gruber, J. Freedman et al. 2011a. Are footfall patterns a function of running surfaces? *Proceedings of the Xth Biennial Footwear Biomechanics Symposium*, Tubingen, Germany, pp. 156–170.

Hamill, J., E. Russell, A.H. Gruber et al. 2009. Extrinsic foot muscle forces when running in varus, valgus and neutral shoes. *Footwear Science* 1(3):153–161.

Hamill, J., E. Russell, A.H. Gruber et al. 2011b. Impact characteristics in shod and barefoot running. *Footwear Science* 3(1):33–40.

Hamill, J., R.E.A. van Emmerik, B.C. Heiderscheit et al. 1999. A dynamical systems approach to the investigation of lower extremity running injuries. *Clinical Biomechanics* 14(5):297–308.

Hasegawa, H., T. Yamaguchi, W.J. Kraemer. 2007. Foot strike patterns of runners at the 15-km point during an elite-level half marathon. *Journal of Strength and Conditioning Research* 21(3):888–893.

Kersting, U., K.D. Newman. 2001. Altered midsole geometry modulates rearfoot kinematics and muscle activation during fatigue in running. *Proceedings of the 5th Symposium on Footwear Biomechanics*, Zurich, Switzerland.

Laughton, C.A., I.S. Davis, J. Hamill. 2003. Effect of strike pattern and orthotic intervention on tibial shock during running. *Journal of Applied Biomechanics* 19:153–168.

Lieberman, D.E., M. Venkadesan, W.A. Werbel et al. 2010. Foot strike patterns and collision forces in habitually barefoot versus shod runners. *Nature* 463:531–535.

McClay, I.S., K. Manal. 1997. Coupling parameters in runners with normal and excessive pronation. *Journal of Applied Biomechanics* 13:109–124.

McClay, I.S., K. Manal. 1998. A comparison of three-dimensional lower extremity kinematics during running between excessive pronators and normal. *Clinical Biomechanics* 13(3):95–203.

Miller, R.H., J. Hamill. 2009. Computer simulation of the effects of shoe cushioning on internal and external loading during running impacts. *Computer Methods in Biomechanics and Biomedical Engineering* 12(4):481–490.

Morio, C., M.J. Lake, N. Gueguen et al. 2009. The influence of footwear on foot motion during walking and running. *Journal of Biomechanics* 42(13):2081–2088.

Nester, C., R.K. Jones, A. Liu et al. 2007. Foot kinematics during walking measured using bone and surface mounted markers. *Journal of Biomechanics* 40:3412–3423.

Nigg, B.M. 2001. The role of impact forces and foot pronation: A new paradigm. *Clinical Journal of Sports Medicine* 11:2–9.

Nigg, B.M., H.A. Bahlsen, S.M. Luethi, S. Stokes. 1987. The influence of running velocity and midsole hardness on external impact forces in heel-toe running. *Journal of Biomechanics* 20(10):951–959.

Nigg, B.M., M. Morlock. 1987. The influence of lateral heel flare of running shoes on pronation and impact forces. *Medicine and Science in Sports and Exercise* 19:294–302.

O'Connor, K., J. Hamill. 2002. Does the heel counter control movement of the rearfoot? *Medicine and Science in Sports and Exercise* 34:5, S26.

Pollard, C.D., I. McClay Davis, J. Hamill. 2004. Influence of gender on hip and knee mechanics during a randomly cued cutting maneuver. *Clinical Biomechanics* 19:1022–1031.

Pratt, D.J. 1989. Mechanisms of shock attenuation via the lower extremity during running. *Clinical Biomechanics* 4:51–57.

Reinschmidt, C., A.J. Van Den Bogert, N. Murphy et al. 1997. Tibiocalcaneal motion during running measured with external and bone markers. *Clinical Biomechanics* 12:8–16.

Roy, J.-P., D. Stefanyshyn. 2006. Shoe midsole longitudinal bending stiffness and running economy, joint energy, and EMG. *Medicine and Science in Sports and Exercise* 38(3):562–569.

Ryan, M.B., G.A. Valiant, K. McDonald et al. 2010. The effect of three different levels of footwear stability on pain outcomes in women runners: A randomized control trial. *British Journal of Sports Medicine* 45(9):715–721.

Squadrone, R., C. Gallozzi. 2009. Biomechanical and physiological comparison of barefoot and two shod conditions in experienced barefoot runners. *Journal of Sports Medicine and Physical Fitness* 49:6–13.

Stackhouse, C.L., I.M. Davis, J. Hamill. 2004. Orthotic intervention in forefoot and rearfoot strike running patterns. *Clinical Biomechanics* 19(1):64–70.

Stacoff, A., J. Denoth, X. Kaelin et al. 1988. Running injuries and shoe construction: Some possible relationships. *International Journal of Sports Medicine* 4:342–357.

Stacoff, A., X. Kaelin. 1988. Pronation and shoe design. In: *Biomedical Aspects of Sport Shoes and Playing Surfaces* (edited by B.M. Nigg and B.A. Kerr), pp. 143–151. Calgary, Alberta, Canada: University of Calgary.

Stacoff, A., Kaelin, X., Stuessi, E. 1991. The effects of shoes on the torsion and rearfoot motion in running. *Medicine and Science in Sports and Exercise* 23(4):482–490.

Stacoff, A., B.M. Nigg, C. Reinschmidt et al. 2000. Tibiocalcaneal kinematics of barefoot versus shod running. *Journal of Biomechanics* 33(11):1387–1395.

Stefanyshyn, D.J., B.M. Nigg. 1997. Mechanical energy contribution to the metatarsophalangeal joint to running and sprinting. *Journal of Biomechanics* 30:1081–1085.

Stefanyshyn, D.J. and Nigg, B.M. 2000. Influence of midsole bending stiffness on joint energy and jump height performance. *Medicine and Science in Sports and Exercise* 32(2):471–476.

Stefanyshyn, D.J., P. Stergiou, B.M. Nigg et al. 2003. Do pronators pronate? *Proceedings of the Sixth Symposium on Footwear Biomechanics*, Calgary, Alberta, Canada, pp. 89–90.

Taunton, J.E., M.B. Ryan, D.B. Clement et al. 2002. A retrospective case-control analysis of 2002 running injuries. *British Journal of Sports Medicine* 36(2):95–101.

Van Woensel, W., P.R. Cavanagh. 1992. A perturbation study of lower extremity motion during running. *International Journal of Sports Biomechanics* 8:30–47.

Williams, K.R., P.R. Cavanagh. 1987. Relationship between distance running mechanics, running economy and performance. *Journal of Applied Physiology* 63(3):1236–1245.

Wright, I.C., R.R. Neptune, A.J. van den Bogert, B.M. Nigg. 1998. Passive regulation of impact forces in heel-toe running. *Clinical Biomechanics* 13(7):521–531.

22 Footwear Influences on Running Biomechanics

Gordon A. Valiant, Allison R. Medellin,
Lorilynn Bloomer, and Sharna M. Clark-Donovan

CONTENTS

22.1 INTRODUCTION

Runners who pronate excessively are often prescribed footwear designed to regulate rearfoot motion to reduce risk of developing a running injury (Grau et al. 2008). Runners who do not pronate excessively during stance are often directed to, or prescribed, footwear that is termed neutral (Wischnia and Frederickson 2004). The designation neutral is presumably based on a rearfoot that is not maximally pronated from a neutral subtalar joint as much as it is for excessive pronators. However, the runner's rearfoot angle is constantly changing while the foot is supporting load on the running surface. Just like a stopped clock is correct two times a day, once in the AM and once in the PM, a runner's subtalar joint typically passes through neutral two times during stance. The supinated subtalar joint passes through neutral during initial pronation following foot strike and again during the rearfoot supination that accompanies hip, knee, ankle, and metatarsophalangeal joint extensions. For their footwear selection, runners would have more clarity if they considered attributes such as fit, cushioning, flexibility, durability, mass, influence on energetics, etc., instead of considering a running shoe designated as neutral. This chapter presents how specific running biomechanics imply specialized footwear needs other than fit and rearfoot motion control and discusses how footwear influences biomechanics of running. Discussion of how foot morphology implies specific footwear needs is presented by Krauss and Mauch (2012).

22.2 FOOTWEAR FLEXIBILITY

22.2.1 Barefoot Running Mechanics

The foot is an inherently flexible structure, and running without shoes provides the ultimate experience of foot extension and flexion. The kinematic responses associated with barefoot running, some of which may be beneficial, differ from those associated with running with shoes. Stacoff et al. (1989) documented less calcaneal eversion and more torsion in the midfoot during barefoot running. There are also loading characteristics unique to barefoot running. For example, involvement of a greater surface area beneath the heel for load support, greater loading beneath the lateral midfoot, and a more rapid onset of pressure through the midfoot following heel strike has been observed in investigations in the Nike Sport Research Lab. Also, forefoot plantar pressure distributions indicate all five metatarsal heads are utilized and more involved in load support, and the hallux and the smaller toes are more active during barefoot push off. An active range of motion of up to 65° at the metatarsophalangeal joint is associated with barefoot running. These findings were implemented in the design of a flexible shoe with a unique combination of flex grooves that contributed to some of the plantar pressure distribution characteristics and foot and ankle kinematics associated with barefoot running. Rearfoot pressures were distributed more uniformly when running in the flexible shoe than when running in the conventional shoe. This distribution is more like the unique distribution associated with barefoot running (Figure 22.1). Lateral midfoot, forefoot, and hallux plantar

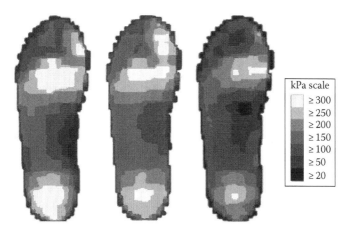

FIGURE 22.1 Peak plantar pressure distributions when running barefoot on grass (left), when running in Nike Free footwear (middle), and when running in footwear with a conventional foam and encapsulated air bladder midsole (right). Maximum pressures are developed at different regions at different times during stance. (Data from Pisciotta, J., Woo, H., Mientjes, M., Bloomer, L., Gregg, M., Nike Sport Research Lab, Beaverton, OR.)

pressure distributions when running in the flexible shoe were also more similar to the barefoot running distributions than to the conventional running shoe distributions.

22.2.2 Benefits Associated with Flexibility

Training with an extremely flexible running shoe has been demonstrated to provide benefits. Brüggemann et al. (2005) demonstrated changes in muscle strength and muscle cross-sectional area in subjects who wore the flexible shoe during their warm-up sessions for a 5 month training period. Beuke and Graumann (2008) also found increases in strength when their subjects wore the extremely flexible running shoe for a portion of a 12 week marathon training program. In addition, Ryan et al. (2009) demonstrated that, when worn as a part of the treatment plan for chronic plantar fasciitis, a very flexible shoe elicited earlier pain relief during the 12 week rehabilitation program.

The lever arm about the ankle in the sagittal plane is reduced with increased shoe flexibility at the metatarsophalangeal joint, reducing the load on the Achilles tendon and triceps surae muscle group (Fuller 1994). When a shoe is flexible about a longitudinal axis, especially in the heel, the lever arm about the ankle joint in the frontal plane is decreased, which can slow the rate of eversion, and which may reduce risk of injury linked to excessive rearfoot motion (Grau et al. 2008). Flexibility that allows frontal plane movement of the forefoot with respect to the rearfoot can also be beneficial in a running shoe. Stacoff et al. (1989) showed a decreased magnitude of pronation when subjects ran wearing shoes that allowed for torsional flexibility. In addition, flexibility of a running shoe midsole can be linked to greater distribution of plantar pressures, which can be associated with better cushioning and comfort (Miller et al. 2000). Conversely, if a shoe is not flexible, discomfort can arise from concentrated regions of high plantar pressure.

While an extremely flexible shoe can elicit training benefits, there are times when a stiffer shoe may be better for performance. Stefanyshyn and Fusco (2004) found that increasing shoe bending stiffness elicited a significant increase in sprinting performance. In addition, Roy and Stefanyshyn (2006) found decreased oxygen consumption rates when subjects ran wearing a stiffer shoe. Thus, footwear flexibility is a consideration for footwear selection and a balance between flexibility and stiffness may be dependent on runners' personal performance needs.

22.3 MIDSOLE CUSHIONING

Running shoe cushioning provides comfort and protection from both transmitted impacts (Valiant 1990) as well as from excessively high localized plantar pressures (Shorten 1989) developed during running, and is thus another consideration when selecting footwear.

22.3.1 Attenuation of Running Impacts

Attenuation of the impact forces associated with running is provided by the compressibility of midsole materials. Softer materials and thicker materials are more compressible and provide more attenuation of impacts than firmer or thinner, less

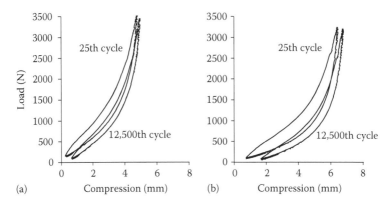

FIGURE 22.2 Response to load-controlled displacement to ~3500 N on a servohydraulic Instron after 25 cycles and 12,500 cycles for a gas-inflated bladder encapsulated in an 11 mm EVA midsole (a) and an 11 mm compression molded EVA midsole (b).

compressible materials (Lafortune et al. 1996). The compressibility of low-density foams makes them common candidates for running shoe midsole material. However, as the cellular walls of foams break down in response to repeated impacts (Brückner et al. 2010), midsoles gradually lose thickness and hence lose some compressibility. Thus, midsole cushioning becomes reduced with wear.

Gas-inflated bladders are also compressible and can be an effective, alternate component for provision of midsole cushioning. A gas-inflated bladder encapsulated in an EVA midsole will not wear out in response to the repeated impacts of running as much as foam midsoles. For example, after 12,500 cycles of load-controlled displacement on a servohydraulic Instron, a foam-encapsulated air cushioning component becomes stiffer and less energy absorbent (Figure 22.2). However, the changes in response to the cycled loading are less than the changes in response of an 11 mm compression molded EVA foam midsole to a similar loading cycle. A loss in resting height of the foam midsole of nearly 2 mm occurs after 12,500 cycles. After 48 h of recovery, there was still a permanent 14% compression set of the foam. This is at least twice the loss in resting height of the combined foam and air after 12,500 cycles. Linear estimates of stiffness obtained from the rising portion of the hysteresis curve at 1800 N, which approximates a typical peak forefoot running load for a 666 N runner, are plotted as a function of number of cycles, on a logarithmic scale, in Figure 22.3. Greatest increases in stiffness occur in the first five compressions. The virgin sample of combined air and foam is stiffer than the virgin foam sample, but at about 20 cycles, average stiffness is similar, and beyond 20 cycles the stiffness of the foam-only midsole is greater than the stiffness of the foam combined with air. In response to the many repeated impacts associated with distance running gaits, air bladders compress and then expand back to resting volume with less breakdown than comparable foam midsoles and thus may provide a more durable, longer lasting alternative cushioning component.

An alternative design to traditional foam midsoles incorporates foam columns inserted beneath the rearfoot between top and bottom plates to reduce stress concentrations and cushion impact forces (Figure 22.4). Different target compliances

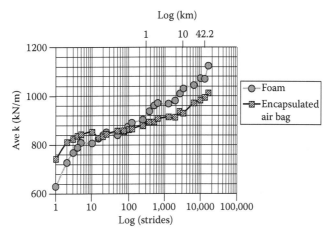

FIGURE 22.3 Changes in average stiffness, associated with 1800 N of compressive load, of 11 mm foam midsoles in response to cycles of load-controlled displacement to 3500 N on a servohydraulic Instron as a function of the number of loading cycles or the number of equivalent marathon strides, plotted on a logarithmic scale.

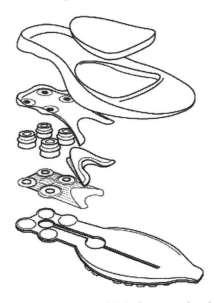

FIGURE 22.4 Sketch of a running shoe midsole incorporating four 22 mm tall × 29 mm diameter (U.S. size M9) compliant foam columns between top and bottom plates for provision of rearfoot cushioning.

for favorably rated cushioning can be achieved from combinations of different foam and plate material and architecture (Valiant et al. 2001). At heel strike, loads are initially opposed by just the lateral columns. Therefore, initial stiffness opposing compression is less than that provided by a larger conventional midsole cross section, so compression of the lateral midsole becomes exaggerated. Since this exaggerated compression requires extra time, the lateral loading dwells longer as evidenced by

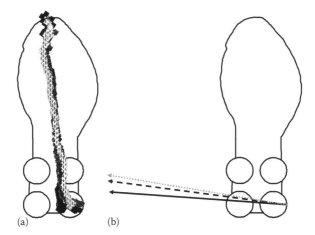

FIGURE 22.5 (a) Example center of pressure at 1 ms intervals when running in footwear, the rearfoot of which includes compliant rearfoot columns (solid black fill), gas-inflated bladder encapsulated in foam (light vertical fill), and homogeneous foam (dark horizontal fill). (b) Initial direction of progression of the center of pressure linearly regressed from the first 4 samples (1000 Hz), averaged for 8 running steps by 21 subjects, when wearing running shoes with compliant rearfoot columns (solid), gas-inflated bladder encapsulated in foam (light dotted), and homogeneous foam (dark dashed).

a delay in the progression of the vertical loading medially to the central rearfoot. In addition to the delayed progression of the center of pressure, the initial progression is also directed more medially or less distally than in other midsole cushioning technologies (Figure 22.5). The extended time to decelerate the impacting rearfoot and compress the columns contributes to cushioning by slowing the rate of increase of vertical impact forces (Figure 22.6).

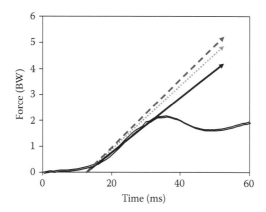

FIGURE 22.6 The thick black line represents the first 60 ms of the vertical component of the ground reaction force, normalized to runner's body weight, typically associated with the impact of a heel strike running gait. Maximum rate of vertical loading, averaged for 21 subjects, 8 trials each, when running in footwear, the rearfoot of which include compliant rearfoot columns (solid), gas-inflated bladder encapsulated in foam (light dotted), and homogeneous foam (dark dashed).

22.3.2   REDISTRIBUTION OF PLANTAR PRESSURES

In addition to attenuating transmitted running impacts, running shoe midsoles are also designed for redistributing localized forces across greater areas beneath the foot. Excessively high plantar pressures associated with some of the smaller metatarsal heads or even the heel pad may be uncomfortable for the runner and may put some of these structures at risk of injury.

Running loads developed on top of soft, compressible foam engage an area greater than the plantar area of the foot. For example, the plantar loads developed when running barefoot across a layer of soft, compressible foam, plotted in Figure 22.7, are redistributed over an area that is greater than the weight-bearing area of the foot when running across just the pressure-measuring transducer. Redistribution across a greater cross-sectional area reduces peak magnitudes of localized load, enhancing comfort and contributing to protection. A further increase in softness and compressibility of underlying foam contributes to engagement of an even greater area for distribution of plantar pressures and greater reductions in peak magnitudes of localized load.

Midsole designs incorporating foam columns in the rearfoot also contribute to redistribution of plantar pressures. Consider a frontal cross section passing between the rearfoot columns corresponding to the center of the heel (Figure 22.8). Pressure distribution beneath the rearfoot is very uniform. The distribution associated with running in other midsole technologies is nonuniform with higher peak pressures beneath the central heel. The greater role in supporting vertical loading by the compressible columns on the outsides of the midsole diminishes the role of the central rearfoot.

22.3.3   INFLUENCE ON RUNNING KINEMATICS

Cushioning provided by running shoe midsoles is another attribute that influences running kinematics. For example, as cushioning decreases with the midsole wear associated with the accumulation of 200 miles of running, maximum ankle dorsiflexion is reduced and plantarflexion at toe-off is increased (Kong et al. 2009). Also, stance time was increased and there was less forward lean of the runners' trunk when running in the worn footwear. Increased knee flexion velocity (Clarke et al. 1983) and increased ankle dorsiflexion velocity (Hardin et al. 2004) during running are also associated with increased running shoe midsole hardness.

The high compressibility of soft shoe midsoles providing enhanced cushioning is often associated with unfavorable contributions to increased rearfoot pronation. Therefore, constructing rear medial midsoles from denser, less compressible foam or other components is the most common design approach for regulation of excessive rearfoot pronation, albeit at the expense of cushioning. However, running shoe midsoles that provide good cushioning characteristics can also be designed to positively influence subtalar joint kinematics by incorporating exceptionally compliant, compressible, rear lateral foam components. In the same way that the compliance of the lateral midsole columns in the example in Figure 22.4 contributes to a reduced rate of impact loading, exaggerated compression of a compliant lateral rear midsole can hold the foot in supination for a longer time, thereby contributing to a reduced

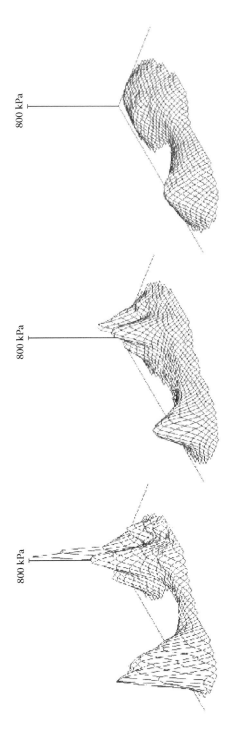

FIGURE 22.7 Peak plantar pressure distribution developed during barefoot running across an EMED platform (novel, gmbh) (left), across a layer of firm foam (middle), and across a layer of soft foam (right). Note that maximum pressures are developed at different regions at different times during stance.

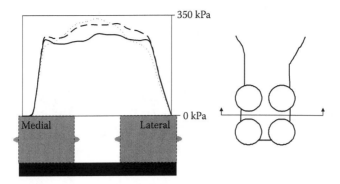

FIGURE 22.8 Peak plantar pressures collected with a 37 sensor pedar insole sampled at 250 Hz (novel, gmbh) plotted at a cross section passing between the columns, the location corresponding to the center of the heel, when running in footwear with compliant rearfoot columns (solid), gas-inflated bladder encapsulated in foam (light dotted), and homogeneous foam (dark dashed). Lateral and medial columns and an outsole thickness included for orientation (n = 19).

rate of rearfoot pronation. The greater the compression of the lateral midsole, the smaller is the moment arm contributing to rearfoot pronation. Isolating the compliant lateral midsole components reduces the influence on compression from adjoining less compressible cushioning components and more effectively reduces the leverage contributing to rearfoot pronation.

Hence, due to the influence of running shoe cushioning on impact transmission, plantar pressure distribution, certain kinematic measures, and subjective preferences of comfort, cushioning is another footwear attribute to consider for the selection of appropriate running shoes.

22.4 RUNNING SHOE OUTSOLE TRACTION AND DURABILITY

Outsole traction requirements are also dictated by running biomechanics. The vertical component of the ground reaction force holds the running shoe outsole against the running surface while the horizontal components contribute to relative motion or slip between shoe outsole and running surface. Therefore, the ratio of horizontal to vertical ground reaction force components developed during running identifies minimum traction coefficients required to prevent slip while runners develop horizontal forces for braking and propulsion. During the early braking phase of running, the ratio of horizontal to vertical forces peaks at 0.6 (Figure 22.9). Since the vertical impact force that rapidly increases following foot strike has a large magnitude relative to the horizontal force component, the braking phase of running generally has lower traction demands on the running shoe outsole than the propulsive phase. During the propulsive phase, the peak traction ratio is 0.7. Thus, coefficients of friction exceeding 0.7 for the fore part of shoe outsoles interacting with running surfaces are typically sufficient to prevent slip from a propulsive force component exerted in a backward direction during pushoff.

Providing adequate traction on wet surfaces is usually the greatest challenge in shoe outsole design. Molding outsoles in synthetic rubber compounds that contribute

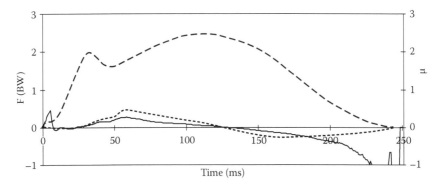

FIGURE 22.9 Vertical component (dashed) and anteroposterior component (dotted) of ground reaction force developed while running at 3.35 m/s over a synthetic rubber running surface. Ratio (μ) of anteroposterior to vertical force components (solid). Note as vertical force approaches zero during unloading when only the distal tip of the shoe may contact the playing surface high ratios with absolute values >1 have less relevance.

to higher friction forces and especially molding rubber compounds in waffle or lug-shaped projections assist in the development of high traction forces on running surfaces that may be wet or slippery.

The wear of running shoe outsoles is also influenced by the biomechanics of running. Wear of running shoe outsoles composed of multiple layers of smooth pattternless rubber was evaluated for wear from a large sample of running field testers. Wearing through different colored layers of rubber after accumulating weeks of running mileage revealed areas of high wear, moderate wear, and minimal wear, as indicated in Figure 22.10. High wear is associated with shear loading beneath the rear lateral outsole at heel strike, and with shear loads developed beneath the central forefoot during propulsion. Moderate wear is associated with the lateral border of the outsole and much of the forefoot region. Minimal wear is associated with the medial rearfoot and with the entire midfoot region except the lateral border.

Hence, an understanding of the regional associations between outsole durability and running shoe outsole design and between outsole traction and design will influence runners' decisions about appropriate outsole design when selecting running footwear.

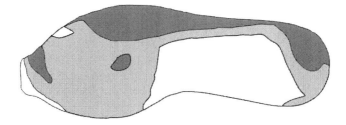

FIGURE 22.10 Average wear of smooth rubber outsoles after equivalent accumulation of running miles by multiple subjects. Regions of high wear (■), moderate wear (▨), and minimal wear (□) denote loss of >1.0, >0.5, and <0.5 mm of outsole thickness, respectively. (Data from Potter, D., Nike Sport Research Lab, Beaverton, OR.)

22.5   RUNNING SHOE INFLUENCE ON ENERGETICS

The association between the elastic characteristics of running shoe midsole materials and the economy of running has been studied. However, storage of strain energy in running shoe midsoles and subsequent reutilization of stored elastic energy with each foot strike has been shown to have a negligible influence on energetics of running. Shorten (1989) demonstrates that a compressed foam running shoe midsole model may store at best 5 J of strain energy, and that this quantity of strain energy may account for only a negligible 0.3 mL O_2/kg/min.

An alternative possibility for positive influence by footwear design on energetics of running has been suggested by Stefanyshyn and Nigg (1997). They propose that athletic footwear can have a larger influence on economy by minimizing energy which is lost as opposed to maximizing stored energy which is returned to the runner with each foot strike. They demonstrate that the metatarsophalangeal joint absorbs 21 J of energy during a running foot step. Since this joint remains in an extended position during support, the lack of flexion prohibits the generation of energy during takeoff. Thus, this joint is only an energy absorber. They then demonstrate that increased bending stiffness of shoe midsoles can contribute to reduction in energy loss at the metatarsophalangeal joint and a potential for influencing energetics of running.

The mass of running footwear rather than the compliance of running footwear may have more of an influence on energetics of running. The energetic cost of carrying extra weight on the feet has been reviewed and discussed and a model has been proposed to explain a portion of the documented effects. A 100 g increase in shoe weight increases the aerobic demands of running by an average of +1% at speeds between 230 and 293 m/min, and this effect may diminish as the velocity of running increases (Frederick et al. 1984). Much of the increase in submaximal energy cost can be accounted for by the increase in potential energy of lifting the extra 100 g vertically 50 cm against gravity during each running stride plus the increase in kinetic energy of accelerating the extra 100 g forward from 0 m/s at toe-off to approximately twice the running speed (Frederick 1985). Hence, mass rather than capability for storage and reutilization of strain energy is the attribute that more appropriately influences economy of running and choice of footwear.

22.6   SUMMARY

The design of running shoes has an influence on many aspects of running biomechanics. Cushioning characteristics influence transmission and attenuation of running impacts as well as the distribution of localized plantar pressures. Kinematics of the ankle, knee, and metatarsophalangeal joints, as well as many other kinematic aspects, are influenced by cushioning characteristics and also by flexibility characteristics of running shoes. Outsole traction requirements and outsole wear are also dictated by biomechanics of running. Runners should consider their running biomechanics as well as their foot morphology when selecting appropriate running footwear. Designating running shoes as neutral does not address any of these influences and only informs the runner that the shoe is not designed to regulate excessive rearfoot pronation. More clarity is afforded to the runner by describing footwear by provision of specific attributes such as cushioning, the shape of the upper for fit, flexibility, mass, durability, and traction or by a combination of attributes.

QUESTIONS

22.1 What are some possible performance benefits of stiffer footwear midsoles? What are some benefits associated with flexible midsoles?

22.2 Name some differences in kinematics and in loading variables between bare-foot running and shod running.

22.3 What is the primary mechanism by which running shoe midsoles attenuate impact force and reduce peak localized plantar pressures?

22.4 What are some of the effects of decreased midsole cushioning on running kinematics?

22.5 What footwear design factor may have the greatest influence on the energy cost of running?

22.6 Considering your own anthropometrics, lower extremity kinematics, and running style, what attributes do you feel would be most important when selecting your own personal running footwear?

REFERENCES

Beuke, B. and Graumann, L. 2008. Nike Free running study. In *Proceedings of 7th ISEA Conference*, pp. 1–5, Biarritz, France.

Brückner, K., Odenwald, S., Schwanitz, S. et al. 2010. Polyurethane-foam midsoles in running shoes—Impact energy and damping. *Procedia Engineering* 2: 2789–2793.

Brüggemann, G.P., Potthast, W., Braunstein, B. et al. 2005. Effect of increased mechanical stimuli on foot muscles functional capacity. In *Proceedings of the International Society of Biomechanics XXth Congress*, p. 553, Cleveland State University, Cleveland, OH.

Clarke, T.E., Frederick, E.C., Cooper, L.B. 1983. Biomechanical measurement of running shoe cushioning properties. In *Biomechanical Aspects of Sport Shoes and Playing Surfaces*, B.M. Nigg and B.A. Kerr (eds.), pp. 25–33. Calgary, Alberta, Canada: University of Calgary.

Frederick, E.C. 1985. The energy cost of load carriage on the feet during running. In *Biomechanics IX-B*, D.A. Winter, R.W. Norman, R.P. Wells, K.C. Hayes, A.E Patla (eds.), pp. 295–300. Champaign, IL: Human Kinetics Publishers.

Frederick, E.C., Daniels, J.R., Hayes, J.W. 1984. The effect of shoe weight on the aerobic demands of running. In *Current Topics in Sports Medicine*, N. Bachl, L. Prokop, R. Suckert (eds.), pp. 616–625. Vienna, Austria: Urban & Schwarzenberg.

Fuller, E.A. 1994. A review of the biomechanics of shoes. *Clinics in Podiatric Medicine and Surgery* 11: 241–258.

Grau, S., Maiwald, C., Krauss, I. et al. 2008. What are causes and treatment strategies for patellar-tendinopathy in female runners? *Journal of Biomechanics* 41: 2042–2046.

Hardin, E.C., Van Den Bogert, A.J., Hamill, J. 2004. Kinematic adaptations during running: Effects of footwear, surface, and duration. *Medicine and Science in Sports and Exercise* 36: 838–844.

Kong, P.W., Candelaria, N.G., Smith, D.R. 2009. Running in new and worn shoes: A comparison of three types of cushioning footwear. *British Journal of Sports Medicine* 43: 745–749.

Krauss, I. and Mauch, M. 2012. Foot morphology. In *Science of Footwear*, R.S. Goonetilleke (ed.), pp. 19–46. Boca Raton, FL: CRC Press.

Lafortune, M.A., Hennig, E.M., Lake, M.J. 1996. Dominant role of interface over knee angle for cushioning impact loading and regulating initial leg stiffness. *Journal of Biomechanics* 29: 1523–1529.

Miller, J.E., Nigg, B.M., Liu, W. et al. 2000. Influence of foot, leg and shoe characteristics on subjective comfort. *Foot and Ankle International* 21: 759–767.

Roy, J.-P.R. and Stefanyshyn, D.J. 2006. Shoe midsole longitudinal bending stiffness and running economy, joint energy, and EMG. *Medicine and Science in Sports and Exercise* 38: 562–569.

Ryan, M., Fraser, S., McDonald, K. et al. 2009. Examining the degree of pain reduction using a multielement exercise model with a conventional training shoe versus an ultraflexible training shoe for treating plantar fasciitis. *Physician and Sportsmedicine* 37: 68–74.

Shorten, M.R. 1989. Elastic energy in athletic shoe cushioning systems. In *XII International Society of Biomechanics Congress Proceedings*, R.J Gregor, R.F. Zernicke, W.C. Whiting (eds.), p. 120, UCLA, Los Angeles, CA.

Stacoff, A., Kaelin, X., Stuessi, E. et al. 1989. The torsion of the foot in running. *International Journal of Sport Biomechanics* 5: 375–389.

Stefanyshyn, D. and Fusco, C. 2004. Increased shoe bending stiffness increases sprint performance. *Sports Biomechanics* 3: 55–66.

Stefanyshyn, D.J. and Nigg, B.M. 1997. Mechanical energy contribution of the metatarsophalangeal joint to running and sprinting. *Journal of Biomechanics* 30: 1081–1085.

Valiant, G.A. 1990. Transmission and attenuation of heel strike accelerations. In *Biomechanics of Distance Running*, P.R. Cavanagh (ed.), pp. 225–247. Champaign, IL: Human Kinetics Books.

Valiant, G.A., Kilgore, B.J., Tawney, J. et al. 2001. Development of a new running shoe cushioning system. In *Proceedings of the 5th Symposium on Footwear Biomechanics*, E. Hennig, A. Stacoff, H. Gerber (eds.), pp. 78–79, Laboratory for Biomechanics, Department of Materials, ETH, Zürich, Switzerland.

Wischnia, B. and Frederickson, R. March 2004. The spring 2004 shoe buyer's guide. In *Runner's World*, pp. 33–73. Emmaus, PA: Rodale, Inc.

23 Effects of Footwear on Muscle Function

George S. Murley and Karl Landorf

CONTENTS

23.1 INTRODUCTION

Footwear is a fundamental piece of clothing that is worn in most communities throughout the world. Footwear protects the foot from extraneous trauma, it can provide comfort, and it is also a fashion item. It addition, certain footwear, for example, running shoes, are purported to provide improvements in performance and reductions in injury. Footwear can have a significant effect on the musculoskeletal system and there has been a substantial amount of research relating to these effects. One area of the musculoskeletal system that footwear has an effect on is *muscle function*. This chapter, therefore, concentrates on the effects of footwear on muscle function. To ensure a comprehensive and transparent review of the literature on this topic, and to minimize bias in the selection of information, we systematically searched the literature to incorporate all relevant studies.

23.2 SYSTEMATIC LITERATURE REVIEW DETAILS

The following section presents details of how we searched for information relating to the effects of footwear on muscle function and the results of the search undertaken. Titles and abstracts from 390 studies were reviewed following a literature search of major databases (MEDLINE and CINAHL®). Of the studies reviewed, 34 investigated the *effect of footwear on lower limb muscle activity during walking or running* and were subsequently included in the evidence presented in this chapter.

The most common styles of footwear investigated were athletic or running style shoes (19 studies), high-heeled or negative-heeled shoes (7 studies), and rocker-sole and unstable style shoes (6 studies). Of these studies, 22 investigated participants during walking and 10 during running (including two that comprised both walking and running).

Muscle activity was generally assessed using two methods: (1) electromyography (EMG) and (2) muscle function magnetic resonance imaging (mfMRI). However, studies investigating the effect of footwear on muscle function have almost exclusively used EMG, except two studies that have used mfMRI. EMG usually involves mounting surface electrodes on the skin overlying individual muscles to sample the electrical activity evoked by motor units during muscular contraction. Occasionally, researchers will use indwelling electrodes to access deeper muscles within the body. In contrast, mfMRI involves the measurement of exercise-induced changes in muscle cell metabolism and muscle fluid uptake following exercise and is reflected by an increase in MRI signal intensity (O'Connor et al., 2006; Murley, 2008).

The advantage of using EMG over other modalities for assessing muscle activation, such as MRI or ultrasound, is the ability to investigate muscle activity simultaneously with dynamic weight-bearing tasks such as walking (Semple et al., 2009). This allows interpretation of muscle activity specific to different phases of the gait cycle. There is a wide range of EMG parameters reported in the literature that relate to muscle activity, including temporal (timing) and intensity variables. Temporal variables include onset, duration, and time to maximum activity. Intensity variables include wavelet analysis, integrated and normalized peak amplitude. Furthermore, EMG signals can be evaluated at a range of different stages of the gait cycle (e.g., preheel strike phase, propulsion phase, etc.), which provides important information on dynamic everyday tasks. (For additional technical information relating to EMG, readers are directed to the work by Cram and Criswell [2011]).

The following section presents a summary of the findings of studies investigating the effect of footwear on lower limb muscle activity. Where possible, the various styles of footwear are broadly classified and discussed according to shoe structure or a specific shoe feature (i.e., high-heeled and negative-heeled shoes). This chapter does not provide a systematic quality assessment (i.e., methodological quality) of the literature; for this, we redirect readers to the work of Murley et al. (2009).

23.3 DISCUSSION

The following section discusses the findings from the studies that were identified in our systematic search of the literature. To aid discussion, we categorized studies into four broad areas, with each representing a different type of footwear, including the following: (1) athletic or running shoes, (2) rocker-sole and unstable style shoes, (3) high-heeled and negative-heeled shoes, and (4) occupational-specific and other shoes. We now discuss the studies relating to each type of footwear.

23.3.1   ATHLETIC OR RUNNING SHOES

Over several decades spanning back to the late 1980s, subtle variations in the properties of athletic footwear, such as alterations in heel counter stiffness (Jorgensen, 1990) and midsole density or stiffness (Komi et al., 1987; Wakeling et al., 2002; von Tscharner et al., 2003; Roy and Stefanyshyn, 2006) have been investigated almost exclusively during running. Various styles of athletic footwear have been used to investigate issues, such as the specific role of lower limb muscles (O'Connor and Hamill, 2004; O'Connor et al., 2006), differences between barefoot and shod running (Komi et al., 1987; Jorgensen, 1990; Ogon et al., 2001; von Tscharner et al., 2003; Divert et al., 2005), attenuation of muscle fatigue in those with overpronating feet (Cheung and Ng, 2010), and tuning of lower limb muscles through alteration in sensory input signals (Boyer and Nigg, 2004). The results of these studies are presented in Table 23.1.

23.3.1.1   Midsole Elasticity/Viscosity/Hardness

Researchers in this area of interest have primarily focused on investigating the potential to "tune" muscle activity by manipulating the impact force. In this instance, the impact force is manipulated through wearing shoes with midsoles of varying hardness, or no shoes at all (i.e., barefoot). Researchers speculate that tuning muscles by altering the input signal (i.e., impact force) could provide insight into the mechanisms that alter muscle activity during running, and as a consequence, may assist our understanding of the causes of lower limb injury.

Two studies used a process related to EMG wavelet analyses to quantify total EMG amplitude (von Tscharner et al., 2003) and EMG frequency characteristics (Wakeling et al., 2002) while comparing various running shoes with different midsole hardness or density. von Tscharner et al. (2003) found significantly greater EMG intensity for tibialis anterior preheel strike and lower intensity postheel strike with running shoes compared to barefoot. Wakeling et al. (2002) investigated the effects of running with a hard and soft insole in a small sample of six male and female runners. They reported significant changes in the intensity ratio between high- and low-frequency bands of selected muscles, although they did not report the post hoc findings for any muscle or shoe effects.

A further two studies investigated differences between barefoot and shod running (Ogon et al., 2001; Divert et al., 2005). Ogon et al. (2001) evaluated erector spinae (L_3) at jogging speeds in 12 healthy volunteers. They reported that jogging barefoot was associated with a significantly earlier onset during heel strike, however, a more prolonged latency to peak acceleration of the lower back during heel strike. The authors hypothesized that these findings reflect the benefits of running shoes in temporally synchronizing destabilizing external forces with stabilizing internal forces around the lumbar spine. Divert et al. (2005) investigated several lower leg muscles in 35 primarily male runners. In contrast

TABLE 23.1
Athletic or Running Shoes

Author/s (Date)	Participant Characteristics Age (Standard Deviation)	Footwear/Test Conditions	Muscles	Task	EMG Variables	Main Findings
Boyer and Nigg (2004)	10 Male subjects who exercised regularly Age: 25 (standard error 4.2) years	(Heel hardness—Asker C) (1) Elastic heeled shoe (52) (2) Viscous heeled shoe (50) (3) Viscoelastic (40) (4) Viscoelastic (55) (5) Viscoelastic (70)	Rec. femoris Bic. femoris Tib. anterior Med. gastroc	Running	Wavelet analysis (intensity derived from time and frequency)	No significant differences between conditions
Cheung and Ng (2009)	20 Novice female runners Age: 25.8 (±3.7) years	(1) Control "cushion" shoe (2) Motion-control shoe	Vast. medialis Vast. lateralis	Running	Onset of amplitude	Vast. medialis—significantly earlier onset relative to vast. lateralis with motion-control shoe compared to control "cushion" shoe
Cheung and Ng (2010)	20 Novice female runners Age: 25.8 (±3.7) years	(1) Neutral shoe (2) Motion-control shoe	Tib. anterior Per. longus	Running	Normalized RMS amplitude and median frequency	Per. longus and tib. anterior—significant positive correlation between RMS amplitude and mileage in the neutral shoe Per. longus—significantly larger drop in median frequency with neutral shoe

Divert et al. (2005)	35 Healthy runners (31 males, 4 females) Age: 28 (±7) years	(1) Barefoot (2) Standard running shoe	Tib. anterior Per. longus Med. gastroc Lat. gastroc Soleus	Running	Mean EMG amplitude	Med. gastroc, lat. gastroc, soleus—significantly greater amplitude with barefoot condition compared to running shoe
Jorgensen (1990)	11 Symptom-free heel strike runners (5 females, 6 males) Age: 25.5 years (range 14–37)[a]	(1) Barefoot (2) Athletic shoe with rigid heel counter (3) Athletic shoe with heel counter removed	Hamstrings[b] Quadriceps[b] Triceps surae[b] Tib. anterior	Running	Normalized EMG amplitude Time to peak amplitude No. of turns in EMG signal	Triceps surae and quadriceps— significantly earlier activity, greater amplitude and no. of turns with heel counter removed compared to shoes with rigid heel counter
Komi et al. (1987)	4 Males with "athletic background" Age: 32 (±9.4) years	(1) Barefoot (2), (3), and (4) 'Jogging shoes" (5) and (6) "Indoor shoes" (indoor shoes comprised harder sole characteristics)	Rec. femoris Vast. medialis Lat. gastroc Tib. anterior	Running	Mean and integrated EMG	No significant differences between conditions
Nigg et al. (2003)	20 Male runners free from serious injury[a]	(1) Shoe with mainly elastic heel (shore C = 45) (2) Shoe with softer more viscous heel (shore C = 26)	Tib. anterior Med. gastroc Vast. medialis Hamstring group[b]	Running	RMS amplitude	No significant differences between conditions
O'Connor and Hamill (2004)	10 Males ("rearfoot strikers") Age: 27 (±5) years	(1) Running shoe—neutral (2) Running shoe—medial wedge (3) Running shoe—lateral wedge (EVA rearfoot wedge tapered by 1 cm across heel of midsole + no heel counter on shoes)	Med. gastroc. Lat. gastroc Soleus Tib. posterior Tib. anterior Per. longus	Running	Integrated and mean EMG EMG onset and offset	No significant differences between conditions

(continued)

TABLE 23.1 (continued)
Athletic or Running Shoes

Author/s (Date)	Participant Characteristics Age (Standard Deviation)	Footwear/Test Conditions	Muscles	Task	EMG Variables	Main Findings
O'Connor et al. (2006)	10 Male "rearfoot strikers" Age: 27 (±5) years	(1) Running shoe—neutral (2) Running shoe—medial wedge (3) Running shoe—lateral wedge (EVA rearfoot wedge tapered by 1 cm across heel of midsole + no heel counter on shoes)	Med. gastroc Lat. gastroc Soleus Tib. posterior Tib. anterior Ext. dig. longus Per. longus	Running	MRI transverse relaxation times (T2). Average of five slices through muscle belly (scan occurred 3.6 ± 0.3 min after run completed)	Tib. anterior—significantly less EMG activity with the neutral compared to varus shoe Soleus—significantly less EMG activity with the neutral compared to the varus and valgus shoe
Ogon et al. (2001)	12 healthy volunteers (7 males, 5 females) Age: 32.9 (±7.9) years	(1) Unshod (2) Shod (running shoes)	Er. spinae (L_3)	Running	Onset latency/ timing	Er. spinae—significantly earlier onset with unshod condition compared to shod following heel strike Er. spinae—significantly shorter latency with shod condition compared to unshod following peak acceleration
Roy and Stefanyshyn (2006)	13 subjects with weekly mileage >25 km Age: 27 (±5.1) years	(1) Unmodified control shoe (2) Modified stiff shoe (3) Modified stiffest shoe	Vast. lateralis Rec. femoris Bic. femoris Med. gastroc Soleus	Running	RMS EMG amplitude from four intervals of stance phase	No significant differences between conditions

Study	Subjects	Footwear conditions	Muscles	Activity	EMG measure	Findings
Serrao and Amadio (2001)	3 "runners"; Age: 24.7 (±3.2) years	(1) Barefoot (2) Individuals' running shoes	Vast. lateralis Med. gastroc	Walking and running	Normalized mean EMG	Vast. lateralis—significantly delayed peak EMG with running shoes compared to barefoot (during walking and running) Med. gastroc—significantly delayed peak EMG with running shoes compared to barefoot (during walking only)
von Tscharner et al. (2003)	40 male "runners"; weekly mileage >25 km[a]	(1) Barefoot (2) Neutral running shoe (3) Pronation control running shoe	Tib. anterior	Running	Wavelet analysis (total EMG intensity)	Tib. anterior—significantly higher and significantly lower EMG intensity with running shoes[a] compared to barefoot during pre- and postheel strike periods of gait cycle
Wakeling et al. (2002)	6 "runners" (3 females) Age: 23.3 (±4.1) years (3 males) Age: 26.0 (±2.5) years	(1) Hard midsole running shoe (2) Soft midsole running shoe	Bic. femoris Rec. femoris Med. gastroc Tib. anterior	Running	Wavelet analysis (low- and high-frequency bands, preheel strike)	Muscles[b]—significantly altered[a] total EMG intensity with different midsole[a] hardness preheel strike

Posterior trunk: Er. spinae—erector spinae; *Anterior compartment thigh*: Rec. femoris—rectus femoris, Vast. lateralis—vastus lateralis, Vast. medialis—vastus medialis; *Posterior compartment leg*: Gastroc.—gastrocnemius, Med. gastroc.—medial gastrocnemius, Lat. gastroc.—lateral gastrocnemius, Flx. d. longus—flexor digitorum longus, Tib. posterior—tibialis posterior, Flx. h. longus—flexor hallucis longus; *Anterior compartment leg*: Tib. anterior—tibialis anterior, Ext. h. longus—extensor hallucis longus, Ext. d. longus—extensor digitorum longus; *Lateral compartment leg*: Per. longus—peroneus longus, Per. brevis—peroneus brevis.

[a] Muscle unspecified.

[b] No further information available.

to Ogon et al. (2001), this investigation reported EMG amplitude rather than temporal characteristics, and participants were assessed while running as opposed to jogging. When running barefoot, participants displayed significantly greater preactivation (EMG amplitude) for medial and lateral gastrocnemius and soleus. The authors suggested that this finding might indicate enhanced storage and restitution of elastic energy in these muscles.

There have been five other studies that have investigated the effect of athletic shoes with variable levels of midsole stiffness or density on similar muscle groups (i.e., thigh and lower leg); however, none reported any significant findings (Komi et al., 1987; Stefanyshyn et al., 2000; Nigg et al., 2003; Boyer and Nigg, 2004; Roy and Stefanyshyn, 2006). The discrepancies between these studies and those discussed earlier may be related to the relatively small sample sizes (leading to issues with statistical power). In addition, incomparable methods, particularly the different approaches to processing the EMG signal and the types of EMG parameters reported makes comparison between these studies difficult.

In summary, for midsole hardness, the findings from the aforementioned studies indicate that lower limb muscle activity can be "tuned" with shoes that have different hardness to accommodate the impact force at heel strike. Furthermore, the material hardness of the midsole may selectively activate specific muscle fiber types, with potential for implications on muscle fatigue and athletic performance.

23.3.1.2 Motion-Control Shoes

One of the key distinguishing features between different styles of modern athletic/running footwear is the density of the midsole. Footwear manufacturers generally produce a "neutral" or "cushioned" midsole for those with high-arched or rigid foot types and then a range of "motion-control" midsoles, with varying densities in the medial and lateral regions of the midsole (e.g., dual density), that aim to reduce pronation in those with flat-arched and hypermobile foot types.

With this in mind, some recent studies have investigated whether motion-control footwear, with a dual-density midsole, reduces lower leg muscle fatigue and restores quadriceps muscle timing in novice female runners with excessive foot pronation (Cheung and Ng, 2009, 2010). These studies found that when compared to a motion-control shoe, the neutral shoe is associated with significant positive gains in root mean square EMG amplitude with increasing running mileage for tibialis anterior and peroneus longus, and a larger decline in median frequency for peroneus longus. This indicates that there is greater fatigue in peroneus longus with a *neutral* shoe as running mileage increases. Another interesting finding from these studies is that motion-control shoes tend to normalize the onset of vastus medialis activation in those with excessive pronation, which may have implications for those with patellofemoral pain syndrome caused by the vastii muscle onset imbalance.

In summary, for motion-control shoes, only two studies have investigated the effect of this type of footwear on lower limb muscle activity. These studies indicate that motion-control shoes have a positive influence on (i.e., reduce) lower leg muscle fatigue and normalize vastii muscle onset in those with excessive foot pronation.

23.3.1.3   Midsole Wedging

Another variable that is thought to influence lower limb motion and muscle activity is the angulation of the shoe midsole. For example, the midsole can be made with a varus or valgus angulation. A varus wedged sole would be expected to reduce demand for muscles that support the medial column of the foot (e.g., tibialis posterior), while increasing demand on muscles that support the lateral column of the foot (e.g., peroneus longus). Perturbing the foot in this way may have clinical implications for individuals with dysfunction of specific muscles or joints (e.g., a varus wedged sole may reduce demand on the inverter musculature in a flexible flatfoot). In contrast, a valgus wedged sole would cause the opposite effect; that is, reduce demand on muscles that support the lateral column of the foot, while increasing demand on muscles that support the medial column of the foot.

With this in mind, two related studies have investigated the roles of the extrinsic foot muscles during running by manipulating the frontal plane angulation of the midsole (O'Connor and Hamill, 2004; O'Connor et al., 2006). The shoes used in these investigations featured (1) a flat midsole (with no angulation), (2) a varus-angled midsole, or (3) a valgus-angled midsole. The shoes with varus- and valgus-angled midsoles were intended to induce foot supination and pronation, respectively, during different stages of the gait cycle while running.

The investigators used EMG to record muscle amplitude and temporal parameters (O'Connor and Hamill, 2004; O'Connor et al., 2006), and muscle function MRI (O'Connor et al., 2006) to assess metabolic activity and workload of lower limb muscles. The EMG analysis indicated that the midsole with a varus angle significantly increased tibialis anterior EMG amplitude compared to the neutral midsole, while the neutral midsole significantly decreased the EMG amplitude for soleus compared to the varus- and valgus-angled midsoles. No significant differences between footwear conditions were detected from the mfMRI analysis.

In summary, for midsole wedging, there were some significant changes reported in these studies, although these findings are unexpected and inconsistent. Therefore, there is insufficient evidence at this stage that midsole wedging has a systematic effect on lower limb muscle activity.

23.3.1.4   Heel Counter

The heel counter (plastic reinforcement placed in the shoe upper around the heel) has been proposed to support the hindfoot and provide shock absorption through confinement of the fat pad of the heel (Jorgensen, 1990). These functions are presumed to alter the demand of lower limb muscles to control motion and attenuate shock. A single investigation of 11 asymptomatic heel-strike runners was undertaken using athletic shoes with and without a heel counter (Jorgensen, 1990). The findings from this study indicated that EMG amplitude for triceps surae and quadriceps occurred significantly later with the heel counter. The authors proposed that this finding could be due to increased shock absorption with the heel counter, leading to less demand for the leg and thigh muscles. Clearly, because there is only one study on this topic, further research is needed to determine the influence of the heel counter on lower limb muscle activity.

In conclusion, various characteristics of athletic or running style shoes influence lower limb muscle function. There is some evidence that altering the configuration

TABLE 23.2
Rocker-Sole and Unstable Style Shoes

Author/s (Date)	Participant Characteristics Age (Standard Deviation)	Footwear/Test Conditions	Muscles	Task	EMG Variables	Main Findings
Bullock-Saxton et al. (1993)	15 Healthy adults (5 men, 10 women) Age: 18–20 years	(1) Barefoot (2) Balance shoes	Glut. maximus Glut. medius	Walking	Mean square value and rate of recruitment	Glut. maximus and medius—significantly greater activity and shortened time to maximum activity with balance shoes, compared to barefoot
Forestier and Toschi (2005)	9 Healthy subjects Age: 37.0 (±12.0) years	(1) Barefoot (2) Ankle destabilization shoe	Med. gastroc. Lat. gastroc. Tib. anterior Per. longus Per. brevis	Walking	Integrated and normalized Onset time (Per. longus and brevis only)	Tib. anterior, Per. brevis, and Per. longus— significantly greater EMG amplitude with destabilization shoe compared to barefoot
Nigg et al. (2006b)	8 healthy subjects (3 females, 5 males) Age: 28.0 (±3.6) years	(1) Unstable shoe (2) Control shoe	Glut. medius Bic. femoris Vast. medialis Med. gastroc. Tib. anterior	Walking	Wavelet analysis (total EMG intensity)	No significant differences between conditions

Study	Subjects	Footwear conditions	Muscles	Activity	Parameters	Results
Peterson et al. (1985)	15 Healthy adult women Age: 25.5 (range) 21–30 years	(1) Rocker shoe with rigid wooden sole and open heel (2) Subjects own athletic shoes made of canvas or leather and laced up	Vast. lateralis Gastroc[a] Soleus	Walking	Timing of muscle activity	No significant differences between conditions
Romkes et al. (2007)	12 Healthy subjects (6 females, 6 males) Age: 38.6 (±13.2) years	(1) Individuals' regular shoes (2) Masai Barefoot Technologies® (MBT-shoes)	Sem. tend. Rec. femoris Vast. lateralis Vast. medialis Med. gastroc. Lat. gastroc. Tib. anterior	Walking	RMS from 16 equal intervals over gait cycle (normalized from barefoot condition)	Tib. anterior, Med. gastroc, Lat. gastroc, Vast. lateralis, Vast. medialis, and Rec. femoris—significantly greater RMS EMG activity with MTB-shoes compared to regular shoes (during part of contact phase) Tib. anterior, Med. gastroc, Lat. gastroc. significantly greater and Rec. femoris significantly lower RMS EMG with MTB-shoes compared to regular shoes (in part of swing phase)
Stoggl et al. (2010)	12 Healthy students (6 males, 6 females) Age: 25 (±2) years	(1) Running shoe (2) Unstable (MBT) shoe	Bic. femoris Vast. medialis Vast. lateralis Med. gastroc Tib. anterior Per. longus	Walking	Integrated EMG, RMS amplitude, median power frequency	No significant differences between conditions

Gluteal region: Glut. maximus—gluteus maximus, Glut. medius—gluteus medius; *Posterior compartment thigh:* Sem. tend.—semitendinosis, Sem. memb.—semitendinosis, Bic. femoris—biceps femoris, Lat. hamst.—lateral hamstring; *Anterior compartment thigh:* Rec. femoris—rectus femoris, Vast. lateralis—vastus lateralis, Vast. medialis—vastus medialis; *Posterior compartment leg:* Gastroc.—gastrocnemius, Med. gastroc.—medial gastrocnemius, Lat. gastroc.—lateral gastrocnemius, Flx. d. longus—flexor digitorum longus, Tib. posterior—tibialis posterior, Flx. h. longus—flexor hallucis longus; *Anterior compartment leg;* Tib. anterior—tibialis anterior, Ext. h. longus—extensor hallucis longus, Ext. d. longus—extensor digitorum longus; *Lateral compartment leg;* Per. longus—peroneus longus, Per. brevis—peroneus brevis.

[a] Muscle unspecified.

and material properties of a shoe midsole can have an effect on lower limb muscle activity. Studies investigating shoes with differing midsole hardness have demonstrated a phenomenon described as "muscle tuning," whereby muscle activity is tuned to minimize soft tissue vibration and reduce fatigue. This has been speculated to be a mechanism that protects the lower limb from injury; however, further research is required to tease out this issue. Motion-control features influence lower leg muscle fatigue and vastii activation patterns in those with excessively pronated feet. Finally, one study found that a heel counter significantly alters lower limb muscle activity.

23.3.2 Rocker-Sole and Unstable Style Shoes

Rocker-sole shoes are designed to assist forward momentum of the body over the foot by having a rounded (convex) sole that is curved in an anterior–posterior direction. Like a rocker-sole shoe, unstable shoes feature a similar anterior–posterior curvature, but they are also curved in a medial–lateral direction. The purpose of including both anterior–posterior and medial–lateral curves is to induce instability via the uneven surface of the sole of the shoe, which supposedly mimics barefoot locomotion (Landry, 2011). It is thought that the uneven surface challenges lower leg muscles that may otherwise be underused when wearing more stable, traditional footwear. Therefore, the unstable shoe design is intended to strengthen lower leg muscles and it is thought that this can reduce pain for a range of musculoskeletal conditions, including knee joint osteoarthritis and low back pain. Recent studies provide limited evidence supporting the use of unstable shoes for these conditions, compared to more supportive, traditional footwear (Nigg et al., 2006a, 2009). The results of the studies included in this section are presented in Table 23.2.

In terms of altering muscle activity, the earliest study of a rocker-sole shoe was conducted by Peterson et al. (1985) on 15 healthy women. The rocker-soled shoe, which featured a rigid wooden sole and open heel, was compared to the participants' own athletic shoes. No significant differences in vastus lateralis and triceps surae EMG muscle timing were detected.

Following on from this work, Bullock-Saxton et al. (1993) investigated the effect of walking with "balance shoes" (i.e., unstable shoes) in 15 healthy participants. They assessed the effect of balance shoes on gluteus maximus and medius and compared the shoes to a barefoot condition. They reported significantly greater gluteus maximus and medius activity and shortened time to maximum muscle activity with balance shoes. While this finding indicates greater work undertaken by the gluteal muscles with balance shoes, it is unclear whether this was due to the effect of the unstable sole or simply because of increased weight of the balance shoe. This is always a significant issue when comparing shoes to a barefoot condition or shoes of differing weight.

More recently, one study (Forestier and Toschi, 2005) used a mechanical destabilization device under the heel of a shoe while walking to induce destabilization of the rearfoot in nine healthy participants. A significant increase in tibialis anterior, peroneus longus, and peroneus brevis EMG amplitude was reported with the destabilization shoe compared to barefoot. Again, this device was compared to a barefoot condition, so the findings may just be related to the weight of the shoe.

Three other studies (Nigg et al., 2006b; Romkes et al., 2006; Stoggl et al., 2010) have compared the effect of unstable shoe design to either participants' own shoes (Romkes et al., 2006) or running shoes (Nigg et al., 2006b; Stoggl et al., 2010). Unfortunately, sample sizes in these studies were generally low, ranging from 8 to 12. Nevertheless, the first study by Romkes et al. (2006) reported that the unstable shoe significantly altered root mean square EMG amplitude of rectus femoris, vastus medialis, vastus lateralis, tibialis anterior, medial, and lateral gastrocnemius within defined intervals of stance and swing phase. The second study by Nigg et al. (2006b) also recorded EMG from gluteus medius, biceps femoris, vastus medialis, medial gastrocnemius, and tibialis anterior and found no significant changes in total EMG amplitude using wavelet analysis. The third study by Stoggl et al. (2010) investigated almost identical muscles to those of Romkes and colleagues; however, in this study the investigators were interested in whether variability of muscle activity was altered, rather than just direction changes in muscle amplitude or frequency. There were no significant differences comparing the unstable shoe to the running shoe condition.

In conclusion, for rocker-sole or unstable shoes, there is limited evidence that this type of footwear has a systematic effect on lower limb muscle activity during walking. The lack of significant findings may be attributed to factors such as the small sample sizes (and possibility of type 2 statistical errors), the style of control shoe used for comparison, and the types of EMG variables evaluated. That is, the unstable shoe design may influence muscle activity, however, previous studies have been limited in their design and have not been able to clearly demonstrate such an effect. There is, therefore, a need for further research on the efficacy of unstable shoe designs and destabilization devices to determine whether they produce predictable and consistent changes in muscle activity.

23.3.3 High-Heeled and Negative-Heeled Shoes

23.3.3.1 High-Heeled Shoes

The high-heeled shoe is a very popular style of fashion footwear. It is a unique style of shoe that is known to be associated with a range of foot deformities, including bunions and claw toes. High-heeled footwear has also been found to produce alterations in lower limb joint mechanics and plantar pressures (Stefanyshyn et al., 2000). While there are several features that characterize the high-heeled shoe such as a narrow and shallow toe-box, the elevation of the heel is a distinguishing feature, which in some shoes can reach 10 cm in height (Stefanyshyn et al., 2000).

The effect of differing heel height on muscle activity in females has been investigated in several studies, with heel heights ranging from 0 to 8 cm (Joseph, 1968; Lee et al., 1990; Stefanyshyn et al., 2000; Lee et al., 2001; Gefen et al., 2002; Park et al., 2010). The results of these studies are presented in Table 23.3. By systematically increasing the heel height, Lee et al. (2001) clearly demonstrated consistent increases in peak EMG for erector spinae activity in the back (i.e., a systematic effect). Increasing heel height is also associated with decreased medial gastrocnemius and tibialis anterior peak EMG amplitude (Lee et al., 1990) and increased rectus femoris, soleus, and peroneus longus root mean square EMG amplitude (Stefanyshyn et al., 2000). In contrast to these findings, one earlier study reported no significant

TABLE 23.3

High-Heeled and Negative-Heeled Shoes

Author/s (Date)	Participant Characteristics Age (Standard Deviation)	Footwear/Test Conditions	Muscles	Task	EMG Variables	Main Findings
Gefen et al. (2002)	8 Female subjects 4 Habitual wearers of high-heeled shoes 4 Habitual wearers of flat-heeled shoes Age: 26.0 (±4) years	(1) Barefoot (2) "Own" footwear[a] (3) Fatiguing exercise 1 (4) Fatiguing exercise 2	Med. gastroc. Lat. gastroc. Soleus Per. longus Tib. anterior Ext. hall. longus	Walking	Normalized EMG median frequency	Med. and Lat. gastroc.—significantly faster decrease in median EMG frequency for Lat. gastroc relative to Med. gastroc. in habitual high-heeled wearers (after fatiguing exercises) compared to low-heeled wearers Per. longus—significantly faster decrease in median EMG frequency for habitual high-heeled wearers

Joseph (1968)	6 Subjects[a]	(1) Low-heeled shoes (heel height 1–2.5 cm) (2) High-heeled shoes (heel height 5.5–8 cm)	Er. spinae Glut. maximus Glut. medius Bic. femoris Hip flexor[b] Soleus Tib. anterior	Walking	Duration of EMG activity Raw EMG amplitude	No significant differences between conditions
Li and Hong (2007)	13 Female subjects Age: 23.1 (±3.9) years	(1) Normal shoes (2) Negative-heeled shoes	Er. spinae Rec. abdominus Bic. femoris Rec. femoris Lat. gastroc. Tib. anterior	Walking	Mean and integrated EMG Duration of EMG	Bic. femoris, Lat. gastroc., and Tib. anterior— significantly greater EMG amplitude; Lat. gastroc and tib anterior—significantly longer duration with negative-heeled shoe compared to normal shoe
Lee et al. (2001)	5 Healthy young women ("in their twenties")[a]	(1) Low-heeled shoes (0 cm) (2) Medium-heeled shoes (4.5 cm) (3) High-heeled shoes (8 cm)	Er. spinae (L_1/L_2) Er. spinae (L_4/L_5) Vast. lateralis Tib. anterior	Walking	Peak and integrated EMG	Er. spinae (L_4/L_5)—significantly greater peak EMG as heel height[a] increased
Lee et al. (1990)	6 Women ("regular wearers of high-heeled shoes") Age range: 20–31 years[a]	(1) Barefoot (2) 2.5 cm heeled shoes (3) 5.0 cm heeled shoes (4) 7.5 cm heeled shoes	Tib. anterior Med. gastroc.	Walking	Normalized peak and mean peak EMG	Med. gastroc and Tib. anterior—significantly lower mean peak EMG with 2.5, 5.0, and 7.5 cm heeled shoes compared to barefoot Med. gastroc—significantly lower mean peak EMG with 2.5 and 5 cm compared to 5.0 and 7.5 cm heeled shoes, respectively Tib. anterior—significantly greater mean peak EMG with 2.5 cm compared to both 5.0 and 7.5 cm

(continued)

TABLE 23.3 (continued)
High-Heeled and Negative-Heeled Shoes

Author/s (Date)	Participant Characteristics: Age (Standard Deviation)	Footwear/Test Conditions	Muscles	Task	EMG Variables	Main Findings
Park et al. (2010)	17 Healthy women Age: 22.1 (±1.2) years	(1) Barefoot (2) 3 cm high-heeled shoe (3) 7 cm high-heeled shoe	Vast. med. obliq. Vast. lateralis	Walking	RMS amplitude	Vast. med. obliq. to Vast. lateralis—significantly reduced ratio with the 7 cm heel compared to both the 3 cm heel and barefoot conditions
Stefanyshyn et al. (2000)	13 Female subjects Age: 40.6 (±8.3) years	(1) Flat shoe (1.4 cm heel height) (2) Low-heeled shoe (3.7 cm) (3) Medium-heeled shoe (5.4 cm) (4) High-heeled shoe (8.5 cm)	Sem. tend. Bic. femoris Rec. femoris Vast. medialis Gastroc[b] Soleus Tib. anterior Per. longus	Walking	RMS EMG amplitude	Per. longus, Rec. femoris, and soleus—significantly greater RMS EMG with higher-heeled shoes compared to lower-heeled shoes[c]

Posterior trunk: Er. spinae—erector spinae; *Anterior trunk:* Rec. abdom.—rectus abdominus; *Gluteal region:* Glut. maximus—gluteus maximus, Glut. medius—gluteus medius; *Posterior compartment thigh:* Sem. tend.—semitendinosis, Sem. memb.—semitendinosis, Bic. femoris—biceps femoris, Lat. hamst.—lateral hamstring; *Anterior compartment thigh:* Rec. femoris—rectus femoris, Vast. lateralis—vastus lateralis, Vast. medialis—vastus medialis; *Posterior compartment leg:* Gastroc.—gastrocnemius, Med. gastroc.—medial gastrocnemius, Lat. gastroc.—lateral gastrocnemius, Flx. d. longus—flexor digitorum longus, Tib. posterior—tibialis posterior, Flx. h. longus—flexor hallucis longus; *Anterior compartment leg:* Tib. anterior—tibialis anterior, Ext. h. longus—extensor hallucis longus, Ext. d. longus—extensor digitorum longus; *Lateral compartment leg:* Per. longus—peroneus longus, Per. brevis—peroneus brevis.

a   No further information available.
b   Muscle unspecified.
c   Several post hoc findings.

changes in various thigh and lower leg muscles comparing heel heights of 1–2.5 to 5.5–8.0 cm (Joseph, 1968) although this study was performed in the 1960s. The reason for this contrary finding may be that the process of collecting and processing EMG data today is vastly progressed compared to when Joseph conducted his research.

The most recent study by Park et al. (2010) investigated barefoot and 3 and 5 cm heel heights in 17 healthy women. They found that increasing heel height decreased vastus medialis oblique EMG amplitude as a ratio of vastus lateralis activity, although this finding was only observed in the nondominant limb. As such, the meaning of this finding is unclear, but the authors suggest that it may assist with understanding potential mechanisms contributing to patellofemoral pain in women.

Finally, a slightly different approach to investigating the effects of high-heeled shoes was undertaken by Gefen et al. (2002). Rather than investigating the immediate effects of heel height on muscle activity, they compared the median EMG frequency of lower limb muscles from habitual and nonhabitual wearers of high-heeled shoes following a fatiguing exercise. When compared to habitual low-heel wearers, habitual high-heel wearers displayed a significantly faster decrease in median frequency for peroneus longus and lateral gastrocnemius after completing a fatiguing exercise. These findings may have been associated with the participants displaying reduced stability, and as a result, lead to an increased risk of accidental injury.

In conclusion, high-heeled shoes have a systematic effect on back and lower limb muscle activity. Even when high-heeled shoes are removed from one's feet, those who wear them regularly display greater fatigue characteristics for some lower leg muscles, and this may predispose them to accidental injury.

23.3.3.2 Negative-Heeled Shoes

In contrast to the high-heeled shoe style discussed earlier, Li and Hong (2007) compared negative-heeled shoes to normal-heeled shoes. They found that the negative-heeled shoes caused significantly greater EMG amplitude for biceps femoris, tibialis anterior and lateral gastrocnemius, and longer EMG duration for lateral gastrocnemius and tibialis anterior. In light of these findings, the authors suggest that negative-heeled shoes could assist with exercise rehabilitation or training programs, where inclined training surfaces are not available. Clearly, further research is needed for this style of shoe.

23.3.4 Occupational-Specific and Other Shoes

Various other footwear styles have been investigated for their effects on lower limb muscle activity. Three studies have investigated specific occupational-related footwear styles, primarily to identify specific shoe characteristics to reduce muscle fatigue and work-related musculoskeletal injury. These studies have featured shoes with various midsole stiffness properties in the following occupations; waiters (Kersting et al., 2005), those wearing clean room boots (rubber boot with polyurethane or polyvinyl chloride [PVC] sole) (Lin et al., 2007) and nurses (Chiu and Wang, 2007). The results of these studies are presented in Table 23.4.

TABLE 23.4

Occupational-Specific and Other Shoes

Author/s (Date)	Participant Characteristics Age (Standard Deviation)	Footwear/Test Conditions	Muscles	Task	EMG Variables	Main Findings
Bohm, and Hosl (2010)	15 Healthy males Age: 29 (±5) years	(1) Stiff shaft boot (2) Soft shaft boot	Sem. tend Vast. lateralis Tib. anterior Med. gastroc Per. longus	Walking	Cocontraction index of muscle amplitude	Vast. lateralis and Sem. tend.—significantly increased cocontraction between muscles in the stiff-shafted boot condition occurred during single leg stance
Bourgit et al. (2008)	12 Healthy females Age: 24 (±4) years	(1) Standard fitness shoe (2) Special 2° dorsiflexion	Glut. max. Vast. lateralis Rec. femoris Bic. femoris	Walking and running	Normalized RMS EMG amplitude	Knee extensors, Bic. femoris, Tib. anterior, and plantar flexors—significantly altered comparing several combinations of shoes for ballistic plantar flexion exercise Plantar flexors—significantly altered comparing several combinations of shoes for walking exercise

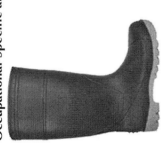

Study	Participants	Footwear conditions	Muscles	Activity	Measure	Results
		(3) Special 4° dorsiflexion (4) Special 10° dorsiflexion	Tib. anterior Lat. gastroc Med. gastroc Soleus			Knee extensors and plantar flexors—significantly altered comparing several combinations of shoes for running exercise
Chiu and Wang (2007)	12 "Healthy" females Age: 23.3 years (±2.1) years	(1) Nursing shoe A (2) Nursing shoe B (3) Nursing shoe C (A, B, and C had different sole, midsole, upper and innersole characteristics)	Bic. femoris Rec. femoris Med. gastroc. Tib. anterior	Walking	Mean normalized EMG amplitude	Med. gastroc—significantly lower EMG amplitude with shoe A and shoe B compared to shoe C (shoe A and B included an arch support design[a])
Cho et al. (2007)	10 Participants over 50 years of age with plantar fasciitis (5 males, 5 females)	Microcurrent shoe (A specially constructed shoe of piezoelectric material that generates a microcurrent during loading)	Tib. anterior Soleus	Walking	Median peak frequency	Tib. anterior—significant reduced median peak frequency pre- and posttesting
Hansen et al. (1998)	8 Healthy women Age: 24 (range: 21–29) years	(1) Clogs without a heel (2) Sports shoe	Er. spinae (L_3)	Walking	RMS amplitude and mean power frequency	No significant differences between conditions
Kersting et al. (2005)	16 Trained waiters (8 males, 8 females) Male age: 27.9 (±2.3) years Female age: 23.9 (±2.0) years	(1) Standard shoe with stiff midsole (2) Neutral shoe with flexible midsole (3) Shoe with soft midsole	Er. spinae (L_3) Med. gastroc Tib. anterior Per. longus	Walking	Integrated EMG (iEMG)	Per. longus and Med. gastroc.—significantly greater iEMG with the shoe with a stiff midsole compared to the soft midsole when walking on a PVC surface Er. spinae—significantly higher activation for the stiff midsole shoe compared to the flexible midsole

(continued)

TABLE 23.4 (continued)
Occupational-Specific and Other Shoes

Author/s (Date)	Participant Characteristics Age (Standard Deviation)	Footwear/Test Conditions	Muscles	Task	EMG Variables	Main Findings
Lin et al. (2007)	12 Healthy female students Age: 24.2 (±1.9) years	(1) "Clean room" boot A (2) "Clean room" boot B (3) "Clean room" boot C Each with different shock-absorbing and elastic properties	Er. spinae Bic. femoris Rec. femoris Gastroc.[b] Tib. anterior	Walking	Mean normalized EMG amplitude	Gastrocnemius[b]—Significantly greater EMG amplitude with boot C and B compared to boot A as a function of time
Sacco et al. (2010)	21 Healthy nondiabetic participants Age: 50.9 (±7.3) years 24 diabetic neuropathic participants Age: 55.2 (±7.9) years	(1) Barefoot (2) Own habitual shoes (sport shoes, loafers, sandals, dress shoes)	Vast. lateralis Lat. gastroc. Tib. anterior	Walking	Time to peak EMG	Lat. gastroc.—significant delay in diabetic participants with wearing habitual shoes compared to barefoot Vast. lateralis—significant delay in nondiabetic participants with wearing habitual shoes compared to barefoot

Posterior trunk: Er. spinae—erector spinae; *Gluteal region*: Glut. maximus—gluteus maximus, Glut. medius—gluteus medius; *Posterior compartment thigh*: Sem. tend.—semitendinosis, Sem. memb.—semitendinosis, Bic. femoris—biceps femoris, Lat. hamst.—lateral hamstring; *Anterior compartment thigh*: Rec. femoris—rectus femoris, Vast. lateralis—vastus lateralis, Vast. medialis—vastus medialis; *Posterior compartment leg*: Gastroc.—gastrocnemius, Med. gastroc.—medial gastrocnemius, Lat. gastroc.—lateral gastrocnemius, Flx. d. longus—flexor digitorum longus, Tib. posterior—tibialis posterior, Flx. h. longus—flexor hallucis longus; *Anterior compartment leg*: Tib. anterior—tibialis anterior, Ext. h. longus—extensor hallucis longus, Ext. d. longus—extensor digitorum longus; *Lateral compartment leg*: Per. longus—peroneus longus, Per. brevis—peroneus brevis.

[a] No further information available.

[b] Muscle unspecified.

The first study investigated three shoes with varying midsole stiffness during walking in 16 trained waiters (Kersting et al., 2005). For peroneus longus and medial gastrocnemius, there was significantly greater integrated EMG (i.e., greater overall muscle effort) with the stiff-midsole shoe compared to the soft-midsole shoe. In addition, erector spinae exhibited significantly greater activation for the stiff-midsole shoe compared to the flexible-midsole shoe.

The second study investigated clean room boots with variable shock-absorbing and elastic properties under different walking conditions (e.g., carrying a load) (Lin et al., 2007). Gastrocnemius EMG amplitude was significantly lower with heavier, more elastic, and shock-absorbing boots when analyzed over time (i.e., after 60 min of walking). This finding indicates that varying the properties of clean room boots can influence muscle activity, but individuals need to be monitored for a long enough period of time (i.e., to induce some fatigue) to detect significant differences.

The third study evaluated three styles of nursing shoes with differing sole, midsole, upper, and innersole characteristics on 12 nurses (Chiu and Wang, 2007). The study found that shoes with "arch support" produced a significant decrease in medial gastrocnemius EMG amplitude. For the findings in these three studies to make clinical sense, EMG changes need to be linked to changes in health status. Ideally, future research needs to factor in symptomatic participants and measure patient reported outcomes, in addition to EMG changes over time.

Several other styles of footwear have been investigated for their effect on lower limb muscle activity; however, these styles cannot readily be classified with those mentioned earlier. These include a study comparing "standard fitness shoes" to a range of "special dorsiflexion" shoes (Bourgit et al., 2008), a "microcurrent" shoe for individuals with heel pain (Cho et al., 2007), clogs and sports shoes (Hansen et al., 1998), barefoot and shod muscle activity in individuals with and without diabetes related neuropathy (Sacco et al., 2010), and a comparison between stiff- and soft-shafted hiking boots (Bohm and Hosl, 2010). The results of these studies are also presented in Table 23.4.

In conclusion, there are too few studies to draw conclusions about the effects of specific occupational shoes on lower limb muscle activity. With the increasing need to create safe, ergonomically sound workplace environments, it is likely that there will be a rapid increase in research exploring the effect of occupational-specific footwear on muscle activity.

23.4   CONCLUSIONS

Footwear is an important piece of clothing that can have a significant effect on the musculoskeletal system, including muscle activity. In this chapter, we have systematically reviewed the literature to comprehensively and transparently cover the effect of footwear on muscle function. While there is a great deal of work yet to be done in this area, there have already been some exciting in-roads made. For example, there is some evidence that muscle activity can be tuned and that indicators of muscle fatigue can be manipulated by varying the composition of the midsole of athletic shoes. In addition, high-heeled shoes have a consistent effect on back and lower limb muscles.

However, clear deficiencies in the literature limit the conclusions that can be made about the wider effects of footwear on muscle function. There is a need for further research of more rigorous methodological quality, including larger sample sizes and greater consensus regarding reporting of electromyographic parameters. In addition, it is still not clear whether changes in muscle function with the use of some types of footwear are consistent and predictable (i.e., systematic), and it is currently not known whether an increase or decrease in EMG variables is beneficial in relation to injury and health status. While it makes intuitive sense that an intervention would be beneficial if it can bring muscle activity closer to that seen in a nonpathological population (measured via EMG), definitive evidence is still lacking.

In closing, therefore, there is some evidence that certain types of footwear can affect muscle function; however, there is insufficient evidence to make conclusions about the effect of footwear on clinically relevant conditions.

QUESTIONS

23.1 Discuss the limitations of the research related to the effects of footwear on lower limb muscle function.

23.2 Suggest ways in which research evaluating the effect of footwear on muscle function could be altered to improve the quality of evidence relating to this area of investigation.

23.3 Summarize the effects of varying the composition of the midsole of athletic shoes on muscle function.

23.4 Is there evidence to support the assertion that current running shoe technologies improve running performance?

23.5 Discuss the evidence related to the effects of high-heeled footwear on lower limb muscle activity.

23.6 Discuss the evidence related to the effects of rocker-sole and unstable footwear on lower limb muscle activity.

REFERENCES

Bohm, H., Hosl, M., 2010. Effect of boot shaft stiffness on stability joint energy and muscular co-contraction during walking on uneven surface. *Journal of Biomechanics* 43, 2467–2472.

Bourgit, D., Millet, G.Y., Fuchslocher, J., 2008. Influence of shoes increasing dorsiflexion and decreasing metatarsus flexion on lower limb muscular activity during fitness exercises, walking, and running. *Journal of Strength and Conditioning Research* 22, 966–973.

Boyer, K.A., Nigg, B.M., 2004. Muscle activity in the leg is tuned in response to impact force characteristics. *Journal of Biomechanics* 37, 1583–1588.

Bullock-Saxton, J.E., Janda, V., Bullock, M.I., 1993. Reflex activation of gluteal muscles in walking: An approach to restoration of muscle function for patients with low-back pain. *Spine* 18, 704–708.

Cheung, R.T.H., Ng, G.Y.F., 2009. Motion control shoe affects temporal activity of quadriceps in runners. *British Journal of Sports Medicine* 43, 943–947.

Cheung, R.T.H., Ng, G.Y.F., 2010. Motion control shoe delays fatigue of shank muscles in runners with overpronating feet. *American Journal of Sports Medicine* 38, 486–491.

Chiu, M.C., Wang, M.J., 2007. Professional footwear evaluation for clinical nurses. *Applied Ergonomics* 38, 133–141.

Cho, M., Park, R., Park, S.H., Cho, Y., Cheng, G.A., 2007. The effect of microcurrent-inducing shoes on fatigue and pain in middle-aged people with plantar fascitis. *Journal of Physical Therapy Science* 19, 165–170.

Cram, J.R., Criswell, E., 2011. *Cram's Introduction to Surface Electromyography*. Jones & Bartlett, Sudbury, MA.

Divert, C., Mornieux, G., Baur, H., Mayer, F., Belli, A., 2005. Mechanical comparison of bare-foot and shod running. *International Journal of Sports Medicine* 26, 593–598.

Forestier, N., Toschi, P., 2005. The effects of an ankle destabilization device on muscular activity while walking. *International Journal of Sports Medicine* 26, 464–470.

Gefen, A., Megido-Ravid, M., Itzchak, Y., Arcan, M., 2002. Analysis of muscular fatigue and foot stability during high-heeled gait. *Gait and Posture* 15, 56–63.

Hansen, L., Winkel, J., Jorgensen, K., 1998. Significance of mat and shoe softness during pro-longed work in upright position: Based on measurements of low back muscle EMG, foot volume changes, discomfort and ground force reactions. *Applied Ergonomics* 29, 217–224.

Jorgensen, U., 1990. Body load in heel-strike running: The effect of a firm heel counter. *American Journal of Sports Medicine* 18, 177–181.

Joseph, J., 1968. The pattern of activity of some muscles in women walking on high heels. *Annals of Physical Medicine* 9, 295–299.

Kersting, U.G., Janshen, L., Bohm, H., Morey-Klapsing, G.M., Bruggemann, G.P., 2005. Modulation of mechanical and muscular load by footwear during catering. *Ergonomics* 48, 380–398.

Komi, P.V., Gollhofer, A., Schmidtbleicher, D., Frick, U., 1987. Interaction between man and shoe in running: Considerations for a more comprehensive measurement approach. *International Journal of Sports Medicine* 8, 196–202.

Landry, S., 2011. Unstable shoe designs: Functional implications. *Lower Extremity Review* 3, 31–36.

Lee, C.H., Jeong, E.H., Freivalds, A., 2001. Biomechanical effects of wearing high-heeled shoes. *International Journal of Industrial Ergonomics* 28, 321–326.

Lee, K.H., Shieh, J.C., Matteliano, A., Smiehorowski, T., 1990. Electromyographic changes of leg muscles with heel lifts in women: Therapeutic implications. *Archives of Physical Medicine and Rehabilitation* 71, 31–33.

Li, J.X., Hong, Y., 2007. Kinematic and electromyographic analysis of the trunk and lower limbs during walking in negative-heeled shoes. *Journal of the American Podiatric Medical Association* 97, 447–456.

Lin, C.-L., Wang, M.-J.J., Drury, C.G., 2007. Biomechanical, physiological and psychophysi-cal evaluations of clean room boots. *Ergonomics* 50, 481–496.

Murley, G.S., 2008. Re: Anomalous tibialis posterior muscle. *Foot* 18, 119–120.

Murley, G.S., Landorf, K.B., Menz, H.B., Bird, A.R., 2009. Effect of foot posture, foot ortho-ses and footwear on lower limb muscle activity during walking and running: A system-atic review. *Gait and Posture* 29, 172–187.

Nigg, B.M., Davis, E., Lindsay, D., Emery, C., 2009. The effectiveness of an unstable sandal on low back pain and golf performance. *Clinical Journal of Sport Medicine* 19, 464–470.

Nigg, B.M., Emery, C., Hiemstra, L.A., 2006a. Unstable shoe construction and reduc-tion of pain in osteoarthritis patients. *Medicine and Science in Sports and Exercise* 38, 1701–1708.

Nigg, B., Hintzen, S., Ferber, R., 2006b. Effect of an unstable shoe construction on lower extremity gait characteristics. *Clinical Biomechanics* 21, 82–88.

Nigg, B.M., Stefanyshyn, D., Cole, G., Stergiou, P., Miller, J., 2003. The effect of material characteristics of shoe soles on muscle activation and energy aspects during running. *Journal of Biomechanics* 36, 569–575.

O'Connor, K.M., Hamill, J., 2004. The role of selected extrinsic foot muscles during running. *Clinical Biomechanics* 19, 71–77.

O'Connor, K.M., Price, T.B., Hamill, J., 2006. Examination of extrinsic foot muscles during running using mfMRI and EMG. *Journal of Electromyography and Kinesiology* 16, 522–530.

Ogon, M., Aleksiev, A.R., Spratt, K.F., Pope, M.H., Saltzman, C.L., 2001. Footwear affects the behavior of low back muscles when jogging. *International Journal of Sports Medicine* 22, 414–419.

Park, K.-M., Chun, S.-M., Oh, D.-W., Kim, S.-Y., Chon, S.-C., 2010. The change in vastus medialis oblique and vastus lateralis electromyographic activity related to shoe heel height during treadmill walking. *Journal of Back and Musculoskeletal Rehabilitation* 23, 39–44.

Peterson, M.J., Perry, J., Montgomery, J., 1985. Walking patterns of healthy subjects wearing rocker shoes. *Physical Therapy* 65, 1483–1489.

Romkes, J., Rudmann, C., Brunner, R., 2006. Changes in gait and EMG when walking with the Masai Barefoot Technique. *Clinical Biomechanics* 21, 75–81.

Roy, J.-P.R., Stefanyshyn, D.J., 2006. Shoe midsole longitudinal bending stiffness and running economy, joint energy, and EMG. *Medicine and Science in Sports and Exercise* 38, 562–569.

Sacco, I.C.N., Akashi, P.M.H., Hennig, E.M., 2010. A comparison of lower limb EMG and ground reaction forces between barefoot and shod gait in participants with diabetic neuropathic and healthy controls. *BMC Musculoskeletal Disorders* 11, 24.

Semple, R., Murley, G.S., Woodburn, J., Turner, D.E., 2009. Tibialis posterior in health and disease: A review of structure and function with specific reference to electromyographic studies. *Journal of Foot and Ankle Research* 2, 24.

Serrao, J.C., Amadio, A.C., 2001. Kinetic and electromyographic adaptations in basefoot locomotion. *Brazilian Journal of Biomechanics* 2, 43–51.

Stefanyshyn, D.J., Nigg, B.M., Fisher, V., O'Flynn, B., Liu, W., 2000. The influence of high heeled shoes on kinematics, kinetics, and muscle EMG of normal female gait. *Journal of Applied Biomechanics* 16, 309–319.

Stoggl, T., Haudum, A., Birklbauer, J., Murrer, M., Muller, E., 2010. Short and long term adaptation of variability during walking using unstable (MBT) shoes. *Clinical Biomechanics* 25, 816–822.

von Tscharner, V., Goepfert, B., Nigg, B.M., 2003. Changes in EMG signals for the muscle tibialis anterior while running barefoot or with shoes resolved by non-linearly scaled wavelets. *Journal of Biomechanics* 36, 1169–1176.

Wakeling, J.M., Pascual, S.A., Nigg, B.M., 2002. Altering muscle activity in the lower extremities by running with different shoes. *Medicine and Science in Sports and Exercise* 34, 1529–1532.

24 Postural Stability Measurement

Implications for Footwear Interventions

Anna Lucy Hatton and Keith Rome

CONTENTS

24.1   INTRODUCTION

It is widely accepted that there is no single test available which has the capacity to measure all aspects of balance performance. Many footwear intervention studies have used a package of balance assessment tools to explore the effects of shoes or insoles on postural stability. Similarly, there appears to be no single stability outcome measure that is sufficiently robust or sensitive to represent overall static or dynamic balance performance. This chapter provides an overview of postural stability assessment techniques and outcome measures commonly used in previous studies exploring the effects of footwear and shoe insole design on static or dynamic balance control; however, this list is not exhaustive. The chapter concludes presenting a selection of current literature reporting the effect of footwear interventions on postural stability in chronic conditions including Parkinson's disease, multiple sclerosis, diabetes, and osteoarthritis. The summary is intended to highlight the relevance of postural stability assessment and footwear in clinical practice.

24.2   DEFINITION OF POSTURAL STABILITY

Within this chapter, the term "balance" will be used to describe the dynamics of body posture, which occurs in response to inertial forces acting on the body, in order to achieve a state of equilibrium between the body and the surrounding environment and to prevent the body from falling. The term "postural stability" refers to an individual's ability to maintain their center of mass (CoM) within specific boundaries of space, commonly known as stability limits. Stability limits are boundaries in which the body can maintain its position without changing the base of support (BOS). During quiet, barefoot, double-limb standing, the stability limit is the area bounded by the two feet on the ground (Shumway-Cook and Woollacott, 1995). Wearing different types of shoes can alter this boundary. As an extreme example, wearing stiletto heels will reduce the stability limits due to the small contact area between the heel and ground; shift the body CoM anteriorly; modify plantar pressure distribution (Snow and Williams, 1994; Snow et al., 1992); and create a smaller critical tipping angle (Tencer et al., 2004), and as such, increase the difficulty of the balance task, demanding greater postural control, relative to wearing a low-heeled or flat shoe.

Postural stability can be affected by other footwear characteristics, including shoe midsole hardness or thickness (Perry et al., 2007; Tsai and Powers, 2008), heel collar height and outer-sole tread (Menant et al., 2008b,c), and the addition of shoe insoles for biomechanical correction (Hertel et al., 2001; Rome and Brown, 2004). Where the design of regular shoes has been modified to create a therapeutic intervention, such as shoes with rocker-bottom soles or negative heels, these alterations can also have implications on postural stability (Albright and Woodhull-Smith, 2009; Landry et al., 2010; Ramstrand et al., 2008, 2010). However, it is also possible that the magnitude to which modifications in shoe design or construction alter postural stability may also be dependent upon an individual's baseline balance ability. Throughout current literature, the effects of footwear on balance performance have been quantified using a range of techniques and outcome measures, which can make comparison between studies difficult.

24.3 FOOTWEAR AND STANDING BALANCE

24.3.1 TRADITIONAL CENTER OF PRESSURE-BASED PARAMETERS

The effects of footwear on static postural stability are commonly evaluated using quantitative posturography, which uses biomechanical instrumentation to measure postural sway, providing information about small body adjustments occurring when standing quietly (Horak, 1997). Center of pressure (CoP) movement, commonly referred to as postural sway, is defined as the small forward and back (AP) and side-to-side (ML) shifts in body position, when standing upright (Shumway-Cook and Woollacott, 1995). The magnitude of CoP movement provides a measure of balance ability (Hasselkus and Shambes, 1975) and the steadiness of an individual. A wide range of traditional CoP-based postural sway parameters, including sway amplitude, variance, and velocity, have been used in previous studies to quantify the effects of different shoe sole profiles (Albright and Woodhull-Smith, 2009; Landry et al., 2010), shoe insoles (Corbin et al., 2007; Percy and Menz, 2001; Priplata et al., 2003; Rome and Brown, 2004; Wilson et al., 2008), and footwear characteristics (Menant et al., 2008b) on standing balance.

Sway parameters measuring the amplitude of CoP movement, such as AP and ML range or the area of CoP excursion, are suggested to relate to the effectiveness of the postural control system (Maki et al., 1990). They may also represent neuromuscular responses to postural imbalances (Winter, 1990), and can predict future risk of falling (Maki et al., 1994; Stel et al., 2003). AP and ML sway range are clinically useful sway parameters when exploring different footwear interventions, as it is only when the CoM moves beyond the limits of stability that the body may fall over, should a postural control mechanism fail.

Measures of sway variance, such as the AP and ML standard deviation (SD), are reported to generate a measure of spatial variability of the CoP about its mean position (Horak, 1997) and may be inherently related to the effects of footwear on functionally relevant aspects of postural stability, including exploratory postural behavior and the number of solutions available to make postural corrections when balance is threatened.

CoP velocity is a time-dependent sway parameter, providing important information about the rate at which the CoP moves within, and about, the base of support and could be a major determinant in balance maintenance (Van Emmerik and VanWegen, 2002). CoP velocity is defined as the total distance covered by the CoP divided by the sampling period (Simoneau et al., 2008). This is considered the most consistent stability parameter and most sensitive for detecting changes in balance, influenced by age and availability of visual information (Prieto et al., 1996; Raymakers et al., 2005); this level of sensitivity may also translate to different types of footwear. CoP velocity is suggested to be proportionate to the level of activity required to regulate postural stability (Maki et al., 1990).

The effects of footwear interventions on postural stability have been explored during both double-(Corbin et al., 2007; Palluel et al., 2008; Rome and Brown, 2004; Wilson et al., 2008) and single-limb (Corbin et al., 2007; Olmsted and Hertel, 2004; Ramstrand et al., 2008; Shimizu and Andrew, 1999) standing balance.

Depending upon the nature of the task, the effect of footwear interventions on postural sway has been measured using static force platforms or dynamic force platforms capable of rotating about a single axis. However, other apparatus including reflective markers (Priplata et al., 2003), the swaymeter (Lord and Bashford, 1996; Menant et al., 2008b), and optical displacement devices (Percy and Menz, 2001) have been used to capture changes in postural sway when wearing different footwear interventions.

Evidence from previous research indicates that traditional CoP-based sway parameters, collected from force platforms, appear to be sufficiently sensitive in detecting changes in postural stability, due to modifications in shoe design (Shimizu and Andrew, 1999) or the presence of shoe insoles (Corbin et al., 2007; Olmsted and Hertel, 2004; Palluel et al., 2008; Rome and Brown, 2004; Wilson et al., 2008). Shimizu and Andrew (1999) concluded that incremental increases in heel height had a detrimental effect on postural stability in healthy, young women. In that study (Shimizu and Andrew, 1999), the length of CoP displacement in both AP and ML planes increased with heel height. Furthermore, as heel elevation increased, the mean position of the CoP was reported to migrate anteriorly and medially toward the first metatarsal head, making it more difficult to balance during single-limb standing (Shimizu and Andrew, 1999).

Custom-made foot orthoses can enhance balance performance in individuals with high- or low-arched feet. Semi-rigid foot orthoses have been shown to improve postural stability during quiet standing in healthy young adults with high-arched feet (Olmsted and Hertel, 2004), while rigid foot orthoses can specifically reduce ML sway in those with low-arched feet (Rome and Brown, 2004).

Novel footwear interventions, including textured shoe insoles, have been investigated regarding their influence on traditional postural sway parameters during quiet standing (Corbin et al., 2007; Palluel et al., 2008; Wilson et al., 2008). Significant differences in CoP velocity and area (Corbin et al., 2007) and AP and ML sway (Palluel et al., 2008) have been observed between textured insole and control conditions in healthy young (Corbin et al., 2007; Palluel et al., 2008) and older (Palluel et al., 2008) people. However, Wilson et al. (2008) reported that wearing textured shoe insoles had no significant effects on AP or ML sway range in healthy middle-aged females. This nonsignificant finding could be attributed to the sensitivity of the sway parameters of interest, or alternatively, represent a true absence of effect of a footwear intervention designed to enhance plantar sensory feedback in a healthy cohort with no plantar sensory impairments.

Traditional CoP-based sway parameters have also been collected from the level of the waist (Lord and Bashford, 1996; Menant et al., 2008b; Percy and Menz, 2001)—a method which also appears to be sensitive in detecting differences in postural stability between shoe conditions. Menant et al. (2008b) concluded that wearing shoes with an elevated heel significantly increased postural sway, relative to a standard shoe in healthy older people. In that study, postural sway data was collected using the swaymeter—a simple device that records body displacement directly at waist level. The swaymeter comprises a rod with a vertically mounted pen at its end, attached to participants by a belt, and extending posteriorly (Figure 24.1) (Lord et al., 2003). Although the swaymeter does not measure CoP movement directly at the

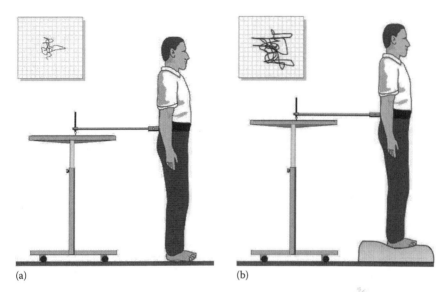

(a)   (b)

FIGURE 24.1 Swaymeter apparatus—(a) sway on ground and (b) sway on foam.

feet, sway measures from this device (Lord et al., 2003) were reported to be strongly associated with CoP measures extracted from force platforms (Sherrington, 2000). Interestingly, Menant et al. (2008b) reported that significant changes in postural stability when wearing shoes with an elevated heel were not detected during other tests of balance including maximal balance range and coordinated stability. This finding may highlight the sensitivity of traditional CoP-based parameters to detect subtle changes in footwear design, over and above functional balance tests.

Percy and Menz (2001) investigated the effect of prefabricated foot orthoses and soft insoles on postural stability in professional soccer players. In that study, postural sway data was also obtained from AP and ML trunk movement, using optical displacement devices (Percy and Menz, 2001). Foot orthoses were observed to have no significant effects on AP or ML sway, although a trend was observed for reduced ML sway during single-limb standing, while wearing prefabricated orthoses (Percy and Menz, 2001).

The effect of footwear interventions on postural stability during standing can be reported using measures of CoP or CoM movement. The CoM is considered to be an imaginary point at which total body mass is concentrated (Lafond et al., 2004). The CoP always oscillates beyond the CoM, in order to maintain the latter in a relatively central position (Winter et al., 1998). This relationship indicates that the CoP controls the CoM, and changes in CoP position will reflect alterations in CoM location (Winter, 2005).

24.3.2 TIME-TO-BOUNDARY

Time-to-boundary (TTB) is a measure which expresses a different, yet related dimension of postural control in comparison to traditional CoP-based measures. TTB estimates the time it would take for the CoP to reach the boundary of the BOS, should the CoP continue on its path at its current velocity (Hertel and Olmsted-Kramer, 2007). This measure is associated with spatiotemporal characteristics of postural control, taking into account the BOS formed by participant's bare feet or footwear, the instantaneous position and velocity of the CoP in AP or ML directions, as required (Hertel and Olmsted-Kramer, 2007). Lower TTB magnitude and variability are indicative of decreased postural control and a more constrained sensorimotor system (McKeon and Hertel, 2008). This outcome measure has been used in footwear intervention studies, exploring the effect of textured shoe insoles on single-limb standing balance in young adults presenting chronic ankle instability (McKeon et al., 2012). Relative to control and sham conditions, when maintaining single-limb stance, textured insoles were shown to significantly decrease TTB magnitude and variability in the mediolateral direction. TTB measures may prove to be more sensitive in detecting postural control deficits associated with wearing different types of footwear than traditional measures because they capture boundary-relevant aspects of postural control; this is especially important when considering shoe features such as heel width and height.

When exploring the effects of footwear interventions on static postural stability, any changes in postural sway between conditions should be interpreted with caution, giving consideration to the context in which balance performance is assessed. Tests of quiet standing are considered sufficiently robust to detect changes in balance control, yet may lack functional relevance as this task rarely occurs in isolation, but rather integrated within activities of daily living (Van Emmerik and Van Wegen, 2002).

24.4   FOOTWEAR AND FUNCTIONAL BALANCE TESTS

A functional approach to balance assessment focuses on an individual's ability to perform tasks relevant to daily life (Shumway-Cook and Woollacott, 1995). This approach has been used in previous studies to quantify the effects of different footwear characteristics on dynamic balance control in healthy young (Olmsted and Hertel, 2004) and older people (Arnadottir and Mercer, 2000; Menant et al., 2008b) and adults recovering from stroke (Ng et al., 2010). It is possible that tests of static postural stability, including single- or double-limb standing balance may not be sufficiently challenging, allowing participants to use alternative motor strategies to maintain upright balance, irrespective of any destabilizing modifications to shoe design. Therefore, the effect of footwear on more dynamic, functional balance tasks has been explored.

24.4.1   Functional Reach Test

The functional reach test assesses the maximal distance one can reach forward beyond arm's length, while maintaining a fixed BOS in the standing position (Duncan et al., 1990). Ng et al. (2010) investigated the effect of footwear on balance in people recovering from stoke using this functional test. However, in that study (Ng et al., 2010), an effect of footwear was not found. The authors postulate that as a single measure of standing balance, the functional reach test may not be sufficiently sensitive to capture the effects of footwear on balance in a cohort of stroke patients with balance impairments at baseline.

Arnadottir and Mercer (2000) also used the functional reach test to investigate the effects of wearing walking shoes (defined as a laced-up, buckled, or Velcro-fastened shoe, with a heel height of 0–2 cm, which included athletic and oxford-type shoes) and dress shoes (defined as a firm-soled, slip-on shoe, with a heel height of at least 4 cm), relative to barefoot, on balance performance in healthy older women. This study reported that older women performed significantly better on the functional reach test when barefoot or wearing walking shoes compared with when they wore dress shoes (Arnadottir and Mercer, 2000). It appears that when barefoot, or wearing low-heeled footwear, older people are capable of reaching further distances, which may be attributed to greater proprioceptive information at the feet and ankles. In comparison, wearing high-heeled dress shoes reduces the size of the BOS and displaces the line of gravity toward the anterior boundary of the BOS—both factors could contribute to relatively poorer performance on the functional reach test (Arnadottir and Mercer, 2000).

24.4.2 MAXIMAL BALANCE RANGE

Maximal balance range requires individuals to lean forward from their ankles without moving the feet or bending at the hips, as far as possible, to the point where balance is just maintained (Figure 24.2) (Lord and Bashford, 1996). Menant et al. (2008b) explored the effect of variations in heel and shoe sole characteristics on maximal AP balance range in healthy older adults. However, in that study, no significant differences were reported for maximal balance range between standard and modified shoe conditions (Menant et al., 2008b).

24.4.3 COORDINATED STABILITY

The coordinated stability test measures an individual's ability to adjust the position of their CoM in a controlled manner, when approaching their limits of stability. Using the swaymeter, individuals are required to move their CoM by bending or rotating

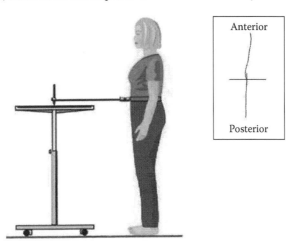

FIGURE 24.2 Maximal balance range.

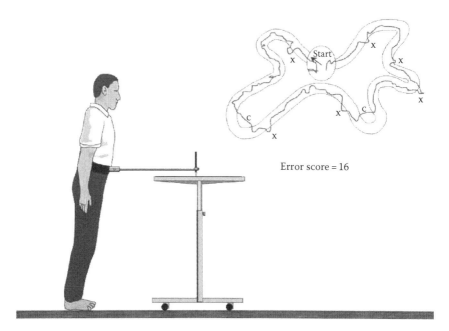

FIGURE 24.3 Coordinated stability with example track and error score.

at the trunk, without moving their feet, in order to guide a pen (attached to the end of the swaymeter rod) around a marked track (Figure 24.3). Errors are scored when the CoM cannot be altered sufficiently, resulting in failure to negotiate a corner of the track. Menant et al. (2008b) also used the coordinated stability balance test to explore the effect of footwear modification on postural stability. Similar to the findings for maximal balance range, no significant differences were reported for coordinated stability between eight different shoe conditions (Menant et al., 2008b); this may be due to the feet-in-place positioning required for the test.

24.4.4 STAR EXCURSION BALANCE TEST

The Star Excursion Balance Test (SEBT) involves a series of lower extremity maximum reaching movements, incorporating single-limb standing balance. Participants stand at the center of a grid placed on the floor, from which eight lines extend at 45° increments. The lines are labeled according to their direction of excursion (relative to the stance leg): anterior, anterolateral, lateral, posterolateral, posterior, posteromedial, medial, and anteromedial. Participants maintain single-limb standing balance, while reaching with the opposite leg to touch as far as possible along one of the eight lines, as instructed. The distance from the center of the grid to the touch point is measured (Figure 24.4).

Olmsted and Hertel (2004) used the SEBT to investigate the effect of custom-made semi-rigid foot orthoses on postural stability in healthy young adults with different types of foot structure. When people with high-arched feet wore semi-rigid foot orthoses, they could reach further in anterolateral, lateral, and posterolateral directions, relative to a no-orthotic condition (Olmsted and Hertel, 2004). The SEBT

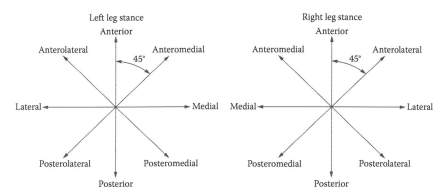

FIGURE 24.4   Directions of the Star Excursion Balance Test, relative to the stance limb.

has also been used to identify significant improvements in medial reach in adults with unilateral chronic ankle instability, after wearing a foot orthotic for 4 weeks (Albright and Woodhull-Smith, 2009). Therefore, the SEBT appears to be sufficiently sensitive in detecting improvements in dynamic postural stability, associated with wearing corrective foot orthoses in both healthy and pathological groups.

24.4.5   Berg Balance Scale

The Berg Balance Scale (BBS) is a performance-based measure of balance that assesses 14 different functional tasks including functional transfers from sitting to standing, unsupported standing under different sensory and BOS conditions, reaching, turning, and stepping. This balance test was originally designed to assess balance performance in older people; however, the functional basis of the BBS components is highly appropriate to the assessment of footwear and postural stability. When considering the test components involving turning, stepping, and transfers, changes in shoe sole material and design may have implications on slip propensity by altering the coefficient of friction between the sole and ground, proprioceptive feedback, or mechanical stability provided to the foot and ankle. The effect of wearing shoes, relative to barefoot, on postural stability in older women attending a day hospital has been investigated using the BBS (Horgan et al., 2009). Horgan et al. (2009) concluded that when older women wore their own usual footwear, they scored significantly higher on the BBS compared to when barefoot; this effect was observed to be greatest in women with the poorest balance ability at baseline. However, this change in BBS score could not be attributed to specific footwear features. While it appears that the BBS is sufficiently sensitive to detect changes in functional balance performance between shod and barefoot conditions, it may not be capable of identifying changes between subtle modifications in footwear design; however, this requires further investigation.

24.4.6   Timed Up and Go Test

The Timed Up and Go (TUG) test is a clinical tool used to assess basic mobility in older people, and essentially assesses the ability to control movement of the center of gravity over a moving BOS. The TUG test measures the time taken to stand up from

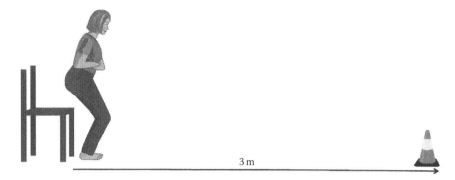

FIGURE 24.5 Timed Up and Go test.

a seated position, walk straight forward for 3 m at a steady, normal pace, turn around, walk back to the chair, and sit back down (Figure 24.5).

The TUG test has shown to be sensitive in detecting changes in balance performance, when healthy older women wore different types of footwear (Arnadottir and Mercer, 2000). These authors hypothesized that wearing shoes with an elevated heel may reduce stability and shock absorption capabilities at the feet and ankle joints, while simultaneously increasing the metabolic cost of walking. As such, these shoe features may prolong the time it takes individuals to successfully complete the TUG test, with larger scores representing poorer walking balance (Arnadottir and Mercer, 2000). In that study, older women were observed to complete the TUG test significantly faster when wearing walking shoes, relative to dress shoes (Arnadottir and Mercer, 2000). Scores for completing the TUG test barefoot were located in-between those reported for walking and dress shoes. This change improvement in TUG test score, when wearing walking shoes, may be clinically relevant. It is possible that differences in TUG test scores between different types of footwear could be attributed to their shock-absorbing properties. Shoes providing greater shock absorption, such as walking shoes, may enable participants to walk faster with greater comfort and reduced impact; this may be especially important for individuals with conditions such as diabetic peripheral neuropathy.

24.5 FOOTWEAR AND REACTIVE BALANCE CONTROL

Compensatory stepping reactions are initiated as a protective mechanism to restore postural control when upright balance is disturbed. The effect of modified footwear on reactive balance control has been explored in children (Ramstrand et al., 2008), young (Albright and Woodhull-Smith, 2009), middle-aged (Ramstrand et al., 2010) and older (Menant et al., 2008b) adults.

Reactive balance control is commonly assessed using a dynamic force platform, which is rotated or tilted in different directions, at varied magnitudes and speeds, to create an unexpected platform perturbation (Albright and Woodhull-Smith, 2009; Ramstrand et al., 2008, 2010). Three different studies have used this apparatus and methodology to investigate the effect of an unstable shoe construction on reactive balance responses in children with developmental disabilities (Ramstrand et al., 2008), healthy young adults (Albright and Woodhull-Smith, 2009), and healthy older women (Ramstrand et al., 2010).

Reactive postural responses to upward and downward force platform rotations have been measured by quantifying the time taken to stabilize the CoP after an unexpected perturbation (Ramstrand et al., 2008), where restabilization was defined as the time when the CoP was maintained within a $5 \times 5\,mm^2$ area for a period of 500 ms (Shumway-Cook et al., 2003). Over three separate testing occasions, time to stabilization during upward platform rotation perturbations, in children with developmental disabilities, was shown to significantly reduce while wearing unstable shoes. Using the same methodology in a group of women aged over 50 years, Ramstrand et al. (2010) reported that time to CoP stabilization significantly reduced over the three testing occasions during downward platform rotations when wearing unstable shoes. In both studies, three testing occasions were conducted, with a minimum of 4 weeks between each. Therefore, the results from these two studies (Ramstrand et al., 2008, 2010) may point to two important factors: (1) long-term wear of unstable shoes may lead to improvements in reactive balance control and (2) time to stabilization is a suitable parameter, capable of detecting the training effect of wearing unstable shoes.

Albright and Woodhull-Smith (2009) also used a dynamic force platform to investigate the effect of five different shoes, including a control, rocker-bottom sole, and negative heel shoe, on postural stability in healthy young adults, in response to a backward platform translation. In comparison to Ramstrand et al. (2008, 2010), a wider range of CoP and CoM sway variables were extracted, including mean sway amplitude, range and variance, functional stability margin, and anterior and posterior peak velocities. Other sway parameters specific to the perturbation event were time to peak, defined as the period between the onset of the perturbation and the maximum peak displacement, and peak duration time, which is a measure of the time period involved in the stopping and reversing of the direction of movement (Albright and Woodhull-Smith, 2009). Both experimental shoe conditions were observed to have destabilizing effects. Relative to the control shoe, rocker-bottom and negative heel shoes both significantly increased movement of the CoP and/or CoM and showed a longer duration of displacement and thus delayed onset of balance recovery. Measures of both CoP and CoM movement appear to be sensitive to detecting changes in reactive postural responses between control and rocker-bottom sole shoes.

Choice-stepping reaction time is another test which assess reactive postural responses, requiring individuals to transfer their bodyweight forward and laterally, in response to a light stimulus, mounted within a nonslip platform comprising four white light panels—two located anterior to the individual (one in front of each foot) and one panel at each side (adjacent to each foot). Each step must be executed in a controlled manner in order to return to a central position and then complete subsequent steps in unexpected directions (Figure 24.6).

This test demands stepping responses similar to those executed when individuals, particularly older people, have to suddenly make changes in their walking path due to hazards (Lord and Fitzpatrick, 2001). Menant et al. (2008b) investigated the effect of different types of footwear on choice-stepping reaction time, defined as the time interval between the light stimulus and the completion of the correct step, in healthy older people. Shorter time intervals represent better reaction times and

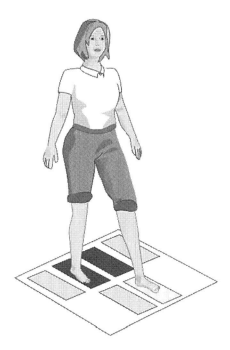

FIGURE 24.6 Choice-stepping reaction time.

greater postural stability. No significant differences in choice-stepping reaction times were reported between a standard shoe and several modified shoes, including an elevated heel, soft sole, hard sole, flared sole, beveled heel, high heeled-collar, and tread sole (Menant et al., 2008b). It is possible that the small modifications made to the shoe sole to create a range of different types of footwear were not sufficient to significantly alter reactive postural responses during sudden stepping in older people. Alternatively, the choice-stepping reaction time may not be sensitive enough to detect changes in reactive balance control, resulting from subtle changes in footwear design.

24.6 FOOTWEAR AND WALKING STABILITY

Previous studies have explored the effects of footwear design and shoe insoles on temporospatial gait variables during walking, as a measure of dynamic stability (Jenkins et al., 2009; Menant et al., 2009; Novak and Novak, 2006; Wilson et al., 2008). In these studies, gait variables of interest have included different combinations of gait velocity, stride length and duration, step length and width, cadence, stance, swing and double-limb support times, stride variability, and step time variability.

The effect of novel shoe insoles on gait patterns have been assessed during unobstructed level-ground walking in healthy and pathological groups (Jenkins et al., 2009; Wilson et al., 2008). Wilson et al. (2008) concluded that wearing textured insoles within standardized shoes had no effect on the BOS (defined as the perpendicular distance from the heel point of one footfall to the line of progression of the opposite foot) in healthy middle-aged females, relative to a control condition.

This variable was considered to be important in maintaining stability during walking. The results from Wilson et al. (2008) suggest that BOS may not be the most sensitive parameter to detect changes in walking stability between shoe insole conditions, and that a single gait parameter does not provide sufficient information about the effect of footwear interventions on dynamic balance control. Alternatively, it is possible that level-ground walking tests may only be sensitive in teasing out changes in gait patterns between different types of footwear interventions in individuals with poor balance at baseline. In comparison, when individuals with Parkinson's disease wore balance-enhancing insoles within their shoes during level-ground walking, Jenkins et al. (2009) observed significant changes in single-limb support time.

Unobstructed walking may not be capable of teasing out the effects of different footwear interventions on walking balance, due to the relatively unchallenging nature of the task. As such, the effects of footwear on more complex gait assessments have been explored, requiring individuals to negotiate uneven terrain comprising inclined wooden platforms (Menant et al., 2008a, 2009, Perry et al., 2008), layers of foam, and artificial grass (Figure 24.7) (Menant et al., 2009); wet, slippery surfaces (Menant et al., 2009; Tsai and Powers, 2008); and narrow balance beams (Robbins et al., 1994).

These methodologies are suggested to simulate daily life, such as stepping over paving curbs or maintaining balance on uneven, or wet slippery surfaces, and

FIGURE 24.7 Irregular surface created using layers of foam, wooden blocks, and artificial grass.

thus environmental conditions during which footwear would normally be worn by individuals in the Western world.

Dynamic balance performance during walking over uneven terrain in different types of shoes (Menant et al., 2008a) and with balance-enhancing insoles (Maki et al., 2008; Perry et al., 2008) has been measured by exploring relationships between the CoM and BOS, commonly referred to as the "stability margin." This measure quantifies the minimum AP or ML displacement of the CoM relative to the BOS during the single-limb support phase of gait, and thus the degree to which the CoM approaches the limits of stability. Larger values of the stability margin in both AP and ML planes indicate greater stability (Maki et al., 2008; Perry et al., 2008). The posterior CoM–BOS margin is calculated as the minimum distance between the CoM AP position and the posterior border of the BOS, normalized to step length. The lateral CoM–BOS margin is calculated as the minimum distance between the CoM ML position and the lateral border of the BOS and normalized to step width (Menant et al., 2008a).

Using this method of assessment, Perry et al. (2008) and Maki et al. (2008) reported a significant improvement in walking stability when older adults with mild age-related loss of foot sensation wore a balance-enhancing shoe insole, relative to a control condition. Menant et al. (2008a) concluded that relative to a standard shoe, wearing shoes with a soft insole led to significantly greater lateral stability margins in young and older adults, during walking. It was suggested that a softer, more malleable shoe insole may lead to restricted ML CoM movement, due to a reduction in joint position sense and poorer mechanical support at the foot (Menant et al., 2008a). Wearing shoes with an elevated heel significantly reduced the posterior CoM–BOS margin, indicating poorer dynamic stability, relative to control. This finding could be attributed to a reduced BOS, smaller critical tipping angle, and anterior shifting of the CoM associated with high-heeled shoes. This evidence suggests the CoM–BOS stability margin is a sensitive measure in detecting changes in dynamic postural stability when wearing different types of footwear or footwear interventions in young (Menant et al., 2008a) and older people (Maki et al., 2008; Menant et al., 2008a; Perry et al., 2008).

Measures of pelvic acceleration in the frontal plane of motion have been shown to detect differences in dynamic stability in young and older people when wearing different types of shoes (Menant et al., 2011). During walking over an irregular surface in shoes with an elevated heel, Menant et al. (2011) observed significant reductions in ML pelvic accelerations in both subject groups. Pelvic acceleration may be a suitable parameter for providing insight into the postural strategies used to counteract lateral instability when walking over an irregular surface or in high-heeled shoes (Menant et al., 2011).

Robbins et al. (1994) investigated the effect of six modified athletic shoes, which differed only in midsole hardness and thickness, on dynamic postural stability in healthy young men when walking over a balance beam at a standardized speed. In this study, the primary outcome measure was the frequency of balance failures (i.e., when individuals fell off the beam). Robbins et al. (1994) reported that midsole hardness was positively related to stability and midsole thickness negatively related to stability. Thus, wearing shoes with a thin and hard midsole led to fewer "balance failures" from the beam and this combination of footwear properties may provide superior stability.

Another measure of dynamic balance performance, associated with footwear studies, is slip events (Tsai and Powers, 2008). The presence of a slip is defined as the forward horizontal displacement of a heel marker, beyond a specified distance, such as 1 cm, following initial heel contact during walking (Tsai and Powers, 2008). Tsai and Powers (2008) reported that wearing shoes with harder soles led to a greater proportion of slip events in healthy young adults, compared to soft sole shoes, during walking on a slippery surface. It was postulated that a hard shoe sole, with limited deformation capacity upon contact with the ground, substantially reduces the friction between the shoe and the floor surface, thus increasing the likelihood of slipping. It is also possible that changes in gait patterns may occur in anticipation of a potential slip event, which may be specific to the type of footwear being worn and their frictional or mechanical properties.

24.7 THERAPEUTIC FOOTWEAR INTERVENTIONS AND POSTURAL STABILITY IN PATHOLOGICAL CONDITIONS

24.7.1 PARKINSON'S DISEASE

The long-term effects of the balance-enhancing insole on dynamic postural stability in Parkinson's disease have been explored (Jenkins et al., 2009). The balance-enhancing insole has been shown to significantly increase single-limb support time during level ground walking in individuals with Parkinson's disease, relative to a flat insole (Jenkins et al., 2009). Considering the construction of the balance-enhancing insole, a flexible raised ridge extends from the first to the fifth metatarsal heads, around the posterior perimeter of the rearfoot. Therefore, as this tubing essentially "cups" and provides a barrier around the heel and lateral periphery of the foot, it is perhaps not unexpected that this insole is capable of improving single-limb support time (Jenkins et al., 2009) and also lateral stability in older people (Maki et al., 2008; Perry et al., 2008). This may suggest the balance-enhancing insole brings about its effect on balance performance through a combination of mechanical and sensory mechanisms. These findings (Jenkins et al., 2009; Maki et al., 2008; Perry et al., 2008) appear to suggest that the balance-enhancing insoles may have an obvious mechanical effect with a putative sensory mechanism.

Novak and Novak (2006) investigated the effect of step-synchronized vibrating insoles on gait parameters in adults with Parkinson's disease. Walking with vibrating insoles (whereby the vibratory stimulus was supra-sensory) has been shown to significantly increase walking speed, cadence, stride duration, and stride length and significantly decrease stride variability. Vibrating insoles were also shown to improve some gait parameters in healthy controls; however, this effect was more marked in the Parkinson's group presenting poorer balance ability at baseline (Novak and Novak, 2006).

24.7.2 MULTIPLE SCLEROSIS

Ramdharry et al. (2006) investigated the effect of contoured dynamic foot orthoses on standing balance and walking speed in adults with multiple sclerosis, at the initial delivery of the intervention and after 4 weeks wear. At both testing occasions, a significant increase in CoP velocity was observed, when the foot orthoses were worn. This destabilizing effect was greatest when standing with eyes closed (Ramdharry et al., 2006).

Kelleher et al. (2010) explored the effect of textured shoe insoles on gait patterns and corresponding lower limb muscle activity in adults with multiple sclerosis. Textured insoles investigated in that study were constructed from coarse sandpaper, fixed to fine leather insoles within participants own shoes. When wearing textured insoles, participants with multiple sclerosis showed a slight increase in walking velocity, but this did not reach the level of statistical significance, and it still remained slower than that of healthy controls. Findings from this study also tentatively suggest that textured insoles may have the capacity to bring accelerating and braking ground reaction forces and measures of lower limb joint kinematics in adults with multiple sclerosis, closer to values observed in healthy adults. During the stance phase of walking, mean EMG activity in lateral gastrocnemius was also shown to increase when wearing the textured insoles (Kelleher et al., 2010). Enhanced plantar tactile stimulation may trigger such kinetic and kinematic responses. However, in participants with multiple sclerosis, consideration must also be given to the presence of lower limb spasticity and potential triggers, such as tactile stimulation, which may exacerbate abnormal joint movement and levels of muscle tone. This evidence suggests that providing enhanced tactile stimulation to the feet may assist in restoring normal gait patterns in adults with neuromuscular disease (Kelleher et al., 2010). However, it remains unknown whether textured insoles are also capable of altering fine motor control (Kelleher et al., 2010); without such evidence it is difficult to postulate whether changes in gait velocity and joint movement in people with multiple sclerosis translate to better balance performance.

24.7.3 DIABETES

Previous studies have demonstrated that modifications to footwear can influence balance and stability in healthy, elderly subjects (Arnadottir and Mercer, 2000; Horgan et al., 2009; Lord and Bashford, 1996; Menant et al., 2008b). However, adjustments made to footwear for the diabetic foot are generally more dramatic and, therefore, are expected to have a greater influence on postural stability. Footwear, orthotics, casts, and braces used for treatment or prevention of diabetic-related plantar ulceration, designed to off-load injured or at-risk foot areas may actually interfere with normal gait and posture and, therefore, stability.

A number of recent studies have reported on using vibrating probes to stimulate the plantar surface of the foot and thereafter improve balance performance in individuals with diabetic neuropathy (Hijmans et al., 2008; Priplata et al., 2006). Vibrating insoles, which apply subsensory mechanical noise to the plantar surface of the feet, have been shown to significantly improve standing balance in adults with neuropathy (Hijmans et al., 2008). Subsensory vibratory stimulation may lead to earlier detection of changes in foot pressure, thus earlier corrective postural responses (Hijmans et al., 2007). The underlying mechanism relating to vibrating footwear interventions centers round the concept of stochastic resonance.

This phenomenon suggests applying subsensory noise can enhance detection, transmission, and flow of weak signals, which may influence sensorimotor function (Gravelle et al., 2002; Priplata et al., 2003). Subsensory noise may augment fine touch sensitivity on the plantar surface of the feet in older adults (Dhruv et al., 2002), thereafter providing enhanced sensory information about the supporting surface for balance control.

Priplata et al. (2006) reported significant differences in a number of balance parameters, when patients with diabetic neuropathy, stroke, and healthy older adults wore vibrating insoles. Research by Hijmans et al. (2008) supports findings that vibrating insoles can improve balance in adults with diabetic neuropathy. That study concluded vibrating insoles had a statistically significant interaction on postural stability during balance tests including a secondary distraction task (Hijmans et al., 2008). Vibrating insoles only had a significant effect on balance performance—in the presence of neuropathic disease and when participants carried out an attention-demanding task. No significant differences in balance performance were observed in healthy controls, suggesting vibratory stimulus may have greater effect in individuals with known balance deficits originating from pathology.

Van Geffen et al. (2007) investigated whether insoles with a low Shore A value (15°), commonly prescribed for patients with a diabetic neuropathy, had a negative effect on postural stability, because these insoles may reduce somatosensory input under the feet. Postural stability was assessed in diabetic patients with a neuropathy and healthy controls while wearing a shoe without insole, on a flat insole with a low Shore A value (15°) and on a flat insole with a higher Shore A value (30°). These assessments were conducted under four different conditions: (1) eyes open, no dual-task; (2) eyes closed, no dual-task; (3) eyes open, dual-task (mental arithmetic); and (4) eyes closed, dual-task. Although diabetic patients were observed to have poorer balance ability compared to healthy controls, the insoles had no significant effects on postural stability in either group. Therefore, prescribing insoles with a low Shore A value (15°), compared to insoles with a higher Shore A value (30°), has no significant negative effect on postural stability in patients with a diabetic neuropathy (Van Geffen et al., 2007).

24.7.4 OSTEOARTHRITIS

Nigg et al. (2006) assessed the effectiveness of unstable shoes in reducing knee pain in adults with knee osteoarthritis, in addition to changes in balance, ankle and knee joint range of motion, and ankle strength compared with a high-end walking shoe as control, over 12 weeks. Nigg et al. (2006) reported that wearing unstable shoes did not significantly improve balance in adults with osteoarthritis. Although wearing the unstable shoe did lead to a significant increase in static balance, this difference was not significant between intervention and control groups.

24.8 CONCLUSION

The effects of footwear interventions on static and dynamic postural stability can be measured using a wide range of apparatus, methodologies, and outcome measures. However, these initial choices, made early in the research or clinical process, can have major implications on the interpretation of postural stability data. Within this chapter, footwear interventions, including novel shoe insoles and modifications to regular shoes, have been shown to significantly alter postural stability in both healthy and pathological groups. These effects may be therapeutically beneficial or lead to postural instability, and as such, could have potentially serious implications on an individual's long-term health and well-being. Therefore, careful consideration should be given to the methods used to investigate the effects of footwear interventions on postural stability, ensuring (1) balance tests are appropriate for specific patient groups to perform, yet sufficiently challenging to tease out the effects of footwear interventions on stability and (2) outcome measures are sensitive enough to detect true changes in balance performance attributed to subtle or gross footwear modification.

QUESTIONS

24.1 How do you define postural stability?

24.2 What footwear characteristics can affect postural stability measures?

24.3 The effects of footwear on postural stability can be quantified by measuring the center of pressure (CoP) or center of mass (CoM) movement. What is the relationship between the CoP and CoM?

24.4 What is the time-to-boundary measurement and what important factors related to postural stability does this parameter capture, over and above traditional CoP-based measurements?

24.5 What features of footwear/shoe insole design may lead to improved performance in functional balance tests?

24.6 What specific changes in gait variables are reported to represent improved dynamic stability when wearing footwear or shoe insoles?

24.7 In which clinical conditions are footwear interventions reported to have a beneficial effect on postural stability?

REFERENCES

Albright B and Woodhull-Smith W (2009) Rocker bottom soles alter the postural response to backward translation during stance. *Gait and Posture* **30** 45–49.

Arnadottir S and Mercer V (2000) Effects of footwear on measurements of balance and gait in women between the ages of 65 and 93 years. *Physical Therapy* **80** 17–27.

Corbin D, Hart J, McKeon P, Ingersoll C, and Hertel J (2007) The effect of textured insoles on postural control in double and single limb stance. *Journal of Sport Rehabilitation* **16** 363–372.

Dhruv N, Niemi J, Lipsitz L, and Collins J (2002) Enhancing tactile sensation in older adults with electrical noise stimulation. *NeuroReport* **13** 597–600.

Duncan P, Weiner D, Chandler J, and Studenski S (1990) Functional reach: A new clinical measure of balance. *Journals of Gerontology: Series A Biological Sciences and Medical Sciences* **45** M192–M197.

Gravelle D, Laughton C, Dhruv N, Katdare K, Niemi J, Lipsitz L, and Collins J (2002) Noise-enhanced balance control in older adults. *NeuroReport* **13** 1853–1856.

Hasselkus B and Shambes G (1975) Aging and postural sway in women. *Journal of Gerontology* **30** 661–667.

Hertel J, Denegar C, Buckley W, Sharkey N, and Stokes W (2001) Effect of rearfoot orthotics on postural sway after lateral ankle sprain. *Archives of Physical Medicine and Rehabilitation* **82** 1000–1003.

Hertel J and Olmsted-Kramer L (2007) Deficits in time-to-boundary measures of postural control with chronic ankle instability. *Gait and Posture* **25** 33–39.

Hijmans J, Geertzen J, Schokker B, and Postema K (2007) Development of vibrating insoles. *International Journal of Rehabilitation Research* **30** 343–345.

Hijmans J, Geertzen J, Zijlstra W, Hof A, and Postema K (2008) Effects of vibrating insoles on standing balance in neuropathy. *Parkinsonism and Related Disorders* **14** S27.

Horak F (1997) Clinical assessment of balance disorders. *Gait and Posture* **6** 76–84.

Horgan N, Crehan F, Bartlett E, Keogan F, O'Grady A, Moore A, Donegan C, and Curran M (2009) The effects of usual footwear on balance amongst elderly women attending a day hospital. *Age and Ageing* **38** 62–67.

Jenkins M, Almeida Q, Spaulding S, Van Oostveen R, Holmes J, Johnson A, and Perry S (2009) Plantar cutaneous sensory stimulation improves single-limb support time, and EMG activation patterns among individuals with Parkinson's disease. *Parkinsonism and Related Disorders* **15** 697–702.

Kelleher K, Spence W, Solomonidis S, and Apatsidis D (2010) The effect of textured insoles on gait patterns of people with multiple sclerosis. *Gait and Posture* **32** 67–71.

Lafond D, Duarte M, and Prince F (2004) Comparison of three methods to estimate the center of mass during balance assessment. *Journal of Biomechanics* **37** 1421–1426.

Landry S, Nigg B, and Tecante K (2010) Standing in an unstable shoe increases postural sway and muscle activity of selected smaller extrinsic foot muscles. *Gait and Posture* **32** 215–219.

Lord S and Bashford G (1996) Shoe characteristics and balance in older women. *Journal of the American Geriatrics Society* **44** 429–433.

Lord S and Fitzpatrick R (2001) Choice-stepping reaction time: A composite measure of falls risk in older people. *Journals of Gerontology: Series A, Biological Sciences and Medical Sciences* **56** M627–M632.

Lord S, Menz H, and Tiedemann A (2003) A physiological profile approach to falls risk assessment and prevention. *Physical Therapy* **83** 237–252.

Maki B, Cheng K, Mansfield A, Scovil C, Perry S, Peters A, McKay S, Lee T, Marquis A, Corbeil P, Fernie G, Liu B, and McIlroy W (2008) Preventing falls in older adults: New interventions to promote more effective change-in-support balance reactions. *Journal of Electromyography and Kinesiology* **18** 243–254.

Maki B, Holliday P, and Fernie G (1990) Ageing and postural control: A comparison of spontaneous- and induced-sway balance tests. *Journal of the American Geriatrics Society* **38** 1–9.

Maki B, Holliday P, and Topper A (1994) A prospective study of postural balance and risk of falling in an ambulatory and independent elderly population. *Journal of Gerontology Series A: Biological Sciences and Medical Sciences* **49** M72–M84.

McKeon P and Hertel J (2008) Systematic review of postural control and lateral ankle instability, Part 1: Can deficits be detected with instrumented testing? *Journal of Athletic Training* **43** 293–304.

McKeon P, Stein A, Ingersoll C, and Hertel J (2012) Altered plantar receptor stimulation impairs postural control in those with chronic ankle instability. *Journal of Sport Rehabilitation* **21** 1–6.

Menant J, Perry S, Steele J, Menz H, Munro B, and Lord S (2008a) Effects of shoe characteristics on dynamic stability when walking on even and uneven surfaces in young and older people. *Archives of Physical Medicine and Rehabilitation* **89** 1970–1976.

Menant J, Steele J, Menz H, Munro B, and Lord S (2008b) Effects of footwear features on balance and stepping in older people. *Gerontology* **54** 18–23.

Menant J, Steele J, Menz H, Munro B, and Lord S (2008c) Optimizing footwear for older people at risk of falls. *Journal of Rehabilitation Research and Development* **45** 1167–1182.

Menant J, Steele J, Menz H, Munro B, and Lord S (2009) Effects of walking surfaces and footwear on temporo-spatial gait parameters in young and older people. *Gait and Posture* **29** 392–397.

Menant J, Steele J, Menz H, Munro B, and Lord S (2011) Step time variability and pelvis acceleration patterns of younger and older adults: Effects of footwear and surface conditions. *Research in Sports Medicine* **19** 28–41.

Ng H, McGinley J, Jolley D, Morris M, Workman B, and Srikanth V (2010) Effects of footwear on gait and balance in people recovering from stroke. *Age and Ageing* **39** 507–510.

Nigg B, Emery C, and Hiemstra L (2006) Unstable shoe construction and reduction of pain in osteoarthritis patients. *Medicine and Science in Sports and Exercise* **38** 1701–1708.

Novak P and Novak V (2006) Effect of step-synchronized vibration stimulation of soles on gait in Parkinson's disease: A pilot study. *Journal of NeuroEngineering and Rehabilitation* **3** 9.

Olmsted L and Hertel J (2004) Influence of foot type and orthotics on static and dynamic postural control. *Journal of Sport Rehabilitation* **13** 54–66.

Palluel E, Nougier V, and Olivier I (2008) Do spike insoles enhance postural stability and plantar-surface cutaneous sensitivity in the elderly? *Age* **30** 53–61.

Percy M and Menz H (2001) Effects of prefabricated foot orthoses and soft insoles on postural stability in professional soccer players. *Journal of the American Podiatric Medical Association* **91** 194–202.

Perry S, Radtke A, and Goodwin C (2007) Influence of footwear midsole material hardness on dynamic balance control during unexpected gait termination. *Gait and Posture* **25** 94–98.

Perry S, Radtke A, McIllroy W, Fernie G, and Maki B (2008) Efficacy and effectiveness of a balance-enhancing insole. *Journal of Gerontology Series A: Biological Sciences and Medical Sciences* **63A** 595–602.

Prieto T, Myklebust J, Hoffmann R, Lovett E, and Myklebust B (1996) Measures of postural steadiness: Differences between healthy young and elderly adults. *IEEE Transactions on Biomedical Engineering* **43** 956–966.

Priplata A, Niemi J, Harry J, Lipsitz L, and Collins J (2003) Vibrating insoles and balance control in elderly people. *Lancet* **362** 1123–1124.

Priplata A, Patritti B, Niemi J, Hughes R, Gravelle D, Lipsitz L, Veves A, Stein J, Bonato P, and Collins J (2006) Noise-enhanced balance control in patients with diabetes and patients with stroke. *Annals of Neurology* **59** 4–12.

Ramdharry G, Marsden J, Day B, and Thompson A (2006) De-stabilizing and training effects of foot orthoses in multiple sclerosis. *Multiple Sclerosis* **12** 219–226.

Ramstrand N, Andersson C, and Rusaw D (2008) Effects of an unstable shoe construction on standing balance in children with developmental disabilities: A pilot study. *Prosthetics and Orthotics International* **32** 422–433.

Ramstrand N, Thuesen A, Nielsen D, and Rusaw D (2010) Effects of an unstable shoe construction on balance in women aged over 50 years. *Clinical Biomechanics* **25** 455–460.

Raymakers J, Samson M, and Verhaar H (2005) The assessment of body sway and the choice of the stability parameter(s). *Gait and Posture* **21** 48–58.

Robbins S, Waked E, Gouw G, and McClaran J (1994) Athletic footwear affects balance in men. *British Journal of Sports Medicine* **28** 117–122.

Rome K and Brown C (2004) Randomized clinical trial into the impact of rigid foot orthoses on balance parameters in excessively pronated feet. *Clinical Rehabilitation* **18** 624–630.

Sherrington C (2000) The effects of exercise on physical ability following fall-related hip fracture. PhD, University of New South Wales, Sydney, New South Wales, Australia.

Shimizu M and Andrew P (1999) Effect of heel height on the foot in unilateral standing. *Journal of Physical Therapy and Science* **11** 95–100.

Shumway-Cook A, Hutchinson S, Kartin D, Price R, and Woollacott M (2003) Effect of balance training on recovery of stability in children with cerebral palsy. *Developmental Medicine and Child Neurology* **45** 591–602.

Shumway-Cook A and Woollacott M (1995) *Motor Control Theory and Practical Applications.* Baltimore, MD: Williams & Williams.

Simoneau E, Billot M, Martin A, Perennou D, and Van Hoecke J (2008) Difficult memory task during postural tasks of various difficulties in young and older people: A pilot study. *Clinical Neurophysiology* **119** 1158–1165.

Snow R and Williams K (1994) High heeled shoes: Their effects on center of mass position, posture, three-dimensional kinematics, rear-foot motion, and ground reaction forces. *Archives of Physical Medicine and Rehabilitation* **74** 568–576.

Snow R, Williams K, and Holmes GJ (1992) The effects of wearing high heels on pedal pressure in women. *Foot and Ankle* **13** 85–92.

Stel V, Smit J, Pluijm S, and Lips P (2003) Balance and mobility performance as treatable risk factors for recurrent falling in older persons. *Journal of Clinical Epidemiology* **56** 659–668.

Tencer A, Koepsell T, Wolf M, Frankenfeld C, Buchner D, Kukull W, LaCroix A, Larson E, and Tautvydas M (2004) Biomechanical properties of shoes and risk of falls in older adults. *Journal of the American Geriatrics Society* **52** 1840–1846.

Tsai Y and Powers C (2008) The influence of footwear sole hardness on slip initiation in young adults. *Journal of Forensic Sciences* **53** 884–888.

Van Emmerik R and Van Wegen E (2002) On the functional aspects of variability in postural control. *Exercise and Sport Science Reviews* **30** 177–183.

Van Geffen J, Dijkstra P, Hof A, Halbertsma J, and Postema K (2007) Effect of flat insoles with different shore A values on posture stability in diabetic neuropathy. *Prosthetics and Orthotics International* **31** 228–235.

Wilson M, Rome K, Hodgson D, and Ball P (2008) Effect of textured foot orthotics on static and dynamic postural stability in middle-aged females. *Gait and Posture* **1** 36–42.

Winter D (1990) Kinesiological electromyography. *Biomechanics and Motor Control of Human Movement*, 2nd edn, Chapter 8. Toronto, Ontario, Canada: John Wiley & Sons, Inc.

Winter D (2005) Kinetics: Forces and moments of force. *Biomechanics and Motor Control of Human Movement*, 3rd edn. Hoboken, NJ: John Wiley & Sons, Inc.

Winter D, Patla A, Prince F, Ishac M, and Gielo-Perczak K (1998) Stiffness control of balance in quiet standing. *Journal of Neurophysiology* **80** 1211–1221.

25 Footwear, Balance, and Falls in the Elderly

Jasmine C. Menant and Stephen R. Lord

CONTENTS

25.1 INTRODUCTION

As the world's population is ageing, falls in the elderly constitute a serious public health problem. It is estimated that one in three people living in the community aged 65 years and over, experience at least one fall per year (Campbell et al. 1990; Tinetti et al. 1988), with devastating human and economic costs to the community (Lord et al. 2007). There are many identified risk factors for falls, including intrinsic risk factors associated with age-related degeneration of the balance and neuromuscular systems (Lord et al. 1991, 1994) and medical conditions (Nevitt et al. 1989) and extrinsic risk factors including environmental factors (Lord et al. 2006; Nevitt et al. 1989).

Most falls occur during motor tasks (Hill et al. 1999), and footwear has been identified as an environmental risk factor for both indoor and outdoor falls (Berg et al. 1997; Connell and Wolf. 1997). By altering somatosensory feedback to the foot and ankle and modifying frictional conditions at the shoe-sole/floor interface, footwear influences postural stability and the subsequent risk of slips, trips, and falls.

Although shoes are primarily devised to protect the foot and facilitate propulsion (McPoil 1988), the influence of fashion on footwear design throughout the ages has compromised the natural functioning of the foot (Thompson and Coughlin 1994). As a result, little is known about what constitutes safe footwear for the elderly when undertaking activities in and around the home (American Geriatrics Society et al. 2001). Because footwear appears to be an easily modifiable falls risk factor, it is imperative to identify the specific shoe features that might facilitate or impair balance in the elderly so as to design targeted fall prevention interventions and provide evidence-based recommendations.

In this chapter, we initially describe the types of footwear commonly worn by the elderly. We then review the evidence pertaining to the effects of specific footwear characteristics on balance and related factors in the elderly. Finally, we highlight studies where footwear has been recognized as a falls risk factor and fall prevention trials that have involved footwear interventions.

25.2 OLDER PEOPLE'S CHOICE OF FOOTWEAR

Several studies have examined the footwear habits of the elderly, through self-reports (Dunne et al. 1993; Kerse et al. 2004; Larsen et al. 2004; Munro and Steele 1999; Paiva de Castro et al. 2010) or by direct assessment of usual footwear (Burns et al. 2002; Horgan et al. 2009; Jessup 2007; Mickle et al. 2010), sometimes using the footwear assessment form (Menz and Sherrington 2000). Slippers appear to be the indoor footwear of choice. Accordingly, approximately 36% of older community-dwellers (Munro and Steele 1999) and nursing home residents (Kerse et al. 2004) and 66% of 44 patients in a subacute aged-care hospital (Jessup 2007) report wearing slippers indoors. Munro and Steele (1999) also noted that 30% of older community-dwellers ($n = 128$) walked within the house barefoot or in socks. As this study was undertaken in Australia, the warm climate might have contributed to the large proportion of people not wearing shoes. The presence of swollen feet may also lead the elderly to wear loose-fitting footwear or no footwear at all (Mickle et al. 2010).

Overall, there is a consensus that many older people, irrespective of their dwelling status, wear poorly fitted shoes (Burns et al. 2002; Finlay 1986; Larsen et al. 2004; Menz and Morris 2005; Paiva de Castro et al. 2010), which may lead to foot problems and pain, and in turn increase the risk of falls (Menz et al. 2006b). Variations in study methodologies might explain why ill-fitting footwear comprise predominantly shoes that were too big in length or width in some studies (Burns et al. 2002; Paiva de Castro et al. 2010), and shoes that were too narrow in others (Menz and Morris 2005). While incorrect shoe length has been significantly associated with ulceration of the foot and with pain (Burns et al. 2002), overly narrow footwear has been also strongly associated with the presence of corns on the toes (Menz and Morris 2005). Interestingly, despite women reporting more foot pain than men, wearing high-heel shoes more than twice a week (10.6% of women) was not associated with foot pain in older Brazilian women ($n = 227$) (Paiva de Castro et al. 2010).

A poor slip-resistant sole, inadequate fastening mechanism, and excessively flexible heel counter are the most frequent detrimental shoe characteristics identified using a modified version of a footwear assessment form (Menz and Sherrington 2000) in aged-care hospital inpatients (Jessup 2007) and outpatients (Horgan et al. 2009). These shoe features are likely to promote slips and trips because they fail to provide foot support. Further, a telephone interview regarding shoes worn at the time of a fall in 652 community-dwellers aged 65 and over found that only 26% of the participants were wearing "sturdy shoes" when they fell (Dunne et al. 1993). These findings, however, may be limited to participants' varying interpretations of what constitutes a sturdy shoe.

In summary, many older people wear inappropriate and/or ill-fitting footwear both inside and outside the home. Shoes are replaced infrequently, possibly due to a lack of knowledge about the importance of safe shoes and/or financial consider-ations (Dunne et al. 1993; Munro and Steele 1999). The choice of footwear might be somewhat dictated by comfort and the need to accommodate painful or swollen feet (Mickle et al. 2010), explaining the tendency for the elderly to wear excessively flexible and/or over-long and wide shoes. The elderly might also favor shoes without fasteners for practical reasons, as they remove the need to bend down to tie laces or fasten straps.

25.3   FOOTWEAR, BALANCE, AND GAIT IN THE ELDERLY

Experimental studies that have examined the effects of specific characteristics of footwear on balance and gait highlight some potential mechanisms as to why certain shoe types are associated with a higher risk of falls in the elderly. These investiga-tions also assist us in identifying footwear properties that may be beneficial to older people's balance and should be considered in the design of safe shoes. This section describes findings related to how various features of footwear can facilitate or impair balance during standing, gait, and undertaking functional tasks.

25.3.1   Barefoot Versus Wearing Shoes

Although it could be assumed that proprioception and plantar sensitivity provide optimal input to the postural control system when the wearer is barefoot versus wear-ing shoes, research studies have provided conflicting evidence. Robbins et al. (1995) showed that when required to estimate the amplitude and the direction of the slope of a weight-bearing surface, the older adults' joint position awareness was 162% lower than that of their younger counterparts when barefoot, possibly due to age-related decline in plantar tactile sensitivity. Wearing running shoes further increased mean estimate error in joint position in both groups, suggesting attenuation of the tactile sensory input through footwear. Robbins et al. further showed that older adults made fewer errors when barefoot compared to when wearing shoes in estimating the maxi-mum supination angle of the soles of their feet when walking along a beam (Robbins et al. 1997). In contrast, Waddington and Adams (2004) reported that older com-munity-dwellers ($n = 20$) were significantly better at discriminating ankle inversion movements when shod compared to when barefoot. However, the fact that their par-ticipants had undergone wobble-board balance training for 5 weeks in self-selected shoes may in part explain these findings.

Given the significant associations between walking barefoot indoors and increased risk of falling (Koepsell et al. 2004; Menz et al. 2006c), it is critical to address the effects of barefoot versus shoe-wearing conditions on balance in the elderly. Interestingly, being barefoot or wearing shoes did not affect standing balance (maintaining balance while standing on a firm or a compliant surface with eyes open or closed) in 30 older adults who had vestibular problems (Whitney and Wrisley 2004). Similarly, Arnadottir and Mercer (2000) did not report any significant differences in functional reach performance in older women ($n = 35$) barefoot compared to when wearing walking shoes. However, these older women took less time and achieved greater self-selected speed in the timed-up and go and 10 m walk tests when wearing shoes, presumably because footwear enhanced plantar shock absorption and therefore improved comfort. In agreement with these later findings, Horgan et al. (2009) noted that older women ($n = 100$) at high risk of falls performed significantly better in a series of balance tests (Berg Balance Scale), when wearing their own usual footwear compared to barefoot.

Contrasting findings were reported by Lord and Bashford (1996) whereby older women ($n = 30$) performed worse in a test of maximal balance range but exhibited less postural sway and better scores in a leaning balance test (coordinated stability) when barefoot than when wearing standard low heels shoes. It should be noted, however, that in this study (Lord and Bashford 1996) participants were novice wearers of a pair of standard low-heel shoes compared to the previous studies (Arnadottir and Mercer 2000; Whitney and Wrisley 2004) in which participants wore their own flat or walking shoes. Furthermore, older community-dwellers required to walk on a 7.8 cm wide beam in various footwear conditions failed the task more frequently when they were barefoot than when wearing shoes (Robbins et al. 1992, 1997), possibly due to decreased function of the toes associated with long-term wearing of shoes (Robbins et al. 1992). Whether walking barefoot or wearing socks over common household surfaces such as polished wooden floors might exacerbate the risk of slipping (Menz et al. 2006c) remains to be examined. The potential risk of stumbling or tripping from excessive slip resistance when walking barefoot or in socks over a carpeted surface also requires further investigation.

The conflicting findings regarding differences in joint position sense and standing balance in the elderly between barefoot and shod conditions may be attributed to methodological differences. However, it appears that wearing shoes enhances walking stability. Furthermore, Burnfield et al. (2004) reported significantly higher plantar pressures in the elderly walking barefoot versus shod, suggesting that the elderly should avoid walking around barefoot as it could increase the risk of foot trauma. Wearing shoes also protects the foot from irregularities in walking surfaces, and is likely to provide more grip than the plantar sole of the foot, reducing the risk of slipping, especially indoors.

25.3.2 SOCKS

Only two studies to date have investigated the slip resistance of commercially available nonslip socks and involved mechanical testing and/or in vivo testing in small samples of young healthy adults. In the first study (Chari et al. 2009), a wet pendulum test revealed that nonslip socks did not provide better slip resistance than compression stockings; however, the stockings showed slippage at the lowest angle of an inclined vinyl-covered platform on which participants stood (Chari et al. 2009). In that experiment, nonslip socks were not as slip resistant as the barefoot condition was. However these findings need to be interpreted with caution given the limited number of participants ($n = 3$) and the static nature of the task. Another investigation (Hübscher et al. 2011) reported that conventional socks and slippers increased the risk of slipping compared to barefoot walking on a linoleum walkway, as indicated by significantly increased absolute deceleration times at heel strike. In contrast, commercially available nonslip socks appeared to provide as much slip resistance as barefoot walking as suggested by similar absolute deceleration times at heel strike. While promising, these results require further validation in samples of the elderly and on additional household walking surfaces likely to be slippery, such as polished floorboards.

25.3.3 TEXTURED INSOLES

The critical contribution of plantar cutaneous sensation to postural control is well documented (Kavounoudias et al. 1998; Perry et al. 2000, 2001). Skin mechanoreceptors within the plantar surface of the foot provide information to the central nervous system about body position to induce postural responses (Kavounoudias et al. 1998). Hence, providing extra tactile sensory input to the plantar surface of the feet has the potential to improve balance control. Mechanical noise applied to the soles of the feet at a subsensory level using vibrating gel-based insoles led to significant reductions in postural sway (Priplata et al. 2003); this effect was more pronounced in the older participants whose threshold of tactile sensitivity would be higher than their younger counterparts (Priplata et al. 2003). Although Suomi and Koceja (2001) reported small but significant reductions in postural sway in older versus younger participants wearing magnetic insoles, participants were not blinded to the insole conditions and the texture of the magnetic and nonmagnetic insoles was different. Further, Hinman (2004) did not observe any significant difference in standing or leaning balance in 56 older community-dwelling people with a history of falls or balance problems who were wearing either some pairs of magnetic insoles (15 magnets with either a gauss rating of 3,900 or 12,000 each) or placebo insoles.

After 4 weeks of wearing textured foot orthotics in standardized shoes, 40 healthy women showed no significant effects of wearing the devices on postural sway (anterior–posterior [AP] and medial-lateral [ML] range of center of pressure [COP] excursions) during standing with their eyes open or eyes closed or on step width during walking at self-selected speed (Wilson et al. 2008). The effects of surfaces of the same textures (1 mm raised circles and 1 mm raised squared pyramid shapes) on postural sway measurements and muscle activity during unperturbed stance were later investigated in 24 young adults (Hatton et al. 2009) and 50 healthy older adults (Hatton et al. 2011). While no significant effect of textured surface versus smooth control surface were noted in the young adults (Hatton et al. 2009), the ML sway range was significantly reduced in the eyes closed condition in the older group, suggesting preliminary evidence of a beneficial effect of specific textured surface on postural stability in older adults (Hatton et al. 2011). Interestingly, when wearing textured insoles with higher nubs (2.5 mm) in their own athletic shoes, without any previous familiarization, young participants exhibited similar COP area and excursion velocity during quiet standing with eyes open and closed, suggesting the beneficial effect of extra tactile sensory input from the textured insoles on postural control when visual input is inhibited (Corbin et al. 2007). Variations in textured patterns (1 mm high nubs [Hatton et al. 2009; Hatton et al. 2011; Wilson et al. 2008] vs. 2.5 mm high nubs [Corbin et al. 2007]), study populations, and the use of insoles (Corbin et al. 2007; Wilson et al. 2008) versus support surfaces (Hatton et al. 2009; Hatton et al. 2011) could account for the different findings between these studies.

Hosoda et al. (1997) found that, contrary to their hypothesis, wearing "health sandals" (textured insoles with small projections), versus slippers (with smooth insoles), increased latency responses to AP perturbations from a motorized balance platform in young adults. In contrast, improvements in measurements of postural sway in ML and AP planes during quiet standing with eyes closed were reported in both young and older adults following five minutes of walking or standing wearing sandals with spike insoles made of semirigid polyvinyl chloride (PVC) (Palluel et al. 2008). These results suggest that spike insoles temporarily enhance postural stability during unperturbed stance in older adults; the authors postulated that the spike insoles possibly stimulated deep mechanoreceptors given that plantar surface threshold sensitivity on several foot locations was similar pre- and post-five minutes of standing in spike insoles (Palluel et al. 2008). Further, Maki et al. (1999) evaluated the effects of facilitating plantar sensation on balance control by providing 7 young (mean age: 26 years) and 14 older (mean age: 69 years) participants insoles with a raised edge at the plantar surface boundaries. Fewer "extra" steps and arm movement reactions were noted in the elderly wearing the modified insoles when stepping in response to unpredictable forward perturbations. Older people wearing the modified insoles also maintained a greater margin of stability relative to the posterior border of the base of support (BOS) during continuous platform perturbations when required to resist the perturbation without stepping.

In summary, the benefits associated with wearing vibrating insoles or insoles that mechanically facilitate plantar tactile sensitivity are likely to be particularly useful to the elderly who suffer from age-related declines in plantar sensitivity or to counteract the detrimental effects on balance of thick, soft sole shoes prescribed to people suffering from ulcers or peripheral neuropathy (Hijmans et al. 2007). However, these postural control-enhancing insoles may not be easily combined with the orthotic devices that some older people wear and their long-term effects have yet to be demonstrated. Considering the limited evidence of the beneficial effects of magnetic insoles on balance and the lack of clear mechanisms for this relationship, magnetic insoles should not be recommended for wear in the elderly.

25.3.4 Heel Height

Research has shown that heel elevation is associated with an increased risk of falling in the elderly (Gabell et al. 1985; Tencer et al. 2004). By elevating and shifting the center of mass (COM) forward, high-heel shoes affect balance control and lead to postural and kinematic adaptations (Snow and Williams 1994). The plantar-flexed ankle position adopted when walking in elevated heel shoes might contribute to larger vertical and horizontal ground reaction forces at heel strike (Ebbeling et al. 1994; Hong et al. 2005; Snow and Williams 1994). Reduced calcaneal eversion (Ebbeling et al. 1994; Snow and Williams 1994) and absence of foot rollover (Ebbeling et al. 1994) in the plantarflexed ankle might in fact prevent the foot from pronating, affecting the foot's natural shock absorption mechanism. In turn, compensation strategies arise at the knee and hip as shown by altered kinematics and kinetics (de Lateur et al. 1991; Ebbeling et al. 1994; Gehlsen et al. 1986; Opila-Correia 1990a,b; Snow and Williams 1994; Soames and Evans 1987). Age and gender interactions appear to lead to different trunk and pelvis kinematics during gait. When wearing shoes with high heels, older women and young men show a flattened lumbar lordosis (de Lateur et al. 1991; Opila-Correia 1990b) while younger women display increased trunk lordosis (Opila-Correia 1990b). Significant increases in forefoot loading during high-heeled gait (Gastwirth et al. 1991; Mandato and Nester 1999), with especially greater pressures in the medial forefoot (Corrigan et al. 1993; Hong et al. 2005; Snow et al. 1992; Speksnijder et al. 2005; Yung-Hui and Wei-Hsien 2005), have been well documented. Such increased pressures might contribute to the development of plantar calluses (Menz et al. 2007), as suggested by significant associations between plantar calluses and wearing shoes with heels higher than 2.5 cm in older women (Menz and Morris 2005).

Changes in spatiotemporal parameters of gait including slower walking speed (Esenyel et al. 2003; Opila-Correia 1990a,b), shorter step or stride length (Ebbeling et al. 1994; Esenyel et al. 2003; Gehlsen et al. 1986; Opila-Correia 1990a), and increased walking cadence (Ebbeling et al. 1994; Gehlsen et al. 1986; Wang et al. 2001) recorded in individuals wearing high-heel shoes compared to low-heel shoes or barefoot suggested a more cautious walking pattern. Raising the COM increases the moment arm of the ML moment of force applied at the COM about the shoe/floor interface, resulting in a smaller ML perturbation required in order for a fall to occur and, thus, a smaller critical tipping angle of the elevated heel shoe (Tencer et al. 2004).

There is some research evidence that experience in walking with elevated heel shoes alters lower limb muscle activity patterns. Lee et al. (1987, 1990), for example, found that men and women wearing high-heel shoes exhibited reduced gastrocnemius muscle activity, possibly because the plantar-flexed position of the ankle alters the length–tension relationship of this muscle. However, while men showed a significant increase in tibialis anterior muscle activity (possibly to counteract a feeling of instability), women, who were regular high-heel shoe wearers, displayed the opposite muscle activity pattern (Lee et al. 1990). Contradicting findings were shown in five young healthy women whose peak tibialis anterior muscle activity did not significantly change when walking in medium- and high-heel shoes compared to low-heel shoes (Lee et al. 2001). The level of experience of these women wearing high-heel shoes, which was not specified, and the imposed walking velocity could have contributed to the differential results reported here. In another study following a fatiguing exercise simulating high-heeled gait, habitual high-heel shoe wearers showed low-level endurance of the peroneus longus muscle and an imbalance in muscle activity between the lateral and medial heads of gastrocnemius (Gefen et al. 2002). This muscle imbalance might increase foot instability as suggested by abnormal lateral movements of the COP under the heel and first metatarsal head observed in habitual high-heel shoe wearers (Gefen et al. 2002).

Although Lindemann et al. (2003) did not find any differences in postural sway or walking velocity in a sample of frail older women ($n = 26$) wearing either 1 or 2 cm heel height tennis shoes, most studies have demonstrated the detrimental effects of wearing elevated heel shoes (>3.5 cm heel height) on stability. Brecht et al. (1995) reported that young women ($n = 27$) maintained significantly better balance on a moving platform when subjected to various accelerations while wearing tennis shoes (1.9 cm mean heel height) compared to cowboy boots (3.7 cm mean heel height). Subsequent research showed that the elderly performed significantly worse in tests of postural sway (Lord and Bashford 1996; Menant et al. 2008b), leaning balance (Arnadottir and Mercer 2000; Lord and Bashford 1996), timed-up and go, and 10 m walk (Arnadottir and Mercer 2000) when wearing elevated heel shoes compared to when barefoot or when wearing low-heel shoes.

We recently conducted a series of gait studies where elevated heel shoes only differed from standard Oxford-type control shoes with regard to heel height (4.5 vs. 2.7 cm) (Menant et al. 2008b, 2009a,b, 2011). We showed that increased heel height led to increased double-support time, foot clearance, and horizontal heel velocity at heel strike, when walking at self-selected speed on linoleum, irregular, and wet surfaces in 26 healthy older adults (Menant et al. 2009a). Further, compared to when wearing the standard shoes, both young and older adults generated smaller vertical and braking loading rates when walking in the elevated heel shoes on even and uneven surfaces (Menant et al. 2008a), and displayed significant reductions in pelvis ML accelerations when walking on level and irregular surfaces (Menant et al. 2011). Interestingly though, when wearing the elevated heel shoes, the elderly did not appear to have more difficulty terminating gait rapidly than when wearing the flatter heel standard shoes (Menant et al. 2009b). Overall, these findings demonstrate that increasing shoe heel height by 4.5 cm without altering heel/surface contact area impairs balance control during walking in the elderly.

The variety of findings pertaining to the effects of high-heel shoes on balance and gait can be attributed to inconsistencies in the choice of footwear. While some studies have compared barefoot to high-heel dress shoe conditions (Gastwirth et al. 1991; Gehlsen et al. 1986; McBride et al. 1991; Opila et al. 1988), others have compared tennis shoes or flat shoes to high-heel shoes with a narrow toe-box (Gastwirth et al. 1991; Gehlsen et al. 1986; Mandato and Nester 1999; McBride et al. 1991; Opila-Correia 1990a,b; Soames and Evans 1987; Wang et al. 2001). Some researchers have used only a shoe heel attached to the heel of the foot of the individual (Corrigan et al. 1993), whereas others have used each individual's dress shoes or have provided a standard dress shoe (Lee et al. 1990; Lord Bashford 1996; McBride et al. 1991; Opila-Correia 1990a,b; Speksnijder et al. 2005). Few studies have managed to isolate the effect of heel height by keeping a shoe of similar design but systematically increasing the heel height (Ebbeling et al. 1994; Eisenhardt et al. 1996; Hong et al. 2005; Lee et al. 1987; Lindemann et al. 2003; Menant et al. 2008a,b, 2009a,b, 2011; Snow and Williams 1994; Yung-Hui and Wei-Hsien 2005). It is thus questionable as to whether some study findings reflect the true effects of heel height or are influenced by other shoe design factors. Nevertheless, recent research provides consistent evidence (Menant et al. 2008a,b, 2009a,b, 2011) that the elderly should be advised against wearing such footwear as it reduces balance and gait stability.

25.3.5 Sole Cushioning Properties

In parallel with the development of running footwear research investigating midsole cushioning for impact forces attenuation (Robbins et al. 1988; Robbins and Gouw 1991; Robbins and Waked 1997), studies have been conducted to investigate the effects of sole and midsole thickness and hardness on stability in the elderly. Two studies by Robbins et al. (1994) involving young and older men (Robbins et al. 1992) demonstrated the detrimental effect that soft and thick shoe midsoles (shore A-15 [for the studies reviewed here, shore-A hardness ranges from shore A-15 for soft soles to shore A-58 for hard soles], 27 mm at the heel and 16 mm under the first metatarsal-phalangeal joint [first MTP joint]) have on balance control, assessed by the frequency of falls from a walking beam. The data also showed that the older men perceived the shoes with soft thick midsoles to be the most comfortable among shoes of varying hardness (shore A-15, A-33, and A-50) and thickness (13 mm at the heel and 6.5 mm under the first MTP joint versus 27 mm at the heel and 16 mm under the first MTP joint) (Robbins et al. 1992), possibly because the soft and thick midsoles enabled even distribution of load across the plantar surface of the foot. This even distribution of load, in turn, was hypothesized to reduce plantar tactile sensory feedback and subsequently impair balance control. The authors also suggested that the midsole mechanical instability generated frontal plane movements at the ankle through material compression. Further work by this research group on age and footwear effects on joint position sense (Robbins et al. 1995) clarified these proposed mechanisms, and concluded that shoes with soft thick soles impair stability by reducing joint position sense.

This concept was then tested in more dynamic conditions requiring young and older men to walk on a beam in shoes of varying midsole hardness and thickness and to estimate the maximum supination angle of the sole of their foot (Robbins et al. 1997). In agreement with previous findings (Robbins et al. 1992), balance was worst in the thick and the soft midsole shoes, especially in the older group. Errors in judgment of foot position were positively correlated with midsole thickness, and negatively correlated with balance and with midsole hardness. Foot position awareness was worse, by approximately 200%, in the older compared to the younger adults in any footwear condition, and the older participants' mean position error was greatest in shoes with the thickest and softest midsoles (Robbins et al. 1992). Sekizawa et al. (2001) later confirmed those findings, reporting worst ankle joint position sense in young males standing in thick sole shoes (50 mm at the heel and 30 mm under the first MTP joint) compared to barefoot.

In an attempt to combine comfort and stability, Robbins et al. (1998) investigated the effects of a soft low-resilience material on postural sway and perceived comfort in 30 young and 30 older adults. The authors hypothesized that, in addition to providing a cushioning sensation, soft low-resilience interfaces would remain compressed after foot strike and prevent excessive frontal plane movement of the foot, as would be expected with high-resilience materials. Results of the study confirmed that in both groups, sway velocity was significantly lower when standing on the thin low-resilience interface compared to the thick high-resilience interface. Overall, there was a trend toward the low-resilience material being more comfortable than the high-resilience material. In accordance with these findings, optimum comfort and stability might be obtained if the soles of the shoe are thin and hard combined with low-resilience insoles. However, no significant differences in measures of postural sway and leaning balance during standing (maximal balance range and coordinated stability) were found between a medium hardness sole (shore A-42) and a hard sole (shore A-58) shoe in a population of 42 older women, leading to the conclusion that the soft soled shoes used in this study might not have been compliant enough to affect balance (Lord et al. 1999). Further, we (Menant et al. 2008b) did not find any difference in tests of postural sway, leaning balance, and choice-stepping reaction time in the elderly ($n = 29$) wearing soft sole shoes (shore A-25) or hard sole shoes (shore A-58) versus shoes with soles of medium hardness (shore A-40).

There is, however, increasing evidence that soft sole shoes might impair balance control during gait. Perry et al. (2007) examined balance control in young people performing tests of rapid unplanned stopping both barefoot and when wearing mid-soles of three different hardnesses (shore A-15, shore A-33, and shore A-50) fixed to their feet. Compared to the hard midsoles, the soft midsoles led to a significant reduction in ML range of COM displacement, to possibly counteract the lack of mechanical support of the material. A reduction in the COM–COP distance together with a significantly greater vertical loading rate in the softer midsoles compared to barefoot during terminal stance demonstrates how softer midsoles may impair balance control in the sagittal plane during stopping. Perry et al. (2007) concluded that soft sole shoes might threaten older peoples' stability, as greater muscular activity is required to maintain stability during stopping in this footwear condition. Likewise, our research recently showed that compared to standard shoes of exact same design but harder sole (shore A-40 hardness), soft sole shoes (shore A-25 hardness) caused a greater COM-lateral BOS margin without concurrent BOS adjustment during even and uneven surfaces walking trials (Menant et al. 2008a), and led to an increase in step width during control, irregular, and wet surface walking trials (Menant et al. 2009a). It might be that compression of the malleable soft sole material countered the inclination of the slanted platforms of the uneven surface or, alternatively, that the restricted ML COM excursions reflected reactive gait adaptations to the uneven surface onto which stepping is unlikely to have been preplanned (Menant et al. 2008a). In contrast, in our second study (Menant et al. 2009a), viewing the level, irregular and wet surfaces would have allowed the participants to undertake anticipatory motor actions to adapt to the footwear worn relative to the type of flooring. Therefore, the participants may have compensated for the potential lack of foot mechanical support and reduced plantar tactile sensory feedback caused by the soft sole shoes by altering their foot placement in the ML plane and thus increasing their step width (Menant et al. 2009a). Although reflecting different mechanisms, the postural adjustments to the soft sole shoes described in our research studies (Menant et al. 2008a, 2009a) indicate that shoes with soles of hardness shore A-25 impair

ML balance control during walking. In these studies (Menant et al. 2008a, 2009a), increasing sole hardness beyond that of a standard shoe (shore A-58 vs. shore A-40) did not appear to affect balance or gait in the young and the elderly. It should be noted that the sole hardness used in these studies for the hard sole shoes and the standard shoes was greater than that used in previous studies (Perry et al. 2001; Robbins et al. 1992, 1994, 1995, 1997, 1998).

In summary, variations in sole or midsole hardness do not appear to significantly alter balance during standing. However, thick and soft sole shoes impair stability during walking by reducing foot position awareness and mechanical stability, and may pose an even greater threat to stability during challenging tasks (Menant et al. 2008a, 2009a; Perry et al. 2007). In contrast, a shoe sole hardness of at least shore A-40 appears to provide optimal stability for the elderly in a range of static and dynamic balance challenging tasks (Menant et al. 2008a, 2009a,b, 2011). Despite this evidence, epidemiological studies have failed to confirm whether sole hardness or thickness influence fall risk in the elderly (Tencer et al. 2004). Because of the constrained nature of balance tests in the investigations conducted by Robbins and colleagues (beam walking) (Grabiner and Davis 1993), further studies are required before definitive recommendations can be made regarding midsole thickness and sole material's resilience.

25.3.6 COLLAR HEIGHT

High-collar shoes were initially investigated in the context of sports injuries to prevent ankle sprains, by providing extra mechanical support around the ankle. Relative to low-collar sports shoes, high-collar sports shoes offer significantly better resistance against inversion (Ottaviani et al. 1995) and reduced ankle inversion angular velocity (Stacoff et al. 1996) in young adults performing various sporting tasks.

In addition to providing greater mechanical stability to the ankle joint, the extra sensory input provided by a high collar is thought to facilitate joint position sense (You et al. 2004) and, in turn, improve ML balance control. In fact, a tactile stimulus applied to the leg of younger, older, and neuropathic participants has been found to reduce body sway during standing (Menz et al. 2006a). Significant improvements in postural sway and leaning balance were also noted in laced boots versus low-collar shoes in a group of 42 women aged 60–92 years old (Lord et al. 1999), while no difference in tests of balance and stepping were found in 29 male and female community-dwellers wearing low-collar shoes versus 11 cm high-collar shoes (Menant et al. 2008b). In contrast, compared to trainers, cowboy boots were found to impair balance control in young women standing on a platform that was translated in the AP direction (Brecht et al. 1995). However, in addition to a higher collar, the boots also had an "inverted" heel of 3.7 cm, which likely contributed to the subject's instability (Figure 25.1).

Our recent work suggests that increasing shoe collar height might enhance older people's AP and ML balance control during ambulation, as young and older adults showed a small but significant increase in COM-lateral BOS margin (Menant et al. 2008a), an increase in step width and double-support time (Menant et al. 2009a), and a reduction in the magnitude of ML pelvis accelerations (Menant et al. 2011) during

FIGURE 25.1 Oxford-type standard shoe with a low collar (left) versus high collar (right), used in studies by Menant and colleagues. (From Menant, J.C. et al., *Arch. Phys. Med. Rehabil.*, 89, 1970, 2008a; Menant, J.C. et al., *Gerontology*, 54, 18, 2008b; Menant, J.C. et al., *Gait Posture.*, 29, 392, 2009a; Menant, J.C. et al., *Gait Posture*, 30, 65, 2009b; Menant, J.C. et al., *Res. Sports Med.*, 19, 28, 2011.)

walking on challenging surfaces in high-collar shoes (11 cm high suede leather collar around the ankle) versus standard shoes. In addition, participants were able to terminate gait significantly faster in the high-collar shoes compared to the standard shoes (Menant et al. 2009b), principally because they reached a stable upright posture faster.

Despite preliminary evidence, more research is recommended to confirm the potential benefits of high-collar shoes on stability during challenging motor tasks, since potential aesthetic concerns of such footwear combined with their lack of suitability for hot climates might deter the elderly, especially women, from wearing such shoes on a regular basis.

25.3.7 Sole Geometry: Flaring

By increasing the BOS, a flared sole might improve ML stability (Menz and Lord 1999) and therefore warrants consideration when designing shoes for the elderly (Helfand 2003). However, to our knowledge, all published investigations before our work had only examined this shoe feature in the context of preventing running injuries (Nigg and Morlock 1987; Stacoff et al. 1996) (Figure 25.2).

We recently demonstrated that a shoe with a 20° sole flare does not enhance balance in young and older adults during standing balance, stepping, and gait tasks compared to a control shoe of similar sole/surface contact area to that of most commercially available footwear (Menant et al. 2008b, 2009a,b). These results suggest that

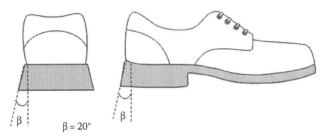

FIGURE 25.2 Shoe with 20° flared sole used in studies by Menant and colleagues. (From Menant, J.C. et al., *Gerontology*, 54, 18, 2008b; Menant, J.C. et al., *Gait Posture*, 29, 392, 2009a; Menant, J.C. et al., *Gait Posture*, 30, 65, 2009b.)

the sole/surface contact area of commercially available Oxford-type shoes is sufficient to provide optimal balance in healthy older adults walking on a range of surfaces.

25.3.8 SLIP-RESISTANT SOLE PROPERTIES

Studies on prospective falls have identified slips and trips as the most commonly reported causes of falls in the elderly (Berg et al. 1997; Hill et al. 1999; Lord et al. 1993). Not wearing shoes indoors is suggested to contribute to indoor slips since walking barefoot or in socks increases the risk of falls in the elderly by more than 10-fold (Koepsell et al. 2004; Menz et al. 2006c). Furthermore, ice- and snow-related slips contribute to a high number of injurious falls in cold climate countries. In a 1-year prospective study, 34% of ice- and snow-slip-related injuries in a Swedish town occurred in adults aged between 50 and 79 years (Bjornstig et al. 1997), and shoes lacking slip-resistant soles likely contributed to these incidents. In light of these data, it seems clear that the elderly would benefit from wearing slip-resistant footwear (Gao and Abeysekera 2004).

Swedish research into the prevention of outdoor winter slips (Gard and Berggard 2001, 2006) evaluated various antiskid devices fixed to the footwear of the elderly who were performing simple walking tasks over five slippery surfaces (ice with sand, ice with gravel, ice with snow, ice with salt, and ice alone). An antiskid device applied to the shoe heel was rated the best in terms of walking safety and balance, time to put on and ease of use, and it did not significantly affect gait and posture as compared to either whole-foot or forefoot-only devices (Gard and Berggard 2001). In addition, compared to whole-foot or toe antiskid devices, a heel device was preferred and perceived as providing the best walking safety and balance by 107 men and women aged 22–80 years (Gard and Berggard 2006).

Using a slip-resistance testing machine, Stevenson et al. showed that none of several rubber nitrile heeled shoes tested could provide a safe friction coefficient for walking over smooth wet surfaces contaminated with detergent or oil. However, they found that roughening of the floor surfaces increased safety when nitrile- or PVC-heeled shoes were used (Stevenson et al. 1989). A series of studies on the slip resistance of various rubber soles on water-wet floors as well as on oil-contaminated surfaces and icy surfaces (Manning and Jones 2001) showed that to reduce the risks of slips, floor polish should be avoided where possible and the roughness of new footwear should be increased by abrading them. Later work from Gao et al. (2004) involving four types of footwear of varying materials and sole tread, hardness, and roughness confirmed a significant positive association between sole roughness and slip resistance.

Further testing showed that a rubber heel with a bevel of about 10°, which provides a greater contact area at heel strike than a square rubber heel, offered better slip resistance over both dry and wet floor surfaces (Lloyd and Stevenson 1989). For the wet floor, a tread pattern reduced the lubricating effect of the water at heel contact, but showed dangerously low coefficients of friction on oily surfaces. Using similar methodology, Menz et al. (2001) reported comparable findings. An Oxford-type shoe (a leather shoe with lacing and a low heel) with various heel configurations provided safe dynamic coefficients of friction on common dry household surfaces, the beveled heel configuration being the most slip resistant.

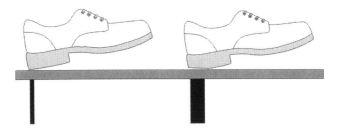

FIGURE 25.3   Effect of a beveled heel (right) versus a squared heel (left) on footwear/ surface contact area at heel strike during walking.

While dress shoes with broad heels reached a significantly greater coefficient of friction than narrow-heeled ones, overall women's dress shoes could not be considered safe regarding slip resistance. Unfortunately, none of the Oxford-type shoes or the dress shoes, even when equipped with a patterned sole, had a safe coefficient of friction on wet oil-contaminated surfaces (Figure 25.3).

In gait studies involving young and older participants walking and stopping rapidly on level dry and wet surfaces, a 10° beveled heel did not appear to affect balance compared to a squared heel, even during tasks performed on the potentially slippery surface (Menant et al. 2009a,b). There are two possible reasons for this discrepancy. First, it is possible that the results of controlled mechanical "bench test" studies do not translate well to gait studies due to individual variations in kinematics and kinetics. Alternatively, the 10° angle of the beveled heel might not have been sufficient to increase slip resistance given that the mean shoe/floor angle displayed by the participants at the time of heel strike was approximately twice as large (Menant et al. 2009a).

Findings from our experiments also revealed that tread sole shoes compared to smooth sole shoes did not improve stability in young and older participants even when they walked or stopped on a wet surface (Menant et al. 2009a,b). It is possible that anticipatory adaptations to the wet potentially slippery surface masked the hypothesized effects of the tread sole shoe compared to the standard shoe (Menant et al. 2009a). In a series of mechanical testing studies, Li and Chen (2004, 2005; Li et al. 2006) found that tread grooves 9–12 mm wide, 5 mm deep, and with oblique or perpendicular orientations relative to the longitudinal axis of the shoe provided higher slip resistance on water-contaminated surfaces than narrower and shallower tread grooves oriented longitudinally. In our studies (Menant et al. 2009a,b), the grooves of the tread sole were 3–8 mm wide and 4 mm deep, and oriented perpendicular and oblique to the longitudinal axis of the shoe, except at the toe and heel regions where they were parallel to the longitudinal axis. It is therefore possible that because of the tread groove width and orientation, the commercial tread used in this study presented suboptimal slip resistance on water-contaminated surfaces, which could explain why it did not perform better than the standard shoe's smooth sole on the wet surface.

Although sufficient slip resistance is required when walking on potential slippery surfaces, excessive foot-floor slip resistance might also place the elderly at increased risk of falling. Connell and Wolf (1997) in fact reported on two near-fall incidents in which the elderly were pivoting and the slip resistance from both their shoe soles and the flooring became too high and resisted the rotation of their lower extremity, resulting in a loss of balance. While the slip resistance of the shoe soles and that of the flooring might have been acceptable if considered individually, they appeared to be too high when combined. Menz et al. (2006c) also recorded that, during their prospective falls study in older retirement village residents, four indoor fallers and one outdoor faller perceived their fall to be caused by their shoe getting "stuck," suggesting cases where excessive slip resistance might have led to trips and/or loss of balance. Too much friction at the shoe/walking surface interface may be hazardous to stability for the elderly who have a shuffling gait, such as those suffering from Parkinson's disease. For these people, a smooth surface may be desirable as shufflers tend to have a very low toe clearance, which may increase the risk of trips when traversing an irregular or highly slip-resistant surface.

In summary, while in vivo research to date has failed to uncover hypothesized beneficial effects of a tread sole on stability during walking and stopping, mechanical testing suggests that Oxford-type shoes equipped with a tread sole appear to provide sufficient slip resistance for walking over dry and water-wet surfaces. Older women should be advised to avoid wearing high-heel dress shoes as, in addition to their known detrimental effects on posture and balance, these shoes do not have a safe coefficient of friction even with a broad heel. To prevent slips, areas contaminated with detergent or oil should be avoided and frequently cleaned. Roughening these surfaces will also offer a greater slip resistance. While providing useful information regarding the safety of footwear/floor interactions, mechanical friction testing has some limitations in that it cannot replicate human behavior in terms of gait biomechanics and psychophysiological factors (Gao and Abeysekera 2004). For example, prior knowledge of a slippery surface leads to postural and temporal gait adaptations, in turn, lowering the required coefficient of friction (Cham and Redfern 2002). In addition, coefficient of friction measurements determined from mechanical testing should be interpreted with caution because of the variety of devices and assessment techniques that have been used. Further clinical investigations are recommended to determine whether it is appropriate to rely on the numerous published mechanical testing studies. Finally, providing that the elderly do not have a shuffling gait where excessive shoe/surface traction might be an issue, they should still be encouraged to wear shoes with slip-resistant soles, as slippery surfaces are not always anticipated in everyday life.

25.4 FOOTWEAR AND FALLS IN THE ELDERLY

25.4.1 FOOTWEAR AS A RISK FACTOR FOR FALLS

While experimental work has focused on investigating the effects of footwear features on balance and gait, retrospective and prospective falls research studies have examined which types and characteristics of shoes are associated with falls in the elderly.

Wearing shoes with slippery soles or slippers was a predisposing factor for falls in a prospective falls study of 96 older community-dwelling people (Berg et al. 1997).

In another prospective falls research study involving 100 older participants, Gabell et al. (1985) identified inadequate footwear as a major contributing factor. Of the 22 falls, 10 occurred while participants were wearing either heavy boots or boots with cutaway heels, slip-on shoes, or slippers. They also found that a history of high-heel shoe wearing in women was a predisposing factor for falling. However, 20 out of the 22 falls reported by Gabell et al. (1985) occurred outdoors that might explain why, contrary to other studies, walking barefoot did not appear to be a major falls risk factor.

In a 2 year prospective investigation of falls in which older community-dwelling fallers ($n = 327$) were matched with nonfallers with similar demographics, Tencer et al. (2004) found that shoes with heels greater than 2.5 cm increased the risk of falls compared to athletic or canvas shoes (odds ratio: 1.9). They also found that falls risk significantly decreased with an increase in median sole/surface area above 74 cm^2 (median sole/surface area for high-heel dress shoes was 49 cm^2). Walking barefoot or wearing stockings increased the risk of falls the most, by up to 11 times compared to walking in athletic or canvas shoes (Koepsell et al. 2004). A similar finding was reported by two additional studies conducted in the community (Larsen et al. 2004) and in retirement villages (Menz et al. 2006c).

In a retrospective study using a footwear assessment form that identifies shoe characteristics relevant to a loss of balance or a fall (Menz and Sherrington 2000), Sherrington and Menz (2003) noted that 75% of a sample of 95 older people (mean age: 78.3 (7.9) years old) who suffered a hip fracture-related fall were wearing improper footwear at the time of the incident. The largest proportion of falls occurred while the elderly were walking inside their homes (48%) and slippers were the most common type of footwear worn (22% of the fall cases). Unsafe features of shoes identified in this study (Sherrington and Menz 2003) included a lack of fixation (63%), excessively flexible heel counter (43%), and an excessively soft sole (20%). The participants who had tripped ($n = 32$) were more likely to have been wearing slippers or ill-fitting shoes without proper fixation. Hourihan et al. (2000) also reported that at the time of a hip-fracture-related fall, 24% of 104 older people were barefoot or in socks, 33% were wearing slippers, and 22% were wearing slip-on footwear. Similarly, analysis of footwear habits among nursing home residents ($n = 606$) revealed a strong association between wearing slippers (as opposed to shoes) and fractures (Kerse et al. 2004). Finally, Keegan et al. (2004) found that slip-on shoes and sandals were associated with a greater risk of suffering a foot fracture from a fall (odds ratio: 2.3, and 3.1, respectively), and that wearing medium/high-heel shoes and narrow shoes increased the risk of fractures at five sites (foot, distal forearm, proximal humerus, pelvis, and shaft of the tibia/fibula) in people aged 45 years and over.

Overall, these findings provide some evidence to suggest that suboptimal footwear, regularly worn by the elderly, increases the risk of falls. The elderly might exacerbate their risk of slipping by walking barefoot, in socks or in shoes without slip-resistant outer soles, or tripping by wearing ill-fitting slippers or shoes lacking fasteners. Wearing shoes not appropriate in size might also lead to foot problems, which, in turn, can place the elderly at an increased risk of falls (Menz et al. 2006b). Indoor footwear, or the lack of it, seems to be more implicated in the etiology of falls than outdoor shoes, possibly because more studies have been conducted in the elderly living in residential aged care, who engage less often in outdoor activities.

25.4.2 Footwear Intervention for Fall Prevention

Two studies have examined the potential of a footwear intervention to prevent falls. The first investigated the effects of an elastometer netting ("Yaktrax Walker") worn around the sole of the shoe, on outdoor slips and falls during winter in North America, in a sample of community-dwelling fallers aged over 65 years old (McKiernan 2005). The relative risks of outdoor slips, falls, and injurious falls for the group wearing the device versus the control group who wore their habitual winter shoes were 0.5, 0.45, and 0.13, respectively. These devices, therefore, may provide a useful and inexpensive solution to the problem of outdoor falls on icy surfaces.

The second study comprised a randomized controlled trial investigating the effects of providing a multifaceted podiatry intervention involving footwear advice and cost subsidy, provision of orthotics, exercise program, and falls advice on falls during a 12 month follow-up period in 305 older people with disabling foot pain (Spink et al. 2011). Of the 153 participants who received the intervention, 103 (67%) were provided with a pair of orthotics and 55% reported wearing them most of the time; 41 (27%) were provided with a footwear voucher and advice and more than a third of them reported wearing the new footwear most of the time. There were 36% fewer falls in the intervention group ($n = 153$) compared to the control group (incidence rate ratio 0.64, 95% confidence interval [CI] 0.45 to 0.91, $p = 0.01$) as well as a trend for fewer fractures from falls. Although these findings suggest that a multifaceted podiatry intervention is effective at reducing falls in the elderly with foot pain, the contribution of footwear to this intervention might have been minor relative to that of the exercise program as only 17% of the intervention participants purchased new footwear. Nevertheless, the authors suggest that the foot orthoses might have contributed to improved balance by enhancing foot stability and plantar sensory feedback as well as reducing plantar pressure and foot pain.

25.5 CONCLUSIONS: OPTIMAL FOOTWEAR FOR THE ELDERLY

The question "what is the safest footwear for the elderly who have fallen or are at risk of falling?" raised by the American and the British Geriatrics Societies and the American Academy of Orthopedic Surgeons in 2001 (American Geriatrics Society et al. 2001) does not have a definitive answer, despite substantial advances in the field of footwear and falls research. There is now sufficient epidemiological evidence to suggest that the elderly should wear appropriately fitted shoes both inside and outside the house, since walking barefoot and in socks indoors are the footwear conditions associated with the greatest risk of falling. Given the high prevalence of foot problems among the elderly (Mickle et al. 2010), there is a need for shoes that can accommodate the extra girth of the forefoot for individuals with hallux valgus, increased depth of the toe box for those with lesser toe deformities, and footwear with compliant material able to stretch around swollen feet (Mickle et al. 2010). Nevertheless, providing that they have adequate fastenings, shoes with a low square heel, a sole of medium hardness (shore A-40), and a high collar provide optimal balance in healthy

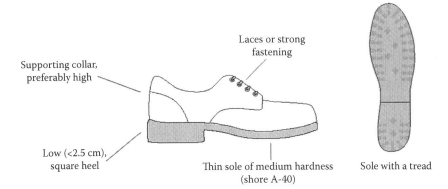

Supporting collar, preferably high

Laces or strong fastening

Low (<2.5 cm), square heel

Thin sole of medium hardness (shore A-40)

Sole with a tread

FIGURE 25.4   Recommended shoe features for the elderly.

older community-dwelling people during functional balance and stepping tests and walking and stopping on a range of surfaces. This shoe type is therefore recommended as safe for wearing in and around the home, on dry and wet linoleum floors, as well as on irregular terrain (Figure 25.4).

In contrast, shoes with an elevated heel are deemed unsafe because of their detrimental effects on balance and gait patterns and because of their reported lack of comfort and stability. The elderly should also be advised not to wear shoes with soft soles (sole hardness less than shore A-33), as these can alter balance control during challenging gait tasks. A tread sole may further prevent slips on wet and slippery surfaces.

Prevention of falls should also include education of the elderly and their carers/family (for those housebound or institutionalized) regarding these footwear recommendations as it is likely that financial and comfort aspects currently outweigh safety considerations when the elderly purchase shoes. Future directions for research should include clinical studies assessing slip-resistant features of the sole that can prevent indoor slipping. Finally, the potential benefits of somatosensory stimulating insoles on postural control should be explored further.

QUESTIONS

25.1 What foot problems have been associated with poorly fitting footwear in the elderly?

25.2 What are the most frequently reported detrimental characteristics identified on the elderly's footwear?

25.3 What is the physiological rationale for providing textured insoles to the elderly?

25.4 What are some kinematic and kinetic consequences of high-heel footwear?

25.5 What is the scientific evidence regarding sole cushioning properties?

25.6 What are the pros and cons of tread sole shoes for the elderly?

25.7 What type or characteristics of footwear have been identified as risk factors for falls in the elderly?

25.8 What are unsafe features of shoes for the elderly?

REFERENCES

American Geriatrics Society, British Geriatrics Society and American Academy of Orthopaedic Surgeons Panel on Falls Prevention. 2001. Guidelines for the prevention of falls in older people. *Journal of the American Geriatrics Society* 49: 664–672.

Arnadottir, S. A. and Mercer, V. S. 2000. Effects of footwear on measurements of balance and gait in women between the ages of 65 and 93 years. *Physical Therapy* 80: 17–27.

Berg, W. P., Alessio, H. M., Mills, E. M., and Tong, C. 1997. Circumstances and consequences of falls in independent community-dwelling older adults. *Age and Ageing* 26: 261–268.

Bjornstig, U., Bjornstig, J., and Dahlgren, A. 1997. Slipping on ice and snow—Elderly women and young men are typical victims. *Accident Analysis and Prevention* 29: 211–215.

Brecht, J. S., Chang, M. W., Price, R., and Lehmann, J. 1995. Decreased balance performance in cowboy boots compared with tennis shoes. *Archives of Physical Medicine and Rehabilitation* 76: 940–946.

Burnfield, J. M., Few, C. D., Mohamed, O. S., and Perry, J. 2004. The influence of walking speed and footwear on plantar pressures in older adults. *Clinical Biomechanics* 19: 78–84.

Burns, S. L., Leese, G. P., and Mcmurdo, M. E. 2002. Older people and ill fitting shoes. *Postgraduate Medical Journal* 78: 344–346.

Campbell, A. J., Borrie, M. J., Spears, G. F., Jackson, S. L., Brown, J. S., and Fitzgerald, J. L. 1990. Circumstances and consequences of falls experienced by a community population 70 years and over during a prospective study. *Age and Ageing* 19: 136–141.

Cham, R. and Redfern, M. S. 2002. Changes in gait when anticipating slippery floors. *Gait and Posture* 15: 159–171.

Chari, S., Haines, T., Varghese, P., and Economidis, A. 2009. Are non-slip socks really 'non-slip'? An analysis of slip resistance. *BMC Geriatrics* 9: 39.

Connell, B. R. and Wolf, S. L. 1997. Environmental and behavioral circumstances associated with falls at home among healthy elderly individuals. Atlanta FICSIT Group. *Archives of Physical Medicine and Rehabilitation* 78: 179–186.

Corbin, D. M., Hart, J. M., Mckeon, P. O., Ingersoll, C. D., and Hertel, J. 2007. The effect of textured insoles on postural control in double and single limb stance. *Journal of Sport Rehabilitation* 16: 363–372.

Corrigan, J. P., Moore, D. P., and Stephens, M. M. 1993. Effect of heel height on forefoot loading. *Foot and Ankle* 14: 148–152.

de Lateur, B. J., Giaconi, R. M., Questad, K., Ko, M., and Lehmann, J. F. 1991. Footwear and posture. Compensatory strategies for heel height. *American Journal of Physical Medicine and Rehabilitation* 70: 246–254.

Dunne, R. G., Bergman, A. B., Rogers, L. W., Inglin, B., and Rivara, F. P. 1993. Elderly persons' attitudes towards footwear—A factor in preventing falls. *Public Health Reports* 108: 245–248.

Ebbeling, C. J., Hamill, J., and Crussemeyer, J. A. 1994. Lower extremity mechanics and energy cost of walking in high-heeled shoes. *Journal of Orthopaedic and Sports Physical Therapy* 19: 190–196.

Eisenhardt, J. R., Cook, D., Pregler, I., and Foehl, H. C. 1996. Changes in temporal gait characteristics and pressure distribution for bare feet versus various heel heights. *Gait and Posture* 4: 280–286.

Esenyel, M., Walsh, K., Walden, J. G., and Gitter, A. 2003. Kinetics of high-heeled gait. *Journal of the American Podiatric Medical Association* 93: 27–32.

Finlay, O. E. 1986. Footwear management in the elderly care programme. *Physiotherapy* 72: 172–178.

Gabell, A., Simons, M. A., and Nayak, U. S. 1985. Falls in the healthy elderly: Predisposing causes. *Ergonomics* 28: 965–975.

Gao, C. and Abeysekera, J. 2004. A systems perspective of slip and fall accidents on icy and snowy surfaces. *Ergonomics* 47: 573–598.

Gao, C., Abeysekera, J., Hirvonen, M., and Gronqvist, R. 2004. Slip resistant properties of footwear on ice. *Ergonomics* 47: 710–716.

Gard, G. and Berggard, G. 2006. Assessment of anti-slip devices from healthy individuals in different ages walking on slippery surfaces. *Applied Ergonomics* 37: 177.

Gard, G. and Lundborg, G. 2001. Test of Swedish anti-skid devices on five different slippery surfaces. *Accident Analysis and Prevention* 33: 1–8.

Gastwirth, B. W., O'brien, T. D., Nelson, R. M., Manger, D. C., and Kindig, S. A. 1991. An electrodynographic study of foot function in shoes of varying heel heights. *Journal of the American Podiatric Medical Association* 81: 463–472.

Gefen, A., Megido-Ravid, M., Itzchak, Y., and Arcan, M. 2002. Analysis of muscular fatigue and foot stability during high-heeled gait. *Gait and Posture* 15: 56–63.

Gehlsen, G., Braatz, J. S., and Assmann, N. 1986. Effects of heel height on knee rotation and gait. *Human Movement Science* 5: 149–155.

Grabiner, M. D. and Davis, B. L. 1993. Footwear and balance in older men. *Journal of the American Geriatrics Society* 41: 1011–1012.

Hatton, A. L., Dixon, J., Martin, D., and Rome, K. 2009. The effect of textured surfaces on postural stability and lower limb muscle activity. *Journal of Electromyography and Kinesiology* 19: 957–964.

Hatton, A. L., Dixon, J., Rome, K., and Martin, D. 2011. Standing on textured surfaces: effects on standing balance in healthy older adults. *Age and Ageing* 40: 363–368.

Hatton, A. L., Dixon, J., Rome, K., and Martin, D. In press. Standing on textured surfaces: Effects on standing balance in healthy older adults. *Age and Ageing*.

Helfand, A. E. 2003. Basic considerations for geriatric footwear. *Clinics in Podiatric Medicine and Surgery* 20: 593–605.

Hijmans, J. M., Geertzen, J. H. B., Dijkstra, P. U., and Postema, K. 2007. A systematic review of the effects of shoes and other ankle or foot appliances on balance in older people and people with peripheral nervous system disorders. *Gait and Posture* 25: 316–323.

Hill, K., Schwarz, J., Flicker, L., and Carroll, S. 1999. Falls among healthy, community-dwelling, older women: A prospective study of frequency, circumstances, consequences and prediction accuracy. *Australian and New Zealand Journal of Public Health* 23: 41–48.

Hinman, M. R. 2004. Effect of magnetic insoles on balance in older adults. *Journal of the American Geriatrics Society* 52: 166.

Hong, W. H., Lee, Y. H., Chen, H. C., Pei, Y. C., and Wu, C. Y. 2005. Influence of heel height and shoe insert on comfort perception and biomechanical performance of young female adults during walking. *Foot and Ankle International* 26: 1042–1048.

Horgan, N. F., Crehan, F., Bartlett, E. et al. 2009. The effects of usual footwear on balance amongst elderly women attending a day hospital. *Age and Ageing* 38: 62–67.

Hosoda, M., Yoshimaru, O., Takayanagi, K. et al. 1997. The effects of various footwear types and materials, and of fixing of the ankles by footwear, on upright posture control. *Journal of Physical Therapy Science* 9: 47–51.

Hourihan, F., Cumming, R. G., Taverner-Smith, K. M., and Davidson, I. 2000. Footwear and hip fracture-related falls in older people. *Australasian Journal on Ageing* 19: 91–93.

Hübscher, M., Thiel, C., Schmidt, J. et al. In press. Slip resistance of non-slip socks—An accelerometer-based approach. *Gait and Posture*.

Hübscher, M., Thiel, C., Schmidt, J., Bach, M., Banzer, W., and Vogt, L. 2011. Slip resistance of non-slip socks—An accelerometer-based approach. *Gait and Posture* 33: 740–742.

Jessup, R. L. 2007. Foot pathology and inappropriate footwear as risk factors for falls in a subacute aged-care hospital. *Journal of the American Podiatric Medical Association* 97: 213–217.

Kavounoudias, A., Roll, R., and Roll, J. P. 1998. The plantar sole is a 'dynamometric map' for human balance control. *Neuroreport* 9: 3247–3252.

Keegan, T. H., Kelsey, J. L., King, A. C., Quesenberry, C. P., Jr., and Sidney, S. 2004. Characteristics of fallers who fracture at the foot, distal forearm, proximal humerus, pelvis, and shaft of the tibia/fibula compared with fallers who do not fracture. *American Journal of Epidemiology* 159: 192–203.

Kerse, N., Butler, M., Robinson, E., and Todd, M. 2004. Wearing slippers, falls and injury in residential care. *Australian and New Zealand Journal of Public Health* 28: 180–187.

Koepsell, T. D., Wolf, M. E., Buchner, D. M. et al. 2004. Footwear style and risk of falls in older adults. *Journal of the American Geriatrics Society* 52: 1495–1501.

Larsen, R. E., Mosekilde, L., and Foldspang, A. 2004. Correlates of falling during 24 h among elderly Danish community residents. *Preventive Medicine* 39: 389–398.

Lee, C.-M., Jeong, E.-H., and Freivalds, A. 2001. Biomechanical effects of wearing high-heeled shoes. *International Journal of Industrial Ergonomics* 28: 321.

Lee, K. H., Matteliano, A., Medige, J., and Smiehorowski, T. 1987. Electromyographic changes of leg muscles with heel lift: Therapeutic Implications. *Archives of Physical Medicine and Rehabilitation* 68: 298–301.

Lee, K. H., Shieh, J. C., Matteliano, A., and Smiehorowski, T. 1990. Electromyographic changes of leg muscles with heel lifts in women: Therapeutic implications. *Archives of Physical Medicine and Rehabilitation* 71: 31–33.

Li, K. W. and Chen, C. J. 2004. The effect of shoe soling tread groove width on the coefficient of friction with different sole materials, floors, and contaminants. *Applied Ergonomics* 35: 499–507.

Li, K. W. and Chen, C. J. 2005. Effects of tread groove orientation and width of the footwear pads on measured friction coefficients. *Safety Science* 43: 391–405.

Li, K. W., Wu, H. H., and Lin, Y. C. 2006. The effect of shoe sole tread groove depth on the friction coefficient with different tread groove widths, floors and contaminants. *Applied Ergonomics* 37: 743–748.

Lindemann, U., Scheible, S., Sturm, E. et al. 2003. Elevated heels and adaptation to new shoes in frail elderly women. *Zeitschrift fur Gerontologie und Geriatrie* 36: 29–34.

Lloyd, D. and Stevenson, M. G. 1989. Measurement of slip resistance of shoes on floor surfaces—Part 2: Effect of a bevelled heel. *Journal of Occupational Health and Safety— Australia New-Zealand* 5: 229–235.

Lord, S. R., Clark, R. D., and Webster, I. W. 1991. Postural stability and associated physiological factors in a population of aged persons. *Journal of Gerontology* 46: M69–M76.

Lord, S. R. and Ward, J. A. 1994. Age-associated differences in sensori-motor function and balance in community dwelling women. *Age and Ageing* 23: 452–460.

Lord, S. R. and Bashford, G. M. 1996. Shoe characteristics and balance in older women. *Journal of the American Geriatrics Society* 44: 429–433.

Lord, S. R., Bashford, G. M., Howland, A., and Munroe, B. J. 1999. Effects of shoe collar height and sole hardness on balance in older women. *Journal of the American Geriatrics Society* 47: 681–684.

Lord, S. R., Menz, H. B., and Sherrington, C. 2006. Home environment risk factors for falls in older people and the efficacy of home modifications. *Age Ageing* 36(S2): ii55–ii59.

Lord, S. R., Sherrington, C., Menz, H. B. and Close, J. C. T. 2007. *Falls in Older People: Risk Factors and Strategies for Prevention*, 2 edn, Cambridge, U.K.: Cambridge University Press.

Lord, S. R., Ward, J. A., Williams, P., and Anstey, K. J. 1993. An epidemiological study of falls in older community-dwelling women: The Randwick falls and fractures study. *Australian Journal of Public Health* 17: 240–245.

Maki, B. E., Perry, S. D., Norrie, R. G., and Mcilroy, W. E. 1999. Effect of facilitation of sensation from plantar foot-surface boundaries on postural stabilization in young and older adults. *Journals of Gerontology Series A-Biological Sciences and Medical Sciences* 54: M281–M287.

Mandato, M. G. and Nester, E. 1999. The effects of increasing heel height on forefoot peak pressure. *Journal of the American Podiatric Association* 89: 75–80.

Manning, D. P. and Jones, C. 2001. The effect of roughness, floor polish, water, oil and ice on underfoot friction: Current safety footwear solings are less slip resistant than microcellular polyurethane. *Applied Ergonomics* 32: 185–196.

Mcbride, I. D., Wyss, U. P., Cooke, T. D. et al. 1991. First metatarsophalangeal joint reaction forces during high-heel gait. *Foot and Ankle* 11: 282–288.

Mckiernan, F. E. 2005. A simple gait-stabilizing device reduces outdoor falls and nonserious injurious falls in fall-prone older people during the winter. *Journal of the American Geriatrics Society* 53: 943–947.

Mcpoil, T. G. 1988. Footwear. *Physical Therapy* 68: 1857–1865.

Menant, J. C., Perry, S. D., Steele, J. R. et al. 2008a. Effects of shoe characteristics on dynamic stability when walking on even and uneven surfaces in young and older people. *Archives of Physical Medicine and Rehabilitation* 89: 1970–1976.

Menant, J. C., Steele, J. R., Menz, H. B., Munro, B. J., and Lord, S. R. 2008b. Effects of footwear features on balance and stepping in older people. *Gerontology* 54: 18–23.

Menant, J. C., Steele, J. R., Menz, H. B., Munro, B. J., and Lord, S. R. 2009a. Effects of walking surfaces and footwear on temporo-spatial gait parameters in young and older people. *Gait and Posture* 29: 392–397.

Menant, J. C., Steele, J. R., Menz, H. B., Munro, B. J., and Lord, S. R. 2009b. Rapid gait termination: Effects of age, walking surfaces and footwear characteristics. *Gait and Posture* 30: 65–70.

Menant, J. C., Steele, J. R., Menz, H. B., Munro, B. J., and Lord, S. R. 2011. Step time variability and pelvis acceleration patterns of younger and older adults: Effects of footwear and surface conditions. *Research in Sports Medicine* 19: 28–41.

Menz, H. B. and Lord, S. R. 1999. Footwear and postural stability in older people. *Journal of the American Podiatric Medical Association* 89: 346–357.

Menz, H. B., Lord, S. R., and Fitzpatrick, R. C. 2006a. A tactile stimulus applied to the leg improves postural stability in young, old and neuropathic subjects. *Neuroscience Letters* 406: 23.

Menz, H. B., Lord, S. T., and Mcintosh, A. S. 2001. Slip resistance of casual footwear: Implications for falls in older adults. *Gerontology* 47: 145–149.

Menz, H. B. and Morris, M. E. 2005. Footwear characteristics and foot problems in older people. *Gerontology* 51: 346–351.

Menz, H. B., Morris, M. E., and Lord, S. R. 2006b. Foot and ankle risk factors for falls in older people: A prospective study. *Journals of Gerontology Series A—Biological Sciences and Medical Sciences* 61: 866–870.

Menz, H. B., Morris, M. E., and Lord, S. R. 2006c. Footwear characteristics and risk of indoor and outdoor falls in older people. *Gerontology* 52: 174–180.

Menz, H. B. and Sherrington, C. 2000. The Footwear Assessment Form: A reliable clinical tool to assess footwear characteristics of relevance to postural stability in older adults. *Clinical Rehabilitation* 14: 657–664.

Menz, H. B., Zammit, G. V., and Munteanu, S. E. 2007. Plantar pressures are higher under callused regions of the foot in older people. *Clinical and Experimental Dermatology* 32: 375–380.

Mickle, K. J., Munro, B. J., Lord, S. R., Menz, H. B., and Steele, J. R. 2010. Foot shape of older people: Implications for shoe design. *Footwear Science* 2: 131–139.

Munro, B. J. and Steele, J. R. 1999. Household-shoe wearing and purchasing habits. A survey of people aged 65 years and older. *Journal of the American Podiatric Medical Association* 89: 506–514.

Nevitt, M. C., Cummings, S. R., Kidd, S., and Black, D. 1989. Risk factors for recurrent non-syncopal falls: a prospective study. *Journal of the American Medical Association* 261: 2663–2668.

Nigg, B. M. and Morlock, M. 1987. The influence of lateral heel flare of running shoes on pronation and impact forces. *Medicine and Science in Sports and Exercise* 19: 294–302.

Opila-Correia, K. A. 1990a. Kinematics of high-heeled gait. *Archives of Physical Medicine and Rehabilitation* 71: 304–309.

Opila-Correia, K. A. 1990b. Kinematics of high-heeled gait with consideration for age and experience of wearers. *Archives of Physical Medicine and Rehabilitation* 71: 905–909.

Opila, K. A., Wagner, S. S., Schiowitz, S., and Chen, J. 1988. Postural alignment in barefoot and high-heeled stance. *Spine* 13: 542–547.

Ottaviani, R. A., Ashton-Miller, J. A., Kothari, S. U., and Wojtys, E. M. 1995. Basketball shoe height and the maximal muscular resistance to applied ankle inversion and eversion moments. *American Journal of Sports Medicine* 23: 418–423.

Paiva De Castro, A., Rebelatto, J. R., and Aurichio, T. R. 2010. The relationship between foot pain, anthropometric variables and footwear among older people. *Applied Ergonomics* 41: 93–97.

Palluel, E., Olivier, I., and Nougier, V. 2008. Do spike insoles enhance postural stability and plantarsurface cutaneous sensitivity in the elderly? *Age* 30: 53–61.

Perry, S. D., Mcilroy, W. E., and Maki, B. E. 2000. The role of plantar cutaneous mechano-receptors in the control of compensatory stepping reactions evoked by unpredictable, multi-directional perturbation. *Brain Research* 877: 401–406.

Perry, S. D., Radtke, A., and Goodwin, C. R. 2007. Influence of footwear midsole material hardness on dynamic balance control during unexpected gait termination. *Gait and Posture* 25: 94–98.

Perry, S. D., Santos, L. C., and Patla, A. E. 2001. Contribution of vision and cutaneous sensation to the control of centre of mass (COM) during gait termination. *Brain Research* 913: 27–34.

Priplata, A. A., Niemi, J. B., Harry, J. D., Lipsitz, L. A., and Collins, J. J. 2003. Vibrating insoles and balance control in elderly people. *The Lancet* 362: 1123.

Robbins, S. E. and Gouw, G. J. 1991. Athletic footwear: Unsafe due to perceptual illusions. *Medicine and Science in Sports and Exercise* 23: 217–224.

Robbins, S., Gouw, G. J., and Mcclaran, J. 1992. Shoe sole thickness and hardness influence balance in older men. *Journal of the American Geriatrics Society* 40: 1089–1094.

Robbins, S. E., Hanna, A. M., and Gouw, G. J. 1988. Overload protection: Avoidance response to heavy plantar surface loading. *Medicine and Science in Sports and Exercise* 20: 85–92.

Robbins, S. and Waked, E. 1997. Balance and vertical impact in sports: Role of shoe sole materials. *Archives of Physical Medicine and Rehabilitation* 78: 463–467.

Robbins, S., Waked, E., Allard, P., Mcclaran, J., and Krouglicof, N. 1997. Foot position awareness in younger and older men: The influence of footwear sole properties. *Journal of the American Geriatrics Society* 45: 61–66.

Robbins, S., Waked, E., Gouw, G. J., and Mcclaran, J. 1994. Athletic footwear affects balance in men. *British Journal of Sports Medicine* 28: 117–122.

Robbins, S., Waked, E., and Krouglicof, N. 1998. Improving balance. *Journal of the American Geriatrics Society* 46: 1363–1370.

Robbins, S., Waked, E., and Mcclaran, J. 1995. Proprioception and stability: Foot position awareness as a function of age and footwear. *Age and Ageing* 24: 67–72.

Sekizawa, K., Sandrey, M. A., Ingersoll, C. D., and Cordova, M. L. 2001. Effects of shoe sole thickness on joint position sense. *Gait and Posture* 13: 221–228.

Sherrington, C. and Menz, H. B. 2003. An evaluation of footwear worn at the time of fall-related hip fracture. *Age and Ageing* 32: 310–314.

Snow, R. E. and Williams, K. R. 1994. High heeled shoes: Their effect on center of mass position, posture, three-dimensional kinematics, rearfoot motion, and ground reaction forces. *Archives of Physical Medicine and Rehabilitation* 75: 568–576.

Snow, R. E., Williams, K. R., and Holmes, G. B., Jr. 1992. The effects of wearing high heeled shoes on pedal pressure in women. *Foot and Ankle* 13: 85–92.

Soames, R. W. and Evans, A. A. 1987. Female gait patterns: The influence of footwear. *Ergonomics* 30: 893–900.

Speksnijder, C. M., Munckhof, R. J. H., Moonen, S. A. F. C. M., and Walenkamp, G. H. I. M. 2005. The higher the heel the higher the forefoot-pressure in ten healthy women. *The Foot* 15: 17–21.

Spink, M. J., Menz, H. B., Fotoohabadi, M. R., Wee, E., Landorf, K. B., Hill, K. D., and Lord, S.R. In press. Effectiveness of a multifaceted podiatry intervention to prevent falls in community-dwelling older people with disabling foot pain: A randomised controlled trial. *British Medical Journal*.

Spink, M. J., Menz, H. B., Fotoohabadi, M. R., Wee, E., Landorf, K. B., Hill, K. D., and Lord, S. R. 2011. Effectiveness of a multifaceted podiatry intervention to prevent falls in community dwelling older people with disabling foot pain: randomised controlled trial. *British Medical Journal* 342: d3411.

Stacoff, A., Steger, J., Stussi, E., and Reinschmidt, C. 1996. Lateral stability in sideward cutting movements. *Medicine and Science in Sports and Exercise* 28: 350–358.

Stevenson, M. G., Hoang, K., Bunterngchit, Y., and Lloyd, D. 1989. Measurement of slip resistance of shoes on floor surfaces—Part 1: Methods. *Journal of Occupational Health and Safety—Australia New-Zealand* 5: 115–120.

Suomi, R. and Koceja, D. M. 2001. Effect of magnetic insoles on postural sway measures in men and women during a static balance test. *Percept Motor Skills* 92: 469–476.

Tencer, A. F., Koepsell, T. D., Wolf, M. E. et al. 2004. Biomechanical properties of shoes and risk of falls in older adults. *Journal of the American Geriatric Society* 52: 1840–1846.

Thompson, F. M. and Coughlin, M. J. 1994. The high-price of high-fashion footwear. *The Journal of Bone and Joint Surgery* 76-A: 1586–1593.

Tinetti, M. E., Speechley, M., and Ginter, S. F. 1988. Risk factors for falls among elderly persons living in the community. *New England Journal of Medicine* 319: 1701–1707.

Waddington, G. S. and Adams, R. D. 2004. The effect of a 5-week wobble-board exercise intervention on ability to discriminate different degrees of ankle inversion, barefoot and wearing shoes: A study in healthy elderly. *Journal of the American Geriatrics Society* 52: 573–576.

Wang, Y., Pascoe, D., Kim, C., and Xu, D. 2001. Force patterns of heel strike and toe off on different heel heights in normal walking. *Foot and Ankle International* 22: 486–492.

Whitney, S. L. and Wrisley, D. M. 2004. The influence of footwear on timed balance scores of the modified clinical test of sensory interaction and balance. *Archives of Physical Medicine and Rehabilitation* 85: 439–443.

Wilson, M. L., Rome, K., Hodgson, D., and Ball, P. 2008. Effect of textured foot orthotics on static and dynamic postural stability in middle-aged females. *Gait and Posture* 27: 36–42.

You, S. H., Granata, K. P., and Bunker, L. K. 2004. Effects of circumferential ankle pressure on ankle proprioception, stiffness, and postural stability: A preliminary investigation. *Journal of Orthopaedic and Sports Physical Therapy* 34: 449–460.

Yung-Hui, L. and Wei-Hsien, H. 2005. Effects of shoe inserts and heel height on foot pressure, impact force, and perceived comfort during walking. *Applied Ergonomics* 36: 355–362.

Part VI

Activity-Specific Footwear

26 Soccer Shoe Design and Its Influence on Player's Performance

Ewald M. Hennig and Katharina Althoff

CONTENTS

26.1 INTRODUCTION

Soccer is the most watched and played sport in the world. In a survey of the International Federation of Association Football (FIFA), 265 million active soccer players were counted (Kunz, 2007). Approximately 15% of these players are women with a strong increase in numbers between 2000 and 2006. Many scientific papers were published on the biological aspects of soccer, including physiological, anthropometrical, and medical research. Shephard (1999) selected 370 scientific papers for a review on the biological and medical aspects of soccer. The review has an emphasis on anthropometric, physiological, and medical aspects of male and female soccer players. Considering the importance of footwear for the game of soccer, it is surprising how little research has been published in this field. Speed performance on the field with fast accelerations in sprinting and rapid cutting movements is becoming increasingly important in the modern game of soccer. Footwear plays an important role in assisting the player to perform fast movements on the field, providing comfort and protecting the foot during kicking. In this chapter, performance and injury protection soccer shoe research is presented. Originating from game analyses and questionnaires about the desired properties of soccer shoes, the specific demands of male and female players are also presented. After a discourse about the importance of shoe comfort, performance-related shoe features are discussed. We show that soccer shoes not only have an influence on traction but can also be constructed to enhance kicking speed and accuracy.

26.2   PREFERRED SOCCER SHOE PROPERTIES: RESULTS FROM QUESTIONNAIRES

Figure 26.1 shows the results of two surveys, performed from our research team in 1998 and 2006. Based on a list of 11 shoe properties, 322 male soccer players were asked to judge the most desirable characteristics that they expect from their shoe. The players ranked the five most important properties from most (1) to less important (5). Shoe features that were not named by the players received a score of 6.

Comfort is by far the most important shoe feature for soccer players. Important are also stability (little movement between foot and shoe), traction (little movement between shoe and playing ground), and ball sensing (touch of the ball). The influence of the soccer shoe on kicking performance like enhancing shooting power, kicking accuracy, or spin production was judged as less important. Surprisingly, injury protection also played a minor role. Comparing the results from 2006 and 1998 the four most desirable features—comfort, stability, traction, and ball sensing—became even more important. The skill level (94 professional and 228 amateur players) had only a minor influence on the ranking of shoe features. Except for shoe weight, the professional players ranked the desirable shoe properties very similar to the judgment of the amateurs. A lower shoe weight is more important for intensive game situations with increased running distances and high-speed requirements as they occur in professional games.

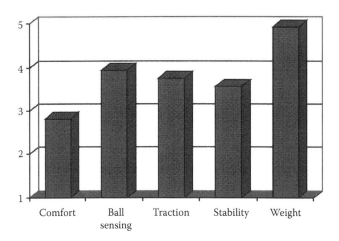

FIGURE 26.1   Ranking of the five most desirable soccer shoe properties (lowest value = best rating).

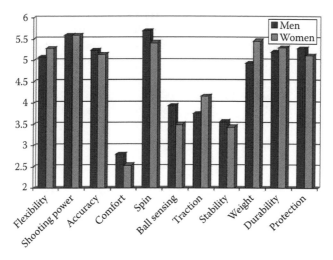

FIGURE 26.2 Ranking of most desirable soccer shoe properties from 322 men and 69 women (lowest value = best rating).

During the 2006 survey, 69 female players were included to judge the most desirable soccer shoe features. As shown in Figure 26.2, the ranking is similar to the men. Comfort and touch of the ball (ball sensing) received better scores from the female players whereas traction and shoe weight were judged less important.

Compared to male soccer players, the women have a three times higher incidence of knee anterior cruciate ligament (ACL) injuries (Agel et al., 2005). Lambson et al. (1996) detected that cleat design has an influence on ACL ruptures. They found a higher ACL injury risk for soccer players wearing an edge cleat design compared to a nonedge design. In spite of a considerably higher injury risk caused by high traction footwear, women still prefer shoes with good traction on the field. Like their male counterparts, they prefer high performance and care less about injury protection.

26.3 GAME OF SOCCER: AN ANALYSIS OF WORLD CUP GAMES

A soccer match is characterized by various situations. On average, a soccer player changes running speed every four seconds (Krustrup et al., 2005; Mohr et al., 2003). Especially, in these situations the shoe-to-ground interaction is important to prevent slipping and thereby improve performance. A detailed analysis of soccer games can provide useful information about the demands of soccer shoes. Therefore, we examined 16 games of the men's world cup 2002 in Korea/Japan and the women's world cup 2003 in the United States (Althoff et al., 2010). Video recordings of the broadcast stations were used to analyze the finals, semifinals, and quarterfinals of each world cup. All actions close to the ball were analyzed and characterized by two experts. Kicking techniques were recorded for all passes and shots. Shots on goal were analyzed in more detail. Based on interviews with established coaches in women soccer, Kirkendall (2007) mentions important differences between the women's and men's game. Men have more endurance, strength, and power, and they are faster. These biological differences lead to a higher pace in the men's game. Furthermore, the coaches noticed skill limitations of the women including the first touch, dribbling, long passes, and goal keeping. The biological and skill differences have a strong influence on the tactics in a game (Kirkendall, 2007). Therefore, the data of the men's and women's world cup were compared to quantify these differences.

Almost 21,000 actions with the ball (passes, shots on goal, dribblings) and 3,500 actions without the ball (tackles) were analyzed with regard to their location on the soccer field. Goalkeeper actions were studied separately. The following results are based on more than 26,000 recorded actions.

26.3.1 Actions with Ball

Shooting, passing, controlling, and dribbling the ball are important skills in soccer. Figure 26.3 shows the total number of these actions for one team in world-class soccer matches. On average, men perform 734 and women 679 actions with the ball. Based on net playing time, the men's teams have 14.4 and the women's teams 11.1 actions per minute. The most frequently performed movements are passing the ball across a short distance (men 293/team, women 243/team) and controlling the ball (men 258/team, women 225/team).

To compare the men and women results, the data in Table 26.1 are normalized to the total number of ball actions. Whereas the men use more short passes, women cover distances more often with long passes. Shots on goal are very important in soccer, but they account for only 2% of all the ball actions.

26.3.2 Dribblings

Dribblings are subdivided into dribbling with feints, dribbling for speed, and dribbling to keep the ball. On average, there are 57 dribblings in a men's game and 44 dribblings in a women's game. Dribblings with feints are performed most frequently. They are

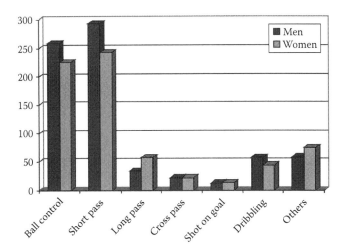

FIGURE 26.3 Actions with the ball for one team in a soccer game.

TABLE 26.1
Distribution of Actions with Ball (%)

Actions with Ball	Ball Control (%)	Short Pass (%)	Long Pass (%)	Cross Pass (%)	Shot on Goal (%)	Dribbling (%)	Others (%)
Men	35	39	5	3	2	8	8
Women	33	36	8	3	2	7	11

normally used to outplay an opponent player. Only small differences between men and women are observed in the use of dribbling for speed and to keep the ball.

26.3.3 TACKLES

198 tackles were counted for the men's and 220 tackles for the women's game. Tscholl et al. (2007) detected 147 tackles per match in top-level tournaments of women. They counted 17.1 sliding tackles per match in the women's and 17.7 sliding tackles in the men's game. The sliding tackling counts from our world cup analyses showed a 20% higher incidence for the men's as compared to the women's games. The higher number of sliding tacklings is an expression of a more powerful and aggressive play in the men's games. Tscholl et al. (2007) mention that in women's soccer, sliding tackles bear the highest risk for injuries.

26.3.4 KICKING TECHNIQUES

Kicking techniques influence kicking velocity and kicking accuracy considerably. Inside kicks are more accurate but slower. Soccer players achieve maximum ball velocity during full instep kicking (Nunome et al., 2002). A soccer game is characterized by passes and shots with different ranges in various situations. Therefore, soccer players try to use the appropriate technique. In the majority of cases, they kick the ball with the inside of the foot (men: 60%, women: 52%). This technique is typically used for short passes and accurate shots, which occur often in a game (Table 26.2).

Compared to men, women kick the ball more often with the instep (men 9%, women 14%). This is likely a consequence of the higher number of long passes in the women's games. Because of a lower muscle mass and strength, women use instep kicks also for distances where men can still use the more accurate

TABLE 26.2
Distribution of Kicking Techniques (%)

Kicking Techniques	Inside (%)	Instep (%)	Outside (%)	Full Instep (%)	Head (%)	Others (%)
Men	60	9	9	6	8	8
Women	52	14	7	8	10	9

inside kicking technique. For example, to pass a ball that covers a distance above 25 m, women use more often the instep as compared to the men (men 66%, women 73%).

26.3.5 SHOTS ON GOAL

For the final eight world cup games, we found that a team shoots on average 12 times on the opponent's goal. These data are very similar for the men's and women's games. However, the women scored 25 goals, whereas the men only achieved 13 goals. Kirkendall et al. (2002) found also a higher success rate in women's soccer. They compared the men's world cup 1998 and the women's world cup 1999 and found a shot-to-goal ratio of 11:1 for the men against 7.6:1 for the women. They also found that a closer distance to goal resulted in a higher probability of scoring. In general, women try to score from a closer distance than men (Figure 26.4), this could be a reason for the higher success rate. However, Kirkendall also showed that women are more successful at all distances. The strikers may benefit from technical weakness of the female goalkeepers (Kirkendall, 2007).

The full instep is the most frequently used technique for shots on goal for both male and female soccer players (Table 26.3). Soccer players can achieve maximum ball velocity with this technique and thereby reduce the time for the goalkeeper to react.

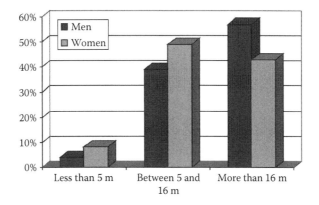

FIGURE 26.4   Distance to goal during shots on goal (in %).

TABLE 26.3
Distribution of Used Kicking Techniques for Shots on Goal (%)

Kicking Techniques for Shots on Goal	Inside (%)	Instep (%)	Outside (%)	Full Instep (%)	Head (%)	Others (%)
Men	9	14	2	60	13	2
Women	15	17	0	46	19	3

Because men achieve higher velocities during maximum full instep kicks, they use this technique more frequently in trying to score a goal. As a consequence of the reduced muscle mass and strength, women are more successful with actions close to the goal, trying to score with headballs, inside, and instep kicks.

26.4 WOMEN'S VERSUS MEN'S SOCCER: CONCLUSIONS FOR FOOTWEAR DESIGN

Game analyses show different movement patterns and ball actions in the women compared to the men's game. Women play more long passes, whereas men pass the ball more often across a shorter distance. Caused by a lower muscle mass and strength, women use different kicking techniques. The total as well as relative number of passes with the instep is much higher in the women's game. Although there are large differences in playing the game, the total distance covered during a game varies only little between men and women. Mohr et al. (2003) and Krustrup et al. (2005) found covered movement distances during a game of 10.3–10.4 km for the women and 10–14 km for the men. However, large differences are found in high intensity game situations for the men with higher running speeds and more explosive movements. Women run 1.32 km (Krustrup et al., 2005) and men 2.43 km (Mohr et al., 2003) at high intensities during a game.

Injury protection is of less importance for both the male and female soccer players (Figure 26.2). However, based on our game analyses results, women's shoes can have protective features without compromising performance. Because of the less aggressive and explosive nature of the women's game, traction demands are lower. High traction was identified as an important factor for the risk of ACL injuries (Lambson et al., 1996). The incidence of ACL ruptures was reported to be three times higher for women (Agel et al., 2005). Stud configuration and stud length should be modified for more protection in shoes for female soccer players. The more frequent use of the instep kicking technique may also have consequences for women soccer shoes. Ball touch (Figure 26.2) was substantially more important for the women. Therefore, a thinner upper material, providing more ball sensation, should be used for women shoes. A sturdy upper material is important for shoes that require good stability and high traction. These demands are lower for the less aggressive play in women's soccer. A thinner upper material may not only provide better touch of the ball but may also improve shoe comfort. Most important, however, is the different foot morphology between men and women (Krauss et al., 2008). For a better fit and comfort, women soccer shoes will have to be built on different lasts, taking anatomical features of the female foot into account.

26.5 FIT AND COMFORT

Probably the most important characteristic for any kind of footwear is fit and comfort. From our questionnaire (Figure 26.2), "comfort" received the highest preference ranking from our male and female soccer players. Early soccer shoes from English factory workers were far from being comfortable. The players used their hard and long laced leatherwork boots with added metal studs for better traction. These hard leather boots often weighed more than 500 g and could become as heavy as 1 kg in wet weather conditions. In these early boots, the lacing area went over the ankle. Modern low cut soccer shoes can have a mass of less than 200 g. Today, carbon outsoles and synthetic materials allow low shoe weights without sacrificing traction and shoe stability. Soccer players prefer shoes that feel like a glove to provide a good touch of the ball during play. Therefore, shoes are sometimes bought too small to get this close fit between shoe and foot. The presence of studs and cleats as part of the outsole may cause high local pressures under the foot during running and other activities on the field. Especially, on a soft ground surface long studs are used for good traction. In these soft ground shoes, typically 6–8 studs are part of the outsole. Such a stud configuration concentrates the acting forces to a few relatively small areas of the foot, potentially creating high local pressures under the foot. To avoid discomfort by high plantar pressures, manufacturers use in-shoe pressure distribution measuring insoles during the design process of their products. The pressure patterns under the foot are determined in soccer specific dynamic loading situations.

For peak pressure reduction, conical-shaped studs and stiff interface plates between the studs and shoe sole are employed to reduce high pressure peaks under the foot. Nevertheless, large differences are still present in commercially available soccer shoes.

Figure 26.5 shows the peak pressure pattern of 18 subjects during a cutting movement on the field in two soccer shoe models with a difference in the heel seat construction. Low pressures under the lateral and high pressures under the medial forefoot are typical for cutting movements in sudden changes of movement direction. Shoe A (light gray) has particularly low peak pressures under the heel. This pressure reduction was achieved by a cup-shaped heel seat construction in the rear of the shoe.

The cupping of the heel construction restricts the displacement of the heel fat pad to the side, which normally reduces the fat pad thickness under the calcaneus in a flat heel construction. Therefore, the natural shock and pressure attenuating

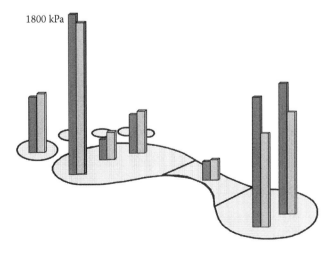

FIGURE 26.5   Peak pressure patterns for cutting movements in two different soccer shoes.

properties of the heel fat pad are used in a clever design approach. Because no additional mechanical protection is needed in the heel seat design, it also helps to reduce shoe weight.

Because a snug fit of the shoe is very important to soccer players, different shoe lasts should be used to accommodate variations in foot anatomy. Based on a cluster analysis from three-dimensional scans of male and female feet, Krauss et al. (2008) concluded that different foot shapes require different shoe lasts for a given shoe size. The authors reported that for both, men and women, shorter feet tend to be more voluminous and longer feet are more narrow and flat. However, to date soccer shoe manufacturer do not offer a choice of soccer shoe constructions for these differences in foot anatomy. In general, the female foot is more slender in the heel and ball region of the foot, and it has a lower toe and instep height (Krauss et al., 2010). Therefore, soccer shoes for women should be constructed by using a gender-specific last. Unfortunately, soccer shoes for women are often only size graded and color adjusted men's shoe constructions.

26.6   TRACTION: ITS INFLUENCE ON PERFORMANCE AND INJURIES

Providing good traction is certainly the most important performance function of soccer shoes. Fast accelerations and decelerations, rapid cut and turn movements are only possible in shoes that provide sufficient traction. Traction on the field is dependent on the playing ground as well as weather conditions. Except for indoor soccer shoes without cleats, outdoor shoes are primarily marketed as hard, firm, and soft ground shoes. For playing on hard fields, the shoes typically have many short studs, often made from hard rubber. These studs are distributed across the entire outsole. Firm ground soccer shoes are the most common type of footwear that players wear on dry soccer grass fields. Most of these shoes have bladed soccer cleats molded into the outsole of the lateral and medial heel and forefoot area as well as the central part of the forefoot. Most soft ground shoes for moist and wet grass field conditions have detachable circular studs that come in different lengths. To quantify the traction properties of soccer shoes by mechanical traction devices is difficult to perform. Even though different turf and surface conditions can be tested, the dynamic behavior of the shoe is almost impossible to simulate. During the contact of the foot with the ground, there are no uniform pressure distributions across the foot (Figure 26.5). Dependent on the kind of movement, plantar pressure patterns will occur that vary with the foot structure, body weight, and the dynamic nature of the movement. Additionally, soccer players will adapt to the properties of their shoes and modify their movement patterns. In a shoe that feels comfortable and provides stability as well as good traction, players will perform activities on the field more powerful and faster. Because of the reduced risk of slipping, the players will move on the field with more confidence.

Based on this mechanism, we developed a functional traction course test (FTC). Because players will adjust their movements to the traction properties of the shoe-to-ground interface conditions, they will run faster through a given slalom course. Our FTC track includes sections for straight accelerations and decelerations as well as a slalom part with multiple cutting and turning movements (Figure 26.6).

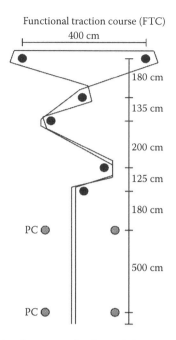

FIGURE 26.6 Functional traction course for determining soccer shoe traction properties.

Using a photocell arrangement, the times for straight acceleration and deceleration as well as the slalom part can be measured. Sterzing et al. (2009b) used
this track and reported results from eight different soccer shoe studies. The
slalom times with cuts and turns were affected most by the shoe to ground interface conditions. Shoes with removed cleats increased the FTC running time by
26% and weather conditions (ice and snow versus dry firm grass) influenced
running times by about 20%. Modifying stud configuration had a smaller effect
(approximately 3%). Even these small changes in traction performance can have
a large effect on playing performance, when players cover the additional distance
needed to get hold of the ball. Too much traction has always been suspected to
be responsible for injuries in soccer. Lambson et al. (1996) performed a 3 year
prospective injury survey with 3119 high school football players. From four shoes
with different cleat designs, they identified one shoe model with an increased
torsional resistance that caused a higher anterior cruciate ligament injury incidence. Apart from this study, there is little evidence that there is a relationship
between traction in soccer and injury risks. Artificial turf has almost completely
replaced natural grass surfaces in American football stadiums. A similar trend is
expected for future field surfaces in soccer. The change in the traction behavior
on artificial turf has always been suspected to increase the injury risk. However,
inconsistent results are reported in the literature. Some studies find increased and
others reduced injury risks, playing on artificial as compared to natural grass
surfaces. Many of the early injury survey studies were not prospective and had a
low number of subjects. Based on two prospective studies, Ekstrand et al. (2006,
2010) concluded that neither male nor female players have an increased overall

injury risk in match and practice situations on artificial turf. However, for the male players Ekstrand et al. (2010) found a reduced risk of quadriceps strains and a higher risk of ankle sprains on artificial turf. Although the overall injury rate is the same on natural and artificial turf, the pattern of injuries may change as a result of different traction conditions. Examining ball skills and movement patterns of male and female elite soccer players on artificial against natural turf, Andersson et al. (2008) found no differences in the total covered distances during a game, the speed profiles of players, and the number of sprints. However, fewer sliding tacklings and more short passes on artificial turf suggest that playing style changed as a consequence of the different traction conditions. As documented by a questionnaire, the players judged their overall performance worse on the artificial turf. The adaptation of the players to different traction conditions on the field may be the main reason why most studies do not find differences in injury rates when comparing different turf or shoe conditions.

26.7 MAXIMUM KICKING VELOCITY

High ball velocities are an important factor for being successful in distance shots on the goal. Although long distance goal kicks do not occur often in a game, they belong to the most important situations during games. Kicking technique and muscular strength influence maximum ball velocities in soccer. However, even soccer shoe construction has an influence on maximum kicking velocity (Hennig and Zulbeck, 1999). Although these authors found statistically significant differences in maximum ball velocities in different shoe models, they could not offer an explanation about the mechanisms behind this phenomenon. From a physics point of view, ball velocity is the result of a simple momentum transfer through the foot. The momentum transfer occurs during a very short time of approximately 10 ms (Nunome et al., 2006) and thus causes high peak reaction forces of approximately 3000 N (Shinkai et al., 2009). Tsaousidis and Zatsiorsky (1996) emphasize that the foot-to-ball interaction is not only an impact situation but is also determined by a throwing-like movement. The authors conclude that more than 50% of the resulting ball speed is not determined by regain of the elastic energy from ball deformation. In an impact situation, the momentum of the foot is influenced by the velocity of the foot and the impacting mass. Therefore, increasing the momentum for higher ball velocities can either be achieved by a high impact velocity or an increased impact mass. The influence of additional shoe weight on the maximum velocity of the kicking foot was reported by Amos and Morag (2002). They found higher foot velocities with lower shoe weight but did not determine maximum ball velocity. Measuring ball velocity in our own studies, we did not find an increase of maximum ball velocity with a lower shoe mass. Therefore, a higher shoe mass probably compensates for the lower foot velocities and creates a similar momentum at impact with the ball. In several studies, Sterzing and Hennig (2008) found an influence of soccer shoe properties on maximum ball speed. Better traction properties of the shoe on the stance leg improved maximum kicking speed. Furthermore, foot protection and toe box height also had a positive influence. The elastic characteristics of the outsole were always suspected to have an influence on maximum ball speed through an elastic rebound of the shoe during ball contact. However, we were not able to find an influence of the outsole properties on ball speed. Contrary to our expectations, we found higher ball velocities during barefoot kicking (Sterzing et al., 2009a). Using high speed video analysis of shod against barefoot kicking, we observed a higher degree of foot plantarflexion in the barefoot condition. More plantarflexion causes a stronger mechanical coupling between the foot and the lower leg, thus resulting in an increased effective impact mass.

26.8   INFLUENCE OF SHOE DESIGN ON KICKING ACCURACY

The results of our questionnaire (Figure 26.2) show that few soccer players believe that a shoe can influence their kicking velocity or kicking accuracy. Kicking accuracy for passing and shots on goal belongs to the most important skills of successful soccer players. Because precise ball placement is an essential part of the game, it is more important than achieving higher ball velocities. This observation is also true for other sports such as tennis. Professional tennis players prefer precision over speed. Therefore, they use high string tensions on their tennis racket. High string tension of a tennis racket improves the accuracy of ball placement but reduces maximum ball speed (Hennig, 2007). We became interested in the question, whether a similar principle as it is known from tennis, can also be applied to soccer footwear.

To explore this question, we performed a study with five commercially available soccer shoes and compared it to barefoot kicking. A custom-made electronic target with a circular diameter of 120 cm was constructed to measure kicking accuracy (Hennig et al., 2009). With 24 subjects, best accuracy kicks were performed from a distance of 10 m to the target center at a height of 115 cm. The subjects were instructed to use their own kicking technique in achieving best accuracy results for the given task. To avoid a possible influence on the results caused by pain during barefoot kicking, elastic socks—providing sufficient comfort—were worn by the subjects. As shown in Figure 26.7, barefoot kicking resulted in a worse precision performance when compared to all other shoe conditions.

Between shoe models, we also found statistically significant differences in kicking accuracy. Based on these results, we performed a series of studies to examine possible influences of footwear construction features on precision performance. From our studies, we found no effect of the friction behavior between the upper shoe and the ball material on accuracy measures. Touch of the ball—or good skin sensation—could also be a factor for better shooting precision. However, based on our data from barefoot kicking we concluded that the feel for the ball is not a likely contributor for influencing kicking accuracy. The difference between the shod and barefoot kicking conditions is the upper material interface. Covering the bony prominences with shoe material, the local pressures are distributed across a larger area and may result in a more homogenous pressure pattern. Therefore, we came up with the hypothesis that shoes creating less pressure inhomogenities across the ball contact area will improve kicking accuracy.

Using a Pedar (Novel Inc.) pressure distribution measuring insole, we tested this hypothesis. We fastened the measuring insole on the upper of two different soccer

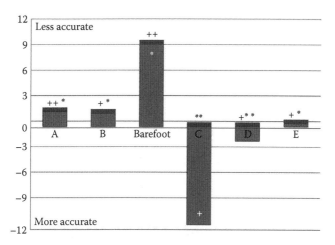

FIGURE 26.7 Deviation from mean accuracy (32.6 cm) in percent (**, ++ $p < 0.01$; *, + $p < 0.05$).

shoe models from our initial accuracy study (Hennig, et al., 2009). Using these two shoes, 20 subjects performed shots for best accuracy on the electronic target. The pressure distribution measuring insoles were adjusted on top of the shoes to the foot anatomy to match the anatomical foot locations of all subjects. At a frequency of 571 Hz, the pressures were recorded between foot and ball during the impact phase. When analyzing the data, the shoe with the better accuracy performance did show a more homogenous pressure pattern (Hennig et al., 2009). In a further study, padding was added to further reduce the pressure gradients, which are caused by the bony prominences of the forefoot. This intervention also improved shooting accuracy. We conclude from our studies that footwear influences kicking precision in soccer. Achieving a more homogenous pressure distribution between the shoe upper and the ball seems to be the key factor for better kicking accuracies.

26.9 CONCLUSIONS

Shoe fit and comfort are the most important characteristics that players expect from their shoes. Because the game of soccer is played differently between women and men, gender-specific footwear should be constructed. For the less aggressive playing style of women, traction demands are lower. Shoes providing less traction and a lower resistance against forefoot rotation may help to reduce the high incidence of anterior cruciate ligaments in female soccer players. For good touch of the ball, a tight fit of the shoe with flexible, soft, and thin upper materials is desirable. Gender differences in foot dimensions have to be considered to guarantee good shoe fit and comfort for women and men. Because mechanical tests cannot mimic traction in playing situations, a functional traction test was found to be successful in the evaluation of soccer shoe traction properties. Increasing maximum ball velocity by shoe design is of a complex nature. Reduced shoe weight increases the maximum velocity of the kicking leg but does not increase the momentum transfer to the ball. Surprisingly, barefoot kicking results in higher ball velocities as kicking with shoes. Shoes prevent the foot from going into full plantarflexion in maximum power kicks. This probably causes a less effective mechanical coupling of the foot to the shank and results in a lower effective impact mass. Kicking precision belongs after traction to the most important factors for being successful in a soccer game. Shoe design influences kicking precision. Barefoot against shod kicking was found to be less accurate. Between shoes, the primary factor for a better precision performance is a more even pressure distribution pattern between ball and shoe. The science of soccer shoe design is still in its infancy. Future challenges will include an optimized combination of desirable shoe features and the design of shoes that will provide good performance but still keep injuries at a low risk.

QUESTIONS

26.1 What are the four most import shoe features for soccer players?

26.2 Which injury occurs much more frequently in female as compared to male soccer players and how can shoes help to reduce these injuries?

26.3 How does the women's game of soccer differ from the men's game?

26.4 What is the basic idea behind a functional traction test for soccer shoes?

26.5 Why is kicking barefoot less accurate than kicking with a soccer shoe?

26.6 What causes kicking barefoot higher ball velocities as compared to shod kicking?

REFERENCES

Agel, J., Arendt, E. A., and Bershadsky, B. (2005). Anterior cruciate ligament injury in national collegiate athletic association basketball and soccer: A 13-year review. *Am J Sports Med*, 33(4), 524–530.

Althoff, K., Kroiher, J., and Hennig, E. M. (2010). A soccer game analysis of two world cups: Playing behavior between elite female and male soccer players. *Footwear Sci*, 2(1), 51–56.

Amos, M. and Morag, E. (2002). Effect of shoe mass on soccer kicking velocity. In *4th World Congress of Biomechanics*. Calgary, Alberta, Canada: Omnipress.

Andersson, H., Ekblom, B., and Krustrup, P. (2008). Elite football on artificial turf versus natural grass: Movement patterns, technical standards, and player impressions. *J Sports Sci*, 26(2), 113–122.

Ekstrand, J., Hagglund, M., and Fuller, C. W. (2010). Comparison of injuries sustained on artificial turf and grass by male and female elite football players. *Scand J Med Sci Sports*, Online publication, April 28.

Ekstrand, J., Timpka, T., and Hagglund, M. (2006). Risk of injury in elite football played on artificial turf versus natural grass: a prospective two-cohort study. *Br J Sports Med*, 40(12), 975–980.

Hennig, E. M. (2007). Influence of racket properties on injuries and performance in tennis. *Exerc Sport Sci Rev*, 35(2), 62–66.

Hennig, E. M., Althoff, K., and Hoemme, A.-K. (2009). Soccer footwear and ball kicking accuracy. *Footwear Science*, 1(Suppl 1), 85–87.

Hennig, E. M. and Zulbeck, O. (1999). The influence of soccer boot construction on ball velocity and shock to the body. In E. M. Hennig and D. J. Stefanyshin (Eds.), *Fourth Symposium on Footwear Biomechanics* (pp. 52–53). Canmore, Canada: University of Calgary.

Kirkendall, D. T. (2007). Issues in training the female player. *Br J Sports Med*, 41(1), i64–i67.

Kirkendall, D. T., Dowd, W. W., and Di Cicco, A. D. (2002). Patterns of successful attacks: A comparison of the men's and women's world cup. *Rev Fútbol Cien*, 1(1), 17–20.

Krauss, I., Grau, S., Mauch, M., Maiwald, C., and Horstmann, T. (2008). Sex-related differences in foot shape. *Ergonomics*, 51(11), 1693–1709.

Krauss, I., Valiant, G., Horstmann, T., and Grau, S. (2010). Comparison of female foot morphology and last design in athletic footwear—Are men's lasts appropriate for women? *Res Sports Med*, 18(2), 140–156.

Krustrup, P., Mohr, M., Ellingsgaard, H., and Bangsbo, J. (2005). Physical demands during an elite female soccer game: Importance of training status. *Med Sci Sports Exerc*, 37(7), 1242–1248.

Kunz, M. (2007). Big count—265 million playing football. *FIFA Mag*, July, 10–15.

Lambson, R. B., Barnhill, B. S., and Higgins, R. W. (1996). Football cleat design and its effect on anterior cruciate ligament injuries. *Am J Sports Med*, 24(2), 155.

Mohr, M., Krustrup, P., and Bangsbo, J. (2003). Match performance of high-standard soccer players with special reference to development of fatigue. *J Sports Sci*, 21(7), 519–528.

Nunome, H., Asai, T., Ikegami, Y., and Sakurai, S. (2002). Three-dimensional kinetic analysis of side-foot and instep soccer kicks. *Med Sci Sports Exerc*, 34(12), 2028–2036.

Nunome, H., Lake, M., Georgakis, A., and Stergioulas, L. K. (2006). Impact phase kinematics of instep kicking in soccer. *J Sports Sci*, 24(1), 11–22.

Shephard, R. J. (1999). Biology and medicine of soccer: An update. *J Sports Sci*, (17), 757–786.

Shinkai, H., Nunome, H., Isokawa, M., and Ikegami, Y. (2009). Ball impact dynamics of instep soccer kicking. *Med Sci Sports Exerc*, 41(4), 889–897.

Sterzing, T. and Hennig, E. M. (2008). The influence of soccer shoes on kicking velocity in full-instep kicks. *Exerc Sport Sci Rev*, 36(2), 91–97.

Sterzing, T., Kroiher, J., and Hennig, E. M. (2009a). Kicking velocity: Barefoot kicking superior to shod kicking ? In T. Reilly and F. Korkusuz (Eds.), *Science and Football VI: The Proceedings of the Sixth World Congress on Science and Football* (pp. 50–56). New York: Routledge.

Sterzing, T., Müller, C., Hennig, E. M., and Milani, T. L. (2009b). Actual and perceived running performance in soccer shoes: A series of eight studies. *Footwear Sci*, 1(1), 5–17.

Tsaousidis, N. and Zatsiorsky, V. (1996). Two types of ball-effector interaction and their relative contribution to soccer kicking. *Hum Mov Sci*, 15, 861–876.

Tscholl, P., O'Riordan, D., Fuller, C. W., Dvorak, J., and Junge, A. (2007). Tackle mechanisms and match characteristics in women's elite football tournaments. *Br J Sports Med*, 41(Suppl 1), i15–i19.

27 Footwear for Preventing Acute Sport-Related Ankle Ligamentous Sprain Injury

Daniel Tik-Pui Fong, Kai-Ming Chan, and Kam-Ming Mok

CONTENTS

27.1 PROBLEM AND PREVALENCE OF ACUTE ANKLE LIGAMENTOUS SPRAIN INJURY IN SPORTS

27.1.1 PROBLEM IN THE PREVIOUS 30 YEARS

All around the world, medical doctors and sports scientists are actively promoting regular physical exercises to gain health benefits. However, in contrary to the promotion of the health benefits, sports also cause injuries (Chan et al. 1993), which are often comparable to traffic accidents, occupational injuries, as well as violence as indicated by the attendances to emergency departments in numerous countries like Sweden (de Loes and Goldie 1988), the United Kingdom (Jones and Taggart 1994), and North Ireland (Abernethy and MacAuley 2003). With the promoted participation in sports and physical activity, a growing exposure to potential injury should increase and thus the rising incidence of sports injury (Shephard 2003).

The first ever epidemiological study of ankle sprains was published in 1977 (Garrick 1977). In this report, the results of a 2 year study of four high schools showed that 14% of the 1176 reported injuries involved the ankle joint, and 85% of these injuries were ankle sprains. The problem was very common in basketball, football, and cross-country, and the typical mechanism was inversion plantar flexion and internal rotation. Since this first report, there were more than 200 epidemiological studies published in the following 30 years, and were systematically reviewed in a recent report (Fong et al. 2007). A total of 227 studies were reviewed, and a total of 201,600 patients with 32,509 ankle injuries were recorded. Ankle was the most commonly injured body sites in 24 of the 70 included sports, and ankle sprains accounted for about 85% of all the ankle injuries. This percentage of ankle sprain injury was comparable to the figures reported 30 years ago. Therefore, together with the increasing trend of sports participants worldwide, the exact number of ankle sprain cases should have dramatically increased over the past 30 years (Figure 27.1).

In 2007, there was a series of 15 injury epidemiology reports on different sports published by a research group in the United States (Hootman et al. 2007). The research group analyzed the data from the injury surveillance system maintained by the National Collegiate Athletic Association, and presented the prevalence and incidence of common injury types from 1988 to 2004. Ankle ligamentous sprain was very often the most common injury types in these sports, with the only exception being ice hockey. Table 27.1 shows the extracted data on ankle sprain percentage, incidence per 1000 athletic-exposure (AE) and rank among all injury types in each of the 15 reports. The percentage of ankle ligamentous sprain injury accounted for as much as 44.1% in game and 29.4% in practice, both in women volleyball. In term of incidence, it accounted for as much as 5.39 and 1.06 per 1000 AE per 1000 athletic-exposure in men football game and men basketball practice, respectively.

27.1.2 CURRENT PROBLEM GLOBALLY

There have been numerous injury epidemiology reports on sports or events involving ankle sprain injuries published from 2007 until now. In soccer, ankle sprain was still the most common injury (Dvorak et al. 2011, Junge and Dvorak 2010a,

FIGURE 27.1   Ankle sprain cases happened in sports events. (From Mok, K.M. et al., *Am. J. Sports Med.*, 39(7), 1548, 2011a, http://ajs.sagepub.com/content/39/7/1548)

Tegnander et al. 2008). The ankle joint was very often the most common injury site, accounting for 22.1%–26.7% of all the injuries (Azubuike and Okojie 2009, Gaulrapp et al. 2010, Kordi et al. 2011, Yard et al. 2008). Among ankle injuries, sprain is the major injury mechanism, causing ligament tear and rupture (Azubuike and Okojie 2009, Le Gall et al. 2008, Oztekin et al. 2009, Yard et al. 2008). In recurrent reinjury, ankle was also the most commonly involved body site, accounting for 28.3% of all the injuries, with 34.9% of them being sprain injuries (Swenson et al. 2009). For referees, ankle sprain and thigh strain were the most common injuries while officiating soccer matches (Bizzini et al. 2009, 2011). For futsal, which is the five-a-side version of soccer, ankle sprain was the second common injuries, account for 10% among the 165 recorded injuries in three consecutive Futsal World Cups in Guatemala 2000, Taipei 2004, and Brazil 2008 (Junge and Dvorak 2010b).

Ankle sprain was also common in basketball, accounting for 23.8% of all the injuries in adolescent (Randozzo et al. 2010) and an incidence rate of 1.12 per 1000 person-hour in the Greek professional league (Kofotolis and Kellis 2007). A report on the U.S. high school basketball leagues showed that ankle injuries accounted for 39.7% of all the injuries, with 44.0% of them being ligamentous sprain

TABLE 27.1

Percentage, Incidence per 1000 Athletic-Exposure and Rank of Ankle Ligamentous Sprain Injury among All Other Injury Types in Game and Practice as Published from a Series of 15 Injury Epidemiology Reports in the United States

Study	Sport	Game			Practice		
		Percentage	Incidence per 1000 AE	Rank	Percentage	Incidence per 1000 AE	Rank
Dick et al. (2007g)	Men baseball	7.4	0.43	2	8.5	0.16	2
Dick et al. (2007b)	Men basketball	26.2	2.33	1	26.8	1.06	1
Agel et al. (2007d)	Women basketball	24.6	1.89	1	23.6	0.95	1
Dick et al. (2007c)	Women field hockey	13.7	0.76	1	15.0	0.37	1
Dick et al. (2007a)	Men football	15.6	5.39	2	11.8–13.9	0.45–1.34	2
Marshall et al. (2007a)	Women gymnastics	16.4	2.48	2	15.2	0.93	1
Agel et al (2007a)	Men ice hockey	4.0	0.65	6	5.5	0.11	3
Agel et al. (2007b)	Women ice hockey	4.2	0.53	4	0	—	—
Dick et al. (2007f)	Men lacrosse	11.3	1.43	1	16.4	0.53	1
Dick et al. (2007d)	Women lacrosse	22.6	1.62	1	15.5	0.51	1
Agel et al. (2007c)	Men soccer	17.0	3.19	1	17.4	0.76	1
Dick et al. (2007e)	Women soccer	18.3	3.01	1	15.3	0.80	2
Marshall et al (2007b)	Women softball	10.3	0.44	1	9.5	0.25	1
Agel et al. (2007e)	Women volleyball	44.1	1.44	1	29.4	0.83	1
Agel et al. (2007f)	Men wrestling	7.5	1.97	2	7.3	0.41	3

(Borowski et al. 2008). It was also often listed in the top three most common injuries in other sports like high school baseball (Collins and Comstock), national fencing competition (Harmer 2008), junior tennis players (Hjelm et al. 2010), cheerleading (Shields et al. 2009), rock climbing (Nelson and McKenzie 2009), and track and field athletes (Malliaropoulos et al. 2009).

In mass sports events, Junge et al. (2009), Laoruengthana et al. (2009) and Owoeye (2010) reported that ankle was the most common injury site and sprain was the most common injury mechanism in Summer Olympic Games 2008 in Beijing, the 37th Thailand National Games 2008 in Phitsanulok, and the 16th National Sports Festival in Nigeria, respectively. From the attendance record in emergency departments, Waterman et al. (2010) reported an incidence rate of 2.15 per 1000 person-years in the United States, and such was highest as it reached 7.20 per 1000 person-years for adolescents between 15 and 19 years old. Fong et al. (2008c) analyzed the attendance record over a 1 year period and reported that 81.3% of the ankle injuries reported in the emergency department were of ligamentous sprain. Among general population, a review paper stated that the incidence of ankle sprain is very high, and the injuries to the lateral ankle ligaments accounted for approximately 25% of all the sports-related injuries (O'Loughlin et al. 2009). The estimated incidence was approximately 5,000 injuries per day in the United Kingdom and 23,000 in the United States.

27.1.3 Need to Prevent the Injury by Proper Prophylactic Footwear

Most athletes are ignorant to acute ankle ligamentous sprain injury, and they do not seek proper medical consultation from orthopedic sports medicine specialists or physiotherapists (Chan et al. 1984). Over 50 years ago, orthopedic surgeons had started suggesting that repetitive ankle sprain injuries may cause long-term chronic ankle instability (Ferran et al. 2009, Freeman 1965). Yeung et al. (1994) suggested that for ankle sprained five times or more, ankle instability become the major sequel in 38% of these patients. In 2004, a group of world renowned orthopedic surgeons concluded in a *World Consensus Conference on Ankle Instability* that there was still no common consensus to properly treat the problem (Chan and Karlsson 2005). Some would prefer operative treatment and some would prefer conservative treatment, and even nowadays, there were still many modalities in both operative (Kramer et al. 2011, Lee et al. 2011, Trc et al. 2010) and conservative treatment (Delahunt et al. 2010, Hubbard and Cordova 2010, Kemler et al. 2011, Sefton et al. 2011, Webster and Gribble 2010). All these modalities achieved different degrees of satisfaction from patients. In other words, there was still no single modality or combination of these modalities which could guarantee a perfect clinical outcome.

Therefore, besides treating the problem, it would also be beneficial to design modalities to prevent the problem. In the understanding of the etiology and mechanism of chronic ankle instability, Dr. Jay Hertel from Pennsylvania proposed a paradigm for the understanding of chronic ankle instability, which consisted of mechanical insufficiencies and functional insufficiencies (Hertel 2002). Recently, Dr. Claire Hiller from Sydney has proposed a modified model from the Hertel model, with chronic ankle instability consisting of mechanical instability, perceived instability, and recurrent sprain (Hiller et al. 2011). To date, there are many research groups in the world looking into the treatment of chronic ankle instability, but comparatively fewer researchers looking at the quantitative biomechanics of lateral ankle sprain, which is the major cause of the development of chronic ankle instability (Fong et al. 2009a). It would be too late to treat the problem after its development; therefore, it would be essential to design new modalities to prevent acute lateral ankle sprain injury. In this chapter, a new invention of an intelligent anti-sprain sport shoe is presented.

27.2 ETIOLOGY AND MECHANISM OF THE INJURY

27.2.1 ANKLE ANATOMY AND BIOMECHANICS AND THE RELATIONSHIP WITH THE INJURY

The human foot is a very complex structure with 26 bones, about 20 musculotendinous units, and over 100 ligaments (Riegger 1988). There are three major articulations: the talocrural joint, which allows plantarflexion and dorsiflexion; the subtalar joint, which allows inversion and eversion; and the distal tibiofibular syndesmosis (Hertel 2002). The fibula extends further to the lateral malleolus than the tibia does to the medial malleolus, thus creating a block to eversion (Attarian et al. 1985); thus, inversion sprains are more common and accounted for 80%–90% of all the ankle sprain injuries. The lateral ligaments at the ankles, namely the anterior talofibular ligaments (ATFL), the calcaneofibular ligament (CFL), and the posterior talofibular ligament (PTFL), support the stability of the ankle. Among these ligaments, ATFL has the lowest ultimate load and is usually ruptured during an inversion sprain (Ferran and Maffuli 2006).

27.2.2 INCORRECT FOOT LANDING AS THE CAUSE OF INJURY

Ankle inversion sprains are caused by a sudden explosive inversion or supination moment at the subtalar joint (Fuller 1999). A joint moment is the result of the magnitude of the vertically projected ground reaction force at foot contact and the perpendicular distance of this force vector from the joint center (moment arm) during a landing. In most situations, such a ground reaction force act on the lateral edge of a slightly inverted ankle to the medial direction. The force vector does not pass through the ankle joint, i.e., there is a long moment arm, and thus creates an explosive inversion moment. In addition, if the ankle joint is plantarflexed with fore foot contact with the ground during touch down, the moment arm is increased and so does the resultant ankle joint moment (Wright et al. 2000). Therefore, incorrect foot landings would be a major factor to cause ankle inversion sprain. Since the joint moment is the product of the force and the moment arm, one may suggest that reducing the force magnitude would be a possible measure. However, in real situation, it would be not be possible to reduce the ground reaction force that much, as the athlete has to land on one foot, and this must involve an impact of two to six times the body weight. Therefore, prophylactic products like high-top shoes, taping, and bracing mainly correct the ankle joint positioning at landing in order to reduce this moment arm and the subsequent ankle joint moment (Surve et al. 1994).

27.2.3 DELAYED REACTION TIME AS ANOTHER ETIOLOGY OF THE INJURY

Another etiology would be the delayed reaction time of the peroneal muscles at the lateral ankle, which functions to initiate ankle eversion, the opposite motion to an inversion ankle sprain injury (Ashton-Miller et al. 1996). It is believed that an ankle sprain injury occurs in 40 ms after landing, as the vertical ground reaction force peaks at around this period of time in landing from a jump (Dufek and Bates 1991). However, the reaction time of the peroneal muscles was reported to be 50 ms or more. For healthy subjects, it ranged from 57 to 69 ms (Karlsson and Andreasson 1992, Vaes et al. 2002). For patients with chronic ankle instability, it reached up to 82–85 ms (Karlsson and Andreasson 1992, Konradsen and Ravn 1991). In a dynamic walking trial, the reaction time is also longer, as reported to be 74 ms (Ty Hopkins et al. 2007). Therefore, it is postulated that the human reflex is not fast enough to accommodate the sudden explosive motion in an inversion ankle sprain injury.

27.2.4 RECENT ADVANCES TO INVESTIGATE THE INJURY MECHANISM

Understanding the injury mechanism is very important before one could suggest preventive measures (Bahr and Krosshaug 2005). The mechanism of inversion ankle sprain injury was reported only qualitatively in the past (Ekstrand and Gilquist 1983, Garrick 1977), even in the recent decade (Andersen et al. 2004, Giza et al. 2003), until the first ever quantitative biomechanics study on an accidental supination ankle sprain incident published in recent years (Fong et al. 2009b). This research group from Hong Kong further adopted a model-based image-matching (MBIM) technique, which was developed by a research group from Oslo for understanding the injury mechanism of anterior cruciate ligament rupture (Koga et al. 2010, Krosshaug and Bahr 2005), to investigate ankle joint kinematics from uncalibrated video sequences (Mok et al. 2011b). They utilized this new advance to systematically investigate ankle joint kinematics from video sequences of injury incidents, and reported two cases during the 2008 Beijing Olympics (Mok et al. 2011a). The peak

FIGURE 27.2 Using model-based image-matching (MBIM) technique, skeleton model was superimposed on the body image in order to measure the ankle joint kinematics. (From Mok, K.M. et al., *Am. J. Sports Med.*, 39(7), 1548, 2011a, http://ajs.sagepub.com/content/39/7/1548)

ankle inversion velocity reached up to 1752 deg/s in a high jump injury and 1397 deg/s in a field hockey injury. The research group is still working on more injury incidents with two or more available cameras capturing video sequences with good quality for the analysis (Figure 27.2).

27.3 ANTI-SPRAIN SPORT SHOE TO PREVENT THE INJURY

27.3.1 PRINCIPLE OF THE ANTI-SPRAIN SPORT SHOE

Since 2006, Chan and his research team (Chan 2006, Chan et al. 2008a) have presented their work on inventing an intelligent sprain-free sport shoe. The invention adopted a three-step mechanism which is similar to the safety airbag feedback system in vehicle. There are sensors in the vehicle to monitor the external crashing forces. There is also a method to identify if there is a hazardous situation. Finally, if such a risk is identified, the airbags inflate and protect the driver and the passengers. In this anti-sprain sport shoe, there are sensors to detect the foot and ankle motion and estimate the ankle supination torque. There is also a mathematical method to identify if such a motion is hazardous. Finally, a correction system would actuate in a very short time to produce a resistive torque to the excessive supination torque, in order to protect the ankle joint from sustaining an ankle sprain injury.

27.3.2 SENSING DEVICE TO MONITOR FOOT AND ANKLE MOTION

One of the two etiologies was suggested to be the incorrect landing, causing excessive and explosive ankle supination torque. Therefore, a sensing device was designed to monitor the foot and ankle motion. Human joint torque, or moment, is usually difficult to measure, and is very often estimated by typical inverse dynamics approach (Eng and Winter 1995, Glitsch and Baumann 1997). This could be done if we have collected the essential anthropometric data, kinematics data, and also kinetics data during a human motion, and this is often done in a laboratory equipped with all the aforementioned devices. Therefore, in order to obtain such data continuous in an outdoor environment, the first step is to devise a mobile in-shoe system to monitor such quantity. A research group from the Netherlands had developed a method to first calculate complete group reaction forces from plantar pressure data (Forner Cordero et al. 2004), and then calculate the joint moment during gait with these group reaction force data (Forner Cordero et al. 2006). However, the purpose of their studies was to get rid of the restriction of force plate positions in the laboratory, and the new method still relied on the kinematics data obtained from a motion analysis system installed in the laboratory. The method was, therefore, still not a feasible one for outdoor setting.

The attempt was successful to first estimate the complete ground reaction forces with only pressure insoles, without relying on the motion analysis system (Fong et al. 2008b), and then developed a mobile in-shoe system for monitoring ankle supination torque during sport motions (Fong et al. 2008a). The system only involved three pressure sensors, mainly at the front part of the plantar region. Unfortunately, the second step to establish a threshold to identify risk of ankle sprain injury was not successful. The research group tried to deliver a supination twisting torque to cadaveric ankle specimens and observe torque magnitude at the start of failure of anterior talofibular ligament as proposed by a previous study (Aydogan et al. 2006), and confirm the minor tear by histological sectioning (Hirose et al. 2004). The tryout was unsuccessful as the start of ligament failure was not so obvious, and there was no way to tell if the scar observed after histological sectioning was caused by the twist or the sectioning procedure.

27.3.3 Mathematical Method to Identify Hazardous Motion

The second idea was more successful—it monitored the foot and ankle kinematics, or motion, rather than the joint torque. Motion sensors at different locations on the foot and ankle were used to collect kinematics data during normal sporting motion and subinjury simulated ankle sprain motion on a pair of mechanical supination sprain simulator (Chan et al. 2008b). The motion sensor used was wearable inertial motion sensor, which is often used to monitor human motion (Fong and Chan 2010). Each motion sensor recorded six parameters, including three-dimensional (3D) linear acceleration and 3D angular velocity. The collected data were used to train up a support vector machine (SVM) model and establish a mathematical identification algorithm, which required only one motion sensor located at any location on the foot segment (Chan et al. 2010). Finally, a method to identify subinjury ankle sprain motion by monitoring only the ankle inversion velocity was established (Chu et al. 2010).

From the investigation, the maximum ankle inversion velocity during common sporting motions ranged from 22.5 to 85.1 deg/s, while the same in simulated ankle sprain motions ranged from 114.0 to 202.5 deg/s (Chu et al. 2010). Our first quantitative analysis reported an inversion velocity of 632 deg/s during a mild ankle sprain injury (Fong et al. 2009b), and our latest analysis on two cases from the Beijing Olympics reported an inversion velocity of 1397 and 1752 deg/s (Mok et al. 2011a). From these quantities, a threshold ankle inversion velocity of 300 deg/s was suggested to identify a hazardous ankle spraining motion.

27.3.4 Correction Device to Prevent Excessive Motion

Besides incorrect landing, another etiology of ankle inversion sprain injury was the delayed peroneal muscle reaction time. Therefore, a corrective device was designed to compensate this deficit. One successful attempt was to deliver electrical stimulus through a pair of electrodes to the peroneal muscle group at the upper lateral shank. This technology was termed functional electrical stimulation (FES) and has been employed in a lot of rehabilitation applications (Kapadia et al. 2011, Sabut et al. 2010). The rationale was that it could generate peroneal muscle contraction and the subsequent ankle joint pronation torque within 21–25 ms, which is the torque latency from the delivery of the signal (Ginz et al. 2004). Therefore, the corrective mechanism could be able to accommodate and resist the sudden ankle torque which happens within 50 ms after the start of ankle twisting.

A biomechanics test was conducted to evaluate its feasibility (Fong et al. 2010). Ten male subjects performed simulated inversion test and supination test on a pair of

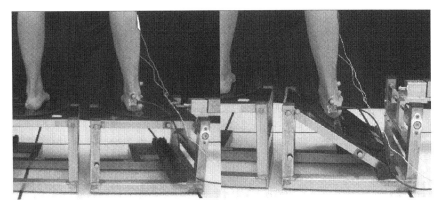

FIGURE 27.3 A mechanical supination sprain simulator to investigate ankle joint motion under sudden inversion. Motion sensor on the heel is to measure the foot tilting velocity.

mechanical sprain simulators (Chan et al. 2008b). In each trial, subject was instructed to stand normally and relax with his weight evenly distributed on both platforms. Without prior notice, one of the platforms fell suddenly for 30°, as trigger by an electronic switch. A myoelectric stimulation device was fabricated to deliver a 130 V square wave signal for 500 ms with different delay time after the electronic trigger (0, 5, 10, and 15 ms). The biomechanical effect was quantified by the maximum heel tilting angle and the maximum heel tilting angular velocity, as determined by a motion analysis system. Results suggested that electrical stimulations with different delay time showed a significant reduction in both the maximum heel tiling angle and angular velocity, and it was concluded effective in biomechanical laboratory trial on simulated spraining motion (Figure 27.3).

27.4 SUMMARY

Injury prevention in sports has been a hot topic since the development of sports medicine services and research. In the past years, most of the modalities were developed based on trial-and-error attempts before understanding well the injury etiology and mechanism. Nowadays, with the advance technique and equipment for human motion analysis, researchers should consider analyzing well the injury biomechanics first, in order to design innovative measures to reduce or prohibit the etiology and mechanism to occur. The measures could be apparels, like the intelligent anti-sprain sport shoe presented in this chapter, and could also be other modalities like exercise intervention, proprioceptive training, landing skill strategy, and even the change of game rules.

We hope that this chapter could inspire researchers to further investigate the science of footwear as the foot and footwear are the contact point to receive ground reaction force to the body in most of the time. Designing proper footwear well could probably adjust the nature of the ground reaction force during human locomotion, and could perhaps reduce the incidence of sports-related injuries.

QUESTIONS

27.1 Why is the ankle sprain problem still significant in sports over the past 30 years, when we have already a handful of newly designed sport shoes like high-top basketball shoe for ankle sprain prevention?

27.2 Why is there always a trade-off between sport performance and injury prevention capacity in a sport shoe?

27.3 Why is ankle sprain injury so common in a person playing sport? Does it happen in animals like horses, tigers, leopards, etc., during running?

27.4 Are sport shoes really protecting the foot and ankle from injuries? Are people playing sports with barefoot sustain more or fewer injuries?

27.5 Why are most sport shoe designed for cushioning? Is the ground reaction force really bad to the human being?

27.6 Can we do anything to help athletes for some sports without shoes (like gymnastics) to prevent ankle sprain injury?

27.7 Can we extend the idea of the intelligent anti-sprain shoe to prevent sprain injury in other joints, like the knee?

27.8 Does the change of game rules help in preventing ankle sprain injury? For example, the rules of netball are different from basketball as it prohibits any player to be with a short distance of the player with the ball. This avoids a lot of body contact; however, ankle sprain injury is more prevalent in netball. What is the reason?

REFERENCES

Abernethy, L., MacAuley, D. 2003. Impact of school sports injury. *British Journal of Sports Medicine*, 37: 354–355.

Agel, J., Dick, R., Nelson, B., Marshall, S.W., Dompier, T.P. 2007b. Descriptive epidemiology of collegiate women's ice hockey injuries: National Collegiate Athletic Association Injury Surveillance System, 2000–2001 through 2002–2003. *Journal of Athletic Training*, 42: 249–234.

Agel, J., Dompier, T.P., Dick, R., Marshall, S.W. 2007a. Descriptive epidemiology of collegiate men's ice hockey injuries: National Collegiate Athletic Association Injury Surveillance System, 1988–1989 through 2002–2003. *Journal of Athletic Training*, 42: 241–248.

Agel, J., Evans, T.A., Dick, R., Putukian, M., Marshall, S.W. 2007c. Descriptive epidemiology of collegiate men's soccer injuries: National Collegiate Athletic Association Injury Surveillance System, 1988–1989 through 2002–2003. *Journal of Athletic Training*, 42: 270–277.

Agel, J., Olson, D.E., Dick, R., Arendt, E.A., Marshall, S.W., Sikka, R.S. 2007d. Descriptive epidemiology of collegiate women's basketball injuries: National Collegiate Athletic Association Injury Surveillance System, 1988–1989 through 2003–2004. *Journal of Athletic Training*, 42: 202–210.

Agel, J., Palmieri-Smith, R.M., Dick, R., Wojtys, E.M., Marshall, S.W. 2007e. Descriptive epidemiology of collegiate women's volleyball injuries: National Collegiate Athletic Association Injury Surveillance System, 1988–1989 through 2003–2004. *Journal of Athletic Training*, 42: 295–302.

Agel, J., Ransone, J., Dick, R., Oppliger, R., Marshall, S.W. 2007f. Descriptive epidemiology of collegiate men's wrestling injuries: National Collegiate Athletic Association Injury Surveillance System, 1988–1989 through 2003–2004. *Journal of Athletic Training*, 42: 303–310.

Andersen, T.E., Floerenes, T.W., Arnason, A., Bahr, R. 2004. Video analysis of the mechanisms for ankle injuries in football. *American Journal of Sports Medicine*, 32: S69–S79.

Ashton-Miller, J.A., Ottaviani, R.A., Hutchinson, C., Wojtys, E.M. 1996. What best protects the inverted weightbearing ankle against further inversion? Evertor muscle strength compares favorably with shoe height, athletic tape, and thee orthoses. *American Journal of Sports Medicine*, 24: 800–809.

Attarian, D.E., McCrackin, H.J., Devito, D.P., McElhaney, J.H., Garrett, W.E. 1985. A biomechanics study of human lateral ankle ligaments and autogenous reconstructive grafts. *American Journal of Sports Medicine*, 13: 377–381.

Aydogan, U., Gilsson, R.R., Nunley, J.A. 2006. Extensor retinaculum augmentation reinforces anterior talofibular ligament repair. *Clinical Orthopaedics and Related Research*, 442: 210–215.

Azubuike, S.O., Okojie, O.H. 2009. An epidemiological study of football (soccer) injuries in Benin City, Nigeria. *British Journal of Sports Medicine*, 43: 382–386.

Bahr, R., Krosshaug, T. 2005. Understanding injury mechanisms: Key component of preventing injuries in sport. *British Journal of Sports Medicine*, 39: 324–329.

Bizzini, M., Junge, A., Bahr, R., Dvorak, J. 2009. Injuries and musculoskeletal complaints in referees—A complete survey in the top divisions of the Swiss football league. *Clinical Journal of Sport Medicine*, 19: 95–100.

Bizzini, M., Junge, A., Bahr, R., Dvorak, J. 2011. Injuries of football referees: A representative survey of Swiss referees officiating at all levels of play. *Scandinavian Journal of Medicine and Science in Sports*, 21: 42–47.

Borowski, L.A., Yard, E.E., Fields, S.K., Comstock, R.D. 2008. The epidemiology of US high school basketball injuries, 2005–2007. *American Journal of Sports Medicine*, 36: 2328–2335.

Chan, K.M. 2006. Ankle injuries in sports—What's new on the horizon? *Journal of Medical Biomechanics*, 21: S6–S7.

Chan, Y.Y., Fong, D.T.P., Chung, M.M.L., Li, W.J., Liao, W.H., Yung, P.S.H., Chan, K.M. 2010. Identification of ankle sprain motion from common sporting activities by dorsal foot kinematics data. *Journal of Biomechanics*, 43: 1965–1969.

Chan, K.M., Fong, D.T.P., Hong, Y., Yung, P.S.H., Lui, P.P.Y. 2008a. Orthopaedic sport biomechanics—A new paradigm. *Clinical Biomechanics*, 23: S21–S30.

Chan, Y.Y., Fong, D.T.P., Yung, P.S.H., Fung, K.Y., Chan, K.M. 2008b. A mechanical supination sprain simulator for studying ankle supination sprain kinematics. *Journal of Biomechanics*, 41: 2571–2574.

Chan, K.M., Fu, F., Leung, L. 1984. Sports injuries survey on university students in Hong Kong. *British Journal of Sports Medicine*, 18: 195–202.

Chan, K.M., Karlsson, J.L. 2005. *ISAKOS – FIMS World Consensus Conference on Ankle Instability*. Hong Kong: CD Concept.

Chan, K.M., Yuan, Y., Li, C.K. et al. 1993. Sports causing most injuries in Hong Kong. *British Journal of Sports Medicine*, 27: 263–267.

Chu, V.W.S., Fong, D.T.P., Chan, Y.Y., Yung, P.S.H., Fung, K.Y., Chan, K.M. 2010. Differentiation of ankle sprain motion and common sporting motion by ankle inversion velocity. *Journal of Biomechanics*, 43: 2035–2038.

Collins, C.L., Comstock, R.D. 2008. Epidemiological features of high school baseball injuries in the United States, 2005–2007. *Pediatrics*, 121: 1181–1187.

Delahunt, E., McGrath, A., Doran, N., Coughlan, G.F. 2010. Effect of taping on actual and perceived dynamic postural stability in persons with chronic ankle instability. *Archives of Physical Medicine and Rehabilitation*, 91: 1383–1389.

Dick, R., Ferrara, M.S., Agel, J., Courson, R., Marshall, S.W., Hanley, M.J., Reifsteck, F. 2007a. Descriptive epidemiology of collegiate men's football injuries: National Collegiate Athletic Association Injury Surveillance System, 1988–1989 through 2003–2004. *Journal of Athletic Training*, 42: 221–233.

Dick, R., Hertel, J., Agel, J., Grossman, J., Marshall, S.W. 2007b. Descriptive epidemiology of collegiate men's basketball injuries: National Collegiate Athletic Association Injury Surveillance System, 1988–1989 through 2003–2004. *Journal of Athletic Training*, 42: 194–201.

Dick, R., Hootman, J.M., Agel, J., Vela, L., Marshall, S.W., Messina, R. 2007c. Descriptive epidemiology of collegiate women's field hockey injuries: National Collegiate Athletic Association Injury Surveillance System, 1988–1989 through 2002–2003. *Journal of Athletic Training*, 42: 211–220.

Dick, R., Lincoln, A.E., Agel, J., Carter, E.A., Marshall, S.W., Hinton, R.Y. 2007d. Descriptive epidemiology of collegiate women's lacrosse injuries: National Collegiate Athletic Association Injury Surveillance System, 1988–1989 through 2003–2004. *Journal of Athletic Training*, 42: 262–269.

Dick, R., Putukian, M., Agel, J., Evans, T.A., Marshall, S.W. 2007e. Descriptive epidemiology of collegiate women's softball injuries: National Collegiate Athletic Association Injury Surveillance System, 1988–1989 through 2003–2004. *Journal of Athletic Training*, 42: 278–285.

Dick, R., Romani, W.A., Agel, J., Case, J.G. Marshall, S.W. 2007f. Descriptive epidemiology of collegiate men's lacrosse injuries: National Collegiate Athletic Association Injury Surveillance System, 1988–1989 through 2003–2004. *Journal of Athletic Training*, 42: 255–261.

Dick, R., Sauers, E.L., Agel, J., Keuter, G., Marshall, S.W., McCarty, K. 2007g. Descriptive epidemiology of collegiate men's baseball injuries: National Collegiate Athletic Association Injury Surveillance System, 1988–1989 through 2003–2004. *Journal of Athletic Training*, 42: 183–193.

Dufek, J.S., Bates, B.T. 1991. Biomechanics factors associated with injury during landing in jump sports. *Sports Medicine*, 12: 326–337.

Dvorak, J., Junge, A., Derman, W., Schwellnus, M. 2011. Injuries and illnesses of football players during the 2010 FIFA World Cup. *British Journal of Sports Medicine*, 45: 626–630.

Ekstrand, J., Gilquist, J. 1983. Soccer injuries and their mechanisms: A prospective study. *Medicine & Science in Sports & Exercise*, 15: 267–270.

Eng, J.J., Winter, D.A. 1995. Kinetic analysis of the lower limbs during walking: What information can be gained from a three-dimensional model? *Journal of Biomechanics*, 28: 753–758.

Ferran, N.A., Maffuli, N. 2006. Epidemiology of sprains of the lateral ankle ligament complex. *Foot and Ankle Clinics*, 11: 659–662.

Ferran, N.A., Oliva, F., Maffulli, N. 2009. Ankle instability. *Sports Medicine and Arthroscopy Review*, 17: 139–145.

Fong, D.T.P., Chan, Y.Y. 2010. The use of wearable inertial motion sensors in human lower limb biomechanics studies: A systematic review. *Sensors*, 10: 11556–11565.

Fong, D.T.P., Chan, Y.Y., Hong, Y., Yung, P.S.H., Fung, K.Y., Chan, K.M. 2008a. A three-pressure-sensor (3PS) system for monitoring ankle supination torque during sport motions. *Journal of Biomechanics*, 41: 2562–2566.

Fong, D.T.P., Chan, Y.Y., Hong, Y., Yung, P.S.H., Fung, K.Y., Chan, K.M. 2008b. Estimating the complete ground reaction forces with pressure insoles in walking. *Journal of Biomechanics*, 41: 2596–2601.

Fong, D.T.P., Chan, Y.Y., Mok, K.M., Yung, P.S.H., Chan, K.M. 2009a. Understanding acute ankle ligamentous sprain injury in sports. *Sports Medicine, Arthroscopy, Rehabilitation, Therapy and Technology*, 1: 14.

Fong, D.T.P., Chu, V.W.S., Chung, M.M.L., Chan, Y.Y., Yung, P.S.H., Chan, K.M. 2010. The effect of myoelectric stimulation on peroneal muscles to resist sudden simulated ankle sprain motions. In *Proceedings of XXVIII International Symposium on Biomechanics in Sports* (p. 207), Marquette, MI.

Fong, D.T.P., Hong, Y., Chan, L.K., Yung, P.S.H., Chan, K.M. 2007. A systematic review on ankle injury and ankle sprain in sports. *Sports Medicine*, 37: 73–84.

Fong, D.T.P., Hong, Y., Yung, P.S.H., Shima, Y., Krosshaug, T., Chan, K.M. 2009b. Biomechanics of supination ankle sprain—A case report of an accidental injury event in laboratory. *American Journal of Sports Medicine*, 37: 822–827.

Fong, D.T.P., Man, C.Y., Yung, P.S.H., Cheung, S.Y., Chan, K.M. 2008c. Sport-related ankle injuries attending an accident and emergency department. *Injury*, 39: 1222–1227.

Forner Cordero, A., Koopman, H.J., van der Helm, F.C. 2004. Use of pressure insoles to calculate the complete ground reaction forces. *Journal of Biomechanics*, 37: 1427–1432.

Forner Cordero, A., Koopman, H.J., van der Helm, F.C. 2006. Inverse dynamic calculations during gait with restricted ground reaction force information from pressure insoles. *Gait and Posture*, 23: 189–199.

Freeman, M.A. 1965. Instability of the foot after injuries to the lateral ligament of the ankle. *Journal of Bone and Joint Surgery—British Volume*, 47: 669–677.

Fuller, E.A. 1999. Center of pressure and its theoretical relationship to foot pathology. *Journal of the American Podiatric Medical Association*, 89: 278–291.

Garrick, J.G. 1977. The frequency of injury, mechanism of injury, and epidemiology of ankle sprains. *American Journal of Sports Medicine*, 5: 241–242.

Gaulrapp, J., Becker, A., Walther, M.N., Hess, H. 2010. Injuries in women's soccer: A 1-year all players prospective field study of the women's Bundesliga (German premier league). *Clinical Journal of Sport Medicine*, 20: 264–271.

Ginz, H.F., Zorzato, F., Iaizzo, P.A., Urwyler, A. 2004. Effect of three anaesthetic techniques on isometric skeletal muscle strength. *British Journal of Anaesthesia*, 92: 367–372.

Giza, E., Fuller, C., Junge, A., Dvorak, J. 2003. Mechanisms of foot and ankle injuries in soccer. *American Journal of Sports Medicine*, 31: 550–554.

Glitsch, U., Baumann, W. 1997. The three-dimensional determination of internal loads in the lower extremity. *Journal of Biomechanics*, 30: 1123–1131.

Harmer, P.A. 2008. Incidence and characteristics of time-loss injuries in competitive fencing: A prospective, 5-year study of national competitions. *Clinical Journal of Sport Medicine*, 18: 137–142.

Hertel, J. 2002. Functional anatomy, pathomechanics, and pathophysiology of lateral ankle instability. *Journal of Athletic Training*, 37: 364–375.

Hiller, C.E., Kilbreath, S.L., Refshauge, K.M. 2011. Chronic ankle instability: Evolution of the model. *Journal of Athletic Training*, 46: 133–141.

Hirose, K., Murakami, G., Minowa, T., Kura, H., Yamashita, T. 2004. Lateral ligament injury of the ankle and associated articular cartilage degeneration in the talocrural joint: Anatomic study using elderly cadavers. *Journal of Orthopaedic Science*, 9: 37–43.

Hjelm, N., Werner, S., Renstrom, P. 2010. Injury profile in junior tennis players: A prospective two year study. *Knee Surgery Sports Traumatology Arthroscopy*, 18: 845–850.

Hootman, J.M., Dick, R., Agel, J. 2007. Epidemiology of epidemiology of collegiate injuries for 15 sports: Summary and recommendations for injury prevention initiatives. *Journal of Athletic Training*, 42: 311–319.

Hubbard, T.J., Cordova, M. 2010. Effect of ankle taping on mechanical laxity in chronic ankle instability. *Foot and Ankle International*, 31: 499–504.

Jones, R.S., Taggart, T. 1994. Sport related injuries attending the accident and emergency department. *British Journal of Sports Medicine*, 28: 110–111.

Junge, A., Dvorak, J. 2010a. Injuries in female football players in top-level international tournaments. *British Journal of Sports Medicine*, 41(Suppl): i3–i7.

Junge, A., Dvorak, J. 2010b. Injury risk of playing football in Futsal World Cups. *British Journal of Sports Medicine*, 44: 1089–1092.

Junge, A., Engebretsen, L., Mountjoy, M.L., Alonso, J.M., Renstrom, P.A., Aubry, M.J., Dvorak, J. 2009. Sports injuries during the Summer Olympic Games 2008. *American Journal of Sports Medicine*, 37: 2165–2172.

Kapadia, N.M., Zivanovic, V., Furlan, J., Craven, B.C., McGillivray, C., Popovic, M.R. 2011. Functional electrical stimulation therapy for grasping in traumatic incomplete spinal cord injury: Randomized control trial. *Artificial Organisms*, 35: 212–216.

Karlsson, J., Andreasson, G.O. 1992. The effect of external ankle support in chronic lateral ankle joint instability: An electromyographic study. *American Journal of Sports Medicine*, 20: 257–261.

Kemler, E., van de Port, I., Backx, F., van Dijk, C.N. 2011. A systematic review on the treatment of acute ankle sprain: Brace versus other functional treatment types. *Sports Medicine*, 41: 185–197.

Kofotolis, N., Kellis, E. 2007. Ankle sprain injuries: A 2-year prospective cohort study in female Greek professional basketball players. *Journal of Athletic Training*, 42: 388–394.

Koga, H., Nakamae, A., Shima, Y., Iwasa, J., Myklebust, G., Engebretsen, L., Bahr, R., Korsshuag, T. 2010. Mechanisms for noncontact anterior cruciate ligament injuries. *American Journal of Sports Medicine*, 38(11): 2218–2225.

Konradsen, L., Ravn, J.B. 1991. Prolonged peroneal reaction time in ankle instability. *International Journal of Sports Medicine*, 12: 290–292.

Kordi, R., Hemmati, F., Heidarian, H., Ziaee, V. 2011. Comparison of the incidence, nature and cause of injuries sustained on dirt field and artificial turf field by amateur football players. *Sports Medicine Arthroscopy Rehabilitation Therapy and Technology*, 3: 3.

Kramer, D., Solomon, R., Curtis, C., Zurakowski, D., Micheli, L.J. 2011. Clinical results and functional evaluation of the chrisman-snook procedure for lateral ankle instability in athletes. *Foot and Ankle Specialist*, 4: 18–28.

Krosshaug, T., Bahr, R. 2005. A model-based image-matching technique for three-dimensional reconstruction of human motion from uncalibrated video sequences. *Journal of Biomechanics*, 38: 919–929.

Laoruengthana, A., Poosamsai, P., Fangsanau, T., Supanpaiboon, P., Tungkasamesamran, K. 2009. The epidemiology of sports injury during the 37th Thailand National Games 2008 in Phitsanulok. *Journal of Medical Association of Thailand*, 92(Suppl): 204–210.

Leanderson, C., Leanderson, J., Wykman, A., Strender, L.E., Johansson, S.E., Sundquist, K. Musculoskeletal injuries in young ballet dancers. *Knee Surgery, Sports Traumatology, Arthroscopy*, 19(9): 1531–1535.

Lee, K.T., Park, Y.U., Kim, J.S., Kim, J.B., Kim, K.C., Kang, S.K. 2011. Long-term results after modified Brostrom procedure without calcaneofibular ligament reconstruction. *Foot and Ankle International*, 32: 153–157.

Le Gall, F., Carling, C., Reilly, T. 2008. Injuries in young elite female soccer players: An 8-season prospective study. *American Journal of Sports Medicine*, 26: 276–284.

de Loes, M., Goldie, I. 1988. Incidence rate of injuries during sport activity and physical exercise in a rural Swedish municipality: Incidence rates in 17 sports. *International Journal of Sports Medicine*, 9: 461–467.

Malliaropoulos, N., Ntessalen, M., Papacostas, E., Longo, U.G., Maffuli, N. 2009. Reinjury after acute lateral ankle sprains in elite track and field athletes. *American Journal of Sports Medicine*, 37L: 1755–1761.

Marshall, S.W., Covassin, T., Dick, R., Nassar, L.G., Agel, J. 2007a. Descriptive epidemiology of collegiate women's gymnastics injuries: National Collegiate Athletic Association Injury Surveillance System, 1988–1989 through 2003–2004. *Journal of Athletic Training*, 42: 234–240.

Marshall, S.W., Hamstra-Wright, K.L., Dick, R., Grove, K.A., Agel, J. 2007b. Descriptive epidemiology of collegiate women's softball injuries: National Collegiate Athletic Association Injury Surveillance System, 1988–1989 through 2003–2004. *Journal of Athletic Training*, 42: 286–294.

Mok, K.M., Fong, D.T.P., Krosshaug, T., Engebretsen, L., Hung, A.S.L., Yung, P.S.H., Chan, K.M. 2011a. Kinematics analysis of ankle inversion ligamentous sprain injuries in sports—Two cases during the 2008 Beijing Olympics. *American Journal of Sports Medicine*, 39(7): 1548–1552.

Mok, K.M., Fong, D.T.P., Krosshaug, T., Hung, A.S. L., Yung, P.S.H., Chan, K.M. 2011b. An ankle joint model-based image-matching motion analysis technique. *Gait and Posture*, 34(1): 71–75.

Nelson, N.G., McKenzie, L.B. 2009. Rock climbing injuries treated in emergency departments in the U.S., 1990–2007. *American Journal of Preventive Medicine*, 37: 195–200.

O'Loughlin, P.F., Murawski, C.D., Egan, C., Kennedy, J.G. 2009. Ankle instability in sports. *The Physician and Sportsmedicine*, 37: 93–103.

Owoeye, O.B. 2010. Pattern and management of sports injuries presented by Lagos state athletes at the 16th National Sports Festival (KADA games 2009) in Nigeria. *Sports Medicine Arthroscopy Rehabilitation Therapy and Technology*, 2: 3.

Oztekin, H.H., Boya, H., Ozcan, O., Zeren, B., Pinar, P. 2009. Foot and ankle injuries and time lost from play in professional soccer players. *Foot*, 19: 22–28.

Randozzo, C., Nelson, N.G., McKenzie, L.B. 2010. Basketball-related injuries in school-aged children and adolescents in 1997–2007. *Pediatrics*, 126: 727–733.

Riegger, C.L. 1988. Anatomy of the ankle and foot. *Physical Therapy*, 68: 1802–1814.

Sabut, S.K., Lenka, P.K., Kumar, R., Mahadevappa, M. 2010. Effect of functional electrical stimulation on the effort and walking speed, surface electromyography activity, and metabolic responses in stroke subjects. *Journal of Electromyography and Kinesiology*, 20: 1170–1177.

Sefton, J.M., Yarar, C., Hicks-Little, C.A., Berry, J.W., Cordova, M.L. 2011. Six weeks of balance training improves sensorimotor function in individuals with chronic ankle instability. *Journal of Orthopaedic and Sports Physical Therapy*, 41: 81–89.

Shephard, R.J. 2003. Can we afford to exercise, given current injury rates? *Injury Prevention*, 9: 99–100.

Shields, B.J., Fernandez, S.A., Smith, G.A. 2009. Epidemiology of cheerleading stunt-related injuries in the United States. *Journal of Athletic Training*, 44: 586–594.

Surve, I., Schwellnus, M.P., Noakes, T., Lombard, C. 1994. A fivefole reduction in the incidence of recurrent ankle sprains in soccer players using the Sport-Stirrup orthosis. *American Journal of Sports Medicine*, 22: 601–606.

Swenson, D.M., Yard, E.E., Fields, S.K. Comstrock, R.D. 2009. Patterns of recurrent injuries among US high school athletes, 2005–2008. *American Journal of Sports Medicine*, 37: 1586–1593.

Tegnander, A., Olsen, O.E., Moholdt, T.T., Engebretsen, L., Bahr, R. 2008. Injuries in Norwegian female elite soccer: A prospective one-season cohort study. *Knee Surgery Sports Traumatology Arthroscopy*, 16: 194–198.

Trc, T., Handl, M., Havlas, V. 2010. The anterior talo-fibular ligament reconstruction in surgical treatment of chronic lateral ankle instability. *International Orthopaedics*, 34: 991–996.

Ty Hopkins, J., McLoad, T., McCaw, S. 2007. Muscle activation following sudden ankle inversion during standing and walking. *European Journal of Applied Physiology*, 99: 371–378.

Vaes, P., Duquet, W., van Gheluwe, B. 2002. Peroneal reaction times and eversion motor response in healthy and unstable ankles. *Journal of Athletic Training*, 37: 475–480.

Vormittag, K., Calonje, R., Briner, W.W. 2009. Foot and ankle injuries in the barefoot sports. *Current Sports Medicine Reports*, 8: 262–266.

Waterman, B.R., Owens, B.D., Davey, S., Zacchilli, M.A., Belmont, P.J. 2010. The epidemiology of ankle sprains in the United States. *Journal of Bone and Joint Surgery (American Volume)*, 92: 2279–2284.

Webster, K.A., Gribble, P.A. 2010. Functional rehabilitation interventions for chronic ankle instability: A systematic review. *Journal of Sport Rehabilitation*, 19: 98–114.

Wright, I.C., Neptune, R.R., van den Bogert, A.J., Nigg, B.M. 2000. The influence of foot positioning on ankle sprains. *Journal of Biomechanics*, 33: 513–519.

Yard, E.E., Schroeder, M.J., Fields, S.K., Collins, C.L., Comstrock, R.D. 2008. The epidemiology of United States high school soccer injuries, 2005–2007. *American Journal of Sports Medicine*, 36: 1930–1937.

Yeung, M.S., Chan, K.M., So, C.H., Yuan, W.Y. 1994. An epidemiological survey on ankle sprain. *British Journal of Sports Medicine*, 28: 112–116.

28 Kinematics Analysis of Walking with Negative-Heeled Shoes on Treadmills

Youlian Hong and Jing Xian Li

CONTENTS

28.1 INTRODUCTION

The strength of the muscles across the ankle joint and the knee joint is important to maintain stability and prevent injury in the knee and ankle. For this reason, patients with ankle and knee joint injuries are rehabilitated to strengthen their muscles around the joints. Many approaches have been developed to enhance muscular strength, one of which is graded walking.[1] The negative-heeled shoes (NHS), also known as earth shoes, were designed to mimic uphill walking so as to build up and exercise the muscles in the trunk and the lower limb. In the normal sports shoes, the heel was approximately 1.5 cm higher than the toe part; these were categorized as normal shoes.[2] In contrast, the NHS have the toe part 1.5 cm higher than the heel, which tilts the foot into about 10° of dorsiflexion (Figure 28.1).

Walking plays an important role in people's daily activity. Previous studies show that the human locomotor pattern has been adaptable to different environments such as uphill and downhill walking, stair ascending and descending, walking with load carriage,[3,4] etc. Factors involved in exploring this topic include changes in walking speed and walking ground.[5] This adaptation is normally achieved by changing the

FIGURE 28.1 A negative-heeled shoe.

pattern of lower-limb motions and trunk postures.[1,3,4] Previous studies in trunk and lower extremity kinematics have shown that gait adaptation on uphill treadmill inclination during walking is characterized by an increasingly flexed posture of the hip, knee, and ankle at initial foot contact as well as a progressive forward tilt of the pelvic and the trunk.[6] An earlier report showed a significant decrease in walking speed and cadence when walking up a steep ramp of greater than $9°$.[7,8] A reduction in step length was also observed during descent walking.[9] Electromyography (EMG) studies regarding muscle activation in uphill treadmill walking and ramp ground walking have found a modifying level of activation of the relevant flexor and extensor muscles in the lower extremity.[1,10–12]

Some studies show the influence of heel height of shoes on locomotor adaptation. It was found that walking in high-heeled shoes produced shorter stride lengths and higher stance time percentage than walking in normal sports shoes.[13] Kinematically, high-heeled gait compared to low-heeled gait was characterized by significantly increased knee flexion at heel strike and during stance phase, and significantly lower knee flexion and hip flexion during swing phase.

Studies that present the kinematics of the trunk and the lower-limb when walking in NHS are limited and the kinematic changes that occur when walking in NHS are not coincidence. The earliest study on NHS stated that there were no changes in the gait pattern of subjects walking barefoot, in tennis shoes, or in NHS.[14] Comparison of the flexion extension of the back, hip, knee, and ankle joints of subjects when standing and walking in NHS, positive heel shoes, and on bare feet suggested that the greatest compensation for heel height occurs distally.[15] Another study[16] reported that walking speed was significantly reduced during walking in NHS as a consequence of a shorter stride length combined with an increased cadence. The walking patterns differed drastically at the ankle joint but no significant difference was found at the knee and hip joints.

Kinematics adaptation in different walking conditions is attributed to the changes of muscle activity in the trunk and the lower extremity. However, for NHS, no study about muscle activity in the trunk and the lower extremity has been reported.

In this light, the purpose of the present investigation is to establish a basic understanding of trunk and lower extremity kinematics adaptation when walking in NHS. It further aims to provide a basis for human locomotion adaptation to NHS.

28.2   MATERIAL AND METHODS

28.2.1   SUBJECTS AND SHOES

A total of 13 female subjects at age 23.08 ± 3.9 years, with body weight of 50.18 ± 5.3 kg, and height of 1.63 ± 0.05 m, volunteered to participate in the study. All subjects were in good health condition. None of them had any experience walking with NHS, and they had no previous histories of muscle weakness or neurological disease, or any drug therapy. All of the subjects were fully informed of the purpose and the procedures involved in the study, to which they gave informal consent according to the university ethics committee's guidelines. To ensure uniformity of the testing conditions, all subjects were provided with two types of shoes, one was the normal sports shoes and the other was the NHS. Both types of shoes were similar in construction and materials except for the heel height. The normal sports shoes tilted the sole into about $10°$ of plantar flexion, while the NHS tilted the sole into $10°$ of dorsiflexion. Since heel heights vary according to the sizes of shoes, in this study, only size 37 (European) shoes were selected.

28.2.2   DATA COLLECTION AND ANALYSIS

Each subject participated in two walking trials, wearing the NHS or normal sports shoes, respectively. The order of the shoes was randomly assigned for each of the trials. In each trial, subject walked on a treadmill at a constant speed of 1.33 m/s, a comfort speed for adults.[17] Before the start of the trial, lightweight spherical (diameter = 2 cm) reflective markers were attached to the right side of the subject at anatomic positions at acromion, greater trochanter, lateral epicondyle of the femur, lateral malleolus, calcaneus, and head of the fifth metatarsal to facilitate the later video digitization. The walking movement was recorded starting from the time the subjects felt comfortable or had achieved a steady stance by a 3-CCD video camera (50 Hz) set at a 1/500 s shutter speed. The camera was positioned laterally to the subject with a distance of 5 m and the lens axis perpendicular to the movement plane. The subjects were not aware of when exactly the data were being measured, in order to minimize possible gait modifications. For each trial, 10 consecutive strides recorded at 1, 4, and 6 min were digitized and analyzed on the video motion analysis system (APAS, United States) using a model of a human body, consisting of the six points covered by reflective markers mentioned earlier. The hip joint angle was determined by the acromion, greater trochanter, and lateral epicondyle of the femur; the knee joint angle was determined by the greater trochanter, lateral epicondyle of the femur, and lateral malleolus; and the ankle joint angle was determined by the lateral epicondyle of the femur, lateral malleolus, and head of the fifth metatarsal. Reference angles for the three joints were the same as those described by Lange et al.[1] For the hip, the thigh and the trunk were aligned vertically equal to $0°$ in standing position, extension was negative, and flexion was positive; the knee joint in full extension was defined as $0°$ and flexion was positive; and the ankle joint in the sole of the foot perpendicular to the shank was defined as $0°$, dorsiflexion was positive, and plantarflexion was negative. Additionally, the sagittal orientation of the trunk segment was measured with respect to the horizontal axis at the anterior direction. Higher values indicated extension and lower values indicated flexion. A Butterworth low-pass filter was used to smooth the position-time data for anatomic landmarks. The maximum and minimum trunk, hip, knee, and ankle joint angles, range of motion (ROM), stride cycle time, cadence, and stride length of each stride were calculated. For each subject the kinematics parameters were averaged across the 10 consecutive strides recorded at each time point and further averaged over the three time points in each shod condition. The parameters were finally averaged across all subjects for each shod condition.

28.2.3 STATISTICAL ANALYSIS

Differences in the measured parameters between NHS and normal sports shoes were examined with a paired sample *t*-test. The 0.05 probability level was used as the criterion for all tests to determine the presence or absence of statistically significant results.

28.3 RESULTS

Table 28.1 shows the mean and standard deviation of each variable and the statistical comparison between the two shod conditions during walking on treadmill. Walking in NHS induced a shorter stride cycle time and stride length compared with in normal sports shoes ($P < 0.01$). Because walking speed was the same, the decrease in stride length when walking in NHS resulted in a consequence of faster cadence ($P < 0.01$).

For the trunk posture, significant differences in the maximum flexion and extension angles were found between the two shod conditions ($P < 0.01$). The ROM of the trunk during walking in NHS tended to be slightly increased, but did not show any statistical significance. Wearing NHS had less impact on trunk posture despite producing a slight backward (0.95°) posture.

At hip joint, maximum flexion and extension angles showed significant differences between the two shod conditions ($P < 0.05$ or 0.01). However, no significant differences were found in ROM. At knee joint, a difference of 1.58° in the maximal extension angles between two shod conditions in stance phase was found with highly significant differences ($P < 0.01$). No significant difference was found between the two shod conditions in the angle at touchdown, the maximal flexion angle during stance and the ROM during stride cycle. At the ankle, the angle at touchdown, the maximum dorsiflexion and plantarflexion angles, and the ROM of the joint all showed significant differences ($P < 0.05$) between the two types of shoes. The difference in maximum dosiflexion during stance phase between two walking conditions was 5.47°. In normal sports shoes, a 3.29° of maximum plantar flexion was observed during stance phase, and no dorsiflexion occurred. Figure 28.2 shows the angular excursions of the hip, knee, and ankle joint angles during a complete stride cycle across subjects with time. Changes in ankle joint show most evident in these joints.

TABLE 28.1

Comparison of Kinematic Variables between Normal Sports Shoes and Negative-Heeled Shoes during Treadmill Walking

Parameter	Normal Sports Shoes		Negative-Heeled Shoes		Statistical Comparison	
	Mean	SD	Mean	SD	T	P
Temporal measurement						
Stride cycle time (s)	1.05	0.04	1.02	0.04	5.16	0.000
Cadence (steps/min)	114.40	1.99	116.80	2.02	−5.36	0.000
Stride length (cm)	140	4.90	137	5.00	4.84	0.000
Trunk position						
Max. flexion angle (°)	80.15	3.67	80.82	3.83	−2.90	0.013
Max. extension angle (°)	88.95	3.78	89.90	3.91	−3.46	0.005
ROM during stride cycle (°)	8.81	1.84	9.07	1.81	−0.91	0.380
Hip joint						
Max. flexion angle (°)	21.87	3.27	20.34	3.27	−2.49	0.028
Max. extension angle (°)	−10.82	4.07	−11.98	3.75	−2.97	0.012
ROM during stride cycle (°)	32.69	5.17	32.32	4.22	0.70	0.495
Knee joint						
Angle at touch-down (°)	7.53	3.39	7.92	3.45	−0.702	0.496
Max. flexion angle during stance (°)	18.66	4.64	19.13	3.34	−0.698	0.498
Max. extension angle during stance (°)	0	0	1.58	1.34	−4.25	0.001
ROM during stride cycle (°)	66.03	2.83	65.28	2.96	0.969	0.352
Ankle joint						
Angle at touch-down (°)	7.89	2.31	13.86	2.46	−18.99	0.000
Max. PF angle during stance (°)	−3.29	2.40	4.85	2.28	−25.19	0.000
Max. DF angle during stance (°)	10.39	2.74	15.86	2.58	−18.71	0.000
ROM during stride cycle (°)	28.01	3.19	28.70	2.98	−2.52	0.027

Max., maximum; PF, plantar flexion; DF, dorsiflexion; ROM, range of motion.

28.4   DISCUSSION

According to the report of Whittle,[17] the normal range of gait parameters for females aged 18–49 years are 0.94–1.66 m/s for speed, 1.06–1.58 m for stride length, and 98–138 steps/min for cadence. These parameters observed in normal sports shoes condition in the present study were 1.33 m/s, 1.40 m, and 114 steps/min, respectively, which were within the ranges provided by Whittle. Furthermore, according to Winter's report,[18] during a completed stride cycle when walking with natural cadence, the ROMs of the hip, knee, and ankle joint were 32.79°, 64.86°, and 29.39°, respectively. The ROMs of walking in NHS observed in the present study were

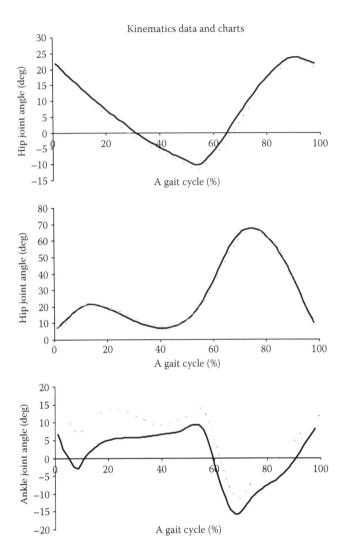

FIGURE 28.2 Lower extremity kinematic data for all individuals' ensemble average during a gait cycle. The horizontal timescale is normalized to a gait cycle from first touchdown to next touchdown of one side foot. Positive angles indicate ankle dorsiflexion, knee flexion, and hip flexion, respectively, and vice versa.

32.32°, 65.28°, and 28.70° for hip, knee, and ankle, respectively, which were closer to the ranges observed by Winter in subjects walking in normal shoes. The data in the present study regarding walking in NHS should represent the normal gait pattern of the female adult population.

Walking speed in the present study was controlled at a constant 1.33 m/s in both walking modes. Compared to normal sports shoes, wearing NHS showed a significantly shorter stride cycle time, associated with faster cadence and shorter stride length. Furthermore, the results obtained were not consistent with the results of the study conducted by De Lateur et al.[15] wherein they compared walking in NHS, barefoot, normal sports shoes, and high-heeled shoes in self-selected walking speeds and found no significant difference in cadence, stride length, and walking speed. The possible reason for this was the different walking speed control between the two studies. Walking speed is determined by stride length and cadence, and all three variables are dependent on each other. When wearing shoes with different heights of heels in free walking, speed, cadence, and stride length could adapt simultaneously, inducing no significant change in each of the three variables. In studying uphill walking, Leroux et al.[6] found that as the treadmill slope increased, the stride length progressively increased and, consequently, the stride cycle duration increased. These authors found that the increase in hip flexion was important, in direct association with the increase in stride length. The increase in hip flexion was an adaptation to the change in treadmill inclination and a way to generate a greater stride length. Different from the study by Leroux et al.,[6,10] our results showed a decrease in hip flexion, which may have been due to the increase in dorsiflexion of the ankle joint and the difference between the tasks of level and uphill walking, which may have consequently resulted in a decrease in stride length and a shorter stride cycle duration. In addition, the maximum extension angle of the knee joint at takeoff when walking in NHS was significantly smaller than when walking in normal sports shoes (Table 28.1), which possibly contributed to the shortening of stride length.[19] In order to keep constant speed, subjects had to increase their cadence to compensate for the decrease in stride length.

While walking in NHS and normal sports shoes, the ankle joint angle at initial contact was at 13.86° and 7.89°, respectively, which implied that the ankle was maintained in dorsiflexion. While walking in normal sports shoes, the initial contact was followed by plantarflexion (−3.29°), by placing the forefoot completely on the ground during the earlier stance phase. The ankle joint changed again to dorsiflexion (10.39°) from mid-stance through takeoff. While walking in NHS, after initial contact, the ankle joint was held in less dorsiflexion (4.85°), and then changed gradually to the greatest dorsiflexion (15.86°) at takeoff. The ROM of the ankle joint was significantly greater in NHS than in normal sports shoes. This might have been caused by the structure of the NHS, where the heel was lower than the toe. When walking in NHS, at initial contact, the ankle could not reach plantarflexion, as it would have been restricted by the lower heel. Consequently, the maximum ankle dorsiflexion angle was significantly increased as compared with walking in normal sports shoes. In contrast, previous research on high-heeled shoes has shown that the ankle joint is maintained in plantarflexion during the whole stance phase.[2,15,20,21]

One of the original concepts of designing NHS was to imitate uphill walking. In order to ascertain whether this aspiration was realized, it is necessary to analyze postural control both in uphill walking and walking in NHS. Leroux et al.[6,10] and Lange et al.[1] reported that increasing the treadmill gradient induced an increasingly flexed posture of the hip, knee, and ankle at initial contact, as well as a progressive forward tilt of the pelvis and the trunk. During uphill walking, after initial contact, the fore part of the foot was placed on the surface of the slope. Similar to walking in NHS, the foot could not retain the plantarflexion position due to the slope of the shoes. Therefore, the ankle joint maintained dorsiflexion for the whole stance and swing phases. In this regard, it could be said that greater ankle joint dorsiflexion would lengthen muscles such as the gastrocnemius and soleus and increase the duration of contract,[6,10] facilitating power generation during the propulsion phase of walking. The findings of the EMG analysis had also confirmed the longer muscle working duration in a gait cycle. Yamamoto and coworkers found that walking exercise in NHS induced an increase in calf blood flow at a moderate speed.[22] However, as the ankle became more dorsiflexed, the length of the moment arm of the Achilles tendon increased, thus promoting the ability of this tendon to produce active propulsion. The opposite result has been determined walking in high-heeled shoes.[2]

After an injury or surgical operation in the gastrocnemius, soleus, or Achilles tendon, the common rehabilitation exercise protocol involves dorsiflexion stretching with a towel or strap, standing on an inclined plane, performing an initial toe raise on a box or step, lifting the body to maximum ankle plantarflexion then lowering it to ankle dorsiflexion, and walking on a treadmill with a slight incline.[23] The purpose of these exercises is to increase the ROM of the ankle joint and to increase the strength of the calf muscle and the Achilles tendon. Similarly, during sports, the highest stress on the Achilles tendon occurs during eccentric contraction of the gastrocnemius and soleus complex; for example, pushing off the weight-bearing foot and simultaneously extending the knee, such as in uphill running. Eccentric exercise is recommended for rehabilitation of those who undergo these stresses in their sports.[24] As described earlier, walking in NHS caused dorsiflexion that could increase the ROM of the ankle joint and might improve the strength of the calf muscle and Achilles tendon. Therefore, walking in NHS might be a viable alternative method to exercising or daily walking during the postsurgical rehabilitation of the calf muscle and the Achilles tendon.

For the knee joint, there were no significant differences in angles, either at initial contact or at maximum flexion, as well as in the ROM between the two types of shoes. The only significant difference was exhibited in the maximum extension angle at takeoff. The results for extension angle in the present study were consistent with the suggestion that extension of the knee contributes to an increase in stride length.[19] For example, in this study, the stride length of walking in normal sports shoes was significantly greater ($P < 0.01$) than that of walking in NHS. During uphill walking, the key mechanism is to lift up the swing leg through a simultaneous increase in hip and knee flexion of that limb.[6] Increased flexion of the knee joint facilitated the swing of the foot to reach the next higher level. Moreover, the more flexed knee joint induced modulation in the timing and duration of activity of the extensors, and contributed to lifting up the whole body to the next higher position during the stance phase. In contrast, level walking does not require the leg to be lifted upward to the same extent during the swing phase. The knee joint would achieve a greater extension angle in NHS than in uphill walking before the foot makes contact with the level ground. The difference between the tasks of level and uphill walking might require different mechanisms of posture control at the knee joint.

The trunk contains the greatest mass of the body. Trunk vertical alignment could also act as an egocentric frame of reference in determining upper and lower limb positions.[6] Trunk orientation, therefore, plays an important role in maintaining balance in walking. The maximum hip flexion angle, extension angle, and ROM, and the maximum trunk flexion angle implied that the whole trunk moved posteriorly, causing the center of gravity to shift backward while wearing NHS, as compared with walking in normal sports shoes. This finding tested our hypothesis and agreed with the results of previous studies,[2,25] which reported that the center of gravity was shifted forward when walking in high-heeled shoes. Leaning the trunk forward facilitated movement of the center of gravity outside the area of support and assisted the lower limbs in generating more momentum during takeoff, both in uphill walking[1,6,10] and level walking.[26,27] From this point of view, walking in NHS induced the upper body to tilt backward, which may cause a disadvantage in the propulsion phase as compared with walking in normal shoes.

From our kinematic analysis, when walking in NHS, the ankle maintained dorsiflexion, which benefited active propulsion. However, the increase in ankle dorsiflexion resulted in the center of gravity shifting backward, which caused difficulty in generating active propulsion. These adverse aspects may compensate each other, so that an adapted walking pattern can be sustained. This posture may be helpful in keeping the upper extremity in an upright position, thereby achieving a more graceful posture. In terms of imitating uphill walking, similar ankle dorsiflexion was found in both walking modes. However, due to the different tasks associated with the two walking modes, the postural adaptation in the knee and hip joints as well as trunk orientation showed considerable differences. Moreover, to know whether postural adaptations show similar patterns in walking in NHS for a relatively long period of time, a follow-up study is needed.

The NHS tilt the foot into about 10° of dorsiflexion. The sole was designed to elevate the forefoot and to mimic uphill walking. Other EMG studies regarding muscle activation in uphill treadmill walking and ramp ground walking have found a modifying level of activation of the relevant flexor and extensor muscles in the lower extremity.[1,10–12] The changes of gait kinematics might relate to the abnormal working style of the lower limb muscles. Further study using EMG methods is suggested.

28.5 CONCLUSION

Walking in NHS alters trunk and lower extremities kinematics. These changes represent an adaptation of human gait by which posture and limb stability are maintained, as the ankle joints are forced into a dorsiflexion posture. To compensate for the changed ankle joint posture, the trunk was slightly postured backward, the hip more extended, and the knee joint more flexed. The negative-heeled gait is a normal gait, but the muscle activities of the lower limbs that are responsible for the gait posture warrant future study.

REFERENCES

1. Lange GW, Hintermeister RA, Schlegel T et al. Electromyographic and kinematic analysis of graded treadmill walking and the implications for knee rehabilitation. *J Orthop Sports Phys Ther* **23**: 294, 1996.
2. Stefanyshyn D, Nigg B, Fisher V et al. The influence of high heeled shoes on kinematics, kinetics, and muscle EMG of normal female gait. *J Appl Biomech* **16**: 309, 2000.
3. Li J, Hong Y, Robinson P. The effect of load carriage on movement kinematics and breathing pattern in children during walking. *Eur J Appl Physiol* **90**: 35, 2003.

4. Hong Y, Li J. Influence of load and carrying methods on gait phase and ground reactions in children's stair walking. *Gait Posture* **22**: 63, 2005.

5. Brandell BR. Functional roles of the calf and vastus muscles in locomotion. *Am J Phys Med* **56**: 59, 1977.

6. Leroux A, Fung J, Barbeau H. Postural adaptation to walking on inclined surfaces: I. Normal strategies. *Gait Posture* **15**: 64, 2002.

7. Prentice SD, Hasler EN, Groves JJ et al. Locomotor adaptations for changes in the slope of the walking surface. *Gait Posture* **20**: 255, 2004.

8. Kawamura K, Tokuhiro A, Takechi H. Gait analysis of slope walking: A study on step length, stride width, time factors and deviation in the center of pressure. *Acta Med Okayama* **45**: 179, 1991.

9. Sun J, Walters M, Svensson N et al. The influence of surface slope on human gait characteristics: A study of urban pedestrians walking on an inclined surface. *Ergonomics* **39**: 677, 1996.

10. Leroux A, Fung J, Barbeau H. Adaptation of the walking pattern to uphill walking in normal and spinal-cord injured subjects. *Exp Brain Res* **126**: 359, 1999.

11. Simonsen EB, Dyhre-Poulsen P, Voigt M. Excitability of the soleus H reflex during graded walking in humans. *Acta Physiol Scand* **153**: 21, 1995.

12. Tokuhiro A, Nagashima H, Takechi H. Electromyographic kinesiology of lower extremity muscles during slope walking. *Arch Phys Med Rehabil* **66**: 610, 1985.

13. Opila-Correia KA. Kinematics of high-heeled gait. *Arch Phys Med Rehabil* **71**: 304, 1990.

14. Mann RA, Hagy JL, Schwarzman A. Biomechanics of the Earth shoe. *Orthop Clin North Am* **7**: 999, 1976.

15. De Lateur BJ, Giaconi RM, Questad K et al. Footwear and posture. Compensatory strategies for heel height. *Am J Phys Med Rehabil* **70**: 246, 1991.

16. Benz D, Stacoff A, Balmer E et al. Walking pattern with missing-heel shoes. Paper presented at *11th Conference of the ESB*, Toulouse France, 1998.

17. Whittle M. *Gait Analysis*, Reed Education and Professional Publishing Ltd, Oxford, U.K., pp. 151–152, 2002.

18. Winter D. *The Biomechanics and Motor Control of Human Gait: Normal, Elderly and Pathological*, University of Waterloo Press, Waterloo, Ontario, Canada, 1991.

19. Valmassy R. *Clinical Biomechanics of the Lower Extremities*, Mosby, Inc., St. Louis, MO, 1996.

20. Meerrfield HH. Female gait patterns in shoes with different heel heights. *Ergonomics* **14**: 411, 1971.

21. Snow R, Williams K. High heeled shoes: Their effect on center of mass position, posture, three-dimensional walking, rearfoot motion, and ground reaction forces. *Arch Phys Med Rehabil* **75**: 568, 1994.

22. Yamamoto T, Ohkuwa T, Itoh H et al. Walking at moderate speed with heel-less shoes increases calf blood flow. *Arch Physiol Biochem* **108**: 398, 2000.

23. Mandelbaum B, Gruber J, Zachazewski J. Achilles tendon repair and rehabilitation, in *Rehabilitation for the Postsurgical Orthopedic Patient,* L Maxey, J Magnusson (eds.), pp. P323–P349, Mosby, Inc., St. Louis, MO, 2001.

24. Brotzman S, Brasel J. Foot and ankle rehabilitation, in *Clinical Orthopaedic Rehabilitation*, S Brotzman (ed.), p. 245, Mosby, Inc., St. Louis, MO, 1996.

25. Ebbeling C, Hamill J, Crussemeyer J. Lower extremity mechanics and energy cost of walking in high-heel shoes. *J Orthop Sports Phys Ther* **19**: 190, 1994.

26. Thorstensson A, Nilsson J, Carlson H. Trunk movement in human locomotion. *Acta Physiol Scand* **121**: 9, 1984.

27. Vogt L, Banzer W. Measurement of lumbar spine kinematics in incline treadmill walking. *Gait Posture* **9**: 18, 1999.

29 Athletic Footwear Research by Industry and Academia

Thorsten Sterzing, Wing Kai Lam, and Jason Tak-Man Cheung

CONTENTS

Athletic footwear research and testing is carried out at industrial and academic institutions with mutual interaction steadily increasing. This chapter illustrates the circumstances and interaction opportunities of the two settings from an industrial perspective. It describes how fruitful interaction can be organized by discussing respective structural characteristics. The continuous demand of functional and commercial footwear innovation serves as background for the introduction of the standard product creation process. Additionally, the important preceding role of academic baseline research to establish solid proof of innovative and functional footwear concepts is described. The key company departments involved in the product creation process, planning, design, and development are introduced by illustrating their respective roles during product creation. Thereby, valuable benefits due to routine and specific interaction with science and research are pointed out. The toolbox of research procedures available for comprehensive footwear evaluation and development is addressed. In order to systematically improve the functional criteria of authentic and advanced performance footwear, this toolbox combines mechanical, biomechanical, athletic performance, and perception procedures, which nowadays are strongly supported by computer simulation approaches. The needs of specific consumer target groups with respect to gender, age, or skill level are referred to, as well as the specific individual service for elite athletes. Future requirements to improve footwear science, research, and testing in industry and academia are proposed. A learning control section and references are provided at the end of the chapter.

29.1 BACKGROUND

29.1.1 ATHLETIC FOOTWEAR

Footwear science addresses aspects of all types of shoes, including dress shoes, working shoes, sport shoes, or medical shoes. However, this chapter focuses on the functional aspects of athletic footwear only, neglecting style and fashion aspects that solely refer to outward appearance issues. Functional aspects of athletic footwear are defined as all shoe modifications having potential to influence objective characteristics as well as subjective perception of human locomotion and human movement. Generally, functional shoe modifications aim to improve three superordinated aspects: comfort, performance, and injury prevention. The inspiring potential of shoe design features to influence these aspects was brought to the scientific community starting from the 1970s and 1980s (Nigg, 2010). The movement mainly originated from three research institutes, the Biomechanics Laboratory at Penn State University (United States), the Nike Research Laboratory (United States), and the Biomechanics Laboratory at the Eidgenoessische Technische Hochschule (Switzerland). The main research focus was on running shoes at that time and initial strong links between the athletic footwear industry and academic institutions were established. Scientific research efforts have been carried out increasingly until today and will go on in the future (Cavanagh, 1980, 1990; Frederick, 1984; Nigg, 1986, 2010). Next to comfort, performance, and injury prevention issues, durability aspects need to be taken into account, which addresses the targeted and expected life cycle period of shoes, but which also refers to gradual alterations of functional footwear properties during regular wear (Wang et al., 2010).

When characterizing target consumers, different skill levels need to be considered. As elite athletes may have highly individual demands of shoe function, personalized footwear accounts for their potentially unique anthropometrics, body composition, and sport-specific techniques. In contrast, common athletes are split into subgroups of gender, age, and skill level to be provided with adequate functional footwear. In addition, cultural and ethnical criteria need to be considered. Males and females differ considerably in their general body biometrics, which are responsible for different athletic performance levels and also specific injury patterns. Thus, it is not surprising that gender-specific biomechanical locomotion patterns were identified (Ferber et al., 2003; Landry et al., 2007a,b). Different biological and physical capabilities were also shown to result in altered playing strategies in ball games (Althoff et al., 2010). Foot morphology was observed to be gender specific already relatively early, and knowledge was enhanced by applying more sophisticated measurement technology in late research (Krauss et al., 2008; Robinson and Frederick, 1990). As foot morphology is also specific to cultures (Mauch et al., 2008), the global challenges for manufacturing well-fitting footwear are obvious. In addition, subjective requirements of different gender and age groups have been addressed in a survey questionnaire about running shoes (Schubert et al., 2011). It should be noted that gender-specific observations have already led to knowledge-based design recommendations on how to create female-specific footwear for running and soccer (Krauss, 2006; Sterzing and Althoff, 2010). Similarly, age-related footwear modifications may be thought of, responding to the development and degeneration of the human body and foot during the human life span.

As athletic footwear is designed for numerous, diverse sports, it needs to account for the various respective requirements. Thereby, each sport can be treated as an individual category, which results in a high number of categories. Integrative approaches identify sports and disciplines that share similar locomotion and movement characteristics, resulting in common overlapping demands. These sports are then grouped into broader and comprehensive categories. Optionally, such integrative approaches may focus on footwear construction aspects or on locomotion type aspects. An integrated comprehensive category for cleated footwear would include sports like soccer, football, rugby, baseball, cricket, and golfing. In contrast, an integrative comprehensive category focusing on locomotion types identifies common important movements for a range of sports and disciplines, such as acceleration, sprinting, deceleration, cutting, turning jumping, or landing. Respective core physiological and biomechanical elements can be extracted and analyzed in order to develop general responding concepts. For instance, basic biomechanical research approaches that address the fundamental circumstances during the push-off and take-off phase of human locomotion are highly beneficial for creating authentic performance track and field footwear. Additionally, gained knowledge is ideally suited to be fed into footwear aiming to provide athletes with a performance edge during various field and court sports. It is noteworthy that such category-based company structures were set up in most footwear companies only during the 1990s. Originally, this was done in an effort to enhance product-related work effectiveness and efficiency.

Athletic footwear also needs to respond to changing environments, for example, changing surface types, as illustrated for tennis and soccer in the following. For tennis, the nature of the game and thus respective footwear requirements differ considerably when playing on hard court in comparison to playing on clay or grass courts. Thus, functionally designed tennis footwear should provide surface specific traction properties providing players with their respective acceleration and deceleration demands. Due to successful development of artificial surfaces and subsequent rule changes by the governing global soccer associations, soccer is nowadays played on natural or on artificial turf. Looking back to the natural grass area, different weather-related surface conditions drove the development of hard, firm, and soft ground stud configurations for responding to hard, dry, and wet surface conditions. For the fairly new artificial turf surfaces, it was shown that footwear construction can adequately respond to this new interface situation by providing players with improved agility running performance and reduced lower extremity loading when wearing adequately designed footwear (Müller et al., 2010b; Sterzing et al., 2010).

29.1.2 Functional Footwear Innovation

Functional design considerations of footwear, have addressed comfort, performance and injury prevention issues ever since. Thereby, innovative structures, materials, smart technologies, and manufacturing methods have been used.

Systematically arranged fixed studs improved functional traction properties of baseball shoes already in the end of the nineteenth century. The general stud concept for cleated footwear to provide functional traction then experienced a significant revision in the midst of the twentieth century. The invention of screw-in studs allowed soccer players to use different stud lengths for coping with temporarily changing surface conditions while maintaining their familiar shoe upper.

Initial efforts to put specific cushioning elements in athletic footwear constructions date back to the beginning of the twentieth century. Rubber and blown-rubber full length insole materials were used in the Converse All-Star vulcanized rubber basketball shoes to provide enhanced cushioning to athletes. It is noteworthy that cushioning is one of the most fundamental footwear concepts trying to accommodate for uneven or hard surfaces and thereby addressing highly desired comfort requirements of athletes. For long, cushioning needs of athletes were only addressed by using different foams as midsole materials. Then, over the years, numerous advanced cushioning concepts were introduced and are strongly associated with respective brands, like the Nike-Air system, working with encapsulated gases, or the Asics-Gel technology making use of the viscoelastic properties of suited materials.

Running shoe construction changed considerably when the first dual density midsoles, originating from collaboration between ASICS and the University of Oregon (United States), were integrated in running shoe designs. Those anisotropic sole units brought a simple but highly influential aspect into running footwear manufacturing.

Geometric modification of running footwear was used to change rearfoot motion characteristics during endurance running. Especially, geometrical varus alignment of running shoes was shown to alter running biomechanics (Milani et al., 1995; Perry and Lafortune, 1995; Van Woensel and Cavanagh, 1992). These early studies analyzed extreme degrees of shoe varus alignment and showed considerably decreased rearfoot motion parameters in runners. Recent research approaches elaborated on the general concept by showing small but systematic biomechanical effects on rearfoot motion control in running shoes with only moderate varus alignment, being suited for everyday use (Brauner et al., 2009; Grau and Horstmann, 2007).

Built-in shoe technology, when considered as a feedback-based mechanical system, was introduced to the athletic footwear market with the Adidas A1. This running shoe obtained and analyzed biomechanical cushioning characteristics during actual running. Based on automated data processing and evaluation, the shoe was capable to adjust its functional cushioning properties at the rearfoot by built-in programming and actuator instrumentation. Smart footwear constructions like this consist of three key components, a sensor unit, a data processing unit, and an actuator unit. Thereby, the process of monitoring, evaluation, and alteration of locomotion patterns is constantly executed throughout exercise. Respective smart shoe designs may have potential to lift footwear to another level in the future, although the ultimate functional and commercial success of these ambitious initiatives has not been proven yet.

In contrast to sophisticated built-in technology footwear, the concept of minimization of footwear, as present with the Nike Free or the Vibram Five Finger, has been showing commercial success. These concepts claim to induce a natural running style to athletes due to their highly flexible midsole and outsole materials as well as their geometrical constructions. However, the terminology of natural running appears to be confusing and will need to be clarified for future discussion of related concepts. When putting barefoot-like running similar to natural running, one neglects that nowadays many people are used to continuous wear of footwear from early childhood on, especially during athletic activities. Thus, it is suggested to distinguish between habitual barefoot runners and habitual shod runners as a first step. The topic of minimization of running footwear has provoked an intense and ongoing public and scientific discussion about the general function and task of athletic footwear for human locomotion.

Generally, functional footwear innovation is an ongoing academic topic and an essential business need for companies in order to maintain or to achieve a certain market position. Thus, anticipation and creation of footwear market demands is a core task for all product-related departments within athletic footwear companies. The innovation process includes all efforts of turning an initial idea into a successful product, ultimately creating commercial value. Thereby, commercial value can be achieved not only by well-selling footwear but also through authentic unique products that are suitable to increase the company's brand reputation. During the process of functional footwear innovation, experimental laboratory and field knowledge is transferred into superior and substantial products by usage of interdisciplinary expertise. This expertise may originate not only from the field of sports science, with its branches of biomechanics, exercise physiology, and psychology but also from the field of engineering, including the areas of mechanics, materials, chemicals, and electronics. The workmanship, knowledge, and experience of podiatrists, shoe technicians, and artisans have to be emphasized as an additional important source for functional footwear innovation. Furthermore, innovative concepts derived from sciences like bionics or biomimics as well as from human motor learning theories are also capable of successfully nurturing product innovation.

29.2 INDUSTRIAL SETTING

29.2.1 COMPANY DEPARTMENTS

Athletic footwear companies require intense and consistent interaction of their specific departments to create well-functioning footwear for the domestic and global markets. A general classification may align departments within the executive level and within the product level (Figure 29.1).

At the executive level, *Sales* is responsible for the development of product distribution channels and the communication on product needs. It explores, establishes, and maintains related business opportunities with wholesale and retail customers, as well as with consumers. *Marketing*, in addition to brand and product promotion efforts, ranging from TV commercials to in-store displays, is responsible for setting the long-term brand and product directions by aligning marketing resources,

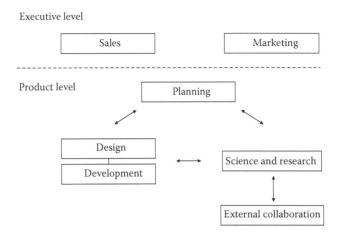

FIGURE 29.1 Company department structure.

competitive advantages, and specific strongholds of the company. Addressing relevant consumer and market necessities, *sales* and *marketing* are responsible to set up the major business directions and to decide on the long-term marketing platform for the company's products. Thereby, an important goal is to establish and preserve an authentic company reputation that ensures positive recognition with respect to product functionality by wholesale and retail customers, as well as by the athletic consumer.

At the product level, all operational steps from initial product planning to final product design and development are carried out in mutual interaction with science and research efforts. Prior to market release, each shoe model ultimately passes through a widely standardized product creation process, the in-line product cycle. The in-line product cycle duration for athletic footwear lasts 18 months, which appears to be a common global standard across athletic footwear companies. The key departments involved in this product creation process are *product planning, product design*, and *product development*, necessarily having well-coordinated interaction all throughout the product creation process. It is highly recommended that *science and research* is involved all throughout the product creation process in order to take over responsibility for all functional aspects of the products by applying objective, valid, and reliable testing procedures. Within the in-line product cycle, *product planning* develops and controls the general product plan, including the product development plan, the product line strategy, the pricing strategy, and also regional product strategies, when applicable. Critically, the product strategy needs to be well aligned with the major business directions and general marketing platform of the company, which is determined at the executive level. Functional considerations of athletic footwear should take a core position within the product strategy as it enhances the potential of footwear products to play substantial roles on the markets, subsequently leading to prolonged and thus commercially more beneficial lifetimes. Therefore, *product planning* should foresee or even try to create future domestic and global market needs by interaction with *product design* and *product development*. Nurturing and supporting these efforts, *science and research* is ideally suited to feed innovative and functional concepts into this process. Thereby, solid scientific proof of such concepts necessarily needs to be established already prior to the initiation of the respective in-line product cycle and should provide the baseline for all subsequent related processes. *Product design* follows up on the set product plan and creates detailed sketches of each footwear product including exact dimensions, color ways, as well as material and manufacturing suggestions. *Product development* then follows up on these sketches by opening respective toolings in order to create prototypes and later on final products in close interaction with the factories. Within a regular in-line product cycle, several prototype rounds are carried out, each allowing for respective testing and subsequent changes to be made.

Continuous interaction following the initial consent about the full product strategy ensures that necessary adjustments can be executed and that the general product plan is viable for all departments involved. *Science and research* takes a specific role prior and within the in-line product cycle as being capable to support the other departments at all stages of the product creation process. Thereby, the genuine scientific knowledge about the functionality of footwear must be already established at the beginning of the in-line product cycle in order to allow handing over the functional technology brief. Located within the company and equipped with the state-of-the-art instrumentation operated by highly educated staff, *science and research* is ideally suited to support numerous related departments of the company by providing functional knowledge and by putting forward innovative functional concepts based on scientific knowledge. Furthermore, it serves as a gateway for valuable external resources due to thoughtfully established scientific collaborations with academic or industrial institutions.

29.2.2 SCIENCE AND RESEARCH INTERACTION

Within the product cycle, the role of the *science and research* department is clearly confined by supporting the interacting departments with educated advice as well as the conduction of product testing. However, additional goals are broader and show closer links to scientific baseline research in human locomotion and human movement. Thus, related timeframes need to be longer than the fixed product cycle schedules of other departments. Short-term (18 months), in-line product cycle related tasks are predominantly linked to the routine commercial business. Main targets are the final monitoring of functional product quality as well as of respective durability issues of products and materials. Midterm tasks (18–36 months) should focus on systematic product creation processes and can well respond to changing footwear needs according to advanced functional performance requirements, to environmental demands, and also to technical requirements put up by governing bodies of sports associations. Thereby, midterm tasks may follow a basic three-phase structure consisting of status quo analysis, knowledge-based prototype testing, and final market comparison (Figure 29.2).

Status quo analyses should be based on comprehensive testing procedures and serve as an intellectual baseline from which knowledge-based prototypes are created which then are subject to further evaluation. Respective findings are then fed into advanced prototypes until the functional quality of the shoes is on the company's

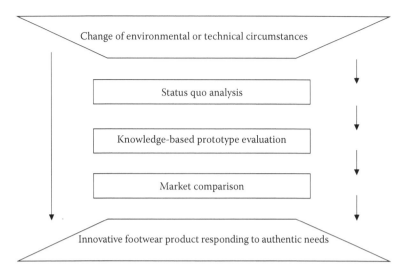

FIGURE 29.2   Midterm innovation structure.

target level. The ultimate goal should always be to establish scientific proof of the functionality of future innovative footwear concepts. Market comparisons are suggested to be carried out at a certain stage to ensure superior function of the respected technology compared to existing products on the market. An illustrative example for a midterm research project is the work on an innovative traction outsole concept for artificial soccer turf shoes (Sterzing et al., 2010). This research effort was carried out in response to the implementation of artificial soccer turf as official playing surface and included the aforementioned three steps, framing a systematic product development effort. The most important benefit of such processes is that solid general footwear manufacturing guidelines can be derived and transferred into a range of shoe models. Furthermore, it is important to understand that additional scientific benefit is gained by providing functional understanding of specific circumstances, which increases the general knowledge of the company.

For the implementation of innovative and advanced authentic performance products, long-term time frames (36 month plus) are recommended. The basic pattern to be followed is to initially use or to generate enhanced understanding of human locomotion and human movement. Based on this, an innovative functional technology idea is needed to be incorporated in athletic footwear design. After initial prototype construction and early evaluation efforts, it is necessary to achieve scientific proof of concept, in a sense of functional validation. Once established, the technology brief can be handed over into the in-line product cycle process (Figure 29.3).

Another important task of *science and research* is to serve as an educative source for company members providing background of general principles of human locomotion and human movement in order to lift up respective company knowledge. Regular educative initiatives in digestible formats provide support and education for staff members from planning, design, and development for an improved understanding of functional athletic footwear needs in general, but also of specific subgroups needs that were addressed earlier. Ideally, planning, design, and development people

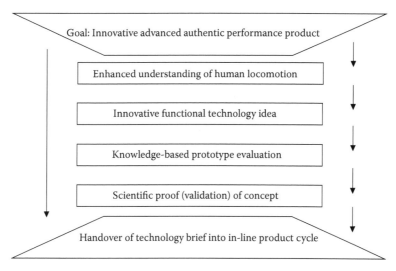

FIGURE 29.3 Long-term innovation structure.

can integrate this knowledge into their daily product-related work. Increased common functional knowledge tremendously facilitates the planning, design, and development efforts during the product creation process and may additionally support sales and marketing of the company's executive level.

29.3 SYSTEMATIC DEVELOPMENT OF ATHLETIC FOOTWEAR

29.3.1 FUNDAMENTAL KNOWLEDGE

To enable comprehensive athletic footwear development based on functional requirements, substantial baseline knowledge about athlete's needs is necessary. Broad game analyses as well as specific locomotion and movement analyses are used to determine objectively the respective sport inherent demands. Thereby, characteristics of single person sports like track and field but also of team sports like various ball games need to be considered. For the latter, frequency and nature of movements need to be obtained, also providing information about their respective importance for game success. Subgroup analyses are recommended to receive more specific information about the effects of different skill level, biometrics, gender, age, playing position, and culture. As locomotion and movement techniques as well as ball game playing patterns may change with time, it is recommended to reconfirm knowledge on a regular basis.

For matching functional athletic footwear design with the subjective needs of athletes, simple interviews and questionnaires are commonly used practices to gather fundamental insight. Thereby, athlete interviews are mainly used when working with elite athletes as they are often provided with individually modified footwear. When general knowledge about athletic footwear demands is required, questionnaires are a useful tool, especially when aiming at gathering substantial information of larger subject groups (Brauner et al., 2012; Schubert et al., 2011). Large-scale surveys allow further analysis of subgroup demands according to gender, age, skill level, or playing position. For the investigation of geographic- and culture-specific requirements, suitable global locations need to be selected for execution of data collection, an ideal opportunity for making use of external collaborations. Well-structured questionnaires provide highly valuable information about numerous footwear aspects like subject characteristics, consumer habits, shoe property importance, desired degree of shoe properties, shoe problems, or foot problems. Such knowledge is an important asset to create a baseline for the planning, design, and development of athletic footwear, ensuring that product creation from the very beginning is aligned with subjective needs of athletes. As athletes' perspective toward footwear requirements may change with time, it is recommended to reconfirm knowledge on a regular basis. In common, game analyses, movement analyses, as well as athlete interviews and questionnaires allow building up a specific footwear feature framework for the different categories; thus, they are crucial tools to start with in the general product creation process.

29.3.2 COMPREHENSIVE FOOTWEAR EVALUATION

Based on earlier works of various researchers, a comprehensive approach incorporating diverse testing procedures is endorsed for the evaluation of athletic footwear. In addition to mechanical, biomechanical, athletic performance, and subjective perception testing procedures, computer simulation testing procedures have to be mentioned (Figure 29.4). In the following the various procedures are characterized and their respective benefits as well as their mutual levels of interaction are described.

Mechanical testing procedures are objective measures to provide information about the material characteristics of athletic footwear. Common mechanical testing procedures for athletic footwear focus on impact, translational and rotational traction, torsion, and flexibility characteristics of footwear. Thereby, materials can be tested as isolated pieces as well as already built-in structures. It is recommended to ultimately test the mechanical function of materials when being built in, as material characteristics may be affected by interfacial and assembly conditions with other components or simply by the manufacturing process. An advantage of mechanical

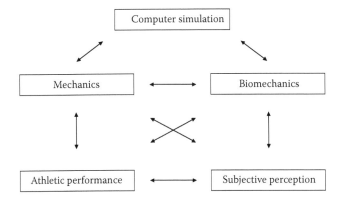

FIGURE 29.4 Comprehensive footwear research and testing structure.

testing is their high reliability and objectivity. A disadvantage is the relatively low validity with respect to actual human locomotion and movement, as commonly used mechanical testing procedures often do not account for the complex loading conditions and do not account for any adaptation processes that are observed during actual human locomotion. In an effort to close this gap, recent research tried to very precisely simulate the biomechanical loading characteristics during foot strike in heel to toe running, when applying mechanical running shoe testing procedures (Heidenfelder et al., 2011). Results indicate the possibility to obtain similar shoe alteration effects for mechanical compared to biomechanical alteration when applying mechanical loading profiles that closely reflect actual biomechanical loading over the suggested life cycle of running shoes. These findings should encourage similar research, potentially leading the path to further valid, subject independent mechanical testing procedures of shoe characteristics in the future. It should be acknowledged that only the recent availability of advanced and also financially affordable mechanical testing instrumentation formed the basis for the progress made.

Biomechanical testing procedures evaluate footwear function by subject involvement. Although field tests are conducted in some occasions, biomechanical footwear evaluation is usually performed in laboratory settings using measurement equipment like force plates, pressure measuring systems, motion analysis systems, muscle electromyography measuring systems, and others (Payton and Bartlett, 2008). In contrast to mechanical procedures, biomechanical testing allows to observe functionally important adaptation mechanisms of subjects due to different footwear or other environmental conditions. Additionally, variability of human locomotion and movement between and within subjects can be analyzed by biomechanical research procedures. For long, variability of locomotion and movement was regarded as nonbeneficial when evaluating athletic footwear function and study protocols were set up in highly standardized manners. Currently, the perspective on human variability in locomotion and movement is shifting. Variability is now regarded to be rather functional in many athletic circumstances and potentially beneficial by itself, thus becoming an increasingly popular research topic with high relevance for the design of athletic footwear. Naturally, these issues have to be addressed by use of biomechanical research procedures.

For athletic performance testing, straightforward measurement protocols evaluating ultimate performance variables are suggested. As athletic performance is a main goal when constructing athletic footwear, the ultimate success margin for the athlete should be objectively proven. Objective benefit of a certain shoe model can be displayed by measuring time, speed, or oxygen consumption during sprinting, agility running, or endurance running (Frederick, 1984; Nigg, 2001; Stefanyshyn and Fusco, 2004; Sterzing et al., 2009). Ball kicking velocity and accuracy in soccer may serve as additional examples for the possibility to quantify athletic performance benefits, obtained by use of adequate footwear (Hennig and Sterzing, 2010; Sterzing and Hennig, 2008). Thereby, the general goal for athletic performance testing with respect to footwear is the objective verification of a performance margin one shoe model might have over another.

Subjective perception testing asks for subjects' individual opinion toward a specific shoe model or single shoe features. It is critical to distinguish two major approaches of subjective testing, the perceived intensity and the preference of these features (NSRL, 2003). The former asks for the plain degree of a certain shoe feature, for example, hard/soft for cushioning or high/low for traction, while not specifically aiming to get a judgment about respective suitability. The latter asks for information which shoe condition is actually perceived to be more suitable and thus preferred. Generally, all shoe properties can be included in subjective perception testing. However, not all shoe properties can be detected by subjects due to neurophysiologic or other constraints, as shown for rearfoot motion (Brauner et al., 2009). In contrast to all other testing methods mentioned, during subjective perception testing subjects themselves are the measurement instrumentation. The human sensory system receives and transmits objective physiological stimuli that are transferred into subjectively interpreted opinions when entering the cerebral level of the human nervous system. Thus, subjective perception testing is highly dependent on subjects' past experiences as shown for running shoe fit testing (Kouchi et al., 2005).

Computer simulation has become an increasingly important research tool in footwear science with continuous advancement driven by numerical techniques and computer technology. The finite element method, as a computational approach, was introduced in the 1970s to analyze characteristics and behavior of plain mechanical structures. The ability to provide realistic simulation even when considering highly sophisticated structural and material as well as specific loading and boundary conditions made the finite element method a versatile tool for biomechanics and footwear applications (Cheung et al., 2009). Computer-aided engineering (CAE) techniques allow rapid change of input parameters to analyze their subsequent effects in a virtual simulation environment without the need of conducting actual experiments. For instance, the effect of contour and material of foot orthoses on plantar peak pressures can be assessed and optimized to a large extent by using finite element simulations without time and cost intensive experiments that coincide with related construction requirements of orthotic prototypes (Cheung and Zhang, 2008). Moreover, finite element analysis provides additional insight into the distribution of internal loads and deformation of biological and mechanical structures or systems being modeled (Cheung et al., 2009). Obtaining such knowledge was formerly usually associated with material destruction or at least complicated invasive experimental methods, sometimes even impossible. Currently, footwear companies increasingly apply CAE approaches for evaluation and design of footwear or its functional components. In fact, computer-aided design (CAD) approaches, only focusing on the appearance and dimension as well as related manufacturing of footwear will no longer be sufficient to fulfill the needs of modern footwear research. Practically, CAE approaches effectively cut down the time and cost of footwear research and development procedures by reducing the frequency of retooling and shoe fabrication for prototyping and mechanical testing, thus ultimately the cost and time of product creation processes.

Comprehensive application of the introduced testing procedures is suggested as respective single findings are not necessarily well aligned. For instance, the sole application of mechanical testing procedures turned out to be insufficient for identification of optimal outsole traction configurations for soccer players. It was shown that an increase of mechanical available traction above a certain level does not result in enhanced biomechanical utilization of traction and enhanced athletic performance

(Luo and Stefanyshyn, 2011; Müller et al., 2010a). The rationale derived from these studies is that optimal levels instead of maximum levels of mechanical traction are beneficial for the athlete, as otherwise athletes adapt their locomotion patterns, probably due to injury prevention considerations. For the dependency of athletic performance and respective subjective perception, controversial findings can be referenced as well. It was shown that subjective perception of athletes can (Sterzing et al., 2009), but does not need to (Sterzing and Hennig, 2008; Sterzing et al., 2011), reflect objective athletic performance measurements.

In common, the introduced comprehensive footwear evaluation and testing approach allows strategic and systematic development of functional and innovative footwear. Mechanical testing provides objective and reliable information about functional and structural material characteristics. Biomechanical procedures deliver objective information about shoe functioning under actual influence of the athlete by consideration of movement adaptation mechanisms and responses of the biological human system. Computer simulation serves as an adjunct to the mechanical and biomechanical testing approaches by identifying suitable engineering designs of different footwear constructions and materials. Athletic performance testing examines the ultimate performance output of athletes related to different footwear conditions. Subjective perception testing shows how footwear performance is perceived by subject groups or individual athletes, which plays a crucial psychological role when trying to achieve maximum performance in competitive sport activities.

29.4 EXTERNAL COLLABORATION

29.4.1 INTERACTION TYPES

A footwear company has various opportunities to interact with external research institutions for enhancement of its scientific research level and to foster functional product innovations. Thereby, external collaborations comprise not only academic institutions but also other private or government bodies and are naturally coordinated by the *science and research* department. Among others, a basic interaction aspect is the mutual transfer of knowledge between the two research partners (Figure 29.5).

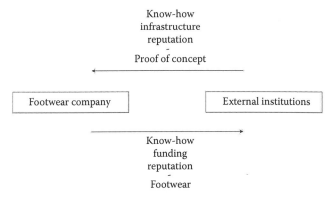

FIGURE 29.5 Interaction of footwear companies and external institutions.

There are various ways for interaction with external collaborators. The company can approach collaborators asking them to carry out specific research tasks according to a set and internally established research protocol. This approach would mean to use human as well as infrastructural resources of an external setting. Another approach is to assign a general and open research topic to external collaborators and ask them to develop a respective research design, which should be subject to mutual finalization prior to be carried out. Here, the external partner is required to provide the research design, results, related interpretation, and future implications for athletic footwear. In addition, external institutions or researchers themselves may approach footwear companies by introducing innovative conceptual ideas and products. In this case, the company needs to evaluate the general quality level and to explore potential research and business benefits of the external collaboration request. On the one hand, the company may benefit from such immediately usable or more readily transferable knowledge with respect to product development. On the other hand, the company is put in a rather passive position, regarding decisions on business opportunities and credibility of research concepts or products without authentic ownership of the research ideas and knowledge. Whereas the first interaction pattern resembles rather an outsourcing of infrastructural and human capacities, the latter two are more useful to bring external technical know-how into the company. However, this know-how needs to be carefully aligned with the company's business, research, and product directions. In order to make efficient use of external knowledge, the company has to ensure that the knowledge provided is fully understood by company's staff members. It is highly recommended to foster continuous mutual interaction between the company and its external research partners. Thereby, ongoing knowledge sharing supports the buildup of scientific knowledge and fundamental understanding within the company, which is of long-term benefit.

29.4.2 GENERAL CONSIDERATIONS

There are also more general opportunities to interact with the public academic domain, such as participation in scientific conferences or open product design competitions. Participation as sponsor in these events may be characterized as taking a rather passive and observational position. In order to enrich the interaction with the academia, it is suggested to take more active roles by sending delegates to suitable conferences and presenting parts of the company's nonconfidential research work, which is beneficial to the scientific reputation of the company and its products. Another frequently used possibility for external interaction is the appointment of internships, master or doctoral theses, as well as postdoctoral positions to suited students or scientists from academic institutions. In case these types of interaction are accurately prepared and executed in a responsible manner, highly valuable results can originate, being of mutual benefit for the company, the respective academic institutions, and the involved individuals.

When collaborating with external academic institutions confidentiality, publication, and intellectual property become an issue. An important goal of universities is to publish scientific knowledge during conferences as well as in scientific journals. However, this may be contradictory to the company's internal goals and policies.

From a company's perspective, publication of knowledge is twofold. A company may want to use publications in order to emphasize the existence of a strong scientific background within the company. Another goal is to show that athletic footwear is truly functional by providing proof of general concepts that are implemented in their footwear. When agreeing to publication, it has to be kept in mind that published knowledge is no longer patentable. Thus, companies will have to file patents of their product designs or concepts prior to publishing related findings either by them or by external research partners. Therefore, a publication delay of scientific findings is usually enforced in order to preserve intellectual knowledge as long as necessary with respect to commercial and business requirements. To some extent, a company may want to use publications in order to show superiority of their products in comparison to their competitors. However, this approach is recommended to be only used with caution, for the following reasons. Such type of knowledge published is commonly regarded to be biased by nature, as a company would only be willing to go public with positive results, dropping those findings showing that competitors' shoes have better function. Therefore, general credibility of these publications appears to be rather low.

For collaborative projects, especially for those involving deliverables with a transfer of technology or concept, which has the potential for commercialization, a negotiation process regarding the share of intellectual property between the company and the external institution needs to be carried out. The chance and result for a successful agreement usually vary from case to case, depending on the nature of the collaborative work and funding support defined by the company as well as individual regulations of academic institutions. Therefore, mutual agreement on intellectual property and related share of benefits should be reached prior to commencement of any collaborative project.

Generally, major goals of companies seeking external collaborations are to make use of the external infrastructure, intellectual input, and reputation of academic research institutions. In order to identify an appropriate academic research institution for such collaboration, it is recommended to have thorough understanding of the resources and strengths of the academic institution. A review of the academic track records and latest published work of the research team as well as on-site visits prior to engagement in research collaboration is of utmost importance for the company to find out the respective dedication of the institution for conducting industrially initiated and mutually beneficial footwear research.

29.5 FUTURE CHALLENGES

Industrial and academic footwear research institutions are challenged by current literature contributions controversially discussing the general concepts and usage of athletic footwear (Lieberman et al., 2010; Nigg, 2001; Richards et al., 2009). It is acknowledged that contradicting findings of footwear research have initiated broad discussions about the value of footwear science and research. An illustrative paradigm is the research area of rearfoot motion. High pronation variable value are associated with an instable foot strike pattern in running, which was linked to overuse injuries in endurance running even though there is still lack of scientific evidence to confirm this theoretical construct. Nevertheless, the assumption that instable rearfoot motion leads to overuse injuries in runners has led to tremendous efforts to construct more stable running shoes accompanied by marketing campaigns educating the running population about the importance of rearfoot stability in endurance running. However, related injury rates were not observed to decrease. It became obvious that prospective scientific research on the epidemiology of running injuries is missing and that current opinion is based on retrospective data only. Thus, injury occurrence and prevention aspects linked to athletic footwear need to be investigated by large scale prospective studies with injury numbers and severity as ultimate outcome variables to be analyzed (Hein et al., 2011; Nigg, 2001). As footwear companies in their daily, business-driven routines often lack respective time needed for such large-scale research efforts, academic institutions should take over and lift related knowledge on a more substantial level.

Another important aspect of future footwear research is the ongoing need for reviewing the current research methods (Oriwol et al., 2009), accompanied by the development of advanced research methods including the critical review of measurement environment and measurement awareness. It is likely that advanced data processing and statistical methods are able to improve the substance of scientific research results.

QUESTIONS

29.1 Explain what is meant by "functional aspects of athletic footwear."
29.2 State five significant athletic footwear concepts brought to the market until today.
29.3 State the four company segments primarily involved in the product creation process.
29.4 State the three phases of a midterm innovation structure, responding to changing environmental or technical circumstances.
29.5 State the four phases of a long-term innovation structure, aiming at the creation of innovative advanced authentic performance products.
29.6 State six footwear-related subgroup criteria for categorization of athletes.
29.7 State six sections of questionnaires when collecting baseline data of athletic footwear requirements.
29.8 State the five research approaches for systematic and comprehensive development of athletic footwear.
29.9 State the two major approaches of subjective perception testing.
29.10 State three benefits of companies due to external collaborations.

REFERENCES

Althoff, K., Kroiher, J., Hennig, E. M. (2010) A soccer game analysis of two world cups: Playing behavior between elite female and male soccer players, *Footwear Science*, 2(1): 51–56.

Brauner, T., Sterzing, T., Gras, N., Milani, T. L. (2009) Small changes in the varus alignment of running shoes allow gradual pronation control, *Footwear Science*, 1(2): 103–110.

Brauner, T., Zwinzscher, M., Sterzing, T. (2012) Basketball footwear requirements are dependent on playing position, *Footwear Science*, 4(3): in press.

Cavanagh, P. (1980) *The Running Shoe Book*, Anderson World, Inc., Mountain View, CA.

Cavanagh, P. (ed.) (1990) *The Biomechanics of Distance Running, Human Kinetics Books*, Champaign, IL.

Cheung, J. T.-M., Yu, J., Wong, D. W.-C., Zhang, M. (2009) Current methods in computer-aided engineering for footwear design, *Footwear Science*, 1(1): 31–46.

Cheung, J. T.-M., Zhang, M. (2008) Parametric design of pressure-relieving foot orthosis using statistics-based finite element method, *Medical Engineering and Physics*, 30: 269–277.

Ferber, R., McClay Davis, I., Williams, D.S. (2003) Gender differences in lower extremity mechanics during running, *Clinical Biomechanics*, 18: 50–357.

Frederick, E. (1984) Physiological and ergonomics factors in running shoe design, *Applied Ergonomics*, 15(4): 281–287.

Grau, S., Horstmann, T. (2007) Entwicklung eines Stabilitaetslaufschuhs zur Praevention von Achillessehnenbeschwerden—Nike Air Cesium, *Sport-Orthopaedie—Sport-Traumatologie*, 23(3): 179–184.

Heidenfelder, J., Sterzing, T., Milani, T. L. (2011) Running shoe properties during age: a comparison of two different approaches, *Footwear Science*, 3(Suppl 1): S70.

Hein, T., Janssen, P., Wagner-Fritz, U., Grau, S. (2011) The influence of footwear and ankle kinematics on the development of overuse injuries. A prospective study, *Footwear Science*, 3(Suppl 1): S72.

Hennig, E. M., Sterzing, T. (2010) The influence of soccer shoe design on playing performance—A series of biomechanical studies, *Footwear Science*, 2(1): 3–11.

Kouchi, M., Mochimaru, M., Nogawa, H., Ujihashi, S. (2005) Morphological fit of running shoes: Perception and physical measurement, *Proceedings of the 7th Footwear Biomechanics Symposium*, Cleveland, OH, pp. 38–39.

Krauss, I. (2006) Frauenspezifische Laufschuhkonzeption, Eine Betrachtung aus klinischer, biomechanischer und anthropometrischer Sicht, *Dissertation, Fakultät für Sozial- und Verhaltenswissenschaften*, Eberhard-Karls-Universität Tübingen, Tübingen, Deutschland.

Krauss, I., Grau, S., Mauch, M., Maiwald, C., Horstmann, T. (2008) Sex-related differences in foot shape, *Ergonomics*, 51(11): 1693–1709.

Landry, S. C., McKean, K. A., Hubley-Kozey, C. L., Stanish, W. D., Deluzio, K. J. (2007a) Neuromuscular and lower limb biomechanical differences exist between male and female elite adolescent soccer players during an unanticipated side-cut maneuver, *American Journal of Sports Medicine*, 35(11): 1888–1900.

Landry, S. C., McKean, K. A., Hubley-Kozey, C. L., Stanish, W. D., Deluzio, K. J. (2007b) Neuromuscular and lower limb biomechanical differences exist between male and female elite adolescent soccer players during an unanticipated run and crosscut maneuver, *American Journal of Sports Medicine*, 35(11): 1901–1911.

Lieberman, D. E., Venkadesan, M., Werbel, W. A., Daoud, A. I., D'Andrea, S., Davis, I. S., Mang'Eni, R. O., Pitsiladis, Y. (2010) Foot strike patterns and collision forces in habitually barefoot versus shod runners, *Nature*, 463: 531–536.

Luo, G., Stefanyshyn, D. (2011) Identification of critical traction values for maximum athletic performance, *Footwear Science*, 3(3): 127–138.

Mauch, M., Mickle, K., Munro, B. J., Dowling, A., Grau, S., Steele, J. R. (2008) Do the feet of German and Australian children differ in structure? Implications for children's shoe design, *Ergonomics*, 51(4): 527–539.

Milani, T. L., Schnabel, G., Hennig, E. M. (1995) Rearfoot motion and pressure distribution patterns during running in shoes with varus and valgus wedges, *Journal of Applied Biomechanics*, 11: 177–187.

Müller, C., Sterzing, T., Lake, M., Milani, T. L. (2010a) Different stud configurations cause movement adaptations during a soccer turning movement, *Footwear Science*, 2(1): 21–28.

Müller, C., Sterzing, T., Lange, J., Milani, T. L. (2010b) Comprehensive evaluation of player-surface interaction on artificial soccer turf, *Sports Biomechanics*, 9(3): 193–205.

Nigg, B. M. (Editor) (1986) *Biomechanics of Running Shoes*, Human Kinetics Books, Champaign, IL.

Nigg, B. M. (2001) The role of impact forces and foot pronation: A new paradigm, *Clinical Journal of Sports Medicine*, 11: 2–9.

Nigg, B. M. (2010) *Biomechanics of Sport Shoes*, University of Calgary, Calgary, Alberta, Canada.

NSRL—Nike Sports Research Lab (2003) Product testing and sensory evaluation, *Nike Sports Research Review*, 2, Beaverton, OR.

Oriwol, D., Sterzing, T., Maiwald, C., Brauner, T., Heidenfelder, J., Milani, T. L. (2009) Pronation velocity values of running shoes are dependent on the mathematical routines applied during data post processing, 22. *Congress International Society of Biomechanics*, Cape Town, South Africa.

Payton, C. J., Bartlett, R. M. (2008) Biomechanical evaluation of movement in sport and exercise, Taylor & Francis Group, New York.

Perry, S. D., Lafortune, M. A. (1995) Influences of inversion/eversion of the foot upon impact loading during locomotion, *Clinical Biomechanics*, 10(5): 253–257.

Richards, C. E., Magin, P. J., Callister, R. (2009) Is your prescription of distance running shoes evidence-based? *British Journal of Sports Medicine*, 43: 159–162.

Robinson, J. R., Frederick, E. C. (1990) Human sexual dimorphism of feet, *American Journal of Human Biology*, 2(2): 195–196.

Schubert, C., Oriwol, D., Sterzing, T. (2011) Gender and age related requirements of running shoes: A questionnaire on 4501 runners, *Footwear Science*, 3(Suppl 1): S148.

Stefanyshyn, D., Fusco, C. (2004) Increased shoe bending stiffness increases sprint performance, *Sports Biomechanics*, 3(1): 55–66.

Sterzing, T., Althoff, K. (2010) Begründung eines Frauenfußballschuhs, *Orthopädieschuhtechnik*, (6): 22–27.

Sterzing, T., Hennig, E. M. (2008) The influence of soccer shoes on kicking velocity in full instep soccer kicks, *Exercise and Sport Sciences Reviews*, 36(2): 91–97.

Sterzing, T., Müller, C., Hennig, E. M., Milani, T. L. (2009) Actual and perceived running performance in soccer shoes: A series of eight studies, *Footwear Science*, 1(1): 5–17.

Sterzing, T., Müller, C., Milani, T. L. (2010) Traction on artificial turf: Development of a soccer shoe outsole, *Footwear Science*, 2(1): 37–49.

Sterzing, T., Müller, C., Wächtler, T., Milani, T. L. (2011) Shoe influence on actual and perceived ball handling performance in soccer, *Footwear Science*, 3(2): 97–105.

Van Woensel, W., Cavanagh, P. R. (1992) Perturbation study of lower extremity motion during running, *International Journal of Sport Biomechanics*, 8: 30–47.

Wang, L., Xian Li, J., Hong, Y., He Zhou, J. (2010) Changes in heel cushioning characteristics of running shoes with running mileage, *Footwear Science*, 2(3): 141–147.

Part VII

Customization

30 Mass Customization and Footwear

Chenjie Wang and Mitchell M. Tseng

CONTENTS

30.1 INTRODUCTION—MASS CUSTOMIZATION AND FOOTWEAR

In this new millennium, two critical driving factors are identified that may dominate the footwear business and initialize a paradigm shift (Boër and Dulio 2007). First, there is an increasing competitive pressure from low labor cost producers, especially from China and the Far East. Those competitors are capable of delivering products of a similar quality, but at a better price. In order to reduce the production cost, major shoe companies are forced to split the production processes, outsource several production steps to the areas with lower labor costs, and organize global production and supply chain network. The low-cost competition also results in a market push toward further diversification and demands for increasingly higher-quality products. Second, there is a trend that the footwear industry is increasingly integrated into the fashion industry, driven by a few multiproduct oligopolies. Consumers now expect shoes to not only offer comfort but also reflect their personal identities. Thus, higher fashion consciousness and personalized shoes have been identified in the footwear consumer market (Piller and Muller 2004). Facing two driving factors of paradigm shift, all shoe providers should develop new business strategies and look for innovative business models in order to maintain a competitive advantage. Shoe producers should spend more effort on product quality improvement, innovation in design and materials, quick response to dynamic fashion trends, and more importantly, the individual consumer's personalized needs and relative supporting service rather than simply offering goods.

Mass customization regards heterogeneous demand among different customers not as a threat, but as a new opportunity for profits. A working definition of mass customization is adopted as "the technologies and systems to deliver goods and services that meet individual customers' needs with near mass production efficiency" (Tseng and Jiao 2001). This definition implies two major goals of mass customization: achieving products of high quality which was defined as satisfying consumers' individual needs and in the meantime keeping the cost as low as close to mass production. In other words, mass customization recognizes each customer as an individual and provides attractive "tailor-made" features which could only have been offered in the pre-industrial craft era, while customers can now afford this kind of product due to the low production cost from modern mass production efficiency and high flexibility. The elicitation of an individual's consumer needs relies on active interactions with the consumer, and the specific information about the consumer's desires is translated into a concrete product or service specification (Zipkin 2001). With an interaction and configuration toolkit, customers are able to express their product requirements and carry out the product configuration process by mapping the requirements into the physical domain of the product, so that customers are considered as "codesigners" (Tseng and Du 1998; Von Hippel 1998). On the other hand, the competitive advantage of mass customization is its capability to combine the efficiency of mass production with the diverse varieties derived from individual consumers' needs and desires, so that the production of mass customized products can still achieve the scale of economy. However, current practices of mass customization in many industries have shown that customers are usually willing to pay a price premium for customization, as the satisfaction of individual needs increase the perceived value of the product (Tseng and Piller 2003; Franke and Schreier 2010; Tseng et al. 2010). Thus, the adopter of mass customization may be able to enjoy a higher profit margin. Therefore, because of the main features in providing customers with more personalized and more profitable products without increasing production costs, mass customization strategy is a good fit to the paradigm shift of the footwear industry.

Mass customization can be a win–win strategy to both the footwear producer and the consumer. With mass customization, customers can have a much larger product range from which to select, and it is easier for them to have shoes which can reflect their individuality. In order to meet the goals, from the producer's perspective, mass customization requires the company to be customer centric, to be able to respond quickly to market, and efficiently manage production and supply chain.

30.1.1 Customer Centric

Mass customization integrates the consumer into the company's value-creation activities. In a traditional footwear business, after a pair of shoes is sold to a consumer, the product life is over from the perspective of the footwear producer, as no further value is generated and returns to the company during the product in use. However, in this new strategy, the consumer takes part in many activities which used to be the domain of the company, such as concept design and product development, which become the resource for future design. The customer's requirements and feedbacks are the company's initiation points of the activities in the value chain, and a long-term relationship of mutual trust is developed between the customer and the company. In many practices, customers are even considered as the "codesigner" more than simply "consumers." The customer can directly interact with the shoe producer to express the requirements or even directly design the shoes themselves (Tseng and Piller 2003). Meanwhile, the company is also able to learn more about the customer's personal preference and serve her needs more accurately than competitors. Thus, the customer-centric company can not only improve customer satisfaction on the product with better fitting to an individual's needs, but also create customers' loyalty so that the customer is more willing to come back in the future to maintain the mutually beneficial relationship with the shoe producer.

30.1.2 Quick Response to Market

With direct communication with the customer and immediate feedback, the shoe providers can respond to the dynamic market more quickly with a mass customization strategy. The knowledge collected from the consumer's codesign can better assist the company to assess and predict the customer's rapidly changing needs, which are the key for next season's new products. In addition, with the customer's online account which is established during online customization, the providers can better employ one-to-one marketing, promoting, and collecting feedback from individual consumers. Thus, analytical knowledge management combined with quick feedback collection can enable the shoe providers to become more reactive to the changes in the shoe market (Boër and Dulio 2007).

30.1.3 Inventory and Supply Chain Management Efficiency

As mass customization is a make-to-order process, the shoes are only made when the purchase order is placed. Thus, the shift from "made-to-stock" to "made-to-order" can significantly improve the production and supply scheduling and reduce the inventory cost and the risks of investment in materials and product development that will not encounter the preference of consumers. Furthermore, mass customization can be conducted online in some cases, so that many providers even sell shoes completely online without any physical stores, such as Zappos. This strategy can further reduce the operation and rental costs by online direct channel sales and increase the profit margin rate (Tseng et al. 2003). Even for those providers still with physical stores, they no longer have to show a full inventory of selection in all stores.

30.2   DIMENSIONS IN CUSTOMIZED FOOTWEAR

Setting the right extent of mass customization is of importance in footwear customization. According to current prevailing practices, mass customization in the footwear business can be divided into three different dimensions—style customization, comfort customization, and function customization (Boër and Dulio 2007).

30.2.1   STYLE CUSTOMIZATION

Style customization in shoes aims at satisfying customers' diverse demands on shoe aesthetics and self-identity reflected by the shoes. Shoe styles have been strongly considered as expressing the wearer's personality and having the capability of transforming them into more handsome, beautiful and confident people. Even for adolescents and young adults, shoes are usually a key signifier of their identities, individualism, ethnicity and personality, and it also happens that their desired shoe styles conflict with what their parents regard as appropriate (Belk 2003). Current trends include antiquated, vintage, and contemporary designs with burnished or tumbled leathers. Causal and formal styles are usually combined and woven leathers are abundant. In addition, graffiti motifs and checkered patterns are increasingly popular in sneakers, and more super-lightweight, air-breathable, waterproof, and abrasion-resistant materials are used in sports shoes (Cantor-Stephens 2007). In style customization, based on the standard lasts and sizes, the customer can choose from diverse style options, including colors, fabrics, leather, and other accessories, within the constraints set by the provider (Piller and Muller 2004).

Style customization should integrate customers' personal preference and current fashion trend. It offers customers the freedom to configure shoes by selecting favorite options among a list of possible variants based the standard shoe design (Boër and Dulio 2007). This strategy has been widely adopted by many large shoe producers, because it requires less effort from both producers and customers. Although style customization involves high costs in design sales, it has little effect on manufacturing and production costs.

The style customization is usually conducted online, and it is a shoe configuration process that is controlled by the customer. Large shoe producers, such as Nike, Adidas, Timberland, and Selve, have different features in style customization, but the basic process for customer participation is common. Customers customize the style through a computer configuration toolkit, usually at the website of the company or sometimes at a computer in a physical retail store. With the toolkit, customers can create their own version of shoes by changing colors or materials on specific parts of the shoes. For instance, a typical configuration toolkit is shown in Figure 30.1, and is developed by Selve, a Munich- and London-based provider of customized footwear. Although customers are limited to a predefined design freedom, they can totally control the configuration process without interception by the producer. However, on the other hand, this strategy also requires that customers know the size of shoes they wear and provide this information during the configuration, as there is no foot measurement in this process. After the customer finishes the configuration and feels satisfied with the customized shoes, a purchase order can be made online,

FIGURE 30.1 Selve style customization toolkit. (Courtesy of Selve, London, U.K.)

which is then transmitted to the manufacturers and initializes a production order. Usually, the style-customized shoes can be delivered to the customer's address within 1 or 2 weeks, or a little bit longer for luxury-class hand-made customized shoes. Currently, many shoe providers are trying to reduce the delivery lead time.

30.2.2 COMFORT CUSTOMIZATION

Comfort customization is to configure a pair of shoes in order to make the consumer feel comfortable and well fit, which is based on the fit of shoes with the foot measurement of the customer (Boër and Dulio 2007). There are two types of approaches for measuring foot shape in order to make the wearer feel comfortable. One is the measurement of static foot shape, including lengths, widths, heights, and girths of feet, which can be measured by laser scanner technologies. The measurement data should match with those of shoes, including the shoe last, the design of the upper, lining, toe shape, insole, outsole etching, and the materials used in fabrication (Witana et al. 2006). Besides the static measurement, it is also important to measure the dynamic shape of the feet in motion, because the foot deforms while walking or running. A popular approach is to analyze the foot motion in the stance phase of the gait cycle, because the nonconformity between the shoe and the foot shape happens during the stance phase (Kimura et al. 2011).

Comfort customization can be addressed by two approaches, custom fit and best-match fit (Boër and Dulio 2007). Custom fit is to exactly and strictly tailor a last and a shoe on the dimensions and morphological data of the consumer's foot. During craftsmen and cobbler times, most shoes were made with a custom fit for each individual customer by carefully measuring the feet of the customer and manually constructing the shoes. A few luxury shoe producers are still working this way, but most ordinary customers cannot afford the expensive price. Best-matched fit is to choose a best-matching last to aim at the privileging comfort, in order to avoid the production of tailor made, individually adjusted lasts. Different from custom fit, the last and all shoe components are predefined in best-matched fit. After measuring a customer's feet parameters, the system selects the one last which is closest to the dimension of measurement. Thus, no individually new last model is produced in best-matched fit. With advanced foot scan technology and sufficient knowledge about foot fit, best-matched fit achieves a compromise between the comfort perceived by the customer and the increased cost in design, sale, and manufacturing. The consumer may not be able to really differentiate the comfort of shoes made by best-match fit and custom fit. As long as the range of the predefined dimensions is large enough, it is believed that no difference is perceived in comfort from a custom fit shoe to a best-matched one. Actually, these two approaches are theoretically the same, as they are both based on optimal approximation, in which best-matched fit approximates to an existing last, while custom fit to a mathematical limit to the real feet.

Comfort customization usually focuses on formal shoes. As this kind of shoes can reflect the owners' identity, people are usually willing to spend more time and money purchasing the fittest ones, just like customized business suits. Large shoe producers who have started comfort customization include the American Otabo, Finnish LeftFoot, and the German Selve. In most practices, customized shoes for comfort fit are sold through real stores, instead of online stores. Sales people use foot scanners to take customer's feet dimensions, store the data, and identify some foot problems, such as different sizes in left and right feet. Then, customers try on the sample shoes selected with best match between customer's foot measurement and the database. In addition, customers can also select different models, outsoles and heel, which may be classified as style customization. After customers finish the configuration and feel satisfied with both the style and comfort, a purchase order is placed in the store and customers may wait for product delivery to their home.

30.2.3 FUNCTION CUSTOMIZATION

Function customization is to make a shoe for each individual consumer by "optimizing" its dimensional parameters, techniques and materials in order to match the use the consumer will make of their shoes (Boër and Dulio 2007). It specifically focuses on the functionality or interfaces of the product, such as selecting speed, precision, power, cushioning, and output devices (Piller and Muller 2004). Due to the intricate and specialized nature, function customization is currently employed by sports shoes manufacturers, such as Adidas, Nike, and ErtlRenz. The shoes for many professional sporting stars are made in this way, and they are usually hand-made individually for each one of them. Designers capture the star's running behaviors and apply specific techniques to protect stars' ankles or improve the sporting performance. In the consumer market, sports shoes have been categorized into different groups according to their diverse functions, such as soccer, running, tennis, golf, etc. The purpose is to maximize the usage utility of the shoes, and provide the best foot protection to customers during sports. However, function customization in the footwear industry should not be limited to sports shoes, and it can be designed for any people who require special functions, such as elderly people. For instance, functional customized shoes may have a huge potential in the foot orthoses and personal health market. It is found that wearing customized shoes can be included in the physical therapy for postsurgical rehabilitation of patients with severe rheumatoid arthritis (Shrader and Siegel 1997). In addition, "Dr Foot" in the United Kingdom offers special insoles to provide extra longitudinal support and cushioning. Therefore, more shoe providers will extend the focus to offer personalized foot health functions through their footwear products.

Most current prevailing sales practices of functional customization are similar to comfort customization. Customers can purchase customized shoes from both online stores and real stores. In online stores, customers usually do not have much freedom, and sometimes they can only select from existing shoe libraries according to purchase navigation. For instance, for Adidas running shoes, customers can only choose shoes based on different running behaviors, including underpronation, normal, and over-pronation. However, such a simple process is reasonable as customers usually do not have much knowledge of technical functionalities. If faced with a complex configuration process, they would rather choose to leave. On the other hand, in the real store, a customer will be invited to a dynamic measurement device, such as the Footscan stage in the Adidas Flagship store, and run on the device to elicit the main features of her running behavior for detailed product configuration. Specific models can be selected for precise sports and uses, such as running, soccer, basketball, and tennis. In addition, customers can also configure the style of the shoes. After a purchase order is placed, the product will be delivered to customer's home in several weeks.

30.2.4 COMBINED STRATEGY AND CHALLENGES

The prevailing practices of shoe customization show that different dimensions of customization strategies are not necessarily mutually exclusive. Instead, companies are taking a combination of these areas to offer the best value within the customer's budget with a combination of different emphases. Thus, the nature of shoe customization is addictive (Boër and Dulio 2007). Style customization has become a popular option provided by many shoe producers, and it is usually employed simultaneously with comfort customization or function customization in one shoe product.

Meanwhile, however, the three dimensions of shoe customization also result in more challenging requirements that a shoe provider should satisfy in order to adopt the paradigm of mass customization successfully. Customer participation in the product configuration process and the resulting large amount of product variants requires for more efficient customer-oriented and product variants management. In addition, a high variety of the products also challenges providers' ability to take advantage of flexible manufacturing and supply management to achieve the economics of scale in order to reduce the production cost. Thus, the shoe providers should apply diverse technical approaches to each step of the value chain in order to address those new challenges.

30.3 TECHNICAL APPROACHES IN SHOE CUSTOMIZATION

From the customer's perspective, the whole shoe customization process can be broken down into four steps: search for favorite "basic" shoes, codesign and configure the shoes, make an order to purchase the customized shoes, and receive the shoes delivered to home soon. To support the customization process, the producer should reconsider the whole value chain from the front end—including product portfolio management, customer interface design, and product configurator—to the back end including flexible manufacturing and supply chain management, as shown in Figure 30.2.

30.3.1 PRODUCT PORTFOLIO MANAGEMENT

Product portfolio management is especially important for style customization, which derives a huge amount of product variety from customer's customization. Customers are provided with a "basic" shoe to customize and a variety of style options and accessories. Therefore, how many varieties should the shoe provider offer to the customer? In the original mindset of mass customization, designers make a large variety of products available and let customers choose from the shelf, but it has been found as unnecessarily wasteful and expensive, which may also lead to mass confusion and constrain customer satisfaction (Huffman and Kahn 1998). Thus, designers must leverage from marketing studies to identify the shoe features preferred by the customer and market segments with higher profit potential. The essential question of the product portfolio is to decide how to offer the right product variety to the right target market (Jiao et al. 2007).

Feasible product portfolios are usually developed with emphasis on portfolio optimization based on measuring customer preference in terms of expected utilities (Jiao and Zhang 2005), in which the objective is to maximize sale profit and market share. Conjoint analysis has been widely employed as an effective means to estimate consumer preference, such as part-worths, importance weights, and ideal points, associated with individual product attributes (Green and Srinivasan 1978). Considering substantial amount of diverse variations in consumer preference, conjoint analysis is usually conducted at the individual level, and the basic steps

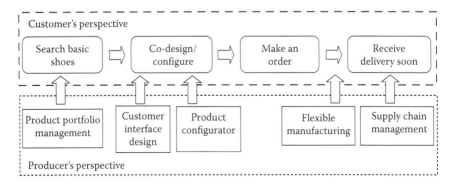

FIGURE 30.2 Technical approaches to support shoe customization.

are introduced as follows (Green and Srinivasan 1990). The first step of conjoint analysis is to select one preference model, such as vector model, ideal point model, or part-worth function model. The mixed model can be developed by allowing some attributes to be treated as the part-worth function model while others follow vector or ideal point models. The second step is to select a data collection method. A popular one is a full-profile method, which describes each option on all of the attributes and measures overall preference judgments directly by behaviorally oriented constructs such as willingness to pay (WTP) and chances of switching to a new brand, usually conducted through online surveys. The third step is to construct a stimulus set, such as fractional factorial design, random sampling from a multivariate distribution, and Pareto-optimal designs. Then, the fourth step is to develop a stimulus presentation method, including verbal description, paragraph description, pictorial or three-dimensional (3D) model presentation and physical products. In addition to rating and ranking scales, the paired comparison is also widely used as measurement scale for the dependent variable, which is the fifth step. Since conjoint analysis is based on regression estimate procedures, its last step is to determine an estimation method to deal with instability of estimated parameters facing various sources of error variance. With the development of information technology, adaptive conjoint analysis is also proposed, which starts with a simple self-explication task through which the respondent's more important attributes are identified (Green and Srinivasan 1990). It is dynamic and the respondent's previous answers are used at each step to generate the next paired comparison questions so that most information can be elicited. In the study of "Design by Customer" (Tseng and Du 1998), adaptive conjoint analysis is applied so that customers can navigate through product families, define their preferences, and then design the products in the sense that they may map their own functional requirements into a physical domain on their own. The technical challenge is at preparation and presentation of product portfolios so that the customers can make the best informed decisions to select a desired one to fulfill their needs.

The other challenge in generating a product portfolio is to tackle the optimal selection of promising products by maximizing the surplus—the margin between the customer-perceived utility and the price of the product—the objective of achieving best profit performance. Green and Krieger (1985) first consider a buyer's problem in which a product line is selected to maximize customers' utilities. They also propose a seller's problem in which the product line is chosen to maximize sellers' return, constrained by each buyer's choice of the most preferred item from the product line. Jiao and Zhang (2005) propose a portfolio selection approach with the goal of maximizing an expected surplus from both the customer and engineering perspectives, in which the manufacturer must determine what combinations of attributes and their levels should be introduced to offer the potential products or be discarded from consideration of product offerings. Such models can integrate marketing analysis inputs with the cost information associated with product development and manufacturing process, and capture the trade-off between the profits derived from offering varieties to the market and cost savings realized by product portfolios that can be produced efficiently within the current company's manufacturing capacities.

30.3.2 CUSTOMER INTERFACE DESIGN

Customer's participation is the key to successful mass customization strategy. In style customization, the customer should input their preference to configure the shoe they like; in comfort and function customization, customers should measure their feet and walking behavior on the foot scanner to capture their foot parameters. When the product is delivered to the customer, another interaction opportunity happens. However, among all of the interactions, the customer's participation in codesigning their own shoe style through a "codesign interface" provided by the producer may be the most important one for the shoe producer, because the customer can easily leave the product without any cost and concerns if she is not satisfied with the process (Franke and Schreier 2010; Franke et al. 2010). How to design the customer interface in order to make the customer interested in shoe customization and enjoy the process is a crucial question.

Web-based mass customization has been employed in most practices of shoe mass customization, which not only allows companies to enhance their interaction with customers, but also increases customers' confidence and educates them about the product they are ordering, which will enhance customers' satisfaction as well (Siddique and Boddu 2004; Franke et al. 2010; Wang and Tseng 2011a). The success of web-based mass customization depends on customers' codesign interface, an ability to allow for learning by doing, to stimulate customers' satisfaction and positive experience, and its integration into the brand concept, as well as its technological capabilities (Franke and Piller 2003). The interface should enable customers to express their requirements and carry out the product development process by mapping the requirements to the physical domain of the product without complexity and confusion (Huffman and Kahn 1998; Tseng and Du 1998; Chen and Wang 2010).

The customer codesign interface should first assist the customer to explore the whole solution space, clearly delineating what will be offered and what will not (Salvador et al. 2009). Conventional design approaches assume that customers know what they want, which, unfortunately, is not true (Tseng et al. 2010). Instead, customers' needs will be clarified or identified through the configuration process, by trying different possibilities, learning from errors, and comparing different solutions, which is an iterative and time-consuming learning process (Salvador et al. 2009). In this process, it is crucial to provide customers with immediate feedback via virtual prototype creation, usually a 3D shoe CAD model, which the customer is able to evaluate and compare with the ideal product in mind.

The customer codesign interface should also balance customers' effort and perceived complexity, to avoid "mass confusion" (Huffman and Kahn 1998). Currently, even a simple product configuration system can provide endless possibilities (Franke and Schreier 2008). It is found that customers can be overwhelmed by the excess variety and external complexity during product configuration (Huffman and Kahn 1998), which is termed "mass confusion." In such cases, customers would postpone buying decisions and consider the product as difficult and undesirable (Salvador et al. 2009). In order to address this problem, it is suggested that the company can provide choice navigation to simplify the ways in which people explore desired product offerings, such as product recommendations based on customers' profile or trial-and-error. The recommendation is generated as the optimal solution based on customers' given information, which would be a good reference for them to further identify and explore their options (Wang and Tseng 2011b; Wang and Tseng 2011c). In addition, an ecosystem design concept is proposed (Tseng et al. 2010; Zhou et al. 2011), in which customers are invited into a virtual reality environment embedded with an affective-related recommendation system supported by the affective-related knowledge database and configuration database. Different from existing personalization recommendation, such systems allow customers to freely configure and experience with the product in a simulated but close-to-practice context and focuses more on customers' needs reflected by the user experience, which may be much closer to customers' real needs.

On the other hand, customer codesign interface should also satisfy customer's hedonic needs as a "creator." Empirically, it is observed that people seem to derive an intrinsic benefit from "doing it themselves," which is also called pride of authorship (Schreier 2006). It describes the output-oriented benefit of having done it oneself, and the positive outcome of such processes constitutes positive feedback, which gives the individual a strong feeling of pride. Unlike satisfaction in high product utility and fit, it indicates a value increment a customer ascribes to a self-designed product, arising purely from the fact that she feels like the originator of the product (Franke et al. 2010; Wang and Tseng 2011a). Due to this psychological effect, customers may overvalue their (often poorly made) creations (Norton 2009), in which case a customer may be willing to accept a lower-quality self-design product even at the same cost of expert design (Ulrich 2011). In other words, customers may appreciate the value not from product utility, but from their participation in the product design process. In order to enhance this psychological effect, customers should invest a large amount of effort in designing the product, or it may not truly make the customer feel she is the product's "originator." Therefore, it may be more challenging to develop a customer codesign interface for such psychological effect than that for high product utility.

To design such an interface, some suggestions are proposed in Franke et al. (2010). First of all, customers can be offered a great deal of design freedom to enable high preference fit and a large degree of decisional control over the process to make the customer feel like they are "the cause" of the outcome of customized shoes. Second, immediate positive feedback can be provided on successful performance during the process. Third, affirmative feedback such as labels and certificates can be provided to the customer to emphasize their role as creator. Thus, designers may actually design customers' enjoyable experience in the shoes' self-design process, rather than just the pair of the shoes alone.

30.3.3 SHOE CONFIGURATOR

It is the shoe configurator that translates customers' design or feet measurement data into a 3D detail model of the shoes. The information generated by the product configurator is essential for shoe production as well as 3D model generation shown to the customer in the codesign interface. To take advantage of economic scale, the configurator generates all customized shoes from a product family with some common structures and product technologies, which form the platform of the family (Erens and Verhulst 1997). In general, a product platform has three aspects: the product architecture, the interfaces, and the standards that provide design rules to which the modules must conform (Baldwin and Clark 2000). It defines the way in which the product elements are arranged into physical components and the way in which components interact, and each function is mapped to the modular component, while standardization and decoupling of the interface are integrated between components (Ulrich and Eppinger 1995).

There are two popular platform-based product configuration approaches based on different product platforms. One common approach is scalable product configuration, in which scaling variables are employed to shrink or stretch the platform in one or more dimensions to satisfy diverse customer needs. This approach is effective in comfort and function customization, in which the lasts can be easily adjusted according to each individual's foot measurement. The other approach is called modular product configuration, from which each product is derived from adding, substituting, and/or removing one or more functional modules (Du et al. 2001). Obviously, this approach is effective in style customization and function customization, in which each changeable component or accessory can be designed as a module. The change of one module does not influence the other parts or the whole design, so that the design cost can be significantly reduced.

In scalable product configuration, the first task is to determine which design parameters take common values, and the second task is to determine the optimal values of common and distinctive variables by satisfying performance and economic requirements (Simpson 2004). Shoe designers can study the foot measurement database to find out which last parameter is usually constant among most customers. The modular product configuration is based on a modular product architecture, which involves one-to-one mappings from functional elements to the physical components of a product. With specified decoupled interfaces between components, each functional element of the product can be changed independently by changing only the corresponding component (Ulrich 1995).

The configurator has become the major approach to link with customers' orders and the components, which organizes the logics of component relationships and essential product parameters. Knowledge management and AI techniques have been widely employed and have shown great promise for automatic product configurations (Simpson 2004; Shooter et al. 2005). A computer-based product platform concept model is introduced in Johannesson and Claesson (2005) to capture both functional behavior and embodiment of design solution, as well as the operative component structure in a configurable system product. This model includes documentation of the design history capturing design motives, decisions and results, embodiment definition of a selected set of design parameters from a function-means-tree, and parameterization of the selections of the design parameters which allow the design solution to appear as different design variants. Unified Modeling Language, UML, has been employed for modeling configuration knowledge bases (Felfernig et al. 2001). The configuration model includes the component model and a set of corresponding functional architectures defining which requirements can be imposed on the product. The configuration model reduces the development time and effort because it can be automatically translated into executable logic representations. In addition, the graph grammar is thought of as an effective method to model the logical organization of product family elements, as well as the mechanism of product variant derivation to satisfy individual customer needs. The advantages of graph grammar include visual representation, formal definition, ease of computational implementation, and extensibility (Du et al. 2002a). A graph rewriting system is proposed in Du et al. (2002b) to support product family design, in which the family graphs performs as the starting graphs for a series of graph operations, and variant graphs are derived by executing predefined rewriting rules in terms of appropriate control structures. It is demonstrated that the graph-based product family design system can provide an interactive environment for customers to make choices among product variants, and negotiations among sales, design, and manufacturing.

30.3.4 FLEXIBLE MANUFACTURING PROCESS

One consequence of mass customization in manufacturing is observed as an exponentially increased number of process varieties, including diverse machines, tools, fixtures, setups, cycle times, and labor (Wortmann et al. 1997), and such process varieties would introduce significant constraints to product planning and control (Kolisch 2000). Thus, a flexible manufacturing process is important to reduce the cost of producing more product variants. From the mass customization perspective, there are two approaches which can be implemented to improve the flexibility of the manufacturing process.

One approach is to implement a manufacturing process family idea. The common components and standardized basic product structure designed for a product family enable similarity in the production process and make process family possible, which comprises a set of similar production processes sharing a common process structure (Martinez et al. 2000). The platform idea can be expanded from product to production process, which implies a focus on commonality of production tools, machines, and assembly lines (Meyer and Lehnerd 1997). In addition, some researchers also present the concept of process configuration which combines the principles of product configuration and process planning, as the commonality across the product variety leads to a number of same or similar productions and operations among process variants (Schierholt 2001). Thus, it is suggested that companies can configure existing operations and manufacturing processes by exploiting similarity among product and process families in order to take advantage of repetitions (Schierholt 2001).

Flexible manufacturing should also rely on the availability of agile manufacturing systems such as a new CAD/CAM manufacturing technology to reduce the response time. It is claimed that the next generation of manufacturing systems must be able to support high levels of flexibility, reconfigurability, and intelligence to allow them to adapt for diverse product variety (Molina et al. 2005). Such manufacturing systems must be rapidly designed, able to convert quickly to new productions, able to adjust capacity quickly, and able to produce increased product varieties in unpredictable quantities (Mehrabi et al. 2000). In order to achieve these objectives, the manufacturing system can be created by incorporating basic process modules which can be rearranged or replaced quickly and reliably. In addition, reconfiguration should allow adding, removing, or modifying specific process capabilities, control and

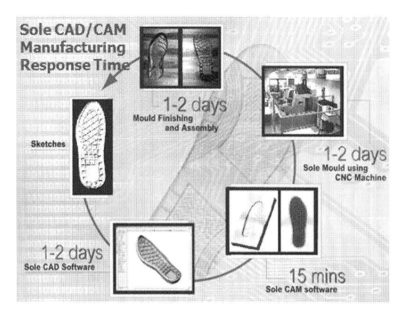

FIGURE 30.3 SoleCAD/CAM manufacturing system. (Courtesy of Hong Kong University of Science and Technology, Hong Kong.)

operation structure to adjust production capacity to respond to diverse product configurations (Mehrabi et al. 2000). For instance, the SoleCAD system developed by HKUST integrates a prototype sole module design and manufacturing system for shoe industry, as shown in Figure 30.3. With the SoleCAD/CAM manufacturing system, the response time from receiving personalized shoe orders to producing a last for shoe production can be reduced to around 5 days. The flexibility of the system also enables designers to easily change the CAD model to produce another personalized last with limited additional cost.

30.3.5 SUPPLY CHAIN MANAGEMENT

In the background of mass customization with the shortened product life cycle and proliferation of product variety, there has been found a tremendous opportunity to enhance supply chain management by including product design as an integral part of supply chain (Tseng et al. 2003). Suppliers can be seen as an extended manufacturing capability. Thus, the company's supply chain should be configured to address customers' requirements for flexibility, agility, cost efficiency, and product variety, and focus more on collaboration with external suppliers or partners (Salvador et al. 2002). For instance, "Earlier Supplier Involvement (ESI)" is proposed as a form of vertical cooperation in which the manufacturer involves suppliers at an early stage of the product development process (Tseng et al. 2003).

Mass customization has challenged the traditional supply chain of many shoe producers. In the traditional supply chain, raw material providers sell to footwear manufacturers who sell to final assembly manufacturers. Then, the products are distributed through wholesalers, distributors, and retailers to the final customers. However, mass customization in the business-to-customer (B2C) commerce background results in an entirely new direct channel opportunity, a process of disintermediation (Tseng et al. 2003). In the direct channel, shoe producers can sell the products directly to the customer, bypassing the traditional intermediaries. It is the electronic commerce that serves as a disintermediator replacing the intermediary links in the traditional supply chain with a direct channel to the customer. The disintermediation of the supply chain leads to better inventory management and production planning by taking advantage of better signals regarding demand levels from consumers directly (Warkentin et al. 2000).

In terms of the EUROShoE project (Boër and Dulio 2007), currently most footwear producers have outsourced the production process to the developing countries, such as China, where the labor cost is much lower than that in the Western countries. In addition, in this era of globalization, a shoe is made by the materials collected and human labor from all over the world. Then, a global production network should be carefully designed and planned by the shoe providers. There are four major location strategies for global production networks that have been identified (Rodrigue et al. 2009). In centralized global production, the production is completed only in one region and then exported to the global market in order to take advantage of economies of scale and address the difficulties in relocation and reproduction. In regional production, the product is manufactured in each region with the size of the production system related to the size of the region market. Such strategy relies more on regional accessibility than economies of scale to reduce high distribution costs. In regional specification, the production is divided into various regions in terms of specific comparative advantages. The manufacture in each region can specialize in producing a specific product and import from other regions what it requires. In vertical transportation integration, different stages of the production are placed at the locations that offer the best comparative advantages. For instance, raw materials can be extracted from the most accessible location, and the manufacturing and assembly can be performed in regions with low labor costs or high levels of expertise. Therefore, shoe providers can select different strategies according to their own production capability and market.

30.4 CONCLUSION

Mass customization can be a competitive factor to help footwear firms face the current competitive global market. Cost pressure and customer preference on product personalization force footwear firms to adopt new strategies, such as mass customization, to increase the level of satisfaction for the customer and manage the cost and complexity in the design and production process. However, is mass customization the panacea to the footwear industry? The answer depends on how the provider restructures the business model and takes advantage of a global production network to reduce the cost. To most shoe providers, individuality is clearly a global trend, so it would be better to consider new strategies as soon as possible in order to survive in this rapidly changing market.

QUESTIONS

30.1 What shoes are being customized in the market? Can you give more examples?

30.2 What is the difference between a pair of "tailor-made" shoes with a mass customized one?

30.3 What is the value derived from the footwear customization? What is the additional cost incurred in the footwear customization?

30.4 What are the key dimensions of the footwear customization? How do they influence the footwear design?

30.5 What is the role of a customer playing in a footwear customization process? How can a customer influence the attributes of the customized shoes?

30.6 How should a conventional design process be changed to realize the mass customization strategy for footwear?

REFERENCES

Baldwin, C. Y. and K. B. Clark. 2000. *Design Rules: The Power of Modularity*, MIT Press, Cambridge, MA.

Belk, R. W. 2003. Shoes and self. *Advances in Consumer Research* **30**: 27–33.

Boër, C. R. and S. Dulio. 2007. *Mass Customization and Footwear: Myth, Salvation or Reality?* Springer, London, U.K.

Cantor-Stephens, F. 2007. Footwear fashion: Sure footed. *Sportswear International* **215**: 86.

Chen, Z. and L. Wang. 2010. Personalized product configuration rules with dual formulations: A method to proactively leverage mass confusion. *Expert Systems with Applications* **37**(1): 383–392.

Du, X., J. Jiao et al. 2001. Architecture of product family: Fundamentals and methodology. *Concurrent Engineering Research and Applications* **9**(4): 309–325.

Du, X., J. Jiao et al. 2002a. Graph grammar based product family modeling. *Concurrent Engineering Research and Applications* **10**(2): 113–128.

Du, X., J. Jiao et al. 2002b. Product family modeling and design support: An approach based on graph rewriting systems. *Artificial Intelligence for Engineering Design, Analysis and Manufacturing: AIEDAM* **16**(2): 103–120.

Erens, F. and K. Verhulst. 1997. Architectures for product families. *Computers in Industry* **33**(2–3): 165–178.

Felfernig, A., G. Friedrich et al. 2001. Conceptual modeling for configuration of mass-customizable products. *Artificial Intelligence in Engineering* **15**(2): 165–176.

Franke, N. and F. T. Piller. 2003. Key research issues in user interaction with configuration toolkits in a mass customization system. *International Journal of Technology Management* **26**(5/6): 578–599.

Franke, N. and M. Schreier. 2008. Product uniqueness as a driver of customer utility in mass customization. *Marketing Letters* **19**(2): 93–107.

Franke, N. and M. Schreier. 2010a. Why customers value self-designed products: The importance of process effort and enjoyment. *Journal of Product Innovation Management* **27**(7): 1020–1031.

Franke, N., M. Schreier et al. 2010b. The "I Designed It Myself" effect in mass customization. *Management Science* **56**(1): 125–140.

Green, P. E. and A. M. Krieger. 1985. Models and heuristics for product line selection. *Marketing Science* **4**(1): 1–19.

Green, P. E. and V. Srinivasan. 1978. Conjoint analysis in consumer research. Issues and outlook. *Journal of Marketing* **55**: 103–123.

Green, P. E. and V. Srinivasan. 1990. Conjoint analysis in marketing: New developments with implications for research and practice. *Journal of Marketing* **54**(4): 3–19.

Huffman, C. and B. E. Kahn. 1998. Variety for sale: Mass customization or mass confusion? *Journal of Retailing* **74**(4): 491–513.

Jiao, J., T. W. Simpson et al. 2007. Product family design and platform-based product development: A state-of-the-art review. *Journal of Intelligent Manufacturing* **18**(1): 5–29.

Jiao, J. and Y. Zhang. 2005. Product portfolio planning with customer–engineering interaction. *IIE Transactions (Institute of Industrial Engineers)* **37**(9): 801–814.

Johannesson, H. and A. Claesson. 2005. Systematic product platform design: A combined function-means and parametric modeling approach. *Journal of Engineering Design* **16**(1): 25–43.

Kimura, M., M. Mochimaru et al. 2011. 3D measurement of feature cross-sections of foot while walking. *Machine Vision and Applications* **22**(2): 377–388.

Kolisch, R. 2000. Integration of assembly and fabrication for make-to-order production. *International Journal of Production Economics* **68**(3): 287–306.

Martinez, M. T., J. Favrel et al. 2000. Product family manufacturing plan generation and classification. *Concurrent Engineering: Research and Applications* **8**(1): 12–22.

Mehrabi, M. G., A. G. Ulsoy et al. 2000. Reconfigurable manufacturing systems: Key to future manufacturing. *Journal of Intelligent Manufacturing* **11**(4): 403–419.

Meyer, M. H. and A. P. Lehnerd. 1997. *The Power of Product Platforms*, Free Press, New York.

Molina, A., C. A. Rodriguez et al. 2005. Next-generation manufacturing systems: Key research issues in developing and integrating reconfigurable and intelligent machines. *International Journal of Computer Integrated Manufacturing* **18**(7): 525–536.

Norton, M. I. 2009. The IKEA effect: When labor leads to love. *Harvard Business Review* **87**(2): 30.

Piller, F. T. and M. Muller. 2004. A new marketing approach to mass customisation. *International Journal of Computer Integrated Manufacturing* **17**(7): 583–593.

Rodrigue, J.-P., C. Comtois et al. 2009. *The Geography of Transport Systems*, Routledge, New York.

Salvador, F., P. M. De Holan et al. 2009. Cracking the code of mass customization. *Mit Sloan Management Review* **50**(3): 71–78.

Salvador, F., C. Forza et al. 2002. Modularity, product variety, production volume, and component sourcing: Theorizing beyond generic prescriptions. *Journal of Operations Management* **20**(5): 549–575.

Schierholt, K. 2001. Process configuration: Combining the principles of product configuration and process planning. *AIE-DAM* **15**(5): 411–424.

Schreier, M. 2006. The value increment of mass-customized products: An empirical assessment. *Journal of Consumer Behaviour* **5**(4): 317–327.

Shooter, S. B., T. W. Simpson et al. 2005. Toward a multi-agent information infrastructure for product family planning and mass customization. *International Journal of Mass Customization* **1**(1): 134–155.

Shrader, J. A. and K. L. Siegel. 1997. Postsurgical hindfoot deformity of a patient with rheumatoid arthritis treated with custom-made foot orthoses and shoe modifications. *Physical Therapy* **77**(3): 296–305.

Siddique, Z. and K. R. Boddu. 2004. A mass customization information framework for integration of customer in the configuration/design of a customized product. *Artificial Intelligence for Engineering Design, Analysis and Manufacturing: AIEDAM* **18**(1): 71–85.

Simpson, T. W. 2004. Product platform design and customization: Status and promise. *Artificial Intelligence for Engineering Design, Analysis and Manufacturing: AIEDAM* **18**(1): 3–20.

Tseng, M. M. and X. Du. 1998. Design by customers for mass customization products. *CIRP Annals—Manufacturing Technology* **47**(1): 103–106.

Tseng, M. and J. Jiao. 2001. Mass customization. In *Handbook of Industrial Engineering*. G. Salvendy (ed.). Wiley, New York, pp. 684–709.

Tseng, M. M., R. J. Jiao et al. 2010. Design for mass personalization. *CIRP Annals— Manufacturing Technology* **59**(1): 175–178.

Tseng, M. M., T. Kjellberg et al. 2003. Design in the new e-commerce era. *CIRP Annals—Manufacturing Technology* **52**(2): 509–519.

Tseng, M. and F. Piller. 2003. *The Customer Centric Enterprise: Advances in Mass Customization and Personalization Consortium*, Springer Verlag, New York.

Ulrich, K. 1995. The role of product architecture in the manufacturing firm. *Research Policy* **24**(3): 419–440.

Ulrich, K. T. 2011. Users, experts, and institutions in design. In K. T. Ulrich (Ed.) *Design: Creation of Artifacts in Society*, University of Pennsylvania, Philadelphia, PA.

Ulrich, K. and S. D. Eppinger. 1995. *Product Design and Development*, McGraw-Hill, New York.

Von Hippel, E. 1998. Economies of product development by users: The impact of "sticky" local information. *Management Science* **44**(5): 629–644.

Wang, C. and M. M. Tseng. 2011a. Comparing four personalization approaches to understand value of personalization. *2011 World Conference on Mass Customization, Personalization, and Co-Creation*. University of California, Berkeley, CA.

Wang, Y. and M. M. Tseng. 2011b. Adaptive attribute selection for configurator design via Shapley value. *Artificial Intelligence for Engineering Design, Analysis and Manufacturing: AIEDAM* **25**(2): 185–195.

Wang, Y. and M. M. Tseng. 2011c. Integrating comprehensive customer requirements into product design. *CIRP Annals—Manufacturing Technology* **60**(1): 175–178.

Warkentin, M., R. Bapna et al. 2000. The role of mass customization in enhancing supply chain relationships in B2C e-commerce markets. *Journal of Electronic Commerce Research* **1**(2): 45–52.

Witana, C. P., S. Xiong et al. 2006. Foot measurements from three-dimensional scans: A comparison and evaluation of different methods. *International Journal of Industrial Ergonomics* **36**(9): 789–807.

Wortmann, J. C., D. R. Muntslag et al. 1997. *Customer Driven Manufacturing*, Chapman & Hall, London, U.K.

Zhou, F., R. J. Jiao et al. 2012. User experience modeling and simulation for product ecosystem design based on fuzzy reasoning petri nets. *IEEE Transactions on Systems, Man and Cybernetics, Part A: Systems and Humans* **402**(1): 201–212.

Zipkin, P. 2001. The limits of mass customization. *Mit Sloan Management Review* **42**(3): 81–87.

31 Strategic Capabilities to Implement Mass Customization of Athletic Footwear
The Example of Miadidas

Frank Piller, Evalotte Lindgens, and Frank Steiner

CONTENTS

31.1 INTRODUCTION

In today's highly competitive business environment, activities for serving customers have to be performed both efficiently and effectively—they have to be organized around a customer-centric supply and demand chain (Piller and Tseng 2010). Since the early 1990s, mass customization has emerged as an idea for achieving precisely this objective. Following Joseph Pine (1993), we define mass customization as "developing, producing, marketing, and delivering affordable goods and services with enough variety and customization that nearly everyone finds exactly what they want." In other words, the goal is to provide customers what they want, when they want it. Hence, companies offering mass customization are becoming customer-centric enterprises (Tseng and Piller 2003), organizing all of their value creation activities around interactions with individual customers.

However, to apply this apparently simple statement in practice is quite complex. As a business paradigm, mass customization provides an attractive business proposition to add value by directly addressing customer needs and in the meantime utilizing resources efficiently without incurring excessive cost. This is particularly important at a time where competition is no longer just based on price and conformance of dimensional quality. When the subject of mass customization is raised, the successful business model of the computer supplier Dell is often cited as one of the most impressive examples. The growth and success of Dell is based on this firm's ability to produce custom computers on demand, meeting precisely the needs of each individual customer and producing these items only after an order has been placed (and paid for), with no finished goods' inventory risk at all. However, beyond Dell, there are many other examples of companies that have employed mass customization successfully (Salvador et al. 2009, see www.configurator-database.com for a broad listing of examples).

As we wear shoes every day, most of us have quite a precise idea of a perfect shoe. But in reality, there often is a trade-off in terms of fit, form, and function. As a result, the footwear sector has been among the industries that have embraced mass customization quite early, with mixed success. Consider the following examples:

- In the mid-1990s, Custom Foot was one of the first companies to embrace mass customization for the footwear industry. Based in Westport, CT, customers could order shoes that promised a perfect fit. Their feet were measured in 14 different ways electronically. Orders were then transmitted to one of a dozen Italian factories. There, custom orders took their place on the assembly line alongside mass-produced shoes. Customers could pick from 150 styles in sizes from women's 4AAAA to men's 16EEEEE, at prices ranging from $100 to $250 (Wong 1996). While in its announcements the company's founder promised to increase the number of stores from 3 in 1996 to 100 by the end of 1997, the company went out of business in 1997. The reason for this failure was not a lack of customer demand, but a lack of stable processes in fulfillment. First, there was a problem with differing cultures. The Italian factories were used to working in batch mode, versus individual production, which caused some problems. For example, the factory workers in Italy wanted to take the usual 4 weeks off in August, when customers in the United States wanted their shoes. But the most serious problem was that the company could not reliably measure a foot and determine what size options would best fit that foot. Subsequently, the system was not capable of producing replicable results.

- More than 10 years later, U.S. sneaker brand Keds took a different approach. In 2009, it launched a custom footwear offering called Kedsstudio.com. The offering focusses on the aesthetic design of the shoes with a retail price of about U.S.$40–60. Users can upload any design or picture on their shoe, which then is printed on the canvas, laser cut, and assembled to a sneaker. Shoes are manufactured with advanced digital printing technology that offers great variety in high quality. Manufacturing takes place in China within 24–48 h, allowing for order lead times for U.S. consumers of 1–2 weeks, depending upon the shipping method selected. To build its mass customization offering, Ked could rely on a new breed of mass customization outsourcing services. It contracted California-based Zazzle Inc. to not just organize its supply chain and custom manufacturing in Asia, but also to connect Keds with creative consumers online. Forty-eight hours after the launch, over 18,000 designs had been published on Zazzle.com. Today, millions of designs have been created. While not all designs are being transferred into an actual order, within a few weeks Keds created more designs with mass customization than its in-house designers in the entire 100 years of the company's existence.

- Selve, a London- and Munich-based manufacturer of women's custom shoes, is a fine example of a company that interacts well with its customers both in traditional stores and online. Selve enables its customers to create their own shoes by selecting from a variety of materials and designs, in addition to a truly custom shoe fit, based on a three-dimensional scan of a customer's feet. Trained consultants provide advice in the company's stores; the online shop offers reorders. Shoes are all made to order in a specialized factory in China and are delivered in about 2 weeks. Customers get this dedicated service for a cost between 200 and 400 Euro, not inexpensive but still affordable for many consumers compared to the price level of a traditional shoe maker (starting at 1000 Euro and more). In its 10 years of existence, Selve could build a growing customer base, and after improving its manufacturing and supply chain processes, also scale up output and sales.

- In 2010, Barcelona, Spain, based brand MUNICH launched a line of custom sports shoe (munichmyway.com) that are not just characterized by a wide variety of designs, but by a local manufacturing approach that strives to relocate manufacturing back to Europe. All custom shoes are made in a small factory in Barcelona, allowing the company to ship shoes quite fast, and also adjust styles and fabric choices rapidly. Despite the higher manufacturing cost, shoes retail for about 120 Euro, which is not much more than the standard shoes of the brand.

What do these examples have in common? Regardless of different product categories, price points, and fulfillment systems, they all have turned customers' heterogeneous needs into an opportunity to create value, rather than regarding heterogeneity as a problem that has to be minimized, challenging the "one size fits all" assumption of traditional mass production. The idea of this chapter is to explore the characteristics of successful mass customization in the footwear industry, using the method of a single case study. We also discuss the mass customization initiative of Adidas AG, one of the largest global sport brands.

The remaining of this chapter is organized as follows: we will start with a brief overview of the mass customization concept and introduce a framework of three strategic capabilities that make mass customization work. We will then discuss the situation of the athletic footwear industry and different approaches to mass customization in this industry. The main part of this chapter will focus on the development of miadidas, the central customization offering of Adidas. The chapter ends with a reflection of the development of mass customization at Adidas.

31.2   MASS CUSTOMIZATION: DEFINITION AND STRATEGIC CAPABILITIES

From a strategic management perspective, mass customization is a differentiation strategy. Referring to Chamberlin's (1962) theory of monopolistic competition, customers gain the increment of utility of a customized good that better fits their needs than the best standardized product attainable would. The larger the heterogeneity of all customers' preferences, the larger is this gain in utility (Kaplan et al. 2007). Davis, who initially coined the term in 1987, refers to mass customization when "the same large number of customers can be reached as in mass markets of the industrial economy, and simultaneously [...] be treated individually as in the customized markets of preindustrial economies" (Davis 1987: 169). Pine (1993: 9) popularized this concept and defined mass customization as "providing tremendous variety and individual customization, at prices comparable to standard goods and services" to enable the production of products and services "with enough variety and customization that nearly everyone finds exactly what they want." A more pragmatic definition was introduced by Tseng and Jiao (2001) who suggested that mass customization corresponds to "the technologies and systems to deliver goods and services that meet individual customers' needs with near mass production efficiency" (Tseng and Jiao, 2001: 658).

More recently, mass customization has been described as a set of organizational capabilities that can enrich the portfolio of capabilities of their organizations (Salvador et al. 2009; Piller and Tseng 2010). Companies that profit from heterogeneities in their customer base successfully have built competences around a set of core capabilities. While specific answers on the nature and characteristics of these capabilities are clearly dependent on industry context or product characteristics, three fundamental groups of capabilities form the ability of a firm to mass customize: solution space development, robust process design, and choice navigation (the derivation of these capabilities builds on work by Salvador et al. (2008, 2009).

31.2.1 SOLUTION SPACE DEVELOPMENT

First and foremost, a company seeking to adopt mass customization has to be able to understand the idiosyncratic needs of its customers. This is in contrast to the approach of a mass producer, where the company focuses on identifying "central tendencies" among its customers' needs, and targets them with a limited number of standard products. Conversely, a mass customizer has to identify the product attributes along which customer needs diverge the most. Once this is understood, the firm knows what is required to properly cover the needs of its customers. Consequently, it can draw up the so-called solution space, clearly defining what it is going to offer and what it is not.

31.2.2 ROBUST PROCESS DESIGN

A second critical requirement for mass customization is related to the relative performance of the supply chain. Specifically, it is crucial that the increased variability in customers' requirements does not lead to significant deterioration in the firm's operations and supply chain (Pine et al. 1993). This demands a robust supply chain design—defined as the capability to reuse or recombine existing organizational and supply chain resources to fulfill differentiated customers' needs. With robust process design, customized solutions can be delivered with near mass production efficiency and reliability.

31.2.3 CHOICE NAVIGATION

Finally, the firm must be able to support customers in identifying their own problems and solutions, while minimizing complexity and burden of choice. When a customer is exposed to too many choices, the cognitive cost of evaluation can easily outweigh the increased utility from having more choices (Huffman and Kahn 1998; Piller 2005). As such, offering more product choices can easily prompt customers to postpone or suspend their buying decisions. Therefore, the third requirement is the organizational capability to simplify the navigation of the company's product assortment from the customers' perspective.

The methods behind these capabilities are often not new. Some of them have been around for many years. However, successful mass customization demands the combination of these methods into capabilities in a meaningful and integrated way. To discuss how this capability framework can be applied in the footwear industry, we will use the case of one of the core offerings of customization in this industry, the mass customization program of Adidas.

31.3 MASS CUSTOMIZATION AT ADIDAS

31.3.1 COMPANY OVERVIEW

The Germany-based Adidas AG is the second largest global supplier of sport goods, with a sales volume of nearly 12 billion Euro across all brands. Among these sales, about 50% account for sport shoes, whereas the rest is generated by apparel and accessories. The company employs about 42,500 members of staff. The core brand adidas had net sales of 8.7 billion Euro in 2010. The production volume of Adidas shoes is about 200 million per year. The company itself today is a pure brand owner, focusing on developing and selling its products, while manufacturing is done by independent suppliers.

The history of Adidas can be traced back to the 1920s when Adolf "Adi" Dassler and his brother began making shoes with the basic materials available after the WW I. At first, slippers were made using old tires to produce the sole. Later those shoes were changed into soccer shoes and gymnastic shoes by adding cleats or nail-on studs (Seifert 2006). The reason for this development was a simple idea to provide every athlete with the best possible equipment. The first time, Dassler shoes were showcased during the Amsterdam Olympics in 1928. The major breakthrough was achieved in the Berlin Olympics in 1936 when a gold medalist wore their shoes. Only 1 year later, the Dassler brothers already manufactured shoes for 11 different kinds of sport.

However, in 1948 both brothers argued on how to proceed with the development of their business, which led to Adolf Dassler's foundation of Adidas, whereas his brother Rudolph established the Puma sports company. From that point of time, Adidas was focusing on performance sport shoes. Its portfolio was complemented by soccer balls in 1963 and apparel in 1967. By then, Adidas was booming and nearly 80% of all the medal-winning athletes were equipped by Dassler's company.

New competitors, such as Nike or Reebok in 1979, started to enter the market focusing on low-quality, fashionable leisure wear targeting teenagers. Those turbulent times were accompanied by the founder's death, which lead to organizational and management changes. After financial troubles, investor Robert Louis-Dreyfus was appointed chairman of the executive board and led the company to an amazing turnaround by focusing on market needs and establishing a strong brand image (Moser et al. 2007). This successful leadership was followed by Herbert Hainer, CEO and Chairman of the Executive board of Adidas since 2001. In the last years, Adidas positioned itself as number two of the world's largest sport manufacturers, a position being reinforced by its acquisition of Reebok in 2006, a former competitor. Today, the company is split into three major divisions that target different customer types: sport performance, sport heritage, and sport style.

31.3.2 INDUSTRY BACKGROUND: REASONS TO ENTER MASS CUSTOMIZATION

The global sports shoe market is characterized by a high level of rivalry between the existing suppliers. Acquisitions such as the one of Reebok by Adidas AG in 2005 or Umbro by Nike in 2007 lead to permanent movements of market shares. Furthermore, low entrance barriers allow new, mostly specialized suppliers to enter the market, even if they are only regionally competitors of the global players. Most companies have outsourced their production facilities to low-wage countries in the Far East. Simultaneously, shortened innovation cycles have led to higher costs in research and development as well as less time for product and quality testing before a product enters the market. From the consumer perspective, the Internet has created an environment where customers are confronted with a huge bandwidth of products as well as better information to compare these (specialized category retailers like Zalando, an online footwear shop, are among the most successful examples of e-commerce). In most cases, a change to another brand evokes no costs. As a result, quality aspects as well as a psychological brand commitment and a perfect fit to customers' needs are becoming key enablers of a successful strategy in the sports shoe industry.

At the same time, the industry moved into a model of high variety and choice. Between 1980 and 2000, the number of styles available in the sport goods market increased by more than 3000% (Cox and Alm 1999). One of the great strengths of any successful footwear company is its ability to create a compelling variety of offerings that excite consumers. But there can be too much of a good thing, as Adidas' competitor Nike has observed: Each quarter, Nike sells about 13,000 different styles of footwear and apparel. Many additional styles make it part way through the process, but do not end up in the final line that goes to market. But each one of these tens of thousands of styles drives costs: costs for design, development, sampling, transportation, storage, and sales. For footwear, 95% of Nike's revenue comes from about 35% of the styles, and for apparel from about 40% (Piller 2007).

Costs of samples to showcase this variety to retailers add up to more than $100 million. In addition, an enormous supply chain complexity has to be handled to plan, distribute, and sell this variety. Still, retail outlets face high overstocks, an increasing fashion risk, and the necessity to provide often large discounts in order to get rid of unwanted products, not to mention lost sales caused by products that have performed better than expected and that are therefore not available in adequate quantities or sizes. Facing these challenges, Adidas decided at the end of the 1990s to "go back to the roots" of its long history and to introduce footwear that is being produced only on-demand according to the exact input of a particular consumer. The motivation to enter mass customization clearly was to find a way to reduce the high complexity of a forecast-based system of high variety.

31.3.3 ADIDAS' MASS CUSTOMIZATION OFFERINGS

Adidas' product development team envisioned already in 1999 that offering custom manufactured is a suitable way of countering the trend of growing heterogeneity of customer needs (Moser et al. 2007). An internal pilot was launched in 2000, called *miadidas*, to evaluate the feasibility of a customized product line within Adidas. The objectives narrowed down the project to emphasize a customized product and to test the demand for customized products. Pilot studies were successful, and the offering was continuously extended to its recent form. Today, Adidas has several product lines for customized goods:

- *miadidas*: The core of Adidas' customization offerings is its original mass customization program, miadidas (Berger and Piller 2003; Seifert 2006). The program is focused on performance shoes and combines fit (measurements), function (climate control, insole), and form (color combination). The program started in 2001 after 2 years of planning and testing. Today, miadidas is offered both online and in selected flagship stores. Custom shoes are available across a number of categories, including soccer, running, basketball, and tennis. miadidas has been positioned as a promise to consumers to achieve the same level of footwear as top athletes would get as part of their custom sponsorship package—building on the company's heritage of being a close partner of sportsmen and women.

- *miteam*: This section of Adidas is targeting sports teams, where a coach, for example, can customize sportswear for all team members. The miteam portfolio does not only offer customized shoes, but also customizable apparel and accessories in addition to shoes.
- *mioriginals*: This offering resembles the programs of Keds, Nike, or Converse and allows consumers to create their own style, based on the selection of style options for predefined color fields of a shoe. While the previous offerings focus on the performance aspect of the product, mioriginals clearly is positioned as a fashion product.
- *micoach*: An interesting addition to the Adidas customization portfolio is micoach, offering users a personalized online training program. Customers can provide input on their training objectives and lifestyle habits, and receive a personalized (however, automatically generated) suggestion for a training program to realize this objective. The site also allows users to monitor their goal achievement. Similar to miadidas, micoach has been positioned in the market as providing average users the same training and coaching know-how as is available to top athletes.

31.4   MIADIDAS: CAPABILITIES AND CHALLENGES

Since the initial launch of miadidas, its fulfillment process has improved continuously. Figure 31.1 shows miadidas' value chain today. The process starts with a configuration process between the consumer and Adidas via some form of consumer interface. A customer order then has to be processed by Adidas' order management system. This process triggers a respective manufacturing process within a corresponding production facility. The process ends with the distribution of the final

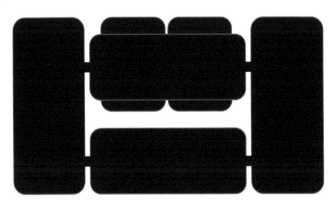

FIGURE 31.1   The miadidas fulfillment process.

product to the end consumer after manufacturing has been completed. In the following, we discuss the miadidas process in greater detail. To structure our arguments, we use the three mass customization capabilities introduced before. We first discuss the development of the solution space at miadidas, then the challenge to establish stable processes, and finally the way how choice navigation works in this case.

31.4.1 SOLUTION SPACE DEVELOPMENT AT MIADIDAS

To define the solution space, a company has to identify those needs where customers are different and where they care about these differences. Matching the options represented by the solution space with the needs of the targeted market segment is a major success factor of mass customization (Hvam et al. 2008). The core requirement at this stage is to access "customer need information," that is, information about preferences, needs, desires, satisfaction, motives, etc., of the customers and users of the product or service offering. Need information builds on an in-depth understanding and appreciation of the customers' requirements, operations, and systems.

For miadidas, this process has been rather simple. First, the decision was made to build the customization activities on existing inline shoes. This allowed for efficiencies in the factory (tooling, production knowledge, inventory of fabrics, and components), but provided a point of reference to consumers on the additional value of a customized product in comparison to a standard shoe. In general, the top-of-the-line variants were selected for each category.

Second, the miadidas team engaged in conventional market research techniques to gather data from representative customers about the scope of the options in a product that should be offered customizable. However, as consumer knowledge about the system had been very low at the launch of the program, Adidas decided to engage in piloting and full-scale prototyping to get deeper insights from potential customers. Participants were recruited during sports events, focusing on people actively doing sports in a serious way (e.g., members of running or soccer clubs, participants at a marathon race, etc.) These pilots revealed some interesting insights:

- 73% of the testers had a very good feeling with the boots.
- Testers would be willing to pay between 10% and 30% more than the regular price for a customized soccer shoe.
- 100% demand for an Adidas customization service in the future.
- 80% understand that the delivery time will take longer (21 days).
- Asked for the criteria that are most important to the customer, fit was mentioned most often by far (68% of the customers interested in custom soccer shoes, 75% for running shoes, respectively), followed by function (14% and 20%), and design (12% and 5%).

As a result, the decision was made to position miadidas as a performance product with the dimensions *mifit* (measurements), *miperformance* (functionality and components), and *midesign* (color options). Until today, miadidas is the only mass customization product by an established sports brand that offers these three options (all competitors, especially NikeID, only offer design customization). Over the past few years, however, the design option became more and more important for two reasons: First, to scale-up the volume of the program, a focus has been placed on online sales. Online sales, however, are not applicable to all customization options. Thus, Adidas was facing a trade-off between keeping its original point of differentiation (fit, form, and function) and the opportunities from using a pure online channel. Second, management had the feeling to serve the "create your own" trend in a better way, focusing on the young, creative consumer that is used to customize her Facebook page and cell phone cover. This trend has become the focus of the mioriginals program.

31.4.2 Robust Process Design at Miadidas

A core idea of mass customization is to ensure that an increased variability in customers' requirements will not significantly impair the firm's operations and supply chain (Pine et al. 1993). This can be achieved through robust process design—the capability to reuse or recombine existing organizational and supply chain resources to deliver customized solutions with high efficiency and reliability. Hence, a successful mass customization system is characterized by stable but flexible responsive processes that provide a dynamic flow of products (Pine et al. 1995; Tu et al. 2001; Salvador et al. 2004; Badurdeen and Masel 2007). Value creation within robust processes is the major differentiation of mass customization versus conventional (craft) customization. Traditional (craft) customizers (like making a custom shoe for a top athlete in the conventional system if Adidas) reinvent not only their products, but also their processes for each individual customer. Mass customizers use stable processes to deliver high-variety goods (Pine et al. 1993), which allows them to achieve "near mass production efficiency" (Tseng and Jiao 2001), but it also implies that the customization options are somehow limited. Customers are being served within a list of predefined options or components, the company's solution space.

The latter principle is illustrated perfectly with the fitting option of miadidas. Traditionally, a custom fit of a shoe is delivered by measuring a customer's feet, creating a custom last representing these shapes, and then producing the shoe around the custom last. In a mass customization setup of footwear, however, in most instances a library of predefined lasts is being used (Boer and Dulio 2007). This library is generally larger than the standard sizing assortment (in the case of Adidas, it represents the entire size range, with three widths per size), and also allows to select a different size for the left and the right foot. Based on the measurement data of the customers, their feet are matched to an existing last. While such a system does not allow the "perfect" fit for all possible shapes, it is "good enough" to present a strong improvement compared to standard sizes available in a regular outlet. Matching a customer order with an existing library of lasts also has two further advantages: Molds and forms required for the production of outsoles can be reduced, which is a major cost advantage. Finally, using pre-defined sizes also allows for stable processes during the sales stage. Customers are enabled to even try the fit of their "custom" shoes by using a sample size in the store. This system improves quality of order taking enormously and transfers the subjective and error-prone process of fitting into a relatively stable and smooth activity.

To manufacture the shoes, Adidas uses the same suppliers for its custom shoes that also produce the corresponding inline products in huge quantities. Adidas was able to achieve this agreement by using its strong position in this market. In addition, for the suppliers flexible manufacturing also became a rather easy option as those companies in general have rather large sample room capacities, with an output of 500,000 pairs a year. For many start-ups in the field of footwear customization, finding a suitable supplier became a long and demanding challenge (as the example of Custom Foot has shown). A global player such as Adidas, on the other hand, is able to bring all manufacturing activities together and reap the benefits of combining mass customization and mass manufacturing.

As soon as a customer order has been submitted to the order management system, the manufacturing of the individualized product has to be triggered. This is an enormously complex process (Figure 31.2), as miadidas shoes have to be manufactured alongside the inline products in the Adidas multiplant network. Adidas manufactures shoes in China, Indonesia, Vietnam, Thailand, and Turkey, and the manufacturing process is different in every plant: For example, the number of local suppliers that is integrated in the respective supply chain varies from factory to factory. Thus, miadidas achieved a large improvement concerning the leverage of existing systems when the supply chain management for miadidas was fully automated in 2006. An ordering system was implemented that automatically allocates orders to production companies. Based on information from planning and development processes, the order management system automatically identifies the manufacturing facility that

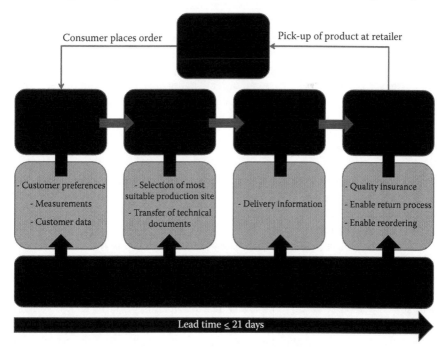

FIGURE 31.2 The miadidas business information process. (Courtesy of Adidas AG, Bavaria, Germany.)

offers the best suitable combination of available resources and capacities. However, the administration of such an automated supply chain faces various problems: An interorganizational information system is required to cope with all different interfaces of the network partners. Furthermore, all information that is relevant for the decision-making processes has to be gathered and made available in a network-wide ERP system. Naturally, all these data have to be updated constantly.

Hence, in an optimal world, designing stable processes for mass customization would start on a greenfield with the development of new products and manufacturing lines as well as new planning resources to enable an automated processing of customer orders. Such a greenfield approach, however, often is difficult for an established company: At the start of the program, volumes are not sufficiently high of to justify all the investments. More importantly, the knowledge and experience for building such a system are not available yet and have to be acquired via processes of organizational experimentation and learning. Hence, in our observation, most mass customization programs are based on the principle of trial-and-error and slow growth. Also within miadidas, sales volumes are not sufficient until today to justify a separate manufacturing line or even dedicated factory. This forces Adidas to provide a mass production system that meets the specific needs of mass customized products. For example, Adidas now has to deal with differences in inventory levels: For the standard inline products, Adidas relies on finished goods inventories. For customized miadidas products, on the other hand, there is no finished goods inventory, but manufacturing demands a larger pool of raw material to cover all variance from the options. Given, however, the relative value of materials compared to the value of the final product, and Adidas' market power toward its suppliers, the material inventory problem is relatively low.

A larger challenge and cost driver is distribution. For miadidas products, distribution has to be organized differently, as customized products at Adidas are manufactured after the customer has placed an order. That means, contrary to buying a pair of shoes at a retailer, miadidas customers cannot take their shoes home directly after the purchase. Instead, customers have to wait for the order to be transferred to the manufacturing facility, for the manufacturing of the shoes and for their distribution, before the product becomes available to them. For miadidas products, this lead time should not exceed 21 days. In consequence, miadidas is using express courier services to distribute the finished goods to the end consumers. From the manufacturing facilities in China, Indonesia, Vietnam, Thailand, and Turkey, the finished products are flown to hubs in the United States, Europe, and Asia. For the U.S. market, miadidas uses FedEx services, whereas DHL services are used in Europe and Asia. The use of express courier services offers an additional benefit to miadidas customers: The courier services offer a tracking system for all transported goods, which allows customers a larger visibility of their order status. On the other hand, distribution via courier services is relatively expensive. Indeed, it is the largest additional cost factor of miadidas compared to the production and distribution of inline shoes. While at the current state of the system a relocation of manufacturing from Asia to Europe is out of consideration, some analysts are already predicting a move from Asia back to Europe or the United States (Gowdner 2011). The additional production cost may be counterbalanced by the savings in time and cost of air delivery.

31.4.3 CHOICE NAVIGATION AT MIADIDAS

Lastly, a mass customizer must support customers in identifying their own needs and creating solutions while minimizing complexity and the burden of choice. The traditional measure for navigating the customer's choice in a mass customization system has been product configuration systems, also referred to as "codesign toolkits" (Franke and Piller 2003, 2004), configurators, choice boards, or customer design systems (Salvador and Forza 2007; Hvam et al. 2008). They are responsible for guiding the user through the elicitation process. Whenever the term "configurator" or "configuration system" is quoted in the literature, for the most part, it is used in a technical sense, usually addressing a software tool. The success of such an interaction system, however, is by no means defined solely by its technological capabilities but also by its integration into the sales environment, its ability to allow for learning, its ability to provide experience and process satisfaction, and its integration into the brand concept. Tools for user integration in a mass customization system contain much more than arithmetic algorithms for combining modular components. In a toolkit, different variants are represented, visualized, assessed, and priced with an accompanying learning-by-doing process for the user. The core idea is to engage customers into fast-cycle, trial-and-error learning processes (von Hippel 1998). Thanks to this mechanism, customers can engage in multiple sequential experiments to test the match between the available options and their needs.

In the miadidas system, there are three possible ways for customers to place an order. Each of these order channels uses a different customer interface: Adidas flagship stores, events, and an online configurator. All three sales channels use a configuration system that helps to visualize the products and that connects the point of sale with the order management system.

31.4.3.1 Retail Stores

The first option is the use of specific Adidas-owned flagship stores. In this case, permanent installations at dedicated retail outlets are chosen. Customers can visit these stores and go through the codesign process with the help of trained sales personnel. This approach offers one major advantage: In the brick and mortar stores, special equipment is available and more detailed customer data can be collected, so that customers can make better use of miadidas' customization offers in terms of fit and functionality.

31.4.3.2 Events

The second sales channel emphasizes special events at selected retailers or wholesale markets. Adidas is able to promote those events to their retail partners by offering them the potential to differentiate themselves from their local competitors by improving the image of their stores. In this case, the retailer takes full responsibility for the marketing of the event and also takes care of the billing process (Berger et al. 2005). Adidas only provides the trained sales personnel for the interaction with the customers for the duration of the event that is usually 2–4 days. This form of event sales can also take place in the context of major sporting events. In this case, Adidas is fully responsible for the whole process, but such events offer the additional benefit of reaching the core target group of miadidas.

31.4.3.3 Online

The third option for placing an order is the Adidas e-commerce platform. This channel is the latest addition to the sales channels of miadidas and will probably become the most important one in the future. In terms of customization of the appearance of the shoes, the online configuration process is mostly identical to the process of placing an order offline. However, the online process cannot offer all options concerning the fit and functionality of the shoes. For reorders, however, this issue can be neglected: Customers can place a first order in an Adidas store and reuse their respective customer data for repeat orders online. That way, mass customization can be an excellent way to generate constant sales based on reorders. On the other hand, the re-buy-situation is not only a chance for miadidas, but it may also turn out to be a strong barrier: Due to the different distribution channels, island solutions have been produced. All distribution partners may not have access to the same pool of data. For example, it is possible to purchase customized shoes in brick and mortar stores and use the same data in e-commerce to reorder the shoes or configure new ones. However, this does not yet work vice versa. This is still an issue that needs effort to facilitate better integration.

The interaction between a manufacturer and the customer that is underlying a codesign process further offers possibilities for building loyalty and lasting customer relationships. Once a customer has successfully purchased an individual item, the knowledge acquired by the manufacturer represents a considerable barrier against any potential switching to other suppliers. Reordering becomes much easier for the customers. The more the customers tell the vendor about their likes and dislikes during the integration process, the better is the chance of a product being created that meets the customers' exact needs at the first try. After delivery of the customized product, feedback from the customer enhances Adidas' knowledge of that customer. The company can draw on detailed information about the customer for the next sale, ensuring that the service provided becomes quicker, simpler, and more focused. The information status is increased and more finely tuned with each additional sale. This data is also used to propose subsequent purchases automatically, once the life of the training shoes is over (for Adidas customers who exercise intensively, this can, in fact, be the case every few months).

When Adidas enters a learning relationship with its customers, it increases the revenues from each customer, because, in addition to the actual product benefits, it simplifies the purchasing decision, so that the customer keeps coming back. Why would a customer switch to a competitor—even one who could deliver a comparable customized product—if Adidas already has all the information necessary for supplying the product? A new supplier would need to repeat the initial process of gathering data from the customer. Moreover, the customer has now learned how self-integration into the process can successfully result in the creation of a product. By aggregating information from a segment of individual customers, Adidas also gains valuable market research knowledge. As a result, new products for the mass market segment can be planned more efficiently, and market research is more effective, because of unfiltered access to data on market trends and customers' needs. This is of special benefit to those companies that unite large-scale make-to-stock production with tailored services. Mass customization can thus become an enabling strategy for higher efficiency of a mass production system. This learning relationship offers new cost-saving potential based on the better access to knowledge about the needs and demands of the customer base (Kotha 1995; Piller and Müller, 2004; Squire et al. 2004).

Benefits of getting access to this knowledge are as follows:

- Reduced or eliminated need for forecasting product demand
- Reduced or eliminated inventory levels of finished goods
- Reduced product returns
- Preventing lost sales if customers cannot find a product in a store that fits their requirements and, thus, allocate the purchasing budget to another item

Choice navigation, however, does not just refer to a technical process of selecting from options or using algorithms to provide the fit. Offering choice to customers in a meaningful way also can become a way for new profit opportunities. Recent research has shown that up to 50% of the additional willingness to pay for customized (consumer) products can be explained by the positive perception of the codesign process itself (Franke and Piller 2004; Schreier 2006; Franke and Schreier 2010; Merle et al. 2010). Product codesigns by customers may also provide symbolic (intrinsic and social) benefits, resulting from the actual process of codesign rather than its outcome.

Adidas tried to address this aspect of choice navigation by various iterations of its online tool, every time improving the user cocreation experience. The largest demonstration of this effect, however, has been the opening of a new kind of performance store that features a very different set-up and sales experience. In 2006, the "mi-innovation center (miC)" opened as part of the Paris flagship store of Adidas at Champs Elysées. The mIC offers consumers customization in technology, style, and design, using many innovative technologies such as a configurator, laser, and infrared technologies, commands generated by gesture translation, a virtual mirror, a digital 3D universe, and radio frequency identification (Kamenev 2006). Until today, however, this concept has not been rolled out to a larger extent and can be considered as a unique experiment.

31.4.4 Alternatives to Mass Customized Footwear Products

Independently from miadidas, Adidas introduced two further radically new product designs that allow for customization in the actual product. We discuss these two alternative approaches at the end of this section to outline further ways of enabling customization from a consumer perspective. The company's underlying rationale is again to reduce risk and uncertainty with regard to product variants, but to do so without the complexity of custom order taking, on-demand manufacturing, and single-piece distribution. The key of these approaches is adaptable, intelligent products (a strategy that dominates customization in many markets, including smart phones, laptops, mattresses, or office furniture).

The first of these products, dubbed *Adidas 1*, is a running shoe that shall provide a huge range of cushioning options in one product (for running shoes, cushioning is a large driver of variety as users demand different degrees of stiffness). To embed flexibility, the shoe is equipped with a sensor, a system to adjust the cushioning, and a microprocessor to control the process. When the shoe's heel strikes the ground, the magnetic sensor measures the amount of compression in its midsole and the microprocessor calculates whether the shoe is too soft or too firm. Then, during the time period when the shoe is airborne, a tiny motor shortens or lengthens a cable attached to a plastic cushioning element, making it more rigid or pliable. Each shoe also has a small user interface that allows for manual adjustments of the product, allowing users to trim the computer's decision to personal taste. According to Adidas, the shoe's range of cushioning options corresponds to at least five previous fixed variants of one shoe, reducing the planning uncertainty by this dimension.

Adidas' second adaptable product, *Tunit*, is a modular system that allows the user to mix and match the various parts of a soccer boot during the usage stage. The boots are sold broken down to three components: upper, chassis (insole), and studs. All components are interchangeable, transferring a sole product into a product system. Introduced for the 2006 World Cup, the product is presently offered in a variety of 10 uppers, 3 chassis, and 3 different sets of studs (uppers and chassis are offered in the standard soccer size range of 16 sizes). The Tunit system changes the conventional way of planning and launching a design variant even more than two other concepts of postponing into the user: First, the underlying form postponement of the Tunit system allows Adidas to plan in a much more flexible manner. The present scope of modules allows configuring 90 different variants. Forecasts have to be performed just on the level of the 10 uppers (colors, fabrics), while studs and insoles are stable modules over many seasons. Conventionally, Adidas would have offered a comparable soccer boot in 28–36 different variants (color–insole–stud combinations), all subject to a detailed forecasting process. In the Tunit system, this risky decision of the final product configuration is postponed into the user domain. Second, the provision of special retail variants or limited editions becomes much more efficient. A new variant now just demands an additional upper. For example, presently Adidas is also offering its standard soccer boots in a "style version" intended for street use. Despite their similarity to the regular version, these streetwear products form an independent product line. With an extension in the range of outsoles and studs, Adidas could eliminate the entire "style" line and include it into the Tunit system. Third, the system provides value to users who do not have to purchase multiple pairs of shoes when playing on different grounds. For Adidas, on the other hand, the opportunity to "upgrade" shoes and sell additional components is a new profit opportunity (with much higher margins).

31.5 DISCUSSION

The development of miadidas reveals a long-learning path and trial-and-error process for implementing and scaling-up a mass customization offering. Mass customization was considered as one possible solution to cope with the high variety and increasing complexity of the inline business. In the initial years, miadidas' real function in the corporate strategy, however, was different. It primarily was a supporting function for the brand image, enhancing also the positioning of other products of the performance category. Besides that, the company could create an image of itself as an innovative player by offering a unique brand experience and improving relationships with its customers. However, serving as a marketing tool or pilot is not a sustainable position for a business unit in a global enterprise.

Therefore, in the later years the miadidas business unit has been reorganized and has been integrated more deeply into the existing processes and routines. Activities such as special retail events became less significant, whereas permanent locations and the online channel became more important (Moser et al. 2007). Most importantly, miadidas can be considered as a knowledge-rich activity that produces information to improve the inline business. The inline (standard) assortment benefits from more accurate forecasts of customer needs and trends as well as more appealing models. At the same time, achievements with regard to design (e.g., modular product architectures), manufacturing technology (flexible printing), logistics (fast distribution system, direct sales), and online sales (e-commerce configurator) served as learning laboratories for the entire company. Several new technologies that first have been introduced to enable miadidas are now supporting efficiency in the inline system.

From our observations of miadidas' development, we conclude that implementing mass customization and, thus, building its underlying three strategic capabilities is by far no easy or straightforward task. Mass customization demands a new customer–company relationship for all members of the organization. This proved to be much more difficult than expected. For the majority of employees at Adidas, the "customer" still is the sales organization, perhaps the retail partner, but—in most instances—not the final consumer. Mass customization, however, demands such a customer-centric perspective.

An indication of this challenge may be the fact that the reordering and learning process (outlined in Section 31.2.3) has not been established in most countries. While rather simple from a technological perspective, and also intuitive from a marketing point of view, enabling a direct relationship with end consumers has proven to be much more difficult than expected. It is the mental gap and the dominance from the existing mass production thinking that is preventing its execution. At the same time, the customization unit within Adidas has to continuously prove its value set and survive as part of a traditional mass producer. Until today, the leadership of this unit masters this challenge very well and has positioned miadidas as an experimental learning space for the entire company (Piller and Walcher 2006). However, the case clearly demonstrates that even for an established and financially successful corporation such as Adidas, implementing mass customization is not easy, despite a suitable market and great customer feedback. Successful mass customization demands building a unique set of strategic capabilities, and changing the corporate mentality to become truly customer centric. But once mastered, it is exactly these challenges that provide the competitive advantage of mass customization.

QUESTIONS

31.1 Why is mass customization considered a suitable strategy specifically for the footwear industry?

31.2 Which three critical capabilities do the literature identify for a successful implementation of mass customization?

31.3 What are the different customization offerings of Adidas? How do these product lines differ in terms of their target groups and customization dimensions?

31.4 How did Adidas derive the solution space for miadidas, that is, how were the customization options chosen?

31.5 Please discuss the trade-off between the custom fit and robust process design that Adidas had to solve.

31.6 How does the cocreation process impact the manufacturer–customer relationship?

31.7 What are the benefits of gaining access to knowledge about individual customer needs?

ACKNOWLEDGMENTS

The documentation of this chapter was possible only due to the continuous interaction between the first author and the miadidas team over the last decade. The authors especially thank Christoph Berger, Director and initiator of the original miadidas program, and Alison Page, his successor and present Director of Customization, for their continuous support and exchange of ideas. For reasons of privacy, all information of this chapter has been taken from public information, company presentations, press statements, and previous publications. The discussion section is solely the opinion of the authors and does not reflect an official opinion of Adidas.

REFERENCES

Badurdeen, F., Masel, D. (2007). A modular minicell configuration for mass customization manufacturing. *International Journal of Mass Customization*, 2(1/2), 39–56.

Berger, C. (2005). Bridging mass customization and mass production at adidas. *Proceedings to the MCPC 2005*, Hong Kong, China.

Berger, C., Moeslein, K., Piller, F., Reichwald, R. (2005). Co-designing the customer interface for customer-centric strategies: Learning from exploratory research. *European Management Review*, 2(3), 70–87.

Berger, C., Piller, F. (2003). Customers as co-designers: The miAdidas mass customization strategy. *IEE Manufacturing Engineer*, 82(4), 42–46.

Boer, C.R., Dulio, S. (2007). *Mass Customization and Footwear: Myth, Salvation or Reality?* Springer, London, U.K.

Chamberlin, E.H. (1962). *The Theory of Monopolistic Competition: A Re-Orientation of Value Theory*, Harvard University Press, Cambridge, MA.

Cox, M., Alm, J. (1998). The right stuff: America's move to mass customization. Federal Reserve Bank of Dallas, 1998 Annual Report, Dallas, TX.

Davis, S. (1987). *Future Perfect*, Addison-Wesley, Reading, MA.

Dulio, S., Boër, C.R. (2004). Integrated production plant (IPP): An innovative laboratory for research projects in the footwear field. *International Journal of Computer Integrated Manufacturing*, 17(7), 601–611

Franke, N., Piller, F. (2003). Key research issues in user interaction with configuration toolkits in a mass customization system. *International Journal of Technology Management*, 26, 578–599.

Franke, N., Piller, F. (2004). Toolkits for user innovation and design: An exploration of user interaction and value creation. *Journal of Product Innovation Management*, 21(6), 401–415.

Franke, N., Schreier, M. (2010). Why customers value self-designed products: The importance of process effort and enjoyment. *Journal of Product Innovation Management*, 27(7), 1020–1031.

Gownder, J.P. (2011). Mass customization is (finally) the future of products. Forrester Report, Cambridge, MA.

Huffman, C., Kahn, B. (1998). Variety for sale: Mass customization or mass confusion? *Journal of Retailing*, 74(4), 491–513.

Hvam, L., Mortensen, N., Riis, J. (2008). *Product Customization*, Springer, New York.

Kamenev, M. (2006). Adidas' high tech footwear. *Business Week*, November 3, 2006.

Kaplan, A.K., Schoder, D., Haenlein, M. (2007). Factors influencing the adoption of mass customization: The impact of base category consumption frequency and need satisfaction. *Journal of Product Innovation Management*, 24(2), 101–116.

Kotha, S. (1995). Mass customization: Implementing the emerging paradigm for competitive advantage. *Strategic Management Journal*, 16(5), 21–42.

Merle, A., Chandon, J., Roux, E., Alizon, F. (2010). Perceived value of the mass-customized product and mass customization experience for individual consumers. *Production and Operations Management*, 19(5), 503–514.

Moser, K., Müller, M., Piller, F.T. (2007). Transforming mass customization from a marketing instrument to a sustainable business model at Adidas. *International Journal of Mass Customization (IJMassC)*, 2(2006) 4, 463–480.

Piller, F. (2005). Mass customization: Reflections on the state of the concept. *International Journal of Flexible Manufacturing Systems*, 16(4), 313–334.

Piller, F. (2007). The consumer decides: Nike focuses competitive strategy on customization and creating personal consumer experiences. February 26, 2007, *MC&OI News* (Online Blog), http://tinyurl.com/5wzvlfn

Piller, F., Tseng, M. (2010). Mass customization thinking: Moving from pilot stage to an established business strategy. In F. Piller and M. Tseng (eds.), *Handbook of Research in Mass Customization and Personalization, Part 1: Strategies and Concepts*, World Scientific Publishing, New York, pp. 1–18.

Piller, F., Walcher, D. (2006). Toolkits for idea competitions: A novel method to integrate users in new product development. *R&D Management*, 36(3), 307–318.

Pine, B.J. (1993). *Mass Customization*, Harvard Business School Press, Boston, MA.

Pine, B.J., Peppers, D., Rogers, M. (1995). Do you want to keep your customers forever? *Harvard Business Review*, 73(2), 103–114.

Pine, B.J., Victor, B., Boynton, A.C. (1993). Making mass customization work. *Harvard Business Review*, 71(5), 108–119.

Salvador, F., de Holan, M., Piller, F. (2009). Cracking the code of mass customization. *MIT Sloan Management Review*, 50(3), 70–79.

Salvador, F., Forza, C. (2007). Principles for efficient and effective sales configuration design. *International Journal of Mass Customization*, 2(1/2), 114–127.

Salvador, F., Rungtusanatham, M., Akpinar, S., Forza, C. (2008). Strategic capabilities for mass customization: Theoretical synthesis and empirical evidence. *Academy of Management Best Paper Proceedings*, Anaheim, CA.

Salvador, F., Rungtusanatham, M., Forza, C. (2004). Supply-chain configurations for mass customization. *Production Planning and Control*, 15(4), 381–397.

Schreier, M. (2006). The value increment of mass-customized products: An empirical assessment. *Journal of Consumer Behavior*, 5(4), 317–327.

Seifert, R.W. (2006) The miadidas mass customization initiative, IMD case study No. POM 249, International Institute for Management Development. Revised version, Lausanne, pp. 1–24.

Squire, B., Readman, J., Brown, S., Bessant, J. (2004). Mass customization: The key to customer value? *Production Planning and Control*, 15(4), 459–471.

Tseng, M., Jiao, J. (2001). Mass customization. In G. Salvendy (ed.), *Handbook of Industrial Engineering*, Wiley, New York, pp. 684–709.

Tseng, M., Piller, F. (2003). The customer centric enterprise. In M. Tseng and F. Piller (eds.), *The Customer Centric Enterprise: Advances in Mass Customization and Personalization*, Springer, New York, pp. 1–18.

Tu, Q., Vonderembse, M.A., Ragu-Nathan, T.S. (2001). The impact of time-based manufacturing practices on mass customization and value to customer. *Journal of Operations Management*, 19(2), 201–217.

von Hippel, E. (1998). Economics of product development by users: The impact of "sticky" local information. *Management Science*, 44(5), 629–644.

Wong, M. (1996). 'Virtual inventory': If shoe fits, order it; new store at megamall doesn't carry any stock. *Star Tribune*, August 6, Page D1.

32 selve Model
Custom Shoes in the Twenty-First Century

Claudia Kieserling

CONTENTS

32.1 BRIEF HISTORY OF SHOE MAKING: FROM SELF-MADE FOOT PROTECTION OVER ELABORATE ARTISANRY TO THE INDUSTRIALIZATION OF FOOTWEAR AND WHERE THE SELVE MODEL STANDS

In the evolution of mankind, footwear served various purposes: protection against cold, heat, and injuries, and along with clothing in general during the various stages of civilization it served numerous cultural expressions.

Primitive people used simply the materials that surrounded them to cover their feet for protection: vegetable substances like bark or reed and untanned animal skins. Along with evolution, the techniques to manufacture got more refined: weaving, stitching, carving, tanning, etc. In addition, the functional demands for footwear got more demanding and the skills to manufacture these goods evolved. Soon, refined footwear became a status for the wealthy and powerful as well as for rituals, rites, and religion.

Further, along with the development of the sense for beauty, footwear became the object of skillful artisans and started to underlie the waves of fashion (Weber, 1982).

When hunters and gatherers, that were self-supporters, settled and developed agriculture, artisans started to develop their skills in various fields and a culture based on division of labor evolved. The shoemaker craft developed differently in various cultures but in any case, shoes were made to order. The purposes, skills, and methods varied but the principle, that shoes were custom made, did not change from the earliest development of crafts and guilds until the beginning of industrialization in the middle of the nineteenth century. In Europe, larger quantities of shoes were manufactured handmade in the eighteenth century and the first trade channels developed to sell pre-manufactured shoes (Schlachter, 1981). Along with the trade, a sizing system developed (which was not necessary with custom-made shoes). With the development of shoe machinery mainly in the United States in the nineteenth century, industrial shoe manufacturing evolved quickly in the United States and Europe and widely spread channels of retail developed.

The advantages of industrial shoe production were (and are) manifold: the entrepreneur has a lucrative field to fulfill a market demand, gives work to a society of laborers and the consumer can choose at a low price from finished products that he can see, test, and try. A democratization of the availability of footwear started.

With the enormous success of industrialization and technology in the nineteenth and twentieth centuries, the system of footwear manufacturing and retail spread across the world. It got more and more refined and nowadays is a highly complex marketing-driven industry.

The shoemaker artisan has survived only as a specialist that serves the elites or as a craftsman in the medical field.

With markets long being saturated, customers becoming more and more demanding, and technology and organizational methods evolving a new field for "mass customized" footwear is opening. Since Joe Pine promoted the idea in the United States in the 1990s and Frank Piller in Europe soon after, selve was one of the very first companies creating a new model for footwear manufacturing and marketing. Today, with footwear artisans almost not existing anymore and customers being highly sophisticated, new markets develop. The traditional system of premanufacturing and traditional retail shows limitations: although the product variety and availability seem limitless, customers leave stores without the shoes they were hoping to find—color, size, style, or fit may not be available at the moment in time the customer is looking for it. This means for the retailer a high risk of stocked goods and lost sales opportunities at the same time. Seeing large traditional retail corporations struggling, this highly successful system is coming to certain boundaries. At the same time, every finished product has to be either discounted or disposed of, while it has already consumed energy and material to be manufactured and transported long distances. It's effect for climate and society (discounted shoes flooding, i.e., African markets, by such killing the local manufacturer/trade) have not been fully considered yet.

The selve model following the Pine and Piller "economies of one" (Pine, 1998) intends to combine the best of two worlds: offering the customer a set of choices they can physically see and try (which they are used to from retail) with the flexibility of the artisan to customize size, fit, style, color, and design details. With the help

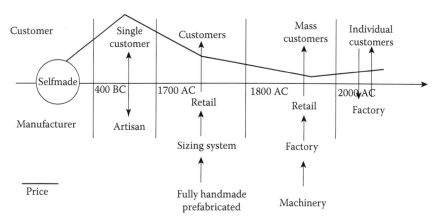

FIGURE 32.1 The customer/manufacturer relation in the history of shoe making.

of modern organization and technology, prices and delivery times are setting new standards for customized footwear and are putting the customer at its center. Eliminating stock and preproduction for using a set of modules in production offers flexibility, timeliness, and affordable prices (Figure 32.1).

32.2 CATEGORIES OF FOOTWEAR IN CONTEMPORARY FOOTWEAR MANUFACTURING AND HOW THEY REACT TO CUSTOMER NEEDS

Traditionally, markets are divided primarily by price or function. The more sophisticated and saturated a market becomes, the more variety and added value driven by marketing become critical.

Manufacturers have no direct contact with customers. Marketing as well as research and development filter influences manufacturers to offer finished goods to customers (Figure 32.2).

Price and function being the main drivers, added value is highly important for marketing positioning in saturated markets. Mass customization and customer codesign options offer important and valuable options for differentiation. Mass customization creates added value in the customer perspective (Figure 32.3).

32.3 CHALLENGES TODAY IN THE FOOTWEAR INDUSTRY IN MARKETING, DESIGN, DEVELOPMENT, AND MANUFACTURING CAPABILITIES IN TERMS OF PERSONNEL, EQUIPMENT, AND RESOURCES

Footwear manufacturing itself has not changed much since the start of industrial manufacturing. From 1840 to 1883, in a rapid succession, shoe machinery was developed mainly in the United States: from cutting machine over stitching machine to lasting machine, but since the introduction of modern chemicals for cementing after 1910 (Schlachter, 1981) as an alternative to stitching for applying outsoles to the

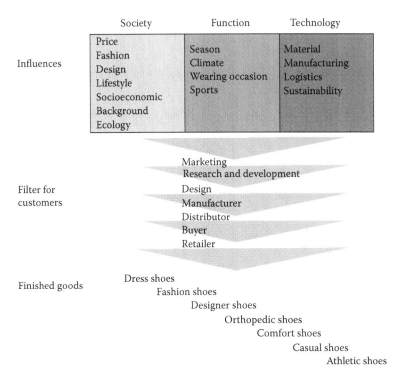

FIGURE 32.2 Customer in the marketing and manufacturing process.

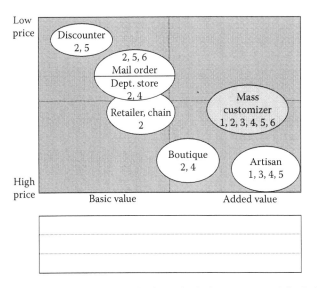

FIGURE 32.3 Mass customization positioning price/value: 1, personal fit; 2, fashion/style; 3, personal style; 4, experience/emotion; 5, availability; 6, additional shopping availability. (From Baumann, F. et al., Mass customization in the footwear industry [In German], Diploma thesis, University of St. Gallen, St. Gallen, Switzerland, p. 63, 1999.)

upper, the developments were various machinery improvements and the automation of some manufacturing processes.

One new technology was PU outsole direct injection, but it has its limitations because it does not allow much variety.

A few innovations have been made on the material side: gore-tex and similar membranes, tanning techniques for special purposes: i.e. waterproof for hiking boots, vegetable tanning for ecological reasons. Professional research in a larger scale is done only in the athletic shoe manufacturing industry, driven by function and performance.

Footwear manufacturing is labor intensive and driven by short life cycles following fashion. With globalized manufacturing and sophisticated and low-cost logistics, it was easy shifting production sites to places of the cheapest labor, since basic skills are relatively easy to train.

After the shoe industry flourished in the United States and Europe in the nineteenth and twentieth centuries, Asia became the biggest manufacturer of footwear, with over 70% of the world shoe production (Boer and Dulio, 2007). Since labor cost in Asia is relatively low and nevertheless the footwear industry is lucrative, i.e., Hong Kong's footwear industry alone earned HK 44.4 billion in 2004, there is little pressure for the industry for innovating the production process. Instead the major changes happened in marketing and logistics. The vertical integration of retail and manufacturing, licensing, mail order, and the increased success of online stores have made footwear a part in the process of an integrated marketing and distribution chain.

Marketing and design are unique to a brand. Manufacturing is an important but interchangeable service provider.

Manufacturing must react quickly to demands from marketing and fashion. Fast training of workers is a key success factor for this. Technicians must be skilled and reliable. Production managers and logistics managers play a key role in the complex process, where timing and cost are critical.

32.4 MODERN APPROACH MADE TO MEASURE SHOES: LOGISTICS, PRACTICALITIES, AND HOW TO SATISFY THE CUSTOMER

selve offers a customer-centric approach in footwear marketing and manufacturing.

A selve market research (174 women in Germany and Switzerland) on customization had the following key findings:

1. *Fit is a critical buying factor*: 70% are interested in personalized fit, accepting a higher price.
2. *Customers are unhappy about the availability in stores*: 84% have experienced that a shoe they intended to buy was not available, 60% experience this often.
3. *Buying shoes is very emotional*: 35% buy just for emotional reasons and another 35% buy to replace other shoes.

Mass customization is the ideal business model to fulfill these needs in saturated, modern markets.

Scanning

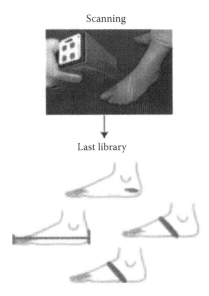

Last library

FIGURE 32.4   Foot scanning and library for match fit manufacturing.

Customers have functional and aesthetic demands and desire for footwear. selve offers a solution space with toolkits and modules to meet these customer demands.

For dress shoes, the most important functional demand is a good fit. selve made a research study on foot measurements, scanned 1000 pairs of women feet, clustered the measurements, and developed its own sizing system. Following these measurements, selve created its own last library for "match fit" manufacturing and last modification for custom lasts (Figure 32.4).

Aesthetic demands and desires are extremely widespread. selve offers a product configurator for customers to design their personal style out of billions of options consisting of last shape, heel height, shoe style, heel shape, leather qualities, leather color, lining options, outsole options, heel cover options, and embellishments (Figures 32.5 and 32.6).

FIGURE 32.5   selve style "Ayleen" variation overview.

FIGURE 32.6   selve style "Ayleen" configurator.

The combination of fit and style is very complex. selve offers skilled consultation for customers to navigate through the options.

Every order is processed within a server-based online system that is connected with the selve-owned factory. The factory stocks lasts, material, and patterns. Custom lasts are manufactured digitally, and custom patterns made accordingly.

Finished shoes are shipped to the selve office and then distributed to the customers.

Every customer's order and contact is stored in the selve database for repeat orders.

32.4.1   On Fit versus Design Customization

Luximon and Goonetilleke (2007) have pointed out that "in spite of a clear indication that design is not the major interest to consumers, companies like Nike and Adidas are offering aesthetic customization." Good fit is very difficult to achieve. The factors that influence the perception of fit are a complex mix of objective and subjective criteria interacting between object (shoe) and subject (wearer). An incomplete list of fit criteria determining and influencing fit is: anatomical shape of foot and last, construction and material selection of the shoe, design of the shoe pattern, climate, season, daytime, biological cycle of the body, and psychological perception.

This highly dynamic and complex mix represents a major challenge for companies: the sales process becomes very expensive and the manufacturing very demanding, thus dramatically increasing the internal cost factors for customization offers (Boer and Dulio, 2007), which explains why big companies have no interest in such difficult ventures. It remains a fact that "Comfort and fit are still neglected by footwear companies" (Luximon and Goonetilleke, 2007), decades after a pioneer of the footwear world Salvatore Ferragamo had that same insight:

> So they (Ferragamos competitors) steal everything ephemeral, everything that tomorrow or next season or next year will be dead and gone, passed into the sphere of the old-fashioned …. Yet they refuse to look at the one unchanging feature of my shoes, the reason why my shoes fit and others do not (Ferragamo, 1957).

The psychological factor is especially difficult to handle, since it is very complex in itself. Especially for women, it may be safe to say that if a shoe is perceived as a very desirable object, fit is willingly sacrificed. The popularity of high heel shoes for women, in spite of its obvious anatomical challenges for the foot, cannot simply be explained by women being less rational than men. Walking in high heel shoes is a rather complex cultural expression: it requires balancing the body on the small and narrow surface of the forefoot and heel. This requires more attention and subtle movements, which is enhancing femininity. In a high heel, the foot cannot roll naturally, which requires steps to be shorter. Further, with lifting the heels the center of gravity of the whole body moves forward which has consequences for the posture. Overall, the liability of balance signals helplessness and fragility, which is a "manipulation of perception" (Tietze, 1988). Nevertheless, there is obviously a pleasure in wearing high heel shoes. And there are ergonomical qualities to high heel shoes (if not exaggerated) that seldom get noted: the muscles of the leg get trained and the pelvis straightens up—both are a good compensation for primarily seated activities. (Tietze, 1988).

Fashion, status, and personal taste are just a few drivers for women effecting the perception of a shoe, making it one of the most difficult products to customize in fit and design.

32.5 OUTLOOK: WHERE IS THE FOOTWEAR INDUSTRY HEADING? WHAT IS THE CENTRAL SUCCESS FACTOR: THE LOWEST MANUFACTURING PRICE OR THE BIGGEST MARKETING POWER? AND WHERE IS THE CUSTOMER IN THESE SCENARIOS?

Customers have needs and desires. Footwear companies make use of these opportunities, some mainly through brand and marketing, others mainly through price. Marketing and branding are not only very powerful but also financially challenging (Figure 32.7).

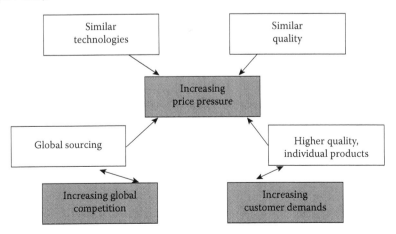

FIGURE 32.7   Three controversial market trends, following Belz et al. (1997).

Big established brands must offer a full range of products to cover the international marketing expenses. Footwear and leather goods manufacturers offer clothes and accessories and vice versa. They cover the full value chain including retail. Retailers are vertically integrated with their own manufacturing structures.

Within this highly sophisticated structures and logistics, the customer is the aim and target. In the steps before the actual sale, he is the subject of marketing and research and can buy or not buy eventually. The generation of consumers that made the current brands successful is used to this process, where they have the final decision but are not being involved in earlier stages.

The generation growing up with smart phones, Internet, and online games are used to configure their online environments and tools. They will expect to be able to configure traditional products as well. They will also have the skills to do so. With technology of product configurators getting more refined and affordable, consumers will want more say—and the "old world" of footwear manufacturing will have to find a way to manufacture based on customer orders.

QUESTIONS

32.1 Will the changing demands of the customers change the footwear industry?

32.2 Will mass customization become an established part of the footwear industry?

32.3 Will the footwear industry embrace the fit challenges of customized footwear on a larger scale?

32.4 Will technologies emerge that will help to cope with the complex challenges of customized fit?

32.5 Will Asia remain the major manufacturing base for footwear or will western countries try to move labor-intensive production back home, to create more jobs for the lower and middle classes?

REFERENCES

Baumann, F., Brunner, D., Garzoli, E., Kieserling, C., 1999, Mass customization in the footwear industry (in German), Diploma thesis, University of St. Gallen, St. Gallen, Switzerland, p. 63.

Belz, C. et al., 1997, *Industry as a Service Economy*, Thexis, St. Gallen, Switzerland, p. 15.

Boer, C., Dulio, S., 2007, *Mass Customization and Footwear: Myth, Salvation or Reality*, Springer, London, U.K., pp. 10, 20.

Ferragamo, S., 1957, *Shoemaker of Dreams*, Harrap & Co, London, U.K., p. 69.

Luximon, A., Goonetilleke, R.S., 2007, Footwear development and foot modeling, in *Handbook of Digital Human Modeling, 2008*, V. Duffy (ed.), Taylor & Francis Group, Boca Raton. FL, p. 19.

Pine, B.J. II, 1998, *Market Strategy of the Future, Preface to Piller* (in German), Hanser, München, Germany, p. 11.

Schlachter, A., 1981, *Shoe, Leather and Shoe Accessories* (in German), Stam, Köln, Germany, pp. 7–9.

Tietze, B., 1988, *Shoes* (in German), Anabas, Giessen, Germany, pp. 96–97.

Weber, P., 1982, z.B. Schuhe, *Shoes Three Millenniums in Picture* (in German), AT Verlag, Stuttgart, Germany, p. 7.

33 Footwear Customization for Manufacturing Sustainability

Claudio R. Boër and Paolo Pedrazzoli

CONTENTS

33.1 INTRODUCTION

To adapt to global competitive pressures, European footwear industry must develop methods and innovative enabling technologies toward a personalized, customer-oriented, and eco-efficient manufacturing (Bischoff et al. 2010). To this end, the vision hereinafter presented strives to put forward a new production paradigm: sustainable mass customization in footwear, presenting customization as one of the main driving forces to achieve effective sustainability. Indeed, manufacturing is growing beyond the economic context, into a social and ecological phenomenon, both from the side of the end user and from the side of governments who, guided by increased customer eco-consciousness, are expected to further introduce eco-taxes, motivating companies to move toward sustainable manufacturing: Manufacturers are thus demanded to merge the need to be reactive toward customer needs and wishes (customized products), with the requisite to be proactive toward ecological and environmental impact (sustainable products).

This chapter first portrays the two concepts of mass customization and sustainability in footwear and then proposes a framework for their actual linking. Eventually, three different, but concurrent EU projects meant to provide solutions to the aforementioned challenges in the field of footwear are presented. The research and experiences discussed in the chapter are carried out in the framework of the last of these aforementioned projects, namely, S-MC-S (sustainable mass customization, mass customization for sustainability—http://web.ttsnetwork.net/SMCS).

33.2 MASS CUSTOMIZATION, SUSTAINABILITY, AND THEIR LINK

The following sections present an overview of the concepts of mass customization and sustainability, meant to provide a common and updated vision of the two paradigms, lowered into the footwear framework. We then highlight the links between them.

33.2.1 MASS CUSTOMIZATION

While taking into account the previous approach to mass customization concept definition (Davis 1987, Pine 1993, Tseng and Jiao 2001), we choose to start from the work done by Frank Piller. Piller (www.mass-customization.de) provided his definition of mass customization, focusing on key concepts that really distinguish mass customization from similar approaches.

> Mass Customization refers to customer co-design process of products and services, which meet the needs of each individual customer with regard to certain product features. All operations are performed within a fixed solution space, characterized by stable but still flexible and responsive processes. As a result, the costs associated with customization allow for a price level that does not imply a switch in an upper market segment (Piller 2004).

This is the definition we choose to start from. Again following Piller's argument and work, this definition can be further interpreted into four statements (Piller 2004).

33.2.1.1 ITEM 1: Customer Codesign

Customers are integrated into value creation by defining and configuring an individual solution. Customization demands that the end user transfers his needs and desires into concrete product specifications. A tool is then needed: whether a paper catalogue, listing variants and combinations, or digital configuration software, the codesign is empowered by a proper tool. The footwear sector offers several examples of web-based tools.

Company	Number of Customizable Shoe Models	Customer Codesign Options
ADIDAS www.miadidas.com	7 Unisex	Depending on the models, ~8–12 shoe parts can be modified; possible to change few materials. Can add a logo (database selection) or a number. Possibility to choose a different size and two different widths for each shoe.
CONVERSE www.converse.com	32 Women 31 Men	10–15 shoe parts can be modified; possibility to choose between colors and patterns. Possibility to add a tag.
FOOTJOY www.footjoy.com	6 Women 15 Men	Three shoe parts can be modified; possibility to add a different tag on each shoe. Possibility to choose a different size and six different widths for each shoe.
KEDS www.kedscollective.com	3 Women 2 Men 2 Kids	Depending on the models, ~10–12 shoe parts can be modified; possible to add a logo, a pattern (database selection) or an uploaded picture.
LEFT www.leftshoecompany.com	112 Men	Possibility to choose up to 17 types of leather and 5 types of outsole. Fitting customization provided trough feet scanning within the shop.
MORGAN MILLER www.morganmillershoes.com	30 Women	Four shoe parts can be modified; possibility to choose between three different shoe widths.
NIKEID nikeid.nike.com	60 Women 100 Men	Depending on the models, ~7–10 shoe parts can be modified; possibility to change sole and insole. Possibility to add six characters on the heel. Possibility to choose a different size and two different widths for each shoe.
OTABO otabo.com	42 Men 3 Women	Two shoe parts can be modified choosing between seven types of leather; possibility to select four different outsoles and to put a tag on the outsole.
PRESCHOOLIANS preschoolians.com	100 Kids	5–10 shoe parts can be modified; possibility to choose between 5 types of sole.
RYZ www.ryz.com	4 Unisex	Four shoe parts can be modified; possibility to upload a picture for sole design.
TIMBERLAND shop.timberland.com	3 Women 2 Men 2 Kids	10 shoe parts can be modified; possibility to put a tag on the tongue and on hangtag (same for both shoes).
VANS shop.vans.com	4 Unisex	10–12 shoe parts can be modified with colors or patterns.

We do not further define here the personalization options of this web-based examples and the related demands to the manufacturing system, as we elaborately discuss the subject in the next sections.

Finally, it is to be pointed out that the codesign phase establishes an individual contact between the manufacturer and the customer, which offers possibilities for building up a long-lasting relationship. Once the customer has successfully purchased an individual shoe, the knowledge acquired by the manufacturer represents a considerable barrier against switching suppliers (Dulio and Boër 2004).

33.2.1.2 ITEM 2: Meeting the Needs of Each Individual Customer

The codesign action aforementioned in ITEM 1 is an action that concretizes the customization potential, expressed by all the possible product configurations (that represent the degree of customization offered by the manufacturer), into a single customized product. Customization offers the increment of utility of a product that fits the needs of the customer better than the standard available. To correctly identify the customization options and dimensions meant to satisfy the additional customer needs is a major success factor of MC. To better express the level of customization offered, we highlight three dimensions: fit, style, or functionality (Boër and Dulio 2007, Piller 2004). *Style (aesthetic design)* relates to modifications aiming at sensual or optical senses, that is, selecting colors, styles, applications, cuts, etc. Many mass customization offerings are based on the possibility to codesign the outer appearance of a product. This kind of customization is often rather easy to implement in manufacturing, demanding a late degree of postponement. *Fit and comfort (measurements)* is based on the fit of a product with the dimensions of the recipient, that is, tailoring a product according to a body measurement or the dimensions of a room or other physical objects. In the case of footwear, this means to measure the two feet in three dimensions and extract the necessary information to choose the best fitting last or even to make the personalized one. It is the most difficult dimension to achieve in both manufacturing and customer interaction, demanding expensive and complex systems to gather the customers' dimensions exactly and transfer them into a product. *Functionality* addresses issues like selecting speed, precision, power, cushioning, output devices, interfaces, connectivity, upgradeability, or similar technical attributes of an offering (Boër and Dulio 2007, Piller 2004).

It is important to highlight that the customer as a single, differently as it happens within the codesign in ITEM 1, does not personally impact the company choices in defining the customization dimension of its product: those are defined by market research, surveys, and anticipation of trends. With reference to the footwear company aforementioned, we can place those examples in a more meaningful context, by stating their capability to cover the customization dimensions just presented.

Company	Customization Dimension Involved
ADIDAS www.miadidas.com	Aesthetic
CONVERSE www.converse.com	Aesthetic
FOOTJOY www.footjoy.com	Aesthetic
KEDS www.kedscollective.com	Aesthetic
LEFT www.leftshoecompany.com	Aesthetic–Fitting–Functional
MORGAN MILLER www.morganmillershoes.com	Aesthetic

(continued)

Company	Customization Dimension Involved
NIKEID nikeid.nike.com	Aesthetic
OTABO otabo.com	Aesthetic–Functional
PRESCHOOLIANS preschoolians.com	Aesthetic–Functional
RYZ www.ryz.com	Aesthetic
TIMBERLAND shop.timberland.com	Aesthetic
VANS shop.vans.com	Aesthetic

33.2.1.3 ITEM 3: Stable Solution Space

The space within which a mass customization offering is able to satisfy a customer's needs must be finite. The term "solution space" represents "the pre-existing capability and degrees of freedom built into a given manufacturer's production system." The concept can be clearly applied to mass production as well. A successful mass customization system is characterized by stable but still flexible and responsive processes. It is the concept of stable processes, used to deliver high variety goods, that allows to achieve "near mass production efficiency." As already mentioned, this also implies that the customization options are limited to certain product features. Customers perform codesign activities within a list of options and predefined components (ITEM 1) that were chosen before his customization activity (ITEM 2). Those options were defined trying to meet the needs of the individual customer, by analyzing the needs of the many. Thus, setting the solution space becomes one of the foremost competitive challenges of a mass customization company.

There is a strong link between the "needs of each individual customer" that defines a potential solution space from the desires and point of view of the customers, the "capability and degrees of freedom built into a given production system" that defines a potential solution space coping with technological and economical consideration of the manufacturer, and the "company specific strategy and policies" that may limit the customization offer due to tactical considerations (this is the case of a shoe company that limits the combination of colors to given preaccepted sets, to preserve the brand style, or do not give the possibility to move along the "aesthetic dimension," again to preserve brand name, but are eager to promote fitting). Thus, the stable solution space is the result of an interaction of those three elements, whose performance indicators may significantly differ from one another (Figure 33.1). Once defined, the stable solution space represents (1) the yet undifferentiated product blueprints (that is the sum of all the potential customization options for the product), (2) the capability and degrees of freedom of the production system, and (3) the adequate supply chain capable to support the product variants.

Figure 33.2 shows the mapping, onto the ITEMs within the MC definition, of the stable solution space: This derives from the desires of the customer (mapped on MC

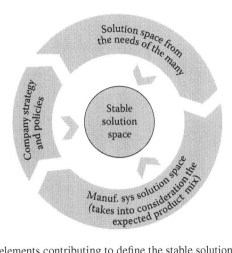

FIGURE 33.1   The elements contributing to define the stable solution space.

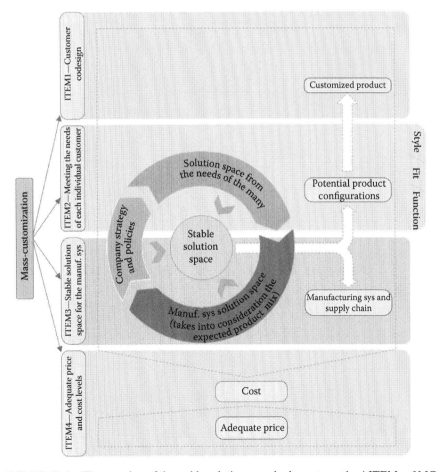

FIGURE 33.2   The mapping of the stable solution space's elements on the 4 ITEMs of MC.

ITEM2 in green) and from the potential solution space coping with technological and economical consideration of the manufacturer (mapped on MC ITEM3 in red). The stable solution space, as aforementioned, represents the potential final product configurations, the production system, and the supply chain meant to manufacture the final product (Figure 33.2). The potential product configurations are the starting points used by customer codesign tools (e.g., the "product configurators" for the personalization available in the websites for shoe personalization aforementioned), in order to define the specification of the final goods. Figure 33.2 also depicts the relation of the SSS with the MC ITEM4: the adequate price (later described).

The implementation of an MC stable solution space (SSS) in the footwear sector, that is a typical mass production sector, requires that the shoe producers change their vision on product, production, and organization (Boër and Dulio 2007). An organizational redefinition is one of the most important issues for the development of a mass customizing company, because different actors (shoe manufacturer, external designers, suppliers, component manufacturers, subcontractors, and customers) are all involved in the shoe life cycle and their relationship needs to be reengineered within an SSS. In *Mass Customization and Footwear: Myth, Salvation or Reality?* (Boër and Dulio 2007) the steps to mass customization implementation and the possible strategies to reach an SSS are described. Tools were also developed to assess each strategy according to the requirements of a specific MC footwear producer. In the EUROShoE project (Dulio and Boër 2004), an integrated production plant (IPP) was designed, developed, and implemented to test several solutions for MC in the footwear sector. This IPP is a sort of laboratory in the field of footwear. The IPP represented a unique implementation that was considered as a "model shoe factory" with very highly integrated and high levels of automation in all its processes. However, to ensure the necessary flexibility and keep the investment costs to a reasonable level, personnel were trained and integrated into the plant in the most critical points. The IPP was able to produce customized shoes with various degrees of customization and it consists of two main sections: the control deck and the shopfloor. The control deck is the equivalent of the sale, administration, production, planning, and design departments of a shoe factory. All its functions are the so-called soft part of a plant. The shopfloor section of the IPP is the "hard part" of the shoe factory that include the cutting, stitching, internal logistic, last making, and making/finishing functions. The IPP plant helped to test the research and development made during the EUROShoE project and certainly contributed to spreading the knowledge of the mass customization paradigm and its application to the footwear sector. Also, several prototypes and demonstrators of various enabling technologies that are needed to implement the paradigm were developed and tested in this IPP-like new scanning technology; new integrated CAD/CAM software to take into consideration the predesign of a shoe collection and the postdesign after the consumer has made his own selection of colors, materials, accessories, etc.; flexible and integrated internal logistic system able to deal with lot size one; flexible cutting and stitching workstations; flexible and fully customized last production; etc.

The IPP, given its outstanding flexibility, represents an SSS that demonstrates a very large customization potential: For example, each of the six footwear companies involved in the project were able to design a customized shoe collection and manufacture within the IPP. The six footwear producers represented six different levels of shoe production from low economic shoes to high luxury brand, from men to women shoes, from everyday walking shoes to high fashion evening shoes, etc.

33.2.1.4   ITEM 4: Adequate Price

Mass customization practice and studies (www.mass-customization.de) shows that consumers are frequently willing to pay a premium price for customization to reflect the increment of utility they gain from a product that better fits their needs than the standard product. Mass customized goods are thus targeting the same market segment that was purchasing the standard goods before, but with adequate price increase.

The SSS is then subject to another constraint, as again shown in Figure 33.2. The number and type of product options, the related manufacturing system, and the adequate supply chain contribute to define the cost for the final customized product. This cost must be compatible with an adequate price so that the customized product does not target a different market segment if compared with the standard one. Again, the EUROShoE project (Boër and Dulio 2007) demonstrated, by relevant consumer analysis, that a premium for customized shoes of 20%–40% is acceptable. It is worthwhile to notice that expressing the premium expressed in percentage allows different footwear sectors to position correctly the target for their product, that is, a 50 Euro shoe will have to cost no more 70 Euro if customized while an upper segment shoe in the price range of 800 Euro, if customized, can be sold for 950 or 1000 Euro.

33.2.2   SUSTAINABILITY

The concept of sustainability, in the way we understand the term now, first hit the ground in 1987, with the Brundtland Report, defining it as "to meet the needs of the present generation without compromising the ability of future generations to meet their own needs" (United Nations General Assembly 1997). Later, as the concept gained popularity and momentum, hundreds of definitions have been proposed, in academic debates and business arenas, referring to a more humane, more ethical, more green, and more transparent way of doing business. Today, the label of "sustainable" is a bottom line requirement: As a matter of fact, sustainability has become a common basic goal for many national and international organizations including industries, governments, NGOs, and universities. However, in spite of the nearly universal recognition that sustainability has received, companies still struggle with the full understanding of the concept and, but just secondly, with the financial viability of the concept. Therefore, the first problem lies with understanding. In the jungle of definitions, and to be able to point out its link with mass customization, we try hereinafter to set some cornerstones, exploring the three sustainability pillars: economical, environmental, and social, always depicted as three overlapping circles. The identification of sustainability indicators has been based on the following criteria:

- Indicators measure aspects that are influenced by the choices made during the design of the SSS.
- According to the SSS definition, indicators are referred not only to the product design, but also to the production system design and the supply chain design.

- A life-cycle perspective has been adopted so as to include in the evaluation also the usage phase and the disposal phase.
- A mass-customized environment has been taken as a reference to select indictors. Even though the link between sustainability and mass customization is sector dependent, the list of indicators has to be as general as possible. Relevance of different indicators could vary depending on the sector.
- Indicators have to be measurable and easily understandable so as to make their use realistic.

In order to identify the individual indicators to be included in the list, a literature review of existing standard has been carried out and new indicators have been introduced when a potential link with mass customization seemed to emerge. The indexes pointed out are by no means expected to be definitive or exhaustive.

33.2.2.1 Environmental Dimension

The environmental dimension of sustainability denominates those aspects that mainly describe environmental performances, in order to minimize the use of hazardous or toxic substances, resources, and energy. These aspects can be summarized as follows: emissions, use of resources, and waste.

	Index	Description	Units
Emissions	Global warming potential (GWP)	The GWP indicator measures the contribution to the global warming caused by the emission of green house gasses in the atmosphere.	kg eq. CO_2
	Photochemical ozone creation potential (POCP)	The POCP indicator calculates the potential creation of tropospheric ozone ("summer smog" or "photochemical oxidation") caused by the release of those gases, which will become oxidants in the low atmosphere under the action of the solar radiation.	g eq. C_2H_4
	Water eutrophication potential (WEP)	WEP indicator measures the contribution to the water eutrophication (enrichment in nutritive elements) of lakes and marine waters caused by the release of polluting substances in the water.	g eq. PO_4^{3-}
	Stratospheric ozone depletion potential (ODP)	The ODP indicator measures the contribution to the depletion of the stratospheric ozone layer caused by gas emissions.	g eq. CFC-11
	Air acidification potential (AAP)	The AAP indicator measures the contribution to the air acidification caused by gas emissions in the atmosphere.	g eq. H^+

(continued)

(continued)

	Index	Description	Units
	Toxicity potential (TP)	The TP is indeed a set of six indicators that measures the relative impact of the emitted substances on specific impact categories (freshwater aquatic environment, marine aquatic environment, freshwater sediment environment, marine sediment environment, terrestrial environment, and humans) due to emission to environmental compartments (air, fresh water, sea water, agricultural, and industrial soil).	kg eq. 1,4-DCB
Use of resources	Land use	The land use reflects the damage to ecosystems due to the effects of occupation and transformation of land.	m^2
	Natural resources depletion (NRD)	NRD indicator measures the depletion of nonrenewable abiotic natural resources (i.e., fossil and mineral resources) as the fraction of the resource reserve used for a single unit out of the solution space weighted by the fraction of the resource reserve that is extracted in 1 year.	$year^{-1}$
	Water depletion (WD)	WD indicator measures the water of any quality (drinkable, industrial, etc.) consumed during the whole life cycle of the product. Water used in closed loop processes are not taken into account.	dm^3
	Energy depletion (ED)	ED indicator measures the energy consumed during the whole life cycle of the product distinguishing between renewable and nonrenewable sources.	kWh
Waste	Waste production (WP)	WP indicator calculates the quantity of waste produced during the whole life cycle of the product.	kg

33.2.2.2 Economic Dimension

The primary objective of a company is ensuring its long-term survival, achieved by satisfying the requirements of its customer at an overall cost that does not exceed the sales revenue. It is also to be taken into account that various components of social and environmental sustainability have a direct impact on economic indexes. For instance, the improvement of environmental parameters can lead to cost reductions thanks to improved efficiency in material and energy use, access to economic incentives, and loans at preferential low interest rates, and decrease of the amount to be paid for environmental compensation taxes. Economical aspects can be summarized as follows: efficiency, investment in technologies and competences, and supply chain risk.

	Index	Description	Details
Efficiency	Unitary cost	The unitary cost indicator measures the expenditure incurred in producing and delivering one product unit, calculated as the average one weighted on the expected product mix.	
	Delivery lead time	The delivery lead time indicator measures the time between the generation of the order and the receipt of the order from the customer.	
	Expected gross profit (EGP)	The EGP indicator measures the difference between the revenues obtained by the yearly product sales (calculated on an expected volume and product mix) and the related costs, before deducting overhead, payroll, taxation, and interest payments.	$EGP = (P - UC) * N$ N is the number of product units expected to be sold UC is the unitary cost. P is the price of the product calculated as the average one weighted on the expected product mix.
	Overall equipment effectiveness (OEE)	The OEE indicator is a qualitative measure of the effectiveness of the manufacturing processes.	$OEE = $ Availability * Performance * Quality Availability is the operating time/ scheduled time (%) Performance is the actual production rate of the equipment/production capability of the equipment (%) Quality is the number of good products/total products made (%)
Investment	R&D investments intensity	The R&D investments intensity indicator provides an allocated measure of the company's R&D investments on the new mass customized product.	R&D intensity = Average yearly R&D investments * Expected sales turnover of the product/expected total sales turnover of the company
SC risks	Supply chain risk	Supply chain risk indicator is a qualitative indicator measuring the risk associated to the provision of components, modules, parts or final products based on the component criticality and the financial reliability of the supplier providing it.	This qualitative indicator takes into account two different factors: *Component criticality*: A qualitative measure of the component availability on the market considering the number of possible alternative suppliers. *Supplier risk*: A qualitative measure of the financial reliability of the supplier that provides the component.

33.2.2.3   Social Dimension

Social dimension of sustainability usually encompass aspects such as social responsibility, health and safety, "polluter pays" principle, and reporting to the stakeholders.
Social aspects can be summarized as follows: working condition and workforce,
product responsibility, and local community benefits.

	Index	Description	Details
Work condition	Injuries intensity	The injuries intensity indicator measures the number of yearly work-related injuries, diseases, and fatalities allocated to the new mass customized product.	Injuries intensity = Average yearly number of injuries * expected sales turnover of the product/ expected total sales turnover of the company
	Employment	The employment indicator measures the percentage of the new employment opportunities created by the introduction of the new mass customized product.	Employment = New employees required by the introduction of the product/number of total employees * 100
	Staff development	The staff development indicator measures the percentage of the company investments in staff development for the production of the new mass customized product.	Staff development = Investments in staff development (i.e., training activities) for the production of the product/total company investments in staff development * 100
	Permanent employment contracts	The permanent employment contracts indicator measures the percentage of the permanent employment positions within the solution space.	Permanent employment = Permanent employment positions within the solution space/total number of employees within the solution space * 100
	Income distribution	The income distribution indicator measures the equity of the employee wage distribution within the solution space.	Income distribution = $(\Sigma IT/IB)/C$ IT = average income of the top 10% employees belonging to the solution space IB = average income of the bottom 10% employees belonging to the solution space C = number of considered companies within the solution space
	Hourly wage	The hourly wage indicator measures the average hourly wage of the employees working within the solution space.	

(continued)

	Index	Description	Details
	Child labor	Child labor indicator measures the number of suppliers within the solution space using child labor.	
Product	Product social features	The product social features indicator measures the number of product features that aim at improving the condition of specific target groups (e.g., product for disabled, elderly, and diabetic people)	
Local community	Charitable contributions intensity	Charitable contributions intensity indicator provides an allocated measure of the expenditures and charitable contributions in favor of the local community.	Charitable contributions intensity = average yearly expenditures * expected sales turnover of the product/expected total sales turnover of the company
	Contribution to local employment	The contribution to local employment indicator measures the percentage of the local supplier within the solution space.	Contribution to local employment = Number of local suppliers within the solution space/total number of suppliers within the solution space * 100

This classification is based on the current state of the art and on the work done within an S-MC-S (Bettoni et al. 2012). The classification, which can be, of course, extended or subject to different means of classification, leads us to identify 27 sustainability indexes.

33.2.3 MASS CUSTOMIZATION AND SUSTAINABILITY

Here we are. Is there a link between mass customization and sustainability indexes improvement?

To tackle this central issue, we must first take a look back to ITEM 4 of the MC definition, to set up a preliminary hypothesis: We are confronting a mass customized shoe with a similar mass produced one, whose similarity makes them comparable and whose final price makes the two shoes to belong to the same market segment. This is an important statement as it endorses a comparison between the two shoes in terms of sustainability indexes. We call SSS_{MC} and SSS_{MP} the SSS related with the mass customized and mass produced shoe. The aforementioned 27 sustainability indexes, $SustIndex_i$ from now on (with i ranging from 1 to 27), define the positioning of the two shoes produced within the two SSSs as far as sustainability is concerned. Index by index, we must look into specific characteristics of the mass customization implementation that makes this MC production system to perform better than the mass production one, in relation to that specific index. In other words, we confront $SustIndex_i(shoe_{MC})$ with $SustIndex_i(shoe_{MP})$, with i ranging from 1 to 27. Clearly, $SustIndex_i(shoe_{MC}) = function(shoe_{MC}, SSS_{MC})$, as the performances of the $SustIndex_i$, referred to a specific mass customized shoe, is definitely depending from the specific configuration chosen and from the SSS that enable that specific shoe to be produced. If we can reasonably state that $SustIndex_i(shoe_{MC})$ outperforms the correspondent $SustIndex_i(shoe_{MP})$, based on the inherent characteristics of SSS_{MC} over SSS_{MP}, then we score a point in presenting customization as one of the main promoter of sustainability.

As an example, let us address the global warming potential (GWP—CO_2 emissions) related to transportations of finished goods, where these goods are shoes with strong customization in the fitting dimension (e.g., left foot). This type of mass customized shoes, taking into consideration the companies that nowadays are producing them, has a very short supply and distribution chain (if compared with mass produced shoes), clearly demonstrating a build-in advantage over the traditional long supply and distribution chains of the corresponding mass produced shoes. This is clearly a qualitative reasoning, limited to a shoe segment that needs the actual calculation of GWP index for a wide range of companies and situations to be confirmed or revised. But, it leads us to think that MC really has an edge over MP in this sustainability index.

While this challenge, related to indexes individual analysis and actual evaluation in real cases, is a current field of research, there is one initial consideration, born from the intrinsic characterization of an MC production system, which contributes to shift the balance in favor of MC. In the footwear sector, an unsold rate of 20% is to be expected. Thus, out of 100 produced shoes to stock, 80 will be sold. In the mass customization case, the shoe is produced just when it is sold. Facing a demand of 80 shoes, 80 will be produced. This clearly reflects in a magnifying effect over each $SustIndex_i(shoe_{MC})$, whose performance is to be increased by an average rate of 20%. This consideration is clearly specific to the shoe sector and represents an average over the whole footwear production.

33.3 R&D PROJECTS ON MASS CUSTOMIZATION AND SUSTAINABILITY

The national Italian research project called "Integrated and automated system for the shoe production - IPP" (Sistema automatizzato ed integrato per la produzione di calzature) paved the way for the research to come, as it developed the design tools and manufacturing processes to set up a stable solution space for a high variety of different type of shoes. The plant was able to produce lot size one and it followed the basic foundation of the mass customized concept: indeed the order for the shoes where done at the shop and distributor level involving directly the consumer for the style issue but not for comfort or fitting. Therefore, a very low level of customization was possible mainly based on the color and some minor modifications of the design like round or straight collar, little accessories, etc.

We have already mentioned one of the first and major projects on mass customization: EUROShoE that was the direct follow-up of the Italian project just described. EUROShoE indeed introduced a full customization of the shoe involving the consumer for all the aspects of style, fit, and comfort and required a complete rethinking of the whole design–selling–manufacturing–distribution–delivery chain. EUROShoE paved the way for a new line of research in footwear and was followed by several projects. Indeed, a national Italian research project IPP mentioned earlier developed the design tools and manufacturing processes to set up an SSS for a high variety of different types of shoes. The plant was able to produce lot size one but did not side along with the very foundation of the mass customized concept: Indeed the order for the shoes where still done at the shop and distributor level without involving directly the consumer. The follow-up project, called CEC-made-Shoe (CEC stands for custom, environment, and comfort—www.cec-made-shoe.com), took the basic results of EUROShoE, as presented earlier in this chapter, and started to introduce the concept of "sustainability" mainly in its environmental aspect: The research focused on the development of more environmental friendly materials and processes for shoes production. The CEC-made-Shoe was an integrated project encompassing from foresight to implementation of the research–industrial innovation value chain, capable of

- Perceiving strategic changes in human, societal, economical needs, and technological potentials as a permanent foresight body
- Making plans and providing resources for developing knowledge concerning products, processes, materials, and organizations and turning such knowledge into industrial technology
- Transferring and implementing it into products, processes, materials, organizations considering their interrelated life cycles, thus developing new generations represented by their respective knowledge
- Contributing through time to innovate consumptions and production paradigms
- Contributing to increase industrial attitude to invest in research as successful projects will prove strategic to tactical rentability
- Complying with and implementing the knowledge-based sustainable development concepts
- Moving the sector from mass production to higher added value and eco-efficient and sustainable products, processes, materials, and organizations

More recently, the project EC-FP7 DOROTHY (http://dorothy.ttsnetwork.net/) is devoted to research methods and technology to develop tools for the design of customer-driven shoes and for the design, configuration, and reconfiguration of flexible multisite, multinational production factories meant to manufacture those customer-driven shoes. In one word, DOROTHY research is focussed to define the optimal SSS for the mass customized shoe in sport-shoes and formal man shoes.

The latest project, EC-FP7 SMCS (http://web.ttsnetwork.net/SMCS), which serves as a framework for the concepts here presented, takes a step further toward the proof of concepts of the new production paradigm, sustainable mass customization. Indeed, it studies the evaluation of mass customization implementation beyond the mere assessment of economic aspects, steering toward the integration of environmental and social consequences into the assessment of the value chain, thus fostering a sustainable approach in MC.

33.4   CONCLUSIONS

This chapter addresses the two concepts of mass customization and sustainability in footwear: It proposes a mapping of the concept of SSS over the four ITEMs of mass customization definition, and points out relevant sustainability indicators useful to evaluate the SSS in terms of economical, ecological, and social performances. Footwear examples are offered, together with the four different but concurrent EU projects, to place those concepts into a meaningful framework and to provide insight in actual implementations. The chapter then focuses on the link between mass customization and sustainability, to discuss whether mass customization can be regarded as one of the main driving forces to achieve effective sustainability. Besides the undisputed commercial and social values, the conclusion is that promising hints are emerging to regard MC also as a major driver for ecological sustainability. It seems, however, difficult to reach the conclusion that MC (or any other production paradigm) has an overall benefit on sustainability as a whole. The value given to commercial, social, or ecological parts of sustainability needs to be addressed or in other words we need to address the function in the relationship $SustIndex_i(shoe_{MC}) = function(shoe_{MC}, SSS_{MC})$.

QUESTIONS

33.1 Does mass customization empower sustainability?

33.2 Does it even make sense to link the two concepts?

33.3 Can we map and formalize mass customization concept so that future implementations can be easily put into existence?

33.4 Which kind of sustainability indicators is relevant to compare MC and sustainability?

REFERENCES

Bettoni, A., Corti, D., Pedrazzoli, P. 2012. Sustainable mass customization—General considerations. *Mass Customization and Personalization Conference*, San Francisco, CA.

Bischoff, R., Bueno, R., Caldeira, J.C. et al. 2010. *Factories of the Future PPP Strategic Multi-Annual Roadmap*. www.manufuture.org

Boër, C.R., Dulio, S. 2007. *Mass Customization and Footwear: Myth, Salvation or Reality?* Springer, New York.

Davis, S. 1987. *Future Perfect*. Addison-Wesley, Boston, MA.

Dulio, S., Boër, C.R. 2004. Integrated Production Plant (IPP): An innovative laboratory for research projects in the footwear field. *International Journal of Computer Integrated Manufacturing*, 17(7): 601–611.

Piller, F. 2004. Mass customization: Reflections on the state of the concept. *International Journal of Flexible Manufacturing Systems*.

Pine, B.J. II. 1993. *Mass Customization*. Harvard Business School Press, Boston, MA.

Tseng, M., Jiao, J. 2001. Mass customization. In: *Handbook of Industrial Engineering*, G. Salvendy (ed.), 3rd edn., Wiley, New York.

United Nations General Assembly. 1997. Report of the World Commission on Environment and Development: Our Common Future. Transmitted to the General Assembly as an Annex to document A/42/427.

Index